Cellular and Molecular Biology of Intermediate Filaments

Cellular and Molecular Biology of Intermediate Filaments

Edited by
Robert D. Goldman
Northwestern University Medical School
Chicago, Illinois

and
Peter M. Steinert
National Cancer Institute
National Institutes of Health
Bethesda, Maryland

Plenum Press • New York and London

Library of Congress Cataloging-in-Publication Data

Cellular and molecular biology of intermediate filaments / edited by
Robert D. Goldman and Peter M. Steinert.
p. cm.
Includes bibliographical references.
ISBN 0-306-43317-6
1. Cytoplasmic filaments. I. Goldman, Robert D., 1939- .
II. Steinert, Peter M.
[DNLM: 1. Cytoskeleton. 2. Molecular Biology. QH 603.C96 C3935]
QH603.C95C45 1990
574.87'34--dc20
DNLM/DLC
for Library of Congress 89-71149
CIP

A Division of Plenum Publishing Corporation
233 Spring Street, New York, N.Y. 10013

Printed in the United States of America

Contributors

Beverly A. Dale, Departments of Oral Biology, Periodontics, Medicine/Dermatology, and Biochemistry, University of Washington, Seattle, Washington 98195

Igor B. Dawid, Laboratory of Molecular Genetics, National Institute of Child Health and Human Development, National Institutes of Health, Bethesda, Maryland 20892

George N. Dessev, Department of Cell, Molecular, and Structural Biology, Northwestern University Medical School, Chicago, Illinois 60611

P. A. M. Eagles, Department of Bio-physics, King's College London, London WC2B 5RL, United Kingdom

R. D. B. Fraser, CSIRO, Division of Wool Technology, Parkville, Victoria 3052, Australia

Giulio Gabbiani, Department of Pathology, University of Geneva, 1211 Geneva 4, Switzerland

H. Gainer, Laboratory of Neurochemistry and Neuroimmunology, National Institute of Child Health and Human Development, National Institutes of Health, Bethesda, Maryland 20892

J. M. Gillespie, CSIRO, Division of Wool Technology, Parkville, Victoria 3052, Australia

Robert D. Goldman, Department of Cell, Molecular, and Structural Biology, Northwestern University Medical School, Chicago, Illinois 60611

Kathleen J. Green, Department of Pathology, Northwestern University Medical School, Chicago, Illinois 60611

Paul V. Haydock, Departments of Periodontics and Medicine/Dermatology, University of Washington, Seattle, Washington 98195

Milan Jamrich, Laboratory of Molecular Genetics, National Institute of Child Health and Human Development, National Institutes of Health, Bethesda, Maryland 20892

Erzsebet Jonas, Laboratory of Molecular Genetics, National Institute of Child Health and Human Development, National Institutes of Health, Bethesda, Maryland 20892

Jonathan C. R. Jones, Department of Cell Biology and Anatomy, Northwestern University Medical School, Chicago, Illinois 60611

Norman Lieska, Department of Anatomy, Texas College of Osteopathic Medicine, Fort Worth, Texas 76107

Marion Lynch, Epithelial Biology Unit, Departments of Dermatology and Pharmacology, The Kaplan Cancer Center, New York University School of Medicine, New York, New York 10016

T. P. MacRae, CSIRO, Division of Wool Technology, Parkville, Victoria 3052, Australia

George S. Michaels, Laboratory of Molecular Genetics, National Institute of Child Health and Human Development, National Institutes of Health, Bethesda, Maryland 20892

Seiji Miyatani, Laboratory of Molecular Genetics, National Institute of Child Health and Human Development, National Institutes of Health, Bethesda, Maryland 20892

W. Michael O'Guin, Epithelial Biology Unit, Departments of Dermatology and Pharmacology, The Kaplan Cancer Center, New York University School of Medicine, New York, New York 10016

H. C. Pant, Laboratory of Preclinical Studies, National Institute on Alcohol Abuse and Alcoholism, National Institutes of Health, Rockville, Maryland 20892

David A. D. Parry, Department of Physics and Biophysics, Massey University, Palmerston North, New Zealand

Linda M. Parysek, Department of Anatomy and Cell Biology, University of Cincinnati Medical School, Cincinnati, Ohio 45267-0521

Barry C. Powell, Department of Biochemistry, University of Adelaide, Adelaide, South Australia 5000, Australia

Katheryn A. Resing, Department of Biochemistry, University of Washington, Seattle, Washington 98195

George E. Rogers, Department of Biochemistry, University of Adelaide, Adelaide, South Australia 5000, Australia

Dennis R. Roop, Departments of Cell Biology and Dermatology, Baylor College of Medicine, Houston, Texas 77030

Thomas D. Sargent, Laboratory of Molecular Genetics, National Institute of Child Health and Human Development, National Institutes of Health, Bethesda, Maryland 20892

Alexander Schermer, Epithelial Biology Unit, Departments of Dermatology and Pharmacology, The Kaplan Cancer Center, New York University School of Medicine, New York, New York 10016

Michael L. Shelanski, Department of Pathology and Center for Neurobiology and Behavior, College of Physicians and Surgeons, Columbia University, New York, New York 10032

Omar Skalli, Department of Pathology, University of Geneva, 1211 Geneva 4, Switzerland

Peter M. Steinert, Dermatology Branch, National Cancer Institute, National Institutes of Health, Bethesda, Maryland 20205; *present address:* Laboratory of Skin Biology, National Institute of Arthritis and Musculoskeletal and Skin Diseases, National Institutes of Health, Bethesda, Maryland 20205

Alasdair C. Steven, Laboratory of Physical Biology, National Institute of Arthritis, Musculoskeletal, and Skin Diseases, National Institutes of Health, Bethesda, Maryland 20892

Marvin H. Stromer, Muscle Biology Group, Department of Animal Science, Iowa State University, Ames, Iowa 50011

Tung-Tien Sun, Epithelial Biology Unit, Departments of Dermatology and Pharmacology, The Kaplan Cancer Center, New York University School of Medicine, New York, New York 10016

Carol M. Troy, Department of Pathology and Center for Neurobiology and Behavior, College of Physicians and Surgeons, Columbia University, New York, New York 10032

Jeffrey A. Winkles, Laboratory of Molecular Genetics, National Institute of Child Health and Human Development, National Institutes of Health, Bethesda, Maryland 20892

Hsi-Yuan Yang, Department of Anatomy and Cell Biology, College of Medicine, University of Illinois, Chicago, Illinois 60612

Robert V. Zackroff, Department of Microbiology, University of Rhode Island, Kingston, Rhode Island 02881

Preface

Research activity on intermediate filaments (IF) has increased dramatically over the past decade. For the most part, this surge of interest is due to their identification as ubiquitous constituents of the cytoskeleton and karyoskeleton (nuclear matrix) of eukaryotic cells and the fact that we know very little regarding their functions. In sharp contrast to the other major cytoskeletal systems, microfilaments and microtubules, IF exhibit a high degree of heterogeneity with regard to their protein subunit composition. Indeed, one can only marvel at the number of different IF polypeptides, their associated proteins (IFAP) and, consequently, the number of genes involved in encoding the multiple constituents of the various IF networks found in different cell types.

The chapters in this book demonstrate how various experimental approaches involving cellular, molecular, biochemical, and immunological methods have been utilized to generate information regarding the structure and function of IF. To this end, we have gathered together chapters from experts in the major fields of IF research. In each chapter, the authors have combined reviews of the available scientific literature with their own ideas on current and future directions for IF research. The chapters have been divided into five major sections which are concerned with the subcellular organization of IF, the molecular structure of IF, the differential expression of IF genes, descriptions of associated proteins involved in the intracellular organization of IF, and finally an analysis of the changes seen in IF in pathological conditions.

Our goal throughout the preparation of this book has been to provide, in a single volume, the basic information necessary to understand the current state of knowledge of the different IF systems. In addition, it is our hope that this volume will serve as a reservoir of background information, which should provide the foundations for stimulating further studies on IF structure and function.

Robert D. Goldman
Peter M. Steinert

Contents

I. The Subcellular Organization of Intermediate Filaments

1. ***Intermediate Filaments: An Overview*** 3
Robert D. Goldman, Robert V. Zackroff, and Peter M. Steinert

1. Historical Overview 3
2. A Comment on Nomenclature 5
3. The Complexity of IF 6
4. A Brief Overview of the Subcellular Organization of IF 7
References 15

2. ***Intermediate (10-nm) Filaments in Muscle*** 19
Marvin H. Stromer

1. Introduction 19
2. Smooth Muscle 19
   2.1. Location of 10-nm Filaments in Cells 19
   2.2. Isolation and Preparation of Desmin and Vimentin 21
   2.3. Molecular Properties of Desmin and Vimentin 22
   2.4. Filament Assembly and Substructure 24
   2.5. IF-Associated Proteins 27
   2.6. Role or Function 28
3. Skeletal Muscle 28
   3.1. Location of 10-nm Filaments in Cells 28
   3.2. Isolation and Preparation of Desmin 29
   3.3. Protein and Filament Properties 30
   3.4. Role or Function 30
4. Cardiac Muscle 30
   4.1. Location of 10-nm Filaments in Cells 30
   4.2. Isolation and Preparation of Desmin 31
   4.3. Role or Function 31
References 32

3. ***Neurofilaments*** 37

P. A. M. Eagles, H. C. Pant, and H. Gainer

1. Introduction 37
2. Neurofilament Arrangement in the Neurons of Vertebrates 38
   2.1. Isolation of Neurofilaments 41
   2.2. Filament Solubility and Structure 42
   2.3. Mammalian Neurofilament Polypeptides 43
   2.4. Reassembly Properties 45
   2.5. Amino Acid Composition and Sequence Studies 45
   2.6. Arrangement of the Triplet Polypeptides within the Neurofilament 48
3. Invertebrate Neurofilaments 52
   3.1. *Myxicola* 53
   3.2. Squid 59
4. Interactions of Neurofilaments 63
5. Posttranslational Modifications of Neurofilaments 66
   5.1. Calcium-Activated Proteolysis 66
   5.2. Phosphorylation of Neurofilaments 70
6. Immunological Probes of Neurofilaments 72
7. Axonal Transport of Neurofilaments 80
8. Ontogeny and Phylogeny of Neurofilaments 81

   References 82

4. ***The Proteins of Hair and Other Hard α-Keratins*** 95

J. M. Gillespie

1. The Nature of Hard α-Keratins 96
   1.1. Solubility 96
   1.2. Chemical Composition of Keratins 97
   1.3. Fine Structure of Hard α-Keratins 99
2. Morphology and Development of the Hair Follicle 100
   2.1. Morphogenesis of the Hair Follicle 100
   2.2. Regulation of Hair Growth Rate 101
   2.3. Development of Fiber Proteins 102
   2.4. Assembly and Cross-Linking of Fiber Proteins 103
3. Constituent Proteins of Intermediate Filaments 104
   3.1. Preparation and Properties of Low-Sulfur Proteins 104
   3.2. Properties of Subunit Polypeptides 105
   3.3. Amino Acid Sequences of Subunit Polypeptides 106
   3.4. Assembly of Subunit Polypeptides 106
4. The Intermediate Filament-Associated Proteins 107
   4.1. The High-Sulfur Proteins 107
   4.2. The High-Tyrosine Proteins 113
5. The Regulation of Keratin Composition 117
   5.1. Molecular Basis of Variability in Keratin Composition 117
   5.2. Genetic Control of Keratin Composition 118

5.3. Nutritional Control of Keratin Composition 122
5.4. Physiological Control of Keratin Composition 124
6. Concluding Remarks 124
References 125

5. ***The Nuclear Lamina: An Intermediate Filament Protein Structure of the Cell Nucleus*** 129
George N. Dessev

1. Introduction 129
2. Definition of NL and Lamin Classification 130
3. Isolation of NL 131
4. Composition of NL and Its Changes during Development 132
5. Biosynthesis of Lamins 133
6. Structure of Lamin Molecules and Their Assembly into Polymers 133
7. Association of NL with Nuclear and Cytoplasmic Structural Elements 134
7.1. NPC 134
7.2. Nuclear Membrane 135
7.3. Cytoskeletal Elements 136
7.4. Chromatin 136
8. Cell Cycle Dynamics of NL 137
9. The Same General Mechanism Is Likely to Be Involved in the Cell Cycle Dynamics of NL and IF 140
10. Cell Cycle Regulation of NL Disassembly 141
References 141

6. ***Interaction of Intermediate Filaments with the Cell Surface*** 147
Kathleen J. Green and Jonathan C. R. Jones

1. Introduction 147
2. IF–Desmosome and Hemidesmosome Interactions 148
3. Morphological Analysis of Desmosome–IF and Hemidesmosome–IF Complexes 148
4. The Biochemical Composition of Desmosomes and Hemidesmosomes 151
5. Putative IF–Plaque Linkers 152
5.1. Desmocalmin 154
5.2. "Band 6" or Desmoplakin IV 155
5.3. Desmoplakin I and II 155
6. IF–Cell Surface Association in Muscle Cells 157
7. Association of IF with Cortical Actin-Containing Structures 158
8. Interaction of IF with the Cell Membrane Complex 160
9. Conclusion 164
References 165

II. The Structure of Intermediate Filaments

7. ***Primary and Secondary Structure of IF Protein Chains and Modes of Molecular Aggregation*** 175
David A. D. Parry

1. Introduction 175
2. Type I–IV IF Protein Chains 176
   2.1. Primary Structure 176
   2.2. Secondary Structure 177
3. Classification of IF Protein Chains 179
4. Structural Features of the Central α-Helix-Rich Domain in IF Protein Chains 182
5. Structure of the Rod Domain of IF Protein Molecules 187
6. Structure of the Terminal Domains of IF Protein Molecules 191
7. Structure of the Four-Chain Aggregates 194
8. Nuclear Lamins: IF-Like Proteins 196
9. Conclusions 198
   References 198

8. ***The Three-Dimensional Structure of IF*** 205
R. D. B. Fraser, T. P. MacRae, and David A. D. Parry

1. Introduction 205
2. Molecular Structure 206
3. Molecular Aggregation 208
   3.1. Enzymatically Derived Fragments 209
   3.2. Ionic Interaction Score 209
   3.3. Oligomers 213
   3.4. Polymers 218
4. Filament Structure 219
   4.1. A Generic Model for IF 219
   4.2. X-Ray Diffraction Studies 219
   4.3. The Surface Lattice 221
   4.4. Disposition of Terminal Domains 223
   4.5. Models Generated Using Interaction Scores 223
   4.6. Radial Distribution of Coiled-Coil Rope Segments 223
5. Concluding Remarks 227
   References 227

9. ***Intermediate Filament Structure: Diversity, Polymorphism, and Analogy to Myosin*** 233
Alasdair C. Steven

1. Introduction 233
2. How Many Different Kinds of IF Are There? 234
   2.1. Tissue-Specific Classification 234
   2.2. IF Types Based on Amino Acid Sequence Homologies 235

2.3. Classification According to Competency for *in Vitro* Self-Assembly or Coassembly ... 235
2.4. Polymorphism of IF That Contain the Same Subunits ... 236
2.5. Other Diffferences: Immunological Specificity, Interspecies Variation, and Distinctions Conferred by Associated Proteins ... 236
3. Structure of Oligomers of IF Proteins ... 237
4. Axial Repeats in IF Structure ... 240
4.1. Regular Distribution of Particular Types of Amino Acid Residues in IF Sequences ... 240
4.2. Axial Repeats from Meridional X-Ray Diffraction ... 241
4.3. Axial Repeats Detected by Electron Microscopy ... 241
4.4. Axial Repeats in Paracrystals ... 244
4.5. Proportionality between the Linear Densities of IF and the Molecular Weights of Their Constituent Subunits ... 245
5. Width and Radial Organization of IF ... 247
5.1. Traditional Estimates of IF Diameter ... 247
5.2. Evaluation of Transverse Traces of Projected Density from Electron Micrographs of Negatively Stained, Rotary-Shadowed, and Unstained IF ... 248
5.3. Radial Density Profiles of IF ... 252
5.4. Filaments within Filaments ... 253
5.5. Interpretation of Images of Negatively Stained IF with and without Proteolytic Modification ... 256
5.6. *In Vitro* Assembly Properties of Proteolytically Modified IF ... 256
5.7. Other Pertinent Observations ... 257
5.8. Current Frame of Reference ... 258
References ... 260

III. Differential Expression of Intermediate Filament Genes

10. ***Hard Keratin IF and Associated Proteins*** ... 267
Barry C. Powell and George E. Rogers

1. Introduction ... 267
2. The Proteins and Genes of Wool ... 269
2.1. Hard Keratin IF ... 269
2.2. Hard Keratin IFAP ... 277
2.3. Hard Keratin IFAP Gene Structure and Organization ... 285
2.4. Evolutionary Expression of Hard Keratin IF and IFAP ... 287
2.5. Trichohyalin in the IRS and Hair Medulla ... 288
3. Differentiation ... 289
3.1. The Structure of Hairs ... 289
3.2. Hair Developmental Program ... 291
3.3. Gene Expression during Differentiation in the Hair Follicle ... 292
References ... 295

11. ***Differentiation-Specific Expression of Keratin Pairs*** 301
W. Michael O'Guin, Alexander Schermer, Marion Lynch, and Tung-Tien Sun

1. Introduction 301
2. Keratin Subfamilies 304
3. Keratin Pairs Defined by Frequent Coexpression 307
4. Patterns of Keratin Pair Expression 311
5. Localization of Various Keratin Pairs 318
6. Structural/Functional Significance of Keratin Pairs 326
7. Conclusions and Perspectives 328
References 330

12. ***Keratins and Their Genes in Xenopus laevis: Structural and Developmental Aspects*** 335
Thomas D. Sargent, Erzsebet Jonas, Milan Jamrich, George S. Michaels, Seiji Miyatani, Jeffrey A. Winkles, and Igor B. Dawid

1. Introduction 335
2. Cytokeratin Gene Sequences from *Xenopus* 336
   2.1. Type I Cytokeratin Sequences 336
   2.2. Type II Cytokeratin Sequences 337
   2.3. Cytokeratin Gene Structure 337
3. Family Relationships among Type I Cytokeratins 340
4. Developmental Aspects 341
   4.1. Distinct Groups of Cytokeratin Genes Are Expressed in Embryonic/Larval and Adult Stages in *Xenopus* 341
   4.2. Correlation between Type I and Type II Cytokeratin Expression 342
   4.3. Developmental Regulation of Cytokeratin Gene Expression 343
5. Conclusions 343
References 343

13. ***Neural Intermediate Filament (Neurofilament) Genes*** 345
Linda M. Parysek

1. Distinctions among NIFP 345
2. NIFP Amino Acid Sequences 346
3. Sequences Related to the Neural IF Multiphosphorylation Site 347
4. NIFP mRNAs 347
5. Introns in NIFP Genes 348
6. NIFP Gene Promoters 349
References 349

14. ***The Structure, Complexity, and Evolution of Intermediate Filament Genes*** 353
Peter M. Steinert and Dennis R. Roop

1. Introduction 353
2. Structure of IF Genes 354

2.1. Complexity ... 354
2.2. IF Gene Linkage and Chromosomal Localization ... 354
2.3. IF Gene Structure ... 356
3. Models for the Evolution of IF Genes ... 356
3.1. Evolution of the Rod Domain ... 356
3.2. Evolution of the End Domains ... 359
3.3. Evolution of Lamin Genes ... 363
4. Conclusions and Future Perspectives ... 363
References ... 364

IV. Accessory Proteins Involved in Regulating the Organization of Intermediate Filaments

15. ***Intermediate Filament-Associated Proteins*** ... 371
Hsi-Yuan Yang, Norman Lieska, and Robert D. Goldman

1. Introduction ... 371
2. IFAP in BHK-21 Cells ... 372
2.1. Criteria for Identification of IFAP in BHK-21 Cells ... 373
2.2. IFAP-300k ... 373
2.3. IFAP-70-280k ... 379
3. Other IFAP ... 380
3.1. Plectin ... 380
3.2. Synemin ... 380
3.3. Paranemin ... 381
3.4. Epinemin ... 382
3.5. p50 ... 382
3.6. NAPA-73 ... 383
3.7. p230 and Spectrin-Related Proteins ... 383
3.8. MAP-2 ... 384
3.9. Desmoplakin 1 and 2 ... 385
3.10. Cellular Expression of the IFAP: A Summary ... 385
4. Speculations Regarding the Possible Functional Significance of IFAP ... 386
References ... 387

16. ***Filaggrins*** ... 393
Beverly A. Dale, Katheryn A. Resing, and Paul V. Haydock

1. Introduction ... 393
2. Isolation and Properties of Filaggrin and Profilaggrin ... 394
2.1. Extraction and Purification ... 394
2.2. Properties of Profilaggrin and Filaggrin ... 395
2.3. Profilaggrin Protein Structure ... 396
2.4. Aggregation with Keratin IF ... 397
3. Profilaggrin mRNA and Gene Structure ... 398
3.1. Use of cDNA Clones ... 398
3.2. Selection and Characterization of Profilaggrin cDNA Clones ... 399

3.3. Size and Expression of Profilaggrin mRNA 399
3.4. Repetitive Structure of the Profilaggrin Gene 399
3.5. Predicted Sequence of the Profilaggrin Repeat Unit 402
4. Filaggrin Synthesis and Relation to Epidermal Cell Differentiation 402
4.1. Introduction 402
4.2. Profilaggrin Phosphorylation 403
4.3. Profilaggrin Phosphatase 403
4.4. Proteolytic Processing 404
5. Synthesis of Profilaggrin and Filaggrin during Fetal Development 405
5.1. Expression in Rodent 405
5.2. Expression in Human Fetal Skin Development 405
6. Filaggrin Alterations Associated with Disorders of Keratinization 407
7. Concluding Remarks and Future Directions 409
References 409

V. Pathological Conditions Relating to Intermediate Filaments

17. ***Expression, Organization, and Involvement of Intermediate Filaments in Disease Processes*** 415
Omar Skalli and Giulio Gabbiani

1. Introduction 415
2. IF as Differentiation Markers of Neoplastic and Nonneoplastic Cells 416
3. Pathological Situations Characterized by Altered IF Expression and/or Organization 421
3.1. Myopathies and Cardiomyopathies 421
3.2. Alcoholic Liver Disease 422
3.3. Atheromatosis 425
3.4. Sarcoidosis 428
3.5. Tumors and Transformed Cells 429
3.6. Aging 431
4. Autoantibodies against IF 431
5. IF as Activators of Inflammatory Reactions 433
6. Conclusions 435
References 436

18. ***The Cytoskeleton in Neurological Disease*** 451
Michael L. Shelanski and Carol M. Troy

1. Introduction 451
2. Neurofibrillary Degeneration 451
2.1. Neurofibrillary Proliferations Composed of 10-nm IF 452
2.2. Biochemical Composition of Neurofibrillary Tangles 452
2.3. Mechanism of Formation of Neurofibrillary Tangles 453
3. Alzheimer's Disease 456
3.1. The Alzheimer Neurofibrillary Tangle 457
3.2. Nonneural Changes in Alzheimer's Disease 458

4. Other Fibrillar Structures 459
5. Inclusions Containing Cytoskeletal Elements 459
5.1. Pick Bodies 459
5.2. Lewy Bodies 460
5.3. Hirano Bodies 460
5.4. Other Inclusions 460
6. Conclusion 461
References 461

Index 467

I

The Subcellular Organization of Intermediate Filaments

1

Intermediate Filaments
An Overview

Robert D. Goldman, Robert V. Zackroff, and Peter M. Steinert

1. Historical Overview

We now know that intermediate filaments (IF) are ubiquitous constituents of virtually all differentiated eukaryotic cells and are present in both the nucleus (as the nuclear lamina) and cytoplasm (as 10- to 15-nm-diameter filaments). However, the realization of this distribution has taken about 100 years. In retrospect, IF networks were first described in the late 1800s–early 1900s by cytologists studying fixed–stained tissue cells. It appears quite likely that many of the fibrillar structures revealed in the cytoplasm with the early silver and gold staining methods were IF. For example, the neurofibrils characteristic of neurons were described early on as an extensive "netlike structure" in the cell body surrounding the nucleus and extending into axons (Wilson, 1928, pp. 40–41). In addition, early cytologists such as Heidenhain and Del Rio depicted various types of columnar epithelial cells as containing extensive arrays of "tonofibrillae." In stratified epithelia, many of the tonofibrillae were described as forming "intercellular plasma bridges" thought to traverse cell membranes (Wilson, 1928, pp. 40–41). These tonofibrils were probably the same fibrous arrays that can now be detected by immunofluorescence methods using antibodies directed against keratin, and the transcellular bridging areas most likely represent the IF bundle–desmosome complexes that typify keratinocytes and other epithelial cells (Jones and Goldman, 1985).

Robert D. Goldman • Department of Cell, Molecular, and Structural Biology, Northwestern University Medical School, Chicago, Illinois 60611. ***Robert V. Zackroff*** • Department of Microbiology, University of Rhode Island, Kingston, Rhode Island 02881. ***Peter M. Steinert*** • Dermatology Branch, National Cancer Institute, National Institutes of Health, Bethesda, Maryland 20205; *present address:* Laboratory of Skin Biology, National Institute of Arthritis and Musculoskeletal and Skin Diseases, National Insitutes of Health, Bethesda, Maryland 20205.

Historically, most of the early work on IF systems had its origins in wool research for obvious economic reasons. In the early 1930s, Astbury and colleagues used x-ray diffraction and discovered the ordered quasi-crystalline array of particles [filaments] in wool/hair that was termed the α-pattern (Astbury and Street, 1931). This was later interpreted in terms of the presence of coiled coils formed by two (or more) α-helices (Crick, 1953). Also in the early 1950s, electron microscopic techniques enabled for the first time high-resolution visualization of the filaments in epithelial and many other types of cells that indicated remarkable similarities in morphological properties. This ultrastructural work was followed rapidly by development of biochemical methods to characterize the wool/hair proteins. Such early data permitted the construction of models of structure, based on the packing of coiled coils (Fraser and MacRae, 1961), that still remain a subject of considerable research activity today.

One of the major problems that slowed progress in IF research until the mid 1970s stemmed from the lack of recognition that all IF were members of the same large family of proteins. As a result, IF were described in many cell types independently and few, if any, comparisons were made with similar structures in other cell types. This resulted in a very large number of descriptive terms originating from such morphological parameters as their size and cell type localization. Based on size, they were variously described as 100 Å, 10-nm, and dece-filaments; 9-nm filaments, 80- to 100-Å filaments, and 110-Å filaments. In many cases, they were named using more descriptive terms, such as neurofilaments and glial filaments, based on the cell type studied. Other descriptive terms included tonofilaments, sarcoplasmic filaments, skeletin filaments, square filaments, round filaments, beta filaments, lentofilaments, protofibrils, cytoplasmic filaments, fine filaments, and filaments (see Eriksson and Thornell, 1979, for a review of nomenclature). Furthermore, proteins comprising different types of IF have been given a variety of names including vimentin, desmin, neurofilament protein or neurofilament triplet protein, keratin, and glial fibrillary acidic protein (see Steinert *et al.*, 1984). In addition, various names such as synemin, filaggrin, and plectin have been given to IF-associated proteins (IFAP; see Chapter 15).

Another source of confusion that persisted for many years stemmed from the idea that IF might be alternate forms of microtubules. This latter argument was based primarily on the apparent proliferation of IF following treatment of cells with agents that depolymerize microtubules (Wisniewski *et al.*, 1968; Holmes and Choppin, 1968; Holtzer *et al.*, 1972, 1976). We now know that in many types of cells, the majority of IF are reorganized and concentrated into large juxtanuclear aggregates or cables when cells are treated with agents such as colchicine and colcemid. Therefore, electron micrographs of thin sections through these latter regions gave the impression that IF were increasing in number as microtubules depolymerized (Goldman, 1971). In addition, actin-containing microfilaments have often been confused with IF in morphological studies (e.g., see Buckley *et al.*, 1978).

The rather rapid progress in the IF field over the past 10 years was initiated at the Cold spring Harbor Cell Motility (1978) meeting, when it became apparent to some of the participants that all of these filamentous types of proteins were members of the same family of cytoskeletal elements and that they were distinctly different from microtubules and microfilaments. From our perspective, three key developments have led to this rather recent stimulation in IF research. First, the demonstration in 1972 that the neurofilaments within the giant axon of *Myxicola* were of the α type (Day and Gilbert, 1972) gave us the first clue that keratin filaments, neurofilaments, and fibroblast IF (vimentin-containing) may be members of the same superfamily (Steinert *et al.*, 1978). A second advance was the development of procedures for the reassembly *in vitro* of native-type filaments from keratinocytes (Steinert *et al.*, 1976) and then from fibroblasts (Starger *et al.*, 1978; Zackroff and Goldman, 1979). The reassembled filaments could re-form without cofactors and energy requirements of any sort, and thus were unlike the other cytoskeletal fibrous proteins including microtubules or microfilaments. Third, the concomitant biochemical characterization of the filament-forming proteins enabled the production of specific antibodies that were then used to fully explore their distribution as well as to probe their dynamic organization in cells.

2. A Comment on Nomenclature

By 1980 it was fashionable to divide IF into five distinct classes depending on their cell type of origin (see Lazarides, 1980, for a review): keratins in epithelial cells; vimentin in cells of mesenchymal origin; desmin in muscle; glial fibrillary acidic protein in astroglia; and neurofilaments in neuronal cells. However, this scheme soon became "leaky" because, for example, some cells contained more than one of these classes of IF, and because a better system soon became apparent, arising from the generation of nucleic acid and protein sequence information from cDNA clones. These sequence data established that all IF-forming proteins possessed a central α-helix-rich rod domain with a conserved secondary structure, but with certain discernible differences that permitted classification into distinct sequence types (Steinert and Parry, 1985; Steinert and Roop, 1988). This same sequence information demonstrated that all IF proteins possessed end domains of highly variable size and chemistry and which could be organized into subdomains, the arrangement of which was also characteristic of each rod domain sequence type (Steinert *et al.*, 1985). In addition, the organization of introns and exons in the genes has been shown to be characteristic of each sequence type. Therefore, the three criteria of rod domain sequences, end domain sequence organization, and gene structure now provide in a straightforward way a classification system for IF proteins and establish rules for admission of new proteins into the IF family. A recent addition has been the nuclear lamins whose relationship to cytoplasmic IF had not been widely appreciated until the availability of sequence information (Fisher *et al.*, 1986; McKeon *et al.*, 1986). Even though lamins form different structures (see Chapter 7), their inclusion in the IF family has stimulated research into their interactions with cytoplasmic IF and the continuity of cytoplasmic and nuclear function (A. Goldman *et al.*, 1986; R. Goldman, *et al.*, 1985, 1986). Currently, there are five types of IF proteins: acidic keratins are type I; neutral–basic keratins are type II; vimentin, desmin, gfap, and/or 57-kDa neuronal IF protein (peripherin; Parysek and Goldman, 1988; Parysek *et al.*, 1988; Leonard *et al.*, 1988) are type III; classical neurofilaments are type IV; and nuclear lamins are type V. We fully expect this list to grow as new sequence information becomes available for IF from simpler organisms and perhaps plants.

While the classification system of the early 1980s is not completely redundant, it should be used only in terms of the newer system. For example, lens epithelia express type III vimentin rather than type I/II keratins as might have been suspected (see Chapter 11). In addition, it appears that some neuronal cell types express both type IV "neurofilaments" as well as the type III IF protein of 57 kDa ("peripherin") (Parysek and Goldman, 1988; Leonard *et al.*, 1988).

Table I. The Complexity of IF

Organizational level	Comments
1. Protein chain	Many chains have been sequenced
	Conserved rod domain drives IF structure
	Hypervariable end domains specify function and properties of chain in IF
2. Two-chain coiled-coil molecule	In-register and parallel arrangement, 50 nm long
	Homodimer or facultative copolymer in types III–V
	Type I/II heterodimer in keratins, although homodimers are possible *in vitro*
	Only indirect evidence for cross-type association
3. Oligomer of two to four molecules	Alignment controversial but probably antiparallel (i.e., IF are nonpolar structures), and either in-register (50 nm) or half-staggered (70 nm)
	Probable smallest particles in "solution" when reversibly disassembled *in vivo* (e.g., phosphorylation)
4. Filament	Polymorphic: can consist of 12–24 molecules wide although some native IF have 16
	Filament may have intermediate levels of protofilamentous and protofibrillar structures
	Good evidence that rod domains form IF "core"
	Circumstantial evidence that protruding end domains define IF function and properties
	Few details have yet been established on structures of either rod or end domains
	Lamins form a structural variant on cytoplasmic IF theme
5. Gene expression	Patterns of expression of proteins now understood
	Exon/intron structures of genes known but location and nature of regulatory sequences not yet known
	Tissue and developmental regulation probably by transcription control or posttranscriptionally (e.g., mRNA stability). Rules not yet known
	May be useful in diagnosis of tumors
6. Cytoplasmic organization	Poorly understood
	Concepts of association with nuclear surface and cell periphery still in early stages of understanding
	Phosphorylation may be important
	Association with other cytoskeletal components such as microtubules and microfilaments poorly understood
7. IFAPs	Proposed regulators of supramolecular organization in cells. May bind IF tightly or loosely across their length or to their ends
	In no case is structure, mechanism of association, or function yet clear
8. Tissue organization	Concept only; proposes a type of "supracellular" organization
	Invoked for perceived cell–cell continuity in epithelia

3. *The Complexity of IF*

Table I summarizes the current status of the IF field of research by simply listing in increasing degrees of molecular order, the extent of our understanding of IF structure. Significantly, the degree of understanding decreases with increasing degrees of complexity down the table. Clarification of those levels of organization, especially those described in parts 5–7, will lead to a better understanding of IF function in cells, apart from their customary structural role. More importantly, we hope the available information will provide a focus for further research in the field.

4. A Brief Overview of the Subcellular Organization of IF

In general, various proportions of type I–IV IF proteins are organized within the cytoplasm into a polymerized system of ~ 10-nm-diameter filaments that course between the nucleus and the cell surface. From a strictly morphological perspective, IF can be very long and they appear to form continuous filamentous arrays from the nuclear region to the region of the plasma membrane. The easiest way to obtain an overview of the IF system is by immunofluorescence microscopy. With this technique, IF appear to radiate from the nuclear region, where they are most concentrated, to the cell surface, where they are least concentrated. This is most readily apparent in various types of cultured cells including fibroblasts (Fig. 1a), epithelial cells (Fig. 1b), and neurons (Fig. 1c).

At the ultrastructural level of resolution, large numbers of IF are seen in the juxtanuclear region (Figs. 2 and 3) where they frequently appear to be anchored to the cytoplasmic surface of the nuclear envelope, either at the outermost membrane or at sites of nuclear pore complexes (Jones *et al.*, 1985; R. Goldman *et al.*, 1985, 1986). From these regions of high concentrations of IF, they radiate toward the cell surface throughout most cytoplasmic areas. In some cases, they are organized in loose arrays of several IF (Fig. 4a). In other instances, they are organized into very tight parallel arrays surrounded by electron-dense material. These latter structures are termed IF bundles or tonofilaments. Such structures are especially common in epithelial cells (Fig. 3; Jones and Goldman, 1985).

Within various cytoplasmic regions, IF are closely associated with organelles such as mitochondria and it has been suggested that IF may be involved in the movement and distribution of these and other organelles (Wang and Goldman, 1978). IF also appear to form associations with the other two cytoskeletal systems. For example, they form close parallel arrays with microtubules in neurons and fibroblasts (Fig. 4b; Goldman and Follett, 1969; Goldman and Knipe, 1973). Although it is widely believed that IF interact with microtubules in a more functional fashion, there is very little convincing biochemical information available bearing on this point. IF also appear to be closely associated with submembranous arrays of actin-containing microfilaments in various types of cells (Fig. 5; Green *et al.*, 1986, 1987).

At the level of the cell surface, IF appear to be attached to desmosomes at the innermost membrane-associated plaque material in epithelial cells (Staehelin, 1974) (Fig. 6), and also with regions of the plasma membrane involved in cell–substrate adhesion such as adhesion plaques or the so-called fibronexus structures that are seen at the cell surface of cultured fibroblasts (see Green and Goldman, 1986).

In summary, there is a large amount of morphological data to suggest that IF form connecting links between nuclear and plasma membrane-associated structures. Furthermore, it is thought that cytoplasmic IF are connected in some fashion to the type V IF system, the nuclear lamins (R. Goldman *et al.*, 1985, 1986; Georgatos and Blobel, 1987a,b). These proteins form an insoluble polymeric network at the inner face of the nuclear envelope membrane to which chromatin appears to be attached (Gerace *et al.*, 1984; see Chapter 5). Thus, the IF family of proteins may act to link the nuclear matrix–chromatin complex with the cytoplasmic IF cytoskeletal system (R. Goldman *et al.*, 1985, 1986). The molecular basis of this latter linkage remains unknown, although some workers have suggested that IF can interact directly with nuclear lamins *in vitro* (Georgatos and Blobel, 1987a,b). Even though there is much work to be done to demonstrate the functional significance of this remarkable cytoskeletal system, it appears, based on

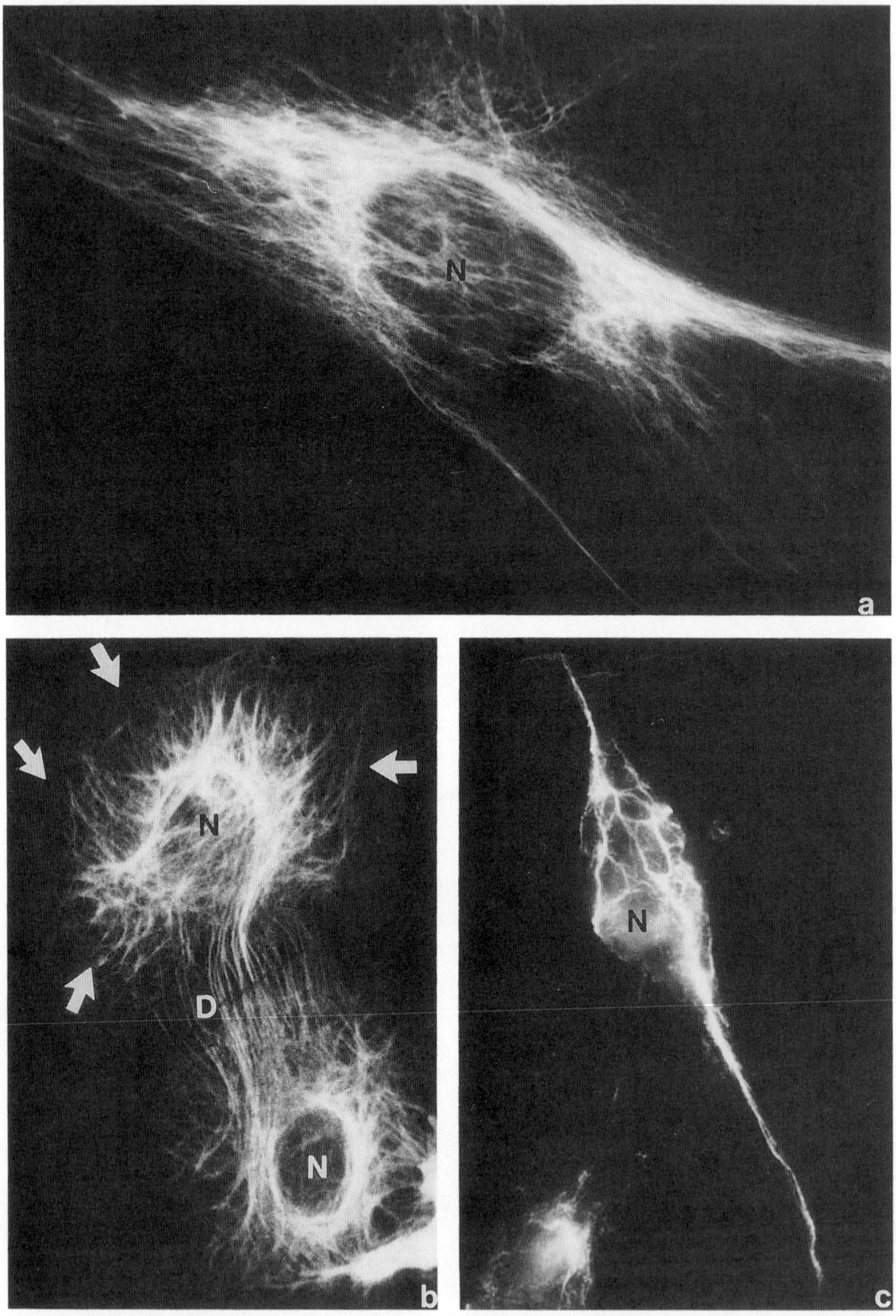

Figure 1. An overview of the IF systems in cultured fibroblasts (a, BHK-21 cell prepared for indirect immunofluorescence with vimentin antibody), epithelial cells (b, mouse epidermal keratinocytes prepared for indirect immunofluorescence with anti-mouse keratin antibody), and nerve cells (c, mouse neuroblastoma cell prepared for indirect immunofluorescence with vimentin antibody). N, nucleus. Note the large accumulation of IF in the juxtanuclear region. The desmosomal region (D) is seen in panel b. (a) ×1063; (b) ×1500; (c) ×1250.

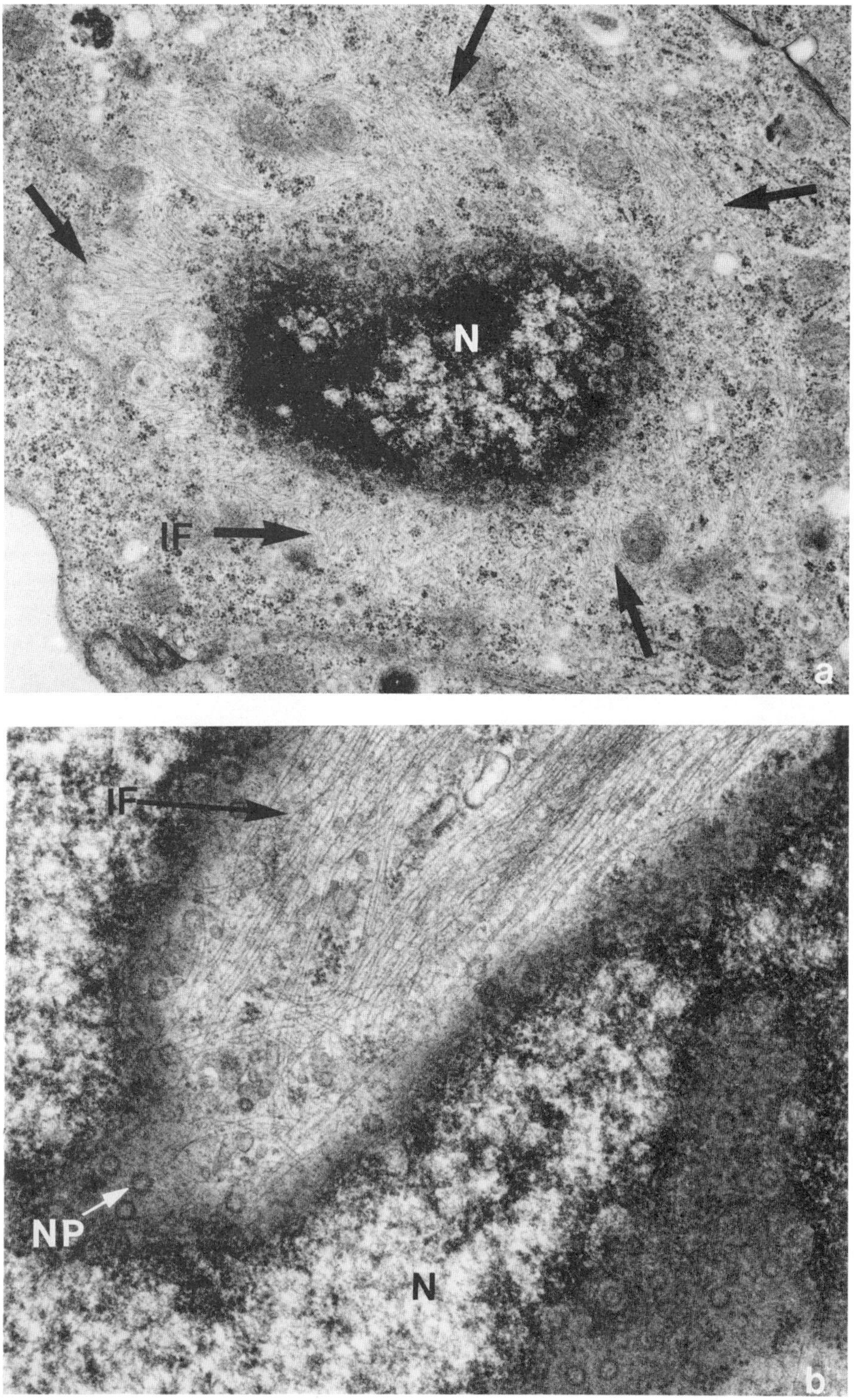

Figure 2. Electron micrographs of thin sections of BHK-21 cells prepared as described in Starger *et al.* (1978). Note IF enrichment in juxtanuclear areas. N, nucleus; NP, nuclear pore complexes.(a) ×20,000;(b) ×34,350.

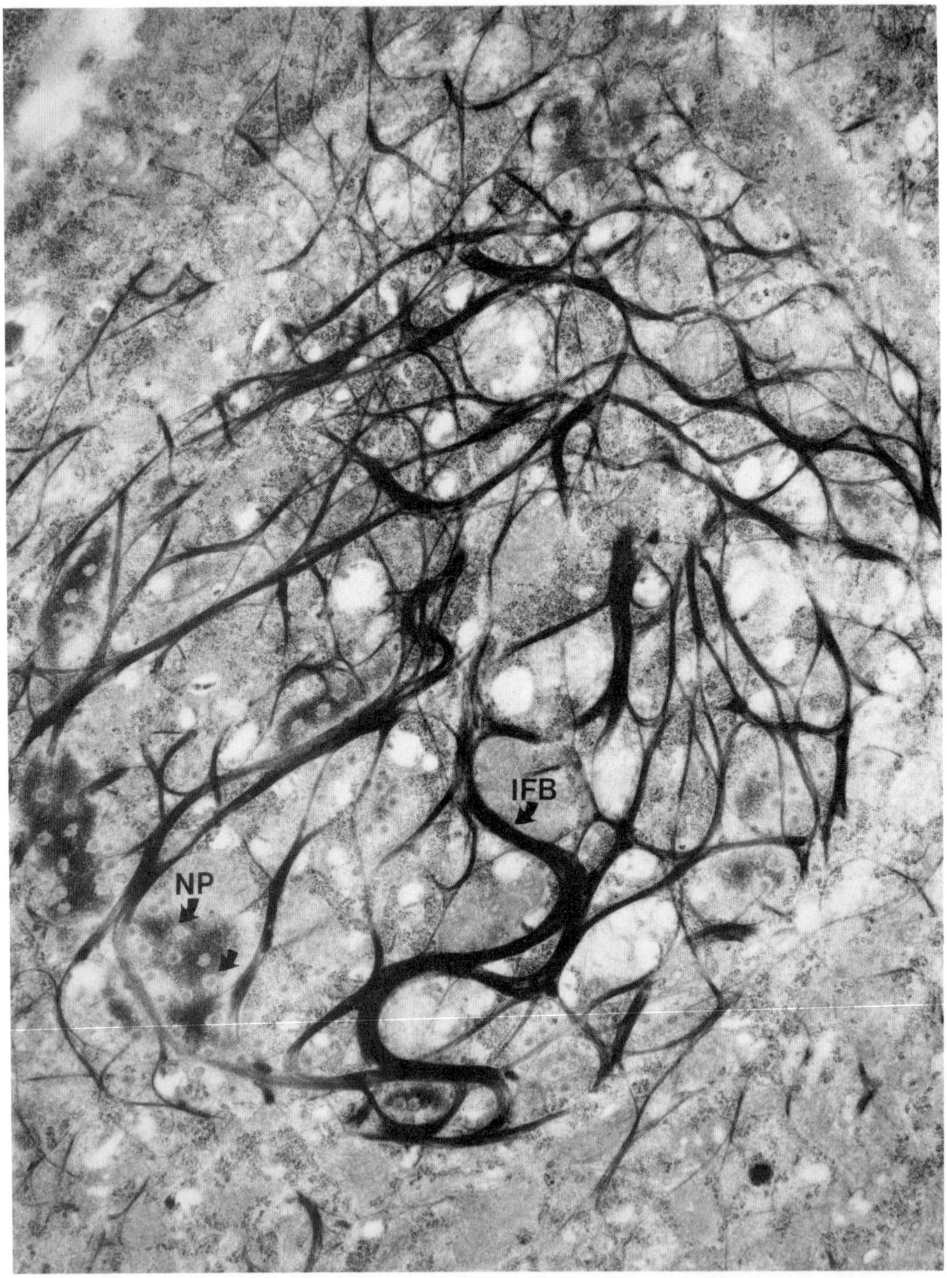

Figure 3. Electron micrographs of a thin section of a mouse epidermal cell showing complex array (cage) of IF bundles (IFB), surrounding the nuclear surface and close associations with nuclear pore complex (NP). Prepared according to Jones *et al.* (1985). ×18,750.

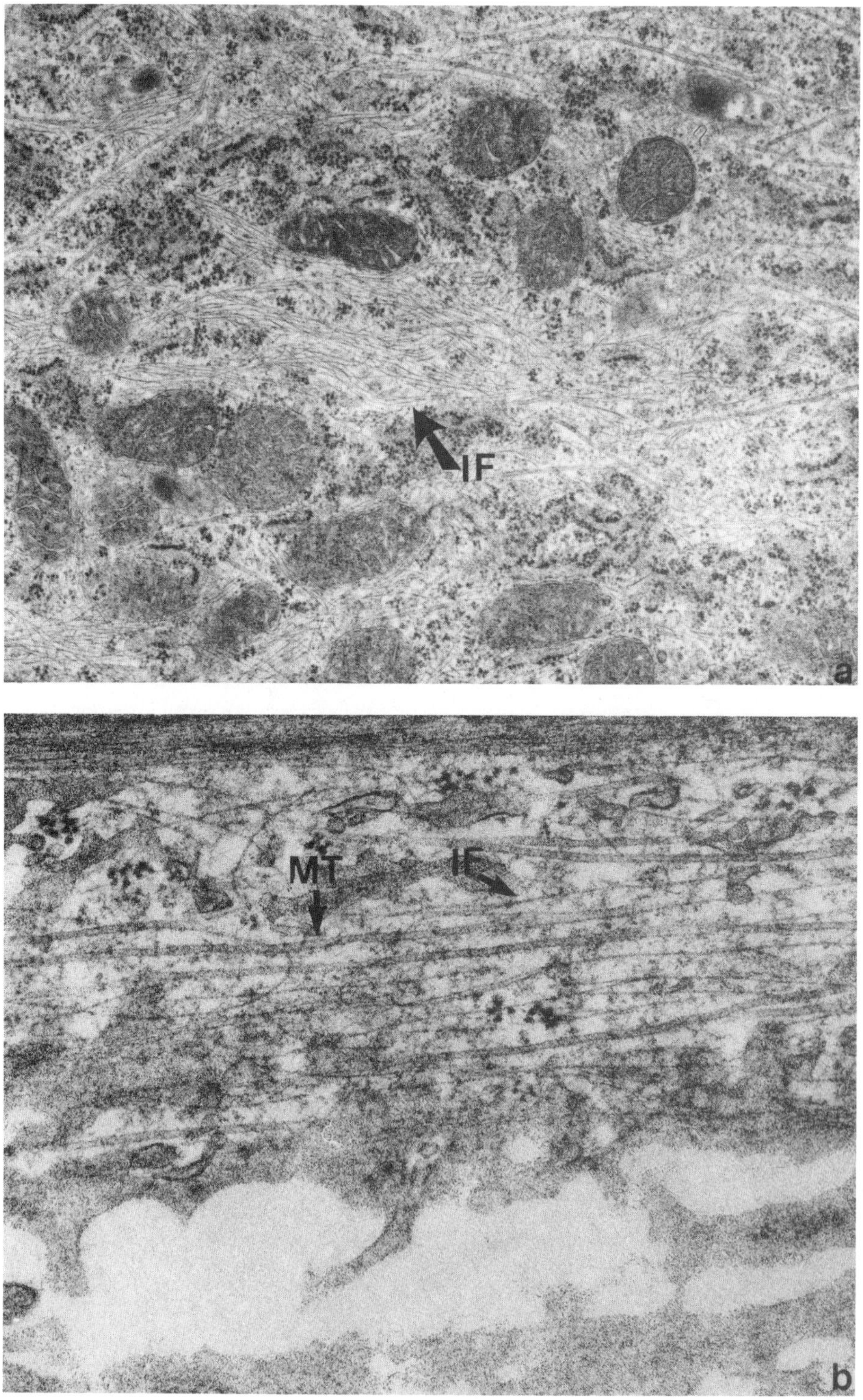

Figure 4. Electron micrographs of thin sections showing different regions of BHK-21 cells. Note loose arrays of IF in a, and parallel arrays of IF and microtubules (MT) in b.(a) ×25,000; (b) ×64,300.

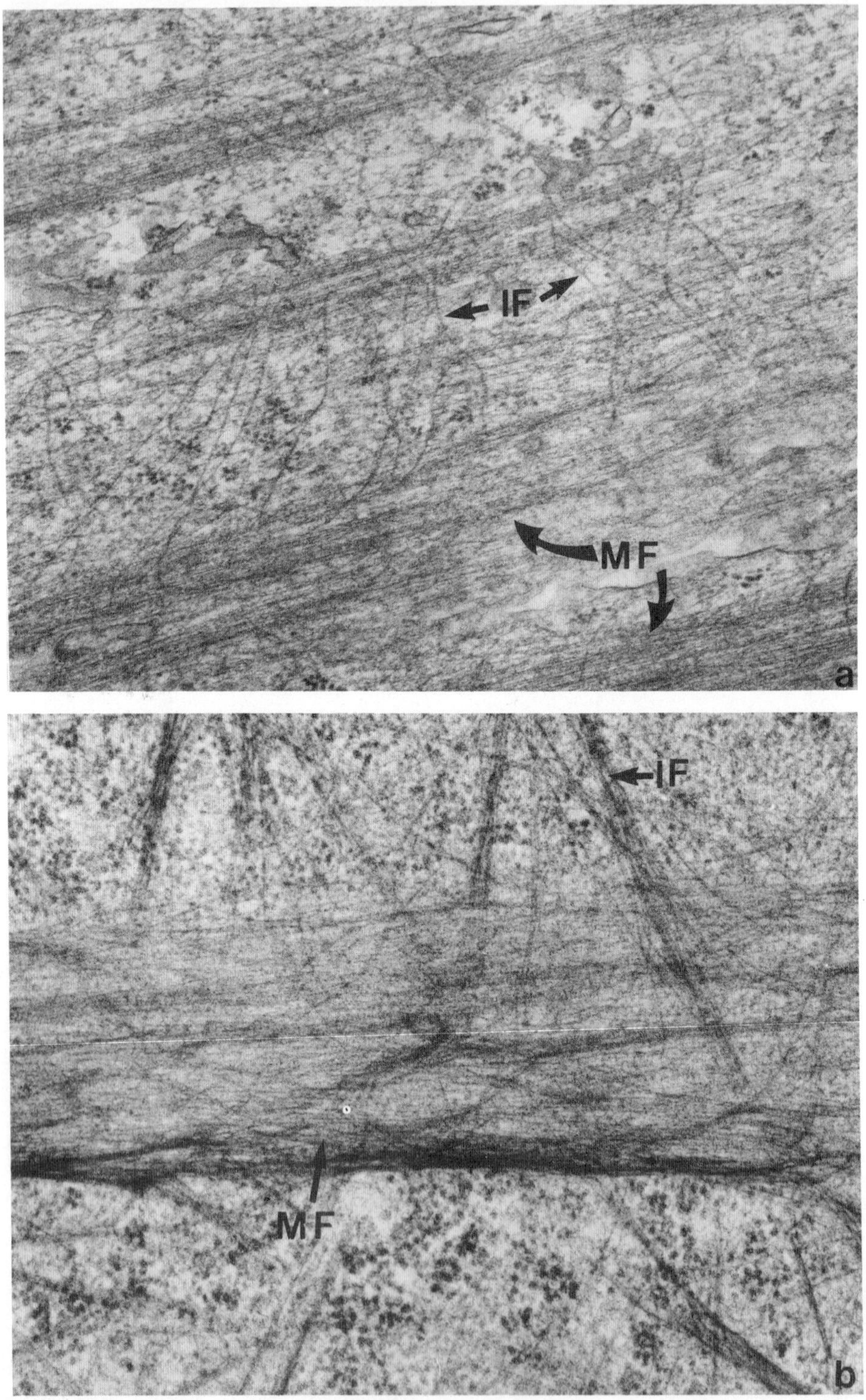

Figure 5. Electron micrographs of thin sections through regions of the submembranous, cortical cytoplasm showing the close relationships between IF and microfilaments (MF). (a) BHK-21 cell; (b) mouse epidermal cell. (a) ×47,000; (b) ×56,000.

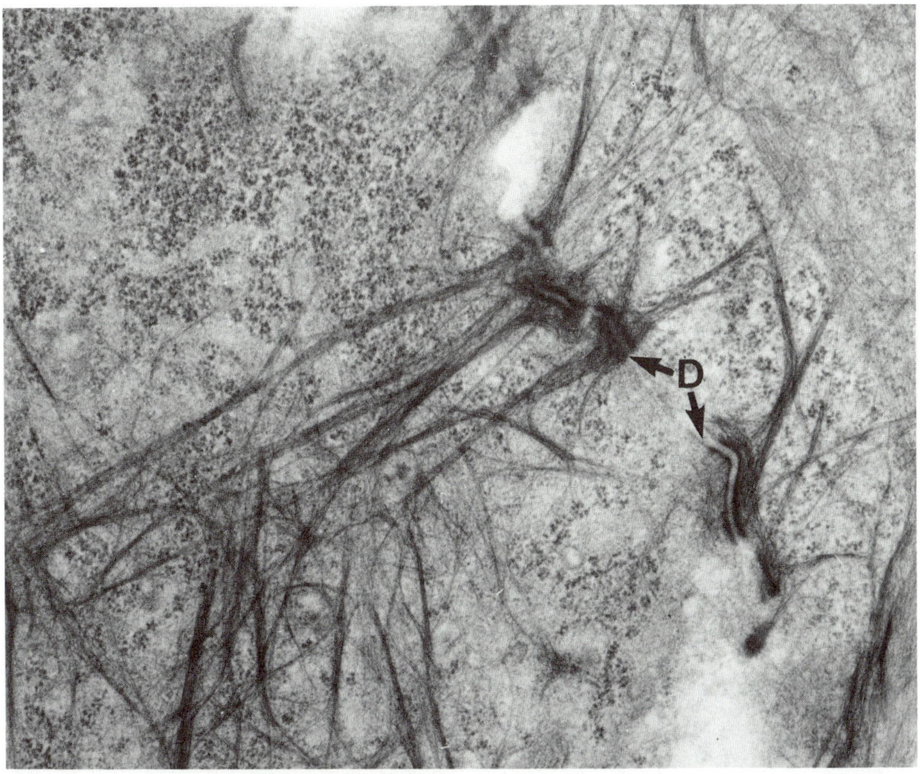

Figure 6. Electron micrograph of a thin section through the cell–cell contact zone of two epidermal keratinocytes. Note associations of IF with desmosomes (D). ×28,000.

morphological evidence alone, that IF may be involved in many physiological activities ranging from signal processing and information transfer from one subcellular compartment to the next, to nuclear centration and anchorage, to the development and maintenance of cell shape.

IF can be isolated and characterized *in vitro* using relatively simple procedures. These methods are based on their relative stability compared to other cytoskeletal elements and various cell structures. This is especially evident in cultured cell lines exposed to detergent (Triton X-100) containing high-ionic-strength buffered solutions, followed by treatment with DNase I to degrade and remove chromatin (Steinert *et al.*, 1982). This latter treatment also appears to have the fortuitous side effect of helping to dissociate actin from the preparation (Hitchcock *et al.*, 1976). The result is a greatly enriched preparation of intact IF and many of their associated proteins and structures, such as the nuclear lamina (R. Goldman *et al.*, 1985; A. Goldman *et al.*, 1986), cell surface-associated structures such as desmosomes (Jones *et al.*, 1982), and IFAPs (see Chapters 15 and 16). IF protein can also be isolated in a soluble state by extraction and homogenization of a tissue in high concentrations of urea (8 M), followed by high-speed centrifugation. Subsequently IF can be reassembled from their constituent soluble subunit proteins by dialysis against buffers of varying ionic strengths depending on the type of IF protein (Steinert, *et al.*, 1982).

Due to their relative stability in *in vitro* preparations, cytoplasmic IF are generally thought to lack the dynamic properties of cytoplasmic microtubules and microfilaments. However, recent studies aimed at determining the mechanisms and factors that regulate the organization and assembly of IF suggests that they are much more dynamic *in vivo* than previously thought. For example, microinjected depolymerized–biotinylated vimen-

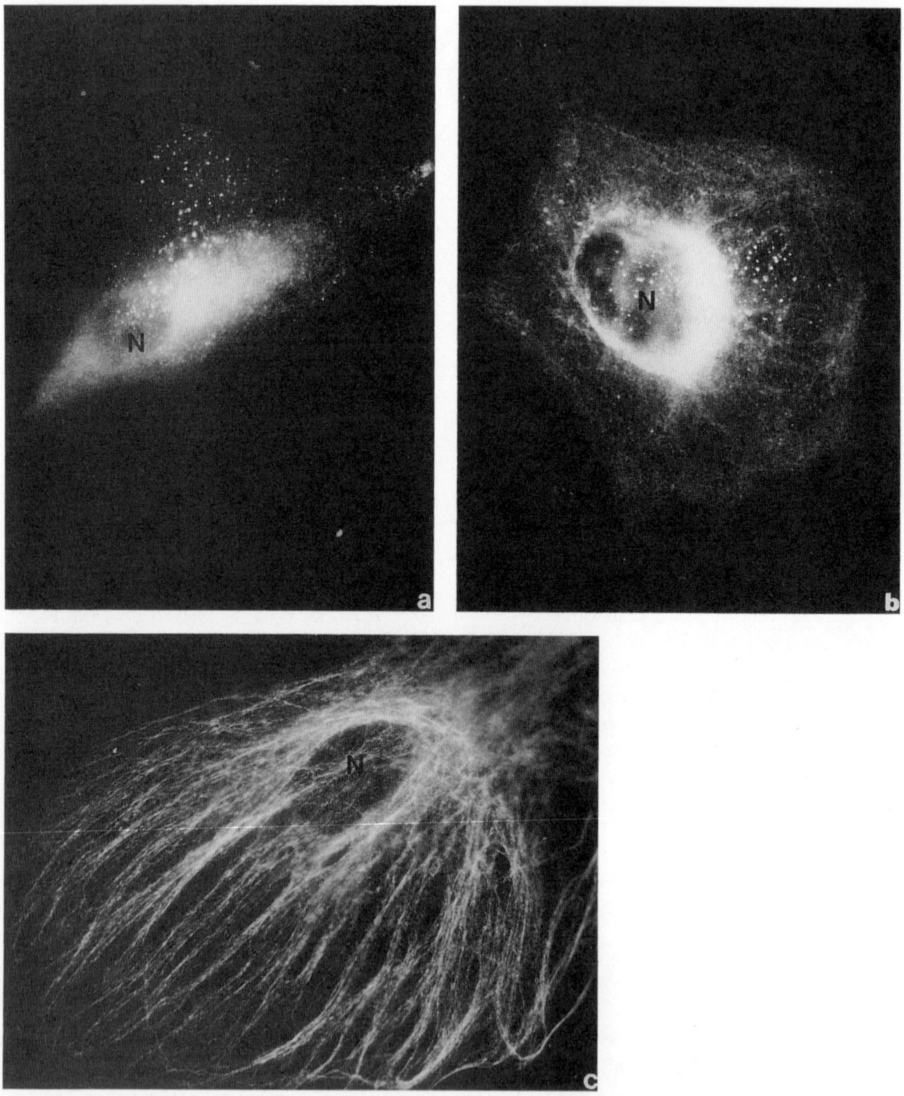

Figure 7. A series of fluorescence micrographs showing the distribution of biotinylated vimentin (using indirect immunofluorescence with antibiotin antibody as described in Vikstrom *et al.* (1989). Panel a shows accumulation of vimentin-enriched "spots" near the nucleus (N) at 30 min postinjection. Panel b shows the location of biotinylated vimentin at 1 hr postinjection; and panel c shows complete IF-like network of biotinylated vimentin at 2 hr postinjection. Double-label fluorescence observations with vimentin antibody have demonstrated that the biotinylated "network" and the endogenous "network" are superimposed. See Vikstrom *et al.* (1989) for details. (a) ×1042; (b) ×1350; (c) ×1065.

tin is incorporated into the endogenous IF network of interphase BHK-21 cells over a period of a few hours (Fig. 7) (Vikstrom *et al.*, 1989). Furthermore, the course of incorporation of the derivatized protein strongly suggests the existence of a juxtanuclear organization center for the incorporation of disassembled vimentin into fully polymerized IF and the polarized growth of IF from this center toward the cell surface (Vikstrom *et al.*, 1989) (see Fig. 7).

Other recent evidence suggests that cytoplasmic IF assembly and disassembly may be regulated by phosphorylation (for a discussion of nuclear lamin phosphorylation, see Chapter 5). *In vitro* experiments have shown that both cyclic AMP-dependent kinase and protein kinase C can phosphorylate type III (vimentin-containing) IF and that the resulting increase in phosphorylation level is coincident with their disassembly (Inagaki *et al.*, 1987, 1988; Geisler and Weber, 1988; Evans, 1988). *In vivo* studies have concentrated on the finding that vimentin phosphorylation increases significantly during mitosis (Evans and Fink, 1982; Evans, 1984; Celis *et al.*, 1983). Recently, it has been shown that this increased level of phosphorylation is related to the presence of specific kinases that are found in mitotic BHK-21 cells and that the resulting hyperphosphorylation is accompanied by IF disassembly during prometaphase–metaphase (Chou *et al.*, 1989). These types of studies demonstrate that IF do have dynamic properties throughout the cell cycle. Undoubtedly these types of investigations will help to reveal the functions of IF, which have until now eluded researchers.

References

Astbury, W. T., and Street, A., 1931, X-ray studies on the structure of hair, wool and related fibres. I, *Philos. Trans. R. Soc.* **230**A:75.

Buckley, I. K., Raju, T. R., and Stewart, M., 1978, Heavy meromyosin labeling of intermediate filaments in cultured connective tissue cells, *J. Cell Biol.* **78:**648–652.

Celis, J. E., Larsen, P. M., Fey, S. J., and Celis, A., 1983, Phosphorylation of keratin and vimentin polypeptides in normal and transformed mitotic human epithelial amnion cells: Behavior of keratin and vimentin filaments during mitosis, *J. Cell Biol.* **97:**1429–1434.

Chou, Y. H., Rosevear, E., and Goldman, R. D., 1989, Phosphorylation and disassembly of intermediate filaments in mitotic cells, *Proc. Natl. Acad. Sci. USA* **86:**1885–1889.

Crick, F. H. C., 1953, The Fourier-transform of a coiled-coil, *Acta Crystallogr.* **6:**685.

Day, W. A., and Gilbert, D. S., 1972, X-ray diffraction pattern of axoplasm, *Biochim. Biophys. Acta* **285:**503–506.

Eriksson, A., and Thornell, L.-E., 1979, Intermediate (skeletin) filaments in heart Purkinje fibers. A correlative morphological and biochemical identification with evidence of a cytoskeletal function, *J. Cell Biol.* **80:**231–247.

Evans, R. M., 1984, Peptide mapping of phosphorylated vimentin, *J. Biol. Chem* **259:**5372–5375.

Evans, R. M., 1988, Cyclic AMP-dependent protein kinase-induced vimentin filament disassembly involves modification of the N-terminal domain of intermediate filament subunits, *FEBS Lett.* **234:**73–78.

Evans, R. M., and Fink, L. M., 1982, An alternation in the phosphorylation of vimentin-type intermediate filaments is associated with mitosis in cultured mammalian cells, *Cell* **29:**43–52.

Fisher, D. Z., Chaudhary, N., and Blobel, G., 1986, cDNA sequencing of nuclear lamins A and C reveals primary and secondary structural homology to intermediate filament proteins, *Proc. Natl. Acad. Sci. USA* **83:**6450–6454.

Fraser, R. D. B., and MacRae, T. P., 1961, Molecular configuration of alpha-keratin, *J. Mol. Biol.* **3:**640–645.

Geisler, N., and Weber, K., 1988, Phosphorylation of desmin *in vitro* inhibits formation of intermediate filaments: Identification of three kinase A sites in the amino terminal head domain, *EMBO J.* **7:**15–201.

Georgatos, S. D., and Blobel, G., 1987a, Two distinct attachment sites for vimentin along the plasma membrane and the nuclear envelop in avian erythrocytes: A basis for a vectorial assembly of intermediate filament, *J. Cell Biol.* **105:**105–116.

Georgatos, S. D., and Blobel, G., 1987b, Lamin B constitutes an intermediate filament attachment site at the nuclear envelope, *J. Cell Biol.* **105:**117–126.

Gerace, L., Comeau, C., and Benson, N., 1984, Organization and modulation of nuclear lamina structure, *J. Cell Sci. Suppl.* 1, 137–160.

Goldman, A. E., Maul, G., Steinert, P. M., Yang, H.-Y, and Goldman, R. D., 1986, Keratin-like proteins which coisolate with intermediate filaments of BHK-21 cells are nuclear lamins, *Proc. Natl. Acad. Sci. USA* **83:**3839–3843.

Goldman, R. D., 1971, The role of three cytoplasmic fibers in BHK-21 cell motility. I. Microtubules and the effects of colchicine, *J. Cell Biol.* **51:**752–762.

Goldman, R. D., and Follett, E. A. C., 1969, The structure of the major cell processes of isolated BHK-21 fibroblasts, *Exp. Cell Res.* **57:**263–276.

Goldman, R. D., and Knipe, C., 1973, Functions of cytoplasmic fibers in non-muscle cell motility, *Cold Spring Harbor Symp. Quant. Biol.* **37:**523–534.

Goldman, R. D., Goldman, A., Green, K., Jones, J., Lieska, N., and Yang, H.-Y., 1985, Intermediate filaments: Possible functions as cytoskeletal connecting links between the nucleus and the cell surface, *Ann. N.Y. Acad. Sci.* **455:**1–17.

Goldman, R. D., Goldman, A. E., Green, K. J., Jones, J. C. R., Jones, S. M., and Yang, H.-Y., 1986, Intermediate filament networks: Organization and possible functions of a diverse group of cytoskeletal elements, *J. Cell Sci. Suppl.* **5:**69–97.

Green, K., and Goldman, R. D., 1986, Evidence for an interaction between the cell surface and intermediate filaments in cultured fibroblasts, *Cell Motil. Cytoskel.* **6:**389–405.

Green, K., Talian, J. C., and Goldman, R. D., 1986, Relationships between intermediate filaments and microfilaments in cultured fibroblasts: Evidence for common foci during cell spreading, *Cell Motil. Cytoskel.* **6:**406–418.

Green, K. J., Geiger, B., Jones, J. C. R., Talian, J. C., and Goldman, R. D., 1987, The relationship between intermediate filaments and microfilaments before and during the formation of desmosomes and adherens-type junctions in mouse epidermal keratinocytes, *J. Cell Biol.* **104:**1389–1402.

Hitchcock, S., Carlson, L., and Lindberg, U., 1976, Depolymerization of F-actin by deoxyibnuclease I, *Cell* **7:**531–542.

Holmes, K. V., and Choppin, P. W., 1968, On the role of microtubules in movement and alignment of nuclei in virus-induced syncytia, *J. Cell Biol.* **39:**526–542.

Holtzer, H., Sanger, J. W., Ishikawa, A., and Strahs, K., 1972, Selected topics in skeletal myogenesis, *Cold Spring Harbor Symp. Quant. Biol.* **37:**549–566.

Holtzer, H., Fellini, S., Rubinstein, N., Chi, J., and Strahs, K., 1976, Cells, myosins and 100A filaments, in: *Cell Motility* (R. D. Goldman, R. D. Pollards, and J. L. Rosenbaum, eds.), Cold Spring Harbor Laboratory, Cold Spring Harbor, N.Y., Book B, pp. 823–839.

Inagaki, M., Nishi, Y., Nishizawa, K., Matsuyama, M., and Sato, C., 1987, Site specific phosphorylation induces disassembly of vimentin filaments *in vitro, Nature* **328:**649–652.

Inagaki, M., Gonda, Y., Matsuyuma, M., Nishizama, K., Nishi, Y., and Sato, C., 1988, Intermediate filament reconstitution *in vitro, J. Biol. Chem.* **263:**5970–5978.

Jones, J. C. R., and Goldman, R. D., 1985, Intermediate filaments and the initiation of desmosome assembly, *J. Cell Biol.* **101:**509–517.

Jones, J., Goldman, A., Steinert, P., Yuspa, S., and Goldman, R., 1982, Dynamic aspects of the supramolecular organization of intermediate filament networks in cultured epidermal cells, *Cell Motil.* **2:**197–213.

Jones, J. C. R., Goldman, A. E., Yang, H.-Y., and Goldman, R. D., 1985, The organizational fate of intermediate filament networks in two epithelial cell types during mitosis, *J. Cell Biol.* **100:**93–102.

Lazarides, E., 1980, Intermediate filaments as mechanical integrators of cellular space, *Nature* **283:**249–256.

Leonard, D. G. B., Gorham, J. D., Cole, P., Green, L. A., and Ziff, E. B., 1988, A nerve growth factor-regulated messenger RNA encodes a new intermediate filament protein, *J. Cell Biol.* **106:** 181–193.

McKeon, F. D., Kirschner, M. W., and Caput, D., 1986, Homologies in both primary and secondary structure between nuclear envelope and intermediate filament proteins, *Nature* **319:**463–468.

Parysek, L. M., and Goldman, R. D., 1988, Distribution of a novel 57kDa intermediate filament (IF) protein in the nervous system, *J. Neurosci.* **8:**555–563.

Parysek, L. M., Chisholm, R. L., Ley, C. A., and Goldman, R. D., 1988, A type III intermediate filament gene is expressed in mature neurons, *Neuron* **1:**395–401.

Staehelin, A. L., 1974, Intercellular functions, *Int. Rev. Cytol.* **39:**191–283.

Starger, J., Brown, W., Goldman, A., and Goldman, R. D., 1978, Biochemical and immunological analyses of rapid purified 10-nm filaments from BHK-21 cells, *J. Cell Biol.* **78:**93–109.

Steinert, P. M., and Parry, D., 1985, Intermediate filaments, *Annu. Rev. Cell Biol.* **1:**41–65.

Steinert, P. M., and Roop, D., 1988, Molecular and cellular biology of intermediate filaments, *Ann. Rev. Biochem.* **57:**575–609.

Steinert, P. M., Zimmerman, S. B., and Idler, W. W., 1976, The self assembly of bovine epidermal keratin filaments *in vitro, J. Mol. Biol.* **108:**547–567.

Steinert, P., Zimmerman, S., Starger, J., and Goldman, R. D., 1978, Ten nanometer filaments of hamster BHK-21 cells and epidermal keratin filaments have similar structure, *Proc. Natl. Acad. Sci. USA* **75:**6098–6101.

Steinert, P., Zackroff, R., Aynardi-Whitman, M., and Goldman, R., 1982, Isolation and characterization of intermediate filaments, *Methods Cell Biol.* **24A:**399–419.

Steinert, P., Jones, J., and Goldman, R. D., 1984, Intermediate filaments, *J. Cell Biol.* **99:**22s–27s.

Steinert, P. M., Parry, D., Idler, W. W., Johnson, L. D., Steven, A., and Roop, D. R., 1985, Amino acid sequences of mouse and human epidermal type II keratins of Mr 67,000 provides a systematic basis for the structural and functional diversity of the end domains of keratin intermediate filament subunits, *J. Biol. Chem.* **260:**7142–7149.

Vikstrom, K. L., Borisy, G. G., and Goldman, R. D., 1989, Dynamic aspects of intermediate filament networks in BHK-21 cells, *Proc. Natl. Acad. Sci. USA* **86:**549–553.

Wang, E., and Goldman, R. D., 1978, Functions of cytoplasmic fibers in intracellular movements in BHK-21 cells, *J. Cell Biol.* **79:**708–726.

Wilson, E., 1928, *The Cell in Development and Heredity,* Macmillan Co., New York.

Wisniewski, H., Shelanski, M. L., and Terry, R. D., 1968, Effects of mitotic spindle inhibitors on neurotubules and neurofilaments in anterior horn cells, *J. Cell Biol.* **38:**224–229.

Zackroff, R. V., and Goldman, R. D., 1979. *In vitro* assembly of intermediate filaments from baby hamster kidney (BHK-21) cells, *Proc. Natl. Acad. Sci. USA* **76:**6226–6230.

2

Intermediate (10-nm) Filaments in Muscle

Marvin H. Stromer

1. Introduction

The purpose of this chapter is to provide an overview of 10-nm filaments in smooth, skeletal, and cardiac muscle with an emphasis on newer developments. The intent was not to provide an exhaustive literature review but, instead, to include a sampling of research from different laboratories. In this way, the reader can easily track research on a particular topic or from a certain laboratory. More detail is provided on 10-nm filaments from smooth muscle because that has been the tissue of choice for the majority of studies on 10-nm filaments from muscle. The sections on 10-nm filaments from skeletal and from cardiac muscle emphasize both unique properties and comparisons with the smooth muscle system.

2. Smooth Muscle

2.1. Location of 10-nm Filaments in Cells

Smooth muscle cells in the digestive tract contain an axial bundle of 10-nm filaments that have been identified by colloidal gold immunocytochemistry as desmin-containing filaments (Stromer and Bendayan, 1988). This bundle extends the length of the cell, partially surrounds the nucleus, and is associated with clusters of mitochondria at the poles of the nucleus. Because an intracellular 10-nm network that is closely associated with the nucleus, mitochondria, and plasma membrane exists in a variety of cells (Goldman *et al.*, 1986), it is likely that a similar axial bundle of 10-nm filaments will be detected in smooth muscle cells from other tissues.

Marvin H. Stromer • Muscle Biology Group, Department of Animal Science, Iowa State University, Ames, Iowa 50011.

Peripheral to this bundle, smooth muscle cell cytoplasm contains many 10-nm filaments that are oriented predominantly parallel to the long axis of the spindle-shaped cell. Exceptions to this orientation can be observed in either intact vascular muscle (Ashton *et al.*, 1975) or saponin-skinned vascular muscle (Garfield and Somlyo, 1985) where some 10-nm filaments appear to loop tangentially near cytoplasmic dense bodies or are obliquely arranged between dense bodies. Tissue freezing followed by deep etching and rotary shadowing has shown that 10-nm filaments also terminate at or near membrane-associated dense bodies (Somlyo and Franzini-Armstrong, 1985). These filaments are most numerous and are readily seen in a zone around cytoplasmic dense bodies in cross sections of smooth muscle cells. Although 10-nm filaments are also visible elsewhere in the cytoplasm, their concentration is usually greatest near the periphery of the cytoplasmic dense body.

The relationship between 10-nm filaments and dense bodies, on the one hand, and thick and thin filaments, on the other, has intrigued many investigators. Cooke and Fay (1972) showed that, both in normal and in EDTA-treated taenia coli strips, unstretched and stretched fibers differed in their distribution of cytoplasmic dense bodies and 10-nm filaments. In unstretched fibers, cytoplasmic dense bodies and 10-nm filaments were uniformly distributed in cell cross sections. After stretching, both the normal fibers and the EDTA-treated fibers, which no longer had thick filaments or thin filament lattices, had dense bodies and 10-nm filaments located in a central zone in cell cross sections. This was interpreted to mean that 10-nm filaments are attached at the ends of fibers and connect cytoplasmic dense bodies into a network that responds to stretch by decreasing the network diameter and thus moves dense bodies toward the cell center. A model involving a cytoskeletal network of 10-nm filaments and dense bodies separate from the actin and myosin filaments was proposed by Small (1977). This model was based on the observation by Small and Sobieszek (1977, 1980) that 10-nm filaments were located between thick and thin filament groups rather than within the myofilament groups. A chainlike arrangement of dense bodies (Fay *et al.*, 1983) and associated 10-nm filaments has been observed in vascular smooth muscle cells (Ashton *et al.*, 1975).

The specific structural relationship between 10-nm filaments and smooth muscle dense bodies has not been resolved. The experiments by Cooke and his collaborators that compared stretched and unstretched smooth muscle cells indicated that some type of connection existed between the 10-nm filaments and both the cytoplasmic and the membrane-associated dense bodies. Campbell *et al.* (1971) suggested that, in embryonic chicken gizzard cells grown in culture, 10-nm filaments appeared to emanate from the substructure of the dense bodies and may have a role in assembly of dense bodies in developing muscle. Ostensibly, the electron-opaque material of the dense body would be added to a scaffold of parallel 10-nm filaments. Isolated dense bodies have "cores" that contain either 10-nm filaments (Cooke and Chase, 1971; Cooke, 1976) or both 10-nm and actin filaments (Nonomura and Ebashi, 1975). Ashton *et al.*, (1975) found that, although 10-nm filament profiles surrounded many cross-sectioned dense bodies, the 10-nm filaments did not enter dense bodies. More recently, Bond and Somlyo (1982) and Somlyo *et al.*, (1984) have described an oblique orientation of 10-nm filaments between adjacent dense bodies rather than forming a link to the next dense body in series. Additional information is needed to determine if 10-nm filaments insert into the ends of dense bodies in various types of smooth muscle cells or if they only connect laterally to dense bodies to form an oblique network. Although such an oblique network would not be parallel to the thick and thin filaments involved in contraction, it has been speculated that such a network might help retain the alignment of the contractile units (Somlyo *et al.*, 1984). Bond and Somlyo (1982) used myosin subfragment 1 to decorate thin filaments that were attached to the sides and ends of dense bodies and showed that the arrowheads always pointed away from the dense bodies. The polarity of thin filaments with respect to the dense body is identical to the polarity of skeletal or cardiac muscle thin filaments with respect to the Z line.

2.2. *Isolation and Preparation of Desmin and Vimentin*

Differences in solubility between 10-nm filaments and thick and thin filaments were first observed by Cooke and Chase (1971). Examination of thin sections of actomyosin-extracted residues of smooth muscle showed that 10-nm filaments and dense bodies persisted but that thick and thin filaments had been removed (Cooke and Chase, 1971; Schollmeyer *et al.*, 1976). The amount of 53-kDa protein seen after electrophoresis on SDS gels was greater in the residue than in the washed muscle fragments, which were the starting material (Cooke, 1976; Schollmeyer *et al.*, 1976). If this residue was treated with 6–8 M urea, both the 10-nm filaments and the 53-kDa band seen in the gels disappeared (Cooke, 1976). This observation, together with the finding that antibodies to the 53-kDa protein bound to native 10-nm filaments (Schollmeyer *et al.*, 1976), indicated that the 53-kDa protein was a major constituent of these filaments.

Extraction of the actomyosin-depleted residue from taenia coli muscle with 1 M acetic acid removed 10-nm filaments and a 53-kDa protein that Small and Sobieszek (1977) called *skeletin*. It is now generally accepted that the name *skeletin* be replaced by the term *desmin* to refer to the principal protein of muscle 10-nm filaments. The acetic acid extraction method also solubilizes the small amounts of actin remaining in the residue. This actin seemingly forms a strong association with the desmin that is difficult to disrupt. Electrophoresis on SDS gels after various column chromatography procedures invariably shows that desmin preparations that were originally extracted with acetic acid still contain actin (Fellini *et al.*, 1978; Hubbard and Lazarides, 1979). Negative staining of the filaments reconstituted from these column fractions showed that both 10-nm and actin filaments were present.

A reliable method for solubilizing and purifying the 10-nm filament protein, desmin, from avian smooth muscle was developed by Huiatt *et al.* (1980). Washed gizzard cell fragments were extracted with a 0.6 M KCl solution to remove actomyosin, and the residue, which was enriched in 10-nm filaments and dense bodies, was solubilized with 6 M urea. The urea extract was first applied to a hydroxyapatite column, and then the appropriate fractions were added to a DEAE-Sepharose CL-6B ion exchange column. This chromatography procedure, which was done in the presence of 6 M urea, results in highly purified desmin fractions that are free of actin. Urea is usually removed by dialyzing against 10 mM Tris-acetate, pH 8.5, 5 mM β-mercaptoethanol. Desmin is soluble under these conditions and, with storage under N_2 and in a toluene atmosphere, may be satisfactorily stored at 2°C for several weeks.

Although smooth muscle cells of the digestive, urogenital, and respiratory tracts contain 10-nm filaments that solely or principally contain desmin, smooth muscle cells in vascular tissue also contain the protein vimentin. It is clear that the vimentin content of blood vessels decreases as distance from the heart increases. The vimentin content of large elastic vessels such as the aorta also is greater than for the muscular vessels (Osborn *et al.*, 1981; Frank and Warren, 1981). The presence of both desmin and vimentin, sometimes in the same cell, in mammalian and avian aorta cells has been verified by immunofluorescence microscopy of frozen sections (Schmid *et al.*, 1982). It is for these reasons that the preferred smooth muscle source for vimentin is the segment of the aorta proximate to the heart (Gabbiani *et al.*, 1981).

Most procedures for the preparation of vimentin are modifications of the desmin procedure. Our laboratory has adapted the desmin procedure of Huiatt *et al.* (1980) to isolate and purify vimentin (Hartzer, 1984). Washed cell fragments are prepared from the tunica media of the aorta. Actomyosin is extracted with a 0.6 M KCl solution, and vimentin is solubilized from the remaining residue with 8 M urea in 14.3 mM β-mercaptoethanol, 20 mM imidazole-HCl, pH 7.0. Vimentin is sequentially purified by batch chromatography in 6 M urea on DEAE-cellulose and by column chromatography in 6 M urea on hydroxyapatite and DEAE-Sepharose CL-6B. Urea can be removed by dialysis against 10 mM Tris-acetate, pH 8.5, 1 mM dithioerythritol. Storage can be done as described for desmin.

2.3. *Molecular Properties of Desmin and Vimentin*

Desmin migrates in SDS gels done by the method of Weber and Osborn (1969) with an apparent molecular weight of 53,000 (Huiatt *et al.*, 1980). Vimentin, however, has a mobility consistent with a molecular weight of 54,000. Izant and Lazarides (1977) showed that the desmin remaining in actomyosin-extracted gizzard residues migrated as two isoelectric variants in a two-dimensional gel electrophoresis system. The two-dimensional gel electrophoresis technique of O'Farrell (1975) was used by Huiatt *et al.* (1980) to demonstrate the existence of α (more acidic, pI 5.4) and β (pI 5.5) isoelectric variants of purified gizzard desmin. Under similar conditions, vimentin also migrates as two slightly more acidic isoelectric variants (pI 5.2 and 5.3). The more acidic of the two vimentin variants is more extensively phosphorylated (Gabbiani *et al.*, 1981; Steinert *et al.*, 1982) and it is generally accepted that this phenomenon also accounts for the two desmin variants. The β variant routinely accounts for about 70–80% of the desmin.

The amino acid composition of desmins from several muscle sources and of vimentin from porcine aorta is shown in Table I. The values indicate that, although similar patterns exist among these proteins, the desmins from diverse species are more closely related to each other than to vimentin. This is consistent with the fact that desmin and vimentin are products of two separate genes (Quax *et al.*, 1984; Capetanaki *et al.*, 1984). The marked similarities in the amino acid contents of desmin from different cell types probably is due to the existence of a single gene for desmin, which has been demonstrated in both the chicken genome (Capetanaki *et al.*, 1984) and the hamster genome (Quax *et al.*, 1984). The very minor differences probably are due to technical variability among the different analyses. Comparison of the amino acid sequences of a 138- to 140-residue fragment from the carboxyl end of chicken gizzard desmin, porcine stomach desmin, and porcine eye lens vimentin (Geisler and Weber, 1981) also showed that the desmins differed in less than 9% of the residues but that porcine desmin and vimentin differed in 36% of the residues.

Smooth muscle desmin has a sedimentation coefficient of 4.6 S, and an ultraviolet circular dichroism spectrum with negative extrema at 208 and 222 nm, typical of a protein containing significant amounts of α helix, and an α helix content of 45% (Huiatt *et al.*,

Table I. Amino Acid Analysis of Muscle Desmins and Vimentin[a]

Amino acid	Porcine cardiac desmin[b]	Porcine skeletal desmin[c]	Avian gizzard desmin[c]	Bovine Purkinje desmin[d]	Porcine aorta vimentin[b]
Asp	9.9	9.7	8.6	10.9	12.4
Thr	5.1	5.3	5.6	6.4	5.4
Ser	7.4	7.1	6.0	6.4	8.5
Glu	18.9	17.6	19.9	21.0	18.0
Pro	2.5	3.6	1.8	2.6	2.6
Gly	5.4	6.4	4.0	5.2	3.1
Ala	9.4	9.2	9.3	9.9	6.6
Cys	0.3	0.3	0.3	0.0	0.3
Val	5.8	6.0	5.2	5.2	5.2
Met	1.8	1.8	2.6	1.9	2.3
Ile	3.9	4.1	4.4	4.3	3.1
Leu	9.8	9.4	10.8	9.4	11.1
Tyr	2.8	2.8	2.9	2.2	2.8
Phe	2.9	3.0	2.8	2.9	2.6
Lys	4.9	5.1	4.9	4.1	5.0
His	1.4	1.4	1.6	1.0	1.5
Arg	8.3	7.7	9.3	6.6	9.4

[a]Values are expressed as mole percent.
[b]From Hartzer (1984).
[c]From O'Shea *et al.* (1979).
[d]From Stigbrand *et al.* (1979).

1980). Vascular vimentin has the same minima in the UV circular dichroism spectrum and virtually the same α helix content, 41% (Hartzer, 1984). Both desmin and vimentin are substrates for several proteases, particularly the calcium-activated protease that is endogenous to many cell types including muscle (O'Shea *et al.*, 1979; Nelson and Traub, 1983).

The complete amino acid sequence of chicken gizzard desmin has revealed additional information about the molecule. The calculated molecular weight is 53,000 based on 463 residues (Geisler and Weber, 1982, 1983). *In vitro* chymotryptic proteolysis of desmin has identified three major domains: an amino-terminal, non-α-helical headpiece of approximately 70 residues that is especially protease-sensitive, a middle rod portion consisting of three α-helical segments connected by short, non-α-helical spacers, and a carboxyl-terminal, non-α-helical tailpiece of about 50 residues (Geisler *et al.*, 1982). The rod portion of desmin contains the single cysteine residue, has a relative mobility on SDS gels of 40 kDa, and contains a sequence that is remarkably similar to other intermediate filament proteins. The highly α-helical rod segment is characterized by a heptad repeat pattern of hydrophobic amino acids, which is needed to form the interchain coiled-coil packed α helices. The hydrophobic residues are usually located in the first and fourth positions in each heptad. Although the sequence homology between one of the α-helical segments of desmin and vimentin has been reported to be 86%, the tailpieces of desmin and vimentin have only 68% homology (Geisler and Weber, 1982). A slightly different structure for the rod domain of intermediate filament protein molecules has been proposed by Parry and Fraser (1985). Four α-helical segments are linked by three non-α-helical spacers. The extent of the homology between the rod domains of desmin, vimentin, and glial fibrillary acidic protein led Parry and Fraser (1985) to classify them as type III chains. On the basis of the nucleotide sequence of cDNA, Quax-Jeuken *et al.* (1983) predicted that hamster vimentin would consist of a nonhelical amino-terminal headpiece of at least 68 residues, two α-helical segments of 135 and 145 residues connected by a 45-residue nonhelical spacer which together form the rod, and a nonhelical carboxyl-terminal tailpiece of 55 residues. Although some details remain to be established, it is clear that both desmin and vimentin consist of α-helical rod domains with extensive sequence homology that are interrupted by two to three (probably two) non-α-helical spacers and a headpiece and a tailpiece where the relationship is evident but where much less extensive homology exists than in the rod domains.

2.4. Filament Assembly and Substructure

The relationship between the 53-kDa desmin monomer and the 10-nm filament has been an active area of investigation. Stromer *et al.* (1979, 1981) observed a helical substructure in 10-nm filaments reconstituted from purified gizzard desmin. If these reconstituted filaments were exposed to either 2 M urea or 8 mM Tris, pH 8.2—two conditions that favor filament disassembly—the predominant structures seen after negative staining were 2–2.5 × 50-nm protofilaments and 3.5–5 × 50-nm subfilaments. Cross-linking the isolated rod domains of gizzard desmin with dimethylsuberimidate and analysis of the products on SDS-polyacrylamide gels showed that tetramers were the predominant species formed (Geisler and Weber, 1982). A similar approach with other cross-linking agents was subsequently used by Pang *et al.* (1983) to demonstrate that the protofilaments of intact smooth muscle desmin and vimentin consisted of tetramers of the respective subunits. The cross-linked protofilament, therefore, consists of a pair of double-stranded coiled coils and, for desmin, has a mobility on SDS gels consistent with a particle size of about 220 kDa.

The orientation of the subunits in the protofilament has been controversial. Geisler and Weber (1982) argued that the two polypeptides in the coiled coil and the two pairs of coiled coils should both be parallel, i.e., with all amino-termini oriented in the same direction. More recently, however, Geisler *et al.* (1985) labeled both ends of desmin rod tetramers with Fab fragments of a monoclonal antibody specific for an epitope near the carboxyl-terminal end of the rod domain. The results showed directly that not only is an antiparallel arrangement much more likely but, based on particle length, the subunits are either aligned or have only a very small stagger. Cleavage of cross-linked desmin protofilaments with *N*-bromosuccinimide also indicated that the dimers are arranged antiparallel (Pang, 1985). Evidence cited by Parry and Fraser (1985) also supports an antiparallel orientation of the two dimers. A monoclonal antibody that recognized an epitope between residues 97 and 255 in the amino-terminal end of thrombin-treated desmin, however, seemed to label 64% of the protofilaments at one end, only 10% at both ends, and 26% were not labeled (Ip, 1988).

The packing of the protofilaments to form 10-nm filaments has been examined by several approaches that involve either electron microscopy of assembly or disassembly products or use of information on the primary and secondary structure of desmin to predict packing geometries. Huiatt *et al.* (1980) demonstrated that when purified, soluble gizzard desmin, which was free of any detectable actin or other proteins, was dialyzed against 150 mM NaCl, 10 mM imidazole-HCl, pH 7.0, 10-nm filaments formed that were morphologically very similar to native 10-nm filaments. This observation clearly showed that desmin, by itself, was capable of assembly and that no other proteins such as actin or accessory proteins were required. Kaufmann *et al.* (1985) found that thrombin removed 67 residues from the amino-terminal end of desmin and that the remaining fragments were not capable of forming filaments. These thrombin fragments could be incorporated into normal filaments only if they were mixed with intact desmin in 7 M urea but not if they were mixed after the urea had been removed by dialysis against 10 mM Tris-HCl, pH 8.2. Removal of the carboxyl-terminal 27 residues with a lysine-specific protease produced a fragment that formed normal filaments when the salt concentration was reduced to 50 mM. Removal of a 4-kDa peptide from the amino-terminus of vimentin also prevented the remainder of the molecule from assembling into filaments (Traub and Vorgias, 1983). Because the isolated rod domains of gizzard desmin do not assemble into filaments (Geisler and Weber, 1982), it has been assumed that the more highly variable, non-α-helical head and/or tail segments are responsible for the assembly into 10-nm filaments. Geisler and Weber (1988) found that phosphorylation of sites in the amino-terminal head domain prevented assembly of desmin into typical, reconstituted filaments. The specific involvement of these two segments in filament assembly presumably will be clarified by future experiments.

Shadowing with heavy metal has also been used to prepare desmin and vimentin filaments for examination in the electron microscope. Henderson *et al.* (1982) and Kaufmann *et al.* (1985) saw a 21-nm axial repeat along reconstituted gizzard desmin filaments transferred to mica by the glycerol spray method. Ip *et al.* (1985a,b) observed a 20- to 22-nm axial repeat along reconstituted vimentin filaments that had been either quick-frozen, etched, and rotary shadowed or negatively stained. Ip *et al.* (1985a) also described aggregates that consisted of two protofilaments of vimentin and that presumably are structurally identical to the subfilaments observed by Stromer *et al.* (1981) in desmin preparations. The overlap of the two protofilaments in the vimentin subfilament and the axial periodicity led Ip *et al.* (1985a) to propose that adjacent protofilaments are staggered by 21 nm as they assemble laterally to form 10-nm filaments.

Electron microscopy of unraveled reconstituted desmin filaments and of early stages of desmin filament assembly has revealed possible intermediates between the protofilament and the 10-nm filament. We have used gentle methods such as dilute buffers to loosen and partly unravel the desmin filament (Stromer, unpublished observations). The maximum number of strands observed is four, but sometimes three or even two are seen. The diameter of these strands is 3.5–4.5 nm, consistent with the diameter of a subfilament. When harsher conditions are used, desmin filaments unravel into more numerous strands, which have diameters very similar to that of a protofilament (see Fig. 3b of Stromer *et al.* 1981). During the early stages of assembly, Milam and Erickson (1984) showed that desmin filaments consisted of two kinds of strands. The smaller strands had the same diameter as the protofilament, whereas the larger strands had diameters twice those of the smaller and seemed to consist of two smaller strands. Milam and Erickson (1984) and Renner *et al.* (1981) have shown, in assembling desmin and vimentin, respectively, the existence of spherical, disklike, or granular particles with 8- to 15-nm diameters. Although the frequency with which such particles are seen is quite variable and thus makes uncertain their identity and function, Potschka (1986) proposed that these particles are an assembly intermediate between laterally packed protofilaments and 10-nm filaments.

The amino acid sequence of chicken gizzard desmin (Geisler and Weber, 1982) has been compared with those of other 10-nm filament proteins, including vimentin. The existence of periodic bands of alternating positive and negative charges on the surface of the coiled coils of desmin and vimentin and their potential role in subunit packing were described by McLachlan and Stewart (1982). It is now generally agreed that the heptad repeats in the rod domain permit formation of the coiled coils. The location of hydrophobic residues in the first and fourth positions of most heptads creates a spiraling band of these residues on the surface of the helix. The concept that the coiled coil forms when two helices are positioned so these two bands interact has been reviewed by Steinert *et al.* (1985). Possible ionic interactions between the two chains in the coiled coil have indicated that the two chains are parallel (Parry and Fraser, 1985). The effect of the amount of stagger between the two coiled coils in a protofilament on the possible ionic interactions between the coils has been studied by Fraser *et al.* (1985).

The concept of how the desmin or vimentin filament is formed is still incomplete, but the following sequence of events can be proposed on the basis of available information. Removal of denaturants such as urea causes the individual peptide chains to assemble into the tetrameric protofilament in a low-ionic-strength, slightly alkaline environment (pH 8–8.5). Increasing the ionic strength and decreasing the pH to or near neutrality promotes the lateral assembly of protofilaments into subfilaments. Although it is possible that a stagger of approximately 21 nm exists between adjacent protofilaments, some investigators have no evidence for a stagger. The subfilaments are assembled to form the 10-nm filament, which, for desmin and vimentin, usually shows a right-handed helical substructure. Measurements of desmin and vimentin 10-nm filaments obtained with the scanning transmission electron microscope (STEM) (Steven *et al.*, 1983) have shown a mass of 36–37 kDa/nm. A 50-nm-long, four-stranded desmin protofilament would have a mass of 4×53 kDa/50 nm or 4.24 kDa/nm. This means that a cross section of the filament should contain eight protofilaments. This proposal also would be consistent with four subfilaments, each composed of two strands of protofilaments.

The diameters of desmin and vimentin intermediate filaments are usually referred to as ranging from 8 to 11 nm. Values reported from several laboratories indicate considerable variability. Diameters measured in STEM micrographs varied from 13.8 to 16.5 nm, but diameters of negatively stained filaments were between 8 and 12 nm (Steven *et al.*, 1982, 1983). It was proposed that the negative stain may have penetrated a peripheral zone of decreased density to produce a smaller diameter than was detected by STEM measurements. Negatively stained cardiac desmin filaments had average diameters of 12.4 nm (Price, 1984). Stromer *et al.* (1987) measured the diameters of reconstituted gizzard desmin filaments 5 min, 1 hr, and 16 hr after assembly initiation. Mean diameters were 14.9 nm at 5 min, had decreased significantly ($p < 0.01$) to 13.6 nm at 60 min, and averaged between 10 and 12 nm after overnight dialysis against filament-forming buffers. Diameters increased as temperature was increased from 2°C to 37°C in the presence of divalent cations or a mixture of a monovalent and divalent cation versus filament-forming buffers that contained only a monovalent cation. It is possible that the specific conditions used in different laboratories to form the filaments, especially the time in dialysis for filament assembly, may explain some of the reported variability in diameter.

The effects of cations and temperature on the kinetics of assembly of gizzard desmin filaments have also recently been studied by a combination of spectrophotometry and electron microscopy (Stromer *et al.*, 1987). Filament-forming buffers that contained calcium or magnesium, or sodium or sodium plus magnesium, were compared at 2, 22, and 37°C. If ionic strength was held constant, the divalent cations were much more active potentiators of assembly than was sodium. When the absorbances at the three temperatures were averaged over the cations, it was clear that the mean absorbance of 22°C was significantly higher than at 2°C and that the mean absorbance at 37°C was significantly higher than at 22°C. A highly significant positive cation–temperature interaction was observed when desmin assembly was done at 37°C in filament-forming buffers that contained either calcium or magnesium. These results indicate the need to specify completely the conditions used to assemble 10-nm filaments so that valid comparisons of morphometric data can be made.

2.5. IF-Associated Proteins

The proximity of 10-nm filaments to dense bodies and to other filaments in smooth muscle cells (Somlyo and Franzini-Armstrong, 1985), the ATP-induced formation of a complex between neurofilaments and microtubules (Runge *et al.*, 1981), the presence of projections at the surface of vimentin filaments (Renner *et al.*, 1981; Franke *et al.*, 1981), and the isolation of a protein associated with desmin and vimentin filaments (Granger and Lazarides, 1980) have created interest in interactions between 10-nm filaments and other cellular components. The isolation (Granger and Lazarides, 1980) and subsequent characterization (Sandoval *et al.*, 1983) of synemin, a 230-kDa protein from chicken gizzard, was the first report of a protein specifically associated with desmin and vimentin filaments. Synemin is phosphorylated and binds both to reconstituted desmin filaments, as demonstrated by antibody localization, and to soluble desmin, as shown by its ability to block the interaction between antidesmin and desmin. Although synemin appears to cross-link vimentin filaments in avian erythrocytes (Granger and Lazarides, 1980), evidence for a similar function in muscle has not been obtained.

A second high-molecular-weight protein, paranemin, has also been associated with 10-nm filaments in embryonic smooth, skeletal, and cardiac muscle cells (Breckler and Lazarides, 1982; Price and Lazarides, 1983). This 280-kDa protein is not present in adult gizzard or skeletal muscle cells but is found, by antibody localization, in adult avian smooth muscle cells of elastic blood vessels. No function has been ascribed to paranemin.

A class of species-distinct, histidine-rich, cationic proteins called filaggrin that is associated with keratinization during epidermal differentiation (Dale *et al.*, 1985) also binds to all classes of 10-nm filaments including desmin and vimentin (Steinert *et al.*, 1981). Filaggrin has been isolated from the stratum corneum of mammalian epidermis, has a molecular weight in SDS-polyacrylamide gel electrophoresis of 48,000 (38,400 determined by sedimentation equilibrium), and bundles 10-nm filaments into macrofibrils. Because filaggrin seemingly is present only in epidermal tissue, the interaction between filaggrin and desmin and vimentin has no discernible physiological significance.

It has also been reported that erythrocyte spectrin mediates the association of reconstituted 10-nm desmin filaments with inside-out vesicles prepared from erythrocyte ghost membranes (Langley and Cohen, 1986). In related experiments, Langley and Cohen also demonstrated that spectrin associated with desmin filaments in a 1 : 230 spectrin : desmin molar ratio.

2.6. *Role or Function*

The specific role of 10-nm filaments in smooth muscle has not been determined. The axial 10-nm filament bundle could maintain the cell shape, serve a support function for the nucleus and associated mitochondria, and may be an originating or anchoring site for the 10-nm filaments. Two hypotheses dominate speculation on the role of these other 10-nm filaments. One hypothesis assigns only a cytoskeletal role and is influenced by the view that 10-nm filaments exist primarily in a domain separate from the actomyosin domain. The second hypothesis recognizes that these domains probably are not separate, and, therefore, both a cytoskeletal role and an undefined, perhaps passive role in contraction may exist. If the role is ultimately determined to be solely cytoskeletal, the ability of smooth muscle to contract to a much greater extent than other muscle types would require a very dynamic cytoskeleton that could adapt to a relatively large range of cell lengths and diameters.

3. *Skeletal Muscle*

3.1. *Location of 10-nm Filaments in Cells*

Immunofluorescence localization of desmin antibodies has shown that desmin filaments are absent from mononucleated myoblasts but, in immature skeletal myotubes, are oriented longitudinally (Bennet *et al.*, 1979). This arrangement of anti-desmin-positive filaments is consistent with the location of 10-nm filaments in developing skeletal muscle observed earlier with the electron microscope (Ishikawa *et al.*, 1968). Antivimentin, however, stained both myoblasts and immature myotubes. Antivimentin staining decreased markedly by 5–6 days in culture when myotubes contained prominent myofibrils and was absent in 10-day cultures. The absence of vimentin in adult skeletal muscle has also been noted by Osborn *et al.* (1982). Antidesmin staining showed both a weak cross-striation and longitudinal filaments between myofibrils in 5- to 6-day cultures. The longitudinal staining structures disappeared, and a strong cross-striation pattern associated with the I–Z region was evident by 10–14 days in culture. A similar rearrangement of desmin filaments from a cytoplasmic to a sarcomeric arrangement has been observed by Gard and Lazarides (1980). Tokuyasu *et al.* (1984) also used immunofluorescence to demonstrate that vimentin and desmin were coincidental in developing myotubes in 7- to 21-day chick embryos but that vimentin diminished during development and was absent at hatching. A subsequent immunoelectron microscopic study (Tokuyasu *et al.*, 1985a) showed that both desmin and vimentin were present in a single 10-nm filament and confirmed the vimentin-to-desmin transition in developing chick embryo muscle. Both Bennett *et al.* (1979) and Tokuyasu *et al.* (1984) noted that desmin staining at the level of the Z line occurred after myofibrils were in lateral registry. On the basis of morphological observations, Kelly (1969) also concluded that myofibrils were present before 10-nm filaments were seen.

The impression of the location of 10-nm filaments in mature skeletal muscle is influenced by the method used to detect the filaments. Lazarides and Granger (1978) prepared Z-line sheets from glycerinated chicken muscle and demonstrated that antidesmin stains the perimeter of the individual Z lines and that anti-α-actinin stains the interior of the Z line. This is consistent with the punctate fluorescence seen between myofibrils at the Z-line level with antidesmin in both isolated chicken skeletal myofibrils (Richardson *et al.*, 1981) and in cryosections of fixed neonatal and adult chicken muscle (Tokuyasu *et al.*, 1984). When seen in the longitudinal view, adult avian or mammalian skeletal muscle stained with antidesmin always shows a concentration of staining between myofibrils at the level of the Z line and often shows a less intense staining across the Z line. This pattern is distinctively different from the uniform and intense staining observed across the Z line per se with anti-α-actinin.

An early immunoelectron microscopy study (Richardson *et al.*, 1981) also showed that desmin filaments were concentrated between myofibrils near the Z line. Tokuyasu *et al.* (1983a) used immunoferritin labeling of thin cryosections of mature skeletal muscle to demonstrate that, in addition to the inter-Z-line location, desmin-containing 10-nm filaments also were seen near mitochondria and nuclei and also between Z lines and the sarcolemma. This labeling of cryosections was corroborated by showing that, in thin sections of plastic-embedded tissue, longitudinal profiles of 10-nm filaments were readily seen at the surface of myofibrils. The results of desmin immunocytochemistry on mature skeletal muscle showed that 10-nm filaments surrounded the myofibril at the Z line and may link mitochondria, nuclei, and myofibrils to the sarcolemma.

A more complex system of 10-nm filaments has been observed in mature rabbit muscle after extraction of most thick and thin filaments with a 0.6 M KI solution (Wang and Ramirez-Mitchell, 1983). Evaluation after either negative staining or critical-point drying revealed a prominent group of four to ten 10-nm filaments at the Z-line site and a smaller group of one to two 10-nm filaments at the M-line site. Both these groups are oriented predominantly perpendicular to the "myofibril" axis. Dense accumulations of filaments at the Z-line sites were connected to successive Z-line sites by about 50–100 10-nm filaments, which were parallel to the "myofibril" and also seemed to be linked perpendicular to the 10-nm filaments at the M-line site. On the basis of the small amount of desmin seen in SDS gels of these myofibril residues (Wang and Ramirez-Mitchell, 1983) and the results of more recent immunolocalization studies, it is likely that the filaments at the Z line probably were composed of desmin but that many of the longitudinal filaments and those at the M-line site were composed of other proteins. Price and Sanger (1984) also saw a prominent array of 10-nm filaments parallel to the myofibril axis near the myofibril surface. These filaments seemingly connected adjacent Z-line sites in 14-day or older embryonic chicken muscle and in KI-extracted adult chicken myofibrils. Identification of these filaments was based solely on their diameters, and, for this reason, the composition is unknown. A more detailed review of the early structural research on 10-nm filaments in embryonic and adult skeletal muscle is contained in Price and Sanger (1983).

3.2. Isolation and Preparation of Desmin

Mammalian skeletal muscle desmin has been isolated from washed porcine myofibrils (O'Shea *et al.*, 1981). Actomyosin was extracted with a 1 M KI solution, and the washed residue was suspended in a 1 M acetic acid solution to release desmin. A desmin-enriched fraction was collected by ammonium sulfate precipitation, was dissolved in 6 M urea, and transferred to a hydroxyapatite column. Final purification was done on a DEAE-Sepharose CL-6B column. Removal of urea by dialysis against 10 mM Tris-acetate, pH 8.5, resulted in a soluble desmin that was equal in stability to smooth muscle desmin.

3.3. Protein and Filament Properties

Skeletal muscle desmin migrated in SDS-polyacrylamide gels with a M_r of 53,000. In two-dimensional electrophoretic gels, DEAE-Sepharose-purified desmin migrated as two variants, a major variant and a minor, more acidic variant. Peptide mapping after protease or cyanogen bromide digestion showed that mammalian skeletal and smooth desmins were virtually identical, and that both were different from but shared many homologous peptides with avian smooth muscle desmin (O'Shea *et al.*, 1981). Purified skeletal muscle desmin formed long reconstituted filaments that averaged 10.4 nm in diameter when dialyzed against a 0.1 M NaCl solution at pH 7.0. These filaments were morphologically indistinguishable from their smooth muscle counterparts.

3.4. Role or Function

Again, the function of 10-nm filaments in skeletal muscle is speculative. Both Tokuyasu *et al.* (1985a,b) and Bennett *et al.* (1979) have argued convincingly that 10-nm desmin filaments probably do not play a role in lateral alignment of myofibrils in developing skeletal muscle. The possibility, however, that the longitudinal orientation of vimentin-containing 10-nm filaments in myoblasts and in immature myotubes could help maintain the bipolar cell shape and assist in the parallel alignment of the elongated myofilaments should be considered. In mature skeletal muscle, the location of 10-nm filaments is also consistent with a cytoskeletal role. The filaments are positioned to link adjacent myofibrils to each other, to mitochondria, nuclei, and the sarcolemma. The nature of these linkages and the means by which this 10-nm filament network adapts during the isovolumetric contraction of muscle undoubtedly will be investigated.

4. Cardiac Muscle

4.1. Location of 10-nm Filaments in Cells

Rash *et al.* (1969, 1970) provided early morphological evidence of the existence of 10-nm filaments in developing cardiac cells. More recently, Forbes and Sperelakis (1983) studied conventionally embedded, thin-sectioned adult mouse ventricle and found that the majority of 10-nm filaments were located in bundles that were aligned with Z lines and were perpendicular to the cell axis. A minority of 10-nm filaments were in smaller, longitudinally oriented bundles directed into desmosomal plaques of the intercalated disk in both mouse and guinea pig hearts.

The use of antibody localization techniques has improved our understanding of the complexity of the 10-nm filament system in cardiac muscle. Myofibril fragments prepared from adult chicken hearts bound fluorescently labeled antidesmin between Z lines of adjacent myofibrils, in the intercalated disk and in bundles between and parallel to the myofibrils (Richardson *et al.*, 1981). Antidesmin staining of cryostat sections of atria and ventricles showed that, in newborn rats, desmin was primarily located in barlike structures in regions of cell junctions and areas of lateral cell adhesion and only faintly associated with Z lines (Carlson *et al.*, 1982). In 4-week-old rats, however, antidesmin staining had shifted to a strong association with both intercalated disks and Z lines. Fuseler and Shay (1982) also observed a transition in arrangement of 10-nm desmin filaments in rat heart myocytes. Myocytes from 9- to 10-day-old hearts contained a complex network of cytoplasmic filaments. Between 12 and 16 days and after the formation of recognizable myofibrils, 10-nm filaments shifted from the cytoplasmic network to longitudinal filaments that outline the myofibril. Desmin foci, which had the same spacing as Z lines, developed from these longitudinal filaments. Cardiac myocytes from neonatal chicks also contained vimentin, but the vimentin was absent in adult chicken cardiac myocytes (Price and Lazarides, 1983; Tokuyasu *et al.*, 1985b). Antidesmin stained both the Purkinje fibers and the ventricular fibers in bovine and chicken hearts (Thornell *et al.*, 1985), but antivimentin stained only the chicken heart Purkinje fibers.

Immunoelectron microscopy has confirmed that desmin is located between Z lines (Richardson *et al.*, 1981) but also has shown more details of the arrangement of longitudinal 10-nm filaments in cardiac cells. Tokuyasu *et al.* (1983b) labeled thin cryosections with antidesmin and found that 10-nm filaments were located in an intermyofibrillar network that included weblike arrays which extended from the Z line toward the center of the sarcomere and bridges between Z lines and mitochondria. Desmin labeling was absent from the fascia adherens of the intercalated disk but was dense near but not on the desmosomal plaques. A follow-up study with improved sample preparation techniques showed that extensive networks of 10-nm filaments exist between chicken cardiac myofibrils (Tokuyasu, 1983).

Current information indicates that 10-nm filaments are rearranged in the myocardium during the transition from embryo to adult and that, in the working myocardium, vimentin is deleted. In the adult avian heart, the 10-nm filament network seems to be more complex than in skeletal muscle. The extent of interspecies variability is, however, unknown.

4.2. Isolation and Preparation of Desmin

Desmin can be isolated from mature mammalian ventricles by preparing well-washed myofibrils and extracting the actomyosin with a 1 M KI solution (Hartzer, 1984). The residue is solubilized in 8 M urea, and the desmin is separated by chromatography on DEAE-cellulose and hydroxyapatite in the presence of 6 M urea. Final purification is done on DEAE-Sepharose CL-6B. Solubility, molecular properties, and filament-forming ability of desmin prepared in this way are essentially identical to those of smooth muscle desmin prepared by the method of Huiatt *et al.* (1980). An alternative method used on bovine Purkinje fibers involves 0.2% SDS and heating at 60°C for an hour to solubilize an actomyosin-extracted myofibril residue (Stigbrand *et al.*, 1979). Gel permeation chromatography in the presence of SDS was used to purify desmin from the mixture. Two-dimensional gels of Purkinje desmin contained at least five isoelectric variants (Thornell *et al.*, 1985) that exist because of differences in the amount of phosphorylation (Kjorell *et al.*, 1987).

4.3. Role or Function

The principal role ascribed to 10-nm filaments is that of a cytoskeletal support that links myofibrils and other organelles to each other and to the sarcolemma. The paucity of myofibrils and the relative abundance of 10-nm filaments in the impulse-conducting Purkinje fibers of birds and larger mammals could indicate an additional but undefined role (Thornell and Eriksson, 1981). An indication that 10-nm filaments in rat heart cells became more numerous after injection with methandrostenolone, an anabolic steroid, suggested that these filaments responded to metabolic changes (Behrendt, 1977). Ferrans and Roberts (1973) also saw increased numbers of 10-nm filaments in hypertrophied human heart cells in patients with obstructive cardiomyopathies. Cardiac hypertrophy, however, induced in the rat by either aortic stenosis (Samuel *et al.*, 1984) or thyroxine injection (Rappaport *et al.*, 1985) caused no changes in the staining pattern seen with fluorescently labeled antidesmin. Tokuyasu *et al.* (1983b) have speculated that branching of cardiac myofibrils and the variable intermyofibrillar space, which frequently contains more mitochondria than skeletal muscle, may necessitate a more complex system of 10-nm filaments in cardiac muscle.

ACKNOWLEDGMENTS. The author's research has been supported by grants from the American Heart Association, Iowa Affiliate, and the Muscular Dystrophy Association. The assistance of Marylou Weigel in the preparation of the manuscript is gratefully acknowledged. This is Journal Paper J-13293 of the Iowa Agriculture and Home Economics Experiment Station, Ames, Iowa 50011, Projects 2361 and 2921.

5. References

Ashton, F., Somlyo, A. V., and Somlyo, A. P., 1975, The contractile apparatus of vascular smooth muscle: Intermediate high voltage stereo electron microscopy, *J. Mol. Biol.* **98:**17–29.

Behrendt, H., 1977, Effect of anabolic steroids on rat heart muscle cells. I. Intermediate filaments, *Cell Tissue Res.* **180:**303–315.

Bennett, G., Fellini, S., Toyama, Y., and Holtzer, H., 1979, Redistribution of intermediate filament subunits during skeletal myogenesis and maturation in vitro, *J. Cell Biol.* **82:**577–584.

Bond, M., and Somlyo, A. V., 1982, Dense bodies and actin polarity in vertebrate smooth muscle, *J. Cell Biol.* **95:**403–413.

Breckler, J., and Lazarides, E., 1982, Isolation of a new high molecular weight protein associated with desmin and vimentin filaments from avian embryonic skeletal muscle, *J. Cell Biol.* **92:**795–806.

Campbell, G., Uehara, Y., Mark, G. and Burnstock, G., 1971, Fine structure of smooth muscle cells grown in tissue culture, *J. Cell Biol.* **49:**21–34.

Capetanaki, Y., Ngai, J., and Lazarides, E., 1984, Characterization and regulation in the expression of a gene coding for the intermediate filament protein desmin, *Proc. Natl. Acad. Sci. USA* **81:**6909–6913.

Carlson, E., Kjorell, U., and Thornell, L.-E., 1982, Differentiation of the myofibrils and the intermediate filament system during postnatal development of the rat, *Eur. J. Cell Biol.* **27:**62–73.

Cooke, P., 1976, A filamentous cytoskeleton in vertebrate smooth muscle fibers, *J. Cell Biol.* **68:**539–556.

Cooke, P., and Chase, R., 1971, Potassium chloride insoluble myofilaments in vertebrate smooth muscle cells, *Exp. Cell Res.* **66:**417–426.

Cooke, P., and Fay, F., 1972, Correlation between fiber length, ultrastructure, and the length–tension relationship of mammalian smooth muscle, *J. Cell Biol.* **52:**105–116.

Dale, B., Resing, K., and Lonsdale-Eccles, J., 1985, Filaggrin: A keratin filament associated protein, *Proc. Natl. Acad. Sci. USA* **82:**330–342.

Fay, F., Fujiwara, K., Rees, D., and Fogarty, K., 1983, Distribution of alpha-actinin in single isolated smooth muscle cells, *J. Cell Biol.* **96:**783–795.

Fellini, S., Bennett, G., Toyama, Y., and Holtzer, H., 1978, Biochemical and immunological heterogeneity of 100Å filament subunits from different chick cell types, *Differentiation* **12:**59–69.

Ferrans, V., and Roberts, W., 1973, Intermyofibrillar and nucleomyofibrillar connections in human and canine myocardium. An ultrastructural study, *J. Mol. Cell. Cardiol.* **5:**247–257.

Forbes, M., and Sperelakis, N., 1983, The membrane systems and cytoskeletal elements of mammalian myocardial cells, in: *Cell and Muscle Motility,* Volume 3 (R. M. Dowben and J. W. Shay, eds.), Plenum Press, New York, pp. 89–155.

Frank, E., and Warren, L., 1981, Aortic smooth muscle cells contain vimentin instead of desmin, *Proc. Natl. Acad. Sci. USA* **78:**3020–3024.

Franke, W., Zerban, H., Grund, C., and Schmid, E., 1981, Electron microscopy of vimentin filaments and associated whisker structures in thin sections and freeze-fractures, *Biol. Cell* **41:** 173–178.

Fraser, R., Mac Rae, T., Suzuki, E., Parry, D., Trajstman, A., and Lucas, I., 1985, Intermediate filament structure: 2. Molecular interactions in the filament, *Int. J. Biol. Macromol.* **7:**258–274.

Fuseler, J., and Shay, J., 1982, The association of desmin with the developing myofibrils of cultured embryonic rat heart myocytes, *Dev. Biol.* **91:**448–457.

Gabbiani, G., Schmid, E., Winter, S., Chaponnier, C., de Chastonay, C., Vandekerckhove, J., Weber, K., and Franke, W., 1982, Vascular smooth muscle cells differ from other smooth muscle cells: Predominance of vimentin filaments and a specific alpha-type actin, *Proc. Natl. Acad. Sci. USA* **78:** 298–302.

Gard, D., and Lazarides, E., 1980, The synthesis and distribution of desmin and vimentin during myogenesis in vitro, *Cell* **19:**263–275.

Garfield, R., and Somlyo, A. P., 1985, Structure of smooth muscle, in *Calcium and Contractility: Smooth Muscle* (A. K. Grover and E. E. Daniel, eds.), Humana Press, Clifton, N.J., pp. 1–36.

Geisler, N., and Weber, K., 1981, Comparison of the proteins of two immunologically distinct intermediate-sized filaments by amino acid sequence analysis: Desmin and vimentin, *Proc. Natl. Acad. Sci. USA* **78:** 4120–4123.

Geisler, N., and Weber, K., 1982, The amino acid sequence of chicken muscle desmin provides a common structural model for intermediate filament proteins, *EMBO J.* **1:**1649–1656.

Geisler, N., and Weber, K., 1983, Amino acid sequence data on glial fibrillary acidic protein (GFA): Implications for the subdivision of intermediate filaments into epithelial and non-epithelial members, *EMBO J.* **2:**2059–2063.

Geisler, N., and Weber, K., 1988, Phosphorylation of desmin in vitro inhibits formation of intermediate filaments; identification of three kinase A sites in the aminoterminal head domain, *EMBO J.* **7:**15–20.

Geisler, N., Kaufmann, E., and Weber, K., 1982, Proteinchemical characterization of three structurally distinct domains along the protofilament unit of desmin 10-nm filaments, *Cell* **30:**277–286.

Geisler, N., Kaufmann, E., and Weber, K., 1985, Antiparallel orientation of the two double-stranded coiled-coils in the tetrameric protofilament unit of intermediate filaments, *J. Mol. Biol.* **182:**173–177.

Goldman, R., Goldman, A., Green, K., Jones, J., Jones, S., and Yang, H.-Y., 1986, Intermediate filament networks: Organization and possible functions of a diverse group of cytoskeletal elements, *J. Cell Sci. Suppl.* **5:**69–97.

Granger, B., and Lazarides, E., 1980, Synemin: A new high molecular weight protein associated with desmin and vimentin filaments in muscle, *Cell* **22:**727–738.

Hartzer, M., 1984, Purification and properties of porcine cardiac desmin and vascular smooth muscle vimentin, Ph.D. dissertation, Iowa State University, Ames (Microfilm No. DA8505825, University Microfilms, Ann Arbor).

Henderson, D., Geisler, N., and Weber, K., 1982, A periodic ultrastructure in intermediate filaments, *J. Mol. Biol.* **155:**173–176.

Hubbard, B., and Lazarides, E., 1979, Co-purification of actin and desmin from chicken smooth muscle and their copolymerization *in vitro* to intermediate filaments, *J. Cell Biol.* **80:**166–182.

Huiatt, T., Robson, R., Arakawa, N., and Stromer, M., 1980, Desmin from avian smooth muscle, *J. Biol. Chem.* **255:**6981–6989.

Ip, W., 1988, Modulation of desmin intermediate filament assembly by a monoclonal antibody, *J. Cell Biol.* **106:**735–745.

Ip, W., Hartzer, M., Pang, Y., and Robson, R., 1985a, Assembly of vimentin in vitro and its implications concerning the structure of intermediate filaments, *J. Mol. Biol.* **183:**365–375.

Ip, W., Heuser, J., Pang, Y., Hartzer, M., and Robson, R., 1985b, Subunit structure of desmin and vimentin protofilaments and how they assemble into intermediate filament, *Ann. N.Y. Acad. Sci.* **455:**185–199.

Ishikawa, H., Bischoff, R., and Holtzer, H., 1968, Mitosis and intermediate-sized filaments in developing skeletal muscle, *J. Cell Biol.* **38:**538–555.

Izant, J., and Lazarides, E., 1977, Invariance and heterogeneity in the major structural and regulatory proteins of chick muscle cells revealed by two dimensional gel electrophoresis, *Proc. Natl. Acad. Sci. USA* **74:**1450–1454.

Kaufmann, E., Weber, K., and Geisler, N., 1985, Intermediate filament forming ability of desmin derivatives lacking either the amino-terminal 67 or the carboxy-terminal 27 residues, *J. Mol. Biol.* **185:**733–742.

Kelly, D., 1969, Myofibrillogenesis and Z-band differentiation, *Anat. Rec.* **163:**403–426.

Kjorell, U., Thornell, L.-E., Lehto, V.-P., Virtanen, I., and Whalen, R. G., 1987, A comparative analysis of intermediate filament proteins in bovine heart Purkinje fibers and gastric smooth muscle, *Eur. J. Cell Biol.* **44:**68–78.

Langley, R., and Cohen, C., 1986, Association of spectrin with desmin intermediate filaments, *J. Cell. Biochem.* **30:**101–109.

Lazarides, E., and Granger, B., 1978, Fluorescent localization of membrane sites in glycerinated chicken skeletal muscle fibers and the relationship of these sites to the protein composition of the Z disc, *Proc. Natl. Acad. Sci. USA* **75:**3683–3687.

McLachlan, A., and Stewart, M., 1982, Periodic charge distribution and the intermediate filament proteins desmin and vimentin, *J. Mol. Biol.* **162:**693–698.

Milam, L., and Erickson, H., 1984, Structural characteristics of the desmin protofilament, *J. Ultrastruct. Res.* **89:**179–186.

Nelson, W., and Traub, P., 1983, Proteolysis of vimentin and desmin by the Ca^{2+}-activated proteinase specific for these intermediate filament proteins, *Mol. Cell. Biol.* **3:**1146–1156.

Nonomura, Y., and Ebashi, S., 1975, Isolation and identification of smooth muscle contractile proteins, in: *Methods in Pharmacology* (E. Daniel and E. Paton, eds.), Plenum Press, New York, pp. 141–162.

O'Farrell, P., 1975, High resolution two-dimensional electrophoresis of proteins, *J. Biol. Chem.* **250:**4007–4021.

Osborn, M., Caselitz, J., and Weber, K., 1981, Heterogeneity of intermediate filament expression in vascular smooth muscle: A gradient in desmin positive cells from the rat aortic arch to the level of the arteria iliaca communis, *Differentiation* **20:**196–202.

Osborn, M., Geisler, N., Shaw, G., Sharp, G., and Weber, K., 1982, Intermediate filaments, *Cold Spring Harbor Symp. Quant. Biol.* **46:**413–429.

O'Shea, J., Robson, R., Huiatt, T., Hartzer, M., and Stromer, M., 1979, Purified desmin from adult mammalian skeletal muscle: A peptide mapping comparison with desmins from adult mammalian and avian smooth muscle, *Biochem. Biophys. Res. Commun.* **89:**972–980.

O'Shea, J., Robson, R., Hartzer, M., Huiatt, T., Rathbun, W., and Stromer, M., 1981, Purification of desmin from adult mammalian skeletal muscle, *Biochem. J.* **195:**345–356.

Pang, Y., 1985, Subunit structure of the protofilament building blocks of intermediate filaments, Ph.D. dissertation, Iowa State University, Ames (Microfilm No. 85-24685, University Microfilms, Ann Arbor).

Pang, Y., Robson, R., Hartzer, M., and Stromer, M., 1983, Subunit structure of the desmin and vimentin protofilament units, *J. Cell Biol.* **97:**226a.

Parry, D., and Fraser, R., 1985, Intermediate filament structure: 1. Analysis of IF protein sequence data, *Int. J. Biol. Macromol.* **7:**203–213.

Potschka, M., 1986, Structure of intermediate filaments, *Biophys. J.* **49:**129–130.

Price, M., 1984, Molecular analysis of intermediate filament cytoskeleton—A putative load-bearing structure, *Am. J. Physiol.* **246:**H566–H572.

Price, M., and Lazarides, E., 1983, Expression of intermediate filament-associated proteins paranemin and synemin in chicken development, *J. Cell Biol.* **97:**1860–1874.

Price, M., and Sanger, J., 1983, Intermediate filaments in striated muscles. A review of structural studies in embryonic and adult skeletal and cardiac muscle, in: *Cell and Muscle Motility,* Volume 3 (R. M. Dowben and J. W. Shay, eds.), Plenum Press, New York, pp. 1–40.

Price, M., and Sanger, J., 1984, Reduction of density and anisotropic distribution of intermediate filaments occur during avian skeletal myogenesis, *Am. J. Anat.* **171:**427–440.

Quax, W., van den Heuvel, R., Egberts, W., Vree, W., Quax-Jeuken, Y., and Bloemendal, H., 1984, Intermediate filament cDNAs from BHK-21 cells: Demonstration of distinct genes for desmin and vimentin in all vertebrate classes, *Proc. Natl. Acad. Sci. USA* **81:**5970–5974.

Quax-Jeuken, Y., Quax, W., and Bloemendal, H., 1983, Primary and secondary structure of hamster vimentin predicted from the nucleotide sequence, *Proc. Natl. Acad. Sci. USA* **80:**3548–3552.

Rappaport, L., Samuel, J., Bertier-Savalle, B., Manotte, F., and Schwartz, K., 1985, Microtubules and desmin filaments during the onset of heart growth in the rat, *Basic Res. Cardiol.* **80:**129–132.

Rash, J., Shay, J., and Biesele, J., 1969, A third class of filaments in early cardiac myocytes, *J. Cell Biol.* **43:**112a.

Rash, J., Biesele, J., and Gey, G., 1970, Three classes of filaments in cardiac differentiation, *J. Ultrastruct. Res.* **33**:408–435.

Renner, W., Franke, W., Schmid, E., Geisler, N., Weber, K., and Mandelkow, E., 1981, Reconstitution of intermediate-sized filaments from denatured monomeric vimentin, *J. Mol. Biol.* **149**:285–306.

Richardson, F., Stromer, M., Huiatt, T., and Robson, R., 1981, Immunoelectron and immunofluorescence localization of desmin in mature avian muscles, *Eur. J. Cell Biol.* **26**:91–101.

Runge, M., Lane, T., Yphantis, D., Lifsics, M., Saito, A., Altin, M., Reinke, K., and Williams, R., Jr., 1981, ATP-induced formation of an associated complex between microtubules and neurofilaments, *Proc. Natl. Acad. Sci. USA* **78**:1431–1435.

Samuel, J., Bertier, B., Bugaisky, L., Marotte, F., Swynghedauw, B., Schwartz, K., and Rappaport, L., 1984, Different distributions of microtubules, desmin filaments and isomyosins during the onset of cardiac hypertrophy in the rat, *Eur. J. Cell Biol.* **34**:300–306.

Sandoval, I., Colaco, C., and Lazarides, E., 1983, Purification of the intermediate filament-associated protein, synemin, from chicken smooth muscle, *J. Biol. Chem.* **258**:2568–2576.

Schmid, E., Osborn, M., Rungger-Brandle, E., Gabbiani, G., Weber, K., and Franke, W., 1982, Distribution of vimentin and desmin filaments in smooth muscle tissue of mammalian and avian aorta, *Exp. Cell Res.* **137**:329–340.

Schollmeyer, J., Furcht, L., Goll, D., Robson, R., and Stromer, M., 1976, Localization of contractile proteins in smooth muscle cells and in normal and transformed fibroblasts, in: *Cell Motility,* Cold Spring Harbor Conferences on cell Proliferation, Volume 3 (R. D. Goldman, T. D. Pollard, and J. Rosenbaum, eds.), Cold Spring Harbor Laboratory, Cold Spring Harbor, N.Y., pp. 361–388.

Small, J., 1977, Studies on isolated smooth muscle cells: The contractile apparatus, *J. Cell Sci.* **24**:327–349.

Small, J., and Sobieszek, A., 1977, Studies on the function and composition of the 10-nm (100 Å) filaments of vertebrate smooth muscle, *J. Cell Sci.* **23**:243–268.

Small, J., and Sobieszek, A., 1980, The contractile apparatus of smooth muscle, *Int. Rev. Cytol.* **64**:241–306.

Somlyo, A. V., and Franzini-Armstrong, C., 1985, New views of smooth muscle structure using freezing, deep-etching and rotary shadowing, *Experientia* **41**:841–856.

Somlyo, A. V., Bond, M., Berner, P., Ashton, F., Holtzer, H., and Somlyo, A. P., 1984, The contractile apparatus of smooth muscle: An update, in: *Smooth Muscle Contraction* (N. L. Stephens, ed.), Dekker, New York, pp. 1–20.

Steinert, P., Contieri, P., Teller, D., Lonsdale-Eccles, J., and Dale, B., 1981, Characterization of a class of cationic proteins that specifically interact with intermediate filaments, *Proc Natl. Acad. Sci. USA* **78**:4094–4101.

Steinert, P., Wantz, M., and Idler, W., 1982, O-Phosphoserine content of intermediate filament subunits, *Biochemistry* **21**:177–183.

Steinert, P., Steven, A., and Roop, D., 1985, The molecular biology of intermediate filaments, *Cell* **42**:411–419.

Steven, A., Wall, J., Hainfeld, J., and Steinert, P., 1982, Structure of fibroblastic intermediate filaments: Analysis by scanning transmission electron microscopy, *Proc. Natl. Acad. Sci. USA* **79**:3101–3105.

Steven, A., Hainfeld, J., Trus, B., Wall, J., and Steinert, P., 1983, The distribution of mass in heteropolymer intermediate filaments assembled in vitro *J. Biol. Chem.* **258**:8323–8329.

Stigbrand, T., Eriksson, A., and Thornell, L., 1979, Isolation and partial characterization of intermediate filament protein (skeletin) from cow heart Purkinje fibers, *Biochim. Biophys. Acta* **577**:52–60.

Stromer, M., and Bendayan, M., 1988, Arrangement of desmin intermediate filaments in smooth muscle cells as shown by high resolution immunocytochemistry, *Cell Motil. Cytoskel.* **11**:117–125.

Stromer, M., Huiatt, T., Richardson, F., and Robson, R., 1979, Studies of disassembly and antibody decoration of synthetic 10-nm desmin filaments from smooth muscle, *J. Cell Biol.* **83**:384a.

Stromer, M., Huiatt, T., Richardson, F., and Robson, R., 1981, Disassembly of synthetic 10-nm filaments from smooth muscle into protofilaments, *Eur. J. Cell Biol.* **25**:136–143.

Stromer, M., Ritter, Y., Pang, Y., and Robson, R., 1987, Effect of cations and temperature on kinetics of desmin assembly, *Biochem. J.* **246**:75–81.

Thornell, L.-E., and Eriksson, A., 1981, Filament systems in the Purkinje fibers of the heart, *Am. J. Physiol.* **241**:H291–H305.

Thornell, L.-E., Eriksson, A., Johansson, B., Kjorell, U., Franke, W., Virtanen, I., and Lehto, V.-P. 1985, Intermediate filament and associated proteins in heart Purkinje fibers: A membrane-myofibril anchored cytoskeletal system, *Ann N.Y. Acad. Sci.* **455**:213–240.

Tokuyasu, K., 1983, Visualization of longitudinally oriented intermediate filaments in frozen sections of chicken cardiac muscle by a new staining method, *J. Cell Biol.* **97**:562–565.

Tokuyasu, K., Dutton, A., and Singer, S., 1983a, Immunoelectron microscopic studies of desmin (skeletin) localization and intermediate filament organization is chicken skeletal muscle, *J. Cell Biol.* **96**:1727–1735.

Tokuyasu, K., Dutton, A., and Singer, S., 1983b, Immunoelectron microscopic studies of desmin (skeletin) localization and intermediate filament organization in chicken cardiac muscle, *J. Cell Biol.* **96**:1736–1742.

Tokuyasu, K., Maher, P., and Singer, S., 1984, Distributions of vimentin and desmin in developing chick myotubes in vivo. I. Immunofluorescence study, *J. Cell Biol.* **98**:1961–1972.

Tokuyasu, K., Maher, P., and Singer, S., 1985a, Distribution of vimentin and desmin in developing chick myotubes in vivo. II. Immunoelectron microscopic study, *J. Cell Biol.* **100**:1157–1166.

Tokuyasu, K., Maher, P., Dutton, A., and Singer, S., 1985b, Intermediate filaments in skeletal and cardiac muscle tissue in embryonic and adult chicken, *Ann N.Y. Acad. Sci.* **455**:200–212.

Traub, P., and Vorgias, C., 1983, Involvement of the N-terminal polypeptide of vimentin in the formation of intermediate filaments, *J. Cell Sci.* **63**:43–67.

Wang, K., and Ramirez-Mitchell, R., 1983, A network of transverse and longitudinal intermediate filaments is associated with sarcomeres of adult vertebrate skeletal muscle, *J. Cell Biol.* **96**:562–570.

Weber, K., and Osborn, M. 1969, The reliability of molecular weight determinations by dodecyl sulfate-polyacrylamide gel electrophoresis, *J. Biol. Chem.* **244:** 4406–4412.

3

Neurofilaments

P. A. M. Eagles, H. C. Pant, and H. Gainer

1. Introduction

The idea of a neurofilamentous skeleton inside nerve cells is not new. Almost 100 years ago Schultze first expressed the view regarding the presence of neurofibrillar networks within neurons. Many workers doubted the existence of these structures because the methods used to visualize them were capricious, different results being found in different laboratories (see review by Parker, 1929). In the 1890s few were able to emulate the techniques of Apáthy whose skills were widely acclaimed. It was not until the development of the novel and reproducible methods for silver staining by Bielschowsky and Ramón y Cajal that neurofibrils were routinely found. With the advent of high-resolution electron microscopic techniques, we now know that the neurofibrillary structures seen by these early microscopists were composed of the 100-Å-diameter intermediate filaments characteristic of nerve cells—neurofilaments (Boycott *et al.*, 1961; Gray and Guillery, 1961). In this review, we discuss the more recent history regarding neurofilaments and focus on current unresolved questions regarding their architecture, enzymatic modification, and development. While being a fairly general account of the neurofilament literature, especially over the past decade, we dwell in depth on key issues—like the nature of neurofilament sidearms, and the phosphorylation dependence of many neurofilament antibodies that will be helpful for both specialist and more general readers.

Abbreviations used in this chapter: Anti-IFA, monoclonal antibody (MAb) that binds to intermediate filaments (Pruss *et al.*, 1981); NFH, the heavy-molecular-weight component of the mammalian triplet with an apparent molecular weight on SDS-PAGE around 200k, also called NF200; NFM, the medium component of the triplet with an apparent molecular weight on SDS-PAGE around 160k, also called NF160; NFL, the light component of the triplet with an apparent molecular weight on SDS-PAGE around 70k, also called NF70.

P. A. M. Eagles • Department of Biophysics, King's College London, London WC2B 5RL, United Kingdom. **H. C. Pant** • Laboratory of Preclinical Studies, National Institute on Alcohol Abuse and Alcoholism, National Institutes of Health, Rockville, Maryland 20892. **H. Gainer** • Laboratory of Neurochemistry and Neuroimmunology, National Institute of Child Health and Human Development, National Institutes of Health, Bethesda, Maryland 20892.

2. Neurofilament Arrangement in the Neurons of Vertebrates

Neurofilaments are found to be major components of the cytoskeleton in many mammalian nerves. Their number and arrangement within the neuron vary depending on their location. In dendrites, neurofilaments are often fewer and less well ordered than in axons where they can be the predominant structural element. Small axons have less of them, larger axons have many (Wuerker and Kirkpatrick 1972). In synapses they form rings, the classical "boutons terminaux" (Gray and Guillery, 1966). Growing nerves exhibit a variety of neurofilament arrangements though they are often observed as small groups, packaged together. Their number increases with development and this increase is thought to stabilize somehow the cytoskeleton overall (Peters and Vaughn, 1967; Peters, 1968; D. Smith, 1973). The filaments have been seen to occupy a central bundle at the core of an elongating axon or growth cone that suggests they are functioning as a strengthening internal scaffold (Yamada *et al.*, 1971), and a central neurofilament bundle is also sometimes highlighted in more mature nerves when treated so as to block axonal transport (e.g., see Tsukita and Ishikawa, 1979).

Studies relating neurofilament content with axon caliber clearly show a direct correlation, with the number of neurofilaments fluctuating in line with the axon diameter (Friede and Samorajski, 1970; Weiss and Mayr, 1971); no such relationship is found for numbers of microtubules, which are constant in number (Tsukita and Ishikawa, 1981). These results show that neurofilaments are not continuous along the axon and are generally taken to mean that the size of the axon is largely determined by factors affecting neurofilaments like their number and arrangement. This in turn could suggest that the transport of neurofilaments may have a role in regulating axon girth as is indicated by the studies of Hoffman *et al.* (1984).

The ratio of microtubules to neurofilaments in the axon varies widely depending on the developmental state (Peters, 1968). In immature axons, microtubules are the major cytoskeletal element and this situation is altered to one where neurofilaments abound as the axon matures and increases in girth (Peters and Vaughn, 1967). These findings are often expressed in the ideas of plasticity and stability within the developing neuron (Lasek, 1981). In the growing cell, plasticity is achieved by the preponderance of microtubules, and as the cell matures, stability of the cytoskeleton results from the specific architecture inherent in the neurofilament network (Willard, 1983). During regeneration of the nerve, there is a reversal of the events leading to maturation, and microtubules once again proliferate.

Measurements of neurofilament density are readily obtained from electron microscopy of axons in cross section. Published micrographs vary widely in their representation of neurofilament arrangement (e.g., Peters, 1968; Bertolini *et al.*, 1970; R. Smith, 1973; Fig. 1); some show dense well-ordered networks while others show sparse colonies with large, apparently empty spaces separating filaments. Despite such differences, a wide variety of studies on many tissues indicate that in neurofilament-rich regions the density can be fairly constant around 100–200 filaments/μm^2 (Norton and Goldman, 1980). At the nodes of Ranvier, there is a dramatic reduction in axon size though neurofilament density remains the same as in internodal regions (Weiss and Mayr, 1971; Tsukita and Ishikawa, 1981), i.e., at the nodes there are fewer filaments.

In growing nerves, the neurofilaments tend to associate into small groups though as the cell matures a situation such as that shown in Fig. 1 is reached where a quite regular

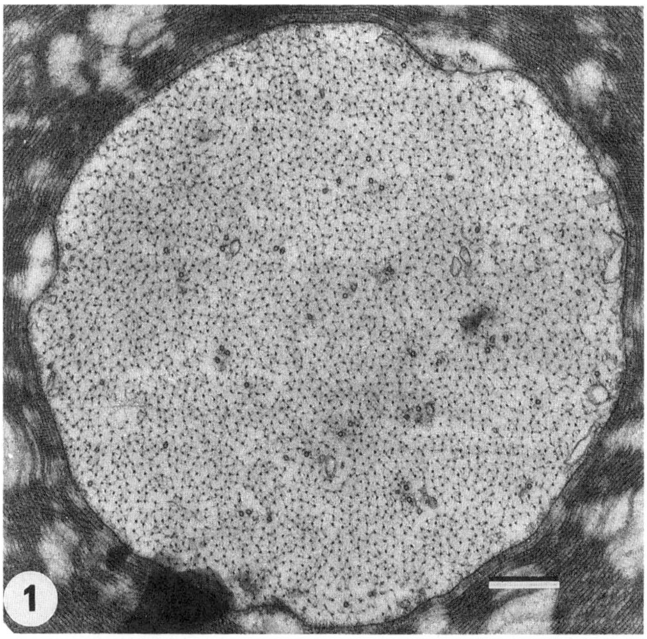

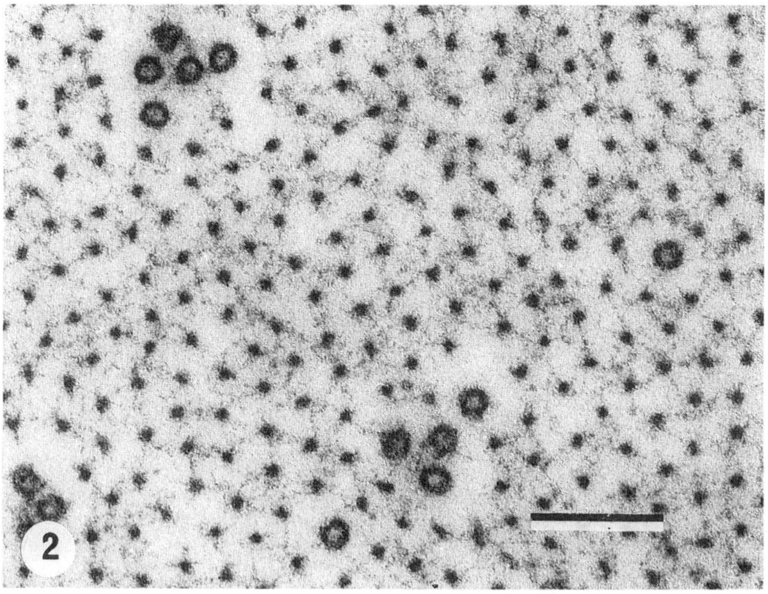

Figure 1. Transverse section of spinal nerve root from cat. The very regular array of neurofilaments with sidearms is clearly seen here. Occasionally, pseudohexagonal packing of the neurofilaments can be observed. Microtubules are often grouped together and small clusters are surrounded by a characteristic halo that appears to exclude neurofilaments, though sidearms can be seen invading this territory. In this section, which is approximately 500 Å thick, the average number of sidearms per neurofilament is 2. The cross-bridges that connect the neurofilament have an average length of 340 ± 100 Å and width of 48 ± 16 Å. In neurofilament-rich areas, the density is around 680 neurofilaments /μm^2. Bar = 0.4 μm.

Figure 2. Transverse section of spinal nerve root from cat. High-power picture from an area shown in Fig. 1. The wispy sidearms of the neurofilaments are clearly seen. Note the regularity in the neurofilament domains and the regions around the microtubule bundles where the neurofilaments seem to be excluded. Bar = 0.12 μm.

neurofilament network can fill the cross section. In such pictures, the neurofilaments are seen to environ microtubule-rich regions and similar relationships are also clearly shown in freeze-fracture studies (Tsukita *et al.*, 1982; Schnapp and Reese, 1982; Hirokawa, 1982). A helical nature to this neurofilament network has so far not been clearly demonstrated as it has with the giant axons of *Myxicola* (Gilbert, 1972) and squid (Metuzals and Izzard, 1969; Lasek *et al.*, 1983). Despite this, there are some indirect indications of intrinsic helicity in the cytoskeleton of cultured mammalian neurons (Shaw and Bray, 1977) and this may well reflect an underlying helical arrangement of the neurofilaments present there.

In cross-sections of nerves, electron microscopy shows the core or backbone of the filament to have a diameter around 100 Å (Fig. 2) and this backbone has been seen to be subdivided into protofilaments (Wuerker, 1970). In longitudinal sections (Fig. 3) the filaments are seen to run for many micrometers along the axon roughly in register, though disease and drug treatment have been shown to perturb this arrangement (see review by Sayre *et al.*, 1985). In nerves like those of Fig. 2, the neurofilaments closely abut one another often producing the effect of a pseudohexagonal lattice in areas. The filaments appear to be held in place by thin projections originally called cross-bridges or sidearms (Wuerker, 1970; Wuerker and Kirkpatrick, 1972) and a number of studies indicate that these structures are composed, at least in part, of polypeptides that originate within the neurofilament. It is likely that the sidearms are highly dynamic, for among other functions they must accommodate local changes in volume as a result of organelle movement.

The sidearms space neighboring filaments apart by between 200 and 400 Å (Hiro-

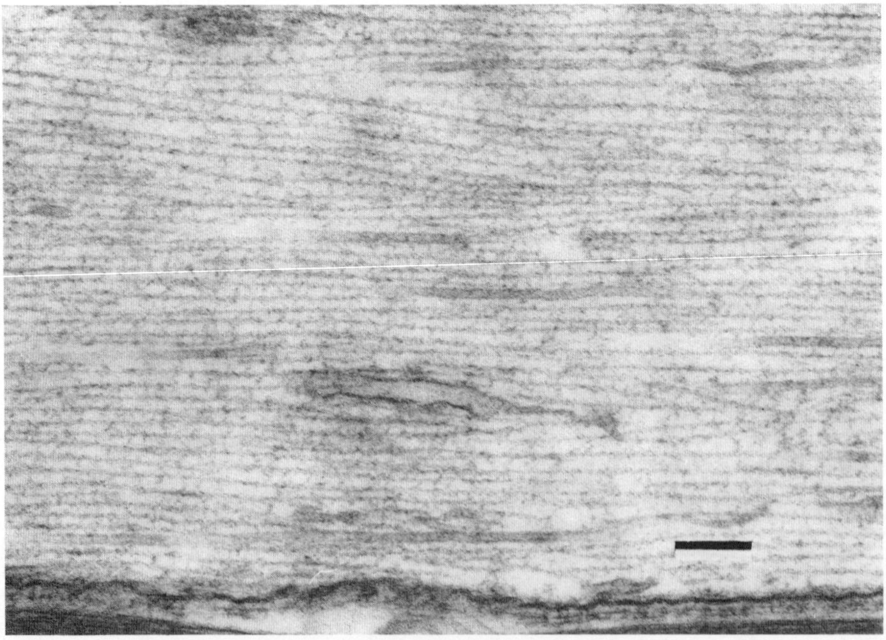

Figure 3. Longitudinal section of spinal nerve root from cat. Neurofilaments lie roughly parallel in such sections with wispy sidearms projecting from the backbone of the filament. At the bottom of the picture is the axolemma and myelin sheath. Bar = 0.2 μm.

kawa, 1982; Schnapp and Reese, 1982) and are themselves 30–50 Å in diameter. They are arranged along the filament with varying degrees of regularity; the distances between them are around 200–500 Å (Hirokawa, 1982; Wuerker, 1970; Yamada *et al.*, 1971; Weiss and Mayr, 1971; Schnapp and Reese, 1982; Hisanaga and Hirokawa, 1988) (Fig. 3). Occasionally, a herringbone pattern in their arrangement can be discerned in longitudinal sections. Physical maltreatment of the nerve can result in the disruption of sidearms, and under these conditions filament densities as high as 1300 neurofilaments/μm^2 can be obtained (unpublished work). It would not be unreasonable to suppose from these observations and from their general appearance that the sidearms have a major role in governing the spatial arrangements of neurofilaments within the neuron.

2.1. *Isolation of Neurofilaments*

The history regarding the isolation and composition of mammalian neurofilaments is tortuous and confusing. It is now clear that some of the earlier studies, reporting neurofilament isolation, in fact were demonstrating the isolation of another class of 100-Å filament, more stable under the isolation conditions used than the neurofilament, which arose from astrocytes (e.g., Shelanski *et al.*, 1971; Davison and Winslow, 1974; Liem *et al.*, 1977; Goldman *et al.*, 1978). The astrocytic filaments are composed of glial fibrillary acidic protein, which has a molecular weight around 50k; thus, for a period of some years the neurofilament subunit was thought to be of this size. A number of other studies (e.g., Hoffman and Lasek, 1975; Schlaepfer and Freeman, 1978; Liem *et al.*, 1978), however, have subsequently shown that the mammalian neurofilament is composed of three major polypeptides with apparent molecular weights around 200k, 160k, and 70k called the triplet (Hoffman and Lasek, 1975), and this is found in nerves of both the central and peripheral nervous systems. The rather complex literature regarding the isolation and identification of the mammalian neurofilament polypeptides has been well reviewed (see Norton and Goldman, 1980; Schlaepfer, 1983; Chiu *et al.*, 1983; Williams and Runge, 1983) and readers are referred to these articles for more detailed discussions.

For the isolation of the mammalian triplet from tissue of the CNS, a range of procedures and modifications to them have been developed by various laboratories (Chiu and Norton, 1982; Chiu *et al.*, 1980; Shecket and Lasek, 1980; Hogue-Angeletti *et al.*, 1982; Brown *et al.*, 1981; Runge *et al.*, 1981c; Liem, 1982). Similar methods are available for neurofilament isolation from the PNS (Schlaepfer and Freeman, 1978; Mori and Kurokawa, 1979, 1980; Carden and Eagles, 1983). The procedure adopted depends to a large extent on the final use of the protein. If one is interested in polypeptides, then a quite crude preparation of filaments will suffice with subsequent purification being done under denaturing conditions in urea or guanidine hydrochloride in order to isolate subunits (Hogue-Angeletti *et al.*, 1982; Moon *et al.*, 1981; Tokutake, 1984; Liem and Hutchison, 1982; Geisler and Weber, 1981). Different approaches are obviously needed if one is studying native filament architecture, when great care must be taken to limit effects of denaturation or disassembly (Schlaepfer and Freeman, 1978; Carden and Eagles, 1983).

Some procedures using CNS tissue yield filaments heavily contaminated by tubulin and associated proteins (Runge *et al.*, 1981c) and this is a direct result of the strong affinity of microtubules for these filaments. The association during isolation was first noticed by Delacourte who subsequently exploited it as a means for isolating large quantities of neurofilaments (Delacourte *et al.*, 1980; Thorpe *et al.*, 1979). Many laboratories have subsequently adopted his procedure as an initial step toward the preparation of pure polypeptides (e.g., see Zackroff *et al.,* 1982; Geisler and Weber, 1981; Lifsics and Williams, 1984).

For the isolation of intact, native filaments, peripheral nerve is often less troublesome to use, because of the lower content of cells bearing other 10-nm filaments, though even here preparations of triplet proteins nearly always have other polypeptides associated with them, normally with molecular weights below 70k. These associated polypeptides are unlikely to be due to proteolysis during preparation because they are present even when protease inhibitors are added. One possibility is that they represent distinct proteins having a strong interaction with the filament, which have copurified. α-Internexin is an example of this (Pachter and Liem, 1985). As some of the associated polypeptides react with neurofilament antibodies, another interesting possibility is that they may in fact be proteolytic products that have arisen as a result of proteolysis occurring to the filament in the intact nerve before isolation (Dahl, 1981; Calvert and Anderton, 1982; Brown *et al.,* 1983; Schlaepfer *et al.,* 1984).

2.2. Filament Solubility and Structure

After purification, neurofilaments are highly insoluble, extremely stable, and very inert structures. They readily entwine forming a thick gel that can exist unchanged for a year or more (Carden and Eagles, 1983). Their inherent stability and insolubility after isolation is presumably a reflection of their role within the nerve where in order for them to maintain a relatively durable cytoskeletal structure they must remain refractory to disassembly under changing physiological conditions.

The solubility of purified filaments depends on pH and ionic strength (Schlaepfer, 1978; Zackroff *et al.,* 1982; Carden and Eagles, 1983). They are least soluble at physiological pH and most soluble at extremes of pH (< pH 4, > pH 9). Filaments also disassemble into soluble units at low ionic strength and these units reassemble when the ionic strength is raised (Zackroff *et al.,* 1982; Carden and Eagles, 1983)—a property that has been exploited in some of the isolation procedures. The precise nature of the structural units that exist under low-ionic-strength conditions is unclear and they are likely to be heterogeneous. Studies show that some must contain at least two sorts of polypeptide (Carden and Eagles, 1986). Varieties of NFM and NFH have also been isolated as soluble components (Cohlberg *et al.,* 1987).

Circular dichroism carried out on intact, isolated neurofilaments indicates a high α-helix content of around 45%. This aggregate value is formed from the individual values of the three major polypeptide components each of which has the capacity to form α helices. The α helices in these filaments can be shown by x-ray diffraction to be arranged in the form of coiled coils that lie roughly parallel to the axis of the filament. They are thought to form the major structural units in the neurofilament core (Wais-Steider *et al.,* 1983).

Negative staining of isolated mammalian neurofilaments shows them to be many micrometers in length; their size appears to vary considerably with the conditions (Schlaepfer, 1977a, 1978). Stain does not readily penetrate the filament's center, suggesting that if it is hollow in nature, as implied by some studies (Schmitt, 1968a; Wuerker, 1970), then the central cavity must be very small and smaller than that which exists in keratin where stain does apparently gain access (Steinert, 1981). It is interesting that in negative stain there is no evidence of the wispy sidearms that are clearly seen in sectioned material from whole nerves and from shadowing of isolated filaments (Ip, 1986; Hisanaga and Hirokawa, 1988). The absence of projections from isolated neurofilaments under negative staining conditions would suggest that the sidearms are either not well contrasted by the technique or that they have collapsed against the filament backbone thus obscuring their detailed structure. Indeed in negative stain the surface of the filament is clearly not well ordered (Schlaepfer, 1977a; Milam and Erickson, 1982; Wais-Steider *et al.*, 1983) and optical diffraction shows no consistent features (unpublished work), which suggests a random orientation of polypeptides on the filament's surface.

Early electron microscopic studies (Wuerker, 1970) indicated that mammalian neurofilaments were composed of four subfilaments or protofilaments, though the number has not been substantiated. The main reason is that protofilaments are hard to visualize by negative staining under normal conditions. Denaturing agents like urea must be used in order to reveal them (Schlaepfer, 1977a) and unfortunately a variety of structures with many ill-defined shapes are produced at the same time.

It is also not clear whether the subfilaments or subunits are helically arranged within the filament. Milam and Erickson (1982) showed by shadowing neurofilaments an axial periodicity of 220–230 Å. As the repeats were almost at right angles to the filament axis, no handedness could be established; thus, if they do in fact reflect a helical period, then this particular helix has a shallow pitch. Perhaps the strongest argument, however, for believing that neurofilaments have some helical symmetry comes from studies where antibodies to individual polypeptides have been used to enhance the underlying periodicities. Decoration of filaments with antibodies to NFH often shows a discontinuous and helical arrangement with a repeat around 1000 Å. Antibodies to NFM also decorate in a periodic and sometimes helical manner, though the helicity is less marked here than with antibodies to NFH (Willard and Simon, 1981; Sharp *et al.*, 1982).

Antibody decoration studies on isolated mammalian neurofilaments were first done by Schlaepfer (1977b) using antisera raised to a 70k protein, and this study, and subsequent work, established NF70 as a neurofilament component. Further investigations with antibodies to each triplet component have shown that all filaments can be decorated (Willard and Simon, 1981; Schlaepfer *et al.*, 1981a; Shaw and Weber, 1981); therefore, all three polypeptides are thought to be arranged together within the neurofilament. Whether a precise ratio is needed for these components within naturally occurring filaments is not too clear at present though it is probably unlikely (Hirokawa *et al.* 1984).

2.3. *Mammalian Neurofilament Polypeptides*

Estimates for the molar ratio of the neurofilament polypeptides vary widely from around 2 : 2 : 1 for NFL : NFM : NFH (Moon *et al.*, 1981) to 9 : 2 : 1 (Mori and Kurokawa, 1980) with numerous intermediate values (Shecket and Lasek, 1980; Chiu *et al.*, 1980; Brown *et al.*, 1981; Chiu and Norton, 1982; Scott *et al.*, 1985). Two problems arise when trying to determine these ratios. First, the characteristics of dye binding by the various polypeptides in SDS are different. Fast green, for example, binds to all three polypeptides in approximately equal amounts (per milligram of protein). Coomassie blue, on the other hand, binds the NFL and NFM components to almost the same extent whereas its association with the NFH component is by comparison reduced to about half—a situation that could lead to a severe underestimate in the amount of the high-molecular-weight component (Chiu and Norton, 1982). The second problem concerns the values to be used for the molecular weights of the two highest triplet components.

The molecular weights of the triplet polypeptides have been the subject of controversy because SDS-PAGE, which was often the sole means of calibration, is now known to overestimate considerably the sizes of NFM and NFH (Kaufmann *et al.*, 1984; Scott *et al.*, 1985) (see Fig. 5). The technique only slightly inflates the size of the NFL component which, based on amino acid analysis, has a weight of around 62k (Geisler *et al.*, 1985a). The aberrant mobility in SDS for the two highest molecular weight components results most probably from reduced binding of SDS to these polypeptides together with peculiar conformational changes (e.g., see Carden *et al.*, 1985). The binding of SDS may be impaired by the extensive network of negative charges on these polypeptides, which is formed in part by large numbers of phosphoserine residues and in part by numerous clusters of glutamic acid residues (Jones and Williams, 1982; Julien and Mushynski, 1982; Geisler *et al.*, 1983). The effect of phosphorylation on the apparent molecular weight of the protein in SDS is easily seen by the significant increase in mobility after enzymatic removal of phosphate (Julien and Mushynski, 1982; Carden *et al.*, 1985). A similar increase in mobility is observed after chemical modification to amino acids in order to reduce the overall charge on the polypeptides (Georges and Mushynski, 1987).

Because of the anomalous behavior of the triplet polypeptides in SDS, more reliable estimates for molecular weights have been sought by using analytical gel filtration and sedimentation equilibrium. Both techniques estimate the molecular weight of the NFL polypeptide as being 60–70k (Kaufman *et al.*, 1984; Lifsics and Williams, 1984), similar to the values obtained by SDS-PAGE and in good agreement with values derived from sequence data; estimates of the weights of the other two triplet components are 30–40% lower than their values derived from SDS-PAGE. Compared with estimates obtained by gel filtration, more reliable values come from sedimentation equilibrium as here no assumptions need be made regarding the conformational state of these proteins, which often behave abnormally on molecular sieves (Lifsics and Williams, 1984).

Sedimentation equilibrium techniques have allowed estimates of around 108k for the molecular weight of NFM (Kaufmann *et al.*, 1984; Cohlberg *et al.*, 1987), in fair agreement with values of 95–100k obtained from available amino acid sequences (Levy *et al.*, 1987; Napolitano *et al.*, 1987; Myers *et al.*, 1987). From the amino acid sequence of NFH (human), its molecular weight is calculated as 110k (Lees *et al.*, 1988), which is almost half the value found by SDS-PAGE. By using sedimentation equilibrium techniques, Kaufmann *et al.* estimated the molecular weight of NFH to be 112k (Kaufmann *et al.*, 1984) and Cohlberg *et al.* produced a value of 136k (Cohlberg *et al.*, 1987).

Many studies have reported on the differences in apparent molecular weights in SDS of the triplet polypeptides (e.g., see Eagles *et al.*, 1980b; Chiu *et al.*, 1980; Shaw *et al.*, 1984) for neurofilaments isolated from various mammalian species though how real these differences are is clearly uncertain. Differential phosphorylation is not the cause (Kaspi and Mushynski, 1985). As extensive structural homology exists between corresponding triplet components from mammalian species, by peptide mapping (Eagles *et al.*, 1980b; Chiu *et al.*, 1980; Brown *et al.*, 1981), antibody labeling (Shaw *et al.*, 1984), and sequencing (Geisler *et al.*, 1985a; Lewis and Cowan, 1985; Julien *et al.*, 1985; Levy *et al.*, 1987; Napolitano *et al.*, 1987; Myers *et al.*, 1987), large variations in polypeptide sizes would be unexpected. Evidence currently available for NFM supports this notion and shows that the protein from mouse has a size of 96k (848 amino acids) (Levy *et al.*, 1987), that from rat a size of 95.6k (844 amino acids) (Napolitano *et al.*, 1987), and that from human a size of 102k (915 amino acids) (Myers *et al.*, 1987).

2.4. *Reassembly Properties*

After isolation in urea, the triplet components can re-form their secondary structure to some extent and in certain cases homopolymeric filaments can be formed that are similar in overall morphology to naturally occurring neurofilaments. Initial studies implicated a role for ATP in the assembly process (Moon *et al.*, 1981; Liem and Hutchison, 1982) though this is now known not to be involved. The best documented example of reassembly is with NFL where circular dichroism studies show that it can renature to contain around 50–79% α helix (Wais-Steider *et al.*, 1983; Lifsics and Williams, 1984). Under suitable conditions of dialysis, many groups have shown that NFL polypeptides, isolated by a variety of procedures, can associate together to form filaments around 100 Å in diameter though of quite variable lengths, O.2 μm to 1 μm (Geisler and Weber, 1981; Moon *et al.*, 1981; Zackroff *et al.*, 1982; Liem and Hutchison, 1982; Gardner *et al.*, 1984; Minami *et al.*, 1984; Lifsics and Williams, 1984; Hisanaga and Hirokawa, 1988). The importance of these experiments is that they demonstrate quite unambiguously that not all of the triplet polypeptides are necessary in order to form a filament. Interestingly, the filaments formed from purified NFL when shadowed reveal a 200 Å periodicity reminiscent of that seen in native neurofilaments (Henderson *et al.*, 1982; Lifsics and Williams, 1984), indicating a similar polypeptide packing arrangement in these two cases. To achieve this similarity in filament size and packing arrangement between the homopolymer of NFL and normal heteropolymeric neurofilaments, NFL must substitute structurally for the other two triplet components but to what extent this occurs is not established.

After isolation, NFM and NFH can also renature and re-form some of their secondary structure. Estimates for the α-helix content vary between 0 and 50% (Lifsics and Williams 1984; Delacourte *et al.*, 1982; Wais-Steider *et al.*, 1983; Cohlberg *et al.*, 1987), and the presence of β structure in the renatured proteins has also been indicated by circular dichroism. Some of the variability in renaturation may be due to the long C-terminal extensions associated with these polypeptides which could interfere with the reassembly process. Despite earlier unsuccessful attempts, NFM has now been shown to re-form into respectable 100-Å filaments (Gardner *et al.*, 1984) though efforts at reassembling the NFH polypeptide into long filaments have so far failed and only short amorphous aggregates have been observed (Tokutake *et al.*, 1984). In the presence of NFL, both NFM and NFH can reassemble into filaments that are morphologically similar to naturally occurring neurofilaments (Geisler and Weber, 1981; Hisanaga and Hirokawa, 1988).

2.5. *Amino Acid Composition and Sequence Studies*

The amino acid composition of the triplet polypeptides is now well established having been determined independently in a number of laboratories (Chiu and Norton, 1982; Hogue-Angeletti *et al.*, 1982; Geisler *et al.*, 1983). The polypeptides are characterized by a high content of acidic residues especially glutamic acid, which amounts to some 20% of the total composition in all three (Fig. 4). This in turn means that the

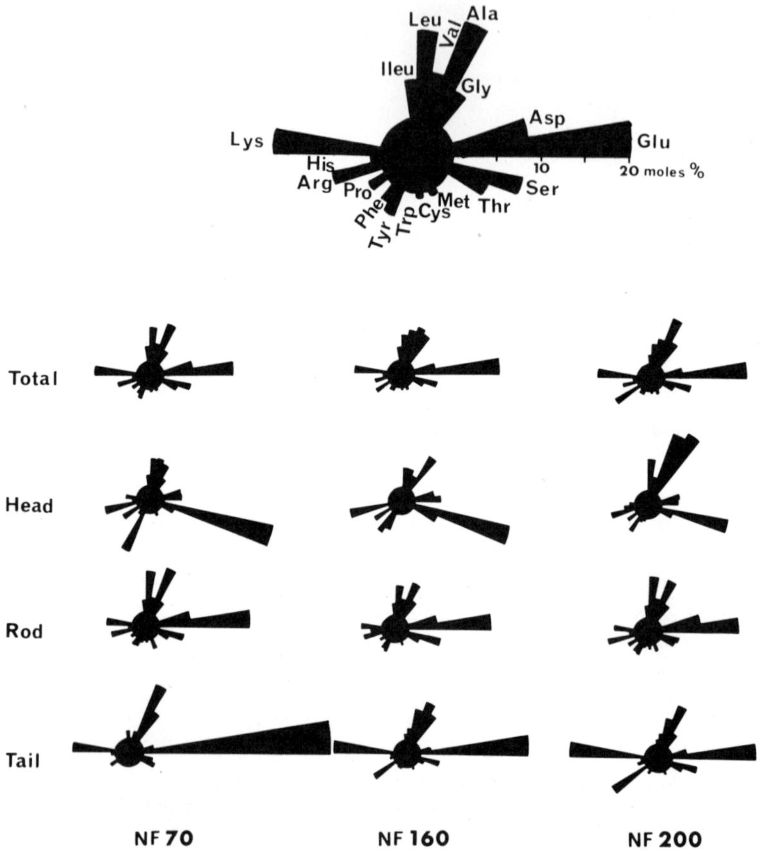

Figure 4. Diagrammatic representation of the amino acid composition of the neurofilament triplet polypeptides and various regions from these polypeptides. The key, which corresponds to the composition of NFL, is shown at the top, where amino acids are grouped according to their properties. For example, basic amino acids like lysine, histidine, and arginine are located on the left of the figure, small aliphatic amino acids like isoleucine, leucine, valine, alanine, and glycine are grouped at the top, and acidic amino acids, aspartic acid and glutamic acid, are found on the right. The amounts of each amino acid in mole percent are indicated by the lengths of the sectors. Therefore, in the example shown, for the amino acid composition of NFL, glutamic acid constitutes about 20% of the total amino acids and aspartic acid constitutes about 8%. The total compositions for NFL, NFM, and NFH are shown in the top row. These data were obtained from the compositions published by Hogue-Angeletti *et al.* (1982). The compositions of the N-terminal head regions of each of the triplet polypeptides are shown in the next row, followed by the rod regions and the tail regions of the polypeptides. These data were obtained from Geisler *et al.* (1983). The composition of the tail region of NFL shown here with the preponderance of glutamic acid, corresponds to the b domain. The similarity in composition of the rod regions from each of the triplet polypeptides is clearly seen and reflects the general characteristics of α helices. The N-terminal regions have a preponderance of serine residues and the C-terminal regions are highly charged with an abundance of lysine and glutamic acid.

isoelectric points for the triplet lie between pH 5 and 6 though quite often the polypeptides streak on isoelectric focusing gels, indicating charge heterogeneity. Lysine is the major basic amino acid and in each is around 10% (Fig. 4). Of the three components, NFH has the highest proline content, around 8%, and this may indicate a distinctly different conformation for parts of this protein compared with the other two.

The formidable and exacting task of sequencing the polypeptides was begun in Weber's laboratory using proteins isolated from pig. From this work the complete sequence of NFL is now available (Geisler *et al.*, 1985a). cDNA clones encoding the NFL polypeptide from mouse (Lewis and Cowan, 1985; Julien *et al.*, 1986) and rat (Julien *et al.*, 1985) have also been described. When compared with the procine polypeptide of equivalent size, analyses of the sequences show that the proteins have considerable similarity, with > 90% homology.

A number of major discoveries have emanated from these studies especially when secondary-structure-prediction methods are applied to the individual sequences. To take NFL first, three distinct regions of the protein can be identified. Initially is the N-terminal region, which covers approximately 80 residues and which is very basic in character. It contains many hydroxy amino acids, and predictions from the sequence indicate a propensity toward forming β structure. Joining the N-terminal sequence is the next region called the rod domain. Approximately 310 amino acids form this particular sequence, the signal feature of which is the heptad repeat of amino acids (A B C, D E F G) where positions A and D are frequently occupied by nonpolar residues. When assembled into an α helix, the nonpolar residues form a seam along one side of the structure. Thus, if these helices are placed in an aqueous environment, there will be a strong thermodynamic tendency for them to associate along this seam. The result is the formation of a coiled coil, for the seam is at an angle to the axis of the helix. It is the rod region of the molecule that is thought therefore to participate in coiled-coil assemblies. In common with other intermediate filaments, the rod may be divided into four helical segments, Ia, Ib, IIa, and IIb, with distinctive regions separating each where interruptions in coiled-coil-forming ability are to be found (Parry and Fraser, 1985; Fraser *et al.*, this volume). The rod of NFL has been isolated by chymotryptic digestion, and circular dichroism shows that it contains between 70 and 80% α helix, supporting the predictions made about the structure for this part of the molecule. Following on from the rod is the C-terminal extension, approximately 17k in molecular weight and peculiar in that nearly half of the residues are glutamic acid (Fig. 4). These residues seem to be confined to the extreme end of the molecule where the structure would look and behave like polyglutamic acid.

From what is known about the sequences of NFM and NFH, they are similar in design (Geisler *et al.*, 1983, 1984, 1985b; Levy *et al.*, 1987; Napolitano *et al.*, 1987; Myers *et al.*, 1987; Lees *et al.*, 1988). They have an α-helical rod domain flanked on one side by a basic N-terminal region and on the other by an acidic C-terminal extension wherein lies the major differences between the triplet components, because the C-terminal regions have vastly different sizes (Fig. 16).

Like the corresponding region in NFL, the C-terminal extensions are highly charged being composed of much lysine (15–26%) though the preponderance of glutamic acid (20–30%) (Fig. 4) probably confers on them a net negative charge. These regions are also rich in proline, which constitutes around 10% of the total amino acids, and this fact together with the low content of hydrophobic amino acids argues against extensive helical and coiled-coil domains being located here at least under physiological conditions. This conclusion is also supported by circular dichroism studies on the isolated regions (Geisler *et al.*, 1983). Details of the region are seen in the complete sequences of NFM from mouse (Levy *et al.*, 1987), rat (Napolitano *et al.*, 1987), and human (Myers *et al.*, 1987) and of NFH from human (Lees *et al.*, 1988) and partial sequences of NFH from rat (Dautigny *et al.*, 1988). Following the rod region is a sequence with a high proline content which leads to a glutamic-acid-rich domain. Then occur a variety of repeats characterized by the motif serine followed by proline with serine often preceded by lysine and proline often followed by glutamic acid or valine. In human NFH the serine–proline motif occurs 41 times.

Knowledge of the amino acid sequences for neurofilament polypeptides has helped in the identification of regions where phosphorylation occurs, where the anti-IFA epitope is located and where Bodian's silver stain is thought to react. Dealing first with phosphorylation, studies show that the two largest neurofilament polypeptides are highly phosphorylated molecules and analyses indicate that the C-terminal extensions contain most of the phosphate. The phosphate on NFH is found mainly as phosphoserine and these residues are unevenly distributed along the length of this region (Julien and Mushynski, 1983; Geisler *et al.*, 1985b; Carden *et al.*, 1985). The phosphate in NFM is thought to be present as phosphothreonine as well as phosphoserine (Julien and Mushynski, 1982; Carden *et al.*, 1985). Isolation and sequencing of the phosphorylated peptides from procine neurofilaments demonstrate that the phosphate is associated with the serine–proline motif (Geisler *et al.*, 1987). This motif occurs in a degenerate sequence that is also highly repetitive (Robinson *et al.*, 1986; Geisler *et al.*, 1987). Moreover, antibodies to phosphorylated epitopes of neurofilaments also bind the sequences (Lee *et al.*, 1988). Thus, it is thought that the motif incorporates the major phosphorylation sites on the molecules.

We now consider the region to which the antibody "anti-IFA" binds. It is well established that this antibody binds not only to each of the triplet components but also to all other intermediate filament polypeptides (Pruss *et al.*, 1981; Pruss, 1985). As it is the rod region of the molecule that is common structurally among these proteins, it was perhaps not too surprising to find that the epitope for the antibody is located here. In fact, studies show that the antibody binds near the C-terminal end of the rod (see Fig. 16) and its epitope has now been defined to within about 20 amino acids (Geisler *et al.*, 1983).

Unlike anti-IFA, Bodian's silver stain is specific in its action for the neurofilament polypeptides and does not stain other intermediate filaments (Gambetti *et al.*, 1981; Phillips *et al.*, 1983). Because of its selective staining, it has been widely used by neuroanatomists to trace axonal pathways. From its selectivity, it might be supposed that Bodian's stain would interact with regions unique to neurofilament proteins and, for the triplet, prime candidates for these regions are the characteristic C-terminal extensions. Recent studies by Gambetti's group have localized the binding region for the stain and it is indeed found in these domains (Autilio-Gambetti *et al.*, 1986).

2.6. Arrangement of the Triplet Polypeptides within the Neurofilament

The most striking difference between neurofilament polypeptides and the polypeptides from other intermediate filaments is their size. It is now known that for the triplet proteins the extra mass is largely accounted for by an increased size of the C-terminal regions, which are thought to reside at the filament's surface. This situation is similar to the arrangement currently believed to occur for other intermediate filaments where the C-terminal domains, though much smaller than for neurofilaments, are exposed near the filament's surface and are thus readily digested off by proteases (Steinert *et al.*, 1983). Therefore, regarding the arrangement and disposition of neurofilament polypeptides, much of the thinking in this area has been influenced by studies on other intermediate filaments about which more is known (Weber and Geisler, 1985). Apart from this, our current understanding of the arrangement of the triplet polypeptides is based on work from sequence analysis, antibody labeling, reassembly of the neurofilament polypeptides, proteolytic digestion, and cross-linking. The majority of the studies to date strongly support the conclusion that the neurofilament can be divided into two quite distinct areas: a backbone, or core, surrounded by a peripheral region. For at least the two largest triplet polypeptides, as much of the protein seems to be exposed peripherally as is sequestered in the backbone and in view of this the peripheral region may largely determine the nature of the interactions that can occur between neurofilaments and other cellular organelles. At present, the major unanswered questions regarding neurofilament architecture revolve around elucidating the conformation and arrangement of the polypeptides in these two distinct areas of the filament.

Like other intermediate filaments, the neurofilament has a backbone composed in large part of α helices. Sequencing studies have shown that rodlike domains, containing this α helix, are present in each of the triplet polypeptides (Geisler *et al.*, 1983), and x-ray diffraction indicates that α helices form coiled coils that are aligned parallel to the sides of the filament (Wais-Steider *et al.*, 1983). The strong implication of these findings is that the rod regions of all three polypeptides are associated with the backbone of the filament.

It was originally suspected that the neurofilament core was composed essentially of one type of polypeptide, NFL. The reasons for this were as follows: first, after isolation NFL could readily be assembled into a filament though no filaments could be obtained, at least at that time, from purified NFM or NFH (Geisler and Weber, 1981). Second, antibodies raised against each of the triplet components labeled the filaments in a characteristic and distinct pattern: antibodies against NFL bound to the backbone of the filament more or less uniformly whereas antibodies directed against NFM and NFH labeled peripheral regions (Willard and Simon, 1981; Sharp *et al.*, 1982). More recently, however, there has been a change in emphasis for the roles of the two higher molecular weight triplet components and the view is now expressed that the rod regions of these polypeptides most probably anchor them as coiled coils located in the backbone (Geisler *et al.*, 1983; Chin *et al.*, 1983; Minami *et al.*, 1984). Part of the reason why NFL is now thought not to be the only component in the backbone, is that purified NFM will itself form long filaments (Gardner *et al.*, 1984).

Some of the most direct evidence for the participation of NFM and NFH in the backbone comes from proteolytic digestion studies using trypsin, chymotrypsin, and V8 protease (Chin *et al.*, 1983; Julien and Mushynski, 1983). Trypsin digestion leaves NFL relatively unimpaired though most of NFH and NFM are cleaved under these conditions. Despite extensive proteolysis, many filaments, apparently quite native-looking, are still present after this treatment (Fig. 5). During digestion, NFH and NFM are cleaved into peptides some of which are highly phosphorylated and soluble during sedimentation, and some of which are enriched in α helix, have molecular weights around 40–50k, and sediment with filaments (see Fig. 5). These latter proteins bind anti-IFA, and thus are most likely to be the remains of the rod regions from the NFH and NFM polypeptides (Eagles and Maggs, 1985). Digestion with chymotrypsin and V8 protease produces a similar picture. Thus, these studies support the idea that the backbone of the neurofilament is composed not only of NFL but also of the rod regions of the two other triplet polypeptides. This is expressed diagrammatically in Fig. 17.

Information on the arrangement of the triplet polypeptides has also been obtained from cross-linking studies. All of the triplet components contain cysteine and some of the cysteine residues are located within the rod regions. Cross-linking of these cysteine residues into disulfide bonds appears to occur naturally in filaments though these bonds are not necessary for filament stability as they can be cleaved with mercaptoethanol

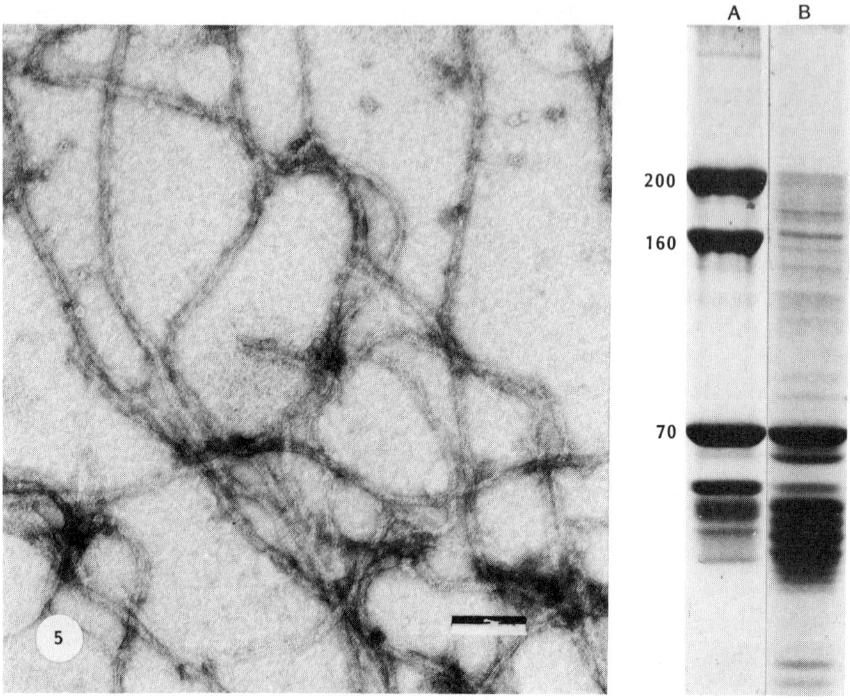

Figure 5. Proteolytic digestion of ox neurofilaments. The neurofilaments were isolated from ox nerve roots and lane A shows their composition. The major polypeptides, NF200, NF160, and NF70, are marked. Digestion with trypsin results in loss of the two highest molecular weight polypeptides with the generation of many fragments that stay attached to filaments and have molecular weights around 40–50k. The gel of these digested filaments is shown in lane B; electron microscopy of the sample that is shown here indicates that filaments are still present after proteolytic digestion and they look surprisingly similar to undigested samples. Data from Chin *et al.* (1983). Bar = 700 Å.

without destroying the filament's integrity (Carden and Eagles, 1983). Chemically cross-linking the cysteines with copper orthophenanthroline has provided useful information on polypeptide disposition, and proteolytic digestion of cross-linked filaments has shown that most of the cysteines that are cross-linked are located within the backbone of the filament (Carden and Eagles, 1983, 1986). A prominent feature of the cross-linking patterns is the self-association of NFL via its single cysteine residue (see Fig. 16). This residue would be located toward the seam of the coiled coil, thus favoring the possibility of these polypeptides being cross linked in pairs. It is highly likely that the polypeptides that cross-link in this way are arranged in a parallel manner with the chains in precise axial register (Parry and Fraser, 1985; Steinert and Parry, 1985). Not all NFL molecules are arranged like this as they do not all cross-link together and it is possible that some of the rod regions of NFL molecules may also lie in close association with the rod regions of NFM or NFH. Support for this proposal comes from studies where cross-links between NFM and NFL can be formed if the packing restraints on these polypeptides are relaxed, which occurs after dissociation of the filament at low ionic strength (Carden and Eagles, 1986).

Less is known about the polypeptides that contribute to the peripheral domains, though by analogy with other intermediate filaments both the N-terminal and C-terminal

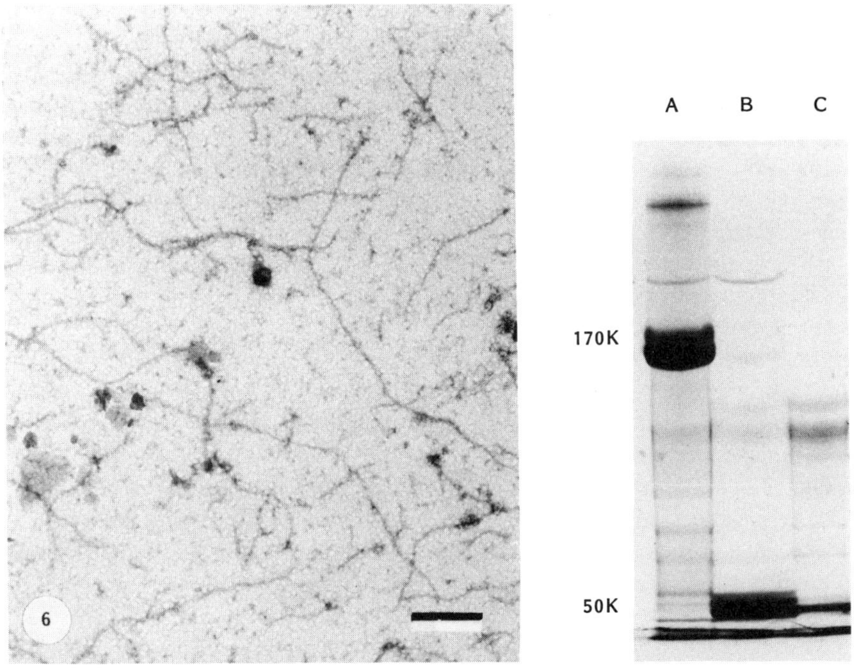

Figure 6. Proteolytic digestion of *Myxicola* neurofilaments. The neurofilaments were initially purified by washing and are shown in lane A. Two major polypeptides, the largest of which has a molecular weight on SDS gels of around 170k, are seen to compose the filaments together with some minor proteins. Digestions of these filaments with chymotrypsin results in cleavage of the major polypeptides. Centrifugation of the sample sediments filamentous material and this is shown in lane B where it is seen to be composed of fragments of around 40–50k. The corresponding supernatant is shown in lane C. Electron microscopy of the digested sample (opposite) shows that after all of the main chains have been cleaved, filaments still exist, and thus these filaments must be made largely of the fragments of 40–50k. These fragments bind anti-IFA and are thought to be analogous to the rod regions of the triplet polypeptides. Essentially similar results are obtained using the calcium-activated protease (Eagles *et al.*, 1981b). Bar = 0.24 μm.

sequences are likely to be involved. The major part of the peripheral region is thought to be formed from the long C-terminal extension of the triplet components and, as previously described, these sequences are singularly rich in charged residues, particularly glutamic acid. This composition together with the observation that most of the phosphate associated with NFM and NFH is also located here means that an extensive network of charges surrounds the filament.

In order to understand better the function of these peripheral regions, they must be isolated and this has been done for the C-terminal regions of NFM and NFH by treating neurofilaments with proteases (Julien and Mushynski, 1983; Chin *et al.*, 1983) and separating the fragments by DEAE chromatography (Chin *et al.*, 1989; see also Milam and Erickson, 1985). Studies show that the C-terminal domains of NFM and NFH, prepared by chymotryptic digestion, behave as discrete highly soluble particles possessing no tendency in either their phosphorylated or dephosphorylated states to form strong interactions. Thus, they are unlikely to be associated together in the intact filament. Equilibrium sedimentation indicates that they have molecular weights around 40k (NFM) and 60k (NFH), considerably lower than the estimates obtained by SDS-PAGE. After denaturation is guanidine hydrochloride, these values are unchanged, suggesting that the molecules are probably monomeric—a surprising situation since their anchoring domains in the filament core are presumably associated in dimeric or tetrameric complexes (Chin *et al.*, 1989). Information on the conformation of these isolated regions is meager. They are not well contrasted by negative staining techniques, but shadowing does reveal something of their structure when a heterogeneous population of elongated molecules can be seen. The dimensions of the particles vary widely though many have diameters around 50 Å with lengths of 200–400 Å (Eagles *et al.*, 1985)—not unlike the dimensions for the sidearm structures that are seen between neurofilaments in intact nerves.

Are the major peripheral C-terminal regions of the neurofilament triplet polypeptides involved in the sidearms? The evidence, though circumstantial, is highly suggestive that they are, at least for NFH and probably also for NFM.

First, proteolytic digestion of filaments causes degradation of both NFM and NFH and results in the release of soluble phsophorylated polypeptides from the filament, the backbone of which seems after this treatment to be morphologically intact. Second, the peptides removed from the filament have properties expected of neurofilament sidearms. For example, they are highly soluble under physiological conditions, are difficult to observe by negative staining, recognizable by shadowing, and of about the right size. Third, the singular sequences obtained for the C-terminal regions establish these domains as highly distinctive structures, which betokens an equally distinctive function. Fourth, NFM and NFH can associate with NFL to form filaments with long whispy projections 500–600 Å in length as seen by shadowing (Hisanaga and Hirokawa, 1988). NFL alone shows none of these projections, which in size and disposition are similar to those found on natural, isolated neurofilaments, themselves resembling sidearms.

One way of proving identity would be to decorate the sidearms with antibodies whose epitopes were clearly defined and located on the C-terminal regions of the polypeptides. It is evident from work so far that antibodies directed against NFH will label the sidearm structures and that antibodies against NFM are occasionally found at the base of the sidearms (Hirokawa *et al.*, 1984), though as yet we do not know where the epitopes for these antibodies lie. As many of the neurofilament antibodies that have been raised so far appear to bind to phosphorylated sites (see Section 5.2 and Matus, 1988), the pictures obtained hitherto by this antibody decoration technique may well reflect the distribution of phosphorylation sites on the sidearm polypeptides. Further elucidation of these most interesting aspects must await more detailed studies.

3. Invertebrate Neurofilaments

Neurofilaments from invertebrates were among the first intermediate filaments to be characterized (Richards *et al.*, 1943; Schmitt, 1950). In this area, studies have been much aided by using the giant axons of squid and the marine worm *Myxicola*. These axons offer a rich source of axoplasm, relatively free of other contaminating tissues and containing abundant neurofilaments. Though despite these considerable advantages, isolation studies were hampered, at least initially, by the presence in axoplasm of a calcium-requiring protease, specific for neurofilaments. Unless great care is exercised to inhibit the protease, a variety of proteolytic breakdown products of neurofilaments may be isolated.

Within the past decade, a number of detailed structural investigations have been done that have resulted in a much better understanding of the overall polypeptide composition and arrangement in squid and *Myxicola* neurofilaments. The current picture that we have of these filaments is that they are in broad outline like their counterparts in mammalian nerves, though their polypeptide construction appears to be somewhat different.

3.1. Myxicola

The marine fan worm *Myxicola infundibulum* while being an essentially sedentary creature is also capable of short bursts of vigorous contractions if disturbed. This escape response has been well studied (Nicol and Whitteridge, 1955; Roberts, 1962) and the fast conduction needed is mediated by a giant axon in the ventral nerve cord. The axon, which may be up to 1.7 mm in diameter, is connected in each segment to numerous unipolar nerve cells. Thus, it forms a syncytium throughout the length of the animal (Nicol, 1948; Nicol and Young, 1946; Wells *et al.*, 1972).

The giant axon in *Myxicola* is singular in its abundance of neurofilaments, which constitute around 70% of the total axoplasmic protein. Microtubules are also present in the axoplasm and they diffract x rays surprisingly well especially as their distribution is sparse. [On average, one microtubule is found for every 200 neurofilaments; see Fig. 12 and Wais-Steider *et al.* (1987).] The peculiarly high proportion of filaments probably accounts for the ease with which this axoplasm can be extracted from the worm without breaking (Gilbert, 1972); calculations show that the breaking strength of this material is comparable to that of wool (Gilbert, 1975a).

Within axoplasm the neurofilaments are arranged on a hierarchy of helical paths. The filaments themselves lie in synchronous waves that represent a series of locally parallel helices whose axes are twisted like the strands in yarn and whose periods are around 10 μm. This twist gives rise to another helix called the segmental helix, which is right-handed and of much larger pitch, approximately 2 mm. The segmental helix in turn begets the cylindrical gel that coils forming a right-handed gel helix when the worm contracts, and uncoils when the worm extends (Gilbert, 1972, 1975a).

Electron microscopy shows that the density of neurofilaments in *Myxicola* axoplasm is around $190/\mu m^2$ and that the filament arrangement is more open with much less cross-bridging than that exhibited by other systems including vertebrate and squid nerves. The filaments have varying diameters (70 to $>$ 200 Å) with either smooth sides or globular projections—the size and appearance depending much on the conditions used (Fig. 8). Fixation of filaments either *in situ* or after removal from axoplasm generally shows spheres (280-Å diameter) or globular structures along the filament (Schmitt, 1950; Gilbert *et al.*, 1975; Krishnan *et al.*, 1979). Negative staining without fixation on the other hand shows filaments of 70- to 80-Å diameter (Gilbert *et al.*, 1975; Wais-Steider *et al.*, 1983) with much smoother sides unless agents like sucrose are added (Fig. 8). In whole axoplasm the various appearances of these neurofilaments are due in part to the attachment of nonneurofilamentous material, though in more purified preparations the conformational state of the sidearm polypeptides, and in addition, local disruptive and disassembly effects, must contribute almost exclusively to the range of views observed.

Myxicola neurofilaments may be purified by a number of procedures. Washing axoplasm under conditions where neurofilaments are stable removes nearly all other components and offers an easy way for rapid purification. Alternatively, column chromatography, cycles of dissociation and association, or cytochrome c precipitation may be used to purify them (Lasek and Hoffman, 1976; Gilbert *et al.*, 1975; Lasek *et al.*, 1979; Eagles *et*

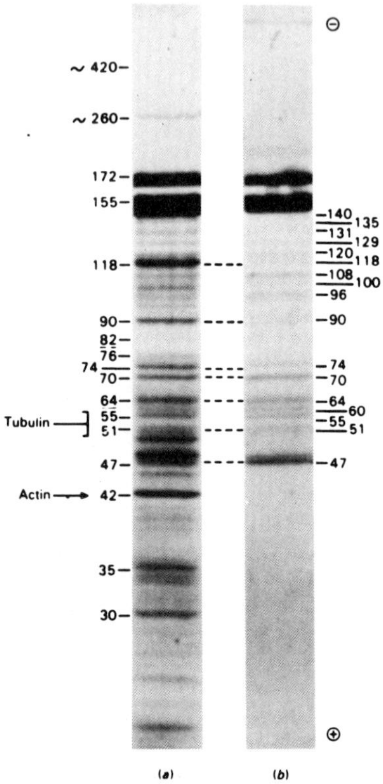

Figure 7. Polyacrylamide-gel patterns of (a) *Myxicola* axoplasm and (b) *Myxicola* neurofilaments. Shown are typical polyacrylamide gels (5–15% acrylamide) of whole axoplasm (a) and of purified neurofilaments (b) that were run in SDS. The neurofilaments were purified by cytochrome c precipitation. The molecular weights of the components are given ($\times 10^{-3}$) together with the regions where actin and tubulin are found. From Eagles *et al.* (1981a).

al., 1981a). Whatever method is used, SDS gel electrophoresis shows a complex polypeptide composition for the purified filaments (see Fig. 7), which consists of two major bands with apparent molecular weights on SDS gels around 170k and 155k together with about 15 minor components (Lasek *et al.*, 1979; Eagles *et al.*, 1981a). Peptide mapping of these components shows that they all contain a peptide that has an apparent molecular weight of around 35k (Fig. 9) situated at one end of the molecule (Eagles *et al.*, 1981a). This end was called the tail end because it appeared to be buried within the filament backbone in a similar manner to the arrangement of the myosin molecules' tail within the myosin filament. (The nomenclature used at this time for *Myxicola* polypeptides is opposite that subsequently employed to describe vertebrate neurofilament polypeptides.) As anti-IFA binds near the 35k fragment (tail region), this region of the molecule is most probably analogous to the rod region of mammalian neurofilaments (Eagles and Maggs, 1985; Eagles *et al.*, 1985). This fact enables the relationship between the *Myxicola* polypeptide and the polypeptides from mammalian neurofilaments to be described diagrammatically as shown in Fig. 16.

The reason for the ladder of polypeptides found in *Myxicola* neurofilaments is unclear. Peptide mapping shows that a simple relationship exists between them and they can be thought of as being derived from the higher molecular weight polypeptides by elimination of fragments from one end of the molecule (Fig. 9 and Eagles *et al.*, 1981a). The ladder is present in axoplasmic samples removed within 10 sec. Therefore, it is likely, but

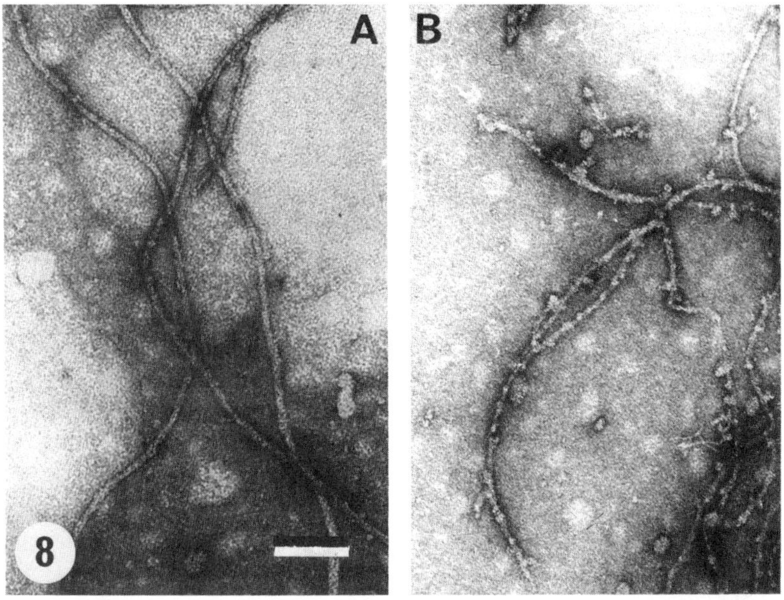

Figure 8. Electron microscopy of Myxicola neurofilaments. (A) Negatively stained *Myxicola* neurofilaments obtained from axoplasm that was homogenized in 50 mM cacodylate buffer pH 6.5 with protease inhibitors. The filaments are long and smooth-sided. Bar = 800 Å. (B) Similar preparation as A but with sucrose added to the buffer. The filaments here have irregular projections on their surface that probably correspond to sidearm domains. Same scale as A.

certainly not proven, that the range of polypeptides that compose it are normal constituents of the filaments *in situ*. How this ladder of polypeptides is formed is unknown though a strong possibility must be that it arises by proteolytic cleavage occurring *in vivo*. It is interesting to note that activation of the naturally occurring, calcium-mediated, protease in axoplasm results in cleavage of the major polypeptides to produce fragments similar in composition to those found in the ladder composing native filaments (Fig. 9). Thus, this protease is a prime contender for the activity. However, why filaments should be degraded in this manner is not at all clear and remains an intriguing puzzle.

The two major bands constituting these filaments are not present in equivalent amounts, the 155k component being about two times more abundant than the higher molecular weight polypeptide. Both proteins have similar amino acid compositions and that for the 155k polypeptide is shown in Fig. 10. The composition is much like that found for the mammalian neurofilament polypeptides (Fig. 4) with glutamic acid composing around 24% of the total composition and with lysine being the most abundant basic amino acid. Both of the *Myxicola* polypeptides are highly phosphorylated with an average content of phosphate around 26 moles/mole (Chin *et al.*, 1989).

The conditions required for disassembly and reassembly of *Myxicola* neurofilaments differ from those found for their mammalian counterparts and unlike these, *Myxicola* filaments dissociate at high ionic strength (> 0.3 M NaCl). Thus, there must be distinct differences in the molecular interactions that hold these two kinds of neurofilaments together. After disassembly, lowering the ionic strength causes filaments to readily assem-

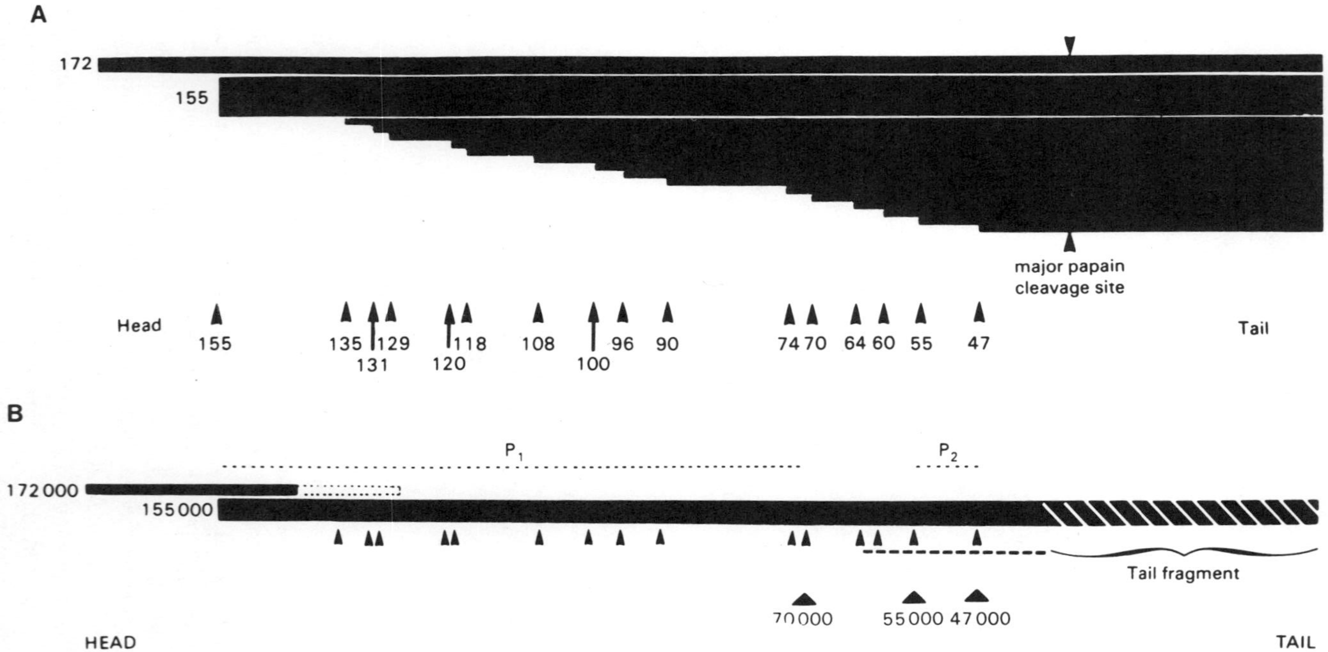

Figure 9. (A) The polypeptide composition of purified *Myxicola* neurofilaments. Diagrammatic representation of the polypeptide composition of purified neurofilaments. Each bar represents a polypeptide the length of which corresponds to its molecular weight (measured from the tail region). The numbers indicate the molecular weight ($\times 10^{-3}$) of the polypeptides. The two arrowheads at the tail represent the sites at which papain cleaves off the 35k fragment. From Eagles *et al.* (1981a). (B) Diagrammatic representation of the location of phosphorylationregions and sites cleaved by the calcium-dependent proteinase on the major neurofilament polypeptides in *Myxicola*. The polypeptides are represented as bars the length of which indicates their molecular weight and the width of which indicates their abundance. The 172k polypeptide is partly shown at the head region; apart from this area it is believed to be identical to the 155k component. The position of the 35k tail-region fragment is hatched. The 15 small arrowheads beneath the 155k polypeptide represent the regions where the minor neurofilament polypeptides terminate. The large triangles indicate the sites where the calcium-dependent proteinase cleaves the molecule. The values 70,000, 55,000, and 47,000 represent the molecular weights of the peptides (measured from the tail) that would be generated by cleavage at these sites. The ^{32}P-labeled regions are marked P_1 and P_2. Most of the label is found in P_1 but the number of residues phosphorylated here have not been determined. From Eagles *et al.* (1981b).

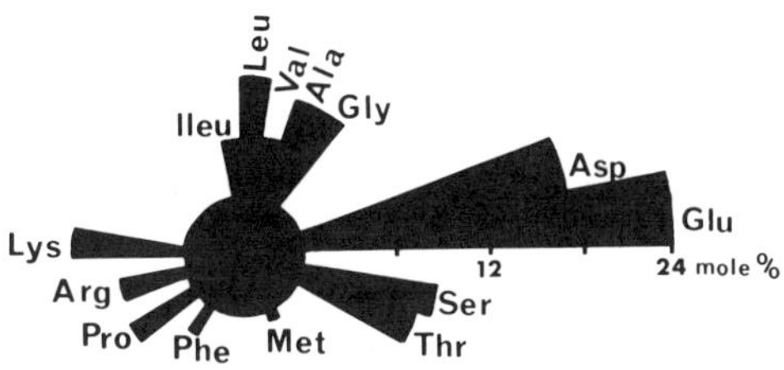

Figure 10. Diagrammatic representation of the amino acid composition of the *Myxicola* 155k neurofilament polypeptide. The key is the same as that used for Fig. 4: the amounts of each amino acid in mole percent are indicated by the lengths of the various sectors. In this polypeptide, glutamic acid thus constitutes around 24% of the total composition. The composition of the larger component of *Myxicola* neurofilaments, having a size of around 172k, is very similar. Eagles and Gilbert (unpublished work).

ble and the processes are quite reversible (Gilbert, 1975b; Gilbert *et al.*, 1975). The unit formed at high ionic strength has been characterized (Bell, 1977). It has a sedimentation value around 6 S and sedimentation equilibrium experiments indicate a mass around 650k. The unit is highly asymmetric and its size is consistent with a structure containing four polypeptide chains, each with a molecular weight of around 160k. Chemical cross-linking investigations strongly support the four-chain model for the unit (Bell, 1978).

The ordered array of neurofilaments in *Myxicola* axoplasm provides an ideal starting material for drawing fibers suitable for x-ray diffraction. By using this material, Day and Gilbert (1972) showed that the major protein component present, the neurofilament, was highly α-helical, with the α helices arranged in coiled coils. Studies with better orientated specimens (Wais-Steider *et al.*, 1987) have shown diffraction occurring from the microtubules as well as neurofilaments, and in these patterns up to 17 meridional or near-meridional reflections, which come from neurofilaments, can be observed (Fig. 12). All of the neurofilament reflections index on a repeat close to 252 Å and some of them, e.g., those at spacings corresponding to 18 and 25.2 Å, are similar in intensity to reflections at the same positions from keratin. To what extent the two surface lattices of these filaments are similar is uncertain.

X-ray diffraction can also be performed on filaments previously digested with proteases. Digestion of filaments with the calcium-requiring protease results in more than half of the polypeptide mass being lost from the filament and the retention of a resistant core containing most of the α helix originally present in the intact filament. X-ray diffraction patterns of these digested filaments show many features of the patterns from undigested, intact molecules (Fig. 13). This would indicate that most of the information present in these patterns comes from the core or backbone of the filament. Presumably, the sidearms, being flexible, do not contribute significantly.

Evidence from electron microscopy indicates that the protofilaments within the neurofilament are arranged helically (Krishnan *et al.*, 1979). Shadowing of filaments also occasionally shows a helical period with a pitch of 300–400 Å and when present this particular helical family is right-handed (Fig. 11). In urea, filaments readily untwist revealing two or more strands (Krishnan *et al.*, 1979) and the subunit structure is further unraveled by shadowing under disassembly conditions in high salt (Fig. 11). The subunits obtained in this way are elongated with lengths around 600–700 Å and widths around 40 Å (Fig. 11). They most probably represent the 6 S particles observed in the analytical untracentrifuge. Under certain conditions of disassembly, apparently spherical particles can be observed (Gilbert *et al.*, 1975; Krishnan *et al.*, 1979), and in some situations almost all the protein is present in this form. In these particles, which are often around 200 Å in diameter, the packing of the subunits must differ from that found in the filament.

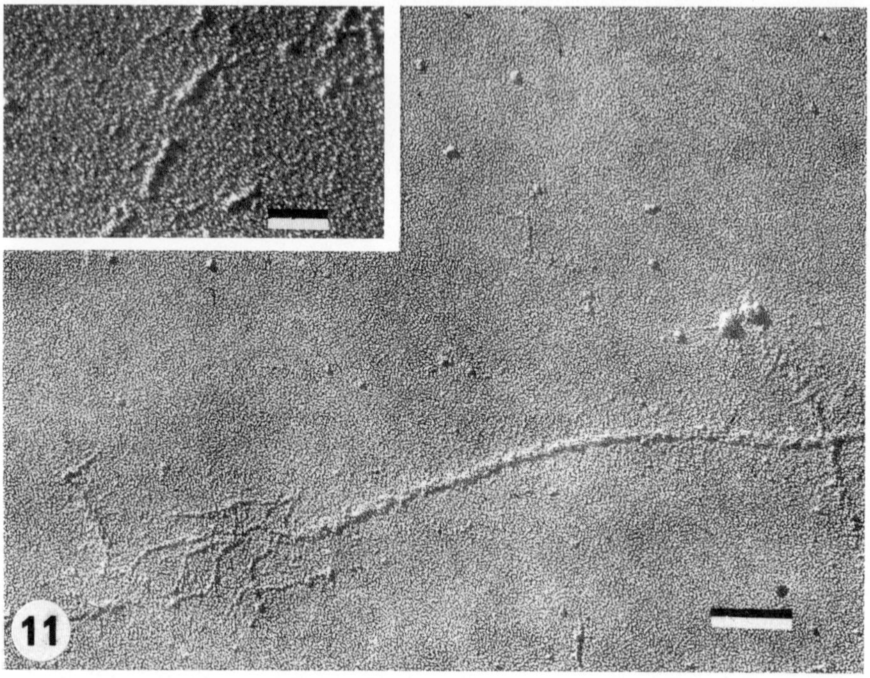

Figure 11. Shadowed *Myxicola* neurofilaments. Shown is an electron micrograph of *Myxicola* neurofilaments that have been shadowed unidirectionally. The filaments were sprayed under conditions causing dissociation of the filaments and a region can be clearly seen where seven or eight subunits have dispersed. The ropelike structure of the filament is evident and occasionally periodicities are seen at spacings around 300–400 Å. Under these dissociating conditions, spherical particles are often observed at the ends of the filaments where they occur in pairs. Bar = 400 Å. (Inset) Shadowed preparation of purified *Myxicola* subunits at higher magnification. The subunits were prepared by dissociating filaments in high salt. Bar = 600 Å.

Interestingly, during disassembly the spherical particles often appear to form first at the ends of filaments (Fig. 11).

The arrangement of polypeptides within the filament has been probed using proteases. Cleavage of filaments with either the calcium-requiring protease or chymotrypsin removes more than half the mass of the major polypeptides while leaving behind an enriched α-helical core that is still filamentous (Fig. 6; Eagles *et al.*, 1980b; Eagles *et al.*, 1985). These results suggested that a large region of the polypeptide is not required for the backbone of these filaments and the region that was removed was therefore thought to be equivalent to the peripheral sidearm domains seen on mammalian neurofilaments in intact nerves (Eagles *et al.*, 1980b). Further work with filaments phosphorylated either by the addition of ATP or by inorganic phosphate (through mitochondrial oxidative phosphorylation) showed that most of the phosphate was located on the nonhelical domains, which were removed by the calcium-mediated protease (Fig. 9 and Eagles *et al.*, 1981b). This observation led to the idea that the sidearms were highly phosphorylated structures (Eagles *et al.*, 1980b), which subsequent studies have shown to be true of mammalian and squid neurofilaments. Sidearm phosphorylation therefore is most probably a common principle of neurofilament architecture.

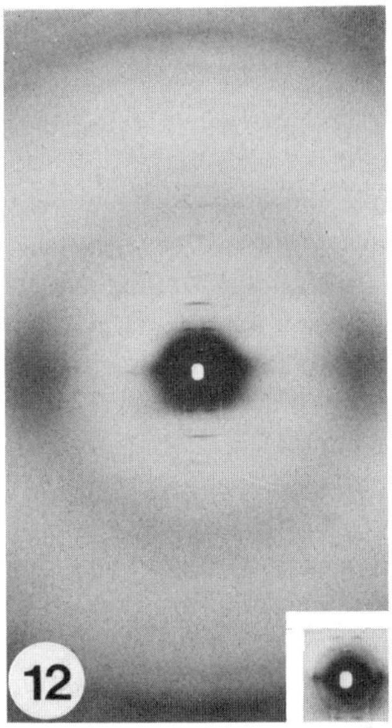

Figure 12. X-ray diffraction pattern from axoplasm of *Myxicola.* The major features come from neurofilaments although some contribution from the microtubules present is also seen. On the equator, reflections at spacings of around 10 Å are present, and on the meridian, arcs at spacings of around 5.1 Å are present. These features arise from the coiled coils of α helices. The meridian also has a number of neurofilament-associated reflections, the strongest of which is at a spacing of around 25 Å. The layer lines are at spacings of around 40 Å and arise from the microtubule lattice. (Inset) A less exposed central region of the pattern showing a reflection around a spacing of 63 Å which comes from the neurofilaments and the J3 reflections that arise from the microtubules. Data from Wais-Steider *et al.* (1987).

3.2. Squid

Since the pioneering studies by Bear *et al.* (1937a,b), the giant axons from the stellar nerves of squid have been used as a rich source of axoplasmic proteins, of which around 13% is neurofilament in origin (Morris and Lasek, 1984). For neurofilament studies the main species used have been *Loligo pealei* (giant axon diameter 75–700 μm), *L. forbesi* (giant axon diameter 0.5–1 mm), and the larger squid *Dosidicus gigas* (giant axon diameter 0.6–1.5 mm) (Arnold *et al.*, 1974). Presumptive neurofilaments were first seen in squid axoplasm by electron microscopy in 1943 (Richards *et al.*, 1943), though the question of whether these filaments were artifacts of fixation, which for a long time had been the subject of much controversy, was not fully resolved until the studies of Schmitt (1950) and Maxfield (1953). By investigating filaments from axoplasm processed under a

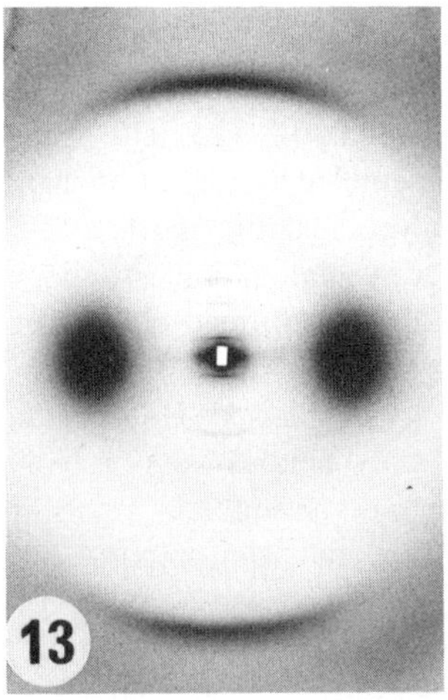

Figure 13. X-ray pattern from neurofilaments after they have been digested with the calcium-activated protease. The reflections coming from the coiled coils are present, indicating that these structures are preserved after digestion. A number of other meridional reflections are also found to be unaltered by this treatment. Under the conditions used here, the protease cleaves off polypeptide regions corresponding to the sidearm domains and leaves mostly intact the α-helical backbone of the filament. Thus, by comparing this picture with that in Fig. 12, it is evident that the sidearm domains do not contribute significantly to the patterns. Data from Wais-Steider *et al.* (1987).

wide variety of conditions, both with and without fixation, stained and unstained, the observations by these workers of filaments being present with diameters around 100 Å in all cases proved beyond a doubt the reality of these structures.

Controversy also surrounded for many years the nature of subunits from which the filaments were made. Initially, it was thought that the squid filaments, and indeed all other neurofilaments, were composed of essentially non-α-helical, globular, subunits (Schmitt, 1968a,b) though this view had to be revised following the discovery that neurofilaments from *Myxicola* could produce the α pattern, characteristic of coiled coils, during x-ray diffraction analysis (Day and Gilbert, 1972). More recently, Wais-Steider *et al.* (1983) showed that purified squid neurofilaments could also produce this α pattern.

Within the axon the neurofilaments form an extensive network, cross-linked by fine projections emanating from the backbone of the filament (Metuzals, 1969; Rice *et al.*, 1980; Hodge and Adelman, 1980). The neurofilamentous network is arranged on a hierarchy of helices, which was first described in detail by Metuzals and Izzard (1969). The overall arrangement seems much like that found in *Myxicola* (Lasek *et al.*, 1983).

Electron microscopy of squid neurofilaments shows that they have diameters ranging from around 70 to 150 Å (Maxfield, 1953; Schmitt, 1950; Davison and Taylor, 1960; Krishnan *et al.*, 1979; Wais-Steider *et al.*, 1983) depending on conditions and preparation technique. In early studies, no provision was made for inhibiting proteases, and some of the filaments thus may have been degraded. (We now know that the presence of a 100-Å-diameter filament is no indication of the intactness of the constituent polypeptides.) Early studies reported that some of the filaments had a beaded appearance (Schmitt, 1950) and more recently clear indications of filamentous material projecting from the backbone have been documented (Gilbert *et al.*, 1975; Metuzals *et al.*, 1983; Lasek *et al.*, 1983; Hodge and Adelman, 1980). Under certain conditions, especially where disassembly is occurring, neurofilaments from squid appear to have a helical substructure like that seen in *Myxicola* (Krishnan *et al.*, 1979).

Various recipes exist in the literature for the isolation of squid neurofilament proteins, though in order to purify intact polypeptides buffers designed to inhibit the calcium-requiring protease are essential and for this task EGTA and leupeptin can be included. The neurofilament proteins may be prepared from cycles of assembly and disassembly, washing axoplasm in buffers designed to maintain neurofilament integrity, sedimentation on sucrose gradients, Millipore filtration, and molecular sieving (Lasek and Hoffman, 1976; Pant *et al.*, 1978; Lasek *et al.*, 1979; Roslansky *et al.*, 1980; Wais-Steider *et al.*, 1983).

That squid neurofilaments could be reversibly dissociated was indicated by the studies of Maxfield (1953) wherein axoplasmic filaments were disassembled and assembled as a result of changing the pH and ionic strength of the buffers. Further studies by Maxfield and Hartley (1957) suggested longitudinal splitting of the filaments during the dissociation process, again as a result of an increase in the pH. Subsequent investigations have confirmed all of these findings (Davison and Taylor, 1960; Gilbert *et al.*, 1975). Filaments have also been reconstituted from proteins extracted in urea and guanidine hydrochloride (Huneeus and Davison, 1970) though it is unclear here to what extent the polypeptides were proteolytically degraded. Based on the ability of squid neurofilaments to disassemble in high salt (> 0.6 M KCl) and to reconstitute when the ionic strength is lowered, the preparation of neurofilaments from squid optic lobe has been achieved (Zackroff and Goldman, 1980).

Neurofilaments from squid axoplasm, when isolated under a variety of conditions designed to prevent proteolysis and by different groups of workers, show as common

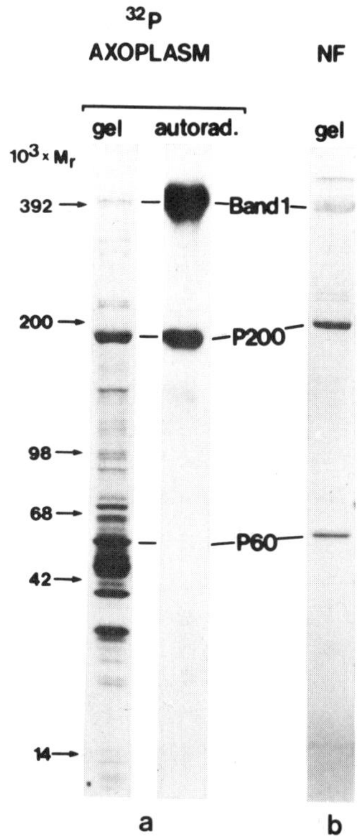

Figure 14. SDS-PAGE of phosphorylated squid axoplasm (a) and purified neurofilaments (b). Axoplasm was phosphorylated with [^{32}P]-Pi and neurofilaments were purified as described by Wais-Steider *et al.* (1983). The gel and corresponding autoradiograph of ^{32}P-phosphorylated axoplasm [(32P) axoplasm] are shown in panel a and the gel of purified neurofilaments (NF) is shown in panel b. A number of minor polypeptides, the identities of which are unclear, are present in the purified preparation. Data from Brown and Eagles (1986).

features three major polypeptide bands with apparent molecular weights on SDS gels around 60k, 200k, and 400k (Fig. 14) (Lasek and Hoffman, 1976; Lasek *et al.*, 1979; Pant *et al.*, 1978; Eagles *et al.*, 1980b; Roslansky *et al.*, 1980; Brown and Eagles, 1986; Gallant *et al.*, 1986). In view of the peculiar behavior of neurofilament polypeptides on SDS gels especially if they are phosphorylated, as is the case with the 200k component and the higher molecular weight component, both of which increase in mobility on dephosphorylation (see Section 5.2; Brown, Wright, and Eagles unpublished work), these values obtained in SDS are most probably gross overestimates. Better values must await further studies.

The composition of filaments isolated from optic lobes differs from that obtained from axoplasmic filaments because the former contains two extra bands with apparent molecular weights around 100k and 74k (Zackroff and Goldman, 1980). It is possible, however, that some of these polypeptides have a nonneuronal origin. A further difference between these filaments is that those from optic lobes have a much greater amount of the 60k polypeptide. Peptide mapping studies show that all three polypeptides from axoplasm have common structural features (Eagles *et al.*, 1980b), and Zackroff *et al.* (1984) have demonstrated that filaments prepared from optic lobes also have extensive homology regarding their constituent polypeptides.

The current picture emerging for the arrangement of polypeptides in squid neurofilaments supports the more general theme regarding neurofilament architecture wherein the lower molecular weight components reside predominantly in the backbone and the high-

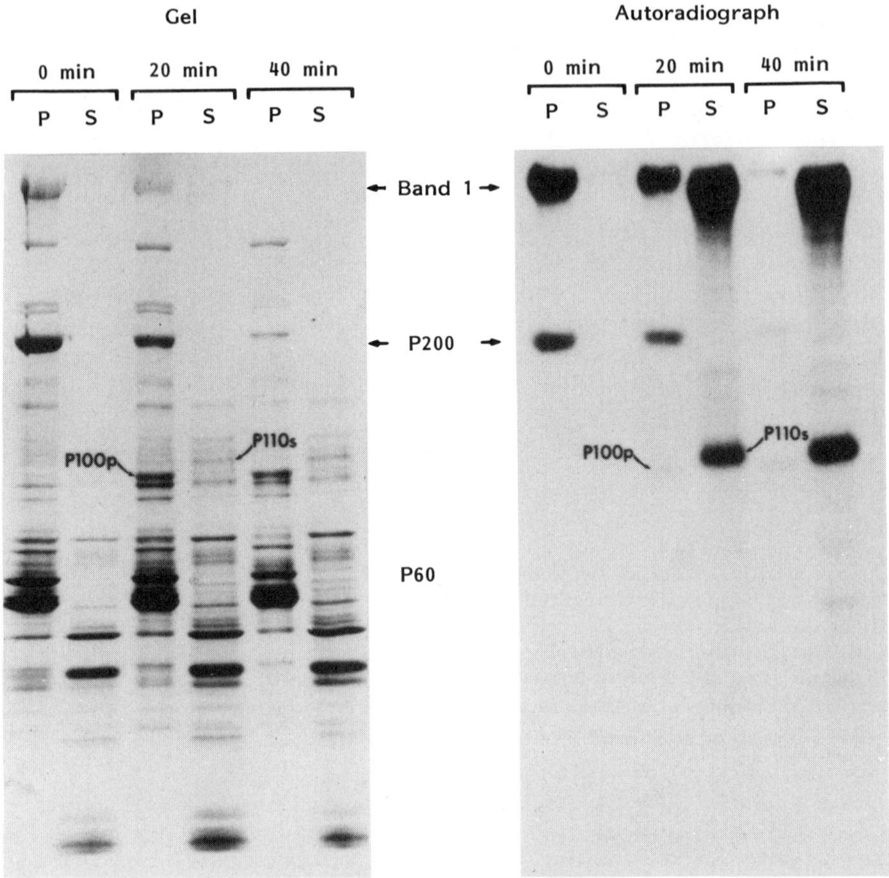

Figure 15. SDS-PAGE of [^{32}P]Pi-phosphorylated squid axoplasm after calcium-dependent proteolysis. Axoplasm was phosphorylated with [^{32}P]Pi and then digested. Proteolysis was terminated at timed intervals. For the control (0 min), tosyl-lysylchloromethane was added before addition of calcium. Axoplasm was then homogenized and separated into pellet (P) and supernatant (S) fractions by centrifugation. Pellet and supernatant samples incubated by 0, 20, and 40 min were analyzed by SDS-PAGE. The Coomassie blue-stained gel and corresponding autoradiograph are shown. Comparison by densitometry of the pellet fractions shows that, after 40 min proteolysis, 90% of Band 1 (HMW) and 85% of P200 have been cleaved, but that 60% of P60 still remains intact. Electron microscopy of the original sample after 40 min proteolysis shows little decrease in overall filament content. P200 is thought to be cleaved into two peptides: P100p, which pellets with filaments, and P110s, which contains the phosphate label and is soluble. P100p binds anti-IFA and presumably contains the rod region of P200, indicating that P110s is the peripheral sidearm domain. Data from Brown and Eagles (1986).

molecular-weight polypeptides have long extensions, peripherally disposed, which project from the filament's surface. The 60k component of squid neurofilaments contains an anti-IFA epitope (Gallant *et al.*, 1986; Brown and Eagles, 1986), and thus presumably contains an α-helical rod region. This view is also supported by the fact that the 60k component after purification will self-assemble to produce smooth filaments similar in structure to the filaments formed from the mammalian polypeptide NFL (Zackroff *et al.*, 1984; Eagles and Wright, unpublished work). The squid 60k component is not phosphorylated by the addition of ^{32}P nor is it very susceptible to degradation by the calcium-requiring protease. The properties of the 200k component, however, are quite different. This component can be heavily labeled with radioactive phosphate both *in situ* and in homogenates. The polypeptide is cleaved after activation of the calcium-requiring protease and many filaments are still present when nearly all of the 200k component has been lost. Two fragments are generated during the process, which have similar apparent molecular weights around 100k (Fig. 15; see also Section 5.1). One of these polypeptide fragments binds to filaments and contains the anti-IFA epitope. Thus, this fragment contains the α-helical rod region. The other polypeptide is highly phosphorylated and solubilized by proteolysis, and thus would presumably form part of the sidearm structure (Figs. 16 and 17) (Gallant *et al.*, 1986; Brown and Eagles, 1986). Regarding the 400k, high-molecular-weight, component associated with these neurofilaments, its structure is unclear. As it can be removed by proteolytic treatment while maintaining a filamentous structure, it is presumably peripheral in nature (Brown and Eagles, 1986). The peculiar staining pattern of this component on gels together with the difficulty in generating peptide maps might indicate that it is cross-linked in some way.

4. *Interactions of Neurofilaments*

In axoplasm, neurofilaments are associated in a complex three-dimensional arrangement not only with themselves but also with other cytoskeletal elements and organelles. Electron micrographs often show clear signs of material that is apparently forming these linkages. The associations may be classified according to whether they bridge the space between contiguous neurofilaments (sidearms), neurofilaments and microtubules or other organelles, or whether they occur in the subaxolemmal space bridging neurofilaments and axolemma. The various interactions are discussed in more detail below.

First, are the sidearms that apparently link neurofilaments. These structures have been described in the foregoing sections. They appear to involve the long C-terminal extensions of the polypeptides which are highly charged and highly phosphorylated. Other proteins, however, may also be involved in their structure. This is indicated by studies of nerves that have been quick-frozen and deep-etched wherein the neurofilament backbone and sidearm often have a knobby or granular appearance owing to attached material (Schnapp and Reese, 1982; Hirokawa, 1982). This material can be removed when the sample is washed in saponin (Hirokawa, 1982).

Neurofilaments that have been purified from *Myxicola* or squid axoplasm or mammalian nerves form highly hydrated gels, providing clear evidence of their potential to form cohesive interactions (Carden and Eagles, 1983; Leterrier and Eyer, 1987). The gels exhibit thixotropy and are stabilized by sucrose and glycerol together with divalent cations like magnesium. As the gel structure depends critically on ionic strength, it is unclear to what extent similar cross-linking interactions occur in axoplasm where the ionic strength is close to 1 M. In axoplasm the ionic environment would substantially modify associations between the charged C-terminal ends of the neurofilament polypeptides, and any interactions they form are likely to be weak though there may be many of them. In fact, isolated C-terminal domains of NFH demonstrably lack a strong affinity for intact neurofilaments, and thus are unlikely to form a very rigid network of sidearms as implied by their electron microscopic appearance (Fig. 2). In view of the high negative charge on the

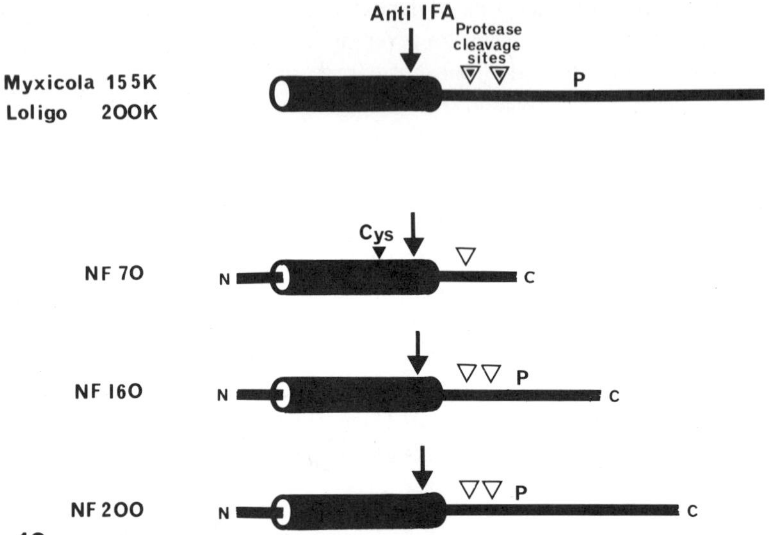

16

Figure 16. Diagrammatic representation of the structure of neurofilament polypeptides from *Myxicola,* squid *(Loligo),* and the mammalian triplet. The highly α-helical rod region is shown here as a cylinder and the projecting regions of the polypeptide, the N-terminal and C-terminal regions, are at the ends. The binding site for the intermediate filament antibody, anti-IFA, is indicated by an arrow and the polypeptides are aligned according to this region. The major phosphorylated regions are designated "P" and the main sites at which proteases cleave are indicated by open arrowheads. For the polypeptides in *Myxicola* and the squid 200k component, cleavage sites for the calcium-activated protease are shown and these separate the highly phosphorylated region from the α-helical and anti-IFA binding region. In the triplet polypeptides, the cleavage sites shown are for trypsin and chymotrypsin. For NF160 and NF200, chymotrypsin-cleavage sites separate the phosphorylated C-terminal regions from the rod domains. The position of the single cysteine residue within the rod region of NF70 is also shown. Adapted from Eagles (1986).

C-terminal domains, they may extend in solution because of their mutual repulsion and in so doing could act as spacers between neighboring neurofilaments. Therefore, neurofilament sidearms may well be more involved with keeping neurofilaments apart than with holding them together.

Links between microtubules and neurofilaments have been observed by a number of groups (e.g., Yamada *et al.,* 1971; Rice *et al.,* 1980; Ellisman and Porter, 1980; Nagele and Roisen, 1982; Hirokawa, 1982; Schnapp and Reese, 1982). Sometimes the association can produce a very regular structure such as that found in tench where nine or ten neurofilaments surround the microtubule (Bertolini *et al.,* 1970). Further evidence for a direct association comes from their coordinated movement during slow axonal transport (Lasek and Hoffman, 1976; Willard, 1983) and it appears that microtubule-associated proteins also move with neurofilaments (Tashiro *et al.,* 1984). An interaction is indicated from various chemical studies. For example, in the initial stages of microtubule purification, neurofilaments are also concentrated (Delacourte *et al.,* 1977; Berkowitz *et al.,* 1977). Whereas there is little doubt on both chemical and physical grounds of an interaction, the nature of the association is unclear. Two prime candidates would be the neurofilament sidearm and the variety of microtubule-associated proteins (MAPs). The possibility

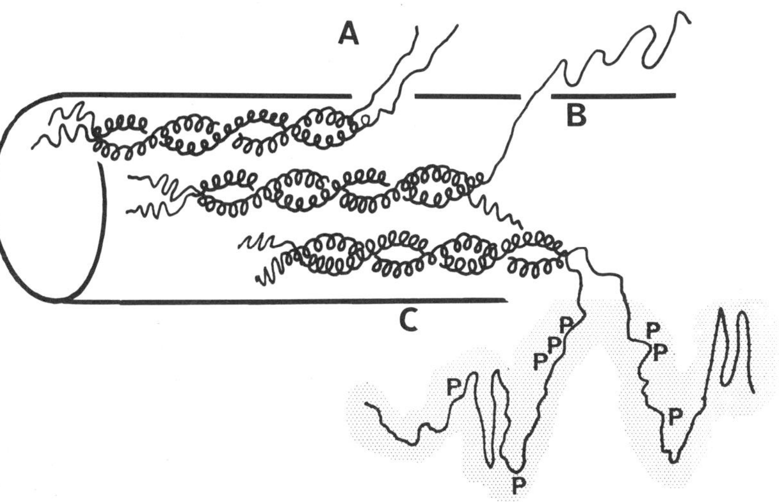

17

Figure 17. Diagrammatic representation of how various neurofilament polypeptides might be arranged when present in filaments. A: example of two polypeptides that are similar in molecular weight with the α helix in the filament backbone arranged in coiled coils with projecting ends disposed peripherally, e.g., NFL polypeptides of the neurofilament triplet. B: example of two polypeptides with dissimilar molecular weights, e.g., *Myxicola* 170k and 50k polypeptides. C: example of two high-molecular-weight polypeptides with coiled coils anchoring them in backbone and long C-terminal extensions that are highly phosphorylated (P) and highly charged (dotted region), e.g., NFH polypeptides of triplet, 170k polypeptides of *Myxicola* neurofilaments, or 200k polypeptides of squid neurofilaments.

that sidearms may be involved is supported by the observation that the structures between neurofilaments are similar in size and shape to those between microtubules and neurofilaments (Hirokawa, 1982) though unfortunately no chemical evidence is available to support this idea. In contrast, studies using purified components clearly show an interaction between MAPs and neurofilaments (Shelanski *et al.*, 1981; Leterrier *et al.*, 1982; Papasozomenos *et al.*, 1983; Aamodt and Williams, 1984a; Williams and Aamodt, 1985). Moreover, MAPs interact mainly with the neurofilament protein NFL (Heimann *et al.*, 1984; Liem *et al.*, 1985; Miyata *et al.*, 1986). The binding of MAPs to neurofilaments helps to explain the observation that these filaments have the ability to stimulate microtubule polymerization (Minami *et al.*, 1982). This rather peculiar result is now attributed in part to MAPs that are present as impurities in many preparations of neurofilaments. [Leterrier *et al.* (1984) have shown that if highly purified neurofilaments containing no MAPs are used for such studies, then the assembly-promoting ability of the filaments is lacking.] On the other hand, neurofilaments may have an intrinsic property of promoting microtubule polymerization and this has been suggested by Minami *et al.* (1984). The property was traced to the NFH component where it was thought to reflect features of the peripheral domains.

Initial studies on the interactions between neurofilaments and microtubules implicated a role for ATP in the process because an increase in viscosity was found for mixtures of neurofilaments and microtubules when ATP was present (Runge *et al.*, 1981b; Williams and Runge, 1983). However, it now seems that the observed viscosity changes are the result of a complex series of events leading to microtubule stabilization by an increased production of GTP—the GTP being formed from the ATP present in the mixture by enzymes that contaminated the preparations (Aamodt and Williams, 1984b).

Two studies (Hirokawa, 1982; Schnapp and Reese, 1982) draw attention to the presence of linkages between neurofilaments and membrane-bound organelles and vesicles. On present evidence, the bridges are unlikely to be involved in vesicle motility *per se,* which seems to be governed by the microtubules. The apparent links may arise from the disruption of neurofilament sidearms that would presumably re-form again when the organelle leaves the zone.

The last of these classes of attachments in which neurofilaments appear to be involved are the linkages between them and the axolemma—connections that span the subaxolemmal space (Hirokawa, 1982). This region, like the subcortical zone of other cells, contains actin. The actin cross-linking protein, fodrin, has also been localized here (Levine and Willard, 1981; Willard, 1983). That neurofilaments might interact directly with fodrin is suggested by two observations. First, neurofilaments and fodrin move coordinately during slow axonal transport (see Willard, 1983), and second, iodinated fodrin binds to neurofilaments *in vitro* and binding is localized to NFL (Frappier *et al.*, 1986).

Apart from the associations described above, there are also intriguing reports of specific neurofilament-associated proteins (Liem *et al.*, 1985) and of neurofilaments binding synapsin (Steiner *et al.*, 1987). Neurofilaments also bind calcium (Abercrombie *et al.*, 1986) and other cations (Lefebvre and Mushynski, 1988). Like many other intermediate filaments, neurofilaments interact with nucleic acids (Traub, 1985). The significance of these findings remains to be established.

Concluding this section, it seems clear that neurofilaments can participate in a number of interactions and so far it looks as though these interactions are mediated by at least two sorts of connections (1) through the long C-terminal extensions of NFH and possibly NFM and (2) through NFL. The stability of the interactions in which these polypeptides participate in axoplasm must be of extreme importance to the cell, and no doubt much future work will focus on this issue.

5. *Posttranslational Modifications of Neurofilaments*

Two posttranslational mechanisms that modify neurofilaments have been extensively studied: calcium-activated proteolysis and phosphorylation. While all neurofilament protein subunits are accessible as substrates for these enzymatic mechanisms, it is the higher molecular weight neurofilament proteins which are most dramatically affected. Although both mechanisms are clearly occurring extensively in neurons, their modes of regulation and functional significance remain a matter largely of speculation. In the sections below, we discuss these mechanisms in some detail.

5.1. *Calcium-Activated Proteolysis*

Calcium-activated neutral proteases (CANPs) are thiol proteases with pH optima around neutrality which are found in the cytosol of a wide variety of tissues and cells, including the nervous system (Ishiura, 1981; Murachi *et al.*, 1981a,b; Schlaepfer and Zimmerman, 1985a,b). In a particular tissue, their activation usually leads to the proteolysis of specific intracellular proteins, with selective physiological consequences. CANPs have also been termed *calcium-dependent neutral proteinases* and *calpains,* and probably represent diverse members of a large gene family. The 80k subunit differs between CANPs, whereas the 30k subunit appears to be highly conserved (Kawasaki *et al.,* 1986). Although the function of the smaller subunit is unclear, at present, it does appear to bind calcium (Emori *et al.,* 1986).

The earliest indication that a calcium-dependent process could disrupt the intracellular organization of the axon came from studies on the squid giant axon (Hodgkin and Katz, 1949; Hodgkin and Keynes, 1956). In these studies, injection of calcium into squid axoplasm caused rapid "liquification" of axoplasm and a similar phenomenon was noted in studies on mammalian nerves (Schlaepfer and Bunge, 1973; Schlaepfer, 1974, 1977c). Although Orrego (1971) found a calcium-dependent proteolytic activity in squid axons, and Guroff (1964) characterized CANP activity in the mammalian nervous system, it was not until the work of Gilbert and his colleagues (Gilbert, 1975b; Gilbert *et al.,* 1975; Anderton *et al.,* 1976) on the giant axons of the fan worm *Myxicola* and the squid *Loligo* that this axonal disruption could be connected to the degradation of neurofilaments.

Following these observations, there have been a large number of reports on both invertebrate and vertebrate neurons showing that CANPs were neutral, thiol proteases that were specifically activated by calcium, and that selectively degraded neurofilament proteins (Gilbert *et al.,* 1975; Schlaepfer and Micko, 1978; Schlaepfer and Freeman, 1980; Pant *et al.,* 1979a; Pant and Gainer, 1980; Eagles *et al.,* 1980b, 1981b; Schlaepfer *et al.,* 1981b; Tashiro and Ishizaki, 1982; Malik *et al.,* 1981, 1986; Nixon, 1986; Nixon *et al.,* 1986; Kamakura *et al.,* 1985, 1986; Zimmerman and Schlaepfer, 1984a,b; Tsung and Lombardini, 1985). In virtually all of the above studies, SDS-PAGE was used to identify the proteins that were proteolyzed in neuronal tissue as a result of calcium treatment. Figure 15 illustrates a typical example of CANP activity. Pant *et al.* (1982) were able to use [^{14}C] casein as an effective substrate to evaluate CANP activities in a variety of squid neural tissues. Because each tissue may contain multiple forms of CANP, it is difficult in most of the studies, where *in situ* preparations or tissue homogenates were used, to evaluate the specific patterns of degradation. However, most of the studies that have analyzed peptide products of ^{32}P-labeled neurofilament proteins (Brown and Eagles, 1986; Eagles *et al.,* 1981b; Gallant *et al.,* 1986; Pant *et al.,* 1986) and ^{125}I- labeled neurofilament proteins (Paggi and Lasek, 1984), or have used immunological assays (Schlaepfer *et al.,* 1984, 1985; Eagles and Maggs, 1985; Brown and Eagles, 1986; Gallant *et al.,* 1986), have suggested that the most sensitive cleavage sites are located as shown in Fig. 16.

Perhaps the least crude of the above type of experiments has been done using purified axoplasm from the squid and *Myxicola* giant axons. By using such preparations, one can at least be sure that the CANPs and protein substrates are only of neuronal origin. Figure 15 illustrates the CANP-induced degradation of squid neurofilament proteins with time. Coomassie blue staining and ^{32}P-autoradiographic analysis of the peptide products show that the 200–220k neurofilament protein is cleaved into two smaller fragments: a 100k protein (100p in Fig. 15) and a 110k protein (110s in Fig. 15). Interestingly, the 100p product cosediments with the intact 220k protein and is reactive with anti-IFA, whereas the 110s product is soluble and contains most of the ^{32}P label. These data, which have been independently generated in two laboratories (Brown and Eagles, 1986; Gallant *et al.,* 1986), form the basis for the diagram in Fig. 16. Thus, it is conceived, from these studies, that the initial CANP cleavage site is between the anti-IFA epitope in the conserved domain and the principal phosphorylated C-terminal end domain of the squid 220k neurofilament protein.

A more complete understanding of CANP enzymology requires the purification of the varied CANPs and kinetic studies using the individual purified neurofilament proteins as substrates. It is known that there are at least two types of CANPs which are extractable from a variety of tissues and which have distinct elution positions in DEAE-cellulose chromatography. One of the CANPs has a high sensitivity to calcium ($< 40\ \mu M$) and the other a low sensitivity ($> 500\ \mu M$) (Mellgren, 1980; Dayton *et al.*, 1981; Kubota *et al.*, 1981; Suzuki *et al.*, 1981; Murachi *et al.*, 1981a,b; Mellgren *et al.*, 1982; Hathaway *et al.*, 1982). Murachi *et al.* (1981b) have shown that of 11 tissues studied that had CANP activity, nervous tissue (brain) was the least abundant in the high-calcium-sensitivity form (referred to by those workers as calpain I, or the U-type). Although some evidence is available for calpain I in brain (see Baudry and Lynch, 1980; Baudry *et al.*, 1981), it is unlikely that this is the type of CANP associated with the neurofilaments in axons. In contrast, calpain II (the low-calcium-sensitivity form of CANP, or the m-type) was quite abundant in brain tissue (Murachi *et al.*, 1981b), and appears to be the type of CANP found in axoplasm (Pant and Gainer, 1980; Pant *et al.*, 1982). It has been speculated that calpain II may serve as a proenzyme form of calpain I (Coolican *et al.*, 1986). One experimental procedure has been found that appears to convert calpain II to calpain I (Dayton *et al.*, 1981; Kubota *et al.*, 1981; Suzuki *et al.*, 1981; Hathaway *et al.*, 1982; Mellgren *et al.*, 1982), and this was a consequence of the affinity purification of calpain II on a casein-CH-Sepharose-4B column in the presence of 24 mM calcium (Suzuki *et al.*, 1981; Kubota *et al.*, 1981). The affinity-purified CANP II was slightly smaller (76k) than the unpurified form (80k), and the authors suggest that this change in calcium sensitivity of the CANP II was caused by a limited proteolysis (autolysis) of the CANP on the column to the CANP I (or calpain I) form of the enzyme. It is well known that CANPs when activated by calcium *in vitro* undergo autolysis to inactive enzymes (Puca *et al.*, 1977; Gilbert *et al.*, 1975; Truglia and Stracher, 1981; Murachi *et al.*, 1981a,b; Kubota *et al.*, 1981; Suzuki *et al.*, 1981; Hathaway *et al.*, 1982; Pant *et al.*, 1982), and that CANP I is more rapidly inactivated than CANP II. It is still unclear why this autolysis is "limited" on the affinity column, but not in the *in vitro* case even in the presence of excess casein substrate. Nevertheless, these data indicate that CANP II can be converted to CANP I by limited proteolysis (i.e., limited autolysis; see also Vitto and Nixon, 1986; Coolican *et al.*, 1986). The problem is to find how this is regulated within the cell. In review, Ishiura (1981) pointed out that although endogenous inhibitors of CANP have been found (Murachi *et al.*, 1981a; Tashiro and Ishizaki, 1982), no endogenous factor has been found that can transform CANP II to CANP I. [Some that have been tried without success are AMP, ADP, ATP, trifluoperazine, calmodulin, and tissue extracts; see also Pant *et al.* (1982).] The only candidate at present is "unphysiological" levels of calcium.

Several investigators have attempted to purify CANPs from nervous tissue and to test the efficacy of these separated forms on various isolated neurofilament subunits (Kamakura *et al.*, 1985, 1986; Malik *et al.*, 1981, 1983, 1984, 1985; Zimmerman and Schlaepfer, 1984a,b; Vitto and Nixon, 1986). Between two and three forms of the CANP activity have been detected in these studies, and each has a different pattern of substrate preference with respect to the mammalian neurofilament triplet subunits. Until these forms of CANP are better characterized with respect to purity (e.g., at a minimum amino acid composition), definitive statements with respect to substrate preference are not possible. Similarly, no clear evaluation of the actual site of hydrolysis (i.e., the amino acids present at the peptide bond being cleaved) is available for the neurally derived CANPs. However, leupeptin, which usually acts as a protease inhibitor of arginyl-directed proteases, is quite effective on neural CANP. This is consistent with recent findings that nonneural CANPs appear to cleave at Arg residues (Traub, 1984) and Arg-Arg residues (Hayashi *et al.*, 1985).

Recent recombinant DNA studies have given some insights into the nature of CANPs. Ohno *et al.* (1984) have sequenced the cDNA of the 80k subunit of chicken CANP (i.e., the calpain II type). This 705-amino-acid protein contains four distinct domains, with the second domain closely resembling papainlike thiol proteases, and the fourth domain homologous with calmodulinlike binding proteins. The authors suggest that this 80k protein evolved from the fusion of genes for proteins of completely different functions (i.e., proteases and calcium-binding proteins) and evolutionary origins. Two groups have sequenced the cDNAs of the 30k subunit from porcine kidney (Sakihama *et al.*, 1985) and from rabbit (Emori *et al.*, 1986). This 266-residue protein has two distinct domains. The first contains polyglycyl sequences and hydrophobic amino acid stretches presumed to be binding sites for either the 80k subunit or plasma membrane, whereas the second domain (C-terminal) is very homologous to the calmodulinlike region of the 80k subunit, and contains four E–F hand structures. This suggests that the fourth domain of the 80k subunit and the second domain of the 30k subunit may have evolved from a similar ancestral calcium-binding protein. Thus, it would appear from these studies that CANPs are structurally organized to interact directly with calcium ions.

Given the widespread occurrence of CANPs, it is perhaps not surprising that they have been implicated in a wide variety of cell functions. Some of these include intracellular enzyme activation (Huston and Krebs, 1968; Nishizuki *et al.*, 1979; Togari *et al.*, 1986; Hamon and Bourgoin, 1979), myoblast differentiation (Kaur and Sanwal, 1981), myofibrillar protein turnover (Ishiura *et al.*, 1978; Kameyama and Etlinger, 1979), myelin (Banik *et al.*, 1985) and neuropeptide (Hirao and Takahashi, 1984) turnover, regulation of steroid (Puca *et al.*, 1977; Vedeckis *et al.*, 1980; Murayama *et al.*, 1984), peptide (Cassell and Glasser, 1982), and glutamate (Baudry and Lynch, 1980; Vargas *et al.*, 1980; Baudry *et al.*, 1981) receptors, and modification of various cytoskeletal proteins (Anderson *et al.*, 1977; Phillips and Jakabova, 1977; Triplett *et al.*, 1972; Wallach *et al.*, 1978; Gilbert *et al.*, 1975; Lucas *et al.*, 1979; Traub and Nelson, 1981; Nelson and Traub, 1982; Collier and Wang, 1982; Siman *et al.*, 1984; Nixon, 1986).

With specific reference to neurofilaments, there are two hypotheses with respect to CANP function. While it is clear that axons contain large amounts of CANP activity (Pant *et al.*, 1982), it is in the CANP II form and is not activated in the axon under normal circumstances. This makes biological sense in the axon, where the neurofilaments are stable structural components, and are degraded only after injury to the nerve when intracellular calcium concentrations could reach millimolar levels. Hence, one hypothesis is that the destruction of neurofilaments preceding nerve degeneration (after injury) is mediated by CANP (Schlaepfer, 1974; Schlaepfer and Micko, 1978). Another hypothesis, originally suggested by Lasek and Hoffman (1976), was that the turnover of neurofilaments in nerve terminals may involve CANP, and that one mechanism regulating the growth of axons might be via the action of CANP in the developing nerve terminals (Lasek and Black, 1977). A problem that remains for the latter proposal is that the axonal

CANP directed at neurofilament degradation requires an unphysiological level (> 500 μM) of calcium ions for activation (see above). Even in the synaptic terminal, where during activity the calcium influx is relatively high, the internal calcium concentration rarely exceeds 10 μM (Llinas and Heuser, 1977). Recent studies on the distribution of CANP and endogenous protein substrates in the squid nervous system showed that there was no endogenous substrate or CANP in the nerve terminal, whereas the immediately contiguous area (i.e., the axon) contained both components abundantly (Pant *et al.*, 1982). These studies were interpreted as support for the idea of a programmed turnover of neurofilaments as they enter the nerve terminals, with a subsequent autolytic inactivation of the CANP (shown to occur with an appropriate time course for axonal CANP; see Pant *et al.*, 1982). A preliminary test of this hypothesis was performed by Roots (1983) who found that injection of leupeptin, which inhibits CANP, into the goldfish optic tectum led to an accumulation of neurofilaments in synaptic terminals *in vivo*. It will be very useful with regard to the above hypotheses to employ recently developed antibodies against CANP (Yoshimura *et al.*, 1984) in functional and morphological studies on neurons.

5.2. *Phosphorylation of Neurofilaments*

The first demonstration that neurofilament proteins (indeed intermediate filament proteins) were highly phosphorylated on serine/threonine residues was reported by Pant *et al.* (1978) for squid axoplasm. Subsequent studies on squid (Pant *et al.*, 1979b,c, 1986; Brown and Eagles, 1986) and *Myxicola* (Eagles *et al.*, 1978, 1981b) axoplasm, as well as mammalian systems (Shecket and Lasek, 1982; Eagles *et al.*, 1980a; Julien and Mushynski, 1981, 1982, 1983), have confirmed and extended these findings. Figure 14 illustrates the pattern of protein phosphorylation in squid axoplasm. This figure not only shows that the neurofilament proteins are the major phosphoproteins present, but also makes the point that the kinase responsible for this phosphorylation is present and abundantly active in pure isolated axoplasm. The key questions regarding neurofilament phosphorylation are: In which domains of the neurofilament protein are the phosphorylated residues located? Which neurofilament protein subunits contain phosphorylated residues, and how many residues are there in each case? What types of kinases are involved? How is phosphorylation of neurofilaments regulated in the neuron? What is the function of neurofilament phosphorylation? An attempt to address these questions is presented below.

Despite the fact that estimates for the phosphate content of the triplet polypeptides vary widely (e.g., Jones and Williams, 1982; Geisler *et al.*, 1985b; Julien and Mushynski, 1982, 1983), there is general agreement that NFH is the most phosphorylated and NFL the least. Some of the variation reported for the phosphate content must be due to the value used for the molecular weights of the polypeptides in the calculation and as uncertainty still exists, the precise molar ratios must await more accurate molecular weight determinations. The most thorough reports on the phosphate contents of neurofilament protein subunits come from Mushynski's laboratory (Julien and Mushynski, 1982, 1983; Georges *et al.*, 1986). In these studies the 68k (NFL), 145k (NFM), and 200k (NFH) neurofilament subunits prepared from rat spinal cord and brain stem contained about 3, 9, and 22 moles of phosphate per mole of polypeptide, respectively (Julien and Mushynski, 1982). In a more recent analysis of purified porcine neurofilament protein subunits, Georges *et al.* (1986) found that the NFL, NFM, and NFH subunits in this species contained about 3, 11, and 22 moles of phosphate per mole of polypeptide, respectively. A particularly important aspect of the latter study was that the authors also analyzed the phosphate contents of the subunits following extensive alkaline phosphatase treatment. They found that such "dephosphorylation" treatments, comparable to that described by Carden *et al.* (1985), did not completely "dephosphorylate" these subunits. The resultant (dephosphorylated) NFL, NFM, and NFH subunits contained 2, 6, and 8 moles of phosphate per mole of polypeptide, respectively. These observations will bear on the analysis of immunological epitopes in neurofilaments to be discussed in a later section. In addition, these data indicate that there is a heterogeneity in the phosphorylated residues at least with respect to accessibility to dephosphorylation, and this (as well as the total phosphate content in each subunit) may account for the greater ^{32}P incorporation in the higher molecular weight subunits in the presence of [^{32}P]-ATP and kinases (Pant *et al.*, 1978, 1986; Eagles *et al.*, 1978, 1980b, 1981b; Brown and Eagles, 1986; Shecket and Lasek, 1982; Julien and Mushynski, 1981).

The above commentary about heterogeneous accessibility of phosphate residues in neurofilament subunits, and therefore possibly heterogeneous turnover, places a limit on our current understanding of the locations of the phosphorylated residues. This is because all the analyses on this issue, to date, have depended upon incorporation of ^{32}P into neurofilament proteins. Whereas there is general agreement that the phosphorylated residues are primarily on the C-terminal tail domains (see Fig. 16), adequate details are unavailable. Julien and Mushynski (1983), studying proteolytic fragments of NF subunits that were labeled with ^{32}P *in vivo,* found that the NFM and NFH ^{32}P labels were located in the tail domains. The NFL data were not conclusive due to poor labeling of this subunit. These data are comparable to those reported earlier by Eagles *et al.* (1981b) for *Myxicola* neurofilaments where the ^{32}P label was associated predominantly with soluble fragments produced by an endogenous calcium-activated protease. Similar data have subsequently been reported for squid neurofilaments (Brown and Eagles, 1986; Gallant *et al.*, 1986; Pant *et al.*, 1986). Particularly pertinent to this point are the findings in squid neurofilaments that activation of the calcium-mediated protease (and therefore neurofilament cleavage) before ^{32}P incorporation greatly enhances the labeling in the 100p peptide fragment (Gallant *et al.*, 1986; Pant *et l.*, 1986), in contrast to the situations when proteolysis follows ^{32}P labeling of the intact neurofilaments (e.g., this is the situation in Fig. 15 where only the soluble, 110s, fragment is labeled). Thus, it would appear that the 100p peptide moiety is more accessible to kinase activity after proteolytic cleavage. To clarify the exact locations of the residues containing phosphate, it will be necessary to perform analyses of the kind used by Georges *et al.* (1986) on a more extensive set of peptide fragments.

Several laboratories have attempted to characterize the types of kinases involved in neurofilament phosphorylation (Shecket and Lasek, 1982; Runge *et al.*, 1981a; Toru-Delbauffe and Pierre, 1983; Pant *et al.*, 1986; Julien *et al.*, 1983; Leterrier *et al.*, 1981). In neurofilament-enriched preparations of bovine (Runge *et al.*, 1981a), guinea pig (Shecket and Lasek, 1982), rat (Julien *et al.*, 1983), and squid (Pant *et al.*, 1986) neural tissues, a protein kinase activity was found that phosphorylated neurofilament subunits. It appears to be an independent kinase and it is not affected by cAMP or the kinase inhibitor, nor by the addition of calcium plus calmodulin. It also appears to prefer casein over histone as a substrate, but it is distinct from known casein kinases I and II (Julien *et al.*, 1983; Pant *et al.*, 1986). In addition to the above independent kinase, one subunit of mammalian neurofilaments has been shown to be a substrate *in vitro* for a cAMP-dependent protein kinase that is associated with microtubules (Letterrier *et al.*, 1981; Julien *et* al., 1983). Peptide mapping studies suggest that these two kinases phosphorylate different amino acids in this neurofilament protein (Julien et al., 1983).

The regulation of neurofilament phosphorylation is unclear. That regulation takes place is apparent from the immunochemical analyses of phosphorylated neurofilament epitopes (see below), which indicate that certain epitopes are not present in neuronal perikarya, but are abundant in axonal processes (e.g., Sternberger and Sternberger, 1983; Carden et al., 1987). The predominance of a cAMP-independent and calcium-independent kinase in neurofilament preparations suggests that cAMP and calcium/calmodulin are not involved though an endogenous protein inhibitor of this kinase, restricted to the neuronal cell body, might play a regulatory role (Pant et al., 1986). The functional significance of neurofilament phosphorylation is likewise unclear. One possibility is that neurofilament phosphorylation may control assembly (Wong et al., 1984), but recent studies have shown that dephosphorylation of neurofilament subunits does not influence their reassembly (Georges et al., 1986). Other suggestions are that neurofilament phosphorylation can influence the interactions of neurofilament with other cytoskeletal structures (Sternberger and Sternberger, 1983; Runge et al., 1981a,b; Minami and Sakai, 1985; Matus, 1988) or can confer resistance to the action of the CANP. Insufficient experimental evidence exists for the support of any of these ideas. Since all classes of intermediate filaments appear to undergo phosphorylation (Lazarides, 1980; Steinert et al., 1982), one would expect that this posttranslational modification is of some significance for cytoskeletal organization.

6. Immunological Probes of Neurofilaments

Antibodies are useful for the analysis of specific intermediate filaments in tissues and cells, and have become valuable tools for diagnostic histopathology (Osborn and Weber, 1983). In addition to antibodies that can selectively react with specific intermediate filament subunits, there is available a monoclonal antibody, anti-IFA, that binds to all intermediate filament proteins (Pruss et al., 1981; Pruss, 1985). The epitope recognized by anti-IFA is located in the highly conserved rod domain of all intermediate filaments, including neurofilaments (Phillips et al., 1983; Geisler et al., 1983; see Fig. 16). Despite its inability to distinguish between intermediate filament proteins, anti-IFA is extremely useful for the identification of any unknown protein as belonging to the intermediate filament class (e.g., see Fig. 18). It is the general experience that anti-IFA is less effective in detecting high-molecular-weight neurofilament proteins than smaller intermediate filament proteins on Western blots. To improve this, Pruss (1985) recommended using pronase in the electrotransfer procedure, ostensibly to improve transfer of the high-molecular-weight proteins to the nitrocellulose. Trypsin treatment of the nitrocellulose following electrotransfer of the protein, however, is a more effective procedure to enhance anti-IFA reactivity (Szaro and Gainer, unpublished data). We believe that the C-terminal end domain of the high-molecular-weight neurofilament proteins is responsible for the apparently lower affinity of anti-IFA, and that the protease simply removes the influence of this domain on the rod domain containing the epitope for anti-IFA.

A most important requirement for neurofilament studies is a means by which the individual subunits can be recognized. The Bodian silver method is a nonimmunological procedure that can distinguish neurofilaments from other intermediate filaments (Gambetti et al., 1981; Autilio-Gambetti et al., 1986). However, this procedure is analogous to the

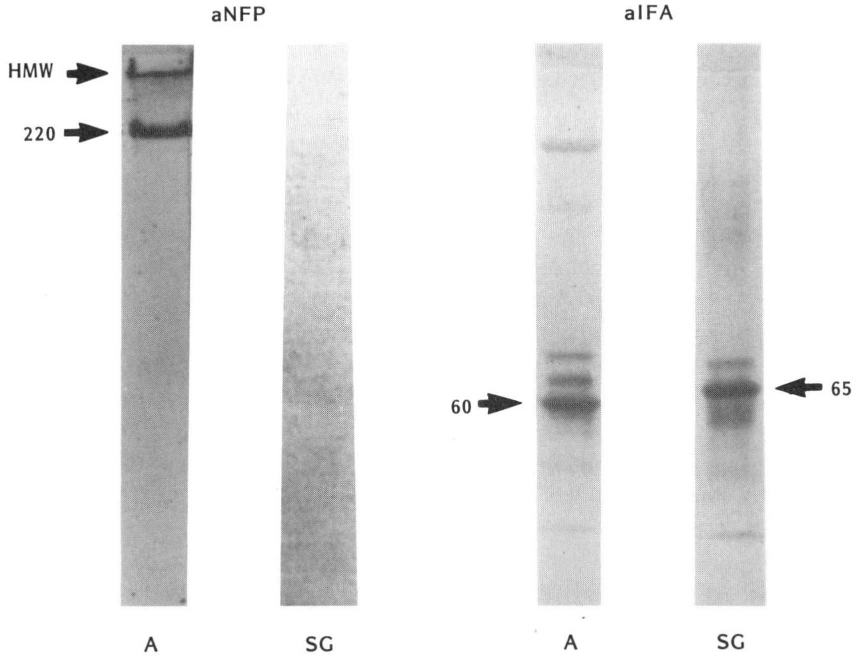

Figure 18. Western blots from squid axoplasm and stellate ganglion. Western blots of squid axoplasm (A) and stellate ganglion cell body (SG) proteins, using aNFP, a monoclonal antibody that cross-reacts with a phosphorylated epitope on P200 (220k in this gel system) and the high-molecular-weight (HMW, Band 1 > 400k) neurofilament proteins, and anti-IFA, a monoclonal antibody that cross-reacts with all intermediate filament proteins (Pruss *et al.*, 1981). Note that the aNFP MAb stained only the HMW and 220k proteins in axoplasm (A), and showed virtually no staining in the stellate ganglion (G). Anti-IFA stained a number of proteins in axoplasm, the major band being the 60k neurofilament protein (note the light band in SG corresponding to the 190k protein in Fig. 20). In the SG sample, anti-IFA intensely stained a 65k band and exhibited a number of stained bands, including a moderately stained 60k band. Data from Cohen *et al.* (1986).

anti-IFA method in that it cannot select between the neurofilament protein subunits themselves. For this purpose, highly specific neurofilament antibodies are necessary and at least initially many of these were obtained fortuitously. Drager *et al.* (1983) using an anti-α-MSH antiserum found that it cross-reacted specifically with neurofilament proteins, and Sternberger *et al.* (1982) making monoclonal antibodies presumably directed at the antigen, rat hypothalamus, produced 37 neuron-specific antibodies many of which were unexpectedly reactive to epitopes on neurofilament proteins (Sternberger and Sternberger, 1983). In addition, serum from normal human subjects has been found to contain autoantibodies to NF200 (Stefansson *et al.*, 1985). At present, however, many specific antibodies against precise neurofilament protein subunits have been generated (Autilio-Gambetti *et al.*, 1981; Vitadello *et al.*, 1986; Brown *et al.*, 1983; Debus *et al.*, 1983; Lee *et al.*, 1982) and used for a variety of experimental purposes including immunoprecipitation (Glicksman and Willard, 1982), immunocytochemical (Sternberger and Sternberger, 1983; Hirokawa *et al.*, 1984; Willard and Simon, 1981; Shaw *et al.*, 1981; Noetzel *et al.*, 1986), and immunochemical (Carden *et al.*, 1985) experiments.

If antibodies (polyclonal or monoclonal) are to be used for molecular identification in the above procedure, it is essential to evaluate first the natures of the antigenic determinants (epitopes) being recognized by the antibodies. This has been particularly important for neurofilament antibodies, since many recognize only phosphorylated sites (Sternberger and Sternberger, 1983; Carden *et al.*, 1985; Matus, 1988). In an exceptionally thorough analysis of a large panel of monoclonal antibodies generated against bovine neurofilaments, Carden *et al.* (1985) found that of 90 MAbs recognizing epitopes in the C-terminal peripheral domain of NF200, all showed reduced affinity for the dephosphorylated form. For the NF150 MAbs, about 50% appeared to be phosphorylation dependent, whereas the NF70 MAbs appeared to be independent of dephosphorylation procedures (but see above concerning the efficacy of dephosphorylation procedures for NF70). In a recent study on the cross-reactivities of MAbs between species (bovine, human, rat, rabbit, and chicken), Lee *et al.* (1986) found extensive conservation of epitopes in core domains and certain phosphorylation sites (particularly in NF200) across species lines. Another interesting conclusion of this study was that core domains are generally poorly immunogenic (in contrast to peripheral end domains), and in addition, when such antibodies are available they perform poorly in immunohistochemistry (Trojanowski *et al.*, 1985).

Some of the above points are illustrated by recent experiments on the squid giant axon (Cohen *et al.*, 1986) in which two MAbs were used; one, aNFP, was a specific MAb directed at a phosphorylated (or dephosphorylation-sensitive) epitope on the 220k (and > 400k, HMW) neurofilament protein, and the other was anti-IFA. Figure 18 shows Western blots, using these antibodies, of proteins in isolated axoplasm from the giant axon (A) and stellate ganglion neuronal perikarya (and glia) preparations (SG). The aNFP detects 220k and HMW proteins in axoplasm, but not immunoreactivity in SG proteins, whereas anti-IFA is effective in both axoplasm and SG blots. Interestingly, the HMW and 220k proteins in axoplasm were poorly reactive against anti-IFA, but the 60k polypeptide in squid axoplasm (Zackroff and Goldman, 1980; Zackroff *et al.*, 1984) was highly reactive, and immunoreactivity toward a 65k protein dominated the SG protein pattern. Figures 19 and 20 show immunohistochemical analyses of the giant axon and stellate ganglion, respectively. In Fig. 19, aNFP stains the axoplasm strongly, but not the glial and connective tissue surrounds, whereas anti-IFA stains the glial sheath intensely and the axoplasm poorly. The weak immunocytochemical staining of axoplasm by anti-IFA (Fig. 19), despite its strong staining of axoplasm on Western blots (Fig. 18), is consistent with previous observations (Trojanowski *et al.*, 1985) that antibodies directed against the core domain perform less effectively than those binding peripheral domains when visualizing neurofilaments by immunohistochemistry.

Figure 20 shows similar experiments in the squid stellate ganglion. The axons (A) in the ganglion behave as the giant axon (Fig. 18) with respect to these two antibodies, whereas the neuronal cell bodies (GC) do not react at all with aNFP. In contrast, the cell bodies do react with anti-IFA showing clusters of immunoreactivity and nuclear membrane labeling within the cell, and intense labeling of glia surrounding both the cell bodies and the axons in the neuropil. Several interpretations of the data are possible. First, the phosphorylated forms of neurofilaments appear restricted mainly to the axon (i.e., aNFP immunoreactivity), consistent with current interpretations of mammalian systems (Sternberger and Sternberger, 1983; see also Lee *et al.*, 1987). Second, the anti-IFA staining in the stellate ganglion shows predominance in the glial elements, suggesting that the 65k major stained band in Fig. 18 is probably glial in origin (note a less intense 60k stained band in the SG in Fig. 18) which is probably the neurofilament polypeptide in neurons and axons lightly

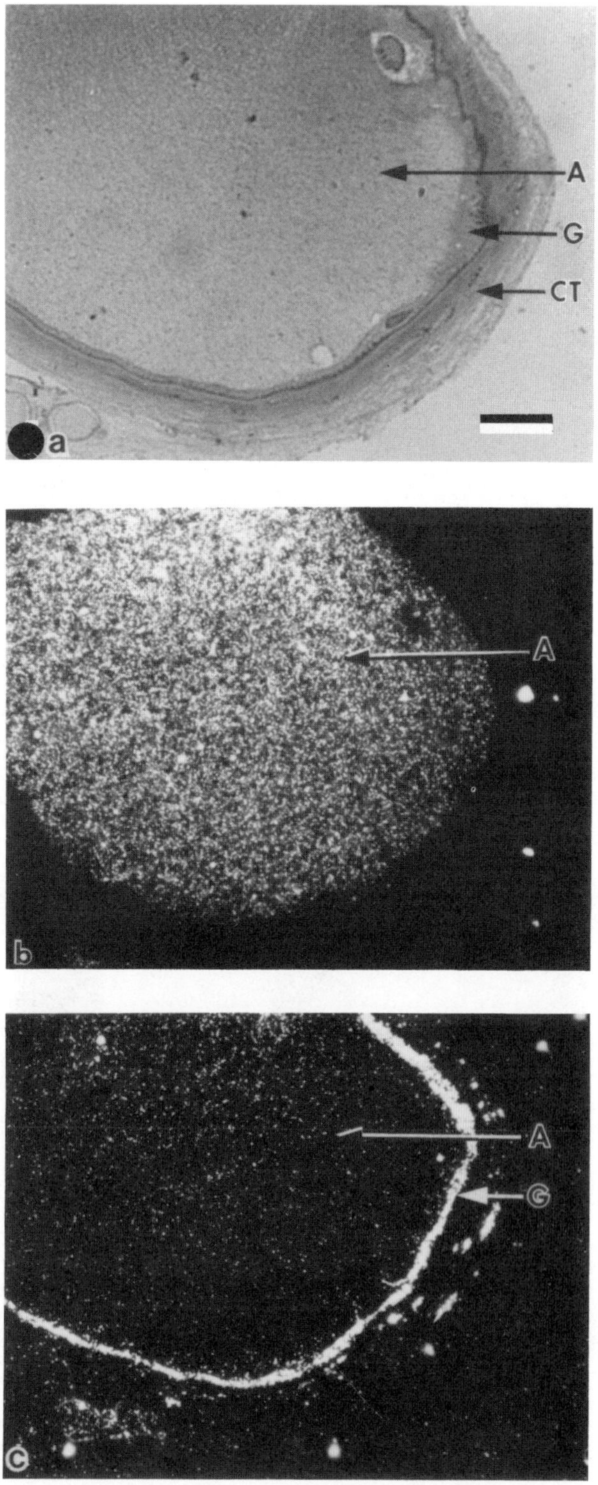

Figure 19. Antibody labeling of sections from squid axons. Light micrographs of tangential semithin sections of the giant axon stained with toluidine blue (a) or labeled with aNFP (b) or anti-IFA (c) MAbs and processed using the silver enhancement procedure. (a) The giant axon preparation consists of three components, the axon (A), the glial cell (G) surrounding the axon, and the outer connective tissue layer (CT) surrounding the glial cell. (b) Intense labeling of the giant axon (A) by the aNFP antibody is seen. No labeling is observed in the glial cell or connective tissue elements. (c) Labeling of the giant axon (A) by anti-IFA is less intense than that seen with the aNFP antibody. In contrast, intense labeling of the glial cell (G) is seen around the giant axon (A). Bar = 24 μm. Data from Cohen *et al.* (1986).

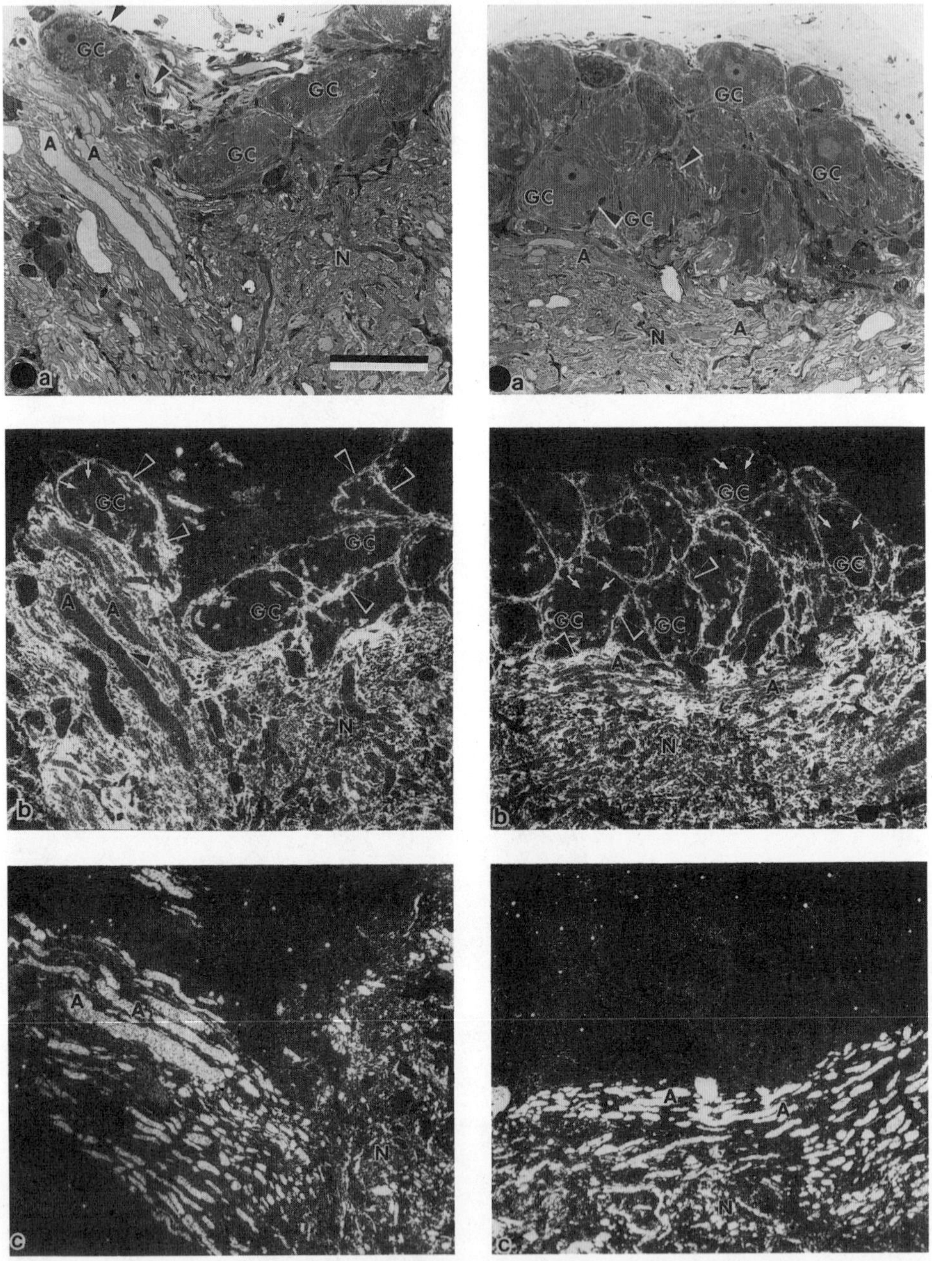

Figure 20. Light micrographs of semithin sections of the stellate ganglion from a region where axons emerge. Sections were stained with toluidine blue (a) or labeled with aIFA (b) or aNFP (c) MAbs and processed using the silver enhancement procedure. (a) Ganglion cells (GC) surround the neuropil (N) and an area containing several axons (A). Arrowheads indicate glial elements. (b) Intense labeling by anti-IFA is seen predominantly in glial elements (arrowheads) surrounding ganglion cells (GC) and axons (A). Labeling is also detected in nuclear membranes (arrows) of ganglion cells. Moderate amounts of labeling are seen in the cytoplasm of ganglion cells and axons. (c) Labeling by aNFP is restricted to axons (A) within the stellate ganglion. No reactivity is seen within ganglion or glial cells. Bar = 0.11 mm. Data from Cohen *et al.* (1986).

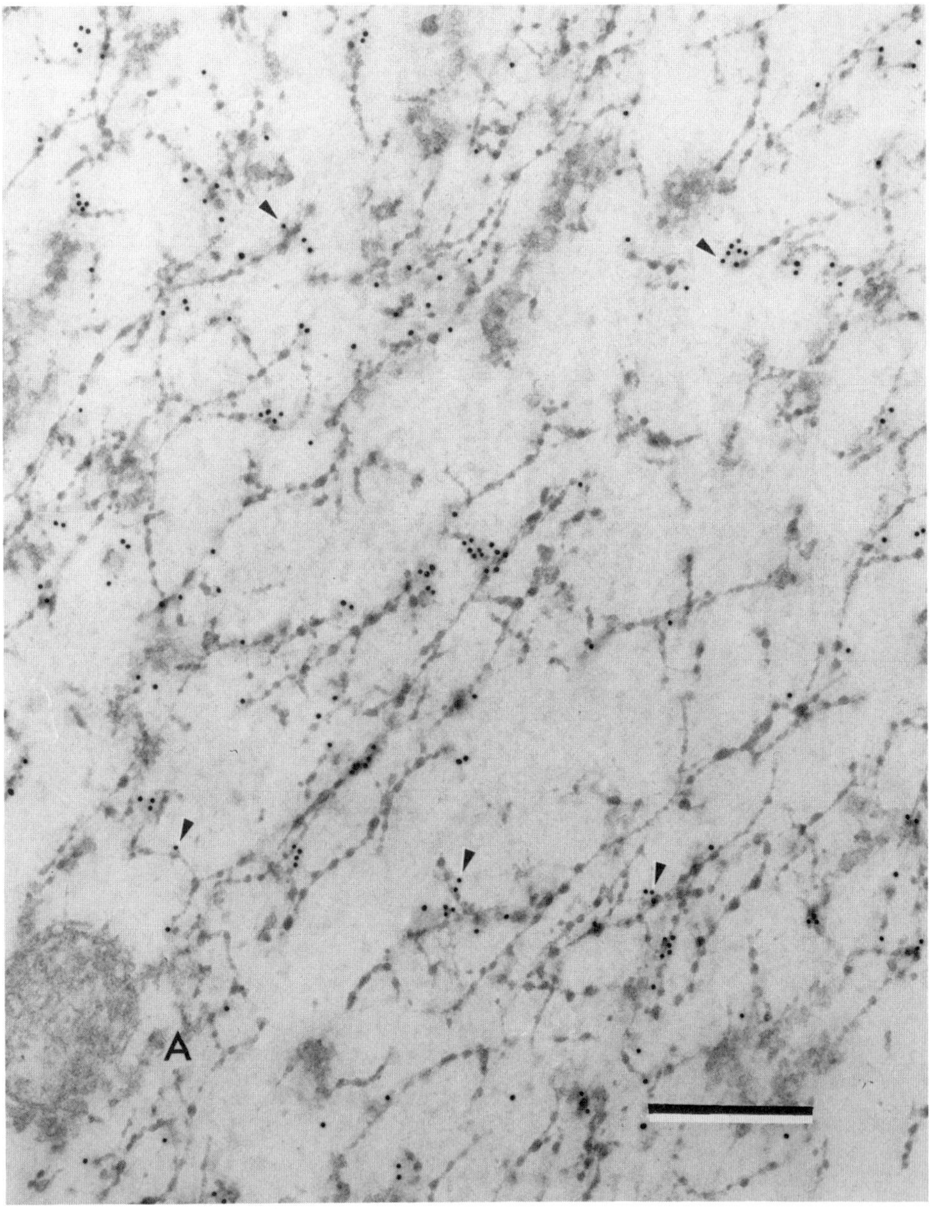

Figure 21. High-magnification electron micrographs of squid giant axon. High-magnification electron micrographs of sections from squid giant axons (A, C) incubated with aNFP, and from a glial cell surrounding the giant axon (B) incubated with anti-IFA. The secondary, goat anti-mouse, antibodies were labeled with 100 Å colloidal gold. In the axon (A, C), neurofilaments appear as straight regions interrupted at intervals by "beadlike" portions (arrowheads). Note extensive labeling of the neurofilaments adjacent to "beadlike" regions. In the glial cell (B), extensive labeling by anti-IFA is seen, whereas comparable sections of glia incubated with aNFP were unlabeled (not illustrated). Bars = 0.4 μm (A), 0.6 μm (B), 0.3 μm (C). Data from Cohen *et al.* (1986).

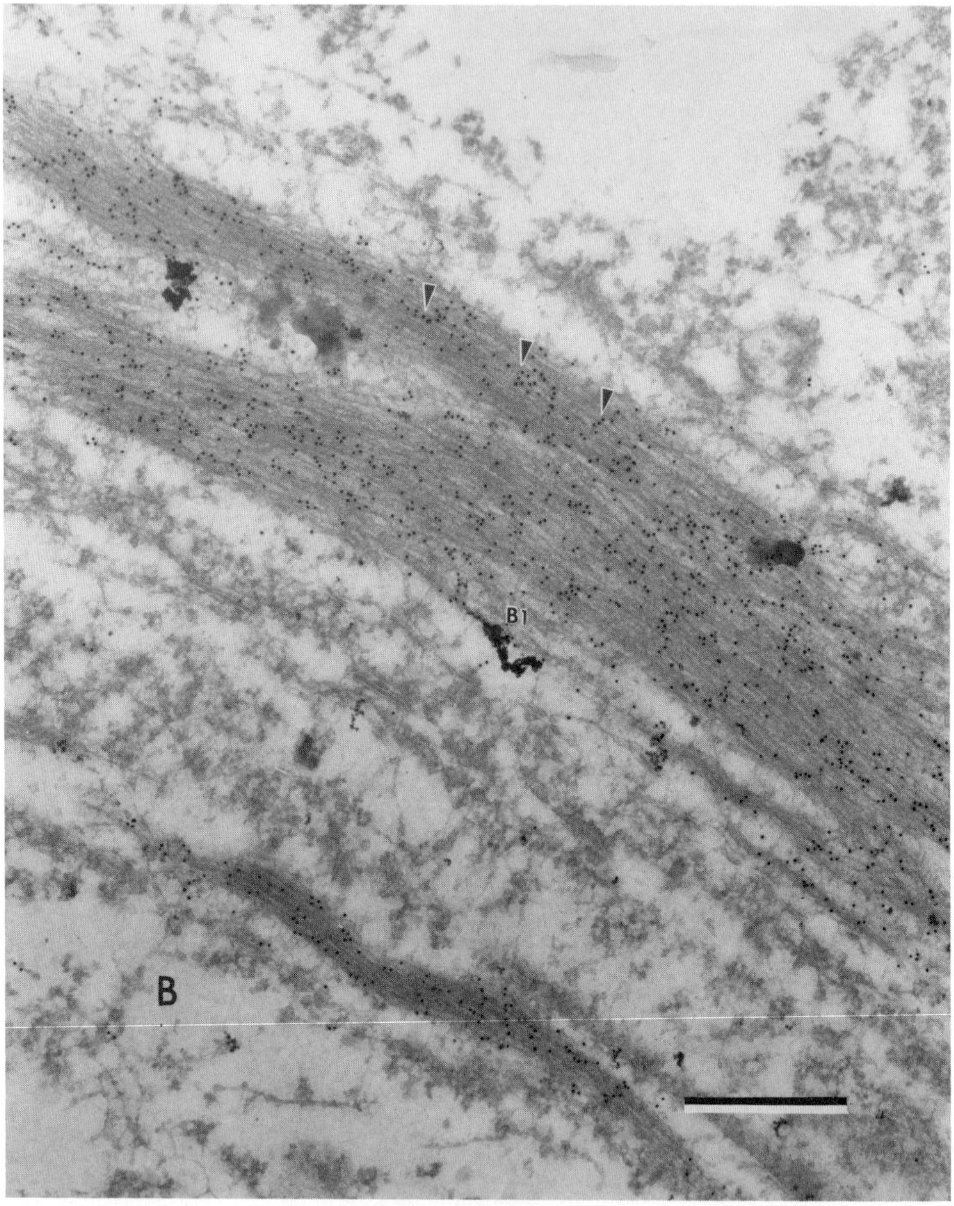

Figure 21. (*continued*)

stained in Fig. 20. Figure 21 shows immunogold labeling at high magnification for axoplasm reacted with aNFP (Fig. 21A,C) and glia labeled with anti-IFA (Fig. 21B). Note that the intense label in axoplasm by aNFP (Fig. 19) is actually distributed in a clustered manner (Fig. 21A,C) along neurofilaments, whereas the equal intense labeling of glia by anti-IFA (Figs. 19 and 20) is evenly distributed across the filaments.

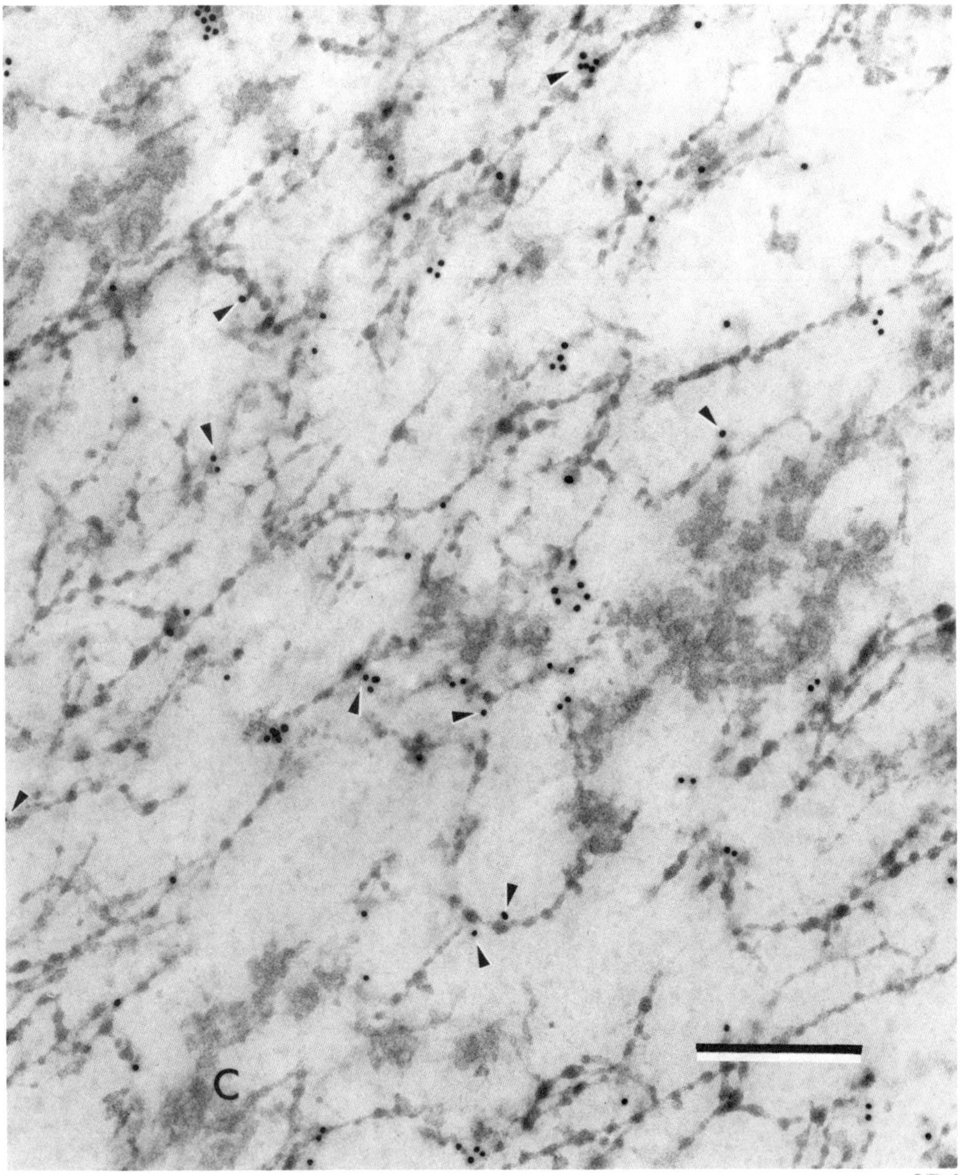

Figure 21. (continued)

Given such immunological information, in combination with the biochemical (phosphorylation and CANP) data discussed earlier, it is possible to construct a hypothesis about neurofilament regulation in the neuron. Figure 22 depicts such a model for the squid stellate ganglion neuron, which is consistent with current views about mammalian neurons (Lasek and Hoffman, 1976; Sternberger and Sternberger, 1983; Oblinger and Schick, 1985).

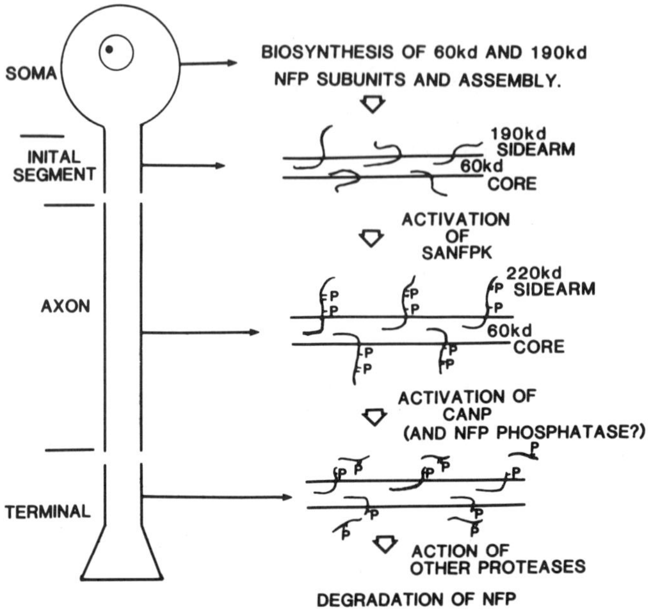

Figure 22. A model for the topographical regulation of neurofilaments in squid stellate ganglion neurons. The 60k and 190k neurofilament protein subunits, representing the putative core and unphosphorylated sidearm components, respectively, in the squid, are synthesized and assembled for axonal transport in the cell body. This is consistent with the ability of anti-IFA to stain islands of smooth intermediate filaments in the cell body, and the absence of aNFP (phosphorylated epitope) staining in the cell body. Since the appearance of aNFP staining is exclusively located in axonal structures both in the stellate ganglion and in the giant axon, we interpret this to mean that activation of the squid axon neurofilament protein kinase (SANFPK) is initiated somewhere in the region of the axonal initial segment (axon hillock), possibly regulated by an inhibitor of the kinase located in the cell body (see Pant *et al.*, 1986). In the giant axon, there is very little of the 190k subunit and virtually all of the high-molecular-weight neurofilament protein subunits are in either the 220k or > 400k form (i.e., the HMW component), both of which stain intensely with aNFP and are heavily phosphorylated by the kinase. Both the 220k and HMW proteins (we believe the HMW is a covalently linked 220k dimer, unique to the giant axon) are greatly modified and rendered unreactive to aNFP by alkaline phosphatase treatment (Cohen *et al.*, 1986). The absence of detectable 60k, 220k, or HMW proteins in the nerve terminal by either immunocytochemical (Fig. 20) or biochemical (Pant *et al.*, 1982) methods is consistent with the proposal that the degradation of neurofilaments occurs at this site, in part, by the activation of calcium-activated neutral protease (CANP) and possibly a neurofilament protein (NFP)-specific phosphatase.

7. *Axonal Transport of Neurofilaments*

Neurofilaments are synthesized and assembled in neuronal perikarya and axonally transported in the slowest component of axonal transport (Hoffman and Lasek, 1975; Lasek and Hoffman, 1976; Black and Lasek, 1980; Lasek, 1981; Willard, 1983). Aside from the vexing question regarding the nature of the motive force that is responsible for slow axonal transport in general and about which little is known, there is also the issue of the structure of the neurofilament lattice that is transported. Pulse–chase axonal transport studies are in general agreement that the newly synthesized neurofilament triplet moves through the axon in a relatively coherent fashion (Black and Lasek, 1980; Lasek, 1981;

Willard, 1983). The question, however, is: to what extent is this moving lattice plastic when regarding the exchange of molecular components, such as neurofilaments, with the stationary phase in the axon? There is insufficient experimental evidence to resolve fully this issue at present (see Willard, 1983, for a detailed discussion of this point). However, some plasticity in the moving neurofilament lattice is indicated by evidence of proteolytic modification of the moving neurofilament proteins (Nixon, 1980, 1983; Nixon *et al.*, 1982). In addition, Nixon and his colleagues (Nixon, 1983; Nixon *et al.*, 1985) have presented some experimental evidence for an axonal transport model in which the moving wave of neurofilament proteins contributes to and serves to maintain a stationary axonal cytoskeletal lattice.

8. *Ontogeny and Phylogeny of Neurofilaments*

In an interesting essay on the function and evolution of neurofilament proteins, Lasek *et al.* (1985) pointed out that the central biological value of neurofilament proteins during evolution was to allow for the construction of giant axons for escape responses (e.g., in the squid, *Myxicola,* and Mauthner cells in vertebrates). These authors further stated the view that the principal current use of neurofilament proteins is to provide the cellular infrastructure necessary to increase axon radial growth and caliber (e.g., in motor and sensory neuron axons). That this function is "useful but nonessential" (a description also applicable to the gene concerned) is indicated by the fact that a whole phylum, the Arthropoda, can produce a nervous system, indeed respectable giant axons, without the presence of neurofilaments. The above paper also contains a survey of neurofilament protein subunits in organisms across a wide range of phyla. Using a reasonably rigorous criterion for neurofilament protein identity, i.e., staining by anti-IFA, more specific antibodies, and the Bodian silver method, the authors concluded that most species have at least one low-molecular-weight neurofilament protein (60–70k) and one or more high-molecular-weight neurofilament proteins ($>$ 100k). Exceptions to the rule are *Myxicola* and lampreys, which only have $>$ 100k subunits, and *Aplysia,* which only has low-molecular-weight subunits.

Since high- and low-molecular-weight neurofilament subunits are encoded by separate genes, Lasek *et al.* (1985) argued that the observed phylogenetic diversity in the neurofilament family is probably due to at least three gene duplications, evolutionary variations in the tail segment's length and base pair compositions, and gene deletions (as in the Arthropoda). These authors also suggested that early (ca. 700–800 million years ago) gene duplication of the ancestral intermediate filament gene gave rise to two completely independent intermediate filament lineages, i.e., the neuronal (neurofilament) and nonneuronal intermediate filament subfamilies. Recent recombinant DNA studies (see Lewis and Cowan, 1986) have confirmed and extended this view. These workers cloned the 68k neurofilament gene in mouse, and determined the positions of the introns. In addition to showing that this gene encodes a 543-amino-acid polypeptide that contains the α-helical rod region common to all intermediate filament proteins, Lewis and Cowan (1986) found three introns all of which were in different positions from the introns found in all other members of the intermediate filament protein family. Desmin, vimentin, and GFAP genes, which represent one subfamily, contain eight introns at identical positions, whereas type I and type II keratin genes have seven and six homologously placed introns, respectively, and one additional intron each in a unique location (see Fig. 3 in Lewis and Cowan, 1986). The distinct positions of the introns in the mouse NF68 gene are homologous to those in the human gene encoding the 150k triplet component recently cloned and sequenced (Lazzarini and Schlaepfer, personal communication). Thus, based on intron locations, there are at least three subfamilies of the intermediate filament gene family: the keratins, nonkeratin, nonneuronal filaments, and the neurofilaments. These data also are consistent with an early divergence of the neurofilament proteins and the nonneuronal and keratin filament proteins from the primordial intermediate filament gene, and the subsequent independent evolution of each group.

Separate genes for the different forms of neurofilament proteins implies that independent gene regulation (transcription) is possible. Indeed, the various forms seem to be differentially expressed during neuronal development (Shaw and Weber, 1982; Sharp *et al.*, 1982; Willard and Simon, 1983; Glicksman and Willard, 1985; Scott *et al.*, 1985; Calvert and Anderton, 1982; Pachter and Liem, 1984) (though as some of these reports employed antibodies that were probably recognizing phosphorylated epitopes, the situation is complex). In general, however, it appears that the NFL (68k) and NFM (150k) forms precede NFH (200k) expression, often by weeks. In some neurons, the NFH product is never elaborated. Glicksman and Willard (1985) speculated that the delay of expression of NFH may be to allow the accomplishment of "certain plastic functions." Presumably then, the introduction of NFH would end this hypothetical "critical period" and allow for the "consolidation" of the axon [presumably correlated with increasing axon caliber; see Hoffman *et al.* (1983)]. The above sequence of neurofilament expression is consistent with the notion expressed earlier that the principal role of neurofilaments, especially NFH (Calvert, 1982), is to increase the radial growth of the axon (Lasek *et al.*, 1985). However, there have been reports of even earlier expressions of the neurofilament proteins in precursors of neurons in postmitotic stages, well before axon formation (Bennett and DiLullo, 1985a,b). What the functions are in these cells of the neurofilament proteins, which are often expressed transiently, is completely unclear at present.

ACKNOWLEDGMENTS. We would like to thank the following people: John Hopkins for providing unpublished micrographs, Figs. 1–3; Lochelle Cohen for assistance and advice about electron microscopy; and Mrs. Carol Eagles for typing the manuscript. P.A.M.E. gratefully acknowledges support from the Foundation for Age Research, the Wellcome Trust, and the Medical Research Council (U.K.).

References

Aamodt, E. J., and Williams, Jr., R. C., 1984a, Microtubule-associated proteins connect microtubules and neurofilaments *in vitro*, *Biochemistry* **23:**6023–6031.

Aamodt, E. J., and Williams, Jr., R. C., 1984b, Association of microtubules and neurofilaments *in vitro* is not mediated by ATP, *Biochemistry* **23:**6031–6035.

Abercrombie, R. F., Gammeltoft, K., Jackson, J., and Young, L., 1986, An intracellular calcium binding site on neurofilament proteins of *Myxicola* giant axon, *J. Gen. Physiol.* **88:**9a.

Anderson, D. R., Davis, J. L., and Carraway, K. L., 1977, Calcium-promoted changes of the human erythrocyte membrane. Involvement of spectrin, trans-glutaminase, and membrane bound protease, *J. Biol. Chem.* **252:**6617–6623.

Anderton, B. H., Bell, C. W., Newby, B. J., and Gilbert, D. S., 1976, Neurofilaments, *Biochem. Soc. Trans.* **4**:544–547.

Arnold, J. M., Summers, W. C., Gilbert, D. L., Manalis, R. S., Daw, N. W., and Lasek, R. J., 1974, A guide to laboratory use of the squid *Loligo pealei*, Marine Biological Laboratory, Woods Hole, Mass.

Autilio-Gambetti, L., Velasco, M. E., Sipple, J., and Gambetti, P., 1981, Immunochemical characterization of antisera to rat neurofilament subunits, *J. Neurochem.* **37**:1260–1265.

Autilio-Gambetti, L., Crane, R., and Gambetti, P., 1986, Binding of Bodian's silver and monoclonal antibodies to defined regions of human neurofilament subunits: Bodian's silver reacts with a highly charged unique domain of neurofilaments, *J. Neurochem.* **46**:366–370.

Banik, N. L., McAlhaney, W. W., and Hogan, E. L., 1985, Calcium stimulated proteolysis in myelin; evidence for a calcium-activated neutral proteinase associated with purified myelin of rat CNS, *J. Neurochem.* **45**:581–588.

Baudry, M., and Lynch, G., 1980, Regulation of hippocampal glutamate receptors: Evidence for the involvement of a calcium-activated protease, *Proc. Natl. Acad. Sci. USA* **77**:2298–2302.

Baudry, M., Bundman, M. C., Smith, E. R., and Lynch, G. S., 1981, Micromolar calcium stimulates proteolysis and glutamate binding in rat brain synaptic membranes, *Science* **212**:937–938.

Bear, R. S., Schmitt, F. O., and Young, J. Z., 1937a, The ultrastructure of nerve axoplasm, *Proc. R. Soc. London Ser. B* **123**:505–519.

Bear, R. S., Schmitt, F. O., and Young, J. Z., 1937b, Investigation on the protein constituents of nerve axoplasm, *Proc. R. Soc. London Ser. B* **123**:520–529.

Bell, C. W., 1977, Hydrodynamic properties of neurofilaments, *J. Physiol. (London)* **266**:83P–84P.

Bell, C. W., 1978, An approach to the structure of neurofilaments using crosslinking reagents and the analytical ultracentrifuge, Ph.D. thesis, University of London.

Bennett, G. S., and DiLullo, C., 1985a, Expression of a neurofilament protein by the precursors of a subpopulation of ventral spinal cord neurons, *Dev. Biol.* **107**:94–106.

Bennett, G. S., and DiLullo, C., 1985b, Transient expression of a neurofilament protein by replicating neuroepithelial cells of the embryonic chick brain, *Dev. Biol.* **107**:107–127.

Berkowitz, S. A., Katagiri, J., Binder, H. K., and Williams, Jr., R. C., 1977, Separation and characterization of microtubule proteins from calf brain, *Biochemistry* **16**:5610–5617.

Bertolini, B., Monaco, G., and Rossi, A., 1970, Ultrastructure of a regular arrangement of microtubules and neurofilaments, *J. Ultrastruct. Res.* **33**:173–186.

Black, M. M., and Lasek, R. J., 1980, Slow components of axonal transport: Two cytoskeletal networks, *J. Cell Biol.* **86**:616–623.

Boycott, B. B., Gray, E. G., and Guillery, R. W., 1961, Synaptic structure and its alteration with environmental temperature: A study by light and electron microscopy of the central nervous system of lizzards, *Proc. R. Soc. London Ser. B* **154**:151–172.

Brown, A., and Eagles, P. A. M., 1986, Squid neurofilaments—Phosphorylation and calcium-dependent proteolysis *in situ*, *Biochem. J.* **239**:191–197.

Brown, B. A., Nixon, R. A., Strocchi, P., and Marotta, C. A., 1981, Characterisation and comparison of neurofilament proteins from rat and mouse CNS, *J. Neurochem.* **36**:143–165.

Brown, B. A., Majocha, R. E., Staton, D. M., and Marotta, C. A., 1983, Axonal polypeptides cross-reactive with antibodies to neurofilament proteins, *J. Neurochem.* **40**:299–308.

Calvert, R., 1982, Is polypeptide heterogeneity the key to neurofilament function? *Nature* **300**:580–581.

Calvert, R., and Anderton, B. H., 1982, *In vivo* metabolism of mammalian neurofilament polypeptides in developing and adult rat brain, *FEBS Lett.* **145**:171–175.

Carden, M. J., and Eagles, P. A. M., 1983, Neurofilaments from ox spinal nerves: Isolation, disassembly, reassembly and crosslinking properties, *Biochem J.* **215**:227–237.

Carden, M. J., and Eagles, P. A. M., 1986, Chemical cross-linking analyses of ox neurofilaments, *Biochem. J.* **234**:587–591.

Carden, M. J., Schlaepfer, W. W., and Lee, V. M.-Y., 1985, The structure, biochemical properties, and immunogenicity of neurofilament peripheral regions are determined by phosphorylation state, *J. Biol. Chem.* **260**:9805–9817.

Carden, M. J., Trojanowski, J. Q., Schlaepfer, W. W., and Lee, V. M.-Y., 1987, Two stage expression of the neurofilament polypeptides during rat neurogenesis with early establishment of adult phosphorylation patterns, *J. Neurosci.* **7**:3489–3504.

Cassell, D., and Glasser, L., 1982, Proteolytic cleavage of epidermal growth factor receptor. A calcium dependent, sulphydryl-sensitive proteolytic system in A431 cells, *J. Biol. Chem.* **257**:9845–9848.

Chin, T.-K., Eagles, P. A. M., and Maggs, A., 1983, The proteolytic digestion of ox neurofilaments with trypsin and alpha-chymotrypsin, *Biochem. J.* **215**:239–252.

Chin, T.-K., Harding, S., and Eagles, P. A. M., 1989, Characterization of two proteolytically derived soluble polypeptides from the neurofilament triplet components NFM and NFH, *Biochem. J.* **264:** 53–60.

Chiu, F.-C., and Norton, W. T., 1982, Bulk preparation of CNS cytoskeleton and the separation of individual neurofilament proteins by gel filtration: Dye binding characteristics and amino acid composition, *J. Neurochem.* **39:**1252–1260.

Chiu, F.-C., Korey, B., and Norton, W. T., 1980, Intermediate filaments from bovine, rat and human CNS: Mapping analysis of the major proteins, *J. Neurochem.* **34:**1149–1159.

Chiu, F.-C., Goldman, J. E., and Norton, W. T., 1983, Biochemistry of neurofilaments, in: *Neurofilaments* (C. A. Morotta, ed.), University of Minnesota Press, Minneapolis, pp. 27–56.

Cohen, R. S., Pant, H. C., House, S., and Gainer, H., 1986, Characterization and distribution of phosphorylated and nonphosphorylated subunits of squid neurofilaments, *16th Annu. Meet. Soc. Neurosci.* **12:**1571 (Part II).

Cohlberg, J. A., Hajarian, H., and Sainte-Marie, S., 1987, Discrete soluble forms of middle and high molecular weight neurofilament proteins in dilute aqueous buffers, *J. Biol. Chem.* **262:**17009–17015.

Collier, N. C., and Wang, K., 1982, Purification and properties of human platelet P235: A high molecular weight protein substrate of endogenous calcium-activated protease(s), *J. Biol. Chem.* **257:**6937–6943.

Coolican, S. A., Haiech, J., and Hathaway, D. R., 1986, The role of subunit autolysis in activation of smooth muscle calcium-dependent proteases, *J. Biol. Chem.* **261:**4170–4176.

Dahl, D., 1981, Isolation of neurofilament proteins and of immunologically active neurofilament degradation products from extracts of brain, spinal cord and sciatic nerve, *Biochim. Biophys. Acta* **668:**299–306.

Dautigny, A., Pham-Dinh, D., Roussel, C., Felix, J. M., Nussbaum, J. L., and Jolles, P., 1988, The large neurofilament subunit (NF-H) of the rat: cDNA, cloning and *in situ* detection, *Bioch. Biophys. Acta* **154:**1099–1106.

Davison, P. F., and Taylor, E. W., 1960, Physical-chemical studies of proteins of squid nerve axoplasm, with special reference to the axon fibrous protein, *J. Gen. Physiol.* **43:**801–823.

Davison, P., and Winslow, B., 1974, The protein subunit of calf brain neurofilament, *J. Neurobiol.* **5:**119–133.

Day, W. A., and Gilbert, D. S., 1972, X-ray diffraction patterns of axoplasm, *Biochim. Biophys. Acta* **285:**503–506.

Dayton, W. R., Schollmeyer, J. V., Lepley, R. A., and Cortex, L. R., 1981, A calcium-activated protease possibly involved in myofibrillar protein turnover: Isolation of a low calcium requiring form of the protease, *Biochim. Biophys. Acta* **659:**48–61.

Debus, E., Weber, K., and Osborn, M., 1983, Monoclonal antibodies specific for glial fibrillary acidic (GFA) protein and for each of the neurofilament triplet polypeptides, *Differentiation* **25:**193–203.

Delacourte, A., Plancot, M.-T., Than, K.-K., Hildebrand, H., and Biserke, G., 1977, Investigation of tubulin fibres formed during microtubule polymerisation cycles, *FEBS Lett.* **77:**41–46.

Delacourte, A., Filliatreau, G., Boutteau, F., Biseike, G., and Schreval, J., 1980, Study of the 10nm-filament fraction isolated during the standard microtubule preparation, *Biochem. J.* **191:**543–546.

Delacourte, A., Dousti, M., and Loucheux-Lefebvre, M.-H., 1982, Characterisation of mammalian neurofilament subunits by circular dichroism, *Biochim. Biophys. Acta* **709:**99–104.

Drager, U. C., Edwards, D. L., and Kleinschmidt, J., 1983, Neurofilaments contain alpha-melanocyte-stimulating hormone (alpha-MSH)-like immunoreactivity, *Proc. Natl. Acad. Sci. USA* **80:**6408–6412.

Eagles, P. A. M., 1986, Neurofilament architecture, in: *Electron Microscopy and Alzheimer's Disease* (J. Metuzals, ed.), San Francisco Press, San Francisco, pp. 6–8.

Eagles, P. A. M., and Maggs, A., 1985, The location of domains of neurofilament polypeptides, *Ann N.Y. Acad. Sci.* **455:**779–781.

Eagles, P. A. M., Bell, C., Maggs, A., Wais, C., and Gilbert, D. S., 1978, Neurofilament structure and its enzymatic modification, in: *The Cytoskeleton and Contractile Networks of Non-muscle Cells,* Cold Spring Harbor Laboratory, Cold Spring Harbor, N.Y., p. 47.

Eagles, P. A. M., Gilbert, D. S., Hopkins, J. M., Maggs, A., and Wais-Steider, C., 1980a, Neurofilament structure and enzymatic modification, *Cell Biol. Int. Rep.* **4:**731.

Eagles, P. A. M., Gilbert, D. S., and Maggs, A., 1980b, Neurofilament structure and enzymic modification, *Biochem. Soc. Trans.* **8:**484–487.

Eagles, P. A. M., Gilbert, D. S., and Maggs, A., 1981a, The polypeptide composition of axoplasm and of neurofilaments from the marine worm *Myxicola infundibulum, Biochem. J.* **199:**89–100.
Eagles, P. A. M., Gilbert, D. S., and Maggs, A., 1981b, The location of phosphorylation sites and calcium dependent proteolytic cleavage sites on the major neurofilament polypeptides from *Myxicola infundibulum, Biochem. J.* **199:**101–111.
Eagles, P. A. M., Hopkins, J. M., Rao, R., and Chin, T.-K., 1985, Neurofilament sidearm, proteolysis and phosphorylation, in: *Proc. 43rd Annu. Meet. Electron Microsc. Soc. Am.*, San Francisco Press, San Francisco, pp. 744–747.
Ellisman, M. H., and Porter, K. R., 1980, Microtrabecular structure of the axoplasmic matrix: Visualisation of cross-linking structures and their distribution, *J. Cell Biol.* **87:**464–479.
Emori, Y., Kawasaki, H., Imajoh, S., Kawashima, S., and Suzuki, K., 1986, Isolation and sequence of cDNA clones for the small subunit of rabbit calcium-dependent protease, *J. Biol. Chem.* **261:**9472–9476.
Frappier, T., Regnorf, F., and Pradel, L. A., 1986, Association of fodrin to the 70kDa neurofilament protein, *Abstracts 15th European Congress on Muscle and Motility,* p. 115.
Friede, R. L., and Samorajski, T., 1970, Axon caliber related to neurofilaments and microtubules in sciatic nerve fibers of rats and mice, *Anat. Rec.* **167:**379–388.
Gallant, P. E., Pant, H. C., Pruss, R. M., and Gainer, H., 1986, Calcium-activated proteolysis of neurofilament proteins in the squid giant neuron, *J. Neurochem.* **46:**1573–1581.
Gambetti, P., Autilio-Gambetti, L., and Papasozomenos, S. C., 1981, Bodian's silver method stains neurofilament polypeptides, *Science* **213:**1521–1522.
Gardner, E. E., Dahl, D., and Bignami, A., 1984, Formation of 10-nanometer filaments from the 150k-dalton neurofilament protein *in vitro, J. Neurosci. Res.* **11:**145–155.
Geisler, N., and Weber, K., 1981, Self-assembly *in vitro* of the 68,000 molecular weight component of the mammalian neurofilament triplet proteins into intermediate-sized filaments, *J. Mol. Biol.* **151:**565–571.
Geisler, N., Kaufmann, E., Fisher, S., Plessmann, U., and Weber, K., 1983, Neurofilament architecture combines structural principles of intermediate filaments with carboxy-terminal extensions increasing in size between triplet proteins, *EMBO J.* **2:**1295–1302.
Geisler, N., Fisher, S., Vandekerckhove, J., Plessmann, U., and Weber, K., 1984, Hybrid character of a large neurofilament protein (NF-M): Intermediate filament type sequence followed by a long and acidic carboxy-terminal extension, *EMBO J.* **3:**2701–2706.
Geisler, N., Plessmann, U., and Weber, K., 1985a, The complete amino acid sequence of the major mammalian neurofilament protein (NF-L), *FEBS Lett.* **182:**475–478.
Geisler, N., Fischer, S., Vandekerckhove, J., Van Damme, J., Plessmann, U., and Weber, K., 1985b, Protein-chemical characterisation of NF-H, the largest mammalian neurofilament component; intermediate filament-type sequences followed by a unique carboxy-terminal extension, *EMBO J.* **4:**57–63.
Geisler, N., Vandekerckhove, J., and Weber, K., 1987, Location and sequence characterization of the major phosphorylation sites of the high molecular mass neurofilament proteins M and H, *FEBS Lett.* **221:**403–407.
Georges, E., and Mushynski, W. E., 1987, Chemical modification of charged amino acid moieties alters the electrophoretic mobilities of neurofilament subunits on SDS-polyacrylamide gels, *Eur. J. Biochem.* **165:**281–287.
Georges, E., Lefebre, S., and Mushynski, W. E., 1986, Dephosphorylation of neurofilaments has no effect on reassembly of subunits, *J. Neurochem.* **47:**477–483.
Gilbert, D. S., 1972, Helical structure of *Myxicola* axoplasm, *Nature New Biol.* **237:**195–198.
Gilbert, D. S., 1975a, Axoplasm architecture and physical properties as seen in the *Myxicola* giant axon, *J. Physiol. (London)* **253:**257–301.
Gilbert, D. S., 1975b, Axoplasm chemical composition in *Myxicola* and solubility properties of its structural proteins, *J. Physiol. (London)* **253:**303–319.
Gilbert, D. S., Newby, B. J., and Anderton, B. H., 1975, Neurofilament disguise, destruction and discipline, *Nature* **256:**586–589.
Glicksman, M. A., and Willard, M., 1982, Separation of neurofilaments form ATPase activity by precipitation with anti-neurofilament antibodies, *J. Neurochem.* **38:**1774–1776.
Glicksman, M. A., and Willard, M., 1985, Differential expression of the three neurofilament polypeptides, *Ann N.Y. Acad. Sci.* **455:**479–491.
Goldman, J. E., Schaumburg, H. H., and Norton, W. T., 1978, Isolation and characterisation of glial filaments from human brain, *J. Cell Biol.* **78:**426–440.

Gray, E. G., and Guillery, R. W., 1961, The basis for silver staining of synapses of the mammalian spinal cord: A light and electron microscope study, *J. Physiol.* (*London*) **157:**581–588.
Gray, E. G., and Guillery, R. W., 1966, Synaptic morphology in the normal and degenerating nervous system, *Int. Rev. Cytol.* **19:**111–182.
Guroff, G., 1964, A neutral, calcium-activated protease from the soluble fraction of the rat brain, *J. Biol. Chem.* **239:**149–155.
Hamon, M., and Bourgoin, S., 1979, Characterization of the calcium-induced proteolytic activation of tryptophan hydroxylase from the rat brain stem, *J. Neurochem.* **32:**1837–1844.
Hathaway, D. R., Werth, D. K., and Haeberle, J. R., 1982, Limited autolysis reduces the calcium requirements of a smooth muscle calcium-activated protease, *J. Biol. Chem.* **257:**9072–9077.
Hayashi, M., Chiomata, M., Nakamura, M., Imahori, K., and Kawashima, S., 1985, Hydrolysis of protamine by calcium-activated neutral protease (CANP), *J. Biochem.* **97:**1363–1370.
Heimann, R., Shelanski, M. L., and Liem, R. K. H., 1984, *In vitro* binding of microtubule associated proteins and neurofilaments, *J. Cell Biol.* **99:**190a.
Henderson, D., Geisler, N., and Weber, K., 1982, A periodic ultrastructure in intermediate filaments, *J. Mol. Biol.* **155:**173–176.
Hiraco, T., and Takahashi, K., 1984, Purification and characterization of a calcium-activated neutral protease from monkey brain and its action on neuropeptides, *J. Biochem.* **96:**775–784.
Hirokawa, N., 1982, Cross-linker system between neurofilaments, microtubules and membranous organelles in frog axons revealed by the quick-freeze, deep-etching method, *J. Cell Biol.* **94:**129–142.
Hirokawa, N., Glicksman, M. A., and Willard, M. B., 1984, Organisation of mammalian neurofilament polypeptides within the neuronal cytoskeleton, *J. Cell Biol.* **98:**1523–1536.
Hisanaga, S., and Hirokawa, N., 1988, Structure of the peripheral domains of neurofilaments revealed by low angle rotary shadowing, *J. Mol. Biol.* **202:**297–305.
Hodge, A. J., and Adelman, Jr., W. J., 1980, The neuroplasmic network in *Loligo* and Hermissenda neurons, *J. Ultra Struct. Res.* **70:**220–241.
Hodgkin, A. L., and Katz, B., 1949, Effect of calcium on the axoplasm of giant nerve fibres, *J. Exp. Biol.* **26:**292–294.
Hodgkin, A. L., and Keynes, R. D., 1956, Experiments on injections of substances into squid giant axons by means of a microsyringe, *J. Physiol.* (*London*) **131:**592–616.
Hoffman, P. N., and Lasek, R. J., 1975, The slow component of axonal transport. Identification of major structural polypeptides of the axon and their generality among mammalian neurons, *J. Cell Biol.* **66:**351–366.
Hoffman, P. N., Lasek, R. J., Griffin, J. W., and Price, D. L., 1983, Slowing of the axonal transport of neurofilament proteins during development, *J. Neurosci.* **3:**1694–1700.
Hoffman, P. N., Griffin, J. W., and Price, D. L., 1984, Control of axonal caliber by neurofilament transport, *J. Cell Biol.* **99:**705–714.
Hogue-Angeletti, R. A., Wu, H.-L., and Schlaepfer, W. W., 1982, Preparative separation and amino acid composition of neurofilament triplet proteins, *J. Neurochem.* **38:**116–120.
Huneeus, F. C., and Davison, P. F., 1970, Fibrillar proteins from squid axons, *J. Mol. Biol.* **52:**415–428.
Huston, R. B., and Krebs, E. G., 1968, Activation of skeletal muscle phosphorylase kinase by calcium. II. Identification of the kinase activation factor as a proteolytic enzyme, *Biochemistry* **7:**2116–2121.
Ip, W., 1986, Structural studies of isolated mammalian neurofilaments, in: *Electron Microscopy and Alzheimer's Disease* (J. Metuzals, ed.), San Francisco Press, San Francisco.
Ishiura, S., 1981, Calcium dependent proteolysis in living cells, *Life Sci.* **29:**1079–1088.
Ishiura, S., Murofuishi, H., Suzuki, K., and Imahori, K., 1978, Studies of a calcium-activated protease from chicken skeletal muscle. I. Purification and characterization, *J. Biochem.* **84:**225–230.
Jones, S. M., and Williams, Jr., R. C., 1982, Phosphate content of mammalian neurofilaments, *J. Biol. Chem.* **257:**9902–9905.
Julien, J.-P., and Mushynski, W. E., 1981, A comparison of *in vitro* and *in vivo*-phosphorylated neurofilament polypeptides, *J. Neurochem* **37:**1579–1585.
Julien, J.-P., and Mushynski, W. E., 1982, Multiple phosphorylation sites in mammalian neurofilament polypeptides, *J. Biol. Chem.* **257:**10467–10470.
Julien, J.-P., and Mushynski, W. E., 1983, The distribution of phosphorylation sites among identified proteolytic fragments of mammalian neurofilaments, *J. Biol. Chem.* **258:**4019–4025.

Julien, J.-P., Smoluk, G. D., and Mushynski, W. E., 1983, Characteristics of the protein kinase activity associated with rat neurofilament preparations, *Biochim. Biophys. Acta* **775:**25–31.

Julien, J.-P., Ramachandran, K., and Grosveld, F., 1985, Cloning of a cDNA encoding the smallest neurofilament protein from the rat, *Biochim. Biophys. Acta* **825:**398–404.

Julien, J.-P., Meyer, D., Flavell, D., Hurst, J., and Grosveld, F., 1986, Cloning and developmental expression of the murine neurofilament gene family, *Mol. Brain Res.* **1:**243–250.

Kamakura, K., Ishiura, S., Suzuki, K., Sugita, H., and Toyokiera, Y. 1985, Calcium activated neutral protease in the peripheral nerve which requires micromolar calcium, and its effect on neurofilament triplet, *J. Neurosci. Res.* **13:**391–403.

Kamakura, K., Ishiura, S., and Sugita, H., 1986, Mu-type calcium-activated neutral protease in the rat peripheral nerve, *J. Neurosci. Res.* **15:**167–173.

Kameyama, T., and Etlinger, J. D., 1979, Calcium-dependent regulation of protein synthesis and degradation of muscle, *Nature* **279:**344–346.

Kaspi, C., and Mushynski, W. E., 1985, Characterization of interspecies size heterogeneity of the large subunit of mammalian neurofilaments, *Ann. N.Y. Acad. Sci.* **455:**794–796.

Kaufmann, E., Geisler, N., and Weber, K., 1984, SDS-PAGE strongly overestimates the molecular masses of the neurofilament proteins, *FEBS Lett.* **170:**81–84.

Kaur, H., and Sanwal, B. D., 1981, Regulation of activity of a calcium-activated neutral protease during differentiation of skeletal myoblasts, *Can. J. Biochem.* **59:**743–748.

Kawasaki, H., Imajoh, S., Kawashima, S., Hayashi, H., and Suzuki, K., 1986, The small subunits of calcium dependent proteases with different sensitivities are identical, *J. Biochem.* **99:**1525–1532.

Krishnan, N., Kaiserman-Abramof, I. R., and Lasek, R. J., 1979, Helical substructure of neurofilaments isolated from *Myxicola* and squid giant axons, *J. Cell Biol.* **82:**323–335.

Kubota, S., Suzuki, K., and Imahori, K., 1981, A new method for the preparation of a calcium-activated neutral protease highly sensitive to calcium ions, *Biochem. Biophys. Res. Commun.* **100:**1189–1194.

Lasek, R. J., 1981, The dynamic ordering of neuronal cytoskeletons, *Neurosci. Res. Progr. Bull.* **19:**7–32.

Lasek, R. J., and Black, M. M., 1977, How do axons stop growing? Some clues from the metabolism of the proteins in the slow component of axonal transport, in: *Mechanisms, Regulations, and Special Functions of Protein Synthesis in the Brain* (E. Roberts, S. Roberts, A. Lajtha, W. H. Gispen, eds.) Elsevier/North-Holland, Amsterdam, pp. 161–169.

Lasek, R. J., and Hoffman, P. N., 1976, The neuronal cytoskeleton, axonal transport and axonal growth, in: *Cell Motility,* Volume 3 (R. Goldman, T. Pollard, and J. Rosenbaum, eds.), Cold Spring Harbor Laboratory, Cold Spring Harbor, N.Y., pp. 1021–1049.

Lasek, R. J., Krishnan, N., and Kaiserman-Abramof, I. R., 1979, Identification of the subunit proteins of 10nm neurofilaments isolated from axoplasm of squid and Myxicola giant axons, *J. Cell Biol.* **82:**336–346.

Lasek, R. J., Metuzals, J., and Kaiserman-Abramof, I. R., 1983, Cytoskeletons reconstituted *in vitro* indicate that neurofilaments contribute to the helical structure of axons, in: *Developing and Regenerating Vertebrate Nervous Sytems,* (P. W. Coates, R. R. Mankwald, and A. D. Kenny, eds.), Liss, New York, pp. 1–18.

Lasek, R. J., Phillips, L., Katz, M. J., and Autilio-Gambetti, L., 1985, Function and evolution of neurofilament proteins, *Ann N.Y. Acad. Sci.* **455:**462–478.

Lazarides, E., 1980, Intermediate filaments as mechanical integrators of cellular space, *Nature* **283:**249–256.

Lee, V., Wu, H. L., and Schlaepfer, W. W., 1982, Monoclonal antibodies recognize individual neurofilament triplet proteins, *Proc. Natl. Acad. Sci. USA* **79:**6089–6092.

Lee, V. M.-Y., Carden, M. J., and Schlaepfer, W. W., 1986, Structural similarities and differences between neurofilament proteins from five different species as revealed using monoclonal antibodies, *J. Neurosci.* **6:**2179–2186.

Lee, V. M.-Y., Carden, M. J., Schlaepfer, W. W., and Trojanowski, J. Q., 1987, Monoclonal antibodies distinguish several differentially phosphorylated states of the two largest rat neurofilament subunits (NF-H and NF-M) and demonstrate their existence in the normal nervous system of adult rats, *J. Neurosci.* **7:**3474–3488.

Lee, V. M.-Y., Otvos, L., Carden, M. J., Hollosi, M., Dietzschold, B., and Lazzarini, R. A., 1988, Identification of the major multiphosphorylation site in mammalian neurofilaments, *Proc. Natl. Acad. Sci. USA* **85:**1998–2002.

Lees, J. F., Shneidman, P. S., Skuntz, S. F., Carden, M. J., and Lazzarini, R. A., 1988, The structure and organization of the human heavy neurofilament subunit (NF-H) and the gene encoding it, *EMBO J.* **7:**1947–1955.

Lefebvre, S., and Mushynski, W. E., 1988, Characterization of the cation-binding properties of porcine neurofilaments, *Biochemistry* **27:**8503–8508.

Leterrier, J.-F., and Eyer, J., 1987, Properties of highly viscous gels formed by neurofilaments *in vitro, Biochem. J.* **245:**93–101.

Leterrier, J.-F., Liem, R. K. H., and Shelanski, M. L., 1981, Preferential phosphorylation of the 150,000 molecular weight component of neurofilaments by a cyclic-AMP-dependent microtubule-associated kinase, *J. Cell Biol.* **90:**755–760.

Leterrier, J.-F., Liem, R. K. H., and Shelanski, M. L., 1982, Interaction between neurofilaments and microtubule-associated proteins: A possible mechanism for intraorganelle bridging, *J. Cell Biol.* **95:**982–986.

Leterrier, J.-F., Wong, F., Liem, R. K. H., and Shelanski, M. L., 1984, Promotion of microtubule assembly by neurofilament-associated microtubule-associated proteins, *J. Neurochem.* **43:**1385–1391.

Levine, J., and Willard, M., 1981, Fodrin: Axonally transported polypeptides associated with the internal periphery of many cells, *J. Cell Biol.* **90:**631–643.

Levy, E., Liem, R. K. H., D'Eustachio, P., and Cowan, N. J., 1987, Structure and evolutionary origin of the gene encoding mouse NF-M, the middle-molecular-mass neurofilament protein, *Eur. J. Biochem.* **166:**71–77.

Lewis, S. A., and Cowan, N. J., 1985, Genetics, evolution and expression of the 68,000mol.wt. neurofilament protein: Isolation of a cloned cDNA probe, *J. Cell Biol.* **100:**843–850.

Lewis, S. A., and Cowan, N. J., 1986, Anomalous placement of introns in a member of the intermediate filament multigene family: An evolutionary conundrum, *Mol. Cell Biol.* **6:**1529–1534.

Liem, R. K. H., 1982, Simultaneous separation and purification of neurofilament and glial filament proteins from brain, *J. Neurochem.* **38:**142–149.

Liem, R. K. H., and Hutchison, S. B., 1982, Purification of individual components of the neurofilament triplet: Filament assembly from the 70000-dalton subunit, *Biochemistry* **21:**3221–3226.

Liem, R. K. H., Yen, S.-H., Loria, C. J., and Shelanski, M. L., 1977, Immunological and biochemical comparison of tubulin and intermediate brain filament protein, *Brain Res.* **132:**167–171.

Liem, R. K. H., Yen, S.-H., Salomon, G. D., and Shelanski, M. L., 1978, Intermediate filaments in nervous tissue, *J. Cell Biol.* **79:**637–645.

Liem, R. K. H., Pachter, J. S., Napolitano, E. W., Chin, S. S. M., Moraru, E., and Heimann, R. 1985, Associated proteins as possible cross-linkers in the neuronal cytoskeleton, *Ann. N.Y. Acad. Sci.* **455:**492–508.

Lifsics, M. R., and Williams, Jr., R. C., 1984, Physiochemical characterisation of the 68000 dalton protein of bovine neurofilaments, *Biochemistry* **23:**2866–2875.

Llinas, R. R., and Heuser, J. E., 1977, Depolarization–release coupling systems in neurons, *Neurosci. Res. Progr. Bull.* **15:**557–687.

Lucas, R. C., Rosenberg, S., Lawrence, J., Shafig, S., and Stracher, S., 1979, The isolation of a cytoskeleton and contractile apparatus from human platelets, *Cytoskel. Struct. Protides Biol. Fluids* **26:**465–470.

Malik, M. N., Myers, L. A., Iqbal, K., Sheikh, A. M., Scotto, L., and Wisniewski, H. M., 1981, Calcium activated proteolysis of fibrous proteins in the central nervous system, *Life Sci.* **29:**795–802.

Malik, M. N., Fenko, M. D., Iqbal, K., and Wisniewski, H. M., 1983, Purification and characterization of two forms of calcium-activated neutral protease from calf brain, *J. Biol. Chem.* **258:**8955–8962.

Malik, M. N., Fenko, M. D., and Wisniewski, H. M., 1984, Purification and partial characterization of two forms of calcium-activated neutral protease from calf brain synaptosomes and spinal cord, *Neurochem. Res.* **9:**233–240.

Malik, M. N., Fenko, M. D., and Wisniewski, H. M., 1985, Three forms of brain calcium-activated neutral protease, *Trans. Soc. Neurochem.* **16:**245.

Malik, M. N., Sheikh, A. M., Fenko, M. D., and Wisniewski, H. M., 1986, Purification and degradation of purified neurofilament proteins by the brain calcium-activated neutral proteases, *Life Sci.* **39:**1335–1343.

Matus, A., 1988, Neurofilament protein phosphorylation—where, when and why, *Trends Neurosci.* **11:**291–292.

Maxfield, M., 1953, Axoplasmic proteins of the squid giant nerve fiber with particular reference to the fibrous protein, *J. Gen. Physiol.* **37:**201–216.

Maxfield, M., and Hartley, Jr., R. W., 1957, Dissociation of the fibrous protein of nerve, *Biochim. Biophys. Acta* **24:**83–87.

Mellgren, R. L., 1980, Canine cardiac calcium-dependent proteases: Resolution of two forms with different requirements for calcium, *FEBS Lett.* **109:**129–133.

Mellgren, R. L., Repetti, A., Muck, T. C., and Easly, J., 1982, Rabbit skeletal muscle calcium-dependent protease requiring millimolar calcium. Purification, subunit structure and calcium-dependent autolysis, *J. Biol. Chem.* **257**:7203–7209.

Metuzals, J., 1969, Configuration of a filamentous network in the axoplasm of the squid (*Loligo pealii L.*) giant nerve fibre, *J. Cell Biol.* **43**:480–505.

Metuzals, J., and Izzard, C. S., 1969, Spatial patterns of threadlike elements in the axoplasm of the giant nerve fiber of the squid (*Loligo pealii L.*) as disclosed by differential interference microscopy and by electron microscopy, *J. Cell Biol.* **43**:456–479.

Metuzals, J., Hodge, A. J., Lasek, R. J., and Kaiserman-Abramof, I. R., 1983, Neurofilamentous network and filamentous matrix preserved and isolated by different techniques from squid giant axon, *Cell Tissue Res.* **228:** 415–432.

Milam, L. M., and Erickson, H. P., 1982, Visualization of a 21-nm axial periodicity in shadowed keratin filaments and neurofilaments, *J. Cell Biol.* **94**:592–596.

Milam, L. M., and Erickson, H. P., 1985, A structural comparison of tryptic fragments of three types of intermediate filaments, *J. Ultrastruct. Res.* **90**:251–260.

Minami, Y., and Sakai, H., 1985, Dephosphorylation suppresses the activity of neurofilaments to promote tubulin polymerization, *FEBS Lett.* **185**:239–242.

Minami, Y., Murofushi, H., and Sakai, H., 1982, Interaction of tubulin with neurofilaments: Formation of networks by neurofilament-dependent tubulin polymerisation, *J. Biochem.* **92**:889–898.

Minami, Y., Endo, S., and Sakai, H., 1984, Participation of 200k or 150k subunit of neurofilament in construction of the filament core with 70k subunit and promotion of tubulin polymerisation by incorporated 200k subunit, *J. Biochem.* **96**:1481–1490.

Miyata, Y., Hoshi, M., Nishida, E., Minami, Y., and Sakai, H., 1986, Binding of microtubule-associated protein 2 and tau to the intermediate filament reassembled from neurofilament 70kDa subunit protein, *J. Biol. Chem.* **261**:13026–13030.

Moon, H. M., Wisniewski, T., Merz, P., DeMartin, J., and Wisniewski, H. M., 1981, Partial purification of neurofilament subunits from bovine brains and studies on neurofilament assembly, *J. Cell Biol.* **89**:560–567.

Mori, H., and Kurokawa, M., 1979, Purification of neurofilaments and their interaction with vinblastine sulphate, *Cell Struct. Funct.* (*Jpn.*) **4**:163–167.

Mori, H., and Kurokawa, M., 1980, Morphological and biochemical characterisation of neurofilaments isolated from rat peripheral nerve, *Biochem. Res.* **1**:24–31.

Morris, J. R., and Lasek, R. J., 1984, Monomer–polymer equilibria in the axon: Direct measurement of tubulin and actin as polymer and monomer in axoplasm, *J. Cell Biol.* **98**:2064–2076.

Murachi, T., Tanaka, K., Hatanaka, M., and Murakami, T., 1981a, Intracellular calcium-dependent protease (calpain) and its high molecular weight and endogenous inhibitor (calpastatin), *Adv. Enzyme Regul.* **19**:407–424.

Murachi, T., Hatanaka, M., Yasumoto, Y., Nakayama, N., and Tamaka, K., 1981b, A quantitative distribution study on calpain and calpastatin in rat tissue and cells, *Biochem. Int.* **2**:651–656.

Murayama, A., Fukai, F., and Murachi, T., 1984, Action of calpain on the basic estrogen receptor molecule of porcine uterus, *J. Biochem.* **95**:1617–1704.

Myers, M. W., Lazzarini, R. A., Lee, V. M.-Y., Schlaepfer, W. W., and Nelson, D. L., 1987, The human mid-size neurofilament subunit: A repeated protein sequence and the relationship of its gene to the intermediate filament gene family, *EMBO J.* **6**:1697–1626.

Nagele, R. G., and Roisen, F. J., 1982, Ultrastructure of a new microtubule–neurofilament coupler in nerves, *Brain Res.* **253**:31–37.

Napolitano, E. W., Chin, S. S. M., Colman, D. R., and Liem, R. K. H., 1987, Complete amino acid sequence and *in vitro* expression of rat NF-M, the middle molecular weight neurofilament protein, *J. Neurosci.* **7**:2590–2599.

Nelson, W. J., and Traub, P., 1982, Purification and further characterization of the calcium activated proteinase specific for the intermediate filament proteins vimentin and desmin, *J. Biol. Chem.* **257**:5544–5553.

Nicol, J. A. C., 1948, The giant nerve fibres in the central nervous system of *Myxicola* (*Polychaeta sabellidae*), *Q. J. Microsc. Sci.* **89**:p.1–45.

Nicol, J. A. C., and Whitteridge, D., 1955, Conduction in the giant axon of *Myxicola infundibulum, Phys. Comp. Oecol.* **4**:101–117.

Nicol, J. A. C., and Young, J. Z., 1946, The giant nerve fibre of *Myxicola infundibulum* (Grube), *Nature* **158**:167–168.

Nishizuki, Y., Takai, Y., Hashimoto E., Kishimoto, A., Kuroda, Y., Sakai, K., and Yamamura, H., 1979,

Regulation and functional compartment of three multifunctional protein kinase systems, *Mol. Cell. Biochem.* **23:**153–165.

Nixon, R. A., 1980, Protein degradation in the mouse visual system. I. Degradation of axonally transported and retinal proteins, *Brain Res.* **200:**69–83.

Nixon, R. A., 1983, Proteolysis of neurofilaments, in: *Neurofilaments* (C. A. Marotta, ed), University of Minnesota Press, Minneapolis, pp. 117–154.

Nixon, R. A., 1986, Fodrin degradation by calcium-activated neutral proteinase (CANP) in retinal ganglion cell neurons and optic glia: Preferential localisation of CANP activities in neurons, *J. Neurosci.* **6:**1264–1271.

Nixon, R. A., Brown, B. A., and Marotta, C., 1982, Posttranslational modification of a neurofilament protein during axoplasmic transport: Implications for regional specialization of CNS axons, *J. Cell Biol.* **94:**150–158.

Nixon, R. A., Logvinenko, K. B., and Fisher, W. H., 1985, Evidence for a stationary non-uniform network of neurofilaments along retinal ganglion cell axons, *Ann N.Y. Acad. Sci.* **455:**797–800.

Nixon, R. A., Quakenbush, R., and Vitto, A., 1986, Multiple calcium-activated neural proteinases (CANP) in mouse retinal ganglion cell neurons: Specifications for endogenous neuronal substrates and comparison to purified brain CANP, *J. Neurosci.* **6:**1252–1263.

Noetzel, M. J., Roots, B. I., and Agrawal, H. C., 1986, Appearance and phosphorylation of the 210kd neurofilament protein in newborn rat brain, spinal cord, and sciatic nerve, *Neurochem. Res.* **11:**363–374.

Norton, W. T., and Goldman, J. E., 1980, Neurofilaments, in: *Proteins in the Nervous System,* 2nd ed. (R. A. Bradshaw and D. M. Schneider, eds.), Raven Press, New York, pp. 301–329.

Oblinger, M. M., and Schick, B., 1985, Posttranslational processing of neurofilament proteins in mammalian sensory neurons occurs over a slow time course *in vivo, J. Cell Biol.* **101:**17a.

Ohno, S., Emori, Y., Imajoh, S., Kawasaki, H., Kisaragi, M., and Suzuki, K., 1984, Evolutionary origin of a calcium-dependent protease by fusion of genes for a thiol protease and a calcium binding protein? *Nature* **312:**566–570.

Orrego, F., 1971, Protein degradation in squid giant axons, *J. Neurochem.* **18:**2249–2254.

Osborn, M., and Weber, K., 1983, Tumor diagnosis by intermediate filament typing: A novel tool for surgical pathology, *Lab Invest.* **48:**372–380.

Pachter, J. S., and Liem, R. K. H., 1984, The differential appearance of neurofilament triplet polypeptides in the developing rat optic nerve, *Dev. Biol.* **103:**200–210.

Pachter, J. S., and Liem, R. K. H., 1985, Alpha internexin, a 66KDa intermediate filament-binding protein from mammalian central nervous tissues, *J. Cell Biol.* **101:**1316–1322.

Paggi, P., and Lasek, R. J., 1984, Degradation of purified neurofilament subunits by calcium-activated neutral protease: Characterization of the cleavage products, *Neurochem. Int.* **6:**589–597.

Pant, H. C., and Gainer, H., 1980, Properties of calcium-activated protease in squid axoplasm which selectively degrades neurofilament proteins, *J. Neurobiol.* **11:**1–12.

Pant, H. C., Schecket, G., Gainer, H., and Lasek, R. J., 1978, Neurofilament protein is phosphorylated in the squid giant axon, *J. Cell Biol.* **78:**R23–R27.

Pant, H. C., Terakawa, S., and Gainer, H., 1979a, A calcium-activated protease in squid axoplasm, *J. Neurochem.* **32:**99–102.

Pant, H. C., Yashioka, T., Tasaki, I., and Gainer, H., 1979b, Divalent cation dependent phosphorylation of proteins in squid giant axon, *Brain Res.* **162:**303–314.

Pant, H. C., Pollard, H. B., Pappas, G. D., and Gainer, H., 1979c, Phosphorylation of specific, distinct proteins in synaptosomes and axons from squid nervous system, *Proc. Natl. Acad. Sci. USA* **76:**6086–6090.

Pant, H. C., Gallant, P. E., Gould, R., and Gainer, H., 1982, Distribution of calcium-activated protease activity and endogenous substrates in the squid nervous system, *J. Neurosci.* **2:**1578–1587.

Pant, H. C., Gallant, P. E., and Gainer, H., 1986, Characterization of a cyclic nucleotide- and calcium-independent neurofilament protein kinase activity in axoplasm from the squid giant axon, *J. Biol. Chem.* **261:**2968–2977.

Papasozomenos, S. C., Binder, L. I., Bender, P. K., and Payne, M. R., 1983, Microtubule-associated protein 2 within axons of spinal motor neurons: Associations with microtubules and neurofilaments in normal and β,β'-iminodipropionitrile-treated axons, *J. Cell Biol.* **100:**74–85.

Parker, G. H., 1929, The neurofibril hypotheses, *Q. Rev. Biol.* **4:**155–178.

Parry, D. A. D., and Fraser, R. D. B., 1985, Intermediate filament structure: Analysis of IF protein sequence data, *Int. J. Biol. Macromol.* **7:**203–213.

Peters, A., 1968, Characterisation of microtubules, neurofilaments and cross-bridges in various neuronal types, *Neurosci. Res. Progr. Bull.* **6:**162–168.

Peters, A., and Vaughn, J. E., 1967, Microtubules and filaments in the axons and astrocytes of early postnatal rat optic nerves, *J. Cell Biol.* **32:**113–119.

Phillips, D. R., and Jakobova, M., 1977, Calcium-dependent protease in human platelets, *J. Biol. Chem.* **252:**5602–5605.

Phillips, L. L., Autilio-Gambetti, L., and Lasek, R. J., 1983, Bodian's silver method reveals molecular variations in the evolution of neurofilament proteins, *Brain Res.* **278:**219–223.

Pruss, R. M., 1985, Efficient detection of intermediate filament proteins using a panspecific monoclonal antibody: AntiIFA, *J. Neuroimmunol.* **8:**293–299.

Pruss, R. M., Mirsky, R., Raff, M. C., Thorpe, R., Dowding, A. J., and Anderton, B. H., 1981, All classes of intermediate filaments share a common antigenic determinant defined by a monoclonal antibody, *Cell* **27:**419–428.

Puca, G. A., Nola, E., Sica, V., and Breciani, F., 1977, Estrogen binding of proteins of calf uterus. Molecular and functional characterization of the receptor transforming factor: A calcium-activated protease, *J. Biol. Chem.* **252:**1358–1370.

Rice, R. V., Roolansky, P. F., Pascoe, N., and Houghton, S. M., 1980, Bridges between microtubules and neurofilaments visualised by stereo electron microscopy, *J. Ultrastruct. Res.* **71:**303–310.

Richards, Jr., A. G., Steinback, H. B., and Anderson, T. F., 1943, Electron microscope studies of squid giant nerve axoplasm, *J. Cell. Comp. Physiol.* **21:**129–143.

Roberts, M. B. V., 1962, The rapid response of *Myxicola infundibulum* (*Grube*), *J. Mar. Biol. Assoc. U.K.* **42:**527–539.

Robinson, R. A., Wion, D., and Anderton, B. H., 1986, Isolation of a cDNA for the rat heavy neurofilament polypeptide (NF-H), *FEBS Lett.* **209:**203–205.

Roots, B. I., 1983, Neurofilament accumulation induced in synapses by leupeptin, *Science* **221:**971–972.

Roslansky, P. F., Cornell-Bell, A., Rice, R. V., and Adelman, Jr., W. J., 1980, Polypeptide composition of squid neurofilaments, *Proc. Natl. Acad. Sci. USA***77:**404–408.

Runge, M. S., El-Magbrabi, M. R., Clans, T. H., Pilkis, S. J., and Williams, Jr., R. C., 1981a, A MAP-2 stimulated protein kinase activity associated with neurofilaments, *Biochemistry* **20:**175–180.

Runge, M. S., Laue, T. M., Yphantis, D. A., Lifsics, M. R., Saito, A., Altin, M., Reinke, K., and Williams, Jr., R. C., 1981b, ATP-induced formation of an associated complex between microtubules and neurofilaments, *Proc. Natl. Acad. Sci. USA***78:**1431–1435.

Runge, M. S., Schlaepfer, W. W., and Williams, Jr., R. C., 1981c, Isolation and characterisation of neurofilaments from mammalian brain, *Biochemistry* **20:**170–175.

Sakihama, T., Kakidani, H., Zenita, K., Yumoto, N., Kikuchi, T., Sasaki, T., Kannagi, R., Nakanishi, S., Ohmori, M., Kohi, T., Titani, K., and Murachi, T., 1985, A putative calcium-binding protein: Structure of the light subunit of porcine calpain elucidated by molecular cloning and protein sequence analysis, *Proc. Natl. Acad. Sci. USA* **82:**6075–6079.

Sayre, L. M., Autilio-Gambetti, L., and Gambetti, P., 1985, Pathogenesis of experimental giant neurofilamentous axonopathies: A unified hypothesis based on chemical modification of neurofilaments, *Brain Res. Rev.* **10:**69–83.

Schlaepfer, W. W., 1974, Calcium-induced degeneration of axoplasm in isolated segments of rat peripheral nerve, *Brain Res.* **69:**203–215.

Schlaepfer, W. W., 1977a, Studies on the isolation and substructure of mammalian neurofilaments, *J. Ultrastruct. Res.* **61:**149–157.

Schlaepfer, W. W., 1977b, Immunological and ultrastructural studies of neurofilaments isolated from rat peripheral nerve, *J. Cell Biol.* **74:**226–240.

Schlaepfer, W. W., 1977c, Structural alterations of peripheral nerve induced by the calcium ionophore A23187, *Brain Res.* **136:**1–9.

Schlaepfer, W. W., 1978, Observations on the disassembly of isolated mammalian neurofilaments, *J. Cell Biol.* **76:**50–56.

Schlaepfer, W. W., 1983, Neurofilaments of mammalian peripheral nerve, in: *Neurofilaments* (C. A. Marotta, ed.), University of Minnesota Press, Minneapolis, pp. 57–85.

Schlaepfer, W. W., and Bunge, R. P., 1973, Effects of calcium ion concentration on the degeneration of amputated axons in tissue culture, *J. Cell Biol.* **59:**456–470.

Schlaepfer, W. W., and Freeman, L. A., 1978, Neurofilament proteins of rat peripheral nerve and spinal cord, *J. Cell Biol.* **78:**653–662.

Schlaepfer, W. W., and Freeman, L. A., 1980, Calcium-dependent degradation of mammalian neurofilaments by soluble tissue factor(s) from rat spinal cord, *Neuroscience* **5:**2305–2314.

Schlaepfer, W. W., and Micko, S., 1978, Chemical and structural changes of neurofilaments in transected rat sciatic nerve, *J. Cell Biol.* **78:**369–378.

Schlaepfer, W. W., and Zimmerman, U. J., 1985a, Calcium activated proteolysis of intermediate filaments, *Ann. N.Y. Acad. Sci.* **455:**552–562.

Schlaepfer, W. W., and Zimmerman, U. J., 1985b, Mechanisms underlying the neuronal response to ischemic injury. Calcium-activated proteolysis of neurofilaments, *Prog. Brain Res.* **63:**185–196.

Schlaepfer, W. W., Lee, V., and Wu, H.-L. 1981a, Assessment of immunological properties of neurofilament triplet proteins, *Brain Res.* **226:**259–272.

Schlaepfer, W. W., Zimmerman, U. J. P., and Micko, S., 1981b, Neurofilament proteolysis in rat peripheral nerve. Homologous with calcium-activated proteolysis of other tissues, *Cell Calcium* **2:**235–250.

Schlaepfer, W. W., Lee, C., Trojanowski, J. Q., and Lee, V. M.-Y., 1984, Persistence of immunoreactive neurofilament protein breakdown products in transected rat sciatic nerve, *J. Neurochem.* **43:**857–864.

Schlaepfer, W. W., Lee, C., Lee, V. M.-Y., and Zimmerman, U. J. P., 1985, An immunoblot study of neurofilament degradation *in situ* and during calcium-activated proteolysis, *J. Neurochem.* **44:**502–509.

Schmitt, F. O., 1950, The structure of the axon filaments of the giant nerve fibers of *Loligo and Myxicola, J. Exp. Zool.* **113:**499–515.

Schmitt, F. O., 1968a, The molecular biology of neuronal fibrous proteins, *Neurosci. Res. Progr. Bull.* **6:**119–144.

Schmitt, F. O., 1968b, Fibrous proteins—Neuronal organelles, *Proc. Natl. Acad. Sci. USA* **60:**1092–1101.

Schnapp, B. J., and Reese, T. S., 1982, Cytoplasmic structure in rapid-frozen axons, *J. Cell Biol.* **94:**667–679.

Scott, D., Smith, K. E., O'Brien, B. J., and Angelides, K. J., 1985, Characterisation of mammalian neurofilament triplet proteins, subunit stoichiometry and morphology of native and reconstituted filaments, *J. Biol. Chem.* **260:**10736–10747.

Sharp, G. A., Shaw, G., and Weber, K., 1982, Immunoelectronmicroscopical localization of the three neurofilament triplet proteins along neurofilaments of cultured dorsal root ganglion neurones, *Exp. Cell Res.* **137:**403–413.

Shaw, G., and Bray, D., 1977, Movement and extension of isolated growth cones, *Exp. Cell Res.* **104:**55–62.

Shaw, G., and Weber, K., 1981, The distribution of the neurofilament triplet proteins within individual neurones, *Exp. Cell Res.* **136:**119–125.

Shaw, G., and Weber, K., 1982, Differential expression of neurofilament triplet proteins in brain development, *Nature* **298:**277–279.

Shaw, G., Osborn, M., and Weber, K., 1981, An immunofluorescence microscopical study of the neurofilament triplet proteins, vimentin and glial fibrillary acidic protein within the adult rat brain, *Eur. J. Cell Biol.* **26:**68–82.

Shaw, G., Debus, E., and Weber, K., 1984, The immunological relatedness of neurofilament proteins of higher vertebrates, *Eur. J. Cell Biol.* **34:**130–136.

Shecket, G., and Lasek, R. J., 1980, Preparation of neurofilament protein from guinea pig peripheral nerve and spinal cord, *J. Neurochem.* **35:**1335–1344.

Shecket, G., and Lasek, R. J., 1982, Neurofilament protein phosphorylation: Species generality and reaction characteristics, *J. Biol. Chem.* **257:**4788–4795.

Shelanski, M. L., Albert, S., DeVries, G. H., and Norton, W. T., 1971, Isolation of filaments from brain, *Science* **174:**1242–1245.

Shelanski, M. L., Leterrier, J.-F., and Liem, R. K. H., 1981, Evidence for interactions between neurofilaments and microtubules, *Neurosci. Res. Progr. Bull.* **19:**32–43.

Siman, R., Baudry, M., and Lynch, G., 1984, Brain fodrin: Substrate for calpain I, an endogenous calcium-activated protease, *Proc. Natl. Acad. Sci. USA* **81:**3572–3576.

Smith, D. E., 1973, The location of neurofilaments and microtubules during the postnatal development of Clarke's nucleus in the kitten, *Brain Res.* **55:**41–53.

Smith, R. S., 1973, Microtubule and neurofilament densities in amphibian spinal root nerve fibres: Relationship to axoplasmic transport, *Can. J. Physiol. Pharmacol.* **51:**798–806.

Stefansson, K., Marton, L. S., Dierpefink, M. E., Molnar, G. K., Schlaepfer, W. W., and Helgason, C. M.,

1985, Circulating autoantibodies to the 200,000-dalton protein of neurofilaments in the serum of healthy individuals, *Science* **228:**1117–1119.

Steiner, J. P., Ling, E., and Bennet, V., 1987, Nearest neighbour analysis for brain synapsin 1, *J. Biol. Chem.* **262:**905–914.

Steinert, P. M., 1981, Intermediate filaments, in: *Electron Microscopy of Proteins I* (J. R. Harris, ed.), Academic Press, New York, pp. 125–166.

Steinert, P. M., and Parry, D. A. D., 1985, Intermediate filaments, *Annu. Rev. Cell Biol.* **1:**41–65.

Steinert, P. M., Wantz, M. L., and Idler, W. W., 1982, O-phosphoserine content of intermediate subunits, *Biochemistry* **21:**177–183.

Steinert, P. M., Rice, R. H., Roop, D. R., Trus, B. L., and Steven, A. C., 1983, Complete amino acid sequence of a mouse epidermal keratin subunit and implications for the structure of intermediate filaments, *Nature* **302:**794–800.

Sternberger, L. A., and Sternberger, N. H., 1983, Monoclonal antibodies distinguish phosphorylated and nonphosphorylated forms of neurofilaments *in situ, Proc. Natl. Acad. Sci. USA* **80:**6126–6130.

Sternberger, L. A., Harwell, L. W., and Sternberger, N. H., 1982, Neurotypy: Regional individuality in rat brain detected by immunocytochemistry with monoclonal antibodies, *Proc. Natl. Acad. Sci. USA* **79:**1326–1330.

Suzuki, K., Tsuji, S., Kubota, S., Kimura, Y., and Imahora, K., 1981, Limited autolysis of calcium-activated neutral protease (CANP) changes its sensitivity to calcium ions, *J. Biochem.* **90:**275–278.

Tashiro, T., and Ishizaki, Y., 1982, A calcium-dependent protease selectively degrading the 160,000 Mr component of neurofilaments is associated with the cytoskeletal preparation of the spinal cord and has an endogenous inhibitory factor, *FEBS Lett.* **141:**41–44.

Tashiro, T., Kurokawa, M., and Komiya, Y., 1984, Two populations of axonally transported tubulin differentiated by their interactions with neurofilaments, *J. Neurochem.* **43:**1220–1225.

Thorpe, R., Delacourte, A., Ayers, M., Bullock, G., and Anderton, B. H., 1979, The polypeptides of isolated brain 10nm filaments and their association with polymerised tubulin, *Biochem. J.* **181:**275–284.

Togari, A., Ichikawa, S., and Nagatsu, T., 1986, Activation of tyrosine hydroxylase by calcium-dependent neutral protease, calpain, *Biochem. Biophys. Res. Commun.* **134:**749–754.

Tokutake, S., 1984, Complete separation of the triplet components of neurofilaments by DE-52 column chromatography depends on urea concentration, *Anal. Biochem.* **140:**203–207.

Tokutake, S., Liem, R. K. H., and Shelanski, M. L., 1984, Each component of neurofilament assembles itself to make component-specific filament, *Biomed. Res.* **5:**235–238.

Toru-Delbauffe, D., and Pierre, M., 1983, A rat brain protein kinase phosphorylating specifically neurofilaments, *FEBS Lett.* **162:**230–234.

Traub, P., 1984, Large scale isolation, purification and partial characterization of the intermediate filament-specific calcium-activated proteinase from porcine kidney and Ehrlich ascites tumour cells: A comparative study, *Arch. Biochem. Biophys.* **228:**120–132.

Traub, P., 1985, Are intermediate filament proteins involved in gene expression? *Ann. N.Y. Acad. Sci.* **455:**68–78.

Traub, P., and Nelson, W. J., 1981, Occurrence in various mammalian cells and tissues of the calcium-activated protease specific for the intermediate-sized filament proteins vimentin and desmin, *Eur. J. Cell Biol.* **26:**61–67.

Triplett, R. B., Wingate, J. M., and Carraway, K. L., 1972, Calcium effects on erythrocyte membrane proteins, *Biochem. Biophys. Res. Commun.* **49:**1014–1020.

Trojanowski, J. Q., Obrocka, M. A., and Lee, V. M.-Y., 1985, Distribution of bovine neurofilament subunits in neurons and neuronal processes: Immunohistochemical studies of bovine cerebellum with subunit specific monoclonal antibodies, *J. Histochem. Cytochem.* **33:**557–563.

Truglia, J. A., and Stracher, A., 1981, Purification and characterization of a calcium dependent sulphydryl protease from human platelets, *Biochem. Biophys. Res. Commun.* **100:**814–822.

Tsukita, S., and Ishikawa, H., 1979, Morphological evidence for the involvement of the smooth endoplasmic reticulum in axonal transport, *Brain Res.* **174:**315–318.

Tsukita, S., and Ishikawa, H., 1981, The cytoskeleton in myelinated axons: Serial section study, *Biomed. Res.* **2:**424–437.

Tsukita, S., Usukura, J., Tsukita, S., and Ishikawa, H., 1982, The cytoskeleton in myelinated axons: A freeze-etch replica study, *Neuroscience* **7:**2135–2147.

Tsung, P. K., and Lombardini, J. B., 1985, Identification of low calcium and high calcium-requiring neutral proteases in rat retina, *Exp. Eye Res.* **41:**97–103.

Vargas, V., Greenbaum, L., and Costa, E., 1980, Participation of cysteine proteinase in the high affinity calcium-dependent binding of glutamate to hippocampal synaptic membranes, *Neuropharmacology* **19:**791–794.

Vedeckis, W. V., Freeman, M. R., Schrader, W. T., and O'Malley, B. W., 1980, Progesterone-binding components of chick oviduct: Partial purification and characterization of a calcium-activated protease which hydrolyses the progesterone receptor, *Biochemistry* **19:**335–343.

Vitadello, M., Triban, C., Fabris, M., Gorio, A., and Schiaffino, S., 1986, Heterogeneity of rat neurofilament polypeptides revealed by a monoclonal antibody, *J. Neurochem.* **46:**665–670.

Vitto, A., and Nixon, R. A., 1986, Calcium-activated neutral proteinase of human brain: Subunit structure and enzymatic properties of multiple molecular forms, *J. Neurochem.* **47:**1039–1051.

Wais-Steider, C., Eagles, P. A. M., Gilbert, D. S., and Hopkins, J., 1983, Structural similarities and differences amongst neurofilaments, *J. Mol. Biol.* **165:**393–400.

Wais-Steider, C., White, N. S., Gilbert, D. S., and Eagles, P. A. M., 1987, X-ray diffraction patterns from microtubules and neurofilaments in axoplasm, *J. Mol. Biol.* **197:**205–218.

Wallach, D. P., Davies, J. A., and Pastan, I. 1978, Purification of mammalian filamin, *J. Biol. Chem.* **253:**3328–3335.

Weber, K., and Geisler, N., 1985, Intermediate filaments: Structural conservation and divergence, *Ann N.Y. Acad. Sci.* **455:**126–143.

Weiss, P. A., and Mayr, R., 1971, Organelles and neuroplasmic (axonal) flow: Neurofilaments, *Proc. Natl. Acad. Sci. USA* **68:**846–850.

Wells, J. A., Besso, J. A., Jr., Boldosser, W. G., and Parsons, R. L., 1972, The fine structure of the nerve cord of *Myxicola infundibulum* (*Annelida, Polychaeta*), *Z. Zellforsch. Mikrosk. Anat.* **131:**141–148.

Willard, M., 1983, Neurofilaments and axonal transport, in: *Neurofilaments* (C. A. Marotta, ed.), University of Minnesota Press, Minneapolis, pp. 86–116.

Willard, M., and Simon, C., 1981, Antibody decoration of neurofilaments, *J. Cell Biol.* **89:**198–205.

Willard, M., and Simon, C., 1983, Modulations of neurofilament axonal transport during the development of rabbit retinal ganglion cells, *Cell* **35:**551–559.

Williams, Jr., R. C., and Aamodt, E. J., 1985, Interactions between microtubules and neurofilaments *in vitro, Ann. N.Y. Acad. Sci.* **455:**462–478.

Williams, Jr., R. C., and Runge, M. S., 1983, Biochemistry and structure of mammalian neurofilaments, in: *Cell and Muscle Motility,* Volume 3 (R. M. Dowber and J. W. Shay, eds.), Plenum Press, New York, pp. 41–56.

Wong, J., Hutchison, S. B., and Liem, R. K. H., 1984, An isoelectric variant of the 150,000-dalton neurofilament polypeptide: Evidence that phosphorylation state affects its association with the filament, *J. Biol. Chem* **259:**10867–10874.

Wuerker, R. B., 1970, Neurofilaments and glial filaments, *Tissue Cell* **2:**1–9.

Wuerker, R. B., and Kirkpatrick, J. B., 1972, Neuronal microtubules, neurofilaments and microfilaments, *Int. Rev. Cytol.* **33:**45–75.

Yamada, M., Spooner, B. S., and Wessells, N. K., 1971, Ultrastructure and function of growth cones and axons of cultured nerve cells, *J. Cell Biol.* **49:**614–635.

Yoshimura, N., Hatanaka, M., Kitahara, A., Kawaguchi, N., and Murachi, T., 1984, Intracellular localization of two distinct calcium-proteases (calpain I and calpain II) as demonstrated by using discriminative antibodies, *J. Biol. Chem.* **259:**9847–9852.

Zackroff, R. V., and Goldman, R. D., 1980, *In vitro* reassembly of squid brain intermediate filaments (neurofilaments): Purification by assembly–disassembly, *Science* **208:**1152–1154.

Zackroff, R. V., Idler, W. W., Steinert, P. M., and Goldman, R. D., 1982, *In vitro* reconstruction of intermediate filaments from mammalian neurofilament triplet polypeptides, *Proc. Natl. Acad. Sci. USA***79:**754–757.

Zackroff, R. V., Hill, W. D., and Tytell, M., 1984, Assembly and organization of neurofilaments in the squid giant axon, *J. Cell Biol.* **99:**323a.

Zimmerman, U. J., and Schlaepfer, W. W., 1984a, Calcium-activated neutral protease (CANP) in brain and other tissues, *Prog. Neurobiol.* **23:**63–78.

Zimmerman, U. J., and Schlaepfer, W. W., 1984b, Multiple forms of calcium-activated protease from rat brain and muscle, *J. Biol. Chem.* **259:**3210–3218.

4

The Proteins of Hair and Other Hard α-Keratins

J. M. Gillespie

The mammalian hard α-keratins constitute a homologous group of epidermal appendages comprising wool, hair, hoof, the horns of cattle, goats, sheep, and rhinoceros, claw, baleen, and the quills of porcupine, echidna, and hedgehog. Although of common embryological origin and sharing a common structure, they cover a very large range in amino acid compositions. Some examples are given in Table I where it can be seen that there is a large variation in the proportion of nearly every amino acid. This precludes a simple compositional definition for hard α-keratins and instead requires a detailed list of chemical and structural characteristics, which can be summarized as follows:

1. Epidermal appendages, usually cystine-rich, which are insoluble in usual protein solvents except at extremes of pH, due primarily to an extensive network of disulfide bonding.
2. After solubilization following fission of disulfide bonds, hard α-keratins yield two or three characteristic and unique families of constituent proteins, named because of peculiarities in composition, low-sulfur, high-sulfur, and high-tyrosine proteins.
3. X-ray diffraction analysis and electron microscopy show that these tissues have a unique arrangement of the constituent proteins, comprising intermediate filaments (IFs) traditionally termed microfibrils, usually aligned in the growth direction, surrounded by a nonfilamentous matrix of IF-associated protein (IFAP).

Most of the information available on the biochemistry of hard α-keratins has come from wool research and apart from some investigations on human hair there has been no corresponding study of other keratins. Most structural studies, in contrast, have been

J. M. Gillespie • CSIRO, Division of Wool Technology, Parkville, Victoria 3052, Australia.

Table I. Amino Acid[a] and Protein[b] Composition of Some Hard α-Keratins

	Horn[c]	Baleen[d]	Quill[e]	Hair[f]	Claw[g]	Quill[h]
Lysine	4.2	4.4	3.0	2.7	2.0	1.9
Histidine	1.0	1.7	0.8	0.9	1.0	0.4
Arginine	7.4	6.0	6.8	5.8	6.7	5.3
Aspartic acid	8.5	7.4	7.6	4.9	5.3	5.3
Threonine	5.1	5.1	5.0	6.8	5.4	3.2
Serine	8.5	7.6	8.6	11.7	9.2	7.3
Glutamic acid	16.2	11.8	12.7	11.4	8.8	7.3
Proline	4.8	8.1	7.3	8.4	7.2	5.5
Glycine	7.1	8.3	9.6	6.4	15.1	22.4
Alanine	7.5	5.8	6.6	4.6	3.8	3.8
Half-cystine	4.6	9.5	7.6	17.8	15.4	10.5
Valine	5.6	5.0	5.3	5.8	3.7	3.8
Methionine	0.7	1.0	0.3	0.6	0.7	0.4
Isoleucine	4.0	3.2	2.9	2.6	2.0	1.6
Leucine	9.3	8.1	8.3	5.8	5.3	7.1
Tyrosine	2.9	4.6	4.7	2.0	6.4	10.4
Phenylalanine	2.8	2.4	2.9	1.6	2.1	3.9
High-sulfur protein	6	13	14	39	29	21
High-tyrosine protein	<1	14	15	<1	24	46
Total IFAP	7	27	29	40	53	67

[a]Expressed as residues/100 residues.
[b]Expressed as g/100 g keratin.
[c]Rhinoceros: *Diceros bicornis*.
[d]Whale: *Balaenoptera musculus*.
[e]Porcupine: *Histrix cristata*.
[f]Human.
[g]Domestic cat.
[h]Echidna: *Tachglossus aculeatus*.

made on porcupine quill because of the superior ordering of its IF–IFAP complex. It is generally tacitly assumed that in dealing with a group of homologous tissues, information gained on one tissue can be legitimately applied to the others. The recent advances in our knowledge on the structure of keratins has come from just such a combined approach in which sequence and structural information on wool proteins was combined with x-ray diffraction data on porcupine quill.

1. The Nature of Hard α-Keratins

1.1. Solubility

Hard α-keratins are insoluble in protein solvents at moderate pH values and can only be appreciably solubilized if covalent bonds, either peptide or disulfide, are broken. At present, the most specific way to achieve solubilization is to break disulfide bonds, using reductants such as thioglycollic acid, β-mercaptoethanol, dithiothreitol, or tributylphosphine. Complete reduction usually requires an excess of reagent, a mildly alkaline pH, and the presence of reagents such as concentrated urea, guanidine hydrochloride, sodium iodine, or lithium bromide to swell the keratin, disrupt membranes, and render the disulfides accessible to the reductant. These dispersing agents are normally needed in any case, because proteins are not readily dissolved by dilute buffers (pH 7–9) from completely reduced or completely reduced and alkylated keratin. Keratin proteins in the reduced state rapidly reoxidize and although some studies have been made in the presence of excess reducing agent, it is customary to block the sulfhydryl groups with charged reagents such as iodoacetic acid or β-bromethylamine, which give the most soluble products.

So far it has been implied that if keratins are fully reduced in the presence of a dispersing agent at mildly alkaline pH, they will be completely solubilized. In practice, this is never realized because keratins contain morphological components such as cuticle and associated membranes, cortical cell membranes, nuclear remnants, and sometimes medulla, which are not soluble or have only a limited solubility in keratin solvents. Consequently, we can only achieve a solubility approaching 80% for wool and hair and 90% for echidna quill. However, many samples of human hair have solubilities around 50% with some as low as 5%. Although the solubility of wool can be drastically reduced by mild physical and chemical treatments leading to the formation of lanthionine and lysino-alanine cross-links, there is no evidence that these cross-links are responsible for the poor solubility of many samples of human hair (Gillespie, 1983). This insolubility is a stumbling block to the study of human hair but can be partly overcome by examining the corresponding proteins of nail or the prekeratins of the hair follicle.

1.2. Chemical Composition of Keratins

1.2.1. The Constituent Proteins

Most of the 100 or so proteins present in extracts of wool and other hard α-keratins can be separated by chemical fractionation (Fig. 1) or by gel electrophoresis (Fig. 2), into three family groups, each of which has a characteristic and unique amino acid composition (Table II). The high-sulfur proteins contain by far the most half-cystine and proline and these amino acids together with threonine and serine constitute about two-thirds of the amino acid residues. The two types of high-tyrosine protein differ in cystine content but both are very rich in glycine and tyrosine with about 70% of the molecule contributed by three or four amino acids. Both the high-sulfur and high-tyrosine proteins contain no methionine, are relatively deficient in lysine and histidine with high-tyrosine proteins also deficient in glutamic acid, alanine, valine, and isoleucine. The low-sulfur proteins in contrast are richest in those amino acids favoring α-helix formation: lysine, aspartic and glutamic acids, and leucine. These protein families can also be distinguished by size; the low-sulfur proteins are the largest with subunit polypeptides in the molecular weight range 44–57k, the high-sulfur proteins are intermediate in size with molecular weights between 10k and 30k, while the high-tyrosine proteins are quite small, all less than 10k.

1.2.2. Morphological Location of the Proteins

The exact location of constituent proteins within keratins was for some time a matter of controversy largely because there had been no direct isolation and analysis of any morphological component. There was, however, a large body of electron microscopic evidence involving the use of specific chemical stains and x-ray diffraction data that linked

1. *Preparation of soluble proteins*
   One gram treated with 100 ml 0.2 M β-mercaptoethanol in 8 M urea at pH 10.5–11 2 hr 40°C or overnight at 20°C, filter or centrifuge, add 0.4 g iodoacetic acid and 0.5 g tris for each 10 ml of recovered extract and maintain the pH between 8 and 9 by adding tris until the nitroprusside test is negative, add a few drops of β-mercaptoethanol to react excess iodoacetate, and dialyze

2. *Fractionation of proteins*
   Add zinc acetate to 0.02 M to dialyzed protein, adjust pH to 5.8–6.0

- Filtrate
  - Add sodium citrate to 0.02 M and dialyze
  - *High-sulfur protein* (SCMKB)
- Precipitate
  - Dissolve in 0.02 M sodium citrate dialyze, centrifuge
    - Supernatant
      - Adjust to 1% protein, add 2 vol acetone, 0.3 vol sat. $(NH_4)_2SO_4$
        - Precipitate
          - *Low-sulfur protein* (SCMKA)
        - Filtrate
          - *Type II high-tyrosine protein* and minor protein constituents
    - Precipitate
      - *Type I high-tyrosine protein*

Figure 1. Scheme for the isolation of the main classes of keratin proteins.

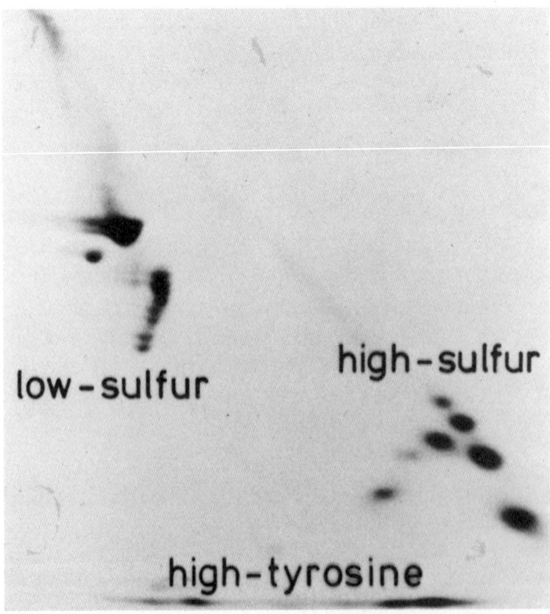

Figure 2. Two-dimensional electrophoresis of S carboxymethyl wool proteins. First dimension (horizontal): PAGE in 8 M urea at pH 8.9. Second dimension (vertical): PAGE in SDS at pH 8.9.

Table II. Amino Acid Compositions[a] of Characteristic Constituent Proteins[b] of Hard α-Keratins

	IFs	IFAPs		
			High-tyrosine protein	
	Low-sulfur protein[c]	High-sulfur protein[d]	Type I[e]	Type II[f]
Lysine	4.1	0.6	0.4	0.3
Histidine	0.6	0.7	1.1	0.2
Arginine	7.9	6.4	5.4	3.8
Aspartic acid	*9.6*	2.0	3.3	2.9
Threonine	4.8	8.9	3.3	1.2
Serine	8.1	13.8	11.9	5.3
Glutamic acid	*16.9*	6.4	0.6	0.7
Proline	3.3	*13.2*	5.3	3.9
Glycine	5.2	6.5	*27.9*	*41.1*
Alanine	7.7	2.1	1.5	0.8
Half-cystine[g]	6.0	*31.4*	6.0	*9.1*
Valine	6.4	3.2	2.1	0.9
Methionine	0.6	0.0	0.0	0.0
Isoleucine	3.8	1.4	0.2	0.1
Leucine	*10.2*	1.9	5.5	5.3
Tyrosine	2.7	0.9	*15.1*	*18.9*
Phenylalanine	2.0	0.6	10.4	5.0

[a]Expressed as residues/100 residues.
[b]S carboxymethyl derivatives.
[c]Isolated from Lincoln wool.
[d]Isolated from raccoon (*Procyon lotor*) hair.
[e]Isolated from Merino wool.
[f]Isolated from echidna (*Tachglossus aculeatus*) quill.
[g]Estimated as S carboxymethylcysteine.

the low-sulfur proteins with the IFs and the high-sulfur proteins with the IFAP (Birbeck and Mercer, 1957; Fraser *et al.*, 1972). Additional support was provided when Fraser *et al.* (1973) showed that the inter-IF volume fraction of a range of keratins agreed well with the mass fraction of their high-sulfur and high-tyrosine protein constituents. Finally, Jones (1976) isolated IFs and showed by amino acid analysis and electrophoresis that they consisted largely, if not entirely, of low-sulfur protein, and Zahn and his colleagues (Thomas *et al.*, 1986) produced short IFs from pure low-sulfur protein.

1.3. Fine Structure of Hard α-Keratins

X-ray diffraction and the electron microscope have been most important tools in gaining an understanding of the ways in which low-sulfur proteins are arranged to form the IFs and of the disposition of the latter with respect to the IFAPs. When the cortex of reduced keratins is stained with a heavy atom stain, cross sections examined in the electron microscope reveal a very characteristic pattern in which the IFs appear as lightly stained circular areas (range 7–10 nm in diameter) surrounded by heavily stained IFAPs (Birbeck and Mercer, 1957; Rogers, 1964). Representative patterns are shown in Fraser *et al.* (1972). The IFs appear to have an incompletely resolved internal structure giving the appearance of a ring and core, but the IFAP is quite featureless. There is some evidence that the inter-IF spacing, and hence the relative proportion of IFAPs, varies between keratins (Rogers, 1964).

All hard α-keratins give a particular x-ray diffraction pattern termed α by Astbury and Street (1931), which serves to distinguish them from the hard keratins of birds and reptiles (Fraser *et al.*, 1972). The interpretation of the α pattern has been progressively refined over the years and the recent papers by Fraser and MacRae (1983) and Parry *et al.* (1987) should be consulted for full details. In summary, the wide-angle reflections (spacings < 2.5 nm) are interpreted as stemming from the presence in the IFs of α-helical proteins arranged in coiled-coil ropes consisting of two segments of low-sulfur protein polypeptide chains. The structural units, consisting of two ropes in a four-chain assembly, are arranged helically to build an annular-shaped IF (Fraser *et al.*, this volume). The microfibrils of all hard α-keratins appear to be built according to this pattern. The low-angle equatorial reflections (spacings > 2.5 nm) give information on the diameter and spacing of the IFs. Over a range of keratins the diameter in the dried state was found to be 7.5 ± 0.3 nm with a spacing that varied from keratin to keratin depending on the content of IFAPs (Fraser *et al.*, 1973; Spei and Zahn, 1979).

2. *Morphology and Development of the Hair Follicle*

2.1. *Morphogenesis of the Hair Follicle*

The hair follicle forms embryonically as a down-growth of the basal layers of the epidermis. As the bud of developing cells enters the dermis, dermal cells form a cluster over which the epidermal cells spread to form the future hair bulb and into which the dermal papilla develops. Essentially then, the hair follicle is a tubular invagination of the epidermis, enclosing a small spike of dermis at the base. This development of the hair follicle is the result of a sequence of interactions between the dermis and epidermis—first a dermal message to the epidermis specifying the nature, size, shape, and distribution pattern of the appendage to be formed, next an epidermal message to the aggregation of dermal cells that form the dermal papilla, and finally a dermal message from the dermal papilla directing the differentiation of the multipotential stem cells of the follicle. The latter are undifferentiated germinative cells and have been traditionally referred to as matrix cells. They have a very high mitotic rate with generation times ranging from 12–13 hr (anagen mouse) to 30–44 hr (sheep). A continued dermal–epidermal interaction is essential for the normal differentiation of the follicle (Malkinson and Griem, 1965; Short *et al.*, 1965; Hardy, 1969; Dhouailly, 1973; Sengel, 1983). It is worth noting that although the dermis has a regional specificity that determines the type of hair or other keratin developing at a particular site, this specificity is not entirely autonomous, as hair type in some animals is also controlled by age, season, and hormonal status.

It is generally considered that the stem cells possess more than one developmental option, are equipotential, and their differentiation into the very dissimilar concentric cellular layers of medulla, cortex, cuticle, and the three layers of the inner root sheath is directed by the dermal papilla (Mercer, 1961; Fraser, 1965). However, Orwin *et al.* (1984) considered that the differentiation path is not irrevocably determined in the follicle bulb but can be influenced during differentiation, for example by diet. Only relatively few (about 20%) of sheep stem cells end up in the wool cortex (Short *et al.*, 1965). These workers showed that an increase in wool fiber mass growth rate, caused by an increased level of nutrition, came from an increased proportion of stem cells differentiating to cortical cells, as well as increases in the rates of mitosis and deposition of keratin.

2.2. *Regulation of Hair Growth Rate*

The growth rate of wool can be regulated by changes in the supply of amino acids available to the follicle, by the hormonal status of the animal, and by the presence in the sheep of a number of chemical substances. Reis (1979) reviewed the effects of amino acids on wool growth and showed that sulfur-containing amino (S-amino) acids play a major role in regulating wool growth rate and large increases can be induced by as little as 1.5–3 g/day of either L-cysteine or DL-methionine. Larger doses of methionine (6 g/day) particularly to sheep on a diet of wheat, imbalanced mixtures of amino acids, or proteins poor in lysine (e.g., zein) caused marked suppression of growth rate. Changes in wool growth rate are regarded as a sensitive indicator of changes in nutritional status (Downes *et al.*, 1976). Dietary deficiency in man is associated with a marked reduction in hair growth rate, an abnormal follicle, and a bulb of reduced diameter. Follicle bulb size has been used as a measure of malnutrition (Bradfield *et al.*, 1967; Friedman and Orraca-Tetteh, 1978). Some unnatural amino acids and amino acid analogues, e.g., mimosine or ethionine, can temporarily stop wool and hair growth and have been investigated as potential biological defleecing agents for sheep (Reis and Panaretto, 1979; Reis *et al.*, 1983).

Endocrine status is an important regulating factor for hair growth; e.g., hypophysectomy, which removes the source of all pituitary hormones, results in the complete suppression of wool growth. Temporary suppression results when sheep are given a single dose of the cortisol hormone analogue, dexamethasone (Wallace, 1979). A comprehensive review of the relation of hormones to hair growth is found in Ebling and Hale (1983).

Suppression of hair growth is often an unpleasant concomitant of anticancer chemotherapy, and certain chemotherapeutic agents, e.g., cyclophosphamide, have been studied as potential biological shearing agents for sheep because of their ability to temporarily suppress wool growth (Reis and Panaretto, 1979). Mouse epidermal growth factor completely suppresses wool growth in sheep but not hair growth in mice (Moore *et al.*, 1982). Many of these chemicals have quite selective actions on wool proteins (Section 5.4). Following suppression of wool growth by dietary, hormonal, or chemical action, there is usually a lag of a week or two followed by regrowth from the follicle at a greater than pretreatment rate.

Although wool growth is probably continuous throughout the life of domestic sheep, this is the exception rather than the rule and most mammals exhibit a periodicity of growth. Such animals have a cyclical process of hair growth (anagen) alternating with periods of degeneration and rest (catagen–telogen). At anagen the germinal cell layers of the follicle bulb differentiate and hair growth recommences. The factors that initiate and terminate these cycles of activity are unknown; however, grafting experiments show that they are specific to the site of origin of the skin (Ebling, 1965; Johnson, 1965; Ebling and Hale, 1983).

2.3. Development of Fiber Proteins

Current ideas concerning the sequence of events in the development of the IF-IFAP composite stem from the histological studies of Rogers (1964), Chapman and Gemmell (1971, 1973), Orwin (1979a,b) and Orwin and Woods (1982) and from the biochemical studies of Downes *et al.* (1963, 1966). Where these workers do not agree I have taken what appears to be the most logical interpretation.

The Merino wool follicle is on average 1700 μm long. In the first 150 μm there is no evidence for IF production, and this zone would be classed as germinative tissue comprising mitotic and differentiating cell zones (Fraser, 1965). IFs become evident thereafter, assembling into bundles, but without IFAP until the 250 μm level when it appears for the first time. The progressive development of fiber bundles is shown in Fig. 3. The bundle arrowed in Fig. 3b at about the 300 μm level already contains 600–700 IFs. The rate of synthesis of the IFAP is very high and at 400 μm the two components have caught up in proportional amount. Synthesis continues to the 600–700 μm level at which point the cells appear filled. The maximum rate of cysteine incorporation is at the 450 μm level while the maximum content of sulfhydryl occurs about 100 μm farther up the follicle and thereafter declines. It is considered that above the 550 μm level, i.e., about one third of the distance up the follicle, oxidation increases in intensity and proceeds concurrently with synthesis. During the latter stages of development, catabolic activity degrades the cell organelles, their associated proteins and nucleic acids. Remnants are concentrated

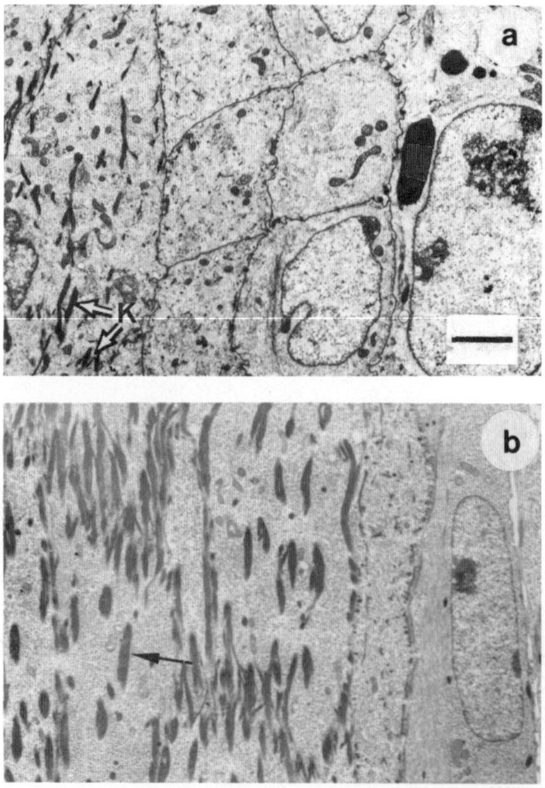

Figure 3. Electron micrographs of longitudinal sections of wool follicles. Arrows indicate keratin filament assemblies. (a) Midzone B; (b) midzone C/low-zone D. Provided courtesy of Dr. D. F. G. Orwin and for (a) by permission of Academic Press. For experimental conditions see Woods and Orwin (1982). Bar = 2.1 μm.

along the boundaries of microfibril bundles where they constitute a zone of weakness (Orwin, 1979b; Spearman, 1982).

The conversion of the stem cell into a mature cortical cell results in a volume increase of about 3-fold but a dry mass increase of about 13-fold, illustrating the large filling by keratin proteins that occurs (Short *et al.*, 1965). There is a 3-fold decrease in cross-sectional area and the cells are transformed from polyhedra to spindles. It is considered that the inner root sheath hardens at a lower level than the cortex and this forms a rigid tube that helps to mold the still plastic fiber. Pressure from dividing cells below and resistance from the follicle wall cause the shape change and are thought to provide the orienting force for the IFs.

It is of interest that the ultra-high-sulfur (UHS) proteins (Section 4.1.4) are not present in urea-soluble fractions from plucked follicle bulbs although the other high-sulfur proteins are. This suggests that they have a late synthesis above 550 μm, and are then rapidly oxidized. MacKinnon (1989) has provided further data showing a late synthesis. High-tyrosine proteins may also have a late synthesis.

2.4. Assembly and Cross-Linking of Fiber Proteins

The steps in the assembly of IFs appear to be as follows (Parry *et al.*, 1987; Conway *et al.*, 1989). One Type II polypeptide chain (wool components 5, 7a, 7b, 7c) and one Type I polypeptide chain (wool components 8a, 8b, 8c-1, 8c-2) line up parallel with their helical portions in register. Strong interactions between the hydrophobic sidechains in the repeating heptapeptides of the helical segments cause the formation of a coiled-coil rope. The exact positional relations of the two chains have not been determined experimentally but possible arrangements have been suggested (Fraser and MacRae, 1983; Woods and Inglis, 1984; Conway *et al.*, 1989; Sparrow *et al.*, 1989). As other IFs have similar structures but entirely different nonhelical N- and C-terminal segments, it is considered that the helical portions of the molecules play the most important part in determining the nature of these aggregation reactions (but see Sauk *et al.*, 1984). Two of these ropes then aggregate in an antiparallel arrangement to form a structural unit composed of four polypeptide chains. Possible staggers in this assembly are discussed in Section III of this volume and by Fraser *et al.* (1988). X-ray diffraction studies suggest that an IF is produced by assembly of these structural units in a helical fashion. There are probably eight structural units per turn of the helix with an axial translation of 19.8 nm between successive units. In this model the IF is an annulus of indefinite length, but presumably no longer than a macrofibril, from which the nonhelical segments project, partly into the annular space and partly to the exterior (Parry *et al.*, 1987; Fraser *et al.*, this volume). Hair IFs as long as 10–15 μm have been observed (Orwin, 1979b; Jones and Pope, 1985).

We cannot say for certain whether IFs contain other proteins. Analyses of isolated IFs (Jones, 1976) show significant departures from the composition expected for a pure low-sulfur protein (Table II). For example, the IFs contain less leucine (8.2 versus 9.1 residues percent) suggesting that they are not 100% low-sulfur protein while the higher proline content (5.2 versus 4.4 residues percent) suggests the presence of IFAPs.

Fraser and MacRae (1980) proposed that after assembly, the IFs are arranged in a lattice and spaced about 9 nm, center to center, by a bridging polypeptide, which may be an IFAP (probably high-sulfur protein) or a projecting segment of the low-sulfur polypeptide chain. In very ordered keratins the IFs have a hexagonal packing and this alone suggests that there must be specific interactions between them. Initially, the space between the IFs is filled with an aqueous medium that is progressively replaced by IFAPs, i.e., by high-sulfur protein in all keratins and high-tyrosine protein in some (Fraser *et al.*, 1972). The extent of replacement is variable depending on the nature of the keratin and the nutritional and physiological status of the animal. This hypothesis provides a simple explanation for the observation that the water content of hydrated specimens is inversely proportional to the IFAP content (Fraser and MacRae, 1980). Although nothing is known of the arrangement of IFAPs or of their interaction with the IFs, the opposite net charges of IF protein and IFAPs could lead to the formation of very tightly bound complexes.

This theory requires the numerous IFAPs to be synthesized outside the developing IF assembly and migrate to the appropriate places in an intricate IF lattice, a process that must be impeded by the insolubility of the IFAPs at physiological pH values and by their propensity for self-aggregation (Gillespie, 1983). This problem could be minimized, if, after synthesis, charged solubilizing groups were added to IFAPs by the formation of mixed disulfides. These disulfides could later be broken by reduction or disulfide–sulfhydryl interchange, after assembly of the IF–IFAP complex. Evidence for blocked sulfhydryl groups lies in the observation of many histologists that much of the sulfhydryl of follicle proteins is masked until comparatively late in development.

Chapman and Ward (1979) alternatively suggested that the IFAP polysomes are bound to the IFs and to the IF bundles so that the IFAPs are synthesized close to their final binding sites. The formation of the lattice would come about by the interaction of IF–IFAP assemblies. However, the composition of IFs (Jones, 1976) does not provide support for this view and in any case it necessitates the rejection of the whole body of evidence supporting differential rates of synthesis.

Finally, the whole structure is consolidated by oxidation of cysteinyl side chains to give cystine cross-links within and between peptide chains. Unless adequate levels of "copper" are available, oxidation is incomplete and a weak distorted hair of abnormal appearance results (Danks, 1983; Gillespie, 1983); however, even in severe cases of copper deficiency in man, mouse, and sheep, newly emergent hair contains constituent proteins that are at least two-thirds oxidized suggesting that cysteinyl residues can differ in their mechanism of oxidation (Section 4.1.6). The sulfhydryl in such hair is remarkably resistant to atmospheric oxidation even over periods of decades. It is not known whether this is due to misalignment of polypeptide chains and the resultant mismatching of cysteinyl residues consequent upon incomplete cross-linking at the appropriate place in the follicle or to the continued need for a copper catalyst to produce the correct oxidizing conditions.

3. Constituent Proteins of Intermediate Filaments

3.1. Preparation and Properties of Low-Sulfur Proteins

For all practical purposes the IFs contain only low-sulfur proteins (Table II), which can be isolated according to the procedure of Fig. 1. Other methods of preparation are summarized by Gillespie (1983). In dilute salt solution, these proteins exist as ill-defined polymers covering a wide span of sizes, but in dilute solutions of SDS, concentrated solutions of urea or guanidine hydrochloride, they dissociate, not to monomers, but to subunit polypeptides of 44–57k (Woods, 1979). All mammalian hard α-keratins yield this class of proteins on solubilization but do so in quite variable amounts. At one extreme, rhinoceros horn is almost entirely composed of low-sulfur proteins while at the other extreme, cat claw and echidna quill probably have less than one-third of their structures contributed by these proteins. Most α-keratins are, however, in an intermediate situation where the IFs and their constituent low-sulfur proteins account for at least 50% of the structure.

Because of the similarity of IFs in all hard α-keratins it has been assumed that their low-sulfur subunit polypeptides are homologous. This has been supported by electrophoretic studies showing that the polypeptides from a wide range of keratins spanned similar ranges of charge and size (Gillespie and Marshall, 1977; Marshall *et al.*, 1985). Furthermore, Hewish *et al.* (1984) by the use of monoclonal antibodies raised against wool polypeptides detected common determinants in a wide range of low-sulfur proteins including those of a monotreme, the echidna.

3.2. Properties of Subunit Polypeptides

When wool low-sulfur protein is electrophoresed in two dimensions in polyacrylamide gels, two well-defined groups of polypeptides are resolved (Fig. 4). The heavier group (57k) comprises components 5, 7a, 7b, and 7c, while the lower molecular weight group (44–48k) contains 8a, 8b, 8c-1, and 8c-2. The subunits appear to be present in equimolar amounts but this is not necessarily true for the corresponding polypeptides of other α-keratins (Gillespie and Marshall, 1977). In the case of human hair and nail, genetic variation has been observed (Marshall, 1983). The isolation and characterization of certain of these polypeptides from wool has been described by Dowling *et al.* (1979). Substantial inter- and intrafamily differences in composition, size, and degree of helicity were found.

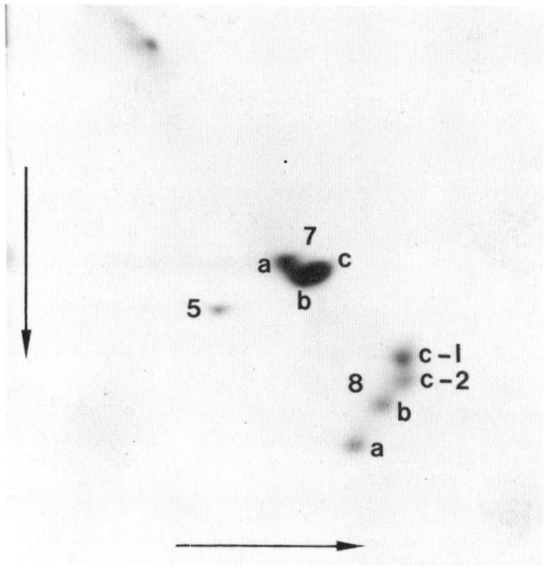

Figure 4. Two-dimensional electrophoresis of S carboxymethyl low-sulfur wool proteins prepared at neutral pH. First dimension (horizontal): PAGE in 8 M urea at pH 8.9. Second dimension (vertical): PAGE in SDS at pH 7.5. Provided courtesy of Mr. L. M. Dowling of this laboratory.

3.3. Amino Acid Sequences of Subunit Polypeptides

An analysis of the amino acid sequences of components 5, 7c, 8a, and 8c-1 has shown that they are built to a common plan, illustrated for 8c-1 in Fig. 5 (Dowling *et al.*, 1983, 1986; Powell and Rogers, 1986). Each component has an N-acetyl amino acid as its N-terminus followed by a nonhelical domain, a rodlike central domain interrupted at three positions by short segments of nonhelical residues, and finally a C-terminal nonhelical domain. In the rod domain the sequences show a succession of similar heptapeptide (heptads) sequences in which the first and fourth residues are usually apolar. The helix-rich segments of the two wool families show considerable homology between themselves and with the helical regions of some other IFs (Dowling *et al.*, 1983; Powell and Rogers, 1986; Conway *et al.*, 1989; Sparrow *et al.*, 1989). The N- and C-terminal nonhelical regions, which contain most of the half-cystine and proline, do not show much homology between wool families and none with the equivalent regions of other IFs (Sparrow *et al.*, 1989)

3.4. Assembly of Subunit Polypeptides

In Section 2.4 the currently accepted mode of aggregation of low-sulfur subunit polypeptides to form the four-chain structural unit was presented without supporting evidence. The molecular weight of the monomeric low-sulfur protein cannot be experimentally determined because any system that dissociates the aggregates also dissociates the monomer to its subunits, consequently the number and nature of the constituent subunits must be arrived at by indirect means (Woods, 1979). Although early interaction studies showed that in solution component 8c-1 interacts specifically with components 5 or 7c, the stoichiometry remained in doubt until Ahmadi *et al.* (1980) cross-linked low-sulfur proteins with dimethyl suberimidate and gained evidence for a four-chain structure of two pairs of unlike subunits. A number of workers carried out cross-linking experiments on helix-rich segments isolated by limited proteolysis of wool and wool low-sulfur proteins and found that the preferred mode of aggregation of the helical segments also favored a four-chain unit (reviewed by Woods and Inglis, 1984). The latter workers isolated a tetrameric helix-rich fragment by limited proteolysis of wool that originated from the N-terminal helical regions and contained two segments from each of the component 7 and 8 families arranged in parallel.

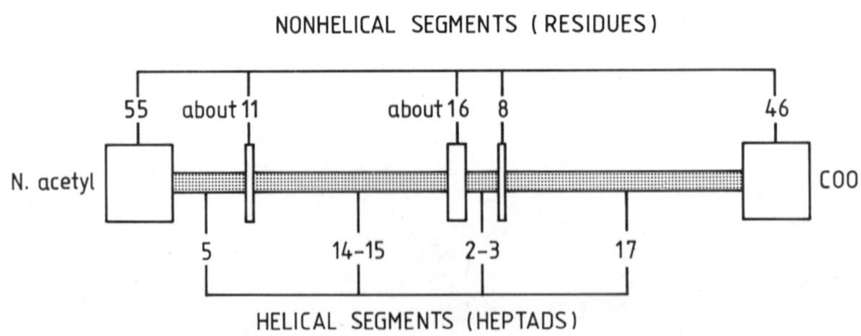

Figure 5. Schematic representation of the general structural features of intermediate filament polypeptide chains, with specific data for wool component 8c-1. Diagram and data courtesy of Mr. L. M. Dowling of this laboratory (Dowling *et al.*, 1986).

4. The Intermediate Filament-Associated Proteins

4.1. The High-Sulfur Proteins

4.1.1. Occurrence and Isolation

Every hard α-keratin contains a heterogeneous cystine-rich protein fraction in relative proportion ranging from below 10% to around 50%. The proteins are very variable in half-cystine content with a maximum observed range from 12.5 residues percent for a wool component up to 35 residues percent for a component of mouse hair. Of the various methods for isolating high-sulfur proteins, the most useful involves reduction followed by alkylation with iodoacetate and fractionation according to the scheme of Fig. 1. Other methods and their associated problems are discussed by Gillespie (1983).

4.1.2. Chemical Composition

The amino acid compositions of high-sulfur proteins isolated from a variety of keratins are shown in Table III. Regardless of their source, whether produced by marsupial or placental mammals or isolated from horn or hair, all these proteins have familial

Table III. Amino Acid Compositions[a] of High-Sulfur Proteins[b] Isolated from a Range of Hard α-Keratins

	Horn		Quill		Baleen	Hair		
	Rhino[c]	Sheep[d]	Porcupine[e]	Echidna[f]	Whale[g]	Sheep[h]	Mouse[i]	Cat[j]
Lysine	1.1	0.8	0.7	0.5	1.8	0.6	0.3	0.4
Histidine	1.1	1.2	1.2	0.5	2.0	0.8	0.3	0.9
Arginine	4.3	6.1	3.5	4.5	4.1	5.9	7.9	6.0
Aspartic acid	5.9	4.9	4.1	3.0	3.3	3.0	1.8	2.6
Threonine	10.4	10.8	7.7	6.9	8.6	10.3	6.8	8.3
Serine	8.8	10.5	9.8	10.4	8.0	12.7	11.7	12.6
Glutamic acid	5.6	6.1	7.2	6.3	6.4	8.4	7.2	6.1
Proline	14.7	13.9	16.5	13.7	15.9	12.5	13.7	11.0
Glycine	6.4	5.1	7.2	10.9	6.9	6.9	6.0	7.4
Alanine	3.3	3.0	4.5	3.6	3.5	2.9	2.2	2.1
Half-cystine[k]	16.9	18.5	22.3	24.2	23.4	18.9	31.9	33.9
Valine	5.5	5.9	6.9	5.1	4.5	5.6	4.2	3.6
Isoleucine	2.9	3.3	2.4	1.1	2.6	3.6	2.0	1.4
Leucine	8.6	5.6	4.0	4.4	4.5	3.9	1.9	2.1
Tyrosine	2.2	1.8	1.2	2.3	2.4	2.1	1.2	1.2
Phenylalanine	2.2	2.6	1.0	2.4	2.0	1.9	0.9	0.6

[a]Expressed as residues/100 residues.
[b]S carboxymethyl derivatives.
[c]*Diceros bicornis.*
[d]Dorset horn.
[e]*Histrix cristata.*
[f]*Tachyglossus aculeatus.*
[g]*Balaenoptera musculus.*
[h]Dorset horn.
[i]Laboratory strain.
[j]Domestic.
[k]Estimated as S carboxymethylcysteine.

resemblances being rich in half-cystine followed in decreasing amount by proline, serine, and threonine. These four amino acids account for at least half the total residues and in the cases of cat hair and raccoon hair (Tables II and III) about two-thirds. Methionine always seems to be absent and these proteins are quite poor in lysine, histidine, and the aromatic amino acids.

4.1.3. Heterogeneity

The high-sulfur proteins provide an example of a group of proteins of apparently similar function that exhibit an extraordinary degree of heterogeneity in both molecular size and chemical composition. This is illustrated by the two-dimensional electrophoresis patterns of the high-sulfur proteins of wool where 20 or more components are separated in addition to the unresolved UHS proteins (Fig. 6). The full extent of heterogeneity is unknown but there are at least 60 well-defined fractions in Merino wool (reviewed by Gillespie, 1983). The system is not as complicated as this as the high-sulfur proteins fall into a limited number of family groups (perhaps six) of which two are fairly minor. There is little sequence homology between the groups (Darskus *et al.*, 1969).

It is instructive to examine the molecular basis of the interfamily heterogeneity by looking at the amino acid compositions and molecular sizes of a number of components or fractions representing the main families of wool high-sulfur proteins (Table IV). The proteins range in size from 98 to 171 residues for sequenced components with an overall molecular mass span from about 11k to 25–28k. Although they all fit the compositional pattern of a high-sulfur protein being rich in half-cystine, proline, and serine, there are quite major differences between them. The half-cystine contents range from 12.5 to about 30 residues percent with large ranges for many other residues, e.g., serine (9–18), glycine (4–8), arginine (4–10), and valine (3–10) and two families contain no lysine or histidine. There is also substantial intrafamily heterogeneity that involves loss of segments and point mutations. This is illustrated for the SCMKB2 family of four major and two minor proteins in Table V.

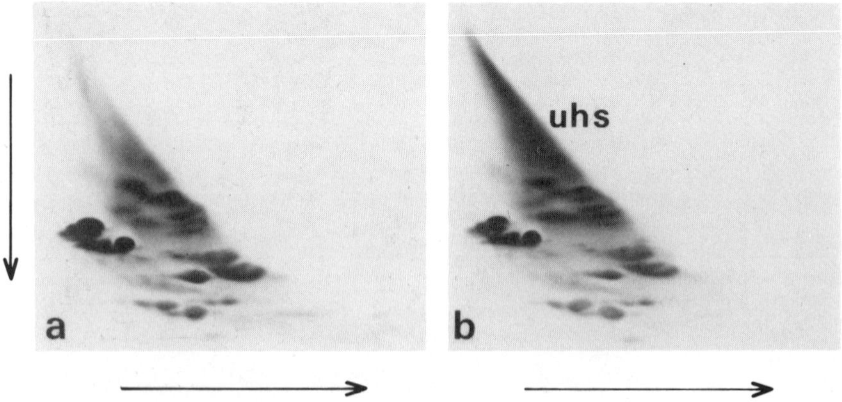

Figure 6. Two-dimensional electrophoresis of S carboxymethyl high-sulfur wool proteins. First dimension (horizontal): PAGE in 3 M urea at pH 2.6. Second dimension (vertical): PAGE in SDS at pH 8.9. High-sulfur proteins from (a) control wool and (b) sulfur-enriched wool.

Table IV. Amino Acid Compositions[a] of High-Sulfur Proteins[b] Isolated from Wool and Mouse Hair

	Wool						Mouse hair
	Protein 2[c]	SCMKBIIIB2[d]	SCMKB2A[e]	SCMKBIIIA3[d]	SCMKBI[f]	UHS fraction[g]	UHS fraction[h]
Lysine	0.6	1.0	0.0	0.0	0.5	0.9	0.4
Histidine	0.7	2.1	0.0	0.0	0.6	1.3	0.2
Arginine	7.3	4.1	3.5	10.2	5.1	6.9	5.4
Aspartic acid	2.6	8.3	0.6	2.4	0.8	0.6	1.0
Threonine	11.3	10.3	9.4	9.4	11.4	11.1	6.7
Serine	17.6	9.3	14.7	9.4	14.8	12.7	13.7
Glutamic acid	7.7	4.1	11.8	7.1	9.8	7.9	9.6
Proline	10.2	13.4	10.6	15.0	12.4	12.8	12.6
Glycine	6.6	5.2	8.8	4.7	6.3	4.2	5.5
Alanine	1.3	3.1	2.9	2.4	3.0	2.0	1.5
Half-cystine[i]	12.5	16.5	22.9	24.4	23.7	29.9	34.9
Valine	9.9	6.2	2.9	7.1	4.2	4.3	3.8
Isoleucine	3.4	4.1	5.3	1.6	2.7	1.7	1.2
Leucine	4.0	7.2	1.8	2.4	1.8	1.3	1.5
Tyrosine	2.8	2.1	3.5	1.6	2.1	1.9	1.9
Phenylalanine	1.5	3.1	1.2	2.4	0.7	0.5	1.1
Residues/mol.	148[j]	98	171	127	NE	NE	NE
Molecular weight ($\times 10^3$)	NE[k]	11	19	16	25–28	NE	19

[a]Expressed as residues/100 residues.
[b]S carboxymethyl derivatives.
[c]Marshall and Gillespie (1981).
[d]Swart *et al.* (1975).
[e]Gillespie (1963), Elleman (1972).
[f]Gillespie (1962).
[g]Gillespie (1983).
[h]Marshall and Gillespie (1976).
[i]Estimated as S carboxymethylcysteine.
[j]Estimated from amino acid analysis.
[k]NE, not estimated.

Table V. Sequence Heterogeneity within the SCMKB2 Family of High-Sulfur Proteins[a]

Component	Number of residues	Sequencing differences from component A	
		Deletions	Replacements
A	171	0	0
A1	171	0	3
B	156	15	8
B1	156	15	9
C	151	20	12
C1	151	20	13

[a]For complete sequences see Lindley (1977).

4.1.4. *The UHS Proteins*

Wool and hairs of the artiodactyl may contain in addition to a normal set of high-sulfur proteins, a variable amount of a distinctive group of very cystine-rich UHS proteins. These are similar in composition to the high-sulfur proteins of hairs such as those of the mouse and cat (cf. column 6 in Table IV with columns 7 and 8 in Table III). Variations in the cystine and high-sulfur protein contents of wool caused by the dietary or physiological manipulation of the sheep (Section 5.3) are due to changes in the level of the UHS proteins. For wool and other artiodactyl hairs, variations in the half-cystine content are linearly related to the content of these proteins (Fig. 7) (Gillespie and Broad, 1972). They can be detected in high-sulfur protein preparations as a peak in moving boundary electrophoresis patterns or as a diagonal wedge of unresolved protein in two-dimensional gels (Gillespie, 1983; Fig. 6). Efforts have been made to isolate homogeneous components

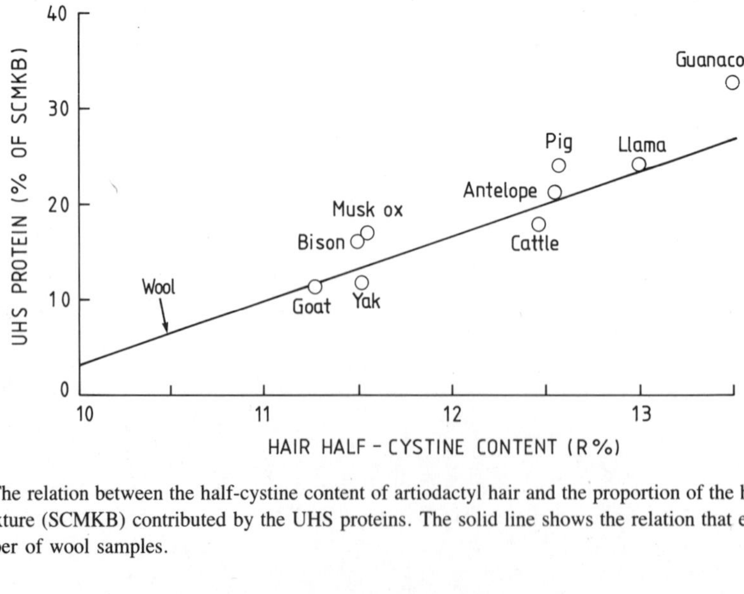

Figure 7. The relation between the half-cystine content of artiodactyl hair and the proportion of the high-sulfur protein mixture (SCMKB) contributed by the UHS proteins. The solid line shows the relation that exists for a large number of wool samples.

from wool and mouse hair but only limited success has been achieved due to the extensive heterogeneity and almost continuous distribution of components (reviewed by Marshall and Gillespie, 1976; MacKinnon, 1989).

4.1.5. *Size and Amino Acid Sequences*

Most is known about the high-sulfur proteins of wool and mohair where a wide variety of techniques have shown that the major components range in size from about 10 to 30k with minor components reaching 100k. The proteins of human, mouse, and guinea pig hairs and certain horny keratins appear to fall within the same size range as those of wool (reviewed by Gillespie, 1983).

Amino acid sequences are only known for the high-sulfur proteins of wool and mohair. For wool, 3 components have been sequenced from the 11,000 family (BIIIB), 11 from the 16,000 family (BIIIA), and 6 from the 19,000 family (SCMKB2) (reviewed by Lindley, 1977). The complete sequences are given by Lindley (1977), Parry *et al.* (1979), and Powell and Rogers (1986) and only the more significant findings will be discussed here. In the native state, all the proteins have half-cystine as the C-terminal amino acid and all except some of the 16,000 family have the blocked N-terminal sequence N-acetylalanine. The 16,000 and 19,000 components but not the 11,000 components contain a repetitive cystine-rich pentapeptide that takes the form -Cys-Cys-X-Pro-Y-. In the major wool protein SCMKB2A where about 85% of the molecule appears to be constructed of this repeating sequence, X is most commonly glutamine, glutamic acid, or arginine (62%) and Y serine or threonine (59%). There are also considerable variations in the type of amino acid in positions 1, 2, and 4; e.g., half-cystine only occurs to the extent of 59% in position 1 and 62% in position 2. These variations in half-cystine in the pentapeptide have structural implications that will be discussed in the next section. The amino acid sequence of one UHS component has been determined by MacKinnon (1989).

4.1.6. *Structure of High-Sulfur Proteins*

The little direct evidence for structure in native high-sulfur proteins in the matrix suggests that they have a globular form with an average diameter of 2.1 nm. It has always seemed unlikely that they would contain much α helix because of their high content of those amino acids (e.g., proline) that do not favor helix formation. Parry *et al.* (1979) pointed out that the regular occurrence of proline in position 4 of the repeating pentapeptide would preclude both α-helix and β-sheet formation. Studies in dilute salt solutions of S carboxymethyl high-sulfur proteins found no evidence for the presence of α helix or β structure but this might be due to the disordering effect of the large numbers of charged carboxymethyl groups. It is difficult to extrapolate the findings to the native state; however, a preparation in the -SH form in 0.2% tributylphosphine—5 M sodium ioide–25% propanol was found to contain about 10% α helix (Marshall and Gillespie, 1978).

Little is known of the arrangement of disulfide bonds in the native high-sulfur proteins although it has been argued on theoretical mechanical grounds that many disulfides must be intrachain for otherwise the IFAPs would resemble a highly cross-linked nonswelling resin. In fact, when keratin are placed in water or formic acid, most of the swelling takes place in the IFAPs (Fraser and MacRae, 1980). Intrachain disulfide bonds have been detected in incompletely reduced high-sulfur wool protein (Marshall and

Gillespie, 1978). Theoretical consideration of the sequences of high-sulfur proteins also led to the view that intra-chain bonding must be a common feature of high-sulfur proteins that contain the pentapeptide repeat. Lindley (1977) produced a theoretical model for SCMKB-111B2, suggesting that it takes up a highly convoluted form in which all except 2 of the 16 half-cystines are intrachain bonded. Parry *et al.* (1979) extended these arguments to possible bonding arrangements for a linear peptide consisting of sequentially arranged pentapeptides (Fig. 8a). Excluding the type of bonding that would form cyclocystine, known to be absent from wool, they suggest two types of bonding. The first involves cysteine #2 of the first pentapeptide bonding to cysteine #1 of the second pentapeptide and so on (Fig. 8b). The second arrangement has cystine #1 of the first pentapeptide bonding to cystine #2 of the second pentapeptide and so on (Fig. 8c). Fraser and MacRae (1980) favor the first alternative. Both arrangements involve convoluted chain conformations. In an ideal situation, with cysteinyl residues filling positions 1 and 2 of the pentapeptides, it is feasible that the protein could be largely intrachain bonded; however, there is always degeneracy and as a consequence there must be many unpaired cysteinyl residues. These, together with those outside the pentapeptides, presumably form interchain disulfides or provide the small amount of sulfhydryl found in normal keratins.

The thiol-disulfide system of the folded pentapeptides appears to be in a cyclic environment and on the basis of studies with model compounds might be expected to have a negative oxidation–reduction potential, e.g., α-lipoic acid −0.42 V (Reed, 1957). In contrast, simple thiol-disulfides have a slightly positive oxidation–reduction potential (+0.1 V; Eldjarn and Pihl, 1957) and this probably applies to noncyclic thiol-disulfides in keratins. Cyclic thiols would be expected to be the easiest to oxidize and this might be accomplished at the normal oxidation–reduction potential of the cortical cell contents without needing a copper catalyst. It may well be that the noncyclic thiols require more vigorous oxidizing conditions that can only be obtained with a copper catalyst and in its absence provide the thiol of copper-deficient wool and hair.

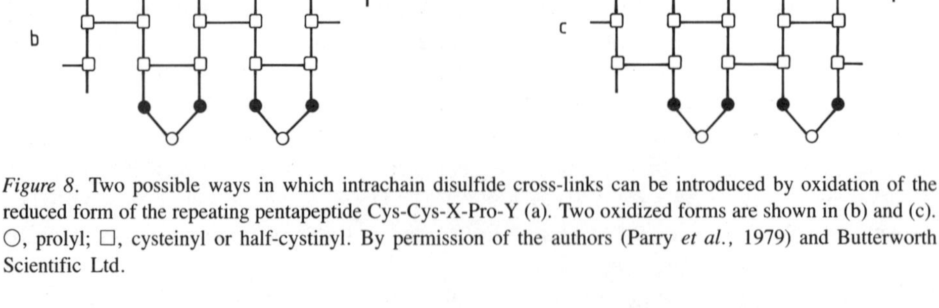

Figure 8. Two possible ways in which intrachain disulfide cross-links can be introduced by oxidation of the reduced form of the repeating pentapeptide Cys-Cys-X-Pro-Y (a). Two oxidized forms are shown in (b) and (c). ○, prolyl; □, cysteinyl or half-cystinyl. By permission of the authors (Parry *et al.*, 1979) and Butterworth Scientific Ltd.

Table VI. The High-Tyrosine Protein Content of a Range of Hard α-Keratins

Keratin type[a]	Total high-tyrosine protein content[b]
Man	<1
Rhinoceros horn[c]	<1
Ox hoof[d]	4
Ox horn[d]	8
Hedgehog quill[e]	13
Sheep[f]	14[k]
Porcupine quill[g]	15
Mouse[h]	16[k]
Rabbit claw[h]	18
Cat claw[i]	24
Echidna quill[j]	46[k]

[a]Hair unless otherwise stated.
[b]g/100 g keratin.
[c]*Diceros bicornis.*
[d]Domestic cattle.
[e]*Erinaceus europaeus.*
[f]Merino.
[g]*Histrix cristata.*
[h]Laboratory strain.
[i]Domestic.
[j]*Tachyglossus aculeatus.*
[k]Maximum observed value.

4.2. The High-Tyrosine Proteins

4.2.1. Occurrence and Isolation

Proteins that are rich in glycine are relatively plentiful in nature, e.g., silk, collagen, and elastin, but it is less common to find ones that are also rich in the aromatic amino acids. They can be found in the beak and claw of birds (Frenkel and Gillespie, 1976) and in insect egg cases (Kawasaki *et al.*, 1971). Such proteins also form part of IFAPs of many α-keratins where they are referred to as high-tyrosine proteins (Gillespie, 1983). These proteins are the most easily solubilized of the wool proteins and quite substantial yields can be obtained by treatment with β-mercaptoethanol alone at pH 9.5. They are normally isolated along with the low-sulfur proteins in the zinc precipitate fraction (Fig. 1) from which they can be recovered as two fractions that differ in solubility and composition. The less soluble group, Type I, is comparatively poor in half-cystine but rich in phenylalanine, while the reverse is true for the Type II fraction (Table II).

The occurrence of high-tyrosine proteins in keratins is quite erratic, seemingly bearing no relation to function. For example, they are undetectable in human and raccoon hairs but present in moderate amounts in wool and mouse hair. They are undetectable in rhinoceros horn, present in moderate amounts in some claws but in extremely large amounts in echidna quill (Table VI). Examples of the amino acid compositions of Type I and Type II high-tyrosine proteins are given in Table II.

4.2.2. Heterogeneity of Composition and Size

All high-tyrosine proteins examined have been found to be heterogeneous in composition and size although the extent and nature of this heterogeneity has only been explored for wool proteins. The complexity in the Type I high-tyrosine protein was first revealed by chromatography on QAE-cellulose at pH 10.5, which gave two major and eight minor fractions apparently separated on the basis of differences in tyrosine content, which ranged from 7 to 20 residues percent (Gillespie, 1983). The main fractions contained 26–33% glycine, 8–12% phenylalanine, 9–16% serine, and 2–6% half-cystine and most were heterogeneous (Fig. 9). Chromatography on DEAE-cellulose at pH 7.5 was used to separate purified components from each of five main QAE-cellulose fractions. Peptide mapping and sequence studies of the purified components indicated that each QAE-cellulose fraction represented a family of proteins with no interfamily but considerable intrafamily homology. The amino acid compositions of representatives of each of the main families (Table VII) illustrate the large interfamily differences that cover the whole range of amino acids. Molecular weight ranged from 7k (D and F) to 8.9k (C). It is of interest that component G/H 2 has a very low content of half-cystine (1.6 residues percent) and is similar in overall composition to the unfractionated Type I high-tyrosine protein isolated from porcupine quill (Table VII).

The nature of the intrafamily heterogeneity is uncertain for although members separate from each other as sharply resolved peaks on chromatography, amino acid and amide analysis and peptide mapping do not show significant differences between them (Gillespie, 1983). In the case of components C2 and C3, amino acid sequencing suggested that they were identical (Marshall *et al.*, 1980). It is possible that this apparent heterogeneity is due to conformational differences, perhaps arising from racemization caused by the conditions of preparation and fractionation, for there is certainly less if pH values are kept below 8

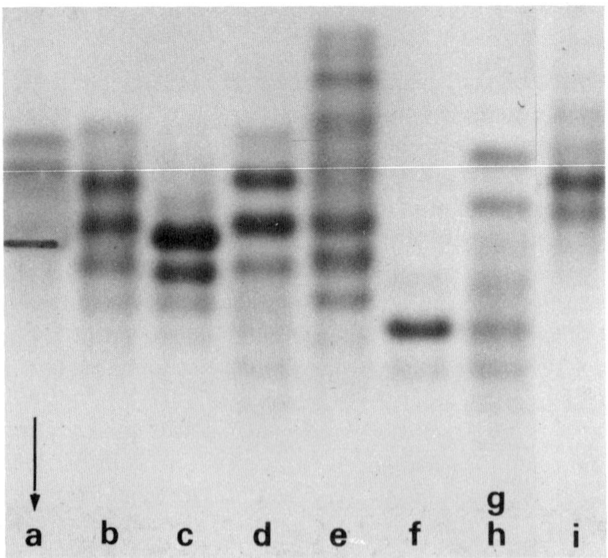

Figure 9. Electrophoresis on cellulose acetate in 8 M urea at pH 8.9 of the major fractions, separated by chromatography on QAE-cellulose, of S carboxymethyl Type I high-tyrosine proteins of wool.

Table VII. Amino Acid Compositions[a] of Type I High-Tyrosine Protein[b] Fractions

	Purified Type I components of wool[c]					Porcupine quill Type I
	C2[d]	D2	F[e]	G/H2	I4	
Lysine	0.2	0.0	0.0	0.6	0.3	0.3
Histidine	1.4	1.4	0.0	1.9	0.2	0.3
Arginine	4.9	7.4	3.4	3.9	4.6	3.9
Aspartic acid	4.3	2.9	1.9	1.9	4.8	3.8
Threonine	4.8	1.5	3.4	1.0	3.3	1.6
Serine	12.6	12.9	12.6	12.8	9.0	6.8
Glutamic acid	0.5	0.2	0.2	1.2	1.8	1.3
Proline	6.6	1.5	6.7	2.4	6.6	3.8
Glycine	30.8	34.0	26.5	35.9	25.8	36.2
Alanine	1.1	0.2	3.1	0.6	0.6	1.0
Half-cystine[f]	5.6	5.4	6.4	1.6	3.3	1.4
Valine	1.6	0.2	4.7	0.4	3.2	1.4
Isoleucine	0.0	0.0	0.0	0.0	1.4	0.3
Leucine	5.7	4.6	3.5	7.7	4.3	6.7
Tyrosine	10.1	13.8	18.0	19.8	20.2	22.5
Phenylalanine	10.0	13.9	9.6	8.3	9.5	8.1

[a]Expressed as residues/100 residues.
[b]S carboxymethyl derivatives.
[c]Gillespie and Frenkel (1974).
[d]Sequenced, Marshall *et al.* (1980).
[e]Sequenced, Dopheide (1973).
[f]Estimated as S carboxymethylcysteine.

(Marshall *et al.*, 1980). It seems that there is a similar phenomenon with the component 8 family of low-sulfur polypeptides (L. M. Dowling, personal communication; cf. Figs. 2 and 4).

The nature and extent of heterogeneity of Type II high-tyrosine proteins are even less well understood. If the proteins are first chromatographed on QAE-cellulose and the fractions electrophoresed, something like 20 components are evident although many are quite minor. Amino acid analysis of the main fractions (Table VIII) indicates that there are quite substantial differences between them in many amino acids although all are rich in glycine, tyrosine, half-cystine, and serine (Marshall *et al.*, 1980). Two homogeneous components (about 8k) isolated from fraction 5 have identical contents of the four major amino acids, but differ significantly in several of the others. This suggests that the Type II proteins, like other wool proteins, will be found to divide into a limited number of multimembered families. The amino acid composition of one of these components is shown in Table VIII) where it can be noted that four amino acids contribute 82% of the residues and that six amino acids are missing.

4.2.3. *Amino Acid Sequence and Structure*

Complete sequences are known for three Type I proteins: components 2 and 3 from family C (Marshall *et al.*, 1980), the single component of family F (Dopheide, 1973), and Type II component 5 (Table IX, A. S. Inglis, F. Casagranda, and C. M. Roxburgh,

Table VIII. Amino Acid Compositions[a] of Type II High-Tyrosine Protein Fractions from Wool[b]

	Fraction number				Purified component 5[c]
	3	4	5	6	
Lysine	0.4	0.4	0.3	0.2	0.0
Histidine	0.6	0.3	0.1	0.1	0.0
Arginine	5.0	5.4	4.8	4.6	4.7
Aspartic acid	3.3	2.0	1.6	1.7	2.2
Threonine	3.9	2.3	0.8	0.9	0.0
Serine	12.0	11.0	10.7	10.1	12.4
Glutamic acid	2.3	1.6	0.6	0.3	0.0
Proline	5.6	3.4	1.5	1.9	1.7
Glycine	27.0	31.6	37.1	37.1	38.8
Alanine	1.4	1.0	0.4	0.4	0.0
Half-cystine[d]	10.0	10.9	10.5	10.1	10.2
Valine	2.0	1.5	0.8	0.8	0.0
Isoleucine	0.9	0.6	0.2	0.0	0.0
Leucine	6.0	6.0	5.9	5.2	6.4
Tyrosine	13.9	17.9	21.1	22.7	20.8
Phenylalanine	5.5	3.8	3.1	3.5	2.7

[a]Expressed as residues/100 residues.
[b]Isolated as described by Marshall *et al.* (1980).
[c]Partial sequence shown in Table IX.
[d]Estimated as S carboxymethylcysteine.

personal communication). A comparison of the sequences of C2, F, and the Type II component shows no evidence of homology although there are some similarities between them. All have tyrosine as the C-terminal amino acid and C2 and F have threonine or serine respectively as the N-terminal amino acid. The N-terminal amino acid of each Type II component is blocked by an N-acetyl group. It is very common for glycine to be followed by tyrosine, serine, or half-cystine in both Type I (30–40% of the sequence) and Type II (about 75%) components. In spite of the high content of glycine, the sequence -Gly-Gly- is relatively uncommon, occurring no more than twice in any component, in marked contrast to the glycine-rich hard keratins of birds and reptiles and the glycine-rich segments of human epidermis (Gregg *et al.*, 1984; Inglis *et al.*, 1988; Steinert *et al.*, 1983). On the basis of amino acid composition, it has been suggested that the glycine–tyrosine-rich proteins of avian beak and claw and reptilian claw might be homologous with the mammalian high-tyrosine proteins (Gillespie *et al.*, 1982a) but sequence data now indicate that little if any homology exists (Gregg *et al.*, 1984; Inglis *et al.*, 1988).

There is no evidence available on the structural arrangement of native high-tyrosine protein. No structure (α or β) has been detected in this laboratory either in solution by optical rotatory dispersion or in films by x-ray diffraction. However, Amiya *et al.* (1980) reported the existence of a β structure in Type I protein films cast from formic acid. The very regular long runs of the repeating sequences -Gly-X- in the Type II fraction are similar to those found in silk and it will be of interest to see what sort of structural model can be proposed for this unusual protein.

Table IX. Amino Acid Sequences[a] of High-Tyrosine Proteins from Wool

Type I	T R F F C C G S Y F P G Y P S Y G T N F
Component C2[b]	H R T F R A T P L N C V V P L G S P L G
	Y G C D G Y S S L G Y G F G G S S F S N
	L G C G Y G G S F Y R P W G S G S G F G
	Y S T Y
Type I	S Y C F S S T V F P G C Y W G S Y G Y P
Component F[c]	L G Y S V G C G Y G S T Y S P V G Y G F
	L G Y S V G C G Y G S T Y S P V G Y G F
	G Y G Y D G G S A F G C R R F W P F A L
	Y
Type II	Ac(G G G G Y)L G C G S Y G Y G G L G C G Y
Component 5[d]	G S C Y G N Y F R R L G C G Y G C G Y G
	Y G S R S L C G S G Y G C G S R P L Y G
	C G Y G C G S G Y G S G F G Y Y

[a]Single-letter code. See IUPAC–IUB (1969).
[b]Gillespie (1983), Dopheide (1973).
[c]Gillespie (1983), Marshall *et al.* (1980).
[d]Marshall *et al.* (1980), A. S. Inglis, F. Casagranda, and C. M. Roxburgh (personal communication). The length and sequence of the parenthetical N-terminal peptide are uncertain.

5. The Regulation of Keratin Composition

5.1. Molecular Basis of Variability in Keratin Composition

Hard α-keratins cover a large range in amino acid composition (Table I). This variability occurs at a number of levels. One structural type, e.g., hair, can differ in composition in samples from different species, different animals within a species, from different sites on one animal, and even different positions on one sample. Substantial variability is also observed between different structural types (e.g., hair and nail) from the same or different species (Marshall, 1986; Marshall and Gillespie, 1989). Examples of such compositional variations are shown for the cystine and tyrosine contents of a wide range of keratins (Fig. 10). It can be seen that half-cystine varies from 5–6 residues percent for rhinoceros horn up to 19 residues percent for raccoon fur. Tyrosine also spans a very wide range from less than 3 residues percent (human hair) up to nearly 9 residues percent (echidna quill). The data presented in Fig. 10 also show that highly significant linear relationships exist between the content of half-cystine and high-sulfur proteins on the one hand and between tyrosine and high-tyrosine proteins on the other. Similar relations exist between proline and high-sulfur proteins and between glycine and high-tyrosine proteins.

The analytical data shown in Fig. 10a also illustrate the very large variations that have been found in the high-sulfur protein content of wool, where the range is at least from 16 to 34%. Examples of the substantial variations in the high-tyrosine protein content of different samples of wool and porcupine quill are illustrated in Fig. 10b. A single echidna quill sampled radially gave high-tyrosine protein contents that ranged from 36 to 45%. Every individual type of keratin critically examined has been found to exhibit such variability. Keratins are thus not materials of unique composition and we can conclude that variations in composition stem from alterations in the ratio of IFAPs to IFs,

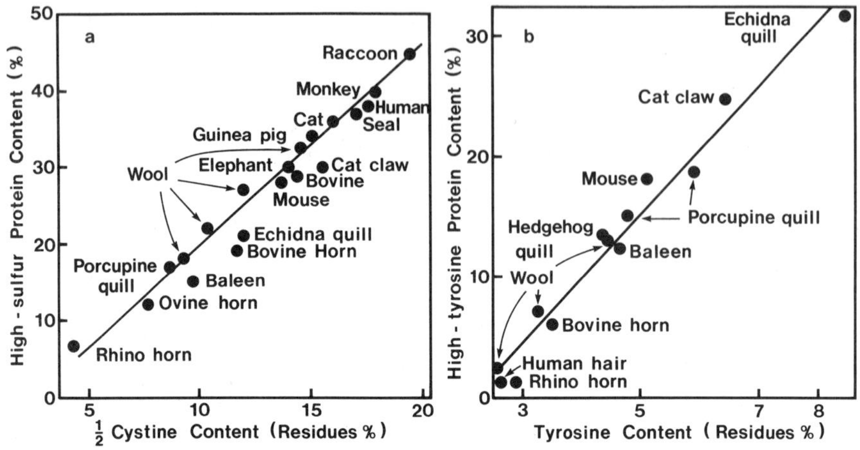

Figure 10. Variability in composition of hard α-keratins. (a) Variability in half-cystine and high-sulfur protein contents. (b) Variability in tyrosine and high-tyrosine protein contents.

from changes in the relative proportions of the two IFAPs, and from variations in their constituent protein components. We know something of the factors regulating composition and these will be examined in turn.

5.2. Genetic Control of Keratin Composition

5.2.1. Genetic Control Over Relative Proportions of Proteins

There is a genetic control over the relative proportions of constituent proteins for although individual keratins do not have a fixed composition, each appears to have its own characteristic compositional range. This can be illustrated by a number of examples. Human and raccoon hairs always contain large amounts of high-sulfur protein but little or no high-tyrosine protein whereas mouse hair is always moderately rich in both protein types. Lincoln and the Felting lustre mutant wool contain very little high-tyrosine protein whereas Merino wool usually contains between 6 and 14%. Rhinoceros horn is always built largely of IFs whereas echidna quill and cat claw have IFAPs as the major structural element. Presumably these specific differences in composition have some relation to function but nothing is known of this.

5.2.2. Species Specificity of Keratin Proteins

The sets of hard keratins produced by one animal species appear to contain nearly identical groups of constituent proteins although the relative proportions of components are usually different. This has been shown for sheep horn, hoof, and wool, rabbit claw and hair, human hair and nail, and echidna quill, claw, and hair (reviewed by Gillespie, 1983). The higher resolving power of two-dimensional electrophoresis in addition to confirming this has shown that as compared with hair, nail contains an additional unique low-sulfur protein component (Marshall, 1983).

A number of workers have studied the species specificity of keratin proteins using single-dimensional electrophoresis but the results have generally been rather disappointing with specificity being demonstrated in some mammalian orders but not others (reviewed by Marshall *et al.*, 1985). The greatly increased degree of discrimination of two-dimensional electrophoresis has changed the situation and precise identification of an animal species now seems possible (Marshall *et al.*, 1985; Marshall, 1985). We can illustrate this with a set of electrophoretic patterns of proteins from the hair of nine species (Fig. 11). A close examination shows that each pattern is different; even the closely related rat and mouse show differences in the location of at least ten proteins. For precise comparison particularly where the patterns are fortuitously similar, e.g., cat and dog (Fig. 11g and h), the samples may need to be mixed and run together to reveal the spots characteristic for each species (Marshall *et al.*, 1985).

5.2.3. Mutations Affecting Keratin Proteins

A large number of abnormalities of hair have been described, sometimes accompanied by neurological abnormalities or multiple malformations, which appear to be due to single gene defects. However, only in a relatively few cases can we find abnormal proteins present and it must be assumed that the abnormal macro- or microappearance of these hairs is caused by a defect in development of the follicle or in the growth, assembly, or cross-linking of the fiber. Trichothiodystrophy (TTD) in man and the naked mutation in the mouse will be used to illustrate mutations affecting hair proteins in different ways.

TTD hair characteristically contains less than 50% of normal cystine content, a change due to a decrease in total high-sulfur protein content, the loss of the UHS fraction, and the replacement of many normal high-sulfur components by new proteins of lower than normal cystine content (Gillespie and Marshall, 1983, 1989; Gillespie *et al.*, 1988). Only one sample of TTD hair has shown changes in IF polypeptides (Van Neste *et al.*, 1987). Electrophoresis of the proteins of this hair sample (Fig. 12) revealed the loss of the UHS fraction and changes in high-sulfur components in regions c–g in the gel. In addition, two IF polypeptides in region a of the gel showed a change in relative abundances and increased charge differences and there were also well-defined changes to the IF polypeptides in region b of the electrophoretic gel pattern.

The effects of the naked mutation on the proteins of mouse hair have been explored by Raphael *et al.* (1984). As had been shown previously by Tenenhouse and Gold (1976), heterozygous mice produce hair of markedly lower than normal high-tyrosine protein content with no change in the high-sulfur proteins and only a small increase in the low-sulfur components (Table X). Not surprisingly, homozygous mice produce hair with even lower high-tyrosine protein content but unexpectedly with only 50% of the normal level of high-sulfur protein. However, the situation is not as simple as this, for it is clear from the amino acid analyses that there must also be changes in the composition of the proteins. For example, proline is essentially constant although a drop in high-sulfur protein content should be accompanied by a fall in this amino acid. Compensating for the apparent drop in the IFAPs, there should be increases in the IF proteins manifested by increases (30–50%) in glutamic acid and leucine. These increases were not found. Unfortunately, the minute amount of hair produced by the homozygote makes study of this interesting mutation very difficult.

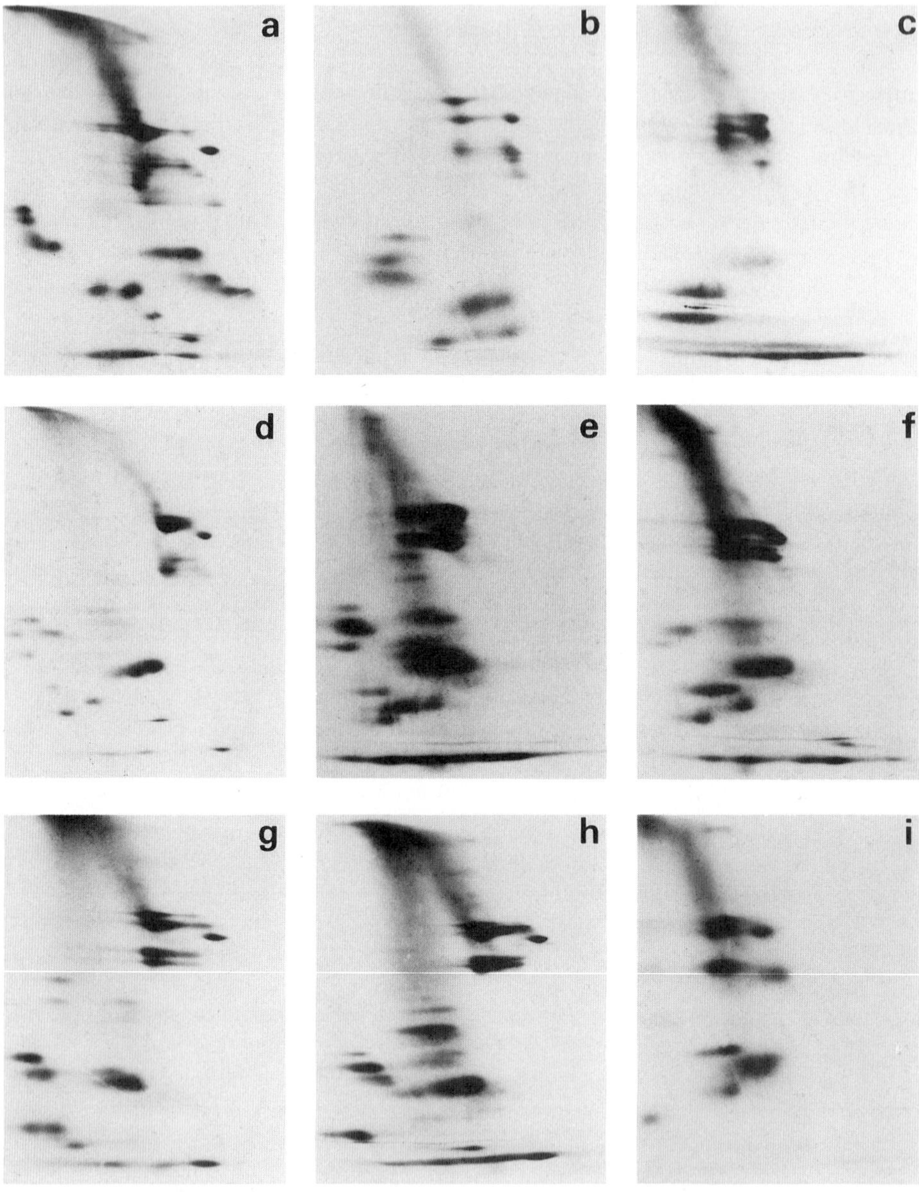

Figure 11. Two-dimensional electrophoresis of S [^{14}C] carboxymethyl proteins of various hairs. First dimension (horizontal): PAGE in 3 M urea at pH 2.9. Second dimension (vertical): PAGE in SDS at pH 8.9. The proteins were visualized by fluorography (Marshall *et al.*, 1985). (a) Sheep; (b) man; (c) kangaroo, *Macropus giganteus;* laboratory strains of (d) guinea pig, (e) mouse, and (f) rat; domestic breeds of (g) dog and (h) cat; (i) elephant, *Elephas maximus*.

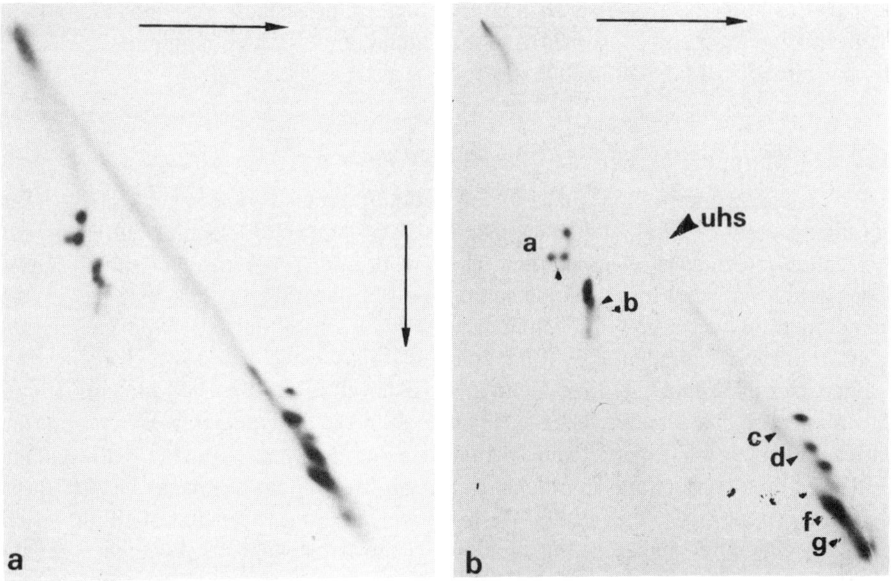

Figure 12. Two-dimensional electrophoresis of S carboxymethyl proteins of human hair. First dimension (horizontal): PAGE in 8 M urea. Second dimension (vertical): PAGE in SDS at pH 8.9. (a) Normal hair from family member; (b) TTD hair.

Table X. Protein[a] and Partial Amino Acid[b] Compositions of Normal and Naked Mutant Mouse Hair

	Mouse type		
		Mutant	
	Control +/+	Heterozygous *N*/+	Homozygous *N*/*N*
Lysine	2.9	3.3	4.3
Aspartic acid	5.4	5.7	6.6
Glutamic acid	12.6	12.9	12.6
Proline	7.2	7.8	7.3
Glycine	10.9	9.0	9.7
Half-cystine	13.9	14.0	8.0
Leucine	6.2	6.2	7.0
Tyrosine	4.7	3.6	3.0
High-sulfur protein[c]	32	32	16
High-tyrosine protein[c]	14	8	4

[a]Expressed as g/100 g hair.
[b]Expressed as residues/100 residues.
[c]Calculated from amino acid composition (Gillespie, 1983).
For complete analyses see Raphael *et al.* (1984).

It has recently been suggested that mutations such as these that affect groups of protein, involve a gene(s) concerned with the regulation of differentiation of follicle stem cells (Gillespie and Marshall, 1989).

5.3. Nutritional Control of Keratin Composition

Reis and Schinckel (1963) made the outstanding discovery that within certain limits the cystine content of wool, and hence the relative proportions of its high-sulfur proteins are susceptible to dietary manipulation. Their work laid to rest the widespread but erroneous belief that wool had a unique composition (discussed by Gillespie, 1983). When cysteine, methionine, or proteins rich in these amino acids are infused into the abomasum of a sheep to avoid degradation by rumen microorganisms, the half-cystine content of the wool increases by as much as 45% due to increased synthesis of the UHS proteins (Fig. 6) and there are also characteristic changes in other amino acids with increases in proline and significant decreases in aspartic acid, alanine, leucine, and phenylalanine (Gillespie and Reis, 1966). Smaller but quite significant changes in wool composition can be effected by changes in normal diet (Reis, 1979). We have evidence in this process of sulfur-enrichment of a reversible rapidly acting biological control mechanism. A number of suggestions have been made of possible modes of action but none appears to have been investigated.

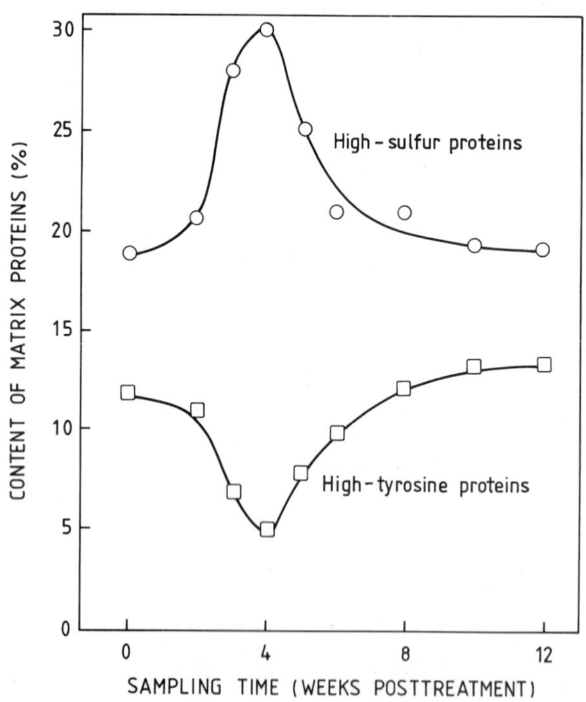

Figure 13. The effect of time after treatment on the protein composition of wool grown after defleecing with epidermal growth factor (27.5 mg).

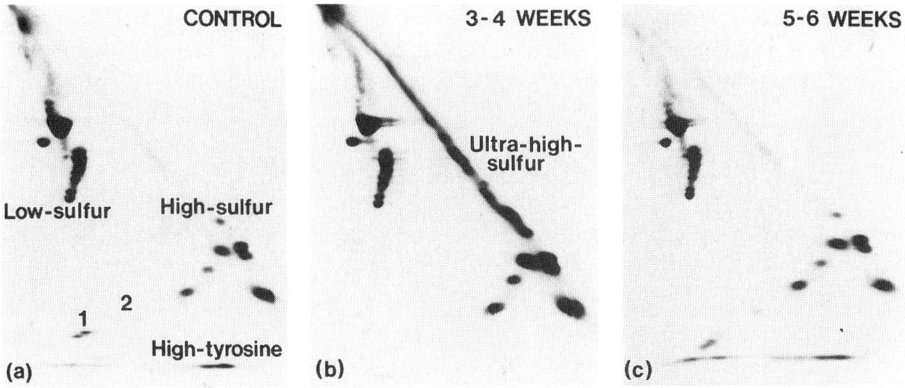

Figure 14. Two-dimensional electrophoresis of S carboxymethyl proteins from wool grown before and at two periods after defleecing with epidermal growth factor (27.5 mg). First dimension (horizontal): PAGE in 8 M urea at pH 8.9. Second dimension (vertical): PAGE in SDS at pH 8.9. (a) Pretreatment wool; (b) regrowth wool, 3–4 weeks; (c) regrowth wool, 5–6 weeks.

1. Mass action may be involved because at constant food intake the UHS protein level in wool is inversely related to wool growth rate. It has been suggested that as cysteine enters the suprabulbar region of the follicle and diffuses laterally, a concentration gradient is set up, the slope depending on the level of cysteine uptake and the rate of overall wool synthesis. As UHS proteins are probably the last to be synthesized, then at low levels of cysteine intake or at high rates of protein synthesis, only traces of cysteine would reach the UHS synthetic region (Gillespie and Reis, 1966).

2. The work of Orwin *et al.* (1984) suggests that the control may be at the level of differentiation of the follicle stem cells. On histological grounds, cortical cells have been divided into *ortho* and *para* classes with a possibly intermediate *meso* type. While the evidence is not ironclad it is probable that *para* cells are much richer in high-sulfur proteins than ortho, with the reverse true for the high-tyrosine proteins. Orwin and coworkers observed that at low levels of sheep nutrition, i.e., low levels of UHS protein production, there was an increase in the relative proportion of *ortho* cells. This suggests that the sulfur-enrichment process does not regulate protein synthesis *per se* but rather the cellular differentiation process. An inverse relation is often observed between the levels of UHS and high-tyrosine proteins (e.g., Figs. 13 and 14), which may also have its basis in an *ortho–para* switch. Unfortunately, no measurement of ratios of cortical cell types has been made during sulfur-enrichment.

3. There may be a direct inductive effect on the protein synthetic apparatus (Gillespie and Reis, 1966) perhaps similar to that operating in developing legume seeds. When the level of sulfate available to the seeds is decreased below optimum, the proportion of S-amino acid-rich proteins synthesized is decreased by a reduction in transcription and the level of mRNA (Beach *et al.*, 1985).

We do not know how generally applicable the findings with sheep are to monogastric animals. Supplementation with S-amino acids or proteins generally causes no change in hair composition, suggesting that most hairs are already at their genetically determined maximum level of high-sulfur proteins and thus can be considered to be already sulfur-enriched. There is general agreement that in man dietary deficiency results in an abnormal follicle and bulb of reduced diameter (Bradfield *et al.*, 1967). There is still disagreement on the effects on composition of hair although sufficient data are available to indicate that in a substantial proportion of cases of protein-calorie malnutrition the half-cystine content is below normal (reviewed by Gillespie, 1968; Friedman and Orraca-Tetteh, 1978). In one such case we have shown that the high-sulfur proteins are changed in overall composition by reduction in the proportion of UHS proteins (Gillespie, 1983).

5.4. Physiological Control of Keratin Composition

We have already mentioned the effects on wool growth rate of various chemicals (Section 2.2); however, in addition to their general effect on follicle activity, they appear to have quite specific effects on the IFAPs (Gillespie, 1983). In illustration we will examine the changes that take place in wool composition following the administration of mouse epidermal growth factor (EGF) to sheep. A dose of 27.5 mg EGF to a Merino sheep resulted in the temporary inhibition of the activity of wool follicles, shedding of the wool, and following a delay of a few weeks regrowth of wool (Moore *et al.*, 1982). The composition of regrowth wool changed progressively with time reaching a maximum divergence from the control in the 3- to 4-week period, followed by progressive return to normal by about 10 weeks (Fig. 13). Over this period the high-sulfur proteins rose from 19% to a maximum of 30% while the high-tyrosine protein content decreased from 12% to 5% and then both slowly returned to pretreatment levels. Two-dimensional electrophoresis (Fig. 14) shows the changes to protein components in wool grown before treatment and at two periods after. In the 3- to 4-week growth period (Fig. 14b), the high-tyrosine proteins were almost completely suppressed, a greater extent than estimated by amino acid analysis, with Type II components the most inhibited. High-sulfur components 1 and 2 were completely suppressed but the UHS proteins were greatly enhanced and instead of appearing as an unresolved wedge, had a well-pronounced banding pattern, suggesting the synthesis of specific components. Figure 14c shows the return to an essentially normal pattern (Gillespie *et al.*, 1982b). Following depilation with cyclophosphamide or *N*-[5(4-aminophenoxy)pentyl]phthalimide, the high-sulfur proteins followed the same pattern of change but after the initial rise they progressively fell below normal and were still as much as 20% below the control level 3 months later (Gillespie *et al.*, 1980).

At present it is not possible to provide a full explanation for the complex changes in wool protein synthesis that follow the temporary cessation of follicle activity. A partial explanation is offered by the observation that regenerating follicles (chemically treated or plucked) produce wool or hair with compositional characteristics similar to that produced in newly operating follicles in young lambs and mice which also progressively change in composition as the animal ages (Gillespie *et al.*, 1980).

6. Concluding Remarks

In spite of the tremendous advances made in α-keratin research over the last 30 years, there are still large gaps in our knowledge, particularly of the biochemistry of the follicle. To single out one area, it is clear that the exact sequence of events in the assembly of the IF–IFAP complex, the interactions between various proteins and assemblies of proteins, and the cross-linking between them are still largely a matter of conjecture. For example, we still do not know whether IF polypeptides are cross-linked internally or to other polypeptides in the IF–IFAP complex. However, some progress is now being made in the prediction of potential cross-links within an IF (Fraser *et al.*, 1988).

Little is known of the structure–function relations of keratins for although many theories linking particular mechanical properties to morphological components have been developed over the years, few have stood the test of time and rapidly increasing knowledge of keratin structure.

To end on a speculative note, the structural complexity of mammalian keratin and the large store of genetic information needed in its synthesis is very intriguing when we consider that a bird can make a functional claw or beak from one, only moderately polymorphic, protein in contrast to the 100 or more proteins needed by a mammal. We can ask the questions, is the mammalian claw functionally superior for its complex structure and if not, what has been its survival value?

References

Ahmadi, B., Boston, N. M., Dobb, M. G., and Speakman, P. T., 1980, Possible four chain repeating unit in the microfibril of wool, in: *Fibrous Proteins: Scientific, Industrial and Medical Aspects,* Volume 2 (D. A. D. Parry and L. K. Creamer, eds.), Academic Press, New York, pp. 161–166.

Amiya, T., Kawaguchi, A., Miyamoto, T., and Inagaki, H., 1980, Ordered structure of high-glycine proteins from reduced Merino wool, *Sen i Gakkaishi* **36:**T479–T483.

Astbury, W. T., and Street, A., 1931, X-ray studies of the structure of hair, wool and related fibers, *Philos. Trans. R. Soc. London Ser. A* **230:**75–101.

Beach, L. R., Spencer, D., Randall, P. J., and Higgins, T. J. V., 1985, Transcriptional and post-transcriptional regulation of storage protein gene expression in sulfur-deficient pea seeds, *Nucleic Acids Res.* **13:**999–1013.

Birbeck, M. S. C., and Mercer, E. H., 1957, The electron microscopy of the human hair follicle, *J. Biophys. Biochem. Cytol.* **3:**203–214.

Bradfield, R. B., Bailey, M. A., and Margen, S., 1967, Morphological changes in human scalp hair roots during deprivation of protein, *Science* **157:**438–439.

Chapman, R. E., and Gemmell, R. T., 1971, Stages in the formation and keratinization of the cortex of the wool fiber, *J. Ultrastruct. Res.* **36:**342–354.

Chapman, R. E., and Gemmell, R. T., 1973, An ultrastructural autoradiographic study of the incorporation of [^{35}S]cystine in the wool fiber cortex, *J. Cell Sci.* **13:**811–819.

Chapman, R. E., and Ward, K. A., 1979, Histological and biochemical features of the wool fiber and follicle, in: *Physiological and Environmental Limitations to Wool Growth* (J. L. Black and P. J. Reis, eds.), University of New England, Publishing Unit, p. 193–208.

Conway, J. F., Fraser, R. D. B., MacRae, T. P., and Parry, D. A. D., 1989, Protein chains in wool and epidermal keratin IF: Structural features and spatial arrangement, in: *The Biology of Wool and Hair* (G. E. Rogers, P. J. Reis, K. A. Ward, and R. C. Marshall, eds.), Chapman & Hall, London, pp. 127–144.

Danks, D. M., 1983, Copper deficiency and the skin, in: *Biochemistry and Physiology of the Skin,* Volume 2 (L. A. Goldsmith, ed.), Oxford University Press, London, pp. 1102–1112.

Darskus, R. L., Gillespie, J. M., and Lindley, H., 1969, The possibility of common amino acid sequences in high-sulfur protein fractions from wool, *Aust. J. Biol. Sci.* **22:**1197–1204.

Dhouailly, D., 1973, Dermo-epidermal interactions between birds and mammals: Differentiation of cutaneous appendages, *J. Embryol. Exp. Morphol.* **30:**587–603.

Dopheide, T. A. A., 1973, The primary structure of a protein component, 0.62, rich in glycine and aromatic amino acids, obtained from wool keratin, *Eur. J. Biochem.* **34:**120–124.

Dowling, L. M., Gough, K. H., Inglis, A. S., and Sparrow, L. G., 1979, Comparison of some microfibrillar proteins from wool, *Aust. J. Biol. Sci.* **32:**437–442.

Dowling, L. M., Parry, D. A. D., and Sparrow, L. G., 1983, Structural homology between hard α-keratin and the intermediate filament proteins, desmin and vimentin, *Biosci. Rep.* **3:**73–78.

Dowling, L. M., Crewther, W. G., and Inglis, A. S., 1986, The primary structure of component 8c-1, a sub-unit protein of intermediate filaments in wool keratin, *Biochem. J.* **236:**695–703.

Downes, A. M., Sharry, L. F., and Rogers, G. E., 1963, Separate synthesis of fibrillar and matrix proteins in the formation of keratin, *Nature* **199:**1059–1061.

Downes, A. M., Ferguson, K. A., Gillespie, J. M., and Harrap, B. S., 1966, A study of the proteins of the wool follicle, *Aust. J. Biol. Sci.* **19:**319–333.

Downes, A. M., Reis, P. J., and Hemsley, J. A., 1976, Proteins and amino acids for wool growth, in: *From Plant to Animal Protein,* Reviews in Rural Science No. 2, University of New England, pp. 143–148.

Ebling, F. J., 1965, Systemic factors affecting the periodicity of hair follicles, in: *Biology of the Skin and Hair Growth* (A. G. Lyne and B. F. Short, eds.), Angus & Robertson, Sydney, pp. 507–524.

Ebling, F. J., and Hale, P. A., 1983, Hormones and hair growth, in: *Biochemistry and Physiology of the Skin,* Volume 1 (L. A. Goldsmith, ed.), Oxford University Press, London, pp. 522–552.

Eldjarn, L., and Pihl, A., 1957, The equilibrium constants and oxidation–reduction potentials of some thiol-disulfide systems, *J. Am. Chem. Soc.* **79:**4589–4593.

Elleman, T. C., 1972, The amino acid sequence of protein SCMK-B2A from the high-sulfur fraction of wool keratin, *Biochem. J.* **130:**833–845.

Fraser, I. E. B., 1965, Cellular proliferation in the wool follicle bulb, in: *Biology of the Skin and Hair Growth* (A. G. Lyne and B. F. Short, eds.), Angus & Robertson, Sydney, pp. 427–445.

Fraser, R. D. B., and MacRae, T. P., 1980, Molecular structure and mechanical properties of keratins, in: *The Mechanical Properties of Biological Material* (J. F. V. Vincent and J. D. Currey, eds.), Society for Experimental Biology Symposium 34, pp. 211–246.

Fraser, R. D. B., and MacRae, T. P., 1983, The structure of the α-keratin microfibril, *Biosci. Rep.* **3:**517–525.

Fraser, R. D. B., MacRae, T. P., and Rogers, G. E., 1972, *Keratins, Their Composition, Structure and Biosynthesis,* Thomas, Springfield, Ill.

Fraser, R. D. B., Gillespie, J. M., and MacRae, T. P., 1973, Tyrosine-rich proteins in keratins, *Comp. Biochem. Physiol.* **44B:**943–947.

Fraser, R. D. B., MacRae, T. P., Sparrow, L. G., and Parry, D. A. D., 1988, Disulfide bonding in α-keratin, *Int. J. Biol. Macromol.* **10:**106–112.

Frenkel, M. J., and Gillespie, J. M., 1976, The proteins of the keratin component of birds' beaks, *Aust. J. Biol. Sci.* **29:**467–479.

Friedman, M., and Orraca-Tetteh, R., 1978, Hair as an index of protein malnutrition, in: *Nutritional Improvement of Food and Feed Proteins* (M. Friedman, ed.), Plenum Press, New York, pp. 132–154.

Gillespie, J. M., 1962, The isolation of the high-sulfur protein SCMKB1, *Aust. J. Biol. Sci.* **15:**572–588.

Gillespie, J. M., 1963, The isolation of the high-sulfur protein SCMKB2, *Aust. J. Biol. Sci.* **16:**241–251.

Gillespie, J. M., 1968, The dietary regulation of the synthesis of hair keratin, in: *Symposium on Fibrous Proteins* (W. G. Crewther, ed.), Butterworths, Australia, pp. 362–363.

Gillespie, J. M., 1983, The structural proteins of hair: Isolation characterization, and regulation of biosynthesis, in: *Biochemistry and Physiology of the Skin,* Volume 1 (L. A. Goldsmith, ed.), Oxford University Press, London, pp. 475–510.

Gillespie, J. M., and Broad, A., 1972, Ultra-high-sulfur proteins in the hairs of the artiodactyla, *Aust. J. Biol. Sci.* **25:**139–145.

Gillespie, J. M., and Marshall, R. C., 1977, Proteins of the hard keratins of echidna, hedgehog, rabbit, ox and man, *Aust. J. Biol. Sci.* **30:**401–409.

Gillespie, J. M., and Marshall, R. C., 1983, A comparison of the proteins of normal and trichothiodystrophic hair, *J. Invest. Dermatol.* **80:**195–202.

Gillespie, J. M., and Marshall, R. C., 1989, Effect of mutations on the proteins of wool and hair, in: *The Biology of Wool and Hair* (G. E. Rogers, P. J. Reis, K. A. Ward, and R. C. Marshall, eds.), Chapman & Hall, London, pp. 257–273.

Gillespie, J. M., and Reis, P. J., 1966, The dietary-regulated biosynthesis of high-sulfur wool proteins, *Biochem. J.* **98:**669–677.

Gillespie, J. M., Frenkel, M. J., and Reis, P. J., 1980, Changes in the matrix proteins of wool and mouse hair following the administration of depilatory compounds, *Aust. J. Biol. Sci.* **33:**125–136.

Gillespie, J. M., Marshall, R. C., and Woods, E. F., 1982a, A comparison of lizard claw keratin proteins with those of avian beak and claw, *J. Mol. Evol.* **18:**121–129.

Gillespie, J. M., Marshall, R. C., Moore, G. P. M., Panaretto, B. A., and Robertson, D. M., 1982b, Changes in the proteins of wool following treatment of sheep with epidermal growth factor, *J. Invest. Dermatol.* **79:**197–200.

Gillespie, J. M., Marshall, R. C., and Rogers, M., 1988, Trichothiodystrophy: Biochemical and Clinical Studies, *Australas. J. Dermatol.* **29:**85–93.

Gregg, K., Wilton, S. D., Parry, D. A. D., and Rogers, G. E., 1984, A comparison of genomic coding sequences for feather and scale keratins: Structural and evolutionary implications, *EMBO J.* **3:**175–178.

Hardy, M. H., 1969, The differentiation of hair follicles and hairs in organ culture, in: *Advances in Biology of the Skin,* IX, *Hair Growth* (W. Montagna and R. L. Dobson, eds.), Pergamon Press, New York, pp. 35–60.

Hewish, D. R., Robinson, C. P., and Sparrow, L. G., 1984, Monoclonal antibody studies of α-keratin low-sulfur proteins, *Aust. J. Biol. Sci.* **37:**17–23.

Inglis, A. S., Gillespie, J. M., Roxburgh, C. M., Whittaker, L. A., and Casagranda, F., 1988, Sequence of a glycine-rich protein from lizard claw: Unusual dilute acid and heptafluorobutyric acid cleavages, in: *Protein: Structure and Function* (J. L. L'Italien, ed.), Plenum Press, New York, pp. 767–774.

IUPAC–IUB Tentative Rules, 1969, A one letter notation for amino acid sequence, *Biochem. J.* **113:**1–4.

Johnson, E., 1965, Inherent rhythms of activity in the hair follicle and their control, in: *Biology of the Skin and Hair Growth* (A. G. Lyne and B. F. Short, eds.), Angus & Robertson, Sydney, pp. 491–505.

Jones, L. N., 1976, Studies on microfibrils from α-keratin, *Biochim. Biophys. Acta* **446:**515–524.

Jones, L. N., and Pope, F. M., 1985, Isolation of intermediate filament assemblies from human hair follicles, *J. Cell Biol.* **101:**1569–1577.

Kawasaki, H., Sato, H., and Suzuki, M., 1971, Structural proteins in silkworm egg-shells, *Insect Biochem.* **1:**130–148.

Lindley, H., 1977, The chemical composition and structure of wool, in: *Chemistry of Natural Protein Fibres* (R. S. Asquith, ed.), Plenum Press, New York, pp. 147–191.

MacKinnon, P. J., 1989, Molecular analysis of the ultra-high-sulfur keratin proteins, Ph.D. Thesis, University of Adelaide.

Malkinson, F. D., and Griem, M. L., 1965, Effects on hair of radiation, alone or in combination with radio-sensitizing or radio-protective agents, in: *Biology of the Skin and Hair Growth* (A. G. Lyne and B. F. Short, eds.), Angus & Robertson, Sydney, pp. 755–779.

Marshall, R. C., 1983, Characterization of the proteins of human hair and nail by electrophoresis, *J. Invest. Dermatol.* **80:**519–524.

Marshall, R. C., 1985, Hair comparison by protein analysis, in: *Proceedings International Symposium on Forensic Hair Comparisons,* U.S. Government Printing Office, Washington, D.C., pp. 71–87.

Marshall, R. C., 1986, Nail, claw, hoof and horn keratin, in: *Biology of the Integument,* Volume 2 (J. Bereiter-Hahn, A. G. Matoltsy, and K. Sylvia Richards, eds.), Springer-Verlag, Berlin, pp. 722–738.

Marshall, R. C., and Gillespie, J. M., 1976, High-sulfur proteins from α-keratins, *Aust. J. Biol. Sci.* **29:** 1–20.

Marshall, R. C., and Gillespie, J. M., 1978, Heterogeneity and incomplete reduction in the high-sulfur proteins of wool, *Aust. J. Biol. Sci.* **31:**219–229.

Marshall, R. C., and Gillespie, J. M., 1981, Changes in wool protein components following chemical defleecing, in: *Proc. Second National Wool Harvesting Research and Development Conference* (P. R. W. Hudson, ed.), Australian Wool Corp., Sydney, pp. 117–121.

Marshall, R. C., and Gillespie, J. M., 1989, Variations in the proteins of wool and hair, in: *The Biology of Wool and Hair* (G. E. Rogers, P. J. Reis, K. A. Ward, and R. C. Marshall, eds.), Chapman & Hall, London, pp. 117–125.

Marshall, R. C., Gillespie, J. M., Inglis, A. S., and Frenkel, M. J., 1980, High-tyrosine proteins of wool. Heterogeneity and biosynthetic regulation, *Proc. Sixth Int. Wool Text. Res. Conf.* **II:**147–158.

Marshall, R. C., Gillespie, J. M., and Klement, V., 1985, Methods and future prospects for forensic identification of hair by electrophoresis, *J. Forensic Sci. Soc.* **25:**57–66.

Mercer, E. H., 1961, *Keratin and Keratinization,* Pergamon Press, New York.

Moore, G. M., Panaretto, B. A., and Robertson, D., 1982, Inhibition of wool growth in Merino sheep following administration of mouse epidermal growth factor and a derivative, *Aust. J. Biol. Sci.* **35:**163–172.

Orwin, D. F. G., 1979a, The cytology and cytochemistry of the wool follicle, *Int. Rev. Cytol.* **60:**331–374.

Orwin, D. F. G., 1979b, Cytological studies on keratin fibres, in: *Fibrous Proteins: Scientific, Industrial and Medical Aspects,* Volume 1 (D. A. D. Parry and L. K. Creamer, eds.), Academic Press, New York, pp. 271–297.

Orwin, D. F. G., and Woods, J. L., 1982, Number changes and developmental potential of wool follicle cells in the early stages of fiber differentiation, *J. Ultrastruct. Res.* **80:**312–322.

Orwin, D. F. G., Woods, J. L., and Ranford, S. L., 1984, Cortical cell types and their distribution in wool fibers, *Aust. J. Biol. Sci.* **37:**237–255.

Parry, D. A. D., Fraser, R. D. B., and MacRae, T. P., 1979, Repeating patterns of amino acid residues in the sequences of some high-sulfur proteins from α-keratins, *Int. J. Biol. Macromol.* **1:**17–22.

Parry, D. A. D., Fraser, R. D. B., MacRae, T. P., and Suzuki, E., 1987, Intermediate filaments, in: *Fibrous Protein Structure* (J. M. Squire and P. J. Vibert, eds.), Academic Press, New York, pp. 193–214.

Powell, B. C., and Rogers, G. E., 1986, Hair keratin: Composition, structure and biogenesis, in: *Biology of the Integument,* Volume 2 (J. Bereiter-Hahn, A. G. Matoltsy, and K. Sylvia Richards, eds.), Springer-Verlag, Berlin, pp. 695–721.

Raphael, K. A., Marshall, R. C., and Pennycuik, P. R., 1984, Protein and amino acid composition on hair from mice carrying the naked (N) gene, *Genet. Res.* **44:**29–38.

Reed, L. J., 1957, The chemistry and function of lipoic acid, *Adv. Enzymol.* **18:**319–347.

Reis, P. J., 1979, Effects of amino acids on the growth and properties of wool, in: *Physiological and Environmental Limitations to Wool Growth* (J. L. Black and P. J. Reis, eds.), University of New England Publishing Unit, pp. 223–242.

Reis, P. J., and Panaretto, B. A., 1979, Chemical defleecing as a method of harvesting wool from sheep, *World Anim. Rev.* **30:**36–42.

Reis, P. J., and Schinckel, P. G., 1963, Some effects of sulfur-containing amino acids on the growth and composition of wool, *Aust. J. Biol. Sci.* **16:**218–230.

Reis, P. J., Tunks, D. A., Rigby, R. D. G., Morton, T. C., and Munro, S. G., 1983, Investigation of some amino acid analogues and metabolites as inhibitors of wool and hair growth, *Aust. J. Biol. Sci.* **36:**157–170.

Rogers, G. E., 1964, Structural and biochemical features of the hair follicle, in: *The Epidermis* (W. Montagna and W. C. Lobitz, eds.), Academic Press, New York, pp. 179–236.

Sauk, J. J., Krumweide, M., Cocking-Johnson, D., and White, J. G., 1984, Reconstitution of cytokeratin filaments in vitro: Further evidence for the role of non-helical peptides in filament assembly, *J. Cell Biol.* **99:**1590–1597.

Sengel, P., 1983, Epidermal-dermal interactions during formation of skin and cutaneous appendages, in: *Biochemistry and Physiology of the Skin,* Volume 1 (L. A. Goldsmith, ed.), Oxford University Press, London, pp. 102–131.

Short, B. F., Wilson, P. A., and Schinckel, P. G., 1965, Proliferation of follicle matrix cells in relation to wool growth, in: *Biology of the Skin and Wool Growth* (A. G. Lyne and B. F. Short, eds.), Angus and Robertson, Sydney, pp. 409–426.

Sparrow, L. G., Dowling, L. M., Loke, V. Y., and Strike, P. M., 1989, Amino acid sequences of wool keratin IF proteins, in: *The Biology of Wool and Hair* (G. E. Rogers, P. J. Reis, K. A. Ward, and R. C. Marshall, eds.), Chapman & Hall, London, pp. 145–155.

Spearman, R. I. C., 1982, The biochemistry of skin disease, *Mol. Aspects Med.* **5:**63–126.

Spei, M., and Zahn, H., 1979, X-ray small angle examination of swollen fiber keratin, *Melliand Textilber.* **60:**523–525.

Steinert, P. M., Steven, A. C., and Roop, D. R., 1983, Structural features of epidermal keratin filaments assembled in vitro, *J. Invest. Dermatol.* **81:**86s–90s.

Swart, L. S., Joubert, F. J., and Parris, D., 1975, Homology in the amino acid sequences of the high-sulfur proteins from keratins, *Proc. Fifth Int. Wool Text. Res. Conf.* **2:**254–264.

Tenenhouse, H. S., and Gold, R. J. M., 1976, Loss of a homologous group of proteins in a dominantly inherited ectodermal malformation, *Biochem. J.* **159:**149–160.

Thomas, H., Conrads, A., Phan, K. H., Van deLocht, M., and Zahn, H., 1986, The in vitro reconstitution of wool intermediate filaments, *Int. J. Biol. Macromol.* **8:**258–264.

Van Neste, D., Gillespie, J. M., and Marshall, R. C., 1987, Heterogeneity of trichothiodystrophy: Preliminary biochemical results, in: *Pediatric Dermatology: Advances in Diagnosis and Treatment* (R. Happle and E. Grosshans, eds.), Springer, Berlin, pp. 170–174.

Wallace, A. L. C., 1979, The effects of hormones on wool growth, in: *Physiological and Environmental Limitations to Wool Growth* (J. L. Black and P. J. Reis, eds.), University of New England Publishing Unit, pp. 257–268.

Woods, E. F., 1979, Microfibrillar proteins of wool: Partial specific volumes and molecular weights in denaturing solvents, *Aust. J. Biol. Sci.* **32:**423–435.

Woods, E. F., and Inglis, A. S., 1984, Organization of the coiled coils in the wool microfibril, *Int. J. Biol. Macromol.* **6:**277–283.

Woods, J. L., and Orwin, D. F. G., 1982, The cytology of cuticle scale pattern formation in the wool follicle, *J. Ultrastruct. Res.* **80:**230–242.

5

The Nuclear Lamina

An Intermediate Filament Protein Structure of the Cell Nucleus

George N. Dessev

1. Introduction

During interphase, eukaryotic cells are physically separated into a nuclear and a cytoplasmic compartment by a complex organelle, the nuclear envelope. The innermost layer of the nuclear envelope is a polymeric protein network (Fawcett, 1966; Fawcett, 1981) termed the nuclear lamina (NL) (Fig. 1), which is tightly associated with the nuclear pore complexes. Apposed to the cytoplasmic surface of the NL is the membrane component of the nuclear envelope, which consists of two lipid bilayers, containing intrinsic membrane proteins. It is likely that the NL is a ubiquitous feature found in all eukaryotic cells, although it is not always equally prominent by electron microscopy.

As a basic element of the nuclear envelope, the NL is believed to be involved in functions of the interphase nucleus, such as structural organization of chromatin, molecular traffic and signal transmission between the nucleus and the cytoplasm. Also, it appears to play a central role in the cell cycle dynamics of the nuclear envelope, which involves growth during interphase and reversible disassembly in mitosis. At present, not all of these possible functions of the NL are supported by hard evidence; most of them remain speculative. Nevertheless, or perhaps because of that, the NL has gained popularity among cell and molecular biologists, who hope that elucidation of its structural and regulatory features may be the key to the mechanisms of a number of nuclear activities.

In this chapter I have tried to briefly outline the information available on the NL; more extensive coverage of different aspects of this subject and more references can be found in a number of reviews (Franke *et al.*, 1981; Hancock and Boulikas, 1982; Hancock, 1982; Gerace and Blobel, 1982; Gerace *et al.*, 1984; Krohne and Benavente, 1986a;

George N. Dessev • Department of Cell, Molecular, and Structural Biology, Northwestern University Medical School, Chicago, Illinois 60611.

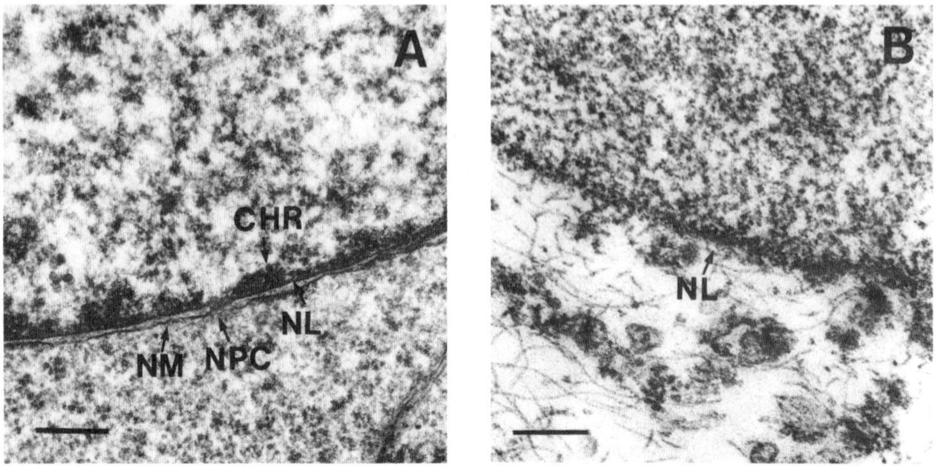

Figure 1. Electron micrographs (thin sections) of (A) an interphase BHK-21 cell, showing the nuclear lamina (NL), double nuclear membrane (NM), nuclear pore complexes (NPC), and condensed chromatin blocks (CHR), attached to the NL, and (B) an interphase BHK-21 cell extracted with nonionic detergent. The latter treatment has removed the nuclear membrane and the NL appears as a single layer. Bars = 0.2 μm.

Gerace, 1986; Franke, 1987; Steinert and Roop, 1988). In addition, I have discussed problems at the borderline of our present knowledge, related to perspectives for future research.

2. *Definition of NL and Lamin Classification*

The NL is defined by (1) location at the nuclear periphery during interphase, (2) insolubility in nonionic detergents at low ionic strength, and (3) insolubility in concentrated solutions of NaCl. The main protein constituents of the NL are a small number of slightly acidic protein species (pI in the range 5–7) with molecular masses between 60 and 75 kDa, termed lamins (Fig. 2C). Two main types of lamins are distinguished: the first includes less acidic lamins A and C and the second the more acidic lamin B. Some functional differences have been suggested for these two types: lamin B serving as a link between the NL and the inner nuclear membrane and lamins A and C being involved in interactions with chromatin (Gerace and Blobel, 1982; Gerace *et al.*, 1984; Gerace, 1986). However, there are examples where a single lamin is sufficient for both purposes (see below). The lamins are immunologically related proteins (Burke *et al.*, 1983; Lehner *et al.*, 1986b; Krohne and Benavente, 1986a). Recently, the primary sequences of a few lamin proteins have been determined, revealing an extensive structural homology not only within the lamin family, but also with the intermediate filament (IF) proteins (McKeon *et al.*, 1986; Fisher *et al.*, 1986; Krohne *et al.*, 1987; Wolin *et al.*, 1987; Stick, 1988). As a result, the lamins have been declared IF type V proteins. Recently, with the discovery of new lamin species, the problem of classification has grown more complex. We expect that in the future, information, based on amino acid sequence, will be used more and more in defining and classifying the lamin proteins.

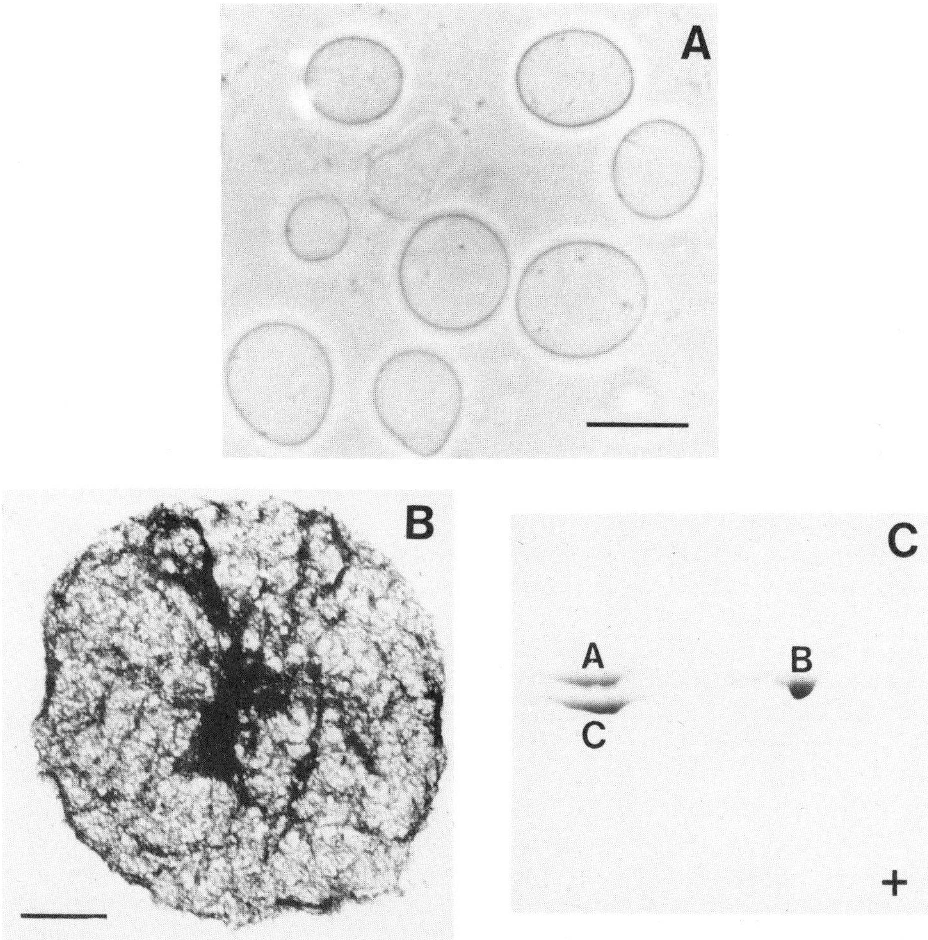

Figure 2. (A) NL structures, isolated from Ehrlich ascites tumor cells after extraction with NP-40, digestion of chromatin at low ionic strength, and extraction with 2 M NaCl (Krachmarov *et al.*, 1986a). Phase contrast. Bar = 5 μm. (B) Electron micrograph of a whole-mount structure. Bar = 1 μm. These preparations contain three main polypeptides (lamins A, B, and C), as demonstrated by two-dimensional IEF/SDS-PAGE in panel C.

3. *Isolation of NL*

NL can be isolated from a number of cell types using nonionic detergents, enzymatic digestion of DNA and RNA, and high-salt extraction (e.g., see Aaronson and Blobel, 1974; Dwyer and Blobel, 1976; Krachmarov *et al.*, 1986a). To obtain relatively pure NL preparations, it is essential to use conditions that prevent aggregation of intranuclear proteins, such as low ionic strength in the absence of divalent cations (Galcheva-Gargova *et al.*, 1982; Krachmarov *et al.*, 1986a), reducing agents (Kaufmann *et al.*, 1983), and low temperature during the digestion step (Evan and Hancock, 1985). Isolated NL structures appear as empty shells (e.g., see Krachmarov *et al.*, 1986a), which retain the shape of the nucleus (Fig. 2A,B). The remarkable stability of isolated NL is difficult to reconcile with its rapid disassembly *in vivo* as a result of what appears to be a low amount of postsynthetic protein modifications (see below). However, one should bear in mind that taking the NL out of its natural environment may induce profound changes in its structure, resulting in a secondary stabilization. For example, NL undergoes a considerable reduction in size upon high-salt treatment (Krachmarov and Dessev, 1988; see also Dessev *et al.*, 1989). Therefore, observations on isolated NL do not necessarily reflect its *in vivo* properties.

A number of unidentified minor polypeptide components have been found in isolated NL preparations (Krachmarov *et al.*, 1986a; Dessev *et al.*, 1989). At present, it is difficult to decide whether these polypeptides are bona fide NL components. However, their study is nevertheless worthwhile, since they may reflect functionally important structural associations of the NL.

4. *Composition of NL and Its Changes during Development*

During the past few years, considerable progress has been made in characterizing the lamins in a number of cell types. Although the lamin family is expected to grow further, sufficient data have already accumulated to permit some assessment of both diversity and degree of conservation of these proteins. Thus, we can consider two groups of cell types, which appear to exhibit clear-cut differences in NL composition. One group includes cells early in the reproductive cycle, such as germ-line cells (oocytes, sperm) (Krohne *et al.*, 1981; Maul *et al.*, 1984; Benavente and Krohne, 1985), embryonic cells (Lehner *et al.*, 1987; Lebel *et al.*, 1987), and relatively undifferentiated somatic cells, such as teratocarcinoma stem cells (Stewart and Burke, 1987), T lymphoblastic cells (Guilly *et al.*, 1987), and lymphocytes (Hornbeck *et al.*, 1988). In these cells, predominantly one lamin has been found. In the course of development and differentiation, new lamin genes become activated and as the cells reach their final differentiated state they acquire a more complex NL composition of two or more lamin species (Stick and Hausen, 1985; Benavente *et al.*, 1985; Lebel *et al.*, 1987).

The composition of the NL also shows variations among adult somatic tissues of different species. For example, two lamins, found in chicken tissues, were initially thought to correspond to mammalian lamin A and B (Shelton *et al.*, 1980). However, it has recently been shown using monoclonal antibodies that the protein, which was considered to be chicken lamin B, is different from rat lamin B, but is immunologically similar to two minor rat polypeptides, which were not previously recognized as lamins. On the other hand, rat lamin B shows immuno-cross-reactivity with a minor chicken polypeptide (Lehner *et al.*, 1986b). There is no doubt that the diversity within the lamin family will increase as new members are described.

One possible, often discussed explanation of these changes is that the NL composition may be important for the functional organization in the nucleus, establishment of differentiated cell functions and differential gene expression. Attractive as it may be, this view is supported by little direct evidence. Moreover, it does not always correlate with the experimental data. For example, in *Xenopus* the oocyte-specific lamin L_{III} is also expressed in adult neurons and muscle cells (Benavente *et al.*, 1985). An alternative explanation of the lamin diversity and developmental changes emerges if we realize that ontogenesis involves turning on and off of large groups of functionally associated genes under the control of master regulatory genes (e.g., see Gehring, 1985), in a sequence that may recapitulate phylogenesis. The changes in the lamin expression may simply result from incorporation of different lamin genes into these genetic blocks, without necessarily having a functional meaning.

It is clear, therefore, that the lamins are both conserved and diverse proteins. While the conserved parts of the lamin molecules are likely to be those involved in their capacity to form polymers, lamin diversity may or may not have a functional significance. Since evolution is where one finds evidence concerning the functional importance of a particular structural property, it will be of great interest to define the molecular features of the lamins that are conserved and those that have been subjected to evolutionary changes. From this point of view, characterizing and sequencing new lamin species from different functionally specialized cell types and from organisms of different evolutionary age is not simply a cataloging work, but is likely to provide important insights into the NL functions and evolution.

5. *Biosynthesis of Lamins*

The lamins are synthesized continuously during interphase (Gerace *et al.*, 1984). In a few cases, precursors of lamins A and B, with slightly lower electrophoretic mobilities than those of the "mature" forms, have been described (Gerace *et al.*, 1984; Laliberte *et al.*, 1984; Dagenais *et al.*, 1985; Lehner *et al.*, 1986a). These precursors are rapidly incorporated into the NL polymer, which is accompanied or followed by their trimming (apparently by a specific proteolytic cleavage) to the sizes of the "mature" lamins. Studies on the distribution of the lamin precursors among different subcellular fractions suggested that the lamins are synthesized on free polysomes (Lehner *et al.*, 1986a). In contrast to the lamins assembled into a polymer, the soluble newly synthesized lamins are not phosphorylated (Ottaviano and Gerace, 1985). Recent studies indicate, however, that the "mature" form of lamin A is a product of proteolysis, which occurs during the fractionation procedures (Lebel and Raymond, 1987). If the proteolysis is suppressed during the cell fractionation, or if the tissue is homogenized directly in trichloroacetic acid, only the higher molecular weight form of lamin A is present. These results argue against the existence of *in vivo* processing of the newly synthesized lamins.

Do lamins incorporate into preexisting NL structures during interphase? Although this may be expected to occur on the basis of the fact that the nuclear size increases as cells grow, to date there is no direct experimental evidence supporting this possibility.

6. *Structure of Lamin Molecules and Their Assembly into Polymers*

Earlier studies employing peptide mapping suggested that the mammalian lamins (lamins A and C in particular) are structurally related polypeptides (Shelton *et al.*, 1980; Kaufmann *et al.*, 1983; Krachmarov *et al.*, 1986a). Although lamin B exhibits a different pattern, all three lamins have been shown to share a common immunological determinant, residing in a 46-kDa chymotryptic peptide (Burke *et al.*, 1983). Further, the lamins appear to be structurally similar to IF proteins by a number of criteria including amino acid composition, immunological cross-reactivity, solubility properties, and ability to form ordered paracrystalline arrays *in vitro* (Goldman *et al.*, 1986). These early data were recently confirmed by determination of the nucleotide sequences of the genes encoding human lamins A and C (McKeon *et al.*, 1986; Fisher *et al.*, 1986), followed by the sequences of a representative of lamin B type, protein L_I (Krohne *et al.*, 1987), another lamin A-like protein (Wolin *et al.*, 1987), and the oocyte-specific lamin L_{III} (Stick, 1988), all from *Xenopus laevis*. These studies have revealed a high degree of homology between the amino acid sequences of both lamins and IF proteins, which extends into similarities in their secondary structures. The main feature of both lamin and IF protein molecules is a central α-helical "rod" domain, 310–360 amino acids long, flanked by nonhelical N-terminal and C-terminal regions of variable length (Steinert and Roop, 1988). Indeed, in agreement with sedimentation data (Havre and Evans, 1983; Aebi *et al.*, 1986), electron microscopy has provided evidence that in solution lamins A and C form rod-shaped dimers about 52 nm long with two globular heads at one end (Aebi *et al.*, 1986). These dimers associate into higher oligomers, forming paracrystalline arrays with an axial periodicity of about 25 nm (Goldman *et al.*, 1986; Aebi *et al.*, 1986), visible as stain-excluding bands, perhaps reflecting a half-staggered arrangement. Under certain conditions, reconstitution experiments with lamins A and C have yielded filaments about 10 nm thick (Aebi *et al.*, 1986), similar to those found in the NL of *Xenopus* oocytes (see below). A similar mechanism of assembly has been suggested for IF type I–IV proteins (reviewed in Steinert and Roop, 1988). It should be noted, however, that the latter do not form paracrystalline arrays.

The filamentous structure of the NL has been noticed in a number of cell types (Dwyer and Blobel, 1976; Franke *et al.*, 1981; Maul and Avdalovic, 1980; Maul, 1977b). Recently, electron microscopic observations have revealed that the NL of *Xenopus* oocytes is made of filaments, approximately 10 nm thick, forming an orthogonal lattice with a crossover spacing of about 52 nm (Aebi *et al.*, 1986), which corresponds to the length of the central α-helical domain in the lamin molecules. When exposed to solutions of low ionic strength, these filaments develop an axial "beading" pattern with a periodicity of about 25 nm, similar to the one observed with IF.

On the basis of a detailed analysis of the available molecular parameters of lamins and IF proteins, a model has been proposed to explain their association into polymeric structures (Parry *et al.*, 1987; see also Parry, this volume). The model places the terminal nonhelical domains outside the backbone formed by the α-helical regions and is in good agreement with the experimental data.

7. Association of NL with Nuclear and Cytoplasmic Structural Elements

The NL is a polyvalent structure that exhibits a number of stable associations with elements of the nuclear envelope, nuclear interior, and cytoplasm, such as nuclear pore complexes (NPC), inner nuclear membrane, and chromatin.

7.1. NPC

Cell fractionation experiments have shown that the NL and NPC are tightly associated and coisolate under conditions that remove cytoplasm, chromatin, lipid membranes, and intranuclear material, termed nuclear matrix (Aaronson and Blobel, 1974; Dwyer and Blobel, 1976). However, in some situations, both *in vivo* and *in vitro,* the NL and NPC show structural independence. It has been found that EDTA destabilizes the NPC, but not the NL, suggesting that divalent cations are essential for their integrity (Bornens and Courvalin, 1978; Krachmarov *et al.*, 1986a). During specific stages of meiosis the NL is transiently disassembled, while the NPC remain intact and associated with the nuclear membrane (Stick and Schwartz, 1983). It is thought that NPC can be inserted into a preexisting NL structure in the absence of cell division, perhaps when a requirement arises for an increased nucleocytoplasmic transport (Maul, 1977a,b). Apart from identifying two proteins with possible NPC location (Gerace *et al.*, 1982; Davis and Blobel, 1986), little is known about the protein composition, molecular structure, and biogenesis of the NPC and about their mechanism of targeting to the nuclear envelope.

7.2. Nuclear Membrane

There are several pieces of evidence related to the interaction between the NL and the lipid-based nuclear membrane. After the nuclear envelope breaks down in mitosis, lamin B remains bound to the vesicles presumably resulting from fragmentation of the nuclear membrane (Gerace and Blobel, 1980). A similar interaction has also been observed in cell-free experiments on reconstitution of the nuclear envelope (Burke and Gerace, 1986). Furthermore, it has been shown that lamin B resists alkaline extraction of total nuclear envelopes, which is supposed to remove peripheral membrane proteins and membrane-enclosed proteins (Lebel and Raymond, 1984). In addition, lamin B can be photoaffinity labeled with the membrane-penetrating compound azidepyrene (Lebel and Raymond, 1984). In these experiments, lamin B behaves as an intrinsic membrane protein. On the basis of these results, it has been proposed that lamin B serves as a link between lamins A/C and the inner nuclear membrane, whereas lamins A and C interact with receptors on chromatin (Gerace and Blobel, 1980; Gerace *et al.*, 1984; Gerace, 1986). However, in a number of cell types, while there is clear morphological evidence for NL–membrane association, only one lamin has been found (Maul, 1980; Maul *et al.*, 1984; Krohne *et al.*, 1981; Benavente and Krohne, 1985; Stewart and Burke, 1987; Lehner *et al.*, 1987; Lebel *et al.*, 1987; Guilly *et al.*, 1987; Hornbeck *et al.*, 1988), eventually performing both functions. The recently published primary structure of a lamin B-type protein from *Xenopus* (L_I) has no long stretches of hydrophobic amino acids, which would constitute membrane-interacting domains (Krohne *et al.*, 1988). The results available so far do not exclude the possibility that lamins A and C could also interact with the nuclear membrane. It has been reported that the mouse lamins A, B, and C incorporate [^{14}C]mannose and therefore are associated with carbohydrate moieties (Maul *et al.*, 1984), which is a feature exhibited by many membrane proteins. It should also be pointed out that the solubilization of lamins A and C has been demonstrated only for mitotic cells (Gerace and Blobel, 1980), where the phosphorylation level of the lamins is substantially increased (Ottaviano and Gerace, 1985). In this context, it would be interesting to study the effect of protein phosphorylation on the association of lamins A and C with elements of the nuclear membrane.

More recent experiments have suggested an indirect mechanism for interaction between the NL and the lipid bilayer through other typical membrane proteins (Senior and Gerace, 1988; Wilson and Newport, 1988).

7.3. Cytoskeletal Elements

There is abundant morphological evidence to suggest that the cytoplasmic IF are associated with the nuclear surface. In a number of physiological situations and in cells treated with microtubule-disrupting drugs, IF retract toward the nucleus, where they form so-called perinuclear caps (Goldman *et al.*, 1986). These results are supported by cell fractionation experiments. In whole-mount specimens of cells, extracted with detergent and digested with DNase/RNase, the cytoskeletal network is seen continuous with the residual elements of the nuclear envelope (Fey *et al.*, 1984). In similarly fractionated preparations, IF coisolate with the remnants of the nuclei and can be seen to terminate on the nuclear surface (Goldman *et al.*, 1986; Katsuma *et al.*, 1987). In NL-enriched preparations from Ehrlich ascites tumor cells, vimentin IF can be visualized in direct contact with the cytoplasmic surface of the NL (Krachmarov *et al.*, 1986a).

Further progress in this problem has been made by studying the *in vitro* reconstitution of bovine vimentin and avian erythrocyte membranes (Georgatos and Blobel, 1987a,b; Georgatos *et al.*, 1987). The results have suggested the existence of two distinct classes of IF binding sites: one located on the nuclear surface and interacting specifically with the C-terminal end domain of vimentin and another, associated with the plasma membrane, recognizing its N-terminal domain. Examination of *in vitro* binding of radiolabeled vimentin to isolated nuclear envelopes has pointed to lamin B as a possible nuclear attachment site for IF. Furthermore, studies with truncated polypeptides (Georgatos *et al.*, 1987) have provided evidence that the plasma membrane binding site of desmin resides within its 67-amino-acid-long N-terminal fragment, which interacts specifically with the plasma membrane protein ankyrin. In similar experiments, interaction of desmin with lamin B has been found to require the presence of 21 amino acid residues located near the C-terminus of its central helical domain. Although these experiments provide important clues concerning the possible mode of interaction between IF and the nuclear envelope, the next essential step should be to find out whether similar interactions take place *in vivo*. There are observations suggesting the extremely dynamic nature of the IF–nuclear association. For example, it has been noted that the nuclei in certain cells can rotate, while the IF system remains immobile (Paddock and Albrecht-Buehler, 1986), and these findings are difficult to reconcile with the existence of rigid bonds between NL and IF. The location of the putative *in vivo* IF binding sites on the nuclear surface is also unclear. As it is difficult to imagine how IF could traverse the double nuclear membrane, it has been suggested that the association occurs around the NPC. However, the NPC are distinct from the NL and have been shown by immunogold labeling not to contain lamins (Burke *et al.*, 1983). All this demonstrates the difficulties that we encounter when attempting to explain the association of IF with the nuclear surface in a conventional way. Clearly, further studies are needed to elucidate the nature of this phenomenon *in vivo*.

7.4. Chromatin

There are numerous indications that in interphase, chromatin is associated with the NL. Displacement of chromatin in an electric field has shown a number of mechanically stable points of attachment to the NL (Skaer *et al.*, 1976). In nuclei exposed to solutions of low ionic strength, fibers of decondensed chromatin appear to terminate on the NL (Puvion-Dutilleul and Puvion, 1980). In meiotic cells, such as oocytes of the surf clam

Spisula solidissima, condensed chromosomes are tightly associated with the NL (Allen, 1953; Maul, 1980). A large proportion of DNA remains associated with the NL and can be visualized as loops radiating from a central core (Hancock and Hughes, 1982; Lebkowski and Laemmli, 1982; Krachmarov *et al.*, 1986b). The NL and DNA are in close contact *in vivo,* since the three lamins can be covalently attached to DNA upon irradiation of living cells (Galcheva-Gargova and Dessev, 1987) or isolated structures (Boulikas, 1986) with UV light, which generates "zero-length" intermolecular cross-links, or by chromium-induced cross-linking (Wedrychowski *et al.*, 1985). Finally, the specific location of the NL in the cell suggests its interaction with chromatin elements.

What are the features that are recognized in the interaction between NL and chromatin? It is unlikely that this association requires specific DNA sequences, since it has been shown that a typical nuclear envelope, consisting of NL, pore complexes, and lipid membranes, is formed around lambda DNA microinjected into *Xenopus* oocytes (Forbes *et al.*, 1983) or in cell-free extracts (Newport, 1987). The kinetics of the nuclear envelope formation in these experiments suggests that DNA has to become associated with proteins in the form of chromatin (Laskey *et al.*, 1977) before it can serve as a template for NL assembly. It is well known that the attachment of chromatin to the nuclear envelope exhibits a specific pattern. Peripheral location of constitutive heterochromatin blocks, containing satellite DNA, which correspond to the centromeric regions of chromosomes, has been demonstrated in mouse cells by *in situ* hybridization (Jones, 1970; Pardue and Gall, 1970). Computer-based three-dimensional reconstitution of the spatial organization of chromosomes in the salivary gland nuclei of *Drosophila* has shown that the chromocenters and the intercalary heterochromatin regions are almost always apposed to the nuclear envelope (Hochstrasser *et al.*, 1986). These data suggest a relation between the higher order structure of chromatin and its association with the NL. In support to this notion, it has been shown that depletion of mitotic cells of lamins by microinjection of lamin-specific antibodies results in profound alterations in the chromatin organization (Benavente and Krohne, 1986). There are also indications that intermediates, such as P1 antigen (Chaly *et al.*, 1984) or a protein termed perichromin (McKeon *et al.*, 1984), may play a role in chromatin–NL association (see also Wilson and Newport, 1988).

At present it is difficult to be more specific about the functional significance of the observations described above. There is little doubt, however, that further research in this exciting field will provide new insight into the structural foundations of the nuclear functions.

8. Cell Cycle Dynamics of NL

As part of the profound structural reorganization that occurs during mitosis, most eukaryotic cells lose their nuclear envelopes. This process involves several distinct events including depolymerization of the NL (Gerace and Blobel, 1980), fragmentation of the nuclear membranes (Chai *et al.*, 1974), and release/disassembly of the NPC (Maul, 1977a,b). The biochemistry of nuclear envelope breakdown has mostly been studied with regard to the nuclear lamins. It has been shown that during mitosis the lamins acquire one or two phosphate groups per molecule, which constitutes a four- to sevenfold net increase in their phosphate content compared to interphase (Ottaviano and Gerace, 1985). The solubilization of the NL during the meiotic breakdown of the nuclear envelope in oocytes

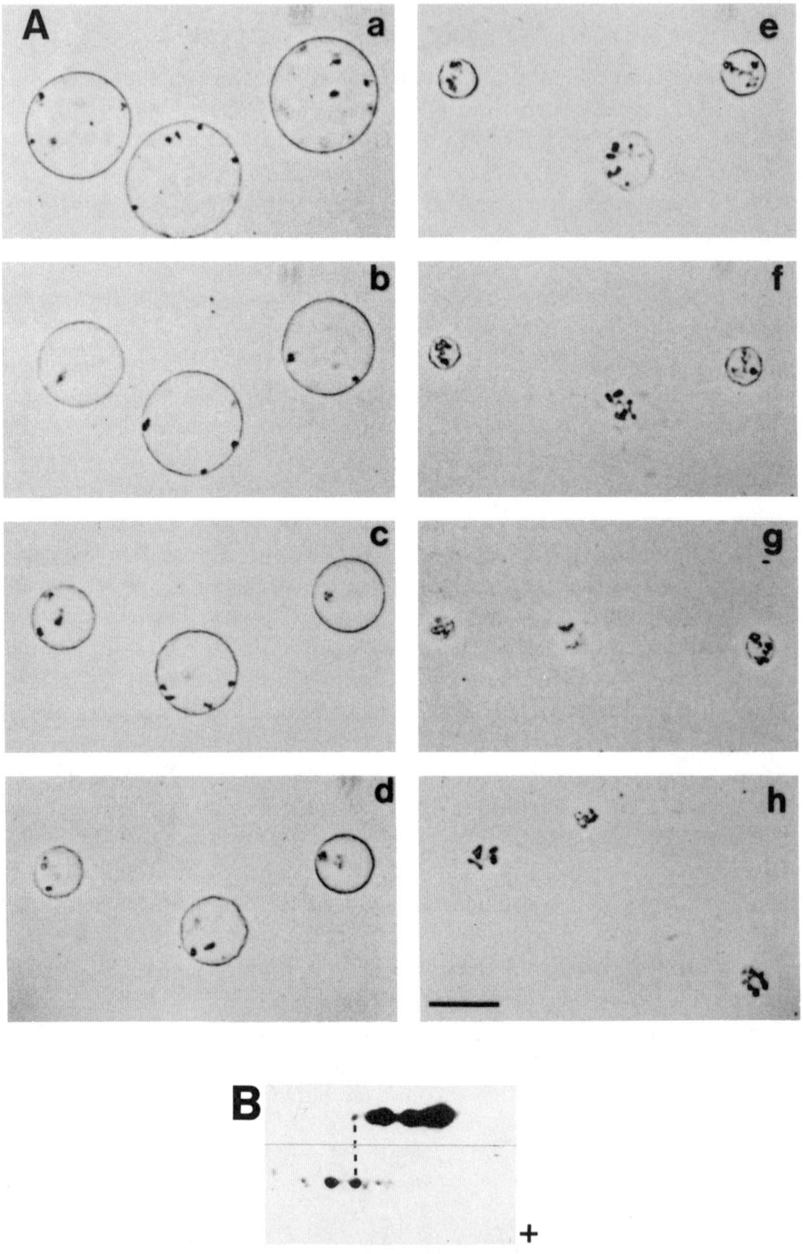

Figure 3. (A) Disassembly of NL in a cell-free system derived from oocytes of the surf clam, *Spisula solidissima*. When detergent-treated oocyte nuclei are exposed to cytosolic extracts from activated *Spisula* oocytes, they undergo a gradual reduction in size and finally disassemble, releasing chromosomes from their attachment to the NL. Eight consecutive stages of disassembly are shown (a–h) at 2, 15, 20, 25, 30, 35, 40, and 45 min, respectively. Bar = 20 μm. (B) In the course of disassembly the single lamin of these cells is extensively phosphorylated and solubilized. As a result of the phosphorylation, its isoelectric point becomes more acid, which is reflected in a shift in two-dimensional IEF/SDS gels (upper panel, autoradiography), compared to the same protein isolated from the polymer NL structure (lower panel, staining with Coomassie blue). The reaction of disassembly was carried out in the presence of [γ-^{32}P]-ATP and the radioactive soluble lamin recovered by immunoprecipitation. From Dessev *et al.* (1989) with permission of Academic Press.

of *S. solidissima,* both *in vivo* and in a cell-free system, has been found to involve extensive phosphorylation of the single lamin of these cells (Fig. 3B) (Dessev and Goldman, 1988; Dessev *et al.*, 1989). Similar observations have been reported for *Xenopus* single oocyte lamin (Krohne and Benavente, 1986b) and for mammalian cell nuclei, incubated in extracts prepared from amphibian oocytes (Miaki-Lye and Kirschner, 1985; Newport and Spann, 1987) or from mitotic cells (Suprynowicz and Gerace, 1986). These results suggest that the mechanism of reversible NL depolymerization involves changes in the phosphorylation level of the lamins.

Evidence for lamin dephosphorylation coincident with their reassembly around chromosomes has been obtained in cell-free conditions (Burke and Gerace, 1986). Since it has been shown that the soluble lamins are reutilized in the new NL at the end of mitosis, it is conceivable that lamin dephosphorylation also occurs *in vivo,* catalyzed by specific, cell cycle-regulated protein phosphatases.

Although a rigorous proof for a causal relationship between protein phosphorylation and NL disassembly is still missing, the idea that the NL dynamics is governed by changes in lamin phosphorylation is in harmony with the present-day concept that this type of postsynthetic modification is employed in regulating many cell functions by modifying the interactions between protein molecules (Flockhart and Corbin, 1982; Hunter and Cooper, 1985; Maller, 1985).

Recent experimental findings suggest, however, that the mechanism of nuclear envelope breakdown may be more complex. A substantial increase in the lamin phosphorylation, but no disassembly of the NL, has been found to occur after stimulation of quiescent BHK cells with insulin (Friedman and Ken, 1988). A similar effect has been described in lymphocytes, where the single lamin type B undergoes a rapid phosphorylation in response to treatment with phorbol esters (Hornbeck *et al.*, 1988). In this case it has been noted that although no nuclear envelope breakdown takes place, the proportion of detergent-insoluble lamin is reduced, suggesting that lamin phosphorylation alters the stability of the NL.

It has been reported that two carboxyl groups of lamin B in mouse cells are reversibly demethylated during mitosis (Chelsky *et al.*, 1987). It is possible, therefore, that postsynthetic modifications other than phosphorylation may be involved in the mitotic dynamics of the NL. Significantly, both demethylation and phosphorylation result in a net increase of the total negative charge of the NL.

The analysis of the molecular changes during the NL disassembly, i.e., how phosphorylation or other modifications of a few sites would weaken the interactions between the helical domains that maintain the backbone of the structure, is a problem that is far from its solution. Still, we could think of the NL as a dynamic system, capable of subunit exchange with the soluble phase at all times in a manner depending on its modification level. It has been shown that during the nuclear envelope breakdown of *Spisula* oocytes in a cell-free system, lamin molecules are continuously released from the NL polymer (Dessev *et al.*, 1989). This process occurs both *in vivo* and *in vitro* (Dessev and Goldman, 1988; Dessev *et al.*, 1989) without disruption of the NL, which undergoes a gradual reduction in size (Fig. 3A) as a result of unknown molecular rearrangements to compensate for the loss of protein. We can imagine that the reverse process will occur during interphase, where the rate of insertion of subunits will be higher than the dissociation rate. This will result in a net mass increase, consistent with nuclear growth (Ottaviano and Gerace, 1985; Wilson and Newport, 1988). In mitosis, the NL undergoes an almost explosive disassembly (Hughes, 1952). The same is true for the meiotic disassembly of the nuclear envelope in *Spisula* oocytes (Allen, 1953; Dessev and Goldman, 1988). We can imagine that upon increasing the concentration of negative charges in the NL, a threshold is eventually reached, beyond which the process becomes avalanche-like. The behavior of the NL during its mitotic disassembly can be formally described by the theory of catastrophes, where relatively small changes can induce dramatic effects (Zeeman, 1976). At this point it appears possible that cooperativity in the modification process (i.e., increase of its efficiency at higher levels of modification) could contribute to the catastrophic character of the NL disassembly.

Phosphopeptide mapping reveals that the lamins are phosphorylated on five to nine different sites both *in vivo* and *in vitro* (Dessev *et al.*, 1988). This, compared with an average of one or two phosphates per molecule found in mitosis (Ottaviano and Gerace, 1985), means that different lamin molecules are phosphorylated on different sites. Such a heterogeneity suggests that what regulates the NL stability may be the concentration of negative charges (phosphates plus carboxyl groups) in the whole structure, or perhaps in some part of it. In other words, there may be a relaxed, rather than a stringent requirement for sequence specificity of the phosphorylated sites. It is relevant to mention that the sites in the desmin molecule, whose phosphorylation prevents the assembly of IF (see below), are located in its N-terminal nonhelical domain (Geisler and Weber, 1988). The phosphorylation sites in keratin are also found on the terminal portions of the molecules (Steinert and Roop, 1988). The locations of the phosphorylation sites in the lamin molecules are not known. Indirect evidence, based on similarity in the phosphopeptide maps of lamins A and C (Dessev *et al.*, 1988), suggests that they may also be in the N-terminal head domains, which are identical for lamins A and C, and not on the C-terminal part, where the two polypeptides differ. It is not known whether this preferential location of phosphate groups results from higher accessibility of the N-terminal parts of the molecules, or from protein kinase specificity.

The factors involved in fragmentation of the nuclear membrane are now known. Although it has been reported that *in vitro* the nuclear membrane disassembly can be uncoupled from that of the NL (Newport and Spann, 1987), the close temporal coordination of these two events *in vivo* suggests a common regulatory mechanism. On the other hand, the NL has been found to disappear in pachytene chicken oocytes, while the nuclear membrane and the pore complexes stay intact (Stick and Schwartz, 1983). Clearly, further experiments are required to elucidate this problem.

9. The Same General Mechanism Is Likely to Be Involved in the Cell Cycle Dynamics of NL and IF

Unlike the lamins, no clear-cut dependence of the IF stability on the cell cycle stage has emerged. However, in some cell types (BHK-21 cells) immunofluorescence observations have revealed dramatic changes in IF organization during the mitotic phase (Jones *et al.*, 1985). These changes include diffuse or spotty fluorescence and short fibers at the spindle poles and are indicative of a certain degree of disassembly.

The phosphorylation level of vimentin has been found to increase in mitosis, together with the appearance of several phosphopeptides absent in interphase vimentin (Evans, 1984). Recent studies have demonstrated that vimentin (Inagaki *et al.*, 1987) and desmin

(Geisler and Weber, 1988), phosphorylated *in vitro* by the catalytic subunit of cAMP-dependent protein kinase, lose their ability to form typical IF. In desmin, 2.3 moles phosphate are incorporated per mole protein and the phosphorylated serines are located in the N-terminal nonhelical domain of the molecule (Geisler and Weber, 1988). Chou *et al.* (1989) have shown that phosphorylation of vimentin and desmin IF by a partly purified kinase, isolated from mitotic cells, correlates with their *in vitro* disassembly. Thus, it appears that not only are the lamins and IF proteins structurally related but, also, their monomer–polymer transition is likely to be controlled by similar mechanisms.

10. Cell Cycle Regulation of NL Disassembly

The NL dynamics can be viewed in the context of the general problem of cell cycle regulation. According to the current concept, the oscillation of the cell between M phase and interphase is governed by a pleiotropic enzymatic system, termed maturation- or M phase-promoting factor (MPF) (Masui and Clarke, 1979; Smith and Eckert, 1971; Wu and Gerhart, 1980; Lohka *et al.*, 1988; Dunphy *et al.*, 1988; Gautier *et al.*, 1988; Arion *et al.*, 1988; for review see Maller, 1985), which becomes activated at the beginning of M phase. It is believed that MPF triggers a cascade of reactions, ultimately leading to M-phase-specific events (breakdown of the nuclear envelope, chromosome condensation, formation of the mitotic spindle, karyokinesis, cytokinesis, and reorganization of cytoskeleton), which occur in a strictly coordinated temporal and topological fashion. At present, little is known about the pathway that leads to nuclear envelope breakdown. Identification of the protein kinase(s) involved in the mitotic phosphorylation of the nuclear lamins will be an essential step toward understanding the cell cycle control over the NL dynamics.

References

Aaronson, R. P., and Blobel, G., 1974, Isolation of a nuclear pore complex in association with a lamina, *Proc. Natl. Acad. Sci. USA* **72:**1007–1011.

Aebi, U., Cohn, J., Buhle, L., and Gerace, L., 1986, The nuclear lamina is a meshwork of intermediate-type filaments, *Nature* **323:**560–564.

Allen, R. D., 1953, Fertilization and artificial activation in the egg of the surf clam, *Spisula solidissima, Biol. Bull.* **105:**213–239.

Arion, D., Meijer, L., Brizuella, L., and Beach, D., 1988, cdc2 is a component of the M-phase-specific histone Hl kinase: Evidence for identity with MPF, *Cell* **55:**371–378.

Benavente, R., and Krohne, G., 1985, Change of karyoskeleton during spermatogenesis of Xenopus: Expression of lamin L_{IV}, a nuclear lamina protein specific for the male germ line, *Proc. Natl. Acad. Sci. USA* **82:**6176–6180.

Benavente, R., and Krohne, G., 1986, Involvement of nuclear lamins in postmitotic reorganization of chromatin as demonstrated by microinjection of lamin antibodies, *J. Cell Biol.* **100:**1847–1854.

Benavente, R., Krohne, G., and Franke, W. W., 1985, Cell type-specific expression of nuclear lamina proteins during development of Xenopus laevis, *Cell* **41:**177–190.

Bornens, M., and Courvalin, J. C., 1978, Isolation of nuclear envelopes with polyanions, *J. Cell Biol.* **76:**191–206.

Boulikas, T., 1986, Protein–protein and protein–DNA interactions in calf thymus nuclear matrix using crosslinking by ultraviolet irradiation, *Biochem. Cell Biol.* **64:**474–484.

Burke, B., and Gerace, L., 1986, A cell free system to study reassembly of the nuclear envelope at the end of mitosis, *Cell* **44:**639–652.

Burke, B., Tooze, J., and Warren, G., 1983, A monoclonal antibody which recognizes each of the nuclear lamin polypeptides in mammalian cells, *EMBO J.* **2**:361–367.

Chai, L. S., Weinfeld, H., and Sandberg, A. A., 1974, Ultrastructural changes on the nuclear envelope during mitosis of Chinese hamster cells: A proposed model of nuclear envelope reformation, *J. Natl. Cancer Inst.* **53**:1033–1050.

Chaly, N., Bladon, T., Setterfield, G., Little, J. E., Kaplan, J., and Brown, D. L., 1984, Changes in distribution of nuclear matrix antigens during the mitotic cell cycle, *J. Cell Biol.* **99**:661–670.

Chelsky, D., Olson, J. F., and Koshland, D. E., Jr., 1987, Cell cycle-dependent methyl esterification of lamin B, *J. Biol. Chem.* **262**:4303–4309.

Chou, Y.-H., Rosevear, E., and Goldman, R., 1989, Phosphorylation and disassembly of intermediate filaments in mitotic cells, *Proc. Natl. Acad. Sci. USA,* **86**:1885–1889.

Dagenais, A., Bibor-Hardy, V., Laliberte, J.-F., Royal, A., and Simard, R., 1985, Detection in BHK cells of a precursor form for lamin A, *Exp. Cell Res.* **161**:269–276.

Davis, L. I., and Blobel, G., 1986, Identification and characterization of a nuclear pore complex protein, *Cell* **45**:699–709.

Dessev, G., Iovcheva, C., Tasheva, B., and Goldman, R., 1988, Protein kinase activity associated with the nuclear lamina, *Proc. Natl. Acad. Sci. USA* **85**:2994–2998.

Dessev, G., and Goldman, R., 1988, Meiotic breakdown of nuclear envelope in oocytes of Spisula solidissima involves phosphorylation and release of nuclear lamin, *Dev. Biol.* **130**:543–550.

Dessev, G., Palazzo, R., Rebhun, L., and Goldman, R., 1989, Disassembly of the nuclear envelope of Spisula oocytes in a cell free system, *Dev. Biol.* **131**:496–504.

Dunphy, W., Brizuella, L., Beach, D., and Newport, J., 1988, The Xenopus cdc2 protein is a component of MPF, a cytoplasmic regulator of mitosis, *Cell* **54**:423–431.

Dwyer, N., and Blobel, G., 1976, A modified procedure for the isolation of a pore complex–lamina fraction from rat liver nuclei, *J. Cell Biol.* **70**:581–591.

Evan, G. I., and Hancock, D. C., 1985, Studies on the interaction of the human c-myc protein with cell nuclei: $p62^{c-myc}$ as a member of a discrete subset of nuclear proteins, *Cell* **43**:253–261.

Evans, R. M., 1984, Peptide mapping of phosphorylated vimentin; evidence for a site-specific alteration in mitotic cells, *J. Biol. Chem.* **259**:5372–5375.

Fawcett, D. W., 1966, On the occurrence of a fibrous lamina on the inner aspect of the nuclear envelope in certain cells of vertebrates, *Am. J. Anat.* **119**:129–146.

Fawcett, D. W., 1981, *The Cell,* W. B. Saunders Co., Philadelphia.

Fey, E. G., Won, K. M., and Penman, S., 1984, Epithelial cytoskeletal frame work and nuclear matrix–intermediate filament scaffold: Three dimensional organization and protein composition, *J. Cell Biol.* **98**:1973–1984.

Fisher, D. Z., Chaudhary, N., and Blobel, G., 1986, cDNA sequencing of nuclear lamin A and C reveals primary and secondary structural homology to intermediate filament proteins, *Proc. Natl. Acad. Sci. USA* **83**:6450–6454.

Flockhart, D. A., and Corbin, J. D., 1982, Regulatory mechanisms in the control of protein kinases, *CRC Crit. Rev. Biochem.* **12**:133–186.

Forbes, D. J., Kirschner, M. W., and Newport, J. W., 1983, Spontaneous formation of nucleus-like structures around bacteriophage DNA microinjected into Xenopus eggs, *Cell,* **34**:13–23.

Franke, W. W., 1987, Nuclear lamins and cytoplasmic intermediate filament proteins: A growing multigene family, *Cell* **48**:3–4.

Franke, W. W., Scheer, U., Krohne, G., and Jarasch, E., 1981, The nuclear envelope and the architecture of the nuclear periphery, *J. Cell Biol.* **91**:39s–50s.

Friedman, D. L., and Ken, R., 1988, Insulin stimulates incorporation of $^{32}P_i$ into nuclear lamins A and C in quiescent BHK-21 cells, *J. Biol. Chem.* **263**:1103–1106.

Galcheva-Gargova, Z., Petrov, P., and Dessev, G., 1982, Effect of chromatin decondensation on the intranuclear matrix, *Eur. J. Cell Biol.* **28**:155–159.

Galcheva-Gargova, Z., and Dessev, G., 1987, Crosslinking of DNA to nuclear lamina proteins by UV-irradiation in vivo, *J. Cell Biochem.* **34**:163–168.

Gautier, J., Norbury, C., Lohka, M., Nurse, P., and Maller, J., 1988, Purified maturation promoting factor contains the product of a Xenopus homolog of the fission yeast cell cycle control gene $cdc2^+$, *Cell* **54**:433–439.

Gehring, W. J., 1985, The homeo box: A key to the understanding of development? *Cell* **40**:3–5.

Geisler, N., and Weber, K., 1988, Phosphorylation of desmin in vitro inhibits formation of intermediate filaments: Identification of three kinase A sites in the aminoterminal head domain, *EMBO J.* **7:**15–20.

Georgatos, S. D., and Blobel, G., 1987, Two distinct attachment sites for vimentin along the plasma membrane and the nuclear envelope in avian erythrocytes: A basis for a vectorial assembly of intermediate filaments, *J. Cell Biol.* **105:**105–115.

Georgatos, S. D., and Blobel, G., 1987b, Lamin B constitutes an intermediate filament attachment site at the nuclear envelope, *J. Cell Biol.* **105:**117–125.

Georgatos, S. D., Weber, K., Geisler, N., and Blobel, G., 1987, Binding of two desmin derivatives to the plasma membrane and the nuclear envelope of avian erythrocytes: Evidence for a conserved site-specificity in intermediate filament–membrane interactions, *Proc. Natl. Acad. Sci. USA* **84:**6780–6784.

Gerace, L., 1986, Nuclear lamina and organization of nuclear architecture, *Trends Biochem. Sci.* **11:**443–446.

Gerace, L., and Blobel, G., 1980, The nuclear lamina is depolymerized during mitosis, *Cell* **19:**277–287.

Gerace, L., and Blobel, G., 1982, Nuclear lamina and the structural organization of the nuclear envelope, *Cold Spring Harbor Symp. Quant. Biol.* **46:**967–978.

Gerace, L., Ottaviano, Y., and Kondor-Koch, C., 1982, Identification of a major polypeptide of the nuclear pore complex, *J. Cell Biol.* **95:**826–837.

Gerace, L., Comeau, C., and Benson, M., 1984, Organization and modulation of nuclear lamina structure, *J. Cell Sci.* Suppl. 1, 137–160.

Goldman, A. E., Maul, G., Steinert, P. M., Yang, H.-Y., and Goldman, R., 1986, Keratin-like proteins that coisolate with intermediate filaments of BHK-21 cells are nuclear lamins, *Proc. Natl. Acad. Sci. USA* **83:**3839–3843.

Goldman, R. D., Goldman, A. E., Green, K. J., Jones, J. C. R., Jones, S. M., and Yang, H.-Y., 1986, Intermediate filament networks: Organization and possible functions of a diverse group of cytoskeletal elements, *J. Cell Sci.* Suppl. 5, 69–97.

Guilly, M. N., Bensussan, A., Bourge, J. F., Bornens, M., and Courvalin, J. C., 1987, A human T lymphoblastic cell line lacks lamins A and C, *EMBO J.* **6:**3795–3799.

Hancock, R., 1982, Topological organization of interphase DNA: The nuclear matrix and other skeletal structures, *Biol. Cell* **46:**105–122.

Hancock, R., and Boulikas, T., 1982, Functional organization in the nucleus, *Int. Rev. Cytol,* **79:**165–214.

Hancock, R., and Hughes, M. E., 1982, Organization of DNA in the interphase nucleus, *Biol. Cell* **44:**201–212.

Havre, P., and Evans, D., 1983, Disassembly and characterization of the nuclear pore complex–lamina fraction from bovine liver nuclei, *Biochemistry* **22:**2852–2860.

Hochstrasser, M., Mathog, D., Gruenbaum, Y., Saumweber, H., and Sedat, J. W., 1986, Spacial organization of chromosomes in the salivary gland of Drosophila melanogaster, *J. Cell Biol.* **102:**112–123.

Hornbeck, P., Huang, K. P., and Paul, W. E., 1988, Lamin B is rapidly phosphorylated in lymphocytes after activation of protein kinase C, *Proc. Natl. Acad. Sci. USA* **85:**2279–2283.

Hughes, A., 1952, *The Mitotic Cycle: The Cytoplasm and Nucleus during Interphase and Mitosis,* Academic Press, New York, p. 79.

Hunter, T., and Cooper, J. A., 1985, Protein tyrosine kinases, *Ann. Rev. Biochem.* **54:**897–930.

Inagaki, M., Nishi, Y., Nishizawa, K., Matsuyama, M., and Sato, C., 1987, Site-specific phosphorylation induces disassembly of vimentin filaments in vitro, *Nature* **328:**649–652.

Jones, J. C. R., Goldman, A. E., Yang, H.-Y., and Goldman, R., 1985, The organizational fate of intermediate filament networks in two epithelial cell types during mitosis, *J. Cell Biol.* **100:**93–102.

Jones, K. W., 1970, Chromosomal and nuclear location of mouse satellite DNA in individual cells, *Nature* **225:**912–915.

Katsuma, Y., Swierenga, S. H. H., Marceau, N., and French, S. W., 1987, Connections of intermediate filaments with the nuclear lamina and the cell periphery, *Biol. Cell* **59:**193–204.

Kaufmann, S. H., Gibson, W., and Shaper, J. H., 1983, Characterization of the major polypeptides of the rat liver nuclear envelope, *J. Biol. Chem.* **258:**2710–2719.

Krachmarov, C., Tasheva, B., Markov, D., Hancock, R., and Dessev, G., 1986a, Isolation and characterization of nuclear lamina from Ehrlich ascites tumor cells, *J. Cell. Biochem.* **30:**351–359.

Krachmarov, C., Iovcheva, C., Hancock, R., and Dessev, G., 1986b, Association of DNA with the nuclear lamina in Ehrlich ascites tumor cells, *J. Cell Biochem.* **31:**59–74.

Krachmarov, C., and Dessev, G., 1988, Reversible contractility of the nuclear lamina, *Comptes rendus de l'Academy Bulgare des Sciences* **41:**119–122.

Krohne, G., and Benavente, R., 1986a, The nuclear lamins: A multigene family of proteins in evolution and differentiation, *Exp. Cell Res.* **162:**1–10.

Krohne, G., and Benavente, R., 1986b, A pool of soluble nuclear lamins in eggs and embryos of Xenopus laevis, in: *Nucleocytoplasmic transport* (R. Peters and M. Trendelenburg, eds.), Springer-Verlag, Berlin, pp. 135–141.

Krohne, G., Dabauvalle, M.-C., and Franke, W. W., 1981, Cell type-specific differences in protein composition of nuclear pore complex–lamina structures in oocytes and erythrocytes of Xenopus laevis, *J. Mol. Biol.* **151:**121–141.

Krohne, G., Wolin, S. L., McKeon, F. D., Franke, W. W., and Kirschner, M. W., 1987, Nuclear lamin L_I of Xenopus laevis: cDNA cloning, aminoacid sequence and binding specificity of a member of the lamin B subfamily, *EMBO J.* **6:**3801–3808.

Laliberte, J.-F., Dagenais, A., Filion, M., Bibor-Hardy, V., Simard, R., and Royal, A., 1984, Identification of distinct messenger RNAs for nuclear lamin C and a putative precursor of nuclear lamin A, *J. Cell Biol.* **98:**980–985.

Laskey, R. A., Mills, A. D., and Morris, N. R., 1977, Assembly of SV40 chromatin in a cell-free system from Xenopus eggs, *Cell* **10:**237–243.

Lebel, S., and Raymond, Y., 1984, Lamin B from rat nuclei exists both as a lamina protein and as an intrinsic membrane protein, *J. Biol. Chem.* **259:**2693–2696.

Lebel, S., and Raymond, Y., 1987, Lamin A is not synthesized as a larger precursor polypeptide, *Biochem. Biophys. Res. Commun.* **149:**417–423.

Lebel, S., Lampron, C., Royal, A., and Raymond, Y., 1987, Lamins A and C appear during retinoic acid-induced differentiation of mouse embryonal carcinoma cells, *J. Cell Biol.* **105:**1099–1104.

Lebkowski, J., and Laemmli, U., 1982, Non-histone proteins and long-range organization of HeLa interphase DNA, *J. Mol. Biol.* **156:**325–344.

Lehner, C. F., Furstenberger, G., Eppenberger, H. M., and Nigg, E. A., 1986a, Biogenesis of the nuclear lamina: In vivo synthesis and processing of nuclear protein precursors, *Proc. Natl. Acad. Sci. USA* **83:**2096–2099.

Lehner, C. F., Kurer, V., Eppenberger, H. M., and Nigg, E. A., 1986b, The nuclear lamin protein family in higher vertebrates: Identification of quantitatively minor lamin protein by monoclonal antibodies, *J. Biol. Chem.* **261:**13293–13301.

Lehner, C. F., Stick, R., Eppenberger, H. M., and Nigg, E. A., 1987, Differential expression of nuclear lamin protein during chicken development, *J. Cell Biol.* **105:**577–587.

Lohka, M. J., Hayes, M. K., and Maller, J. L., 1988, Purification of maturation-promoting factor, an intracellular regulator of early mitotic events, *Proc. Natl. Acad. Sci. USA* **85:**3009–3013.

McKeon, F. D., Tuffanelli, D. L., Kobayashi, S., and Kirschner, M. W., 1984, The redistribution of a conserved nuclear envelope protein during the cell cycle suggests a pathway for chromosome condensation, *Cell* **36:**83–96.

McKeon, F. D., Kirschner, M. W., and Caput, D., 1986, Homologies in both primary and secondary structure between nuclear envelope and intermediate filament proteins, *Nature* **319:**463–468.

Maller, J. L., 1985, Regulation of amphibian oocyte maturation, *Cell Differ.* **16:**211–221.

Masui, Y., and Clarke, H. J., 1979, Oocyte maturation, *Int. Rev. Cytol.* **57:**185–282.

Maul, G. G., 1977a, Nuclear pore complexes: Elimination and reconstitution during mitosis, *J. Cell Biol.* **74:**492–500.

Maul, G. G., 1977b, The nuclear and the cytoplasmic pore complex: Structure, dynamics, distribution and evolution, *Int. Rev. Cytol.* Suppl. 6, 75–186.

Maul, G. G., 1980, Determination of newly synthesized and phosphorylated nuclear proteins in mass-isolated germinal vesicle of *Spisula solidissima, Exp. Cell Res.* **129:**431–438.

Maul, G. G., and Avdalovic, N., 1980, Nuclear envelope proteins from *Spisula solidissima* germinal vesicles, *Exp. Cell Res.* **130:**229–240.

Maul, G. G., Baglia, F. A., Newmeyer, D. D., and Ohlsson-Wilhelm, B. M., 1984, The major 67,000 molecular weight protein of the clam oocyte nuclear envelope is lamin-like, *J. Cell Sci.* **67:**69–85.

Miake-Lye, R., and Kirschner, M. W., 1985, Induction of early mitotic events in a cell-free system, *Cell* **41:**165–175.

Newport, J., 1987, Nuclear reconstitution in vitro: Stages of assembly around protein-free DNA, *Cell* **48:**205–217.

Newport, J., and Spann, T., 1987, Disassembly of the nucleus in mitotic extracts: Membrane vesicularization, lamin disassembly, and chromosome condensation are independent processes, *Cell* **48:**219–230.

Ottaviano, Y., and Gerace, L., 1985, Phosphorylation of the nuclear lamins during interphase and mitosis, *J. Biol. Chem.* **260:**624–632.

Paddock, S. W., and Albrecht-Buehler, G., 1986, Distribution of microfilament bundles during rotation of the nucleus in 3T3 cells treated with monensin, *Exp. Cell Res.* **163:**525–538.

Pardue, M. L., and Gall, J. G., 1970, Chromosome localization of mouse satellite DNA, *Science* **168:**1356–1358.

Parry, D. A. D., Conway, J. F., Goldman, A. E., Goldman, R. D., and Steinert, P. M., 1987, Nuclear lamin proteins: Common structures for paracrystalline, filamentous and lattice forms, *Int. J. Biol. Macromol.* **9:**137–145.

Puvion-Dutilleul, F., and Puvion, E., 1980, New aspects of intranuclear structures following partial decondensation of chromatin: A cytochemical and high-resolution autoradiographic study, *J. Cell Sci.* **42:**305–321.

Senior, A., and Gerace, L., 1988, Integral membrane proteins specific to the inner nuclear membrane and associated with the nuclear lamina, *J. Cell Biol.* **107:**2029–2036.

Shelton, K. R., Higgins, L. L., Cochran, D. L., Ruffolo, J. J., Jr., and Egle, P. M., 1980, Nuclear lamins of erythrocyte and liver, *J. Biol. Chem.* **255:**10978–10983.

Skaer, R. J., Whytock, S., and Emmines, J. P., 1976, Intracellular electrophoresis of chromatin of living cells, *J. Cell Sci.* **21:**470–496.

Smith, L. D., and Eckert, R. E., 1971, The interaction of steroids with Rana pipiens oocytes in the induction of maturation, *Dev. Biol.* **25:**233–247.

Steinert, P. M., and Roop, D. R., 1988, Molecular and cellular biology of intermediate filaments, *Annu. Rev. Biochem.* **57:**539–625.

Stewart, C., and Burke, B., 1987, Teratocarcinoma stem cells and early mouse embryos contain only a single major lamin polypeptide closely resembling lamin B, *Cell* **51:**383–392.

Stick, R., 1988, cDNA cloning of the developmentally regulated lamin L_{III} of Xenopus laevis, *EMBO J.* **7:**3189–3197.

Stick, R., and Hausen, P., 1985, Changes in the nuclear lamina composition during development of Xenopus laevis, *Cell* **41:**191–200.

Stick, R., and Schwartz, H., 1983, Disappearance and reformation of the nuclear lamina structure during specific stages of meiosis in oocytes, *Cell* **33:**949–958.

Suprynowicz, F. A., and Gerace, L., 1986, A fractionated cell-free system for analysis of prophase nuclear disassembly, *J. Cell Biol.* **103:**2073–2981.

Wedrychowski, A., Steven Ward, W., Schmidt, W. N., and Hnilica, L. S., 1985, Chromium-induced cross-linking of nuclear proteins and DNA, *J. Biol. Chem.* **260:**7150–7155.

Wilson, K. L., and Newport, J., 1988, A trypsin-sensitive receptor on membrane vesicles is required for nuclear envelope formation in vitro, *J. Cell Biol.* **107:**57–68.

Wolin, S. L., Krohne, G., and Kirschner, M. W., 1987, A new lamin in Xenopus somatic tissues displays strong homology to human lamin A, *EMBO J.* **6:**3809–3818.

Wu, M., and Gerhart, J. C., 1980, Partial purification and characterization of the maturation-promoting factor from eggs of Xenopus laevis, *Dev. Biol.* **79:**465–477.

Zeeman, E. C., 1976, Catastrophe theory, *Sci. Am.* **234:**(4)65–83.

6

Interaction of Intermediate Filaments with the Cell Surface

Kathleen J. Green and Jonathan C. R. Jones

1. *Introduction*

The molecular architecture of the cytoskeleton–cell membrane complex has been a subject of intense investigation for more than a decade (for reviews see Weatherbee, 1981; Geiger, 1983; Jacobson, 1983; Bennett *et al.*, 1986). In general, the ultimate goal of these studies has been to understand how cytoskeletal proteins and membrane constituents act together in the formation and maintenance of cell shape, in directed cell movement, and in the transduction of signals across the plasma membrane. Although we are far from understanding these basic cellular functions, we have made some progress in identifying molecules that comprise the cytoskeleton–cell membrane complex. This is particularly so with respect to the actin-containing cytoskeleton, where several systems [the red blood cell (e.g., Bennett, 1985; Marchesi, 1985; Beaven *et al.*, 1985; Byers and Branton, 1985; Shen *et al.*, 1986), intestinal microvillus (e.g., Matsudaira and Burgess, 1979; Glenney *et al.*, 1982; Mooseker *et al.*, 1984), and focal contacts/adherens junctions (e.g., Burridge *et al.*, 1987; Geiger *et al.*, 1987)] have provided information regarding molecules involved in linking actin and actin-containing microfilaments (MF) with the cell membrane. Microtubule (MT)–membrane interactions have also been studied, although not as extensively as the MF system (see Dentler, 1981; Murray, 1984; Horst *et al.*, 1987). Until recently, less has been reported regarding interactions between intermediate filaments (IF) and the cell surface; with the exception of IF–desmosome interactions, this topic is often omitted in cytoskeleton–cell surface reviews. However, within the last several years a growing body of literature addressing an IF–cell surface association has developed.

IF are now well established as a major constituent of cytoskeletal networks (for

Kathleen J. Green • Department of Pathology, Northwestern University Medical School, Chicago, Illinois 60611. ***Jonathan C. R. Jones*** • Department of Cell Biology and Anatomy, Northwestern University Medical School, Chicago, Illinois 60611.

recent reviews, see Traub, 1985; Steinert and Roop, 1988). They are integrated into a complex system of cytoplasmic components by interacting with MT (Goldman *et al.*, 1980; Shelanski *et al.*, 1981; Bloom and Vallee, 1983), cell organelles (Wang and Goldman, 1978; Hirokawa, 1982; LeTerrier *et al.*, 1982), MF (Schliwa and van Blerkom, 1981; Hirokawa *et al.*, 1982; Celis *et al.*, 1984; Green *et al.*, 1986, 1987), the nuclear surface (Small and Celis, 1978; Lehto *et al.*, 1978, Goldman *et al.*, 1985, 1986), and IF themselves, via a panoply of putative IF-associated proteins (IFAP) (reviewed in Chapter 15, this volume). Their interaction, directly or indirectly, with the plasma membrane would establish a structural link between the nucleus and the cell surface that would have important implications for the organization of the cytoplasm and cellular communication (Goldman *et al.*, 1985). Using a variety of experimental techniques, investigators have addressed the possibility that IF–cell surface associations occur *in vivo,* and have utilized *in vitro* methods to identify putative linker molecules. The purpose of this review is to bring together the most recent reports that have contributed to the growing body of literature concerning IF–cell surface interactions, along with an update on the established fields of IF–desmosome and IF–muscle cell membrane interactions. We will start with a discussion of the desmosome and the hemidesmosome.

2. *IF–Desmosome and Hemidesmosome Interactions*

The desmosome is considered to be involved in intercellular adhesion while the hemidesmosome appears to play an important role in the adherence of basal cells to the underlying connective tissue (for a historical perspective of early studies on intercellular junctions, see Farquhar and Palade, 1963; Staehelin, 1974). IF of a number of different types are found in association with desmosomes. IF composed of keratin subunits are associated with the typical spot desmosomes or macula adherens of epithelial cells. Desmin-containing IF are associated with desmosome-like junctions present in cardiac myocytes (Kartenbeck *et al.*, 1983; Thornell *et al.*, 1985) and vimentin IF are associated with desmosomal plaques of human meningiomal cells and arachnoidal tissue (Kartenbeck *et al.*, 1984; Cowin *et al.*, 1985a; Moll *et al.*, 1986). Although not definitively proven, it is generally considered from their ultrastructural appearance that keratin-type IF interact with hemidesmosomes (see below).

The following is a brief summary of what is known about the structure and composition of desmosomes and hemidesmosomes. We then discuss the molecules that may be involved in the association of IF with these two cell junction types.

3. *Morphological Analysis of Desmosome–IF and Hemidesmosome–IF Complexes*

The spot desmosome has a characteristic ultrastructural appearance. It is a symmetrical junction in which the plasma membranes of two adjacent cells are separated by an intercellular space of about 30 nm (Fig. 1a; Farquhar and Palade, 1963; Staehelin, 1974; Arnn and Staehelin, 1981). The intercellular region of the desmosome contains dense material and often a dense central stratum can be observed running along the middle (Fig. 1a; Rayns *et al.*, 1969; Hull and Staehelin, 1979). Electron-dense plaques occur on the cytoplasmic surfaces of the desmosome (Fig. 1a). These plaque structures appear to be

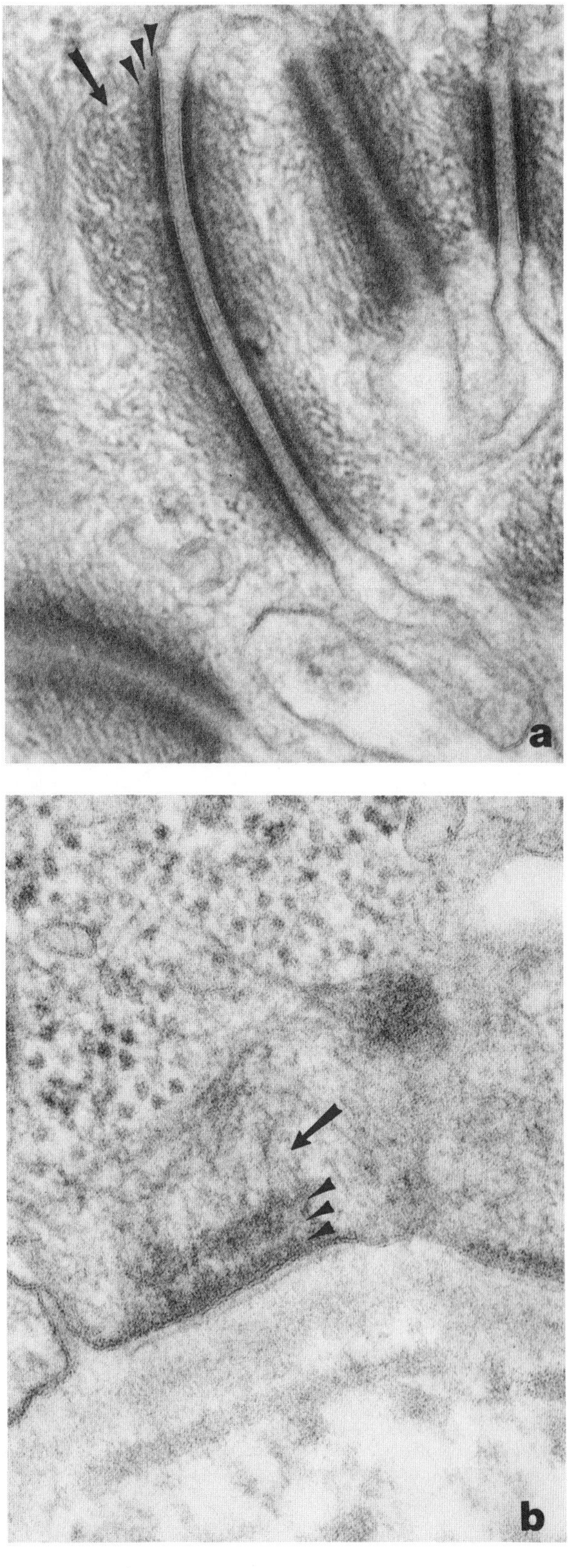

Figure 1. Electron micrographs of desmosomes (a) and a hemidesmosome (b) in bovine tongue epithelial cells. Note that IF are associated with both these junction types (arrows in a and b). The plaques of both the desmosome and the hemidesmosome are composed of three distinct regions (arrowheads in a and b). 66,000×. [As modified from Fig. 2, Jones *et al.* (1986c), *Cell Motil. Cytoskel.* **6:**560–569.]

composed of at least three different zones or regions (Jones *et al.*, 1986c). Immediately subjacent to the plasma membrane is an extremely electron-dense region (Fig. 1a). A less dense area separates the latter region from a layer containing fibrillar material to which bundles of IF appear to attach (Fig. 1a). IF do not appear to terminate at the plaque. Rather, Kelly (1966) has shown by stereo electron microscopy that the majority of IF loop through the fibrillar material of the desmosomal plaque. Furthermore, in freeze-fractured preparations of desmosomes a set of fine "traversing" filaments can be seen extending between IF and the plasma membranes of the desmosome (Kelly and Shienvold, 1976). There are also suggestions in the literature that these traversing filaments go through the membrane and connect with large desmosomal intramembrane globular elements (Fig. 2; Kelly and Kuda, 1981; Pirbazari and Kelly, 1985). The nature of the traversing filaments is unknown (see below) although it has been proposed that they are composed of IF protofilaments that "unravel" from the IF that loop through the desmosomal plaque area (Leloup *et al.*, 1979).

It has been suggested that the plaque of the desmosome may act as some type of organizing center or nucleation site for the IF that associate with it. This is based on an

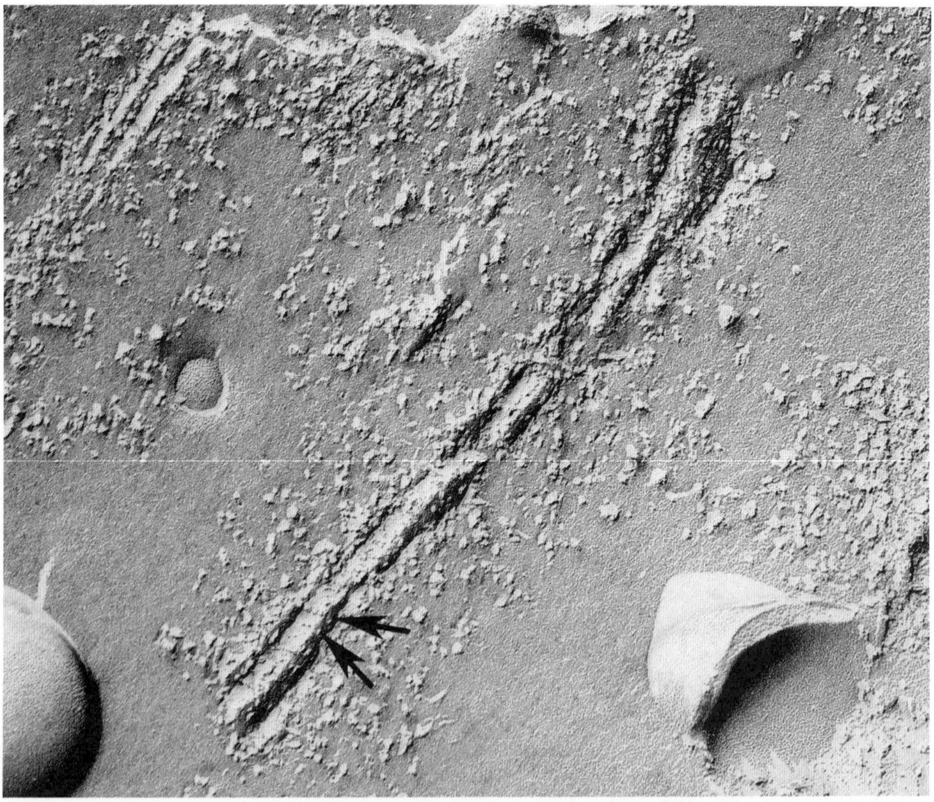

Figure 2. An enriched preparation of desmosomes isolated from bovine tongue epithelium according to Skerrow and Matoltsy (1974) with certain modifications detailed in Mueller and Franke (1983) was processed for freeze-fracture analysis. Note that in the intercellular space, a linear array of globular elements appears to span from one membrane to the other (arrows). These may be related to the filamentous structures spanning the extracellular space described by Pirbazari and Kelly (1985). 80,000×.

analysis of desmosome formation in a number of different systems including early chick blastoderm (Overton, 1962), developing teleost embryos (Lentz and Trinkaus, 1971), mouse epidermis (Krawczyk and Wilgram, 1973), and cultured cells (Dembitzer *et al.*, 1980; Bologna *et al.*, 1986). These authors noted that during desmosome formation the plaque structure of the desmosome appears at the cell surface first and only later do IF become associated with it. Recent studies on desmosome formation in cultured cells by Duden and Franke (1988) would appear to be consistent with this hypothesis. However, other workers have proposed that IF may be involved in the initiation of desmosome assembly in cultured epithelial cells, contrary to the above reports (Jones and Goldman, 1985) (see below for a discussion of this point). Whether the desmosomal plaque is involved in either IF assembly or organization of IF networks remains to be demonstrated definitively.

The hemidesmosome, like the desmosome, possesses a cytoplasmic plaque composed of three zones or regions (Fig. 1b; Jones *et al.*, 1986c). In newt epidermis, IF appear to loop past the hemidesmosomal plaque in a manner similar to that seen in the case of the desmosome (Kelly, 1966; Kelly and Kuda, 1981). However, there are some suggestions in the literature that IF may terminate at the hemidesmosomal plaque in the case of mammalian hemidesmosomes (Jones *et al.*, 1986c). Traversing filaments that appear to link IF to the plasma membrane have been observed in freeze-fracture preparations of hemidesmosomes (Kelly and Shienvold, 1976; Kelly and Kuda, 1981). As in the case of the desmosome, the composition of the traversing filaments remains unknown.

4. The Biochemical Composition of Desmosomes and Hemidesmosomes

A number of procedures for the isolation of desmosomes with or without their attached bundles of IF have been described (Skerrow and Matoltsy, 1974; Drochmans *et al.*, 1978; Jones *et al.*, 1982; Jones and Goldman, 1985). The procedure of Skerrow and Matoltsy (1974) has proved invaluable in the characterization of the polypeptide composition of desmosomes prepared from both bovine tongue and muzzle epithelia (Gorbsky and Steinberg, 1981; Cohen *et al.*, 1983; Mueller and Franke, 1983; Cowin and Garrod, 1983; Kapprell *et al.*, 1985, 1988; Tsukita and Tsukita, 1985; J. C. R. Jones *et al.*, 1986b, 1987; J. C. R. Jones, 1988; S. M. Jones *et al.*, 1988). The desmosomes in these preparations possess much of their *in situ* morphology. However, this procedure results in the isolation of desmosomes without their attached IF bundles. Immunoblotting analysis of such preparations using keratin antibodies reveals that they are not devoid of IF protein. Indeed, it has been argued that the fibrillar material that is found on the outermost surface of the plaques of isolated desmosomes comprises remnants (protofilaments?) of the bundles of IF that were associated with the desmosomes prior to isolation (Skerrow and Matoltsy, 1974; Gorbsky and Steinberg, 1981).

Characterization of the components of desmosomes has involved an immunological approach. Antibody probes have been generated against a number of the polypeptides present in isolated bovine epithelial desmosome preparations (e.g., see Franke *et al.*, 1982; Cowin and Garrod, 1983; Cohen *et al.*, 1983; Mueller and Franke, 1983; Tsukita and Tsukita, 1985; Jones *et al.*, 1986a,b, 1987; Schmelz *et al.*, 1986a,b; Penn *et al.*, 1987; Jones, 1988; Kapprell *et al.*, 1988). All desmosomes possess a high-molecular-weight polypeptide (M_r 250k) termed desmoplakin I (Mueller and Franke, 1983; Franke *et al.*, 1983; Cowin *et al.*, 1985b). In stratified epithelia a related polypeptide of 220k (desmoplakin II) has also been characterized (Franke *et al.*, 1982; Mueller and Franke, 1983; Cowin *et al.*, 1985b). The desmoplakins appear to reside in the plaque of the desmosomes as determined by immunogold electron microscopy (Fig. 3a; Mueller and Franke, 1983; Franke *et al.*, 1983; Jones and Goldman, 1985; Jones *et al.*, 1986b; Miller *et al.*, 1987). Two polypeptides in this same molecular weight range, a 200k protein (D1 antigen) and a 240k calmodulin binding polypeptide (desmocalmin), have also been identified as being plaque components of bovine muzzle epidermal desmosomes (Tsukita and Tsukita, 1985; Franke *et al.*, 1987). Certain other plaque components of desmosomes termed desmoplakin III or plakoglobin (83k) (Gorbsky *et al.*, 1985; Cowin *et al.*, 1986) and desmoplakin IV or "band 6" (75k) (Kapprell *et al.*, 1988) have been described. Plakoglobin appears to be a component not only of desmosomes but also of other junction types possessing symmetrical plaques (Cowin *et al.*, 1986) whereas band 6 is located in the plaque of desmosomes only in stratified and complex epithelia (Kapprell *et al.*, 1988). Bovine epithelial desmosomal glycoproteins have been studied by several groups and a number of glycosylated polypeptides in the molecular weight range 100–165k have been identified (Gorbsky and Steinberg, 1981; Cohen *et al.*, 1983; Cowin and Garrod, 1983, 1984a,b; Jones *et al.*, 1986a,b). Several recent studies have suggested that certain glycoproteins may be restricted to the intercellular region of the desmosomes while others appear to be transmembranous (Fig. 3b; Schmelz *et al.*, 1986a,b; Jones *et al.*, 1986b; Miller *et al.*, 1987; Steinberg *et al.*, 1987).

The biochemical composition of hemidesmosomes remains obscure, in part because there is, as yet, no procedure for the isolation of hemidesmosomes. Some workers have suggested that hemidesmosomal plaques contain desmoplakin (Mueller and Franke, 1983; Franke *et al.*, 1983; Cowin *et al.*, 1985b; Miller *et al.*, 1987). However, this has been questioned (Jones *et al.*, 1986c). Indeed, in our hands antibody preparations directed against the desmoplakins, D1 antigen, and all the major glycosylated desmosomal components fail to recognize hemidesmosomes (Jones, unpublished results). Sakai *et al.* (1986) have shown that antibodies directed against collagen type VII appear to recognize certain fibrils that anchor hemidesmosomes into the underlying connective tissue (Fig. 4; Ellison and Garrod, 1984). Furthermore, autoantibodies in the serum of bullous pemphigoid patients recognize plaque components of hemidesmosomes (Mutasim *et al.*, 1985; Westgate *et al.*, 1985; Regnier *et al.*, 1985; Jones *et al.*, 1986c). Recently, Klatte *et al.* (1989) have characterized 180k and 240k polypeptides present in bovine corneal epithelium that are recognized by bullous pemphigoid autoantibodies. Immunogold localization suggests that both antigens are present on the hemidesmosomal plaque. Furthermore, these authors have identified a monoclonal antibody directed against a 125k polypeptide which localizes to the extracellular surface of the hemidesmosome between the plasma membrane and the lamina densa.

5. *Putative IF–Plaque Linkers*

The following discussion will relate primarily to desmosomes since less is known regarding putative linkers in the hemidesmosome. It should be mentioned here, however, that Jones *et al.* (1986c) and Klatte *et al.* (1989) have suggested that the antigens recognized by autoantibodies in the serum samples of bullous pemphigoid patients may be involved in IF–plaque interactions in hemidesmosomes.

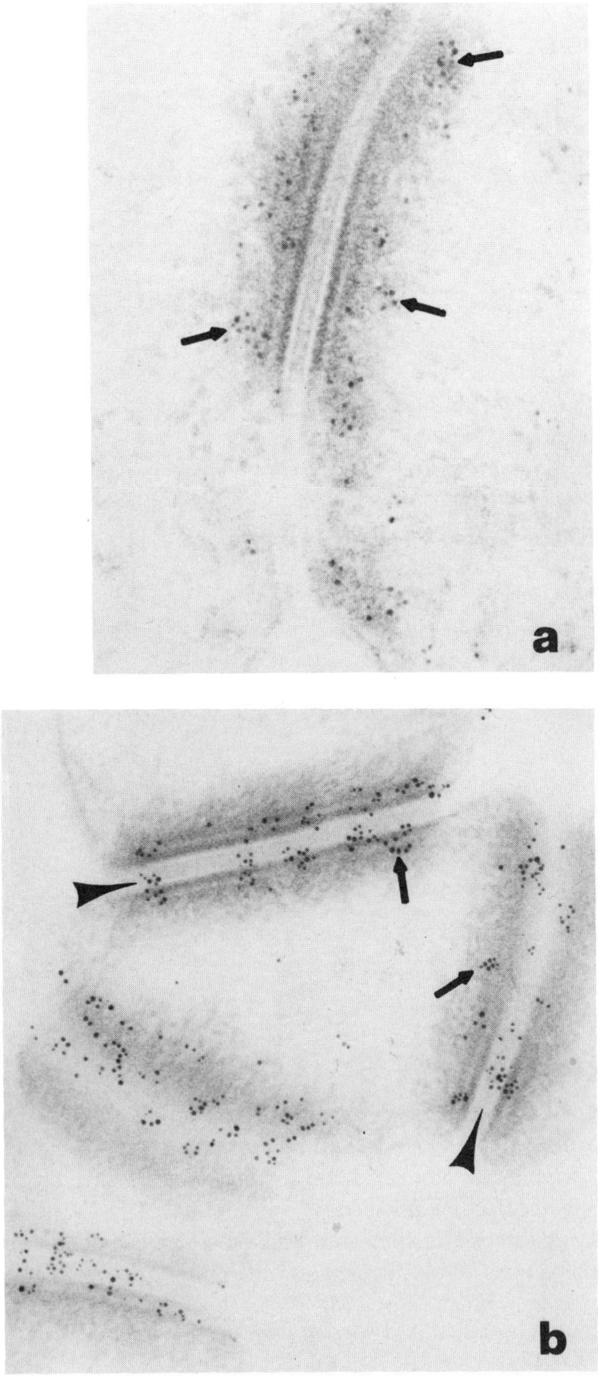

Figure 3. Thin sections of Lowicryl K4M-embedded bovine tongue mucosa were processed for indirect immunogold localization of desmoplakin (a) and the 160/165k (desmoglein) glycoprotein components of desmosomes (b). Note that in (a) gold particles are located in the region of the desmosomal plaque with which IF associate (arrows). In (b) gold particles occur both along the intercellular space of the desmosome (arrowheads) and over the plaque region (arrows), supporting the view that the 160/165k glycoproteins are transmembranous (Schmelz *et al.*, 1986b). 58,000×.

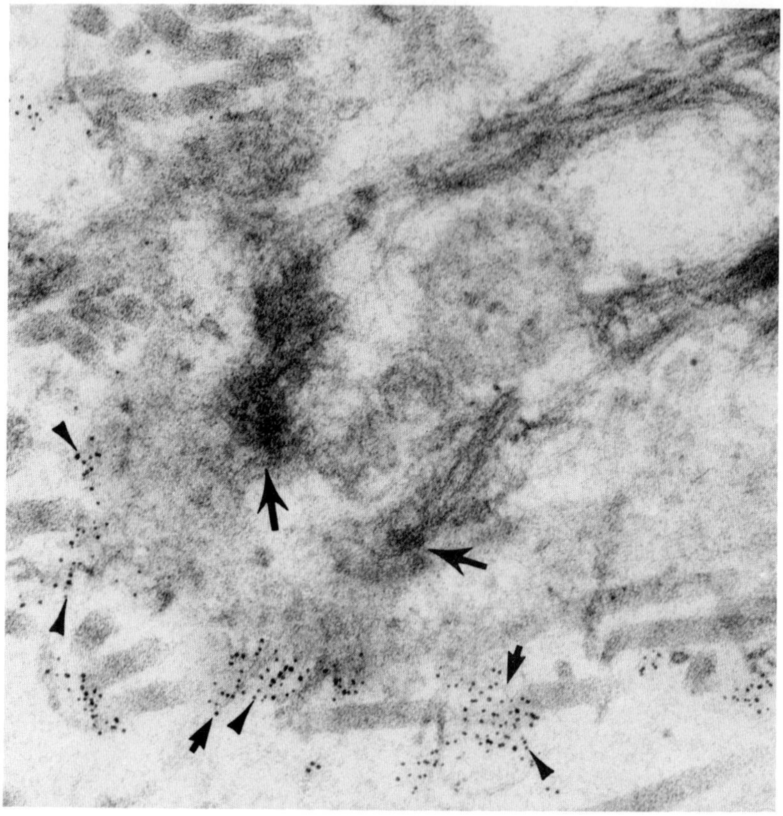

Figure 4. A thin section of Lowicryl K4M-embedded bovine tongue mucosa prepared for indirect immunogold localization using a monoclonal antibody preparation directed against collagen VII (a gift from Dr. Robert Burgeson). Gold particles are located in a region immediately underlying the lamina densa of the dermal–epidermal border (arrowheads). Large arrows indicate the position of two hemidesmosomes. Note that gold particles are associated with anchoring fibrils (small arrows) (Sakai *et al.*, 1986). 49,500×.

5.1. *Desmocalmin*

Desmocalmin is a calmodulin-binding, high-molecular-weight protein associated with bovine muzzle epidermal desmosomes (Tsukita and Tsukita, 1985). Desmocalmin resembles an elongated rod in rotary-shadowed preparations of the purified molecule. Immunoelectron microscopy reveals that it is located on the cytoplasmic surfaces of desmosomes (Tsukita and Tsukita, 1985). Tsukita and Tsukita (1985) examined the interactions of desmocalmin with purified keratin IF. Their results indicate that desmocalmin binds IF in a Mg^{2+}-dependent manner. Interestingly, desmocalmin appears to bind to the ends of keratin IF (Tsukita and Tsukita, 1985). These results led these authors to speculate that desmocalmin contributes to an IF–membrane association in desmosomes.

5.2. "Band 6" or Desmoplakin IV

Recently, Kapprell *et al.* (1988) have shown that the desmosomal plaque protein band 6 (M_r 75k) binds cytokeratins and therefore may be an IF–plaque linker. These authors used an assay system in which various ^{125}I-labeled cytokeratins were incubated with desmosomal polypeptides that had been separated by SDS-PAGE and transferred to nitrocellulose. Their results indicate that type I keratins in particular bind to band 6 protein. However, band 6 appears to be found exclusively in stratified and complex epithelial desmosomes. So, although it is possible band 6 may contribute to IF–cell surface interactions *in vivo* in stratified tissues, in other cell types IF must interact with desmosomes via different polypeptide(s).

5.3. Desmoplakin I and II

Desmoplakin I and II may play a role in the association of IF with the desmosomal plaque. A number of pieces of evidence are consistent with this idea. Immunogold ultrastructural localization reveals that the desmoplakins are located in the region of the plaque to which IF attach (Fig. 3a; Jones and Goldman, 1985; Jones *et al.*, 1986b,c; Miller *et al.*, 1987). Furthermore, it is known that the desmoplakins are glycine rich (Kapprell *et al.*, 1985). It has been proposed that certain IFAP in the skin called filaggrins may induce bundling of IF because of an interaction between high glycine-rich portions of the filaggrins (Dale *et al.*, 1978, 1985) with high glycine-rich portions of the keratins (Steinert and Roop, 1988). Some investigators have suggested by analogy that desmoplakins may mediate attachment of IF to the desmosomal plaque via interaction of its glycine-enriched domains with those of IF (Kapprell *et al.*, 1985). cDNA clones encoding the desmoplakins have recently been identified (Green *et al.*, 1988; 1990). Analysis of the predicted amino acid sequence of the desmoplakins suggests a model by which the C-terminal domain might interact with the rod domain of IF by virtue of a common periodicity in charged residues (Green *et al.*, 1990). Future functional studies utilizing mammalian expression systems will be needed to provide evidence for such binding *in vivo*.

Jones and Goldman (1985) suggested that desmoplakin possesses the properties of an IFAP. The evidence for this was based on a study of desmosome formation in cultured keratinocytes. Keratinocytes can be maintained in tissue culture for prolonged periods in medium containing low levels of Ca^{2+} (Hennings *et al.*, 1980). Under these conditions the cells do not form desmosomes. However, in keratinocytes as well as other cultured epithelial cells, if the Ca^{2+} level in the culture medium is increased to 1.2 mM, desmosome formation is initiated (e.g., Jones *et al.*, 1982; Hennings and Holbrook, 1983; Watt *et al.*, 1984; Jones and Goldman, 1985; Mattey and Garrod, 1986; Green *et al.*, 1987). Jones and Goldman (1985) and Green *et al.* (1987) analyzed keratinocytes both prior to and following the initiation of desmosome assembly by immunofluorescence and immunoblotting using antibody preparations directed against desmoplakin and keratin IF. In cells maintained in low levels of Ca^{2+}, both keratin and desmoplakin appear to be concentrated in the perinuclear region. These investigators reported that upon the initiation of desmosome assembly, the keratin IF are redistributed to areas of cell–cell contact in concert with desmoplakin. Furthermore, immunoblotting results revealed that desmoplakin is associated with IF even in the absence of formed desmosomes (Jones and Goldman, 1985). These results are consistent with those presented by Pasdar and Nelson (1988a,b) and Jones and Grelling (1989) who studied desmosome assembly in the simple epithelial MDCK cell line and human keratinocytes, respectively. Jones and Goldman (1985) speculated that the interaction of desmoplakin and its associated IF with the cell surface initiated desmosome formation, a view supported by Pasdar and Nelson (1988b). This model has been questioned by Duden and Franke (1988). These investigators provided an alternative interpretation of their cell culture data, suggesting instead that new desmosomes are formed *de novo* at the cell surface as previously suggested by others (Overton, 1962; Lentz and Trinkaus, 1971; Krawczyk and Wilgram, 1973; Dembitzer *et al.*, 1980; Bologna *et al.*, 1986). In this model, desmosomal plaque formation is followed by association with IF, without the contribution of preformed cytoplasmic IF–desmoplakin complexes. Regardless of the validity of either model, desmoplakin I and II remain the best candidates for IF–desmosome plaque linkers. Definitive evidence for such an association remains to be shown, however, and recent studies have been unable to demonstrate direct binding of desmoplakins and IF by a variety of *in vitro* assays (Kapprell *et al.*, 1988; O'Keefe *et al.*, 1989).

It is not known whether the association of desmoplakins (DP) and IF revealed in the above studies is a direct one. Other proteins may be involved in a desmoplakin–IF association. In addition, the relative roles of DPI and DPII remain unclear. Like band 6, DPII is apparently found only in stratified and complex epithelia; therefore, DPI is a more likely candidate for interacting with IF in simple epithelial tissues. It also seems probable that desmoplakin does not directly link IF with the desmosomal membrane. Recent immunoelectron microscopic localization data appear to suggest that desmoplakin is restricted to the cytoplasmic surface of the desmosomal plaque and does not reside in the electron-dense zone immediately subjacent to the desmosomal membrane (Fig. 3a; Jones and Goldman, 1985; Jones *et al.*, 1986a,b; Miller *et al.*, 1987). Thus, the association of desmoplakin with the desmosomal membrane may itself be mediated by certain transmembranous glycoproteins such as the 165k glycoprotein recently characterized by Schmelz *et al.* (1986a,b). This latter glycoprotein appears to contribute to the desmosomal plaque (Schmelz *et al.*, 1986a,b; Miller *et al.*, 1987; Jones, 1988). It is tempting to speculate that the traversing filaments described by Kelly and his co-workers (Kelly and Shienvold, 1976) may be composed of a mixture of a number of plaque components including desmoplakin (and desmocalmin?) that associate with transmembrane glycoproteins. The latter may then associate with the adhesion molecules of the desmosome [e.g., the desmocollins (Cowin *et al.*, 1984b), a 140k desmosome-associated glycoprotein (Jones *et al.*, 1986b), and a cadherin/uvomorulin-like molecule (Jones, 1988)]. It is possible that the latter molecules are the transmembrane globular elements observed in freeze-fracture images of desmosomes (Fig. 2; Kelly and Shienvold, 1976). Indeed, we should soon be in a position to identify the series of molecules that link the IF system of one cell with its neighbor via the desmosome. This would appear to be a prerequisite to an understanding of the functions of the IF–desmosome complex in epithelial tissues. The possibility that the IF–desmosome complex not only acts to transduce stretching and/or constricting forces from one cell to another but may also be involved in the transmission of signals controlling cell function, as has been proposed by Goldman *et al.* (1985), should prove an interesting topic for investigation.

6. *IF–Cell Surface Association in Muscle Cells*

Discrete sites of IF attachment with a well-defined morphology such as the desmosome or hemidesmosome are rare. The plasma membrane-associated dense bodies or "attachment plaques" of smooth muscle cells provide one example (see Burnstock, 1970; Small and Sobieszek, 1980; Cooke, 1983). These densely staining zones were described in the early literature by Pease and Molinari (1960) and others and are found as a discontinuous layer beneath the plasma membrane. They appear to be associated with both desmin-containing (Lazarides and Hubbard, 1976) [or skeletin-containing (Small and Sobieszek, 1977)] IF as well as actin-containing MF (Cooke and Chase, 1971; Uehara *et al.*, 1971; Ashton *et al.*, 1975; Small and Sobieszek, 1977). IF are also associated with similar dense bodies in the cytoplasm of smooth muscle cells; however, some investiga-

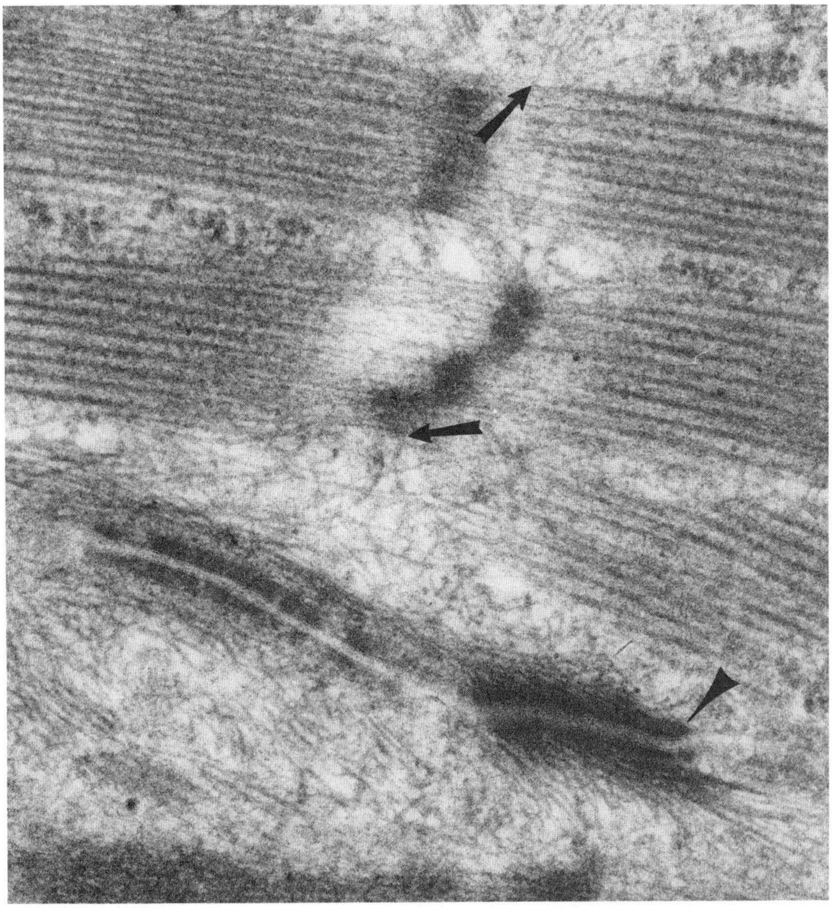

Figure 5. An electron micrograph of rabbit cardiac muscle cells. The arrowhead marks the desmosomal plaque region to which IF are attached. Arrows indicate regions in which individual IF interact with the Z lines of the myofibrils. These IF link adjacent myofibrils and attach the peripheral myofibrils to the sarcolemma. 55,000×.

tors suggest that these cytoplasmic bodies are not associated with MF (Cooke, 1976). Extraction of smooth muscle cells with high-salt-containing buffers leaves a desmin-containing filamentous network still attached to cytoplasmic and membrane-associated dense bodies (Cooke and Chase, 1971). Antibodies directed against α-actinin and tropomyosin appear to react with dense bodies at the cell surface as well as in the cytoplasm (Schollmeyer *et al.*, 1976).

It has been suggested that smooth muscle dense bodies may be analogous to striated muscle Z lines and dense material on the sarcolemma (e.g., Lazarides, 1980). Immunocytochemical techniques have been used to localize desmin, in addition to α-actinin and actin in the Z lines and subsarcolemma densities (Lazarides and Hubbard, 1976; Lazarides and Granger, 1978; Bennett *et al.*, 1978; Campbell *et al.*, 1979). These and other studies have shown that in skeletal myofibrils, parallel IF can be seen to link Z disks within a myofibril, and transverse IF link adjacent myofibrils at the level of the Z disks (Richardson *et al.*, 1981; Price and Sanger, 1983). These transverse IF also extend to the sarcolemma and other membrane systems. In cardiac muscle cells, transverse IF link adjacent Z disks and span the region from the Z disks of peripheral myofibrils to the sarcolemma (Fig. 5; Ferans and Roberts, 1973; Thornell *et al.*, 1978; Fuseler *et al.*, 1981; Forbes and Sperelakis, 1983). Longitudinal (parallel) IF also come together at the Z disk (Thornell *et al.*, 1978). It has been proposed, therefore, that in striated muscle, IF may play a role in integrating various components of the contractile apparatus by linking individual myofibrils laterally at their Z disks as well as linking Z disks to the plasma membrane (Lazarides, 1980; Fuseler *et al.*, 1981; Price and Sanger, 1983).

7. *Association of IF with Cortical Actin-Containing Structures*

IF also appear to be associated with actin-containing structures on or near the cell membrane in fibroblast cells (Fig. 6). These structures include focal contacts or adhesion plaques (Talian *et al.*, 1982; Bershadsky *et al.*, 1987), stress fibers (Green *et al.*, 1986), and phalloidin binding IF focal centers (IFFC) recently described in chick embryo fibroblasts (Green *et al.*, 1986). In the latter case, electron-dense fibrillar nodules that are outpocketings of substrate-associated stress fibers appear to act as foci for IF (vimentin- or desmin-containing) as demonstrated by double-label immunofluorescence microscopy and electron microscopy (Green *et al.*, 1986). It is not known whether these structures contain α-actinin or tropomyosin; if they do, it is possible that they may in some way be analogous to dense bodies of smooth muscle cells.

Several investigators have observed that cytochalasin B and D, alone or in combination with colchicine, cause a redistribution of keratin-containing IF of several epithelial cell types into a netlike array of fibers interrupted by periodic foci, which correspond with accumulations of actin-containing material (Knapp *et al.*, 1983a,b; Celis *et al.*, 1984; Keski-Oja and Atitalo, 1985; Knapp and Bunn, 1987; Green *et al.*, 1987). Knapp *et al.* (1983a,b) have reported that this keratin IF array is stabilized by membrane-associated focal centers. Furthermore, Keski-Oja and Alitalo (1985) have demonstrated that p36, a major cell-surface-associated substrate of tyrosine protein kinases, colocalizes with the drug-induced keratin/actin-containing focal centers in PtK_2 cells. The significance of this is unknown, although the authors speculated that a functional association among actin, keratin, and p36 may exist.

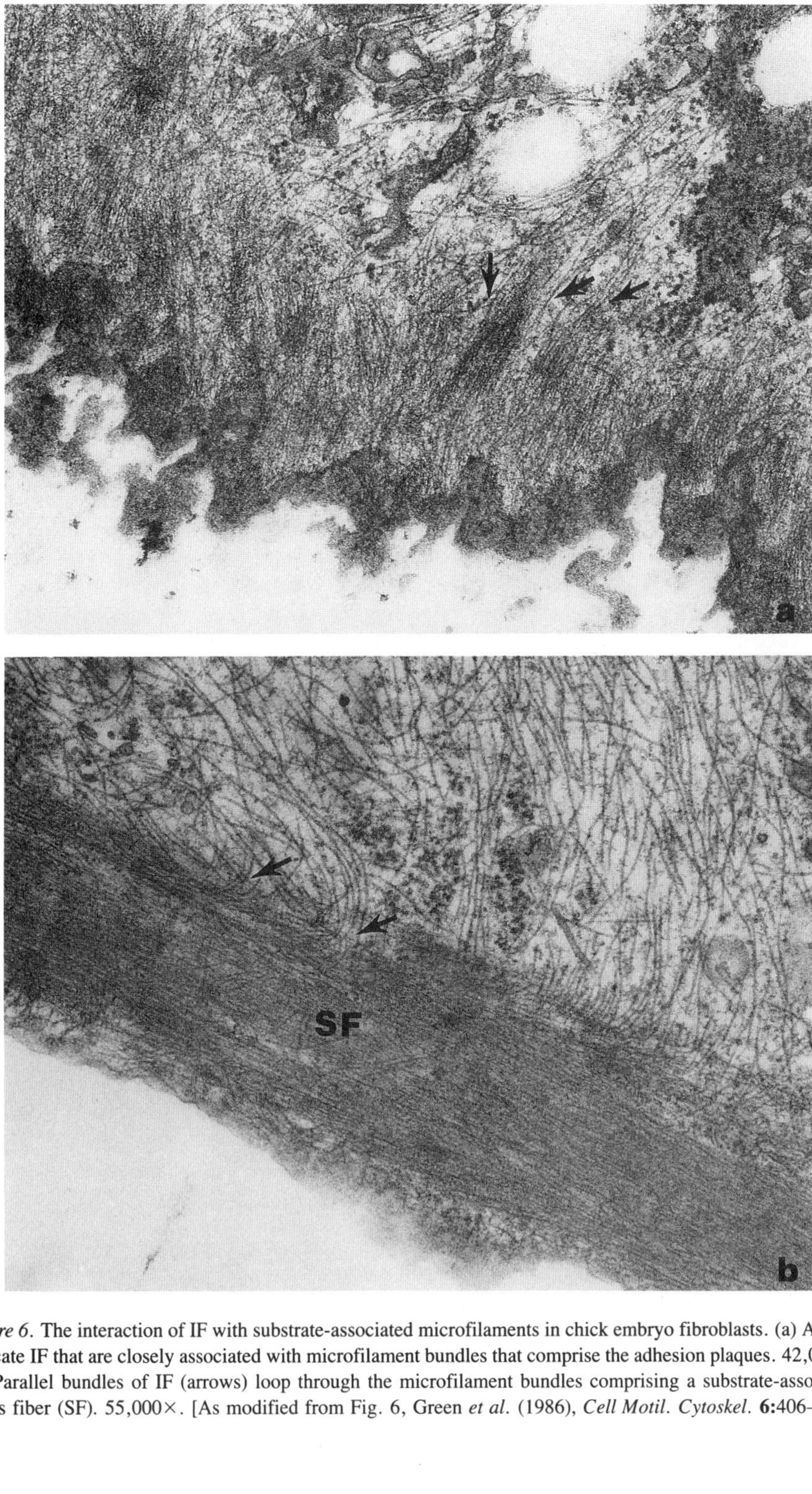

Figure 6. The interaction of IF with substrate-associated microfilaments in chick embryo fibroblasts. (a) Arrows indicate IF that are closely associated with microfilament bundles that comprise the adhesion plaques. 42,000×. (b) Parallel bundles of IF (arrows) loop through the microfilament bundles comprising a substrate-associated stress fiber (SF). 55,000×. [As modified from Fig. 6, Green *et al.* (1986), *Cell Motil. Cytoskel.* **6:**406–418.]

8. Interaction of IF with the Cell Membrane Complex

A more direct interaction between IF and the cell surface has been reported by several investigators. In some cases, investigators have implied a close association between IF and cell surface-related proteins, and in some cases with membrane lipids themselves. In other cases, the molecules involved have not yet been identified. Granger and colleagues (Granger and Lazarides, 1982; Granger *et al.*, 1982) have reported an interaction between IF and the cell surface of nucleated avian erythrocytes. Unlike mammalian erythrocytes, these cells contain nuclei, MT, and vimentin-containing IF. The investigators used a combination of biochemical and morphological techniques to show that vimentin IF, as well as an associated protein called synemin, interact not only with the nuclear surface in these cells, but also with the plasma membrane. They were able to attach erythrocytes to glass coverslips, sonicate the cells to remove nuclei, and by immunofluorescence and rotary shadowing techniques show that IF remained attached to the adhering membranes (Fig. 7). Further work demonstrated that both synemin and vimentin are present in these attached filaments, and that synemin is present as periodically spaced foci, with the periodicity changing throughout development of the erythro-

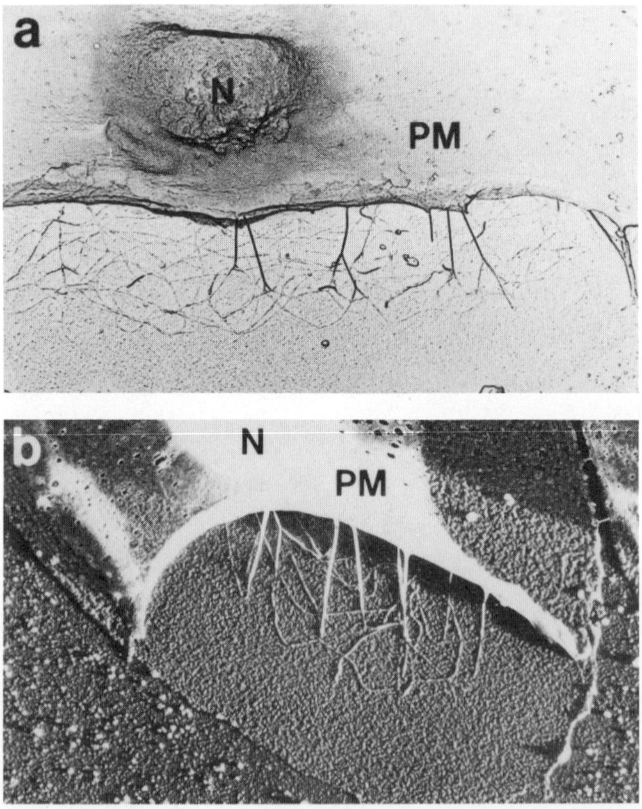

Figure 7. Pt–Pd replicas of critical-point-dried partially disrupted chicken erythrocytes. Note the IF that traverse the cytoplasm and appear to link areas of plasma membrane (PM) on opposite halves of the erythrocyte. N marks the nucleus. 12,500×. [As modified from Fig. 5, Granger and Lazarides (1982), *Cell* **30:**263–275.]

cyte (Granger and Lazarides, 1982). Centonze *et al.* (1986) have described a similar relationship of vimentin IF with the surface-associated cytoskeleton of the amphibian erythrocyte.

An association between vimentin-containing IF and lens cell membranes has also been reported (Bradley *et al.*, 1979; Ramaekers *et al.*, 1982). In addition to ultrastructural evidence, Ramaekers *et al.* (1982) showed that newly synthesized vimentin became associated with purified plasma membranes added to an *in vitro* mixture of reticulocyte lysate containing lens polyribosomes. In the same vein, Traub *et al.* (1985) reported binding of lipids to vimentin during its purification from Ehrlich ascites tumor cells. They pursued this with several approaches for studying a possible interaction between vimentin and lipids *in vitro* (Perides *et al.*, 1986; Traub *et al.*, 1986, 1987). Most recently, Traub *et al.* (1987) employed the technique of gel permeation chromatography to show that the α-helical rod domain in particular appears to interact with neutral lipids. The authors suggested that such an association may facilitate the interaction of intermediate filaments with membrane systems *in vivo*.

Green and Goldman (1986) presented a variety of structural evidence that an association with the cell membrane complex exists in chick embryo fibroblasts and BHK-21 cell fibroblasts. In substrate-attached cells and cells removed from their substrate in the absence of trypsin, IF remained in close proximity with the plasma membrane. However, tryptic removal of cells from their substrate resulted in a retraction of IF from the cell surface, indicating that an alteration of some cell surface molecule may have affected the attachment of IF to the surface. In cells extracted to remove most cytoplasmic constituents including MT and MF, IF remained in association with both the nuclear ghost and the plasma membrane complex (Fig. 8). Finally, the authors showed that IF, in addition to MF, were closely associated with regions of the cell surface known as the "fibronexus" (Singer, 1979) where fibronectin-containing extracellular matrix is found on the outer plasma membrane (Fig. 9).

In this regard, Culp *et al.* (1979) reported that IF in addition to other cytoskeletal proteins, fibronectin, and extracellular matrix proteins were present in the substrate-attached material (SAM) that remained after removal of mouse fibroblast cells from their tissue culture dish. This result is consistent with a close association between IF and the cell surface, particularly at adhesion sites. However, these experiments have been difficult to interpret, since cells may break open during the procedure, allowing cytoplasmic constituents to be released onto the substrate.

Biochemical analysis of detergent-resistant cytoskeletal preparations containing predominantly IF demonstrated the presence of IF subunit polypeptides and several other components, among them the extracellular matrix protein, fibronectin (Lehto *et al.*, 1981; Green and Goldman, 1986). Whether or not fibronectin was present because of an interaction, direct or indirect, with IF has not been demonstrated. Several molecules have been implicated as linkers between cytoplasmic IF and the cell membrane. One of these is a 140k cell surface component IF found in detergent-resistant cytoskeletons of cultured human fibroblasts (Lehto, 1983). The possible relationship between this glycoprotein and the recently described class of "fibronectin receptor" proteins is not known.

Lehto and Virtanen (1983) characterized a cytoskeleton-associated 230k polypeptide found in bovine lens which becomes rapidly reorganized to a dense plaque underneath the perinuclear cap of IF upon treatment of cells with demecolcine. This protein was found to be antigenically related to α-spectrin. Other investigators have noted a possible rela-

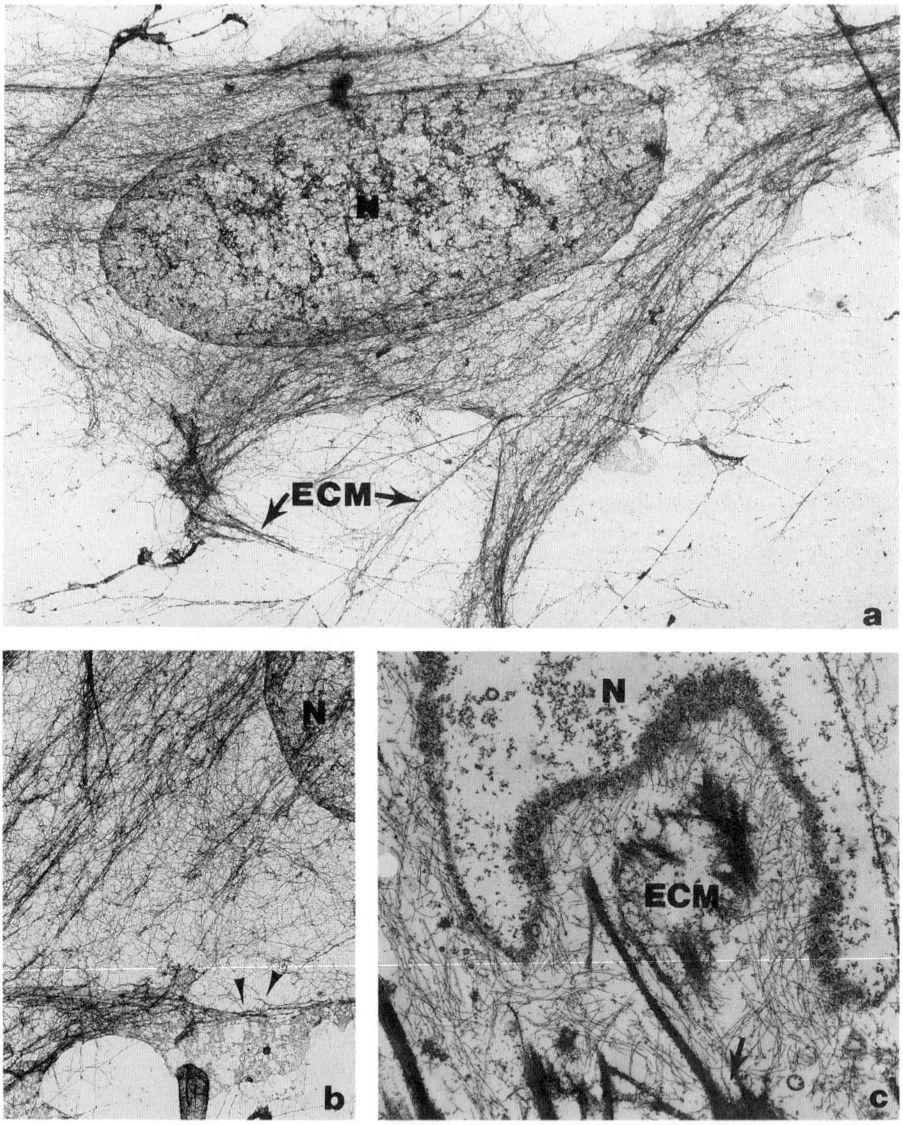

Figure 8. Analysis of chick embryo fibroblast cells, lysed under conditions in which the IF cytoskeleton remains intact [see Figs. 3 and 4, Green and Goldman (1986), *Cell Motil. Cytoskel.* **6**;389–405]. In (a) a critical-point-dried whole cell mount shows maintenance of cell shape and IF distribution from the nucleus (N) to the cell surface. Cell surface-associated extracellular matrix material is designated "ECM." 5500×. In the whole mount shown in (b), IF closely approach the cell surface, which can be identified by plasma membrane remnants (arrowheads). 11,000×. An electron micrograph of a lysed, embedded, and sectioned fibroblast, stained with tannic acid, is shown in (c). Note that IF extend from the nuclear ghost (N) to electron-dense fibrous material (ECM). The close association with the ECM is denoted by the arrow. 14,500×.

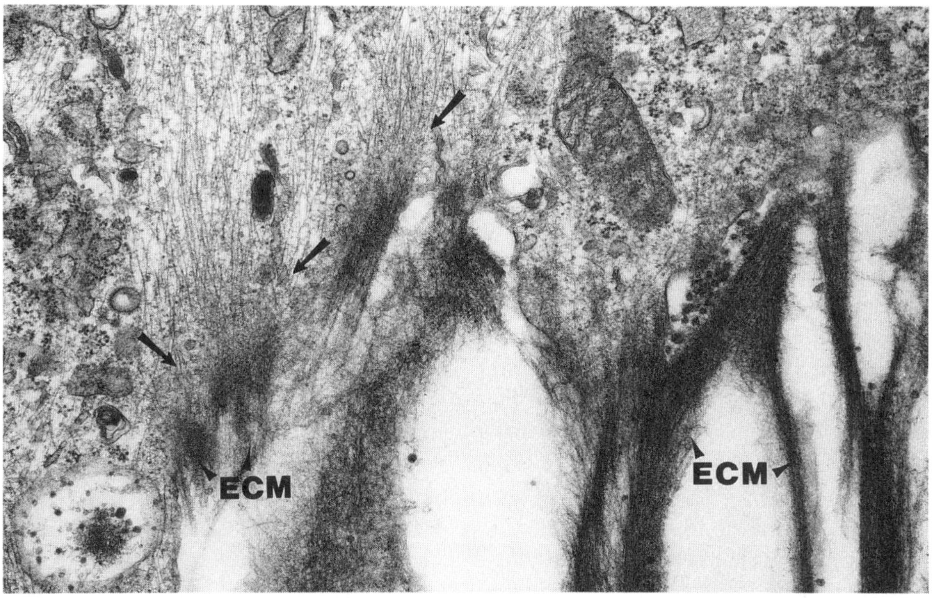

Figure 9. An electron micrograph of a chick embryo fibroblast shows bundles of IF (arrows) closely approaching the cell surface extracellular matrix (ECM) or "fibronexus." 35,000×. [As modified from Fig. 2, Green and Goldman (1986), *Cell Motil. Cytoskel.* **6**:389–405.]

tionship between spectrin and IF. Mangeat and Burridge (1984) demonstrated that microinjection of an antispectrin antibody that causes immunoprecipitation of nonerythrocyte spectrin in living fibroblasts also resulted in an aggregation and altered distribution of vimentin-containing IF. Granger and Lazarides (1984) pointed out that in most cases where an IF–cell surface association is implicated, the particular membranes in question are lined with spectrin or related proteins. It is interesting to note that Hirokawa *et al.* (1982, 1983) observed structural links between actin and IF in mouse intestinal brush border, and suggested that these filamentous linkers may belong to the fodrin–spectrin–TW260/240 family of proteins. Herrmann and Wiche (1987) suggested on the basis of *in vitro* binding studies that the IFAP plectin (otherwise known as IFAP-300K; Yang *et al.*, 1985) is able to interact with spectrin, suggesting the possibility that these molecules are involved in the formation or maintenance of a subplasmalemmal network comprising IF, spectrin, and plectin. It is interesting to note that plectin has also been observed near the cell periphery in some cell types, and immunoelectron microscopy has revealed the presence of plectin in the desmosomal plaque (Wiche *et al.*, 1983). The authors suggested that plectin may play a universal role in anchoring IF.

The possibility that spectrin can directly associate with IF *in vitro* was tested by Georgatos and Marchesi (1985) and by Langley and Cohen (1986, 1987). In the former case, radioiodinated lens vimentin in an unpolymerized form was incubated with various preparations of human erythrocyte membrane vesicles. It was demonstrated that vimentin binds in a saturable manner to IOV (inside-out membrane vesicles) which are stripped of actin and spectrin. Furthermore, it was discovered that removal of protein 4.1 or reconstitution with purified spectrin did not affect the saturable binding of vimentin to these vesicles. However, removal of ankyrin significantly lowered the binding. Ankyrin is a protein that links the erythrocyte actin–spectrin complex to the membrane. Further work by these investigators suggested that vimentin IF subunits associate with the human erythrocyte plasma membrane (Georgatos *et al.*, 1985) and that vimentin and desmin IF associate with nucleated avian erythrocyte plasma membranes (Georgatos and Blobel, 1987a; Georgatos *et al.*, 1987) via their N-terminal head domains. Interestingly, further evidence suggested that the C-terminal tail domains of desmin and vimentin bind to the nuclear envelope protein lamin B (Georgatos and Blobel, 1987b; Georgatos *et al.*, 1987). A model was proposed whereby vectorial assembly of IF occurs from the nuclear envelope toward the cell surface where further growth is prevented by attachment of IF to the plasma membrane skeleton (Georgatos and Blobel, 1987a).

Langley and Cohen (1986), on the other hand, have reported evidence for an association between erythrocyte spectrin and filamentous desmin IF using two *in vitro* assays. Spectrin was shown to promote the interaction of desmin filaments via multiple lateral associations with IOV preparations by electron microscopy. In addition, a sedimentation assay was used to demonstrate that when increasing concentrations of spectrin were incubated with desmin filaments, spectrin cosedimented with the filaments in an apparently saturable manner. However, the fact that spectrin binds to desmin with a low molar ratio suggests that spectrin might be binding to a contaminant (IFAP?) in the IF preparations. More recently, Langley and Cohen (1987) have published *in vitro* data to suggest that spectrins from different sources, i.e., erythrocytes and brain, bind with different affinities to vimentin IF and neurofilaments. That is, vimentin IF bound more erythrocyte than brain spectrin, while neurofilaments bound more brain than erythrocyte spectrin. The authors speculated that this may reflect an important regulatory function with regard to IF–cell surface interactions, especially in cells where multiple spectrin and IF types are present in a single cell type. Together these data imply that IF can bind in more than one manner to molecules that have been demonstrated to be a part of the cell-membrane complex: in a lateral fashion to spectrin, and in an end-on fashion to ankyrin. The recent localization of spectrin to the z lines of muscle also suggests further that a desmin–spectrin interaction may contribute to the linking up of Z lines by IF (Nelson and Lazarides, 1983).

9. Conclusion

The results of the studies reviewed here suggest that IF, whether they are composed of keratin, vimentin, or desmin, interact with structures and molecules associated with the plasma membrane. Currently, the evidence for an *in vivo* association between IF and the cell surface comes primarily from morphological studies. The biochemical nature of such an interaction remains relatively obscure. However, *in vitro* reconstitution experiments using IF, either in subunit or filament form, with various cell surface molecules are now allowing identification of possible IF–cell surface linker molecules. As reviewed above, such *in vitro* studies have recently suggested that IF or IF subunits can interact with ankyrin and spectrin.

Molecular biological techniques should prove invaluable for evaluating the functional role, and the *in vivo* role in particular, of putative IF–cell surface proteins. The introduction of antisense copies of putative "linker" genes or copies that have been altered by site-specific mutagenesis may help demonstrate that the molecule plays a role in the association of IF with the cell surface and may also provide clues to the functions of IF–cell surface complexes. In addition, bacterial expression systems will allow production of large quantities of putative IF–cell surface linker molecules or portions thereof. *In vitro* studies might then be used to determine whether these bacterially expressed polypeptides bind to IF and/or other cell surface molecules. Using a combination of structural, biochemical, immunological, and molecular approaches, we should gain significant insights into our understanding of IF–cell surface interactions, and be in a position to determine the role that such associations play in cellular morphogenesis and signal transduction.

ACKNOWLEDGMENTS. The authors thank Drs. Bruce L. Granger and Elias Lazarides for contributing the micrographs seen in Fig. 7, originally published in *Cell* **30:**263. We are also grateful to Dr. Robert Decker and Marlene Decker for Fig. 5, and Manette McReynolds for Fig. 2. Thanks go also to Walter Glogowski for help in preparing figures and to Carole Becker for typing the manuscript.

J. C. R. J. and K. J. G. are recipients of Basil O'Connor Starter Scholar Research Awards from the March of Dimes. J. C. R. J. is supported by the National Institutes of Health (GM38470 and AR36740) and is a Junior Faculty Research Awardee of the American Cancer Society (JFRA-232). K. J. G. is supported by the National Institutes of Health (HD24430), the Illinois Chapter of the American Cancer Society, and the Council for Tobacco Research, U.S.A.

References

Arnn, J., and Staehelin, L. A., 1981, The structure and function of spot desmosomes, *Dermatology* **20:**330–339.

Ashton, F. T., Somlyo, A. V., and Somlyo, A. P., 1975, The contractile apparatus of vascular smooth muscle: Intermediate high voltage stereo electron microscopy, *J. Mol. Biol.* **98:**17–29.

Beaven, G. H., Jean-Baptiste, L., Ungewickell, E., Baines, A. J., Shahbakhti, F., Pinder, J. C., Lux, S. E., and Gratzer, W. B., 1985, An examination of the soluble oligomeric complexes extracted from the red cell membrane and their relationship to the membrane skeleton, *Eur. J. Cell Biol.* **36:**299–306.

Bennett, G. S., Fellini, J. A., and Holtzer, H., 1978, Immunofluorescent visualization of 100 Å filaments in different cultured chick embryo cell types, *Differentiation* **12:**71–81.

Bennett, V., 1985, The membrane skeleton of human erythrocytes and its implications for more complex cells, *Annu. Rev. Biochem.* **54:**273–304.

Bennett, V., Cohen, C. M., Lux, S. E., and Palek, J., eds., 1986, Membrane Skeletons and Cytoskeleton–Membrane Associations, *UCLA Symp. Mol. Cell. Biol.* **38.**

Bershadsky, A. D., Tint, I. S., and Svitkina, T. M., 1987, Association of intermediate filaments with vinculin-containing adhesion plaques of fibroblasts, *Cell Motil. Cytoskel.* **8:**274–283.

Bloom, G. S., and Vallee, R. B., 1983, Association of microtubule-associated protein 2 (MAP 2) with microtubules and intermediate filaments in cultured brain cells, *J. Cell Biol.* **96:**1523–1531.

Bologna, M., Allen, R., and Dulbecco, R., 1986, Organization of cytokeratin bundles by desmosomes in rat mammary cells, *J. Cell Biol.* **102:**560–567.

Bradley, R. H., Ireland, M., and Maisel, H., 1979, The cytoskeleton of chick lens cells, *Exp. Eye Res.* **28:**441–453.

Burnstock, G., 1970, Structure of smooth muscle and its innervation, in: *Smooth Muscle* (E. Bulbring, A. J. Brading, A. W. Jones, and T. Tomita, eds.), Arnold, London, pp. 1–69.

Burridge, K., Molony, L., and Kelly, T., 1987, Adhesion plaques: Sites of transmembrane interaction between the extracellular matrix and the actin cytoskeleton, *J. Cell Sci. Suppl.* **8:**211–229.

Byers, T. J., and Branton, D., 1985, Visualization of the protein associations in the erythrocyte membrane skeleton, *Proc. Natl. Acad. Sci. USA* **82:**6153–6157.

Campbell, G. R., Campbell, J. C., Stewart, U. G., Smalll, J. V., and Andersen, P., 1979, Antibody staining of 10nm (100 Å) filaments in cultured smooth cardiac and skeletal muscle cells, *J. Cell Sci.* **37:**303–322.

Celis, J. E., Small, J. V., Larsen, P. M., Fey, S. J., De Mey, J., and Celis, A., 1984, Intermediate filaments in monkey kidney TC7 cells: Focal centers and interrelationship with other cytoskeletal systems, *Proc. Natl. Acad. Sci. USA* **81:**1117–1121.

Centonze, V. E., Ruben, G. C., and Sloboda, R. D., 1986, Structure and composition of the cytoskeleton of nucleated erythrocytes: III. Organization of the cytoskeleton of *Bufo maranis* erythrocytes as revealed by freeze-dried platinum carbon replicas and immunofluorescence microscopy, *Cell Motil. Cytoskel.* **6:**376–388.

Cohen, S. M., Gorbsky, G., and Steinberg, M. S., 1983, Immunochemical characterization of related families of glycoproteins in desmosomes, *J. Biol. Chem.* **258:**2621–2627.

Cooke, P., 1976, A filamentous cytoskeleton in vertebrate smooth muscle fibers, *J. Cell Biol.* **68:**539–556.

Cooke, P., 1983, Organization of contractile fibers in smooth muscle, in: *Cell and Muscle Motility,* Volume 3 (R. Dowben and J. Shaw, eds.), Plenum Press, New York, pp. 57–77.

Cooke, P. H., and Chase, R. H., 1971, Potassium chloride-insoluble myofilaments in vertebrate smooth muscle cells, *Exp. Cell Res.* **66:**417–425.

Cowin, P., and Garrod, D. R., 1983, Antibodies to epithelial desmosomes show wide tissue and species cross reactivity, *Nature* **302:**148–150.

Cowin, P., Mattey, D., and Garrod, D., 1984a, Distribution of desmosomal components in the tissues of vertebrates, studied by fluorescent antibody staining, *J. Cell Sci.* **66:**119–132.

Cowin, P., Mattey, D., and Garrod, D., 1984b, Identification of desmosomal surface components (desmocollins) and inhibition of desmosome formation by specific Fab', *J. Cell Sci.* **70:**41–60.

Cowin, P., Franke, W. W., Grund, C., Kapprell, H.-P., and Kartenbeck, J., 1985a, The desmosome–intermediate filament complex, in: *The Cell in Contact* (G. Edelman and J. P. Thiery, eds.), Wiley, New York, pp. 427–460.

Cowin, P., Kapprell, H.-P., and Franke, W. W., 1985b, The complement of desmosomal plaque proteins in different cell types, *J. Cell Biol.* **101:**1442–1454.

Cowin, P., Kapprell, H.-P., Franke, W. W., Tamkun, J., and Hynes, R. O., 1986, Plakoglobin: A protein common to different kinds of intercellular adhering junctions, *Cell* **46:**1063–1073.

Culp, L. A., Murray, B. A., and Rollins, B. J., 1979, Fibronectin and proteoglycans as determinants of cell–substratum adhesion, *J. Supramol. Struct.* **11:**401–427.

Dale, B. A., Holbrook, K. A., and Steinert, P. M., 1978, Assembly of stratum corneum basic protein and keratin filaments in macrofibrils, *Nature* **276:**729–731.

Dale, B. A., Resing, K. A., and Lonsdale-Eccles, J. D., 1985, Filaggrin: a keratin filament associated protein, *Ann. N.Y. Acad. Sci.* **455:**330–342.

Dembitzer, H. M., Herz, F., Schermer, A., Wolley, R. C., and Koss, L. G., 1980, Desmosome development in an in vitro model, *J. Cell Biol.* **85:**695–702.

Dentler, W. L., 1981, Microtubule–membrane interactions in cilia and flagella, *Int. Rev. Cytol.* **72:**1–47.

Drochmans, P., Freudenstein, C., Wanson, J.-C., Laurent, T., Keenan, W., Stadler, J., Leloup, R., and Franke, W. W., 1978, Structure and biochemical composition of desmosomes and tonofilaments isolated from calf muzzle epidermis, *J. Cell Biol.* **79:**427–443.

Duden, R., and Franke, W. W., 1988, Organization of desmosomal plaque proteins in cells growing at low calcium concentrations, *J. Cell Biol.* **107:**1037–1063.

Ellison, J., and Garrod, D. R., 1984, Anchoring filaments of the amphibial epidermal–dermal junction traverse the basal lamina entirely from the plasma membrane of hemidesmosomes to the dermis, *J. Cell Sci.* **72:**163–172.

Farquhar, M. G., and Palade, G. E., 1963, Junctional complexes in various epithelia, *J. Cell Biol.* **17:**375–412.

Ferrans, V. J., and Roberts, W. C., 1973, Intermyofibrillar and nuclear–myofibrillar connections in human and canine myocardium: An ultrastructural study, *J. Mol. Cell. Cardiol.* **5:**247–257.

Forbes, M. S., and Sperelakis, N., 1983, The membrane system and cytoskeletal elements of mammalian myocardial cells, in: *Cell and Muscle Motility,* Volume 3 (R. Dowben and J. Shaw, eds.), Plenum Press, New York, pp. 89–155.

Franke, W. W., Moll, R., Schiller, D. L., Schmid, E., Kartenbeck, J., and Mueller, H., 1982, Desmoplakins of epithelial and myocardial desmosomes are immunologically and biochemically related, *Differentiation* **23:**115–127.

Franke, W. W., Moll, R., Mueller, H., Schmid, E., Kuhn, C., Krepler, R., Artlieb, U., and Denk, H., 1983, Immunocytochemical identification of epithelium-derived human tumors with antibodies to desmosomal plaque proteins, *Proc. Natl. Acad. Sci. USA* **80:**543–547.

Franke, W. W., Cowin, P., Schmelz, M., and Kapprell, H.-P., 1987, The desmosomal plaque and the cytoskeleton, *Ciba Symp.* **125:**26–44.

Fuseler, J. W., Shay, J. W., and Feit, H., 1981, The role of intermediate filaments (10 nm) in the development and integration of the myofibrillar contractile apparatus in the embryonic mammalian heart, in: *Cell and Muscle Motility,* Volume 1 (R. Dowben and J. Shaw, eds.), Plenum Press, New York, pp. 205–259.

Geiger, B., 1983, Membrane–cytoskeleton interaction, *Biochim. Biophys. Acta* **737:**305–341.

Geiger, B., Volk, T., Volberg, T., and Bendori, R., 1987, Molecular interactions in adherens-type contacts, *J. Cell Sci. Suppl.* **8:**251–272.

Georgatos, S. D., and Blobel, G., 1987a, Two distinct attachment sites for vimentin along the plasma membrane and the nuclear envelope in avian erythrocytes: A basis for vectorial assembly of intermediate filaments, *J. Cell Biol.* **105:**105–115.

Georgatos, S. D., and Blobel, G., 1987b, Lamin B constitutes an intermediate filament attachment site at the nuclear envelope, *J. Cell Biol.* **105:**117–125.

Georgatos, S. D., and Marchesi, V. T., 1985, The binding of vimentin to human erythrocyte membranes: A model system for the study of intermediate filament–membrane interactions, *J. Cell Biol.* **100:**1955–1961.

Georgatos, S. D., Weaver, D. C., and Marchesi, V. T., 1985, Site specificity in vimentin–membrane interactions: Intermediate filament subunits associate with the plasma membrane via their head domains, *J. Cell Biol.* **100:**1962–1967.

Georgatos, S. D., Weber, K., Geisler, N., and Blobel, G., 1987, Binding of two desmin derivatives to the plasma membrane and the nuclear envelope of avian erythrocytes: Evidence for a conserved site-specificity in intermediate filament–membrane interactions, *Proc. Natl. Acad. Sci. USA* **84:**6780–6784.

Glenney, J. R., Osborn, M., and Weber, K., 1982, The intracellular localization of the microvillus 110K protein, a component considered to be involved in side-on membrane attachment of F-actin, *Exp. Cell Res.* **138:**199–205.

Goldman, R. D., Hill, B. F., Steinert, P., Aynardi Whitman, M., and Zackroff, R. V., 1980, Intermediate filament–microtubule interactions: Evidence in support of a common organization center, in: *Microtubules and Microtubule Inhibitors* (M. De Brabander and J. De Mey, eds.), Elsevier/North-Holland, Amsterdam, pp. 91–102.

Goldman, R. D., Goldman, A., Green, K., Jones, J., Lieska, N,. and Yang, H.-Y., 1985, Intermediate filaments: Possible functions as cytoskeletal connecting links between the nucleus and the cell surface, *Ann. N.Y. Acad. Sci.* **455:**1–17.

Goldman, R. D., Goldman, A. E., Green, K. J., Jones, J. C. R., Jones, S. M., and Yang, H.-Y., 1986, Intermediate filament networks: Organization and possible functions of a diverse group of cytoskeletal elements, *J. Cell Sci. Suppl.* **5:**69–97.

Gorbsky, G., and Steinberg, M. S., 1981, Isolation of the intercellular glycoproteins of desmosomes, *J. Cell Biol.* **90:**243–248.

Gorbsky, G. J., Cohen, S. M., Shida, H., Giudice, G. J., and Steinberg, M. S., 1985, Isolation of the non-glycosylated proteins of desmosomes and immunolocalization of a third plaque protein: Desmoplakin III, *Proc. Natl. Acad. Sci. USA* **82:**810–814.

Granger, B. L., and Lazarides, E., 1982, Structural association of synemin and vimentin in avian erythrocytes revealed by immunoelectron microscopy, *Cell* **30:**263–275.

Granger, B. L., and Lazarides, E., 1984, Expression of the intermediate filament-associated protein synemin in chicken lens cells, *Mol. Cell. Biol.* **4:**1943–1950.

Granger, B. L., Repasky, E. A., and Lazarides, E., 1982, Synemin and vimentin are components of intermediate filaments in avian erythrocytes, *J. Cell Biol.* **92:**299–312.

Green, K. J., and Goldman, R. D., 1986, Evidence for an interaction between the cell surface and intermediate filaments in cultured fibroblasts, *Cell Motil. Cytoskel.* **6:**389–405.

Green, K. J., Talian, J. C., and Goldman, R. D., 1986, Relationship between intermediate filaments and microfilaments in cultured fibroblasts: Evidence for common foci during cell spreading, *Cell Motil. Cytoskel.* **6:**406–418.

Green, K. J., Geiger, B., Jones, J. C. R., Talian, J. C., and Goldman, R. D., 1987, The relationship between intermediate filaments and microfilaments before and during the formation of desmosomes and adherens type junctions in mouse epidermal keratinocytes, *J. Cell Biol.* **104:**1389–1402.

Green, K. J., Goldman, R. D., and Chisholm, R. L., 1988, Isolation of cDNAs encoding desmosomal plaque proteins: Evidence that bovine desmoplakins I and II are derived from two mRNAs and a single gene, *Proc. Natl. Acad. Sci. USA* **85:**2613–2617.

Green, K. J., Parry, D. A. D., Steinert, P. M., Virata, M. L. A., Wagner, R. M., Angst, B. D., and Nilles, L. A., 1990, Structure of the human desmoplakins: implications for function in the desmosomal plaque, *J. Biol. Chem.* (in press).

Hennings, H., and Holbrook, K. A., 1983, Calcium regulation of cell–cell contact and differentiation of epidermal cells in culture. An ultrastructural study, *Exp. Cell Res.* **143:**127–142.

Hennings, H., Michael, D., Cheng, C., Steinert, P., Holbrook, K. A., and Yuspa, S. H., 1980, Calcium regulation of growth and differentiation of mouse epidermal cells in culture, *Cell* **19:**245–254.

Herrmann, H., and Wiche, G., 1987, Plectin and IFAP-300K are homologous proteins binding to microtubule-associated proteins 1 and 2 and to the 240 kilodalton subunit of spectrin, *J. Biol. Chem.* **262:**1320–1325.

Hirokawa, N., 1982, Cross-linker system between neurofilaments, microtubules and membranous organelles in frog axons revealed by quick-freeze, deep etching method, *J. Cell Biol.* **94:**129–142.

Hirokawa, N., Tilney, L. B., Fujiwara, K., and Heuser, J. E., 1982, The organization of actin, myosin, and intermediate filaments in the brush border of intestinal epithelial cells, *J. Cell Biol.* **94:**425–443.

Hirokawa, N., Cheney, R. E., and Willard, M., 1983, Location of a protein of the fodrin–spectrin–TW260/240 family in the mouse intestinal brush border, *Cell* **32:**953–965.

Horst, C. J., Forestner, D. M., and Beharse, J. C., 1987, Cytoskeletal–membrane interactions: A stable interaction between cell surface glycoconjugates and doublet microtubules of the photoreceptor connecting cilium, *J. Cell Biol.* **105:**2973–2987.

Hull, B. E., and Staehelin, L. A., 1979, The terminal web: A reevaluation of its structure and function, *J. Cell Biol.* **81:**67–82.

Jacobson, B. S., 1983, Interaction of the plasma membrane with the cytoskeleton: An overview, *Tissue Cell* **15:**829–852.

Jones, J. C. R., 1988, Characterization of a 125K glycoprotein associated with bovine epithelial desmosomes, *J. Cell Sci.* **89:**207–216.

Jones, J. C. R., and Goldman, R. D., 1985, Intermediate filaments and the initiation of desmosome assembly, *J. Cell Biol.* **101:**506–517.

Jones, J. C. R., and Grelling, K. A., 1989, Distribution of desmoplakin in normal cultured human keratinocytes and in basal cell carcinoma cells, *Cell. Motil. Cytoskel.* **13:**181–194.

Jones, J. C. R., Goldman, A. E., Steinert, P. M., Yuspa, S., and Goldman, R. D., 1982, Dynamic aspects of the supramolecular organization of intermediate filament networks in cultured epidermal cells, *Cell Motil.* **2:**197–213.

Jones, J. C. R., Yokoo, K. M., and Goldman, R. D., 1986a, Further analysis of pemphigus autoantibodies and their use in studies on the heterogeneity, structure, and function of desmosomes, *J. Cell Biol.* **102:**1109–1117.

Jones, J. C. R., Yokoo, K. M., and Goldman, R. D., 1986b, A cell surface desmosome associated component: Identification of a tissue-specific cell adhesion molecule, *Proc. Natl. Acad. Sci. USA* **83:**7282–7286.

Jones, J. C. R., Yokoo, K. M., and Goldman, R. D., 1986c, Is the hemidesmosome a half desmosome? An immunological comparison of mammalian desmosomes and hemidesmosomes, *Cell Motil. Cytoskel.* **6:**560–569.

Jones, J. C. R., Vikstrom, K. L., and Goldman, R. D., 1987, Evidence for heterogeneity in the 160/165K glycoprotein components of desmosomes, *J. Cell Sci.* **88:**513–520.

Jones, S. M., Jones, J. C. R., and Goldman, R. D., 1988, Fractionation of desmosomes and comparison of the polypeptide composition of desmosomes prepared from two bovine epithelial tissues, *J. Cell. Biochem.* **36:**223–236.

Kapprell, H.-P., Cowin, P., Franke, W. W., Ponstingl, M., and Opferkuch, H. J., 1985. Biochemical characterization of desmosomal proteins isolated from bovine muzzle epidermis: Amino acid and carbohydrate composition, *Eur. J. Cell Biol.* **36:**217–229.

Kapprell, H.-P., Owaribe, K., and Franke, W. W., 1988, Identification of a basic protein of M_r 75,000 as an accessory desmosomal plaque protein in stratified and complex epithelia, *J. Cell Biol.* **106:**1679–1692.

Kartenbeck, J., Franke, W. W., Moser, J. G., and Stoffels, U., 1983, Specific attachment of desmin filament to desmosomal plaques in cardiac myocytes, *EMBO J.* **2:**735–742.

Kartenbeck, J., Schrechheimer, K., Moll, R., and Franke, W. W., 1984, Attachment of vimentin filaments to desmosomal plaques in meningiomal cells and arachnoidal tissue, *J. Cell Biol.* **98:**1072–1081.

Kelly, D. E., 1986, Fine structure of desmosomes, hemidesmosomes and an adepidermal globular layer in developing newt epidermis, *J. Cell Biol.* **28:**51–72.

Kelly, D. E., and Kuda, A. M., 1981, Traversing filaments in desmosomal and hemidesmosomal attachments: Freeze fracture approaches toward their characterization, *Anat. Rec.* **199:**1–14.

Kelly, D. E., and Shienvold, F. L., 1976, The desmosome: Fine structural studies with freeze fracture replication and tannic acid staining of sectioned epidermis, *Cell Tissue Res.* **172:**309–323.

Keski-Oja, J., and Alitalo, K., 1985, Reorganization of plasma membrane-associated 36,000 dalton protein upon drug-induced redistribution of cytokeratin, *Exp. Cell Res.* **158:**86–94.

Klatte, D. H., Kurpakus, M. A., Grelling, K. A., and Jones, J. C. R., 1989, Immunochemical characterization of hemidesmosomal components and their expression in cultured epithelial cells, *J. Cell. Biol.* (in press).

Knapp, L. W., and Bunn, C. L., 1987, The experimental manipulation of keratin expression and organization in epithelial cells and somatic cell hybrids, *Curr. Top. Dev. Biol.* **22:**69–96.

Knapp, L. W., O'Guinn, W. M., and Sawyer, R. H., 1983a, Rearrangement of the keratin cytoskeleton after combined treatment with microtubule and microfilament inhibitors, *J. Cell Biol.* **97:**1788–1794.

Knapp, L. W., O'Guinn, W. M., and Sawyer, R. H., 1983b, Drug-induced alteration of cytokeratin organization in cultured epithelial cells, *Science* **219:**501–503.

Krawczyk, W. S., and Wilgram, G. F., 1973, Hemidesmosome and desmosome morphogenesis during epidermal wound healing, *J. Ultrastruct. Res.* **45:**93–101.

Langley, R. C., Jr., and Cohen, C. M., 1986, Association of spectrin with desmin intermediate filaments, *J. Cell Biochem.* **30:**101–109.

Langley, R. C., Jr., and Cohen, C. M., 1987, Cell type-specific association between two types of spectrin and two types of intermediate filaments, *Cell Motil. Cytoskel.* **8:**165–173.

Lazarides, E., 1980, Intermediate filaments as mechanical integrators of cellular space, *Nature* **283:**249–256.

Lazarides, E., and Granger, B. L., 1978, Fluorescent localization of membrane sites in glycerinated chicken skeletal muscle fibers and the relationship of these sites to the protein composition of the Z-discs, *Proc. Natl. Acad. Sci. USA* **75:**3683–3687.

Lazarides, E., and Hubbard, B. D., 1976, Immunological characterization of the subunit of the 100 Å filaments from muscle cells, *Proc. Natl. Acad. Sci. USA* **73:**4344–4348.

Lehto, V.-P., 1983, 140,000 dalton surface glycoprotein: A plasma membrane component of the detergent-resistant cytoskeletal preparation of cultured human fibroblasts, *Exp. Cell Res.* **143:**271–286.

Lehto, V.-P., and Virtanen, I., 1983, Immunolocalization of a novel, cytoskeleton-associated polypeptide of M_r 230,000 daltons (p230), *J. Cell Biol.* **96:**703–716.

Lehto, V.-P., Virtanen, I., and Kurki, P., 1978, Intermediate filaments anchor the nuclei in nuclear monolayers of cultured human fibroblasts, *Nature* **272:**175–177.

Lehto, V.-P., Vartio, T., and Virtanen, I., 1981, Fibronectin remains in the cytoskeletal preparations of cultured human fibroblasts, *Cell Biol. Int. Rep.* **5:**417–426.

Leloup, R., Lawet, L., Ronveaux, M. F., Drochmans, P., and Wanson, J. C., 1979, Desmosomes and desmogenesis in the epidermis of calf muzzle, *Biol. Cell* **34:**137–152.

Lentz, T. L., and Trinkaus, J. P., 1971, Differentiation of the junctional complex of surface cells in the developing Fundulus blastoderm, *J. Cell Biol.* **48:**455–472.

LeTerrier, J.-F., Liem, R. K. H., and Shelanski, M. L., 1982, Interactions between neurofilaments and microtubule-associated proteins: A possible mechanism for intraorganellar bridging, *J. Cell Biol.* **95:**982–986.

Mangeat, P. H., and Burridge, K., 1984, Immunoprecipitation of nonerythrocyte spectrin within live cells following microinjection of specific antibodies: Relation to cytoskeletal structures, *J. Cell Biol.* **98:**1363–1377.

Marchesi, V. T., 1985, Stabilizing infrastructure of cell membranes, *Annu. Rev. Cell Biol.* **1:**531–561.

Matsudaira, P. T., and Burgess, D. R., 1979, Identification and organization of the components in the isolated microvillus cytoskeleton, *J. Cell Biol.* **83:**667–673.

Mattey, D. L., and Garrod, D. R., 1986, Calcium-induced desmosome formation in cultured kidney epithelial cells, *J. Cell Sci.* **85:**95–111.

Miller, K., Mattey, D., Measures, H., Hopkins, C., and Garrod, D., 1987, Localization of the protein and glycoprotein component of bovine nasal epithelial desmosomes by immunoelectron microscopy, *EMBO J.* **6:**885–889.

Moll, R., Cowin, P., Kapprell, H.-P., and Franke, W. W., 1986, Desmosomal proteins: New markers for identification and classification of tumors, *Lab. Invest.* **54:**4–25.

Mooseker, M. S., Bonder, E. M., Conzelman, K. A., Fishkind, D. J., Howe, C. L., and Keller, T. C. S., III, 1984, Brush border cytoskeleton and integration of cellular functions, *J. Cell. Biol.* **99:**104s–112s.

Mueller, H., and Franke, W. W., 1983, Biochemical and immunological characterization of desmoplakins I and II, the major polypeptides of the desmosomal plaque, *J. Mol. Biol.* **163:**647–671.

Murray, J. M., 1984, Disassembly and reconstitution of a membrane–microtubule complex, *J. Cell Biol.* **98:**1481–1487.

Mutasim, D. F., Takahashi, Y., Labib, R. S., Anhalt, G. J., Patel, H. P., and Diaz, L. A., 1985, A pool of bullous pemphigoid antigen(s) is intracellular and associated with the basal cell cytoskeleton–hemidesmosome complex, *J. Invest. Dermatol.* **84:**47–53.

Nelson, W. J., and Lazarides, E., 1983, Expression of the β subunit of spectrin in nonerythroid cells, *Proc. Natl. Acad. Sci. USA* **80:**363–367.

O'Keefe, E. J., Erickson, H. P., and Bennett, V., 1989, Desmoplakin I and desmoplakin II, *J. Biol. Chem.* **264:**8310–8318.

Overton, J., 1962, Desmosome development in normal and reassociating cells in the early chick blastoderm, *Dev. Biol.* **4:**532–548.

Pasdar, M., and Nelson, W. J., 1988a, Kinetics of desmosome assembly in Madin–Darby canine kidney epithelial cells: Temporal and spatial regulation of desmoplakin organization and stabilization upon cell–cell contact. I. Biochemical analysis, *J. Cell Biol.* **106:**677–685.

Pasdar, M., and Nelson, W. J., 1988b, Kinetics of desmosome assembly in Madin–Darby canine kidney epithelial cells: Temporal and spatial regulation of desmoplakin organization and stabilization upon cell–cell contact. II. Morphological analysis, *J. Cell Biol.* **106:**687–695.

Pease, D. C., and Molinari, S., 1960, Electron microscopy of muscular arteries; pial vessels of the cat and monkey, *J. Ultrastruct. Res.* **3:**447–468.

Penn, E. J., Hobson, C., Rees, D. A., and Magee, A. I., 1987, Structure and assembly of desmosome junctions: Biosynthesis, processing, and transport of the major protein and glycoprotein components in cultured epithelial cells, *J. Cell Biol.* **105:**57–68.

Perides, G., Scherbarth, A., Kuhn, S., and Traub, P., 1986, An electron microscopic study of the interaction in vitro of vimentin intermediate filaments with vesicles prepared from Ehrlich ascites of tumor cell lipids, *Eur. J. Cell Biol.* **41:**313–325.

Pirbazari, M., and Kelly, D. E., 1985, Analysis of desmosomal intramembrane particle populations and cytoskeletal elements: Detergent extraction and freeze-fracture, *Cell Tissue Res.* **241:**341–351.

Price, M., and Sanger, J. W., 1979, Intermediate filaments connect Z-discs in adult chicken muscle, *J. Exp. Zool.* **208:**263–269.

Price, M. G., and Sanger, J. W., 1983, Intermediate filaments in striated muscle, in: *Cell and Muscle Motility,* Volume 3 (R. Dowben and J. Shay, eds.), Plenum Press, New York, pp. 1–40.

Ramaekers, F. C. S., Dunia, I., Dodemont, H. J., Benedetti, E. L., and Bloemendal, H., 1982, Lenticular intermediate-sized filaments: Biosynthesis and interaction with plasma membrane, *Proc. Natl. Acad. Sci. USA* **79:**3208–3212.

Rayns, D. G., Simpson, L. O., and Ledingham, J. M., 1969, Ultrastructure of desmosomes in mammalian intercalated discs: Appearances after lanthanum treatment, *J. Cell Biol.* **42:**322–326.

Regnier, M., Vaigot, P., Michel, S., and Prunieras, M., 1985, Localization of bullous pemphigoid antigen (BPA) in isolated keratinocytes, *J. Invest. Dermatol.* **85:**187–190.

Richardson, F. L., Stromer, M. H., Huiatt, T. W., and Robson, R. M., 1981, Immunoelectron and immunofluorescence localization of desmin in mature avian muscles, *Eur. J. Cell Biol.* **26:**91–101.

Sakai, L. Y., Keene, D. R., Morris, N. P., and Burgeson, R. E., 1986, Type VII collagen is a major structural component of anchoring fibrils, J. Cell Biol. **103:**1577–1586.

Schliwa, M., and van Blerkom, J., 1981, Structural interactions of cytoskeletal components, *J. Cell Biol.* **90:**222–235.

Schmelz, M., Duden, R., Cowin, P., and Franke, W. W., 1986a, A constitutive transmembrane glycoprotein of M_r 165,000 (desmoglein) in epidermal and non-epidermal desmosomes. I. Biochemical identification of the polypeptide, *Eur. J. Cell Biol.* **42:**177–183.

Schmelz, M., Duden, R., Cowin, P., and Franke, W. W., 1986b, A constitutive transmembrane of glycoprotein of M_r 165,000 (desmoglein) in epidermal and non-epidermal desmosomes. II. Immunolocalization and microinjection studies, *Eur. J. Cell Biol.* **42:**184–199.

Schollmeyer, J. E., Furcht, L. T., Goll, D. E., Robson, R. M., and Stromer, M. H., 1976, Localization of contractile proteins in smooth muscle cells and in normal and transformed fibroblasts, in: *Cell Motility*, Vol. A (R. D. Goldman, T. Pollard, J. Rosenbaum, eds.), Cold Spring Harbor Laboratories, pp. 361–388.

Shelanski, M. L., Leterrier, J.-F., Liem, R. K. H., 1981, Evidence for interactions between neurofilaments and microtubules, *Neurosci. Res. Progr. Bull.* **19:**32–43.

Shen, B. W., Josephs, R., and Steck, T. L., 1986, Ultrastructure of the intact skeleton of the human erythrocyte membrane, *J. Cell Biol.* **102:**997–1006.

Singer, I. I., 1979, The fibronexus: A transmembrane association of fibronectin-containing fibers and bundles of 5nm microfilaments in hamster and human fibroblasts, *Cell* **16:**675–685.

Skerrow, C. J., and Matoltsy, A. G., 1974, Isolation of epidermal desmosomes, *J. Cell Biol.* **63:**515–523.

Small, J. V., and Celis, J. E., 1978, Direct visualization of the 10nm (100 Å)-filament network in whole and enucleated cultured cells, *J. Cell Sci.* **31:**393–409.

Small, J. V., and Celis, J. E., 1978, Direct visualization of the 10nm (100 Å)-filament network in whole and enucleated cultured cells, *J. Cell Sci.* **31:**393–409.

Small, J. V., and Sobieszek, A., 1977, Studies on the function and composition of the 10nm (100 Å) filaments of vertebrate smooth muscle, *J. Cell Sci.* **23:**243–268.

Staehelin, L. A., 1974, Structure and function of intercellular junctions, *Int. Rev. Cytol.* **39:**191–278.

Steinberg, M. S., Shida, H., Giudice, G. J., Shida, M., Patel, N. H., and Blaschuk, O. W., 1987, On the molecular organization, diversity and functions of desmosomal proteins, *Ciba Symp.* **125:**3–17.

Steinert, P. M., and Roop, D. R., 1988, Molecular and cellular biology of intermediate filaments, *Ann. Rev. Biochem.* **57:**593–625.

Talian, J., Khan, H., and Goldman, R., 1982, Microfilaments, intermediate filaments and adhesion plaques during movement of cultured fibroblasts, *J. Cell Biol.* **95:**295a.

Thornell, L. E., Eriksson, A., Stigbrand, T., and Sjostrom, M., 1978, Structural proteins in cow Purkinje and ordinary ventricular fibers—A marked difference, *J. Mol. Cell. Cardiol.* **10:**605–616.

Thornell, L. E., Eriksson, A., Johansson, B., Kjorell, U., Franke, W. W., I. Virtanen, I., and Lehto, V.-P., 1985, Intermediate filaments and associated proteins in heart Purkinje fiber: A membrane myofibril anchored cytoskeletal system, *Ann. N.Y. Acad. Sci.* **455:**213–240.

Traub, P., 1985, *Intermediate Filaments*, Springer-Verlag, Berlin.

Traub, P., Perides, G., Scherbarth, A., and Traub, U., 1985, Tenacious binding of lipids to vimentin during its isolation and purification from Ehrlich ascites tumor cells, *FEBS Lett.* **193:**217–221.

Traub, P., Perides, G., Schimmel, H., and Scherbarth, A., 1986, Interaction in vitro of non-epithelial intermediate filament proteins with total cellular lipids, individual phospholipids and a phospholipid mixture, *J. Biol. Chem.* **261:**10558–10568.

Traub, P., Perides, G., Kuhn, S., and Scherbarth, A., 1987, Efficient interaction of nonpolar lipids with intermediate filaments of the vimentin type, *Eur. J. Cell Biol.* **43:**55–64.

Tsukita, S., and Tsukita, S., 1985, Desmocalmin: A calmodulin-binding high molecular weight protein isolated from desmosomes, *J. Cell Biol.* **101:**2070–2080.

Uehara, Y., Campbell, G. R., and Burnstock, G., 1971, Cytoplasmic filaments in developing and adult vertebrate smooth muscle, *J. Cell Biol.* **50:**484–497.

Wang, E., and Goldman, R. D., 1978, Functions of cytoplasmic fibers in intracellular movements in BHK-21 cells, *J. Cell Biol.* **79:**708–726.

Watt, F. M., Mattey, D. L., and Garrod, D. R., 1984, Calcium-induced reorganization of desmosomal components in cultured human keratinocytes, *J. Cell Biol.* **99:**2211–2215.

Weatherbee, J. A., 1981, Membranes and cell movement: Interactions of membranes with the proteins of the cytoskeleton, *Int. Rev. Cytol. Suppl.* **12:**113–176.

Westgate, G. E., Weaver, A., and Couchman, J. R., 1985, Bullous pemphigoid antigen localization suggests an intracellular association with hemidesmosomes, *J. Invest. Dermatol.* **84:**218–224.

Wiche, G., Krepler, R., Artlieb, U., Pytela, R., and Denk, H., 1983, Occurrence and immunolocalization of plectin in tissue, *J. Cell. Biol.* **97:**887–901.

Yang, H.-Y., Lieska, N., Goldman, A. E., and Goldman, R. D., 1985, A 300,000-mol-wt. intermediate filament-associated protein in baby hamster kidney (BHK-21) cells, *J. Cell Biol.* **100:**620–631.

Yang, H.-Y., Lieska, N., and Goldman, R. D., 1989, Intermediate filament associated proteins (IFAPS), in: *Cellular and Molecular Biology of Intermediate Filaments* (R. D. Goldman and P. M. Steinert, eds.), Plenum, New York, pp. 371–391.

II

The Structure of Intermediate Filaments

7

Primary and Secondary Structure of IF Protein Chains and Modes of Molecular Aggregation

David A. D. Parry

1. Introduction

Five classes of intermediate filaments (IF) have been defined on the basis of the cell type from which the filaments were initially isolated and characterized. The first of these classes comprises the keratins, a heterogeneous family of protein chains with molecular weights in the range 40 to 70k; they are expressed in most epithelia. Vimentin chains have a molecular weight of 53k, and are expressed in cells of mesenchymal origin and in cell lines established *in vitro*. The third class comprises desmin chains of molecular weight 53k; they are expressed in smooth, cardiac, and skeletal myogenic cells. Glial fibrillary acidic protein chains, found in glial cells and astrocytes, have molecular weights of 50k and form the fourth class. The fifth class of IF protein contains neurofilament chains with molecular weights of about 62, 98, and 112k; these are expressed in varying amounts in different neuronal tissues. The limitations of this type of classification have become apparent with the observation that some cells are capable of expressing vimentin in addition to their "normal" IF protein species.

Even a cursory inspection of these data shows that there are significant differences in the molecular weights of the IF protein chains and in the number of chain variants expressed in a particular cell type. More detailed studies have shown that other important differences exist, including immunological properties, solubility, and the mode of self-assembly. In contrast, however, IF from all sources have a common morphology and, as will be shown shortly, extensive regions of chain with a high degree of sequence homolo-

David A. D. Parry • Department of Physics and Biophysics, Massey University, Palmerston North, New Zealand.

gy. One of the fundamental problems of IF research is thus to understand how the similarities and differences can be incorporated within a single model.

This chapter will be concerned with establishing the structure of individual IF protein molecules and their primary modes of aggregation. In the following chapter, Fraser *et al.* will develop this theme and show how the molecules aggregate in a variety of closely related polymorphic forms to produce intact filaments compatible with the experimental observations outlined here.

2. Type I–IV IF Protein Chains

Until the early 1980s, research on the hard keratins of wool and quill and on the 10-nm-diameter IF was undertaken quite independently. The groups concerned had relatively little knowledge, or indeed interest, in the achievements of the other. This situation changed almost overnight, primarily as a result of the significant homology observed between segments of wool keratin, desmin, vimentin, and neurofilaments. The hard keratins, as well as those of epidermal origin, thus became recognized as an important class of IF. Structural work on hard keratin (the only source of IF in which the natural orientation is high) was no longer seen as relevant only to those working in the field of wool research. Similarly, the elegant chemical characterization of IF proteins and their reassembly *in vitro* into native-type filaments had major implications for those involved in characterizing the highly cross-linked, heterogeneous, and largely intractable proteins found in wool. The merger of these two areas of research has proved extraordinarily fruitful, and rapid progress has resulted. The amino acid sequence data and their analyses represented the starting point upon which much of the progress on structural aspects of IF has subsequently become possible.

2.1. Primary Structure

Amino acid sequences, either in whole or part, are now known for about 50 IF protein chains. These include several representative sequences from each of the five classes of IF. A current catalogue of the primary structures of these proteins is as follows:

1. Keratins: mouse epidermal keratin of molecular weights 47k (M47K; Singer *et al.*, 1986), 50k (M50K; Steinert and Roop, unpublished), 50 and two 52k (Knapp *et al.*, 1987), 55k (M55K; Steinert and Roop, unpublished), 59k (M59K; Steinert *et al.*, 1983; Krieg *et al.*, 1985), 60k (M60K; Steinert *et al.*, 1984), and 67k (M67K; Crewther *et al.*, 1983; Steinert *et al.*, 1985a); human epidermal keratin of molecular weights 40k (H40K or K19; Eckert, 1988), 46k (H46K or K18; Romano *et al.*, 1986), 47k (H47K or K16; Raychaudury *et al.*, 1986; for nomenclature see Moll *et al.*, 1982, and Quinlan *et al.*, 1985; also Fig. 1), 50k (H50K or K14; Hanukoglu and Fuchs, 1982; Marchuk *et al.*, 1984, 1985), 55k (H55K or K7; Glass *et al.*, 1985), 56k (H56K or K6a, K6b; Hanukoglu and Fuchs, 1983; Tyner *et al.*, 1985), 56.5k (H56.5k or K10; Zhou *et al.*, 1988), 58k (H58K or K5; Lersch and Fuchs, 1988), 65k (H65K or K3; Klinge *et al.*, 1987), and 67k (H67K or K1; Steinert *et al.*, 1985a); bovine epidermal keratin of molecular weights 50k (B50K, major keratin polypeptide VII equivalent to chain 16 of human epidermis, Fig. 1; Jorcano *et al.*, 1984), 54k (B54K, major keratin polypeptide VIb equivalent to chain 14 of human epidermis, Fig. 1; Jorcano *et al.*, 1984; Rieger *et al.*, 1985), and 40k (B40K,

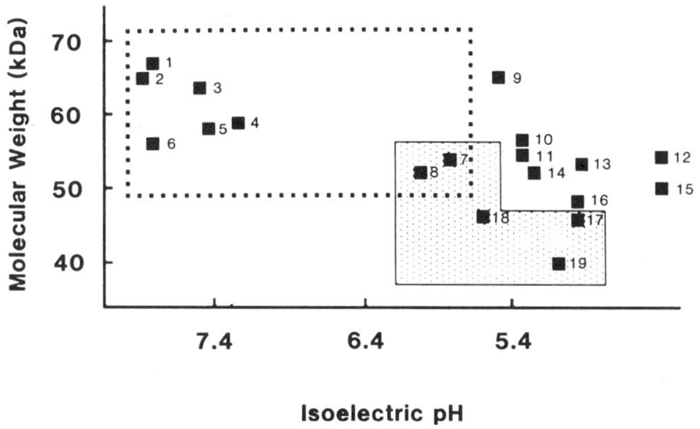

Figure 1. The nomenclature pertaining to the cytokeratin polypeptide chains in human tissues is related to the position of the chains on a two-dimensional gel electrophoresis pattern. The type II "basic" cytokeratins (dotted box, 1–8) have higher isoelectric pHs and generally higher molecular weights than the Type I "acidic" cytokeratins (9–19). Mesothelial keratins are indicated by the shaded box. There is now some evidence that 2, 11, and 12 may not be unique keratin chains.

equivalent to human cytokeratin 19; Bader *et al.*, 1986); frog epidermal keratin of molecular weights 47k (XK70 and DG81; Winkles *et al.*, 1985; Jonas *et al.*, 1985), 51k (XL51; Hoffmann and Franz, 1984) and 64k (XL64—three chains; Hoffmann *et al.*, 1985); wool keratins of molecular weights 48k (component 8c-1; Dowling *et al.*, 1983, 1986; Crewther *et al.*, 1983) and 54k (component 7c; Sparrow and Inglis, 1980; Dowling *et al.*, 1983; Crewther *et al.*, 1983).

2. Vimentin: hamster eye lens vimentin (HELV; Quax-Jeuken *et al.*, 1983; Quax *et al.*, 1983), porcine eye lens vimentin (PELV; Geisler and Weber, 1981; Geisler *et al.*, 1982b), and chicken vimentin (Zehner and Paterson, 1985).

3. Desmin: porcine stomach desmin (PSD; Geisler and Weber, 1981; Geisler *et al.*, 1982a), chicken gizzard desmin (CGD; Geisler and Weber, 1982), and hamster desmin (HD; Quax *et al.*, 1984).

4. Glial fibrillary acidic protein: bovine (BGFAP; Hong and Davison, 1981), mouse (MGFAP; Lewis *et al.*, 1984), and porcine (PGFAP; Geisler and Weber, 1982, 1983).

5. Neurofilament protein: chains of molecular weight 62k (NF-L; Geisler *et al.*, 1982b, 1983, 1985c; Julien *et al.*, 1985, 1987; Lewis and Cowan, 1986), 98k (NF-M; Geisler *et al.*, 1984; Myers *et al.*, 1987; Napolitano *et al.*, 1987; Levy *et al.*, 1987), and 112k (NF-H; Geisler *et al.*, 1985a; Shneidman *et al.*, 1988; Lees *et al.*, 1988). The neurofilaments were obtained from pigs, humans, rats, and mice.

2.2. Secondary Structure

The secondary structure of most proteins can be classified as α-helical, extended β, turn, or random-coil. Visual inspection of an amino acid sequence, however, rarely allows the chain conformation to be deduced and hence algorithms have been developed to allow some quantitation of the most likely secondary structure for any particular portion of sequence. These algorithms use as their data base the conformational information obtained from the determination of the crystal structures of globular proteins. Even though it is widely recognized that the best of these predictive schemes, such as those due to Chou and Fasman (1978) and Garnier *et al.* (1978), provide only a best estimate of the probable secondary structure, no better methods are currently available; they are thus used extensively by workers in the field. It is worth noting, however, that the success rate for predicting α helix is considerably greater than for β or β-turn structures. Part of the reason for this may lie with the fact that the latter structures occur in short pieces that are frequently well separated from one another along the length of the sequence, whereas lengths of α helix are often long by comparison. Recently, Cohen and Parry (1986) have shown that an additional and previously unrecognized feature of amino acid sequences indicates an α-helical secondary structure. This feature is a heptad pattern of residues of the type $(a\text{-}b\text{-}c\text{-}d\text{-}e\text{-}f\text{-}g)_n$ where *a* and *d* are generally apolar residues. Such a pattern formed the basis of the proposal by Crick (1953) of a coiled-coil structure for α-fibrous proteins. While a heptad structure is not found for all α-helical structures in every type of protein, it has now become increasingly recognized as an important structural motif, and its presence provides strong evidence for α helices packed together in bundles.

Within this framework the sequences of all IF protein chains have been analyzed and common features sought. In every case the sequences can be divided into a central domain rich in predicted α-helical structure and N- and C-terminal domains of variable secondary structure (Geisler *et al.*, 1982a, 1983; Geisler and Weber, 1982; Hanukoglu and Fuchs, 1982, 1983; Steinert *et al.*, 1983, 1984, 1985a,b; Crewther *et al.*, 1983; Parry and Fraser, 1985; Steinert and Parry, 1985; Conway and Parry, 1988; Steinert and Roop, 1988). The central domain contains 308–315 residues of which approximately 90% (i.e., 284 residues) is predicted to be α-helical. Four α-helix-rich segments within this domain have been specified and these are known as segments 1A, 1B, 2A, and 2B; all of them have a heptad pattern of residues. This provides strong support for an α-helical designation (Cohen and Parry, 1986). Short regions without a heptad repeat link the helical segments to one another: L1 links segments 1A to 1B, L12 links segment 1B to 2A, and L2 links segment 2A to 2B (Fig. 2). The portions of the α-helix-rich domain defined by segments 1A–L1–1B and 2A–L2–2B are also known as segments 1 and 2, respectively; each

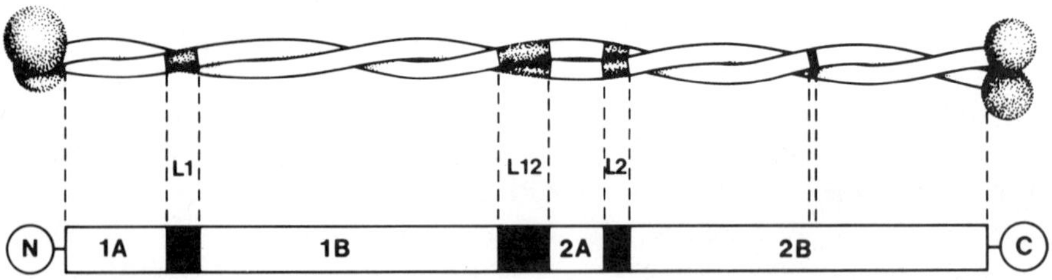

Figure 2. Each IF protein molecule has an N- and a C-terminal non-coiled-coil domain, and a rod domain containing four coiled-coil segments designated 1A, 1B, 2A, and 2B. Each of these segments is a left-handed two-stranded coiled-coil rope. The segments of coiled-coil rope are linked by L1, L12, and L2, which, unlike 1A, 1B, 2A, and 2B, do not have a regular heptad structure. The continuity of the coiled-coil rope is thus broken at these link segments. A break in the phasing of the heptad repeat also occurs at a point close to the center of segment 2B and this again represents a point of discontinuity in the rope structure. The relative phasing of the coiled-coil segments is unknown and thus the choice shown in the figure is arbitrary. From Parry and Fraser (1985) with permission of Butterworths Press.

contains about the same number of residues and is thus of similar size. The concept of an α-helix-rich domain is readily compatible with x-ray, infrared, and physicochemical data that had previously been shown to indicate that IF had a significant α-helix content (Fraser *et al.*, 1976; Steinert *et al.*, 1978; Renner *et al.*, 1981; Steinert, 1981).

The predicted structure of the N- and C-terminal domains varies widely between IF chains and it is apparent that there are few if any common features. Structural subdomains, however, are apparent in certain classes of chain as will be seen in Section 6 (Steinert *et al.*, 1985a,b). It is clear from this work that the range of molecular weights shown by IF protein chains is accounted for primarily by the variation in the sizes of the terminal domains.

3. Classification of IF Protein Chains

S-carboxymethylated derivatives of wool keratin, when enzymatically degraded, yielded fragments enriched in α-helical structure. Two such fragments, designated Type I and Type II, were distinguishable chemically even though they were clearly homologous (Crewther *et al.*, 1978; Gough *et al.*, 1978). Subsequent determination of the amino acid sequence of an "acidic" human epidermal keratin (K14; Hanukoglu and Fuchs, 1982; Marchuk *et al.*, 1984, 1985) revealed that part of that sequence (segment 1B) had a remarkable similarity to the Type I fragment isolated from wool keratin. Furthermore, the 1B segment of a "basic" human epidermal keratin of molecular weight about 56k (K6a; Hanukoglu and Fuchs, 1983) showed a striking resemblance to the Type II fragment from wool characterized by Crewther and colleagues. As a result of these observations, the "acidic" class of keratin chains ($pI < 5.5$) from both hard and soft keratins have been designated Type I chains; the "neutral–basic" class of keratin chains ($pI > 6$) are known as Type II chains (see Fig. 1). The sequence homology is not restricted to the α-helical portions of this central domain. Indeed, small changes in the sequences and the lengths of segments L1 and L12, two of the nonhelical portions present, are as characteristic of chain type as are features of the four α-helix-rich segments (Parry and Fraser, 1985). On the basis of the degree of homology of both the helical and nonhelical segments in the central domain (Table I), all of the keratin sequences subsequently determined have fallen naturally into the Type I [component 8c-1, M50K, M52K, M55K, M59K, H40K (K19), H46K (K18), H47K (K16), H56.5K (K10), H50K (K14), B50K, M52K, B54K, XK70, DG81, XL70, XL81, XL51] or Type II classification [component 7c, M60K, M67K, H55K (K7), H56K (K6a,b), H58K (K5), H65K (K3), H67K (K1), XL64]. Furthermore, on the basis of sequence homology the Type I and Type II chains can each be subdivided in such a way that Type Ia and IIa chains constitute those from the "hard" α-keratins and Type Ib and IIb chains constitute those from the "soft" epidermal α-keratins (Conway and Parry, 1988).

Amino acid sequences of desmin, vimentin, and glial fibrillary acidic protein have also been completed, and while these display homology to both the Type I and Type II keratin chains, it is clear that they are much more similar to one another than to the keratins (Table I). For this reason, this group of IF protein chains (PSD, CGD, HD, HELV, PELV, BGFAP, MGFAP, and PGFAP) have been designated Type III.

The remaining IF protein chains constitute those isolated from neurofilaments. A number of these have now been sequenced *in toto* and sufficient data are available from

Table I. Sequence Homology between Segments 1A, 1B, and 2 in Various IF Protein Chains

Protein	8c-1(I)	M59K(I)	H50K(I)	7c(II)	M67K(II)	H56K(II)	CGD(III)	HELV(III)	MGFAP(III)	PNF-L(IV)	PNF-M(IV)
8c-I(I)	—	6.1×10^{-14}	3.3×10^{-14}	—	5.3×10^{-5}	—	2.6×10^{-4}	6.6×10^{-4}	2.8×10^{-5}	2.0×10^{-3}	5.9×10^{-6}
	—	1.4×10^{-37}	$<10^{-39}$	2.3×10^{-9}	7.4×10^{-7}	3.2×10^{-6}	2.5×10^{-7}	3.1×10^{-7}	3.3×10^{-6}	1.6×10^{-6}	1.3×10^{-4}
	—	1.2×10^{-35}	$<10^{-39}$	5.8×10^{-6}	3.5×10^{-4}	5.0×10^{-11}	6.3×10^{-8}	5.3×10^{-5}	6.5×10^{-10}	2.1×10^{-10}	6.3×10^{-9}
M59K(I)	6.1×10^{-14}	—	3.4×10^{-28}	—	6.1×10^{-7}	—	2.2×10^{-5}	3.4×10^{-7}	1.1×10^{-5}	4.6×10^{-4}	3.3×10^{-7}
	1.4×10^{-37}	—	$<10^{-39}$	1.1×10^{-5}	1.7×10^{-2}	6.9×10^{-2}	5.8×10^{-10}	9.0×10^{-10}	7.5×10^{-5}	1.8×10^{-8}	1.3×10^{-5}
	1.2×10^{-35}	—	$<10^{-39}$	9.2×10^{-6}	3.2×10^{-3}	2.7×10^{-6}	1.6×10^{-12}	3.2×10^{-9}	6.0×10^{-11}	7.1×10^{-11}	2.1×10^{-16}
H50K(I)	3.3×10^{-14}	3.4×10^{-28}	—	—	6.5×10^{-6}	—	6.6×10^{-6}	4.0×10^{-6}	5.3×10^{-4}	3.2×10^{-3}	1.7×10^{-5}
	$<10^{-39}$	$<10^{-39}$	—	4.3×10^{-5}	9.7×10^{-3}	2.1×10^{-3}	4.9×10^{-8}	2.5×10^{-9}	1.1×10^{-6}	9.1×10^{-8}	1.1×10^{-4}
	$<10^{-39}$	$<10^{-39}$	—	6.0×10^{-6}	1.1×10^{-4}	8.2×10^{-13}	2.8×10^{-14}	7.3×10^{-13}	2.3×10^{-16}	8.8×10^{-12}	5.2×10^{-17}
7c(II)	—	—	—	—	—	—	—	—	—	—	—
	2.3×10^{-9}	1.1×10^{-5}	4.3×10^{-5}	—	1.7×10^{-18}	1.8×10^{-22}	3.4×10^{-9}	6.0×10^{-6}	1.5×10^{-10}	3.8×10^{-11}	3.2×10^{-6}
	5.8×10^{-6}	9.2×10^{-6}	6.0×10^{-6}	—	2.6×10^{-37}	$<10^{-39}$	2.3×10^{-9}	6.6×10^{-9}	2.1×10^{-12}	2.3×10^{-5}	4.3×10^{-11}
M67K(II)	5.3×10^{-5}	6.1×10^{-7}	6.5×10^{-6}	—	—	—	1.0×10^{-8}	4.0×10^{-12}	2.1×10^{-12}	1.6×10^{-8}	1.6×10^{-8}
	7.4×10^{-7}	1.7×10^{-2}	9.7×10^{-3}	1.7×10^{-18}	—	$<10^{-39}$	2.8×10^{-5}	9.2×10^{-7}	6.1×10^{-6}	5.0×10^{-3}	5.5×10^{-2}
	3.5×10^{-4}	3.2×10^{-3}	1.1×10^{-4}	2.6×10^{-37}	—	$<10^{-39}$	7.2×10^{-16}	5.4×10^{-14}	8.5×10^{-8}	7.2×10^{-9}	1.5×10^{-11}
H56K(II)	—	—	—	—	—	—	—	—	—	—	—
	3.2×10^{-6}	6.9×10^{-2}	2.1×10^{-3}	1.8×10^{-22}	$<10^{-39}$	—	2.9×10^{-7}	2.6×10^{-4}	2.6×10^{-6}	2.9×10^{-5}	1.7×10^{-4}
	5.0×10^{-11}	2.7×10^{-6}	8.2×10^{-13}	$<10^{-39}$	$<10^{-39}$	—	3.2×10^{-22}	4.8×10^{-23}	1.1×10^{-15}	1.9×10^{-12}	7.3×10^{-20}
CGD(III)	2.6×10^{-4}	2.2×10^{-5}	6.6×10^{-6}	—	1.0×10^{-8}	—	—	8.4×10^{-21}	1.1×10^{-19}	1.5×10^{-7}	2.8×10^{-13}
	2.5×10^{-7}	5.8×10^{-10}	4.9×10^{-8}	3.4×10^{-9}	2.8×10^{-5}	2.9×10^{-7}	—	7.9×10^{-37}	1.0×10^{-26}	8.2×10^{-14}	1.4×10^{-8}
	6.3×10^{-8}	1.6×10^{-12}	2.8×10^{-14}	2.3×10^{-9}	7.2×10^{-16}	3.1×10^{-22}	—	$<10^{-39}$	$<10^{-39}$	$<10^{-39}$	$<10^{-39}$
HELV(III)	6.6×10^{-4}	3.4×10^{-7}	4.0×10^{-6}	—	4.0×10^{-12}	—	8.4×10^{-21}	—	1.6×10^{-23}	2.0×10^{-10}	3.4×10^{-12}
	3.1×10^{-7}	9.0×10^{-10}	2.5×10^{-9}	6.0×10^{-6}	9.2×10^{-7}	2.6×10^{-4}	7.9×10^{-37}	—	1.3×10^{-23}	8.2×10^{-9}	6.0×10^{-10}
	5.3×10^{-5}	3.2×10^{-9}	7.3×10^{-13}	6.6×10^{-9}	5.4×10^{-14}	4.8×10^{-23}	$<10^{-39}$	—	$<10^{-39}$	$<10^{-39}$	$<10^{-39}$
MGFAP(III)	2.8×10^{-5}	1.1×10^{-5}	5.3×10^{-4}	—	2.1×10^{-12}	—	1.1×10^{-19}	1.6×10^{-23}	—	1.1×10^{-12}	3.9×10^{-12}
	3.3×10^{-6}	7.5×10^{-5}	1.1×10^{-6}	1.5×10^{-10}	6.1×10^{-6}	2.6×10^{-6}	1.0×10^{-26}	1.3×10^{-23}	—	3.3×10^{-15}	1.5×10^{-6}
	6.5×10^{-10}	6.0×10^{-11}	2.3×10^{-16}	2.1×10^{-12}	8.5×10^{-8}	1.1×10^{-15}	$<10^{-39}$	$<10^{-39}$	—	1.8×10^{-26}	$<10^{-39}$
PNF-L(IV)	2.0×10^{-3}	4.6×10^{-4}	3.2×10^{-3}	—	1.6×10^{-8}	—	1.5×10^{-7}	2.0×10^{-10}	1.1×10^{-12}	—	8.6×10^{-15}
	1.6×10^{-6}	1.8×10^{-8}	9.1×10^{-8}	3.8×10^{-11}	5.0×10^{-3}	2.9×10^{-5}	8.2×10^{-14}	8.2×10^{-9}	3.3×10^{-15}	—	1.8×10^{-13}
	2.1×10^{-10}	7.1×10^{-11}	8.8×10^{-12}	2.3×10^{-5}	7.2×10^{-9}	1.9×10^{-12}	$<10^{-39}$	$<10^{-39}$	1.8×10^{-26}	—	$<10^{-39}$
PNF-M(IV)	5.9×10^{-6}	3.3×10^{-7}	1.7×10^{-5}	—	1.6×10^{-8}	—	2.8×10^{-13}	3.4×10^{-12}	3.9×10^{-12}	8.6×10^{-15}	—
	1.3×10^{-4}	1.3×10^{-5}	1.1×10^{-4}	3.2×10^{-6}	5.5×10^{-2}	1.7×10^{-4}	1.4×10^{-8}	6.0×10^{-10}	1.5×10^{-6}	1.8×10^{-13}	—
	6.3×10^{-9}	2.1×10^{-16}	5.2×10^{-17}	4.3×10^{-11}	1.5×10^{-11}	7.3×10^{-20}	$<10^{-39}$	$<10^{-39}$	$<10^{-39}$	$<10^{-39}$	—

NF-L, NF-M, and NF-H chains to show that the model established for other IF chains, i.e., N- and C-terminals of variable structure enclosing a near-conserved domain rich in α-helical structure, is still appropriate. This domain in neurofilament chains is homologous to that in Type I, II, and III chains. Further, the N-terminal domains are similar in size to those of other IF chains but the C-terminal domains vary greatly from 149 residues in NF-L to about 500 residues in NF-M and about 700–800 residues in NF-H. In the latter two cases, the C-terminal domains are very significantly larger than found for any other IF protein chain (Table II). From homology studies (Table I) it can be seen that although the neurofilament chains are most similar to the Type III chains, the degree of homology between the NF-L and NF-M chains is greater than that between either of them and desmin, vimentin and glial fibrillary acidic protein. It is also noteworthy that the neurofilament chains differ from the Type III chains in the lengths of segments L1 and L12 (Table II). For these and several other reasons, neurofilament chains are classified as Type IV (Conway and Parry, 1988). Furthermore, the NF-L, NF-M, and NF-H neurofilament chains are clearly distinguishable within the Type IV classification on the basis of homology in both the rod domain and the H1 and H2 subdomains (Conway and Parry, 1988).

Although the division of IF protein chains has been made on the basis of homology within the central α-helix-rich domain, there is an additional reason to believe that the choice made is justifiable. Keratins from both wool and epidermis require both Type I and Type II chains for filament formation and thus the structural unit in the filament is an obligate heteropolymer (Ia, IIa chains for hard α-keratin, Ib and IIb chains for epidermal α-keratin). In contrast, Type III chains—desmin, vimentin, or glial fibrillary acidic protein—can form filaments by themselves without any other chain type present. The molecules forming the filaments are thus homopolymers. Interestingly, however, Type III chains from one source, say desmin, can also copolymerize with different Type III chains (i.e., vimentin or glial fibrillary acidic protein) to form filaments *in vitro* (e.g., see Steinert *et al.*, 1981). There is no evidence, however, of Type III chains copolymerizing with Type I, II, or IV chains to form filaments either *in vivo* or *in vitro*. When NF-L, NF-M, and NF-H chains are present simultaneously in a neuronal tissue, immunoelectron

Table II. Domain and Segment Sizes in Type I–V IF Protein Chains

	Number of residues				
Domain/segment	I	II	III	IV	V
N	55–143	151–179	65–101	88–101	30
1A	35	35	35	35	35 (+6)[a]
L1	11–15	10–14	8	10	0
1B	101	101	101	101	143 (+5)[a]
L12	16	17	16	17 or 22[b]	19
2A	19	19	19	19	19
L2	8	8	8	8	8
2B	121	121	121	121	121
C	46–112	88–162	52–56	149–759	185–277

[a]The 11 residues equivalent to segment L1 have been assigned to segments 1A and 1B since the entire length of segment 1 is predicted to be α-helical and contains a continuous heptad repeat with only a phase change at the previous site of L1.

[b]NF-L and NF-M have 17 residues in segment L12 whereas NF-H chains contain 22 residues.

microscopy has revealed that each neurofilament contains all three chain species (Shaw and Weber, 1981). Although both homopolymer and heteropolymer molecular structures are consistent with these observations, it appears most likely that all three neurofilament chains form homopolymeric IF molecules and that these aggregate further into intact IF containing some appropriate mixture of the triplet proteins (Zackroff and Goldman, 1980; Shaw and Weber, 1982).

4. Structural Features of the Central α-Helix-Rich Domain in IF Protein Chains

Individual α helices, such as those in the central domain of IF protein chains, are generally unstable in an aqueous environment. Chain aggregation, however, will enable a stable structure to be formed and the manner in which this occurs *in vivo* is largely specified by the nature of the sequences of the α-helical segments. In this section some of the characteristics of the segments in the central α-helix-rich domain will be described; in the following sections it will be shown how these influence the molecular structure and its mode of aggregation.

Those segments predicted to be α-helical (segments 1A, 1B, 2A, and 2B) have lengths that are conserved absolutely; those predicted to act as linking segments (segments L1, L12, but not L2) show some variation (Fig. 3). The central, and the longest of the linking segments is L12; it is 16 residues long for Type I and III chains, 17 residues long for Type II, IV-L, and IV-M chains, and 22 residues long for Type IV-H chains (Table II). The sequences of these regions in the different chain types do not favor the formation of α helix but a short piece of the form (apolar-hydrophilic)$_4$ would be compatible with an extended β structure. Segment L1 is 8 residues long for Type III chains, 10 residues long

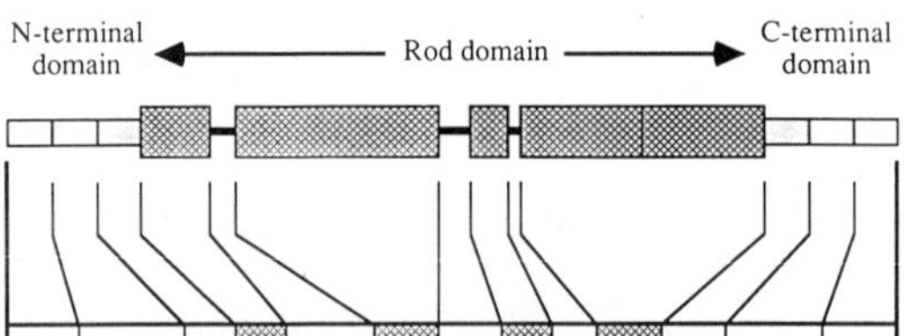

	E1	V1	H1	1A	L1	1B	L12	2A	L2	2B	H2	V2	E2
Ia	?	?	≥7	35	11	101	16	19	8	121	≥0	?	?
Ib	0-10	65-140	7	35	11-14	101	16	19	8	121	0	0-110	5-30
IIa	?	?	≥32	35	10	101	17	19	8	121	≥10	?	?
IIb	5-70	20-160	36	35	11-14	101	17	19	8	121	20	25-125	21
III	19-81		20	35	8	101	16	19	8	121	58	0	0
IV-L	0	0	91	35	10	101	17	19	8	121	81	60-70	
IV-M	0	0	102	35	10	101	17	19	8	121	≥88	?	?
IV-H	0	0	97	35	10	101	22	19	8	121	>110	?	?
V	0	0	30	41	0	148	19	19	8	121	179	?	?

Figure 3. Mammalian IF protein chains may be divided into five types on the basis of the degree of homology in the rod domain. Keratins contain only Type I and Type II chains (Ia, IIa for hard α-keratins and Ib, IIb for epidermal keratins); desmin, vimentin, and glial fibrillary acidic protein are all Type III chains; neurofilament protein chains are designated Type IV-L, IV-M, and IV-H and the nuclear lamin chains are Type V chains. The number of residues in the various segments are indicated for each chain type. The N- and C-terminal domains show substructure (E1, V1, H1, and H2, V2, E2, respectively) with a degree of bilateral symmetry about the rod domain. E1 and E2 are generally basic subdomains at the extreme ends of the IF protein chain, V1 and V2 are subdomains of considerable variability in size and character, and H1 and H2 are subdomains of high homology. From Conway and Parry (1989) with permission of Springer-Verlag.

Table III. Percentages of Acidic and Basic Residues[a] in the N-Terminal and C-Terminal Domains, and in the Rod Domain Segments 1A, 1B, and 2 of Type I–IV IF Protein Chains

Domain or segment	Residue type	IF chain type I	II	III	IV
N-terminal domain	Acidic	1.9	2.8	5.7	6.9
	Basic	4.5	7.3	14.0	11.6
Segment 1A	Acidic	19.9	12.4	19.0	18.6
	Basic	18.8	19.0	14.3	14.3
Segment 1B	Acidic	20.8	21.8	26.1	25.8
	Basic	14.7	14.3	16.8	19.2
Segment 2	Acidic	18.7	19.6	20.8	20.0
	Basic	12.7	16.6	16.2	17.4
C-terminal domain	Acidic	6.4	0.8	15.6	30.9
	Basic	11.5	6.8	16.8	10.5

[a]Expressed as a percentage of the number of residues in the domain or segment. Acidic residues are aspartic and glutamic acid; basic residues are arginine, histidine, and lysine.

for Type IV chains, 10, 11, 12, or 14 residues long for Type II chains, and 11, 14, or 15 residues for Type I chains (Table II).

It follows that any structure formed must be able to tolerate relatively large variations in the size of the link L1. In contrast, link L2 is conserved in length for all chain types; it is also predicted to be α-helical even though it lacks the heptad characteristic of the other α-helical segments.

Although helical segments 1A, 1B, and 2 generally have a net negative charge, there are important differences between the various chain types, most particularly between Type I and Type II chains (Table III). Segment 1A is near neutral for Type I chains but is basic for Type II chains. Segment 2, although acidic for both chain types, has an extra four to five negative charges in the Type I chain compared to its Type II counterpart. These differences largely explain the observed division of keratin chains into the "acidic" (Type I) and "neutral–basic" (Type II) groupings (Steinert *et al.*, 1984). In passing, it should be noted that the N- and C-terminal domains of keratin chains are net basic though the degree to which these are phosphorylated *in vivo* may alter the balance to neutral or even slightly acidic.

The heptad pattern of residues in the α-helix-rich domain is maintained without break in segments 1A, 1B, and 2A. However, toward the center of segment 2B the phasing of the pattern undergoes a discontinuity. In effect, this break or "stutter" may be considered as the insertion of an additional four residues, or, conversely, as the deletion of three residues from an otherwise continuous heptad pattern. This feature is maintained absolutely in all IF protein chains and must represent an important structural feature, the full significance of which has yet to be ascertained. The short link segment L2, which is believed to be α-helical but which lacks the heptad repeat, may be considered as part of a heptad pattern extending from the beginning of segment 2A to the end of segment 2B (noting, of course, the discontinuity in phasing just discussed). The phasing of the heptad pattern across the link L2 would, however, suffer a discontinuity equivalent either to the insertion of five residues or the deletion of two residues in an otherwise continuous pattern. The possible effect on the molecular structure of the two discontinuities in segment 2 will be discussed later.

From a comparison of the "average" IF sequence with the actual IF sequences, it is possible to assess the positions in the heptad repeat that are most highly conserved. The results of this study (Parry and Fraser, 1985; Conway and Parry, 1988) showed that positions *a* and *d* were most highly conserved (or conservatively substituted) followed by positions *e* and *g*. Least conserved positions were *b, c,* and *f*. It was also noted that the apolar residues leucine and isoleucine in positions *a* and *d* were not symmetrically distributed (Parry and Fraser, 1985).

Homology searches on the IF protein sequences have revealed that there are two regions that are largely conserved between chain types. One of these constitutes the last four heptads of the central domain, i.e., at the C-terminal end of segment 2B, and has a sequence

$$\mathrm{EYQ.LL}\genfrac{}{}{0pt}{}{\mathrm{D}}{\mathrm{N}}\mathrm{VK.}\genfrac{}{}{0pt}{}{\mathrm{R}}{\mathrm{A}}\mathrm{L}\genfrac{}{}{0pt}{}{\mathrm{D}}{\mathrm{E}}\mathrm{.EIATYR}\genfrac{}{}{0pt}{}{\mathrm{K}}{\mathrm{R}}\mathrm{LLEGE}\genfrac{}{}{0pt}{}{\mathrm{E}}{\mathrm{D}}\mathrm{.R}\begin{matrix}\mathrm{L}\\\mathrm{M}\\\mathrm{I}\end{matrix}$$

The other has a length of two heptads, a sequence

$$\mathrm{LNDRFASYI}\genfrac{}{}{0pt}{}{\mathrm{D}}{\mathrm{E}}\mathrm{KVRFLE}$$

and is found toward the N-terminal end of the central domain in segment 1A. The structural significance of these observations, however, remains to be established (though see Parry *et al.*, 1987, for one possibility). The position of the conserved regions at the ends of the α-helical-rich domain suggests that they may play a role in providing regions of particular regularity and stability [cf. the sequence (glycine–imino–imino)$_5$ at the C-terminal end of the triple helical region of the collagen α chains] or in establishing key interactions in the molecular assembly.

It has been observed by Huber (1979) and Burley and Petsko (1985) that the presence of the aromatic residues, tyrosine, phenylalanine, and tryptophan appears to be associated with regions of particularly high chain stability in globular proteins. Thus, the occurrence of these residues in IF protein chains may also provide an indication of such regions. Parry and Fraser (1985) have shown that aromatic residues are conserved at both termini of segments 1B and 2B and may be expected to contribute to the stability of these regions (Table IV). It is also worth noting that the two highly conserved regions in the IF protein sequences contain three of the six aromatic residues that are conserved in all four chain types.

The fast Fourier transform (FFT) technique has been used to determine whether any amino acids, or amino acid groupings, have a regular linear distribution along the sequences of the IF proteins. This technique (Parry, 1975, 1979; Fraser and MacRae, 1976; McLachlan and Stewart, 1976) involves the Fourier transformation of a one-dimensional array that represents the characteristic of the amino acid sequence being investigated. Thus, the amino acids of interest in a sequence (say, the negatively charged ones) are assigned a value of unity and all of the other terms in the linear array representing the sequence are equated to zero. This array is usually embedded in a larger array filled with zeros, thus enabling the Fourier transform to be sampled at finer intervals than would otherwise be possible. This allows the true period to be determined with somewhat greater

Table IV. Conservation of Aromatic Residues (Tyrosine and Phenylalanine) in Type I–IV IF Protein Chains[a]

Segment	Residue number[b]	Chain type I	II	III	IV
1A	11		x	x	x
	14[c]	x	x	x	x
	20		x	x	
1B	1	x			
	4[c]	x	x	x	x
	5	x			
	7		x		
	39	x	x		
	43	x	x		
	57		x		
	60			x	
	68		x	x	
	87[c]	x	x	x	x
	92		x		
2A	13	x	x	x	
2B	1[c]	x	x	x	x
	2	x			
	5		x	x	
	29			x	
	61			x	
	68			x	
	93[c]	x	x	x	x
	110[c]	x	x	x	x

[a]Tryptophan residues are conserved in segment 1A at residue 35 (Type I) and 32 (Type II) and in segment L2 at residue 8 (Types I, III, and IV).
[b]Residue numbering commences at the beginning of each segment.
[c]Conserved in all chain types.

accuracy. If the Fourier intensities are scaled in an appropriate way (McLachlan and Stewart, 1976), the probability of a period with a Fourier intensity I being statistically significant is given by the expression exp($-I$). The probability of an intensity I occurring anywhere in the sequence is 0.5 N exp($-I$) where N is the length of the amino acid sequence being considered. This method has the limitation that the statistics rely on the fact that n (the number of amino acids with the property currently being investigated), as well as N must be large. This condition is generally satisfied in the studies on the IF protein chains.

Structural periodicities in the amino acid sequences have been sought for various amino acid groupings including acidic, basic, and apolar. Significant periods of 3.5 (7/2) and 2.33 (7/3) residues have been found for the apolar residues in segments 1A, 1B, and 2B and this reflects the heptad repeat with the apolar residues occurring alternately three and four residues apart in positions *a* and *d*. In addition to the heptad period, a highly significant periodicity has been found in the linear distribution of the acidic residues (aspartic and glutamic acid) and the basic residues (arginine, histidine, and lysine) in both segments 1B and 2. In segment 1B the period for the acidic and the basic residues is 9.56 and 9.59

Table V. Structural Periodicity in the Linear Disposition of Acidic and Basic Residues in Segments 1B and 2 in Type I–IV Protein Chains[a]

		Segment 1B				Segment 2			
		Acidic		Basic		Acidic		Basic	
Type	Chain	Period	Intensity	Period	Intensity	Period	Intensity	Period	Intensity
I	8c-1	9.29	9.55	9.49	6.45	9.89	5.38	9.75	2.93
	M59K	9.29	8.82	9.50	7.80	9.85	5.54	9.78	4.27
	H50K	9.29	9.47	9.58	8.93	9.92	4.95	9.85	2.07
	M50K	9.25	8.75	9.66	5.54	9.91	5.83	9.87	1.62
	M55K	—	—	—	—	9.78	7.39	9.83	1.66
	XK70	9.42	7.44	9.68	9.40	9.80	4.65	9.94	5.17
	DG81	9.33	7.11	9.60	9.25	9.94	5.73	9.99	4.08
II	7c	9.66	6.37	9.54	10.42	9.87	4.67	9.83	7.63
	M67K	9.71	4.95	9.44	5.73	9.91	7.37	9.78	4.86
	H56K	9.57	6.66	9.55	7.63	9.89	7.71	9.77	5.52
	H67K	9.83	5.36	9.53	4.97	9.88	8.04	9.73	3.54
	M60K	9.62	5.15	9.55	7.63	9.87	7.67	9.78	6.19
III	CGD	9.73	5.94	9.57	4.84	9.83	6.34	9.87	6.91
	HELV	9.78	4.08	9.66	4.27	9.79	4.16	9.89	6.83
	MGFAP	9.39	6.84	9.62	4.07	9.78	5.47	9.89	5.92
IV	PNF-L	9.73	3.51	9.83	3.02	9.73	3.77	9.87	6.01
	PNF-M	10.12	5.94	9.69	3.98	9.68	3.27	9.87	3.32
	PNF-H	—	—	—	—	—	—	—	—
I		9.31 ± 0.06	8.52 ± 1.03	9.59 ± 0.08	7.90 ± 1.60	9.87 ± 0.06	5.64 ± 0.88	9.86 ± 0.08	3.11 ± 1.41
II		9.68 ± 0.10	5.70 ± 0.77	9.52 ± 0.05	7.28 ± 2.11	9.88 ± 0.02	7.09 ± 1.37	9.78 ± 0.04	5.55 ± 1.52
III		9.63 ± 0.21	5.62 ± 1.41	9.62 ± 0.05	4.39 ± 0.40	9.80 ± 0.03	5.32 ± 1.10	9.88 ± 0.01	6.55 ± 0.55
IV		9.93 ± 0.28	4.73 ± 1.72	9.76 ± 0.10	3.50 ± 0.68	9.71 ± 0.04	3.52 ± 0.35	9.87 ± 0.00	4.67 ± 1.90
I, II, III, and IV		9.56 ± 0.25	6.62 ± 1.84	9.59 ± 0.10	6.50 ± 2.27	9.84 ± 0.07	5.76 ± 1.46	9.84 ± 0.07	4.62 ± 1.88

[a]Overall periodicities in segments 1B and 2 are 9.578 ± 0.187 and 9.839 ± 0.070 respectively.

residues, respectively (Table V); the periods differ in phase, however, by about 180° (Parry *et al.*, 1977). Thus, segment 1B may be visualized as alternating stripes of negative and positive charge with a period of about 9.58 residues (1.42 nm). The period for the acidic and the basic residues in segment 2 is 9.84 residues (1.46 nm) in both cases and, once again, the periods differ in phase by 180° (McLachlan and Stewart, 1982; Crewther *et al.*, 1983; Dowling *et al.*, 1983; Parry and Fraser, 1985). It is important to note that the period in segment 2 is continuous from one end to the other and does not suffer a phase change across the link segment L2 or at the center of segment 2B (Parry and Fraser, 1985). This observation is a very important one and implies that segment 2 is effectively a single piece of structure in spite of the two heptad discontinuities. It is also quite clear that the slightly different periods found in segments 1B and 2, which differ at the 99.9% significance level, will favor the aggregation through ionic interactions of segment 1B with itself (parallel or antiparallel) and of segment 2 with itself (parallel or antiparallel). However, 1B–2 interactions could also be favored (Parry and Fraser, 1985), if, for example, segment 2 was supercoiled about segment 1B in such a way that the axial projection of the two periods was the same. This could be achieved if segment 2 was supercoiled about an undistorted segment 1B with a pitch length of about 50 nm, thus making approximately one-third of a turn. Since this type of distortion is minimal, it seems probable that 1B–2 interactions could be as important as 1B–1B and 2–2 interactions.

The data described in this section have detailed the important features of the central α-helix-rich domain in IF protein chains. It is now important to use these data to establish the next level of structure in which the chains aggregate to form IF protein molecules.

5. *Structure of the Rod Domain of IF Protein Molecules*

Crick (1953) suggested that a unique meridional reflection with a spacing of 0.51 nm observed in the x-ray diffraction patterns of orientated α-fibrous proteins (which include IF) could be explained in terms of a regular close packing of lengths of α helix. Indeed, straight segments of right-handed α helix with about 3.6 residues per turn are able to pack closely, provided that their axes are inclined at about 20° to one another. If the α helices have a heptad repeat with apolar residues alternating three and four residues apart (i.e., in positions *a* and *d*), then an apolar stripe would be formed that would be inclined in a left-handed manner around the axes of the individual helices. Long-range close packing could thus be achieved if the α helices coil around one another in a left-handed manner. This coiled-coil rope structure has a regular knobs-into-holes packing of apolar side chains and is stabilized by the apolar interactions between chains. Much of the central α-helix-rich domain in the IF protein chains has such a heptad repeat and it seems highly probable that a coiled-coil structure of the general type discussed by Crick will be present in IF. It was noted earlier that the apolar residues in positions *a* and *d* of the heptad pattern were the most highly conserved in the IF chain sequences, a feature now seen to be readily consistent with the proposed coiled-coil structure.

Four questions, however, remain to be answered:

- How many chains are present in the IF molecules?
- What relative orientations do the chains have (parallel or antiparallel)?

- What is the relative axial displacement of the chains?
- What is the chain composition of molecules from different cell types?

Considerable progress has been made in the past few years in answering these questions and much of the remainder of this section will be concerned with these particular aspects of IF structure.

Early physicochemical studies had indicated that IF molecules in keratins contained three chains (Crewther and Harrap, 1967; Crewther and Dowling, 1971; Skerrow *et al.*, 1973; Steinert, 1978a; Steinert *et al.*, 1980), and on this basis it seemed probable that the molecules were three-stranded coiled-coil ropes. More recently, however, strong evidence has been presented for a four-chain aggregate consisting of a pair of two-stranded molecules (Ahmadi and Speakman, 1978; Ahmadi *et al.*, 1980; Gruen and Woods, 1983; Woods and Inglis, 1984; Parry *et al.*, 1985). Data from desmin (Geisler and Weber, 1982) and vimentin (Pang *et al.*, 1983) have also indicated a four-chain structure for these IF. In the case of the keratins, the evidence has been adduced from a variety of methods including cross-linking studies on wool (Ahmadi *et al.*, 1980) and on its proteolytic fragments (Ahmadi and Speakman, 1978; Woods and Gruen, 1981; Gruen and Woods, 1983), molecular weight studies of a chymotryptic digestion product (Woods and Gruen, 1981), and sequence analyses of α-helix-rich segments (Crewther *et al.*, 1978; Gough *et al.*, 1978). *In vitro* reassembly of a four-chain unit (Gruen and Woods, 1983) together with data on the molecular weight, cross-linking, and sequence (Woods, 1983; Woods and Inglis, 1984) provided convincing evidence for a four-chain unit consisting of a pair of two-stranded ropes. The modes of aggregation of molecules in the four-chain aggregate will be discussed in Section 7.

The relative orientation of the two chains and their relative axial register in the IF protein molecule have been established by a variety of methods. From purely theoretical considerations, the amount of coiled-coil structure, as judged by the extent of heptad overlap, is maximized when the chains are parallel and in axial register. In this situation, two 1A segments will be paired, two 1B segments will be paired, and so on. Furthermore, such an arrangement would result in a minimization of the potentially disruptive effects of the link segments and the heptad discontinuities in segment 2. If the chains were antiparallel, the degree of heptad overlap would be reduced because of the asymmetric distribution of the coiled-coil segments. The nonalignment of the link segments and the stutters in the heptad pattern would also tend to increase the amount of structural discontinuity. Direct evidence for parallel chains in the IF molecules comes from the work of Woods and Inglis (1984) on wool keratin and Parry *et al.* (1985) on epidermal keratin. These workers produced enzymatic digests that contained two-chain particles consisting mainly of segment 2B, and four-chain particles consisting primarily of segment 1B. Sequence studies of these particles showed that the chains cannot be antiparallel. Thus, IF protein molecules are two-chain parallel structures. The precise axial alignment of the two chains could not be determined from these studies but the stagger was deduced to be small. Two pieces of data, however, indicate that the chains are in exact axial register. The first of these comes from a theoretical analysis of the possible number of interchain ionic interactions calculated as a function of chain polarity and relative axial displacement. The results clearly show that these interactions are maximized for a parallel in-register arrangement of IF protein chains (Table VI). Many of these ionic interactions involve residues in positions *e* and *g*, which are the most highly conserved positions in the heptad

Table VI. Interchain Ionic Interactions as a Function of Relative Axial Stagger between Similarly Directed IF Protein Chains[a]

Stagger (heptads)	8c-1/ 7c	M59K/ M67K	CGD/ CGD	HELV/ HELV	MGFAP/ MGFAP	PNF-L/ PNF-L	PNF-M/ PNF-M	PNF-L/ PNF-M	LMN A/ LMN A
−2	2	0	−4	0	−8	1	5	3	8
−1	2	5	2	5	1	−1	−3	−1	−4
0	10	9	14	16	18	12	14	12	36
1	2	5	2	5	1	−1	−3	−2	−4
2	4	0	−4	0	−8	1	5	3	8

[a]8c-1 and 7c are Type I and Type II chains from wool keratin; M59K and M67K are Type I and Type II chains from mouse epidermal keratin; CGD, HELV, and MGFAP are Type III chains from chicken gizzard desmin, hamster eye lens vimentin, and mouse glial fibrillary acidic protein; PNF-L and PNF-M are Type IV chains from the light and medium components of porcine neurofilaments; LMN A is a Type V chain from the human nuclear lamina.

pattern after the residues in positions *a* and *d*. The second piece of data is obtained from studies on the *in vitro* oxidation of cysteine residues in IF chains through the use of 1,10-phenanthroline cupric ion complexes (Quinlan and Franke, 1982, 1983) to provide disulfide-bonded dimers. Characterization of these dimers has subsequently been shown by Parry *et al.* (1985) to be consistent only with an in-register chain arrangement. These data, therefore, provide strong support that IF protein molecules contain a pair of parallel chains in exact axial register.

The chain composition of IF molecules in desmin, vimentin, and glial fibrillary acidic protein is unequivocal. Only Type III chains are necessary for filament formation *in vivo* and *in vitro*. The molecules must therefore be homodimers. Neurofilaments contain only Type IV chains; however, as stated earlier, the molecules may be either homodimers (NF-L/NF-L, NF-M/NF-M, or NF-H/NF-H) or heterodimers (NF-L/NF-M, NF-L/NF-H, or NF-M/NF-H). *In vitro* evidence has been presented showing that the neurofilament chains of a single species are able to form homopolymeric IF (e.g., see Zackroff and Goldman, 1980; Shaw and Weber, 1982). However, it has not yet been established that neurofilament molecules *in vivo* are homopolymers but it is a reasonable hypothesis that this is so. The keratins are more complex than the neurofilaments and their chain composition must be considered in terms of the rules that govern their expressions in epithelial tissue (Moll *et al.*, 1982; Woodcock-Mitchell *et al.*, 1982; Wu *et al.*, 1982; Eichner *et al.*, 1984; Sun *et al.*, 1984; Steinert and Parry, 1985; Steinert *et al.*, 1985b). These rules include the observations that all keratin chains may be classified as Type I or Type II, that both Type I and Type II chains are coexpressed in all epithelia, and that both chain types are necessary for filament formation both *in vitro* and *in vivo*. Further, any given epithelium expresses only a few chains at any particular stage of development and in most cases a particular pair of Type I and Type II chains is expressed. The fundamental question, therefore, is whether the four-chain aggregate in keratins contains a Type I homodimer and a Type II homodimer or whether all of the molecules are identical, i.e., Type I/Type II obligate heterodimers. Much of the physicochemical evidence appears to be compatible with either possibility though the latter option is now clearly favored (Crewther *et al.*, 1983; Woods and Inglis, 1984; Quinlan *et al.*, 1984; Parry *et al.*, 1985; Steinert *et al.*, 1989; Sparrow *et al.*, 1989). Consideration, however, must also be given to the possibility that both homodimers *and* heterodimers exist *in vivo*. Such a possibility, if confirmed, would not be unprecedented in fibrous protein structures: $\alpha\alpha$ homodimers and $\alpha\beta$ heterodimers exist in native tropomyosin molecules, and $[\alpha 1(\mathrm{I})]_3$ homotrimers and $[\alpha 1(\mathrm{I})_2\alpha 2]$ heterotrimers occur in native Type I collagen molecules.

From consideration of all these points, it seems certain that the α-helix-rich domains from two IF protein chains aggregate in a parallel mode and in axial register to form the "rod" domain of the IF protein molecule. The coiled-coil segments 1A, 1B, 2A, and 2B contain about 5, 14–15, 2–3, and 17–18 heptads, respectively, and have lengths of about 5.2, 15, 2.8, and 18 nm, respectively, assuming that the average axial rise per residue is 0.1485 nm. If the linking segments L1, L12, and L2 also have the same axial rise per residue, then the overall length of both segment 1 and segment 2 will be about 21–22 nm and that of the rod domain as a whole will be about 46–47 nm. Indeed, the lengths of the IF molecules as determined by electron microscopy do lie in the range 40–50 nm, with a best estimate of about 48 nm (Steinert *et al.*, 1980; Steinert, 1981; Geisler *et al.*, 1982a, 1985b; Quinlan *et al.*, 1984; Ip *et al.*, 1985). This suggests that the coiled-coil segments in the rod domain are linearly arranged and that they do not form a compact bundle of oppositely directed segments. It also suggests that there is poor preservation in the electron microscope of the N- and C-terminal domains of the IF molecules. These observations are consistent with the results of Steinert (1978a,b) who isolated a pair of α-helix-rich fragments from bovine epidermal keratin. One of these (known as Particle 1) was 40 nm long and 2 nm in diameter. The second (known as Particle 2) was 18 nm long and 2 nm wide. From these data, Steinert predicted that epidermal keratin molecules contained two α-helix-rich segments, each about 18 nm long, separated by non-α-helical structure. N- and C-terminal domains of less regular structure were also predicted. Other support for this model of IF molecular structure comes from the work of Crewther and colleagues, who isolated two α-helical-rich fragments from enzymatically degraded *S*-carboxymethylated wool keratin. Oriented fibers of these fragments gave x-ray diffraction patterns showing a 0.51-nm meridional reflection characteristic of a coiled-coil rope and a low-angle meridional pattern indexing on a 16-nm period (Suzuki *et al.*, 1973). Since the purified fragments were about 100–110 residues long and corresponded largely to the heptad-containing 1B segments of Type I and Type II chains, it follows that the length of the fragments will be close to 16 nm, i.e., the period in the x-ray pattern.

From all these considerations, it seems reasonable that the rod domain of IF protein molecules is largely a coiled-coil rope stabilized by apolar interactions (Fig. 4a). However, its continuity is interrupted by irregularities in the vicinity of links L1 and L12. It is also likely that the molecular structure will be affected by the discontinuities in the heptad phasing at link L2 and at a point toward the center of segment 2B. McLachlan and Karn (1983) have suggested that a similar feature observed in the sequence of nematode myosin might be accounted for by a localized alteration in the pitch length of the coiled-coil rope. This explanation may also be valid for the IF proteins. It is possible that the two sites of heptad discontinuity may represent points at which there is a sharp kink in the direction of the axis of the coiled-coil rope (Fig. 4b,c). Indeed, some major disturbance of the geometry of the coiled-coil rope would be required at the link L2 if the integrity of the apolar-stabilized structure is to be maintained (Fig. 4c).

Having established the main features of the rod domain of IF molecules, it is now important to consider the role of the N- and C-terminal domains and also their structural characteristics.

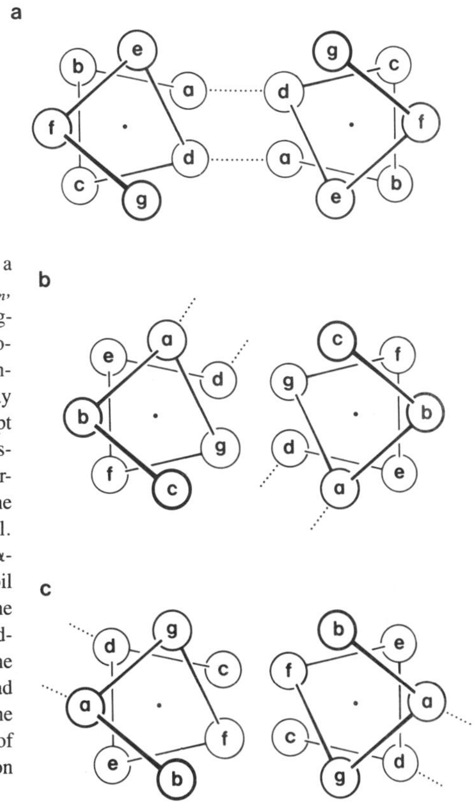

Figure 4. Segments 1A, 1B, 2A, and 2B each have a heptad pattern of residues of the form $(a\text{-}b\text{-}c\text{-}d\text{-}e\text{-}f\text{-}g)_n$, where *a* and *d* are commonly apolar residues. In a regular coiled-coil rope, residues in positions *a* and *d* provide stability through the formation of hydrophobic interactions, as indicated in panel (a). About halfway along segment 2B, the heptad pattern suffers an abrupt phase change. This would lead to the situation illustrated in (b) unless the axis of the coiled-coil rope undergoes a change in direction or the local geometry of the α-helical chains is significantly modified from normal. Furthermore, if the link L2, which is predicted to be α-helical, is considered as part of a continuous coiled-coil rope from the beginning of segment 2A through to the end of segment 2B, then the continuation of the coiled-coil geometry would result in the apolar stripe in the surface of each α-helical strand facing outwards instead of inwards (c). A major discontinuity of some sort in the coiled-coil structure must be postulated in the vicinity of link L2. From Parry and Fraser (1985) with permission of Butterworths Press.

6. *Structure of the Terminal Domains of IF Protein Molecules*

The N- and C-terminal domains in IF protein chains vary greatly with respect to size, volume, charged residue content, and chemical character (Tables II and III; Parry and Fraser, 1985). Type I IF protein chains have the smallest N-terminal domains (the smallest currently known is 55 residues for component 8c-1 from wool keratin). However, the C-terminal domain of the bovine keratin chain with a molecular weight of about 40k (Bader *et al.*, 1986) contains only nine residues. The largest N-terminal domain so far characterized (179 residues) is for a Type II human epidermal keratin of molecular weight 67k (K1). However, the Type IV neurofilament chains, NF-M and NF-H, have massive C-terminal domains containing approximately 500 and 700–800 residues, respectively (Geisler *et al.*, 1984, 1985a; Lees *et al.*, 1988; Shneidman *et al.*, 1988). Excluding the Type IV chains, the volumes of the N- and C-terminal domains lie in the range 7–21 and 6–17 nm^3, respectively (Parry and Fraser, 1985), with the N-terminal domain always appearing to be a little larger than its C-terminal counterpart both in volume and in size.

The total content of charged residues in both the N- and C-terminal domains varies appreciably from one chain type to another though, excluding the effect of phosphorylation, the domains are net basic in all cases other than for the C-terminal domain of Type IV neurofilament chains (Table III). The percentage of charged residues in the N-terminal domain varies from about 6% (Type I) to 20% (Types III and IV) but the variation in the C-terminal domain is even more pronounced—8% (Type II), 18% (Type I), 32% (Type III), and > 40% (Type IV).

The chemical character of the terminal domains also varies markedly among the IF chains. In hard keratins these domains are characterized by the presence of many cysteine residues that form disulfide bonds both within the terminal domains and with the cysteine-rich matrix proteins in which the IF are embedded. The C-terminus of component 8c-1 has an imperfect triplet repeat of the form (Pro-Cys-X) (Dowling *et al.*, 1986), which suggests that a degree of regularity in the disulfide bond disposition may occur. The structures of the terminal domains in wool keratin are not known but it is clear that a disulfide-bonded stabilized aggregate provides an ideal method to obtain an insoluble, hard, and rigid structure (Steinert and Parry, 1985; Steinert *et al.*, 1985b; see also Fraser *et al.*, 1988, for a discussion of the likely disposition of disulfide bonds in α-keratin).

The terminal domains in some of the epidermal keratin chains are also most unusual as they contain segments with highly elevated contents of glycine and serine residues. This frequency results in a number of repeating sequences, some as long as 28 residues (Steinert *et al.*, 1983; Parry and Fraser, 1985; see also Jorcano *et al.*, 1984). The nature of the repeats which contain but a few amino acid types is rarely unique and often alternative ways exist in which a variety of repeats of different lengths may be specified. Those segments rich in glycine adopt an irregular and highly convoluted structure that is both flexible and highly apolar (Conway *et al.*, 1988; Mack *et al.*, 1988; Conway and Parry, 1989; Steinert *et al.*, 1989). It has been cogently argued that such sequences are ideally suited for the epidermis whose role is to provide a flexible, protective barrier between an animal and its environment (Steinert and Parry, 1985).

The C-terminal domains in neurofilament proteins (Type IV chains), besides being unique in size, have extensive regions of sequence containing predominantly glutamic acid residues. This feature accounts for both the high charged residue content and the net acidic character of the chains noted earlier.

The variability in the size, volume, and chemical character of the terminal domains is significant both structurally and functionally. The "internal" volume of the IF may be as low as 15 nm^3/molecule (see model illustrated in Fig. 9 of Fraser *et al.*, 1985) whereas in a two-chain IF molecule the total volume of the terminal domains is in the range 26–76 nm^3 (Parry and Fraser, 1985). It follows that much of the terminal domains must lie on the surface of the IF and be available to interact with IF-associated proteins (IFAP) or other IF. Direct evidence for the external location of the ends has been obtained by Steinert *et al.* (1983, 1985a). In that work they were able to cleave glycine-rich fragments from mouse epidermal keratin IF without affecting the apparent structural integrity of the filament. Since the sequences of the peptides liberated showed that they originated from both the N- and C-terminal domains, it seems highly probable that the terminal domains have a peripheral location both *in vitro* and *in vivo*.

The predicted secondary structure of the terminal domains yields no common feature among the different IF protein chains. In general, there is little predicted α-helical structure, the only exceptions being the C-terminal domains of Type III IF chains and part of the C-terminal domains of some Type I IF chains. These regions of α-helical structure do not have the heptad repeat characteristic of the coiled-coil segments in the rod domain. The terminal domains, which contain mostly β, turn, and coil structure, probably adopt a compact structure with many reversals of chain direction. In contrast, more elongated structures (or bridges) are frequently seen linking parallel neurofilaments in the axon and these, almost certainly, comprise in part the large negatively charged C-terminal domains of the NF-H chains (Willard and Simon, 1981; Sharp *et al.*, 1982; Debus *et al.*, 1982; Hirokawa *et al.*, 1984). Indeed, it is predicted that any highly charged protein domain should adopt a conformation with an appreciable axial ratio since this tends to maximize the surface area of the domain and allows the charged side chains to be hydrated.

Studies of the terminal domains reveal a distinct subdomainal structure with a marked degree of bilateral symmetry (Fig. 3). Commencing from the amino-terminal end of the chain, the subdomains in the N-terminal and the C-terminal domains are designated E1, V1, H1 and H2, V2, and E2, respectively. In Type II chains the H1 and H2 subdomains are 36- and 20-residue sequences adjacent to the rod domain and are homologous in all chains of that type. H1 has a sequence

PVCSPGGIQEVTINQSLLTPLNL.IDP.IQKVK.EE

On the basis of the current sequence data, however, mammalian H2 sequences fall into two classes:

SGEC.PNVSV.VSTSHTSMS [H65K (K3), M67K, H67K (K1)]
NGEGVG.VNISVVQSTVSSG [H55K (K7), H56K (K6a,b), H58K(K5), M60K]

though it remains to be seen whether this is general or merely a feature of the limited data base available. Limited homology does occur between these two 20-residue H2 subdomains. V1 and V2 subdomains lie on the distal sides of the H1 and H2 subdomains, respectively; they have variable lengths and amino acid compositions and are often highly elevated in a few particular amino acids, such as glycine and serine. A feature of these V subdomains is the presence of repeating sequences (Steinert *et al.*, 1983; Jorcano *et al.*, 1984; Parry and Fraser, 1985). The E1 and E2 subdomains are short, generally basic regions at the amino- and carboxy-terminal ends of the chain. The Type I IF chains either lack or have minimal H1 and H2 subdomains and this generally accounts for the lower molecular weight of a Type I chain compared to its coexpressed Type II chain counterpart (Steinert and Parry, 1985; Conway and Parry, 1988; Steinert and Roop, 1988). Although three Type I IF protein chains (H50K, M50K, M55K) have essentially identical C-terminal domains, these bear no relation to those in other Type I chains. It has also been noted by Winkles *et al.* (1985) that a short piece of C-terminal sequence $\genfrac{}{}{0pt}{}{\mathrm{E\,VH}}{\mathrm{DTR}}\mathrm{DG}\genfrac{}{}{0pt}{}{\mathrm{K}}{\mathrm{E}}\mathrm{VVS}$ is homologous in Type I and Type III chains. Within chains of a particular type, variation in molecular weight can be largely explained by differences in the size of the V subdomains. Type III chains also have H1 and H2 subdomains; the former is 20 residues long and has a sequence

VDFSLADA.NTEFK.TRTNE

whereas the latter effectively encompasses the entire C-terminal domain. Type IV chains as a whole have short H1 and H2 subdomains (9 and 13 residues, respectively; Conway and Parry, 1988) though there is strong evidence of long and highly homologous H1 and H2 subdomains in the NF-L, NF-M, and NF-H chains considered as separate subclasses (Conway and Parry, 1988). A pattern is thus beginning to emerge of classes of terminal domain within particular chain types (Jorcano *et al.*, 1984) though more data are required to establish this point.

Studies on the folding characteristics of proteins, both fibrous and globular, suggest that the N-terminal and the C-terminal domain of each IF protein chain will fold separately into a distinct structural entity. Each two-stranded IF protein molecule will therefore have two globular heads at its amino- *and* its carboxy-terminal end. High-resolution electron microscopy, in conjunction with conventional three-dimensional reconstruction techniques should allow this prediction to be tested experimentally.

7. *Structure of the Four-Chain Aggregates*

It has already been established that IF molecules contain two parallel chains in axial register and that four-chain aggregates can be isolated from intact IF or from proteolytically degraded fragments of IF. The manner in which the pair of IF molecules are arranged in these aggregates, with particular regard to their relative polarity and their relative axial shift, is therefore of some importance structurally.

Segments 1B and 2 both contain a regular, linear disposition of acidic and basic residues with a period of about 1.4–1.5 nm (Parry *et al.*, 1977; McLachlan and Stewart, 1982; Dowling *et al.*, 1983; Crewther *et al.*, 1983). It is a reasonable assumption that molecular aggregation will largely be specified by ionic interactions of the type 1BU–1BU, 1BU–1BD, 2U–2U, 2U–2D, 2U–1BU, and 2U–1BD though, of course, the contribution to filament stability provided by specific hydrogen-bonding and nonspecific van der Waals interactions will clearly be very important in the final stages of docking. The terms "U" and "D" refer to up-pointing and down-pointing segments, respectively. Only five molecular arrangements are worthy of serious consideration: in two of these the molecules are parallel to one another (1BU–1BU and 2U–2U, Fig. 5a; 1BU–2U, Fig. 5b) and in the remaining three the molecules are antiparallel [1BU–2D (twice), Fig. 5c; 1BU–1BD, Fig. 5d; 2U–2D, Fig. 5e]. Furthermore, in some of these arrangements the molecules are largely overlapped (Fig. 5a,c) and in the others the molecules are approximately half-staggered (Fig. 5b,d,e). The lengths of the four-chain aggregates are thus either about 47 or 70 nm; both lengths have been observed experimentally in *in vitro* studies (Ahmadi *et al.*, 1980; Potschka *et al.*, 1984; Quinlan *et al.*, 1984; Ip *et al.*, 1985; Geisler *et al.*, 1985b).

Interestingly, a study of β-keratin, prepared by stretching α-keratin in steam, has direct relevance to the discussion of chain polarity in IF. Fraser *et al.* (1969) showed that the β-keratin crystallites contained the same number of chains in cross section as did the IF. Thus, each β crystallite appears to be derived from an individual IF. Since the β crystallites were shown to contain oppositely directed chains, it was argued that IF will also contain oppositely directed (antiparallel) chains (Crewther *et al.*, 1983). Since this does not occur at the molecular level, it follows that pairs of molecules must be antiparallel (Crewther *et al.*, 1983; Fraser and MacRae, 1983).

This conclusion is also consistent with the interpretation of the x-ray diffraction data from hard α-keratin in terms of perpendicular diads relating pairs of molecules (Fraser and MacRae, 1983). Further theoretical evidence favoring an antiparallel arrangement of molecules in the four-chain aggregates comes from studies on the intermolecular ionic interactions (Fraser *et al.*, 1985), though parallel arrangements were not excluded by these

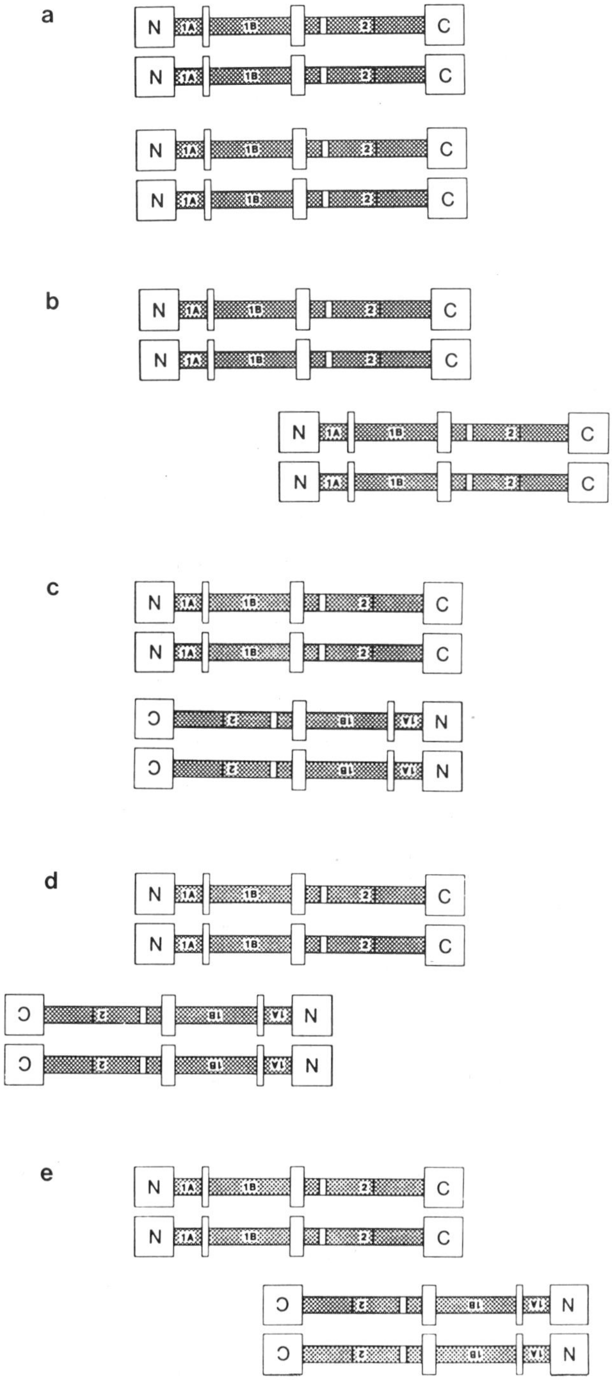

Figure 5. The two-stranded, parallel-chained, IF protein molecules that form four-chain aggregates may in theory be arranged either parallel to one another [(a) and (b)] or antiparallel [(c), (d), and (e)]. In each case the axial alignment of the molecules is governed by the maximization of ionic interactions (salt bridges) between the regularly disposed acidic and basic residues in segments 1B and 2. Excluding the N- and C-terminal domains, the lengths of the four-chain aggregates are either about 45–48 nm [(a) and (c)] or about 1.5 times this length, i.e., 67–70 nm [(b), (d), and (e)]. In these latter cases the molecules are approximately half staggered with respect to one another.

studies. Direct experimental support for an antiparallel arrangement of molecules, however, came from the work of Geisler *et al.* (1985b). They used a monoclonal antibody against the 2B segment of desmin to show that the 48-nm-long aggregates seen in the electron microscope were labeled at both ends (Fig. 5b). Evidence in favor of a second mode of aggregation, i.e., a half-staggered antiparallel arrangement of molecules, comes from studies on α-helix-rich particles derived from both hard and soft (epidermal) α-keratins (Woods and Inglis, 1984; Parry *et al.*, 1985). In these studies, tetramers of 1B segments and dimers of 2B segments were isolated and characterized. These results are consistent with both the antiparallel (Fig. 5d) and parallel models (Fig. 5a) though the former has been favored not only for the reasons stated earlier but also from the fact that only 2B dimers, and not tetramers, were isolated experimentally. The data as a whole thus provide convincing support for the concept of antiparallel molecules that are either half-staggered or almost completely overlapped with respect to their neighbors in the intact IF.

8. *Nuclear Lamins: IF-Like Proteins*

Other IF-like proteins recently sequenced *in toto* are lamins A and C, two of the three major components of mammalian nuclear envelope (McKeon *et al.*, 1986; Fisher *et al.*, 1986) and the lamin B chain from *Xenopus* (Krohne *et al.*, 1987). Detailed analysis of lamins A and C by McKeon *et al.* (1986) and by Parry *et al.* (1986) has revealed that although there are similarities between the lamins and the IF proteins, there are also significant differences (Fig. 6, Table VII). The similarities include the generally similar lengths, disposition, and structures of the segments in the rod and terminal domains, a highly conserved sequence in segment 1A and at the C-terminal end of segment 2B, and the presence of a highly significant 10-residue period in the linear distribution of the acidic and the basic residues in the long coiled-coil segments. The major differences between lamin and the IF proteins can be summarized as follows: (1) segment 1B in lamin is 42 residues (6 heptads) longer than in the IF proteins though the N- and C-terminal portions of segment 1B are homologous in both protein types (Table VII); (2) the lamin proteins have a higher charged residue content in the rod domain that the IF proteins; (3) a continuous heptad substructure exists for the lamins from the N-terminal end of segment 1A to the C-terminal end of segment 1B though a discontinuity in phasing does occur in

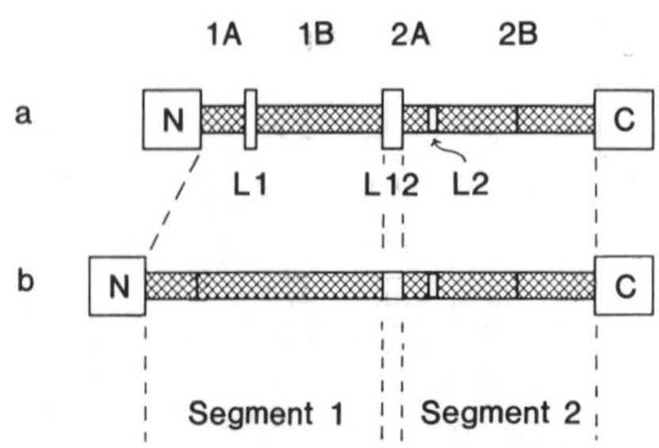

Figure 6. Schematic representation of the structure of (a) Type I–IV and (b) Type V IF protein chains. N and C represent the non-α-helical end domains that are often basic in character. The rod domain lying between N and C has a high predicted α-helix content and also contains a heptad substructure (shown cross-hatched). Link segments L1 and L12 in Type I–IV IF chains are non-α-helical in contrast to those in Type V chains. Breaks in heptad phasing occur at link L2 and at the center of segment 2B in (a) and at the previous site of link L1, also L12, and again at L2 and the center of segment 2B in (b). A well-defined linear periodicity occurs in the disposition of the acidic and the basic residues in segments 1B and 2 in (a) and over the entire rod domain in (b). From Parry *et al.* (1986) with permission of *Biochemical Journal*.

Table VII. Domain and Structure Comparison of IF Protein Chains and Lamins A and C

Domain/ segment	IF proteins		Lamins A and C
N	20–179 residues, basic nonhelical		30 residues, basic nonhelical
1A	35 residues, α-helical, continuous heptad structure contains highly conserved sequence		35 residues, α-helical, continuous heptad structure extending into L1, contains the highly conserved sequence found in IF proteins, 10-residue periodicity in the linear disposition of the acidic and basic residues (Table I) continuous with that in segment 1B
L1	8–15 residues, nonhelical		11 residues, α-helical with a break in heptad continuity that nonetheless allows segment 1 (1A–L1–1B) to be designated a single block of structure
1B	101 residues, α-helical, acidic, continuous heptad structure, regular linear disposition of acidic and basic residues with 9.55-residue period		143 residues, α-helical, neutral, continuous heptad structure extending back into L1, 10-residue periodicity in the linear disposition of the acidic and basic residues continuous with that in segment 1A (Table I)
	• N-terminal 42 residues	$\equiv$	• N-terminal 42 residues
	• Central 28 residues	$\not\equiv$	• Central 70 residues (note the insert of six heptads)
	• C-terminal 31 residues	$\equiv$	• C-terminal 31 residues
L12	16–17 residues, nonhelical		19 residues, α-helical
2	148 residues, α-helical, heptad structure with breaks in continuity at L2 and at the center of 2B, highly conserved C-terminal four heptads, regular linear disposition of acidic and basic residues with 9.85-residue period		148 residues, α-helical, heptad structure with breaks in continuity at L2 and at the center of 2B, same highly conserved C-terminal four heptads as found in IF proteins, 10-residue period in the linear disposition of the acidic and basic residues (Table I)
C	20–800 residues, basic nonhelical		277 residues (lamin A), 185 residues (lamin C), basic nonhelical

the α-helical heptad containing segment L1: in the IF proteins segment L1 represents a non-α-helical, less regular piece of structure; and (4) the number of interchain ionic interactions for a homodimeric lamin molecule is substantially greater than for any of the IF molecules.

The sequence relationship between lamin and IF proteins is clearly of great interest and suggests that they share many structural features as well as a common ancestor. Indeed, lamin chains have been designated Type V IF protein chains. The differences between the proteins, however, imply that the aggregation of molecules differs in significant respects. More detailed structural studies will be required to clarify these points but a recent study of the various *in vivo* and *in vitro* assemblies of lamin proteins (Aebi *et al.*, 1986; Parry *et al.*, 1987) has revealed that the predominant mode of assembly involves an antiparallel half-stagger arrangement of molecules. A similar structural pattern has also been established for other IF proteins.

9. Conclusions

The past few years has seen a rapid advance in knowledge about the structure of IF proteins. Many amino acid sequences have been completed and, from these, a picture has emerged of a common, heptad-containing, α-helix-rich domain of length 46–47 nm flanked by N- and C-terminal domains of distinctive composition and structure. These end domains provide the specificity of the IF and define its function *in vivo*. The α-helix-rich segments in the IF protein chains aggregate to form a coiled-coil rope containing two, in-register, parallel strands. Since the strands are in register, the radius of the coiled coil varies with the sizes of the stabilizing apolar residues in positions *a* and *d* of the heptad pattern. A dominant periodicity found in the linear disposition of the acidic residues and the basic residues in the two major segments of coiled coil strongly suggested that IF molecules aggregate through periodic ionic interactions. Indeed, theoretical and experimental work has now shown that two common modes of molecular aggregation occur *in vitro*. These arrangements, which are stabilized by salt bridges between the regularly disposed acidic and basic residues in segments 1B and 2, comprise a pair of antiparallel molecules that are (1) almost completely overlapped or (2) approximately half-staggered with respect to one another. The presence of substructure within the terminal domains indicates that a portion of these structures may also be involved in specifying the axial and lateral organization of the molecules in the filament.

Searches for sequence homology have revealed that IF protein chains may be divided into five types (Types I–V) on the basis of the extent of similarity within both the helical and nonhelical segments of the rod domain. Keratins contain Type I and Type II chains and both are required for filament formation. Further subdivision is possible on the basis of homology: Type Ia and IIa chains occur in hard α-keratins and Type Ib and IIb chains occur in soft epidermal α-keratins. Desmin, glial fibrillary acidic protein, and vimentin filaments contain only Type III chains. Neurofilaments contain only Type IV chains but in this case three distinct chains exist within the Type IV classification. The nuclear lamina contains Type V chains.

The molecular structure of IF proteins has two features that are incompletely understood. One of these relates to the coiled-coil pitch length of the portion of the IF molecule comprising the rod domain and the other relates to the interpretation of the heptad discontinuities in segment 2. Much more remains to be established. In particular, the structure of the terminal domains is worthy of more detailed study than has yet been undertaken since it is these domains that are important in specifying the function of the intact IF. The main role of the rod domains, which are packed in the body of the IF, is clearly to provide the spatial framework upon which the terminal domains are able to function appropriately.

References

Aebi, U., Cohn, J., Buhle, L., and Gerace, L., 1986, The nuclear lamina is a meshwork of intermediate-type filaments, *Nature* **323:**560–564.

Ahmadi, B., and Speakman, P. T., 1978, Suberimidate crosslinking shows that a rod-shaped, low cystine, high-helix protein prepared by limited proteolysis of reduced wool has four protein chains, *FEBS Lett.* **94:**365–367.

Ahmadi, B., Boston, N. M., Dobb, M. G., and Speakman, P. T., 1980, Possible four-chain repeating unit in the microfibril of wool, in: *Fibrous Proteins: Scientific, Industrial and Medical Aspects,* Volume 2 (D. A. D. Parry and L. K. Creamer, eds.), Academic Press, New York, pp. 161–166.

Bader, B. L., Magin, T. M., Hatzfeld, M., and Franke, W. W., 1986, Amino acid sequence and gene organisation of cytokeratin No. 19, an exceptional tail-less intermediate filament protein, *EMBO J.* **5:**1865–1875.

Burley, S. K., and Petsko, G. R., 1985, Aromatic–aromatic interaction: A mechanism of protein stabilization, *Science* **229:**23–28.

Chou, P. Y., and Fasman, G. D., 1978, Prediction of the secondary structure of proteins from their amino acid sequence, *Adv. Enzymol.* **47:**45–148.

Cohen, C., and Parry, D. A. D., 1986, α-helical coiled-coils—A widespread motif in proteins, *Trends Biochem. Sci.* **11:**245–248.

Conway, J. F., and Parry, D. A. D., 1988, Intermediate filament structure: 3. Analysis of sequence homologies, *Int. J. Biol. Macromol.* **10:**79–98.

Conway, J. F., and Parry, D. A. D., 1989, Structure and spatial organisation of intermediate filament and nuclear lamin molecules, in: *Cytoskeletal and Extracellular Proteins: Structure, Interactions and Assembly,* Springer Series in Biophysics (U. Aebi and J. Engel, eds.), Volume 3, Springer-Verlag, Berlin, pp. 140–149.

Conway, J. F., Fraser, R. D. B., MacRae, T. P., and Parry, D. A. D., 1988, Protein chains in wool and epidermal keratin IF: Structural features and spatial arrangement, in: *Biology of Wool and Hair* (G. E. Rogers, P. J. Reis, K. A. Ward, and R. C. Marshall, eds.), Chapman & Hall, London, pp. 127–144.

Crewther, W. G., and Dowling, L. M., 1971, The preparation and properties of large peptides from the helical regions of the low-sulphur proteins of wool, *Appl. Polym. Symp.* No. 18, 1–20.

Crewther, W. G., and Harrap, B. S., 1967, The preparation and properties of a helix-rich fraction obtained by partial proteolysis of low-sulphur S-carboxymethylkerateine from wool, *J. Biol. Chem.* **242:**4310–4319.

Crewther, W. G., Inglis, A. S., and McKern, N. M., 1978, Amino acid sequences of α-helical segments from S-carboxymethylkerateine A. Complete sequence of a type II segment, *Biochem. J.* **173:**365–371.

Crewther, W. G., Dowling, L. M., Steinert, P. M., and Parry, D. A. D., 1983, The structure of intermediate filaments, *Int. J. Biol. Macromol.* **5:**267–274.

Crick, F. H. C., 1953, The packing of α-helices: Simple coiled-coils, *Acta Crystallogr.* **6:**689–697.

Debus, E., Flugge, G., Weber, K., and Osborn, M., 1982, A monoclonal antibody specific for the 200 kilodalton polypeptide of the neurofilament triplet, *EMBO J.* **1:**41–46.

Dowling, L. M., Parry, D. A. D., and Sparrow, L. G., 1983, Structural homology between hard α-keratin and the intermediate filament proteins, desmin and vimentin, *Biosci. Rep.* **3:**73–78.

Dowling, L. M., Crewther, W. G., and Inglis, A. S., 1986, The primary structure of component 8c-1, a subunit protein of intermediate filaments in wool keratin, *Biochem. J.* **236:**695–703.

Eckert, R. L., 1988, Sequence of the human 40-kDa keratin reveals an unusual structure with very high sequence identity to the corresponding bovine keratin, *Proc. Natl. Acad. Sci. USA* **85:**1114–1118.

Eichner, R., Bonitz, P., and Sun, T. T., 1984, Classification of epidermal keratins according to their immunoreactivity, isoelectric point and mode of expression, *J. Cell Biol.* **98:**1388–1396.

Fisher, D. Z., Chaudhary, N., and Blobel, G., 1986, cDNA sequencing of nuclear lamins A and C reveals primary and secondary structural homology to intermediate filament proteins, *Proc. Natl. Acad. Sci. USA* **83:**6450–6454.

Fraser, R. D. B., and MacRae, T. P., 1976, The molecular structure of feather keratin, Proc. 16th Ornithological Congress, Canberra, Aust. Acad. Sci., pp. 443–451.

Fraser, R. D. B., and MacRae, T. P., 1983, The structure of the α-keratin microfibril, *Biosci. Rep.* **3:**517–525.

Fraser, R. D. B., MacRae, T. P., Parry, D. A. D., and Suzuki, E., 1969, The structure of β-keratin, *Polymer* **10:**810–826.

Fraser, R. D. B., MacRae, T. P., and Suzuki, E., 1976, Structure of the α-keratin microfibril, *J. Mol. Biol.* **108:**435–452.

Fraser, R. D. B., MacRae, T. P., Suzuki, E., and Parry, D. A. D., 1985, Intermediate filament structure: 2. Molecular interactions in the filament, *Int. J. Biol. Macromol.* **7:**258–274.

Fraser, R. D. B., MacRae, T. P., Sparrow, L. G., and Parry, D. A. D., 1988, Disulphide bonding in α-keratin, *Int. J. Biol. Macromol.* **10:**106–112.

Garnier, J., Osguthorpe, D. J., and Robson, B., 1978, Analysis of the accuracy and implications of simple methods for predicting the secondary structure of globular proteins, *J. Mol. Biol.* **120:**97–120.

Geisler, N., and Weber, K., 1981, Comparison of the proteins of two immunologically distinct intermediate sized filaments by amino acid sequence analysis: Desmin and vimentin, *Proc. Natl. Acad. Sci. USA* **78:**4120–4123.

Geisler, N., and Weber, K., 1982, The amino acid sequence of chicken muscle desmin provides a common structural model for intermediate filament proteins, *EMBO J.* **1:**1649–1656.

Geisler, N., and Weber, K., 1983, Amino acid sequence data on glial fibrillary acidic protein (GFA): Implications for the subdivision of intermediate filaments into epithelial and non-epithelial members, *EMBO J.* **2:**2059–2063.

Geisler, N., Kaufmann, E., and Weber, K., 1982a, Proteinchemical characterization of three structurally distinct domains along the protofilament unit of desmin 10 nm filaments, *Cell* **30:**277–286.

Geisler, N., Plessmann, U., and Weber, K., 1982b, Related amino acid sequences in neurofilaments and non-neuronal intermediate filaments, *Nature* **296:**448–450.

Geisler, N., Kaufmann, E., Fischer, S., Plessmann, U., and Weber, K., 1983, Neurofilament architecture combines structural principles of intermediate filaments with carboxy-terminal extensions increasing in size between triplet proteins, *EMBO J.* **2:**1295–1302.

Geisler, N., Fischer, S., Vanderkerckhove, J., Plessmann, U., and Weber, K., 1984, Hybrid character of a large neurofilament protein (NF-M): Intermediate filament type sequence followed by a long and acidic carboxy-terminal extension, *EMBO J.* **3:**2701–2706.

Geisler, N., Fischer, S., Vanderkerckhove, J., van Damme, J., Plessmann, U., and Weber, K., 1985a, Protein-chemical characterization of NF-H, the largest mammalian neurofilament component: Intermediate filament-type sequences followed by a unique carboxy-terminal extension, *EMBO J.* **4:**57–63.

Geisler, N., Kaufmann, E., and Weber, K., 1985b, Antiparallel orientation of the two double-stranded coiled-coils in the tetrameric protofilament unit of intermediate filaments, *J. Mol. Biol.* **182:**173–177.

Geisler, N., Plessmann, U., and Weber, K., 1985c, The complete amino acid sequence of the major neurofilament protein (NF-L), *FEBS Lett.* **182:**475–478.

Glass, C., Kim, K. H., and Fuchs, E., 1985, Sequence and expression of a human type II mesothelial keratin, *J. Cell Biol.* **101:**2366–2373.

Gough, K. H., Inglis, A. S., and Crewther, W. G., 1978, Amino acid sequences of α-helical segments from S-carboxymethyl-kerateine-A. Complete sequence of a type I segment, *Biochem. J.* **173:**373–385.

Gruen, L. C., and Woods, E. F., 1983, Structural studies on the microfibrillar proteins of wool, *Biochem. J.* **209:**587–595.

Hanukoglu, I., and Fuchs, E., 1982, The cDNA sequence of a human epidermal keratin: Divergence of sequence but conservation of structure among intermediate filament proteins, *Cell* **31:**243–252.

Hanukoglu, I., and Fuchs, E., 1983, The cDNA sequence of a type II cytoskeletal keratin reveals constant and variable domains among keratins, *Cell* **33:**915–924.

Hirokawa, N., Glicksman, M. A., and Willard, M. B., 1984, Organization of mammalian neurofilament polypeptides within the neuronal cytoskeleton, *J. Cell Biol.* **98:**1523–1536.

Hoffmann, W., and Franz, J. K., 1984, Amino acid sequence of the carboxy-terminal part of an acidic type I cytokeratin of molecular weight 51 000 from *Xenopus laevis* epidermis as predicted from the cDNA sequence, *EMBO J.* **3:**1301–1306.

Hoffmann, W., Franz, J. K., and Franke, W. W., 1985, Amino acid sequence microheterogeneities of basic (type II) cytokeratins of *Xenopus laevis* epidermis and evolutionary conservativity of helical and non-helical domains, *J. Mol. Biol.* **184:**713–724.

Hong, B., and Davison, P. F., 1981, Isolation and characterization of a soluble, immunoactive peptide of glial fibrillary acidic protein, *Biochim. Biophys. Acta* **670:**139–145.

Huber, R., 1979, Conformational flexibility and its functional significance in some protein molecules, *Trends Biochem. Sci.* **4:**271–276.

Ip, W., Hartzer, M. K., Pang, S. Y. Y., and Robson, R. M., 1985, Assembly of vimentin *in vitro* and its implications concerning the structure of intermediate filaments, *J. Mol. Biol.* **183:**365–375.

Jonas, E., Sargent, T. D., and Dawid, I. B., 1985, Epidermal keratin gene expressed in embryos of *Xenopus laevis, Proc. Natl. Acad. Sci. USA* **82:**5413–5417.

Jorcano, J. L., Rieger, M., Franz, J. K., Schiller, D. L., Moll, R., and Franke, W. W., 1984, Identification of two types of keratin polypeptides within the acidic cytokeratin subfamily I, *J. Mol. Biol.* **179:**257–281.

Julien, J. P., Ramachandran, K., and Grosveld, F., 1985, Cloning of a cDNA encoding the smallest neurofilament protein from the rat, *Biochim. Biophys. Acta.* **825:**398–404.

Julien, J. P., Grosveld, F., Yazdanbaksh, K., Flavell, D., Meijer, D., and Mushynski, W., 1987, The structure of a human neurofilament gene (NF-L): A unique exon–intron organization in the intermediate filament gene family, *Biochim. Biophys. Acta* **909:**10–20.

Klinge, E. M., Sylvestre, Y. R., Freedberg, I. M., and Blumenberg, M., 1987, Evolution of keratin genes: Different protein domains evolve by different pathways, *J. Mol. Evol.* **24:**319–329.

Knapp, B., Rentrop, M., Schweizer, J., and Winter, H., 1987, Three cDNA sequences of mouse type I keratins, *J. Biol. Chem.* **262:**938–945.

Krieg, T. M., Schafer, M. P., Cheng, C. K., Filpula, D., Flaherty, P., Steinert, P. M., and Roop, D. R., 1985, Organisation of a type I keratin gene. Evidence for evolution of intermediate filaments from a common ancestral gene, J. Biol. Chem. **260:**5867–5870.

Krohne, G., Wolin, S. L., McKeon, F. D., Franke, W. W., and Kirschner, M. W., 1987, Nuclear lamin LI of *Xenopus laevis:* cDNA cloning, amino acid sequence and binding specificity of a member of the lamin B subfamily, *EMBO J.* **6:**3801–3808.

Lees, J. F., Shneidman, P. S., Skuntz, S. F., Carden, M. J., and Lazzarini, R. A., 1988, The structure and organisation of the human heavy neurofilament subunit (NF-H) and the gene encoding it, *EMBO J.* **7:**1947–1955.

Lersch, R., and Fuchs, E., 1988, Sequence and expression of a type II keratin, K5, in human epidermal cells, *Mol. Cell. Biol.* **8:**486–493.

Levy, E., Liem, R. K. H., D'Eustachio, P., and Cowan, N. J., 1987, Structure and evolutionary origin of the gene encoding mouse NF-M, the middle-molecular-mass neurofilament protein, *Eur. J. Biochem.* **166:**71–77.

Lewis, S. A., and Cowan, N. J., 1986, Anomalous placement of introns in a member of the intermediate filament multigene family: An evolutionary conundrum, *Mol. Cell. Biol.* **6:**1529–1534.

Lewis, S. A., Balcarek, J. M., Krek, V., Shelanski, M., and Cowan, N. J., 1984, Sequence of a cDNA clone encoding mouse glial fibrillary acidic protein: Structural conservation of intermediate filaments, *Proc. Natl. Acad. Sci. USA* **81:**2743–2746.

Mack, J. W., Torchia, D. A., and Steinert, P. M., 1988, Solid-state NMR studies of the dynamics and structure of mouse keratin intermediate filaments, *Biochemistry* **27:**5418–5426.

McKeon, F. D., Kirschner, M. W., and Caput, D., 1986, Homologies in both primary and secondary structure between nuclear envelope and cytoplasmic intermediate filament proteins, *Nature* **319:**463–468.

McLachlan, A. D., and Karn, J., 1983, Periodic features in the amino acid sequence of nematode myosin rod, *J. Mol. Biol.* **164:**605–626.

McLachlan, A. D., and Stewart, M., 1976, The 14-fold periodicity in α-tropomyosin and the interaction with actin, *J. Mol. Biol.* **103:**271–298.

McLachlan, A. D., and Stewart, M., 1982, Periodic charge distribution in the intermediate filament proteins desmin and vimentin, *J. Mol. Biol.* **162:**693–698.

Marchuk, D., McCrohon, S., and Fuchs, E., 1984, Remarkable conservation of structure among intermediate filament genes, *Cell* **39:**491–498.

Marchuk, D., McCrohon, S., and Fuchs, E., 1985, Complete sequence of a gene encoding a human type I keratin: Sequences homologous to enhancer elements in the regulatory region of the gene, *Proc. Natl. Acad. Sci. USA* **82:**1609–1613.

Moll, R., Franke, W. W., Schiller, D., Geiger, B., and Krepler, R., 1982, The catalog of human cytokeratins: Patterns of expression in normal epithelia, tumors and cultured cells, *Cell* **31:**11–24.

Myers, M. W., Lazzarini, R. A., Lee, V. M. Y., Schlaepfer, W. W., and Nelson, D. L., 1987, The human mid-size neurofilament subunit: A repeated protein sequence and the relationship of its gene to the intermediate filament gene family, *EMBO J.* **6:**1617–1625.

Napolitano, E. W., Chin, S. S. M., Colman, D. R., and Liem, R. K. H., 1987, Complete amino acid sequence and *in vitro* expression of rat NF-M, the middle molecular weight neurofilament protein, *J. Neurosci.* **7:** 2590–2599.

Pang, Y. Y. S., Robson, R. M., Hartzer, M. K., and Stromer, M. H., 1983, Subunit structure of the desmin and vimentin protofilament units, *J. Cell Biol.* **97:**266a.

Parry, D. A. D., 1975, Analysis of the primary sequence of α-tropomyosin from rabbit skeletal muscle, *J. Mol. Biol.* **98:**519–535.

Parry, D. A. D., 1979, Determination of structural information from the amino acid sequences of fibrous proteins, in: *Fibrous Proteins: Scientific, Industrial and Medical Aspects,* Volume 1 (D. A. D. Parry and L. K. Creamer, eds.), Academic Press, New York, pp. 393–427.

Parry, D. A. D., and Fraser, R. D. B., 1985, Intermediate filament structure: 1. Analysis of IF protein sequence data, *Int. J. Biol. Macromol.* **7**:203–213.

Parry, D. A. D., Crewther, W. G., Fraser, R. D. B., and MacRae, T. P., 1977, Structure of α-keratin: Structural implication of the amino acid sequences of the type I and type II chain segments, *J. Mol. Biol.* **113**:449–454.

Parry, D. A. D., Steven, A. C., and Steinert, P. M., 1985, The coiled-coil molecules of intermediate filaments consist of two parallel chains in exact axial register, *Biochem. Biophys. Res. Commun.* **127**:1012–1018.

Parry, D. A. D., Conway, J. F., and Steinert, P. M., 1986, Structural studies on lamin: Similarities and differences between lamin and intermediate filament proteins, *Biochem. J.* **238**:305–308.

Parry, D. A. D., Conway, J. F., Goldman, A. E., Goldman, R. D., and Steinert, P. M., 1987, Nuclear lamin proteins: Common structures for paracrystalline, filamentous and lattice forms, *Int. J. Biol. Macromol.* **9**: 137–145.

Potschka, M., Winkler, H., and Wicke, G., 1984, Charge effects in sedimentation of intermediate filament subassemblies, 8th Int. Biophys. Congr. Bristol, p. 212.

Quax, W., Egberts, W. V., Hendriks, W., Quax-Jeuken, Y., and Bloemendal, H., 1983, The structure of the vimentin gene, *Cell* **35**:215–223.

Quax, W., van der Heuvel, R., Egberts, W. V., Quax-Jeuken, Y., and Bloemendal, H., 1984, Intermediate filament cDNAs from BHK-21 cells: Demonstration of distinct genes for desmin and vimentin in all vertebrate classes, *Proc. Natl. Acad. Sci. USA* **81**:5970–5974.

Quax-Jeuken, Y. E. F. M., Quax, W. J., and Bloemendal, H., 1983, Primary and secondary structure of hamster vimentin from the nucleotide sequence, *Proc. Natl. Acad. Sci. USA* **80**:3548–3552.

Quinlan, R. A., and Franke, W. W., 1982, Heteropolymer filaments of vimentin and desmin in vascular smooth muscle tissue and cultured hamster kidney cells demonstrated by chemical crosslinking, *Proc. Natl. Acad. Sci. USA* **79**:3452–3456.

Quinlan, R. A., and Franke, W. W., 1983, Molecular interactions in intermediate-sized filaments revealed by chemical cross-linking. Heteropolymers of vimentin and glial filament protein in cultured human glioma cells, *Eur. J. Biochem.* **132**:477–484.

Quinlan, R. A., Cohlberg, J. A., Schiller, D. L., Hatzfeld, M., and Franke, W. W., 1984, Heterotypic tetramer (A_2D_2) complexes of non-epidermal keratins isolated from cytoskeletons of rat hepatocytes and hepatoma cells, *J. Mol. Biol.* **178**:365–388.

Quinlan, R. A., Schiller, D. L., Hatzfeld, M., Achtstatter, T., Moll, R., Jorcano, J. L., Magin, T. M., and Franke, W. W., 1985, Patterns of expression and organization of cytokeratin intermediate filaments, *Ann. N.Y. Acad. Sci.* **455**:282–306.

Raychaudury, A., Marchuk, D., Lindhurst, M., and Fuchs, E., 1986, Three tightly linked genes encoding human type I keratins: Conservation of sequence in the 5′-untranslated leader and 5′-upstream regions of coexpressed keratin genes, *Mol. Cell Biol.* **6**:539–548.

Renner, W., Franke, W. W., Schmid, E., Geisler, E., Weber, K., and Mandelkow, E., 1981, Reconstitution of intermediate-sized filaments from denatured monomeric vimentin, *J. Mol. Biol.* **149**:285–306.

Rieger, M., Jorcano, J. L., and Franke, W. W., 1985, Complete sequence of a bovine type I cytokeratin gene: Conserved and variable intron positions in genes of polypeptides of the same cytokeratin subfamily, *EMBO J.* **4**:2261–2267.

Romano, V., Hatzfeld, M., Magin, T. M., Zimbelmann, R., Franke, W. W., Maier, G., and Ponstingl, H., 1986, Cytokeratin expression in simple epithelia I. Identification of mRNA coding for human cytokeratin No 18 by a cDNA clone, *Differentiation* **30**:244–253.

Sharp, G., Shaw, G., and Weber, K., 1982, Immunoelectron microscopic localization of the 3 neurofilament triplet proteins along neurofilaments of cultured dorsal root ganglion neurons, *Exp. Cell Res.* **137**:403–414.

Shaw, G., and Weber, K., 1981, Distribution of neurofilament triplet proteins within individual neurons, *Exp. Cell Res.* **136**:119–126.

Shaw, G., and Weber, K., 1982, Differential expression of neurofilament triplet proteins in brain development, *Nature* **298**:277–279.

Shneidman, P. S., Carden, M. J., Lees, J. F., and Lazzarini, R. A., 1988, The structure of the largest murine neurofilament protein (NF-H) as revealed by cDNA and genomic sequences, *Mol. Brain Res.* **4**: 217–231.

Singer, P. A., Trevor, K., and Oshima, R. G., 1986, Molecular cloning and characterization of the endo B cytokeratin expressed in preimplantation mouse embryos, *J. Biol. Chem.* **261**:538–547.

Skerrow, D., Matoltsy, A. G., and Matoltsy, M. N., 1973, Isolation and characterization of the helical regions of epidermal prekeratin, *J. Biol. Chem.* **248**:4820–4826.

Sparrow, L. G., and Inglis, A. S., 1980, Characterization of the cyanogen bromide proteins of component 7c, a major microfibrillar protein from wool, Proc. 6th Int. Wool Text. Res. Conf. Pretoria, Vol. II, pp. 237–246.

Sparrow, L. G., Dowling, L. M., Loke, V. Y., and Strike, P. M., 1989, Amino acid sequences of wool keratin IF proteins, in: *Biology of Wool and Hair* (G. E. Rogers, P. J. Reis, K. A. Ward, and R. C. Marshall, eds.), Chapman & Hall, London, pp. 145–155.

Steinert, P. M., 1978a, Structure of the three-chain unit of the bovine epidermal keratin filament, *J. Mol. Biol.* **123:**49–70.

Steinert, P. M., 1978b, Structural features of the α-type filaments of the inner root sheath cells of the guinea pig hair follicle, *Biochemistry* **17:**5045–5052.

Steinert, P. M., 1981, Intermediate filaments (IF), in: *Electron Microscopy of Proteins,* Volume 1 (J. R. Harris, ed.), Academic Press, New York, pp. 125–166.

Steinert, P. M., and Parry, D. A. D., 1985, Intermediate filaments: Conformity and diversity of expression and structure, *Annu. Rev. Cell Biol.* **1:**41–65.

Steinert, P. M., and Roop, D. R., 1988, Molecular and cellular biology of intermediate filaments, *Annu. Rev. Biochem.* **57:**593–625.

Steinert, P. M., Zimmerman, S. B., Starger, J. M., and Goldman, R. D., 1978, Ten-nanometer filaments of hamster BHK-21 cells and epidermal keratin filaments have similar structures, *Proc. Natl. Acad. Sci. USA* **75:**6098–6101.

Steinert, P. M., Idler, W. W., and Goldman, R. D., 1980, Intermediate filaments of BHK-21 cells and bovine epidermal keratinocytes have similar ultrastructures, *Proc. Natl. Acad. Sci. USA* **77:**4534–4538.

Steinert, P. M., Idler, W. W., Cabral, F., Gottesman, M. M., and Goldman, R. D., 1981, *In vitro* assembly of homopolymer and copolymer filaments from intermediate filament subunits of muscle and fibroblastic cells, *Proc. Natl. Acad. Sci. USA* **78:**3692–3696.

Steinert, P. M., Idler, W. W., Aynardi-Whitman, M. A., Zackroff, R. V., and Goldman, R. D., 1982, Heterogeneity of intermediate filaments assembled *in vitro, Cold Spring Harbor Symp. Quant. Biol.* **46:** 465–474.

Steinert, P. M., Rice, R. H., Roop, D. R., Trus, B. L., and Steven A. C., 1983, Complete amino acid sequence of a mouse epidermal keratin subunit and implications for the structure of intermediate filaments, *Nature* **302:**794–800.

Steinert, P. M., Parry, D. A. D., Racoosin, E. L., Idler, W. W., Steven, A. C., Trus, B. L., and Roop, D. R., 1984, The complete cDNA and deduced amino acid sequence of a type II mouse epidermal keratin of 60 000 Da: Analysis of sequence differences between type I and type II keratins, *Proc. Natl. Acad. Sci. USA* **81:**5709–5713.

Steinert, P. M., Parry, D. A. D., Idler, W. W., Johnson, L. D., Steven, A. C., and Roop, D. R., 1985a, Amino acid sequences of mouse and human epidermal keratins of M_r 67 000 provide a systematic basis for the structural and functional diversity of the end domains of keratin intermediate filament subunits, *J. Biol. Chem.* **260:**7142–7149.

Steinert, P. M., Steven, A. C., and Roop, D. R., 1985b, The molecular biology of intermediate filaments, *Cell* **42:**411–419.

Steinert, P. M., Torchia, D. A., and Mack, J. W., 1989, Structural features of keratin intermediate filaments, in: *Biology of Wool and Hair* (G. E. Rogers, P. J. Reis, K. A. Ward, and R. C. Marshall, eds.), Chapman & Hall, London, pp. 157–167.

Sun, T. T., Eichner, R., Schermer, A., Cooper, D., Nelson, W. G., and Weiss, R. A., 1984, Classification, expression, and possible mechanisms of evolution of mammalian epithelial keratins: A unifying model in: *Cancer Cells 1, The Transformed Phenotype,* Cold Spring Harbor Laboratory, Cold Spring Harbor, N.Y., pp. 169–176.

Suzuki, E., Crewther, W. G., Fraser, R. D. B., MacRae, T. P., and McKern, N. M., 1973, X-ray diffraction and infrared studies of an α-helical fragment from α-keratin, *J. Mol. Biol.* **73:**275–278.

Tyner, A. L., Eichman, M. J., and Fuchs, E., 1985, The sequence of a type II keratin gene expressed in human skin: Conservation of structure among all intermediate filament genes, *Proc. Natl. Acad. Sci. USA* **82:** 4683–4687.

Willard, M., and Simon, C., 1981, Antibody decoration of neurofilaments, *J. Cell Biol.* **89:**198–205.

Winkles, J. A., Sargent, T. D., Parry, D. A. D., Jonas, E., and Dawid, I. B., 1985, Developmentally regulated cytokeratin gene in *Xenopus laevis, Mol. Cell. Biol.* **5:**2575–2581.

Woodcock-Mitchell, J., Eichner, R., Nelson, W. G., and Sun, T. T., 1982, Immunolocalization of keratin polypeptides in human epidermis using monoclonal antibodies, *J. Cell Biol.* **95:**580–588.

Woods, E. F., 1983, The number of polypeptide chains in the rod domain of bovine epidermal keratin, *Biochem. Int.* **7**:769–774.

Woods, E. F., and Gruen, L. C., 1981, Structural studies on the microfibrillar proteins of wool: Characterization of the α-helix-rich particle produced by chymotryptic digestion, Aust. J. Biol. Sci., **34**:515–526.

Woods, E. F., and Inglis, A. S., 1984, Organization of the coiled-coils in the wool microfibril, *Int. J. Biol. Macromol.* **6**:277–283.

Wu, Y. J., Parker, L. M., Binder, N. E., Beckett, M. A., Sinard, J. H., Griffiths, C. T., and Rheinwald, J. G., 1982, The mesothelial keratins: A new family of cytoskeletal proteins identified in cultured mesothelial cell and nonkeratinizing epithelia, *Cell* **31**:693–703.

Zackroff, R. V., and Goldman, R. D., 1980, *In vitro* reassembly of squid brain intermediate filaments (neurofilaments): Purification by assembly–diassembly, *Science* **208**:1152–1155.

Zehner, Z. E., and Paterson, B. M., 1985, The chicken vimentin gene: Aspects of organization and transcription during myogenesis, *Ann. N.Y. Acad. Sci.* **455**:79–94.

Zhou, X.-M., Idler, W. W., Steven, A. C., Roop, D. R., and Steinert, P. M., 1988, The complete sequence of the human intermediate filament chain keratin 10, *J. Biol. Chem.,* **263**:15584–15589.

8

The Three-Dimensional Structure of IF

R. D. B. Fraser, T. P. MacRae, and David A. D. Parry

1. Introduction

Although it is only comparatively recently that intermediate filaments (IF) have been recognized as a class of cytoskeletal structures, it is now more than 50 years since information was first collected on the three-dimensional structure of individual members of the class. For example, in discussing Speakman's observations on the sorption of water by wool and his own x-ray studies of wool, Astbury (1933) stated that "A wool fibre . . . becomes as much as about 18% greater in diameter but only about 1% longer. This transformation is easily explained from the results of x-ray analysis, which shows that the regain water is for the most part spread over the surfaces of the tiny crystalline aggregates which are much longer than they are thick. Because the crystals are so very tiny, perhaps not more than some twenty times thicker than the water molecules themselves. . . ."

It was in this same era that Astbury and co-workers (Astbury and Street, 1931; Astbury and Woods, 1933) analyzed the high-angle x-ray diffraction data from α-keratins and concluded that the characteristic α pattern originated from a polypeptide chain conformation that was folded, rather than extended as in silk. The search for the precise nature of this conformation has since attracted widespread interest and has now been solved in general terms (Bragg *et al.*, 1950; Pauling and Corey, 1951, 1953; Crick, 1953a,b), but the problem of visualizing the precise nature of the coiled-coil rope structure remains to be solved.

Using as a basis the account of present-day knowledge of the molecular structure of IF molecules determined primarily from chemical and physicochemical studies (Parry, this volume), we will now examine critically the physical evidence concerning the three-dimensional structure of the molecule, of molecular aggregates, and of native IF.

R. D. B. Fraser and T. P. MacRae • CSIRO, Division of Wool Technology, Parkville, Victoria 3052, Australia. ***David A. D. Parry*** • Department of Physics and Biophysics, Massey University, Palmerston North, New Zealand.

2. Molecular Structure

The coiled-coil rope structure, proposed independently by Crick (1952, 1953a,b) and by Pauling and Corey (1953) to explain the high-angle x-ray diffraction pattern obtained from α-keratin, accounts satisfactorily for the simultaneous appearance of prominent meridional reflections of spacing 0.1485 and 0.5165 nm (Fraser *et al.*, 1964a) but the determination of the physical parameters of the rope is complicated by the fact that the diffraction pattern is overlaid by an elaborate interference pattern associated with higher levels of organization (Fig. 1a). The general character of the pattern was found to be consistent with the coiled-coil model (Fraser and MacRae, 1961; Fraser *et al.*, 1964a) and in the latter study the then surprising result was obtained that the width at half height, measured in a direction perpendicular to the equator, of the near-equatorial layer line associated with the pitch of the coiled coil was around 0.15 nm^{-1}. This implied that the coiled coils were interrupted along their length on average at intervals of around 6.7 nm. However, we now know that the heptad repeat in the rod domain (and hence the coiled-coil structure) is interrupted at links L1, L12, L2 and at the center of segment 2B (Fig. 2). The resulting five segments contain a total of 276 residues (Parry, this volume) giving a mean length of $(276 \times 0.1485)/5 = 8.2$ nm. Any irregularities will further reduce the effective length as judged from the layer line width. The x-ray data are thus in good agreement with the expectation from sequence studies.

The determination of the pitch of the coiled coil and the number of strands in the rope is complicated both by the interference pattern mentioned earlier and by the influence of the material in which the coiled coil is embedded. However, Fraser *et al.* (1965) compared the diffraction patterns obtained from α-keratin with those obtained from various muscles

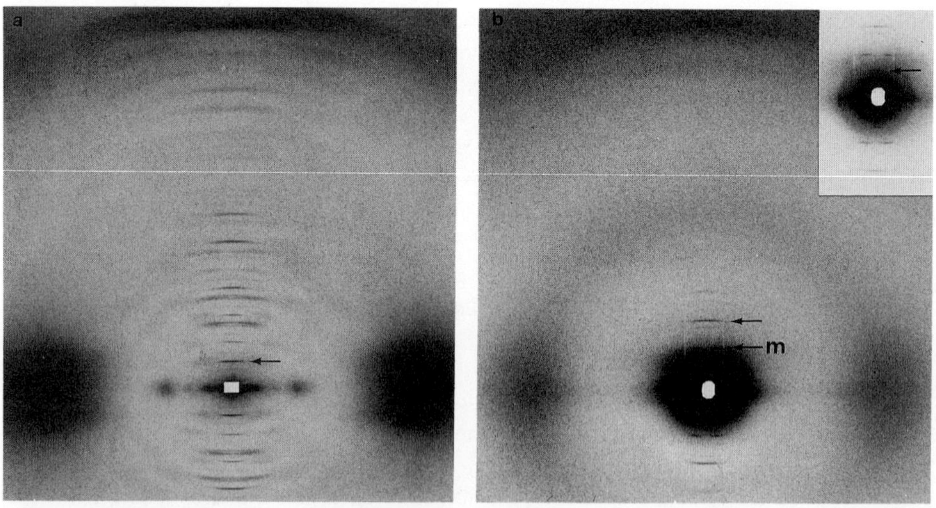

Figure 1. X-ray diffraction patterns obtained from (a) porcupine quill *(Atherurus africanus)* and (b) axoplasm of *Myxicola*, a marine worm (courtesy of Dr. Peter Eagles). The pattern contains reflections from both microtubules (e.g., m→) and IF (→). The spacing of the arrowed IF reflection is 2.52 nm, and the inset shows the 6.3-nm reflection.

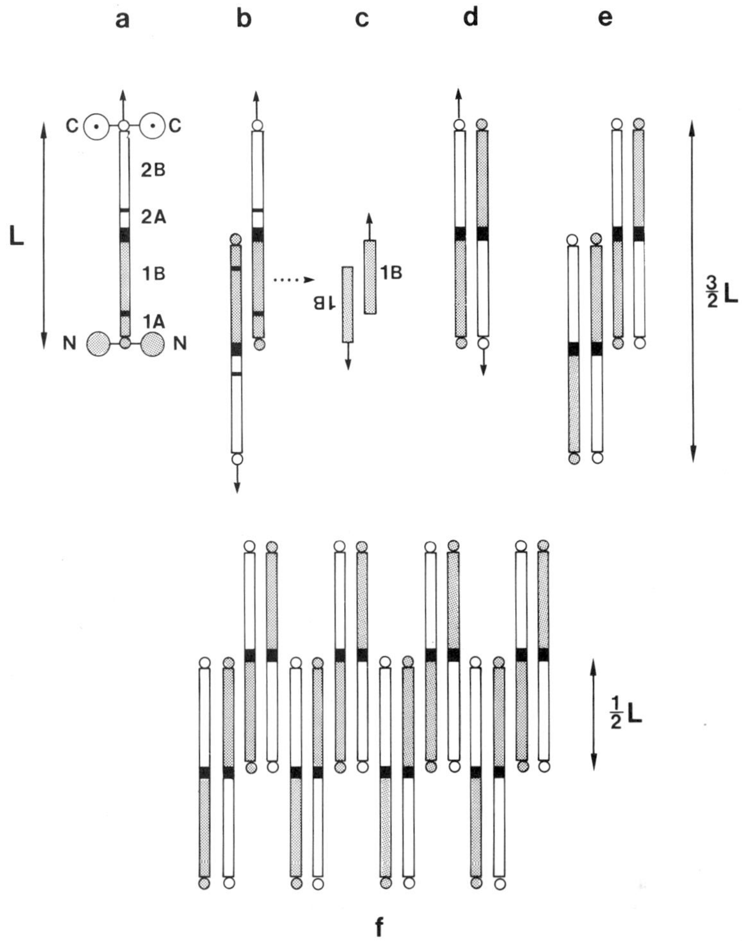

Figure 2. (a) Schematic representation of the IF protein molecule. The central rod domain comprises four segments of two-strand rope 1A, 1B, 2A, and 2B linked by three nonrope segments (black). The circles at either end of the rod segment represent portions of the N- and C-terminal domains that may be involved in filament assembly. The remainders of the N- and C-terminal domains are shown separately for each chain (larger circles) since it is not known whether they interact. (b) A type of association presumed to exist in IF on the basis of the isolation, after partial hydrolysis, of particles containing four 1B chain segments as in (c). The axial stagger shown in these and subsequent illustrations is arbitrary; evidence concerning the actual stagger is discussed in the text. (d) A type of association deduced from electron microscope studies of chymotryptically modified desmin (Geisler *et al.*, 1985). (e) A type of association, which combines (b) and (d), observed in vimentin (Ip *et al.*, 1985). (f) An oligomeric assembly formed by the lateral aggregation of particles of the type shown in (e). Such assemblies can ring around to form short tubes (Ip *et al.*, 1985).

and muscle proteins and concluded that "in the light of the similarity between paramyosin and α-keratin in the dry state, the present work favours a two-chain structure, although this cannot be taken as proved until fully swollen specimens can be prepared and tested." At first, these results appeared to be in conflict with data obtained from chemical and physicochemical studies, which suggested a three-strand rope (Crewther and Dowling,

1971; Skerrow *et al.*, 1973; Steinert, 1978; Steinert *et al.*, 1980) but they are in accord with more recent studies that indicate clearly that two-strand rather than three-strand ropes are present (Ahmadi and Speakman, 1978; Geisler and Weber, 1982; Gruen and Woods, 1983; Parry *et al.*, 1985). Once the number of strands in the coiled-coil rope is decided, it is possible to derive a value for the pitch of the coiled coil from the measured displacement from the equator of the near-equatorial layer lines associated with the coiled coil. The results obtained by Fraser *et al.* (1965) yielded a range of 14–17 nm for the pitch, the lower end of which is similar to values observed in other α-fibrous proteins (Parry, 1975). A knowledge of the pitch of the coiled coil is important in speculations about packing and interactions and toward that end a method has been devised by Fraser *et al.* (1989) that allows the pitch length to be assessed in terms of the number of apolar, ionic, and hydrophilic interactions that can be made between interacting coiled coils. Preliminary results suggest that the coiled-coil pitch length may be about 14 nm for the 1B segments of desmin and hard α-keratin (Fraser *et al.*, 1989).

High-angle x-ray diffraction patterns of the α type have also been recorded from epidermis (Rudall, 1952; Fraser *et al.*, 1972) and from various types of neurofilaments (Wais-Steider *et al.*, 1983) but the quality of the data obtained has not permitted the various coiled-coil parameters to be determined.

3. Molecular Aggregation

In vitro studies of molecular aggregation have proved to be a valuable means of providing information on the three-dimensional arrangement of the molecules in naturally occurring assemblies of fibrous proteins (Fraser and MacRae, 1973a). Electron microscopic studies have been particularly valuable in this regard and the results obtained with IF are discussed in detail by Steven (this volume). Experience gained with other fibrous proteins shows, however, that various polymorphic modes of aggregation are possible *in vitro* and that only some of these are relevant to the native structure (Fraser and MacRae, 1973a). In this work we discuss certain modes of aggregation that seem to be relevant to the structure of native IF, and the attempts that have been made to relate these to the amino acid sequence.

In Fig. 2a the IF molecule is represented diagrammatically as a rod domain comprising two-stranded coiled-coil segments 1A, 1B, 2A, 2B connected by short non-coiled-coil linking segments. At each end of the rod domain a small circle is used to represent symbolically parts of the nonhelical N- and C-terminal domains that may be involved in assembly of the IF (Steinert, 1978, 1981; Kaufmann *et al.*, 1985) though there is increasing evidence (summarized by Conway and Parry, 1988) that it is primarily the N-terminal domains or parts thereof (rather than the C-terminal domains) that are involved in IF assembly. The circles may represent only a few residues or may, in some instances, represent a substantial part of the terminal domain. The portions of the terminal domains not involved in assembly are represented separately for each chain, since it is not known whether they interact with one another. In the cases of desmin and vimentin, the two N- and C-terminal domains will be the same since they are homodimers (Steinert *et al.*, 1981), but if, as has been generally assumed, the α-keratin molecule is a heterodimer (see Parry, this volume, for a discussion of this point) they will be different.

3.1. Enzymatically Derived Fragments

Helix-rich particles can be isolated from wool extracts (Crewther and Dowling, 1971) and chemical and physicochemical studies have established that the major component consists of four-chain fragments derived from the 1B coiled-coil segment (Ahmadi and Speakman, 1978; Crewther *et al.*, 1978; Gough *et al.*, 1978; Woods, 1983; Woods and Inglis, 1984). Similar results have been obtained with epidermal proteins (Parry *et al.*, 1985). Of the two possible ways of bringing the 1B segments in two molecules into approximate register, that shown in Fig. 2b has been favored (Fraser and MacRae, 1983; Crewther *et al.*, 1983) leading to an antiparallel association. The stagger shown in Fig. 2b and c is completely arbitrary and information about the actual stagger has been sought in two ways. Oriented films of a four-chain particle consisting of 1B segments yielded a typical α-type x-ray diffraction pattern associated with the coiled-coil conformation and in addition exhibited a sharp meridional low-angle pattern indicating an axial repeat of 16 nm that probably corresponds closely to the length of the particle (Suzuki *et al.*, 1973). Since the 1B chain segment contains 101 residues, the predicted length of the 1B molecular segment is $101h_{cc}$, where h_{cc}, the mean axial rise per residue in the coiled-coil conformation, has been estimated to be about 0.1485 nm (Fraser *et al.*, 1964a). The predicted length of segment 1B is thus 15 nm. Some uncertainty exists about the integrity of the coiled conformation at the extreme ends of the segments adjacent to the cleavage sites but it seems likely that the relative stagger between the oppositely directed molecular segments cannot be large, and probably less than about 15 residues. Similar particles isolated from epidermal keratin digests were estimated to have a length of about 18 nm from physicochemical and electron microscope studies (Steinert, 1978; Steven, this volume).

3.2. Ionic Interaction Score

Hulmes *et al.* (1973) devised a method to search for potentially favorable apolar and ionic molecular interactions in collagen. This technique has also been applied to IF proteins (Fraser *et al.*, 1985). In this latter study the sequences of pairs of the long rod domain segments were aligned (Fig. 3a) in parallel (U, U) or antiparallel (U, D) orientations and a count made of the number of possibilities for favorable interactions between negatively charged residues (Asp, Glu) and positively charged residues (Arg, Lys) located within the range ± 1 residue translations. The stagger was then varied systematically and the results for the interactions (1BU, 1BD) and (1BU, 2D) in the IF proteins from wool, epidermis, and desmin are illustrated in Figs. 4 and 5. In the case of wool and epidermis, a heterodimer structure for the molecule was assumed. The pronounced fluctuations in the score for ionic interaction potential can be related directly to the periodic distribution of charged residues in these segments (Parry *et al.*, 1977; McLachlan and Stewart, 1982; Crewther *et al.*, 1983; Parry and Fraser, 1985), and the generally lower scores for mixed segments can be traced to the slight difference in period in segments 1B and 2 (9.55 and 9.84 residues, respectively; see Parry and Fraser, 1985). Scores were also calculated for hydrogen-bond and apolar interactions but it was only for ionic interactions that pronounced maxima were observed. The scores for ionic interactions are particularly relevant to the present discussion since they are believed to play a key role in the alignment and registration of molecules during filament formation (Fraser *et al.*, 1981). The highest ionic

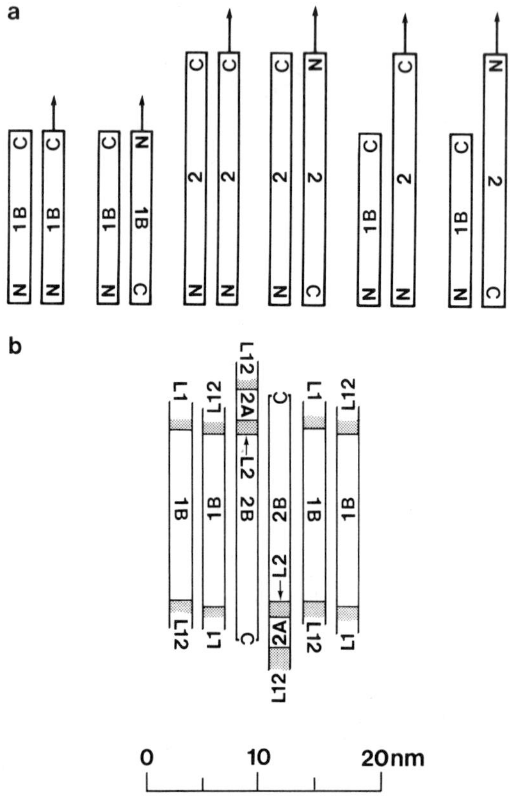

Figure 3. (a) Possible interactions between the long rod domains in IF molecules. These are designated (1BU, 1BU), (1BU, 1BD), (2U, 2U), (2U, 2D), (1BU, 2U) and (1BU, 2D), respectively. In each case the segments are shown in the situation defined as zero stagger and arrows indicate the direction of positive stagger. Δz is defined as the relative axial stagger and is conveniently measured in multiples of h_{cc}, the mean axial translation in the coiled coil. (b) A combination of staggers, derived from Figs. 4 and 5, and the data of Fraser *et al.* (1985), which would be compatible with ring formation as in Fig. 2f.

interaction score for (1BU, 1BD) in desmin (Fig. 4c) is 56 at Δz(1BU, 1BD) = 3 and related peaks are present in the histograms for wool and epidermis IF protein at Δz(1BU, 1BD) = 5 and 4, respectively, where Δz is expressed in units of h_{cc}, the mean axial translation per residue in the coiled coil. Comparable maxima for these latter two materials occur at Δz(1BU, 1BD) = −5 (Fig. 4a,b). It seems reasonable to suppose therefore that the relative stagger between the two 1B molecular segments in Fig. 2b and c is close to ± 5, in good agreement with the results from the x-ray study, discussed earlier.

The method of Hulmes *et al.* (1973) takes no account of the azimuthal distribution of the residues on the surface of the molecule and so the interaction score includes some interactions that are stereochemically not feasible. In spite of this inherent simplicity, the technique has proved remarkably successful for a variety of fibrous proteins, but clearly a more complete analysis, with due allowance for azimuthal aspects of the interacting structure, is now required.

Toward that end, the methodology has recently been extended by Fraser *et al.* (1989) to allow the interface between the coiled-coil molecules to be studied using either radial projections (Fig. 6a) or a projection onto the contact plane (Fig. 6b). In searching for favorable interactions, it is necessary to vary the pitch length of the coiled coils, the relative axial stagger between interacting segments, and the azimuthal coordinates of the interacting coiled-coil ropes. The latter parameters must be varied independently but it is reasonable to suppose that the coiled coils are related by a parallel diad if they are

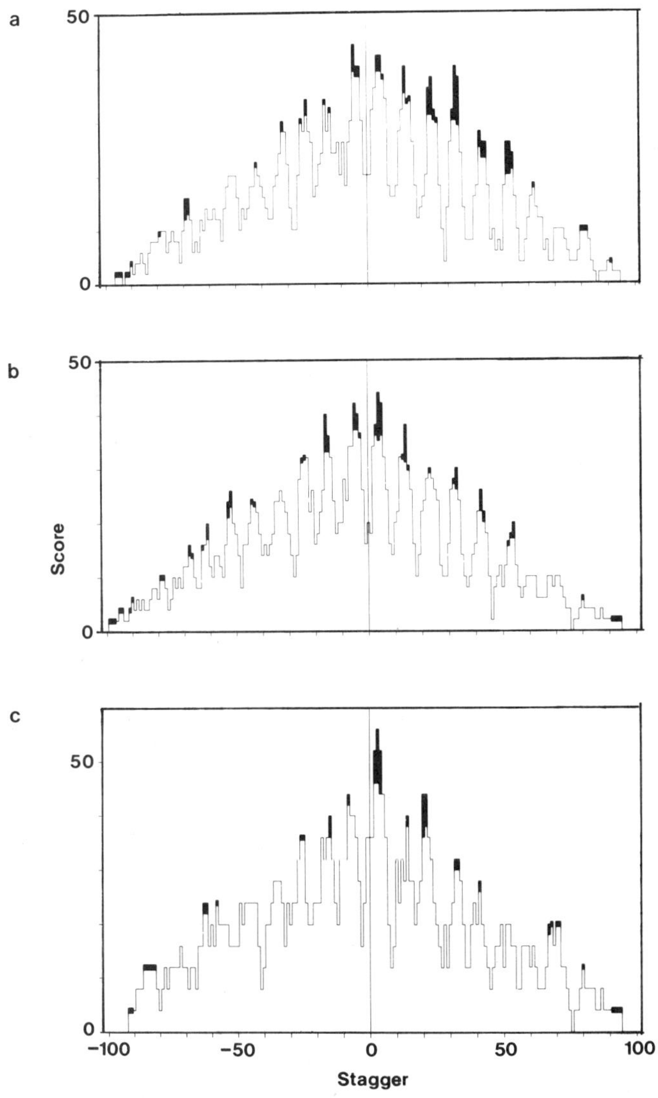

Figure 4. Comparison of ionic interaction scores between the two 1B rod-domain segments of IF molecules for (1BU, 1BD). Where score exceeds the 0.975 quantile calculated for random distributions, the amount by which it exceeds it is shaded. (a) Wool IF molecules, (b) epidermal IF molecules, (c) desmin molecules (Fraser *et al.*, 1985).

similarly directed, or by a perpendicular diad if they are oppositely directed. Ionic, apolar, and hydrogen-bond interactions can all be calculated using this method and suitable cutoff criteria (in terms of the distance of separation of the residues involved) can be placed on each interaction type. Such methods have been used previously by Trus and Piez (1976) and Hofmann *et al.* (1978) to study molecular interactions in collagen. An example of a simple two-dimensional analysis is given in Fig. 7 for antiparallel 1B segments from chicken gizzard desmin. An examination of the distribution of residues on the surface of

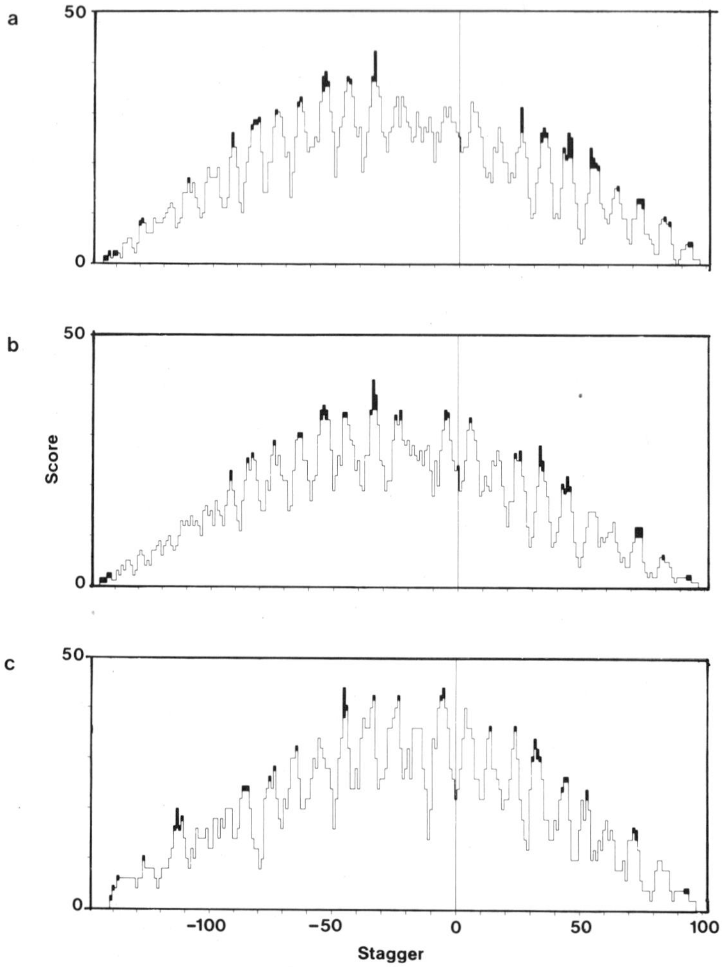

Figure 5. Ionic interaction scores for (1BU, 2D). Details as in Fig. 4.

the 1B rope segment of chicken gizzard desmin, using the contact plane method and allowing the possibility that interactions can be made between appropriate residues less than 0.5 nm apart on the contact plane (Fig. 6b), reveals a highly favorable mode of aggregation for a pair of 1B rope segments related by a perpendicular diad and a pitch length of 14 nm, a relative stagger of Δz(1BU, 1BD) = +2, and an azimuth around $-5°$ (Table I, Fig. 7). This may be compared with the result obtained from the one-dimensional search, which gave the principal maximum at Δz(1BU, 1BD) = 3 (Fig. 4c). A highly favorable interaction was also found for wool 1B rope segments at Δz(1BU, 1BD) = -6, which again compares favorably with results of the one-dimensional search, which yielded a major maximum at Δz(1BU, 1BD) = -5 (Fig. 4a). This latter result is in good accord with the x-ray data discussed earlier. Other experimental and theoretical evidence supporting the conclusion that the relative axial stagger between 1B rope segment is small has been summarized by Parry (this volume).

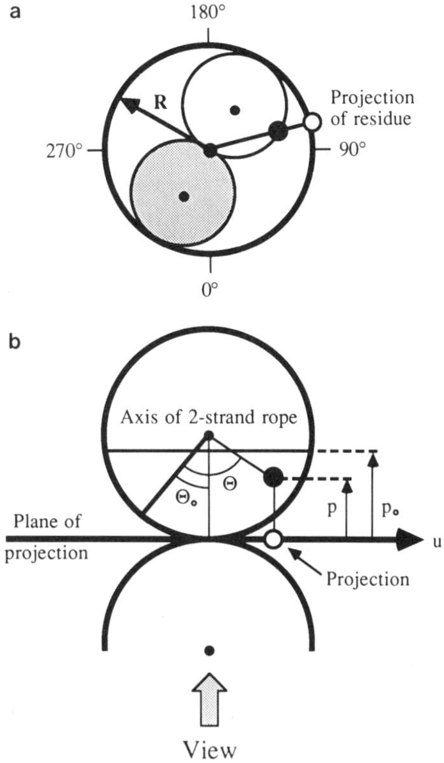

Figure 6. (a) Schematic representation of a two-stranded coiled coil in section. The radii of the coiled coil and the constituent α helices are R and $R/2$, respectively. The positions of the residues (filled circles) making up the coiled-coil structure may be projected outwards onto a cylindrical surface to generate a "radial projection." It is convenient to omit residues in the *a* and *d* positions since they are localized close to the axis of the coiled-coil rope and are unlikely to be involved in interrope interactions. Such radial interactions are useful in studying the surface distribution of residues and searching for potentially favorable stagger/azimuth combinations. Individual stagger/azimuth combinations can be studied in detail by projecting the residues onto the "contact plane" as in (b). Only residues for which the distance from the contact plane (p) is less than p_o (an arbitrary cutoff value based on physical considerations of proximity between residues that could interact) are included. Potential interactions can then be classified as apolar (i.e., between any two of leucine, valine, isoleucine, methionine, phenylalanine, alanine, and tyrosine), ionic (i.e., between aspartic acid or glutamic acid and arginine or lysine), or hydrogen bonding (i.e., between any two of asparagine, glutamine, serine, threonine, arginine, or tyrosine).

3.3. *Oligomers*

There have been many observations of particles with lengths in the range 45–55 nm (e.g., see Steven, this volume). Also, Geisler *et al.* (1985) have demonstrated that a four-chain particle approximately 48 nm in length, obtained from desmin treated with chymotrypsin, consisted of two molecular rod domains in approximate register with an antiparallel orientation as in Fig. 2d. This type of association gives a substantially lower ionic interaction score than a parallel orientation since there are peaks of height 46 at Δz(1U, 1U) = ± 5 and height 66 at Δz(2U, 2U) = ± 5 for the parallel case in desmin (Fraser *et al.*, 1985) giving a total 112 compared to peak scores of 44 ($\Delta z = -5, -45$) and 42 ($\Delta z = -23, -33$) for (1BU, 2D) giving totals of 88 or 84 respectively for possible antiparallel cases. The chains in the chymotryptically excised particle contain an additional 26 residues at the N-terminal end and 8 residues at the C-terminal end that are not part of the rod domain and it is possible that specific interactions between these projections and the rod domain stabilize the antiparallel rather than the parallel form. Alternatively, the failure to predict the observed type of association may simply be a reflection of the fact that the scoring scheme does not incorporate any allowance for azimuthal orientation and so identifies necessary rather than sufficient conditions for favorable interaction. Furthermore, the positive effect that helix dipole moments have on the stability of an antiparallel assembly of coiled-coil segments (compared to a parallel one) has not been included in the ionic interactions study discussed earlier and will undoubtedly provide a further signifi-

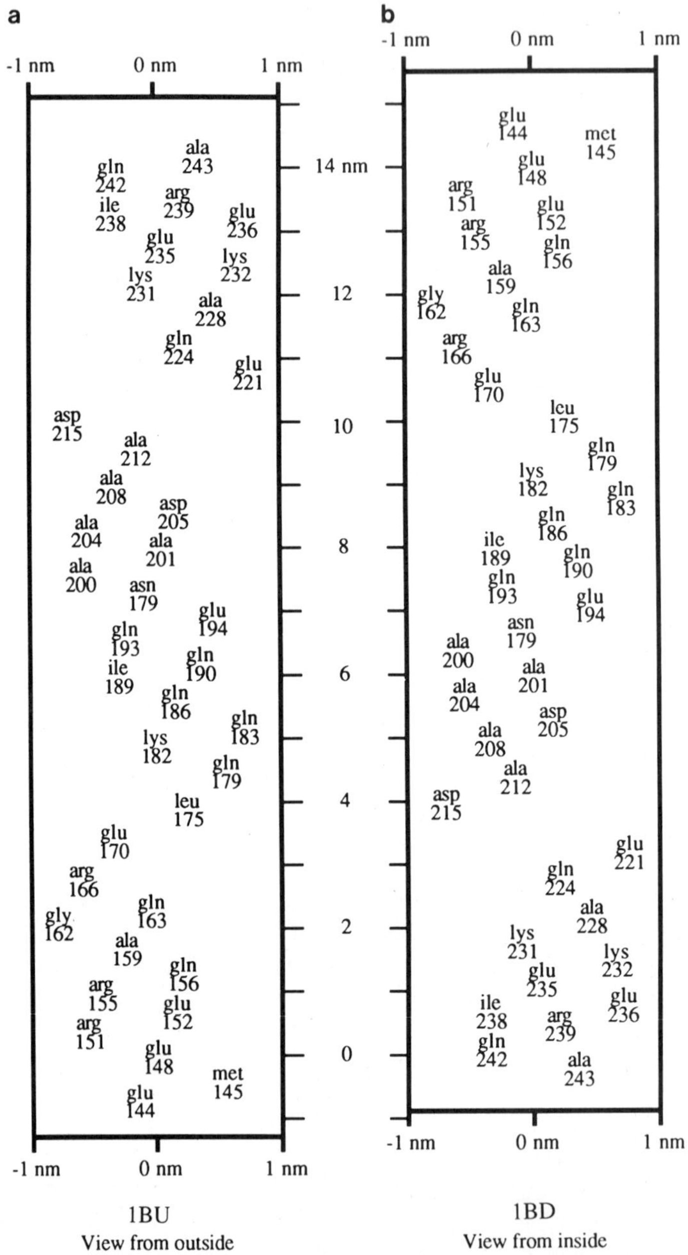

Figure 7. Projections of the positions of the residues in the 1B rope segment of chicken gizzard desmin (residues 143–243) that lie in the outer positions of a two-stranded coiled-coil rope (i.e., in positions *b, c, e, f,* and *g*), onto the "contact plane" (Fig. 6b). The relative stagger and azimuth correspond to the case discussed in the text. (a) The upwardly directed rope segment; (b) the downwardly directed rope segment. The rope segments are assumed to be related to one another by a perpendicular diad. Superposition of (a) and (b) with the illustrated stagger yields favorable apolar, ionic, and hydrophilic interactions (Table I). The value of p_o chosen was 0.5 nm. Favorable interactions are listed in Table I. More generalized and detailed calculations using full three-dimensional data with appropriate cutoff values for the various interaction types are in progress (Fraser *et al.*, 1989).

Table I. Favorable Interactions between Two Coiled-Coil Ropes Consisting of Chicken Gizzard Desmin 1B Chain Segments[a]

1BU	1BD	Type
Met 145	Ala 243	Hydrophobic
Glu 148	Arg 239	Ionic
Arg 151	Gln 242	Hydrogen bond
Arg 155	Glu 235	Ionic
Gln 163	Gln 224	Hydrogen bond
Arg 166	Arg 219	Hydrogen bond
Glu 170	Arg 219	Ionic
Leu 175	Ala 212	Hydrophobic
Lys 182	Asp 205	Ionic
Leu 189	Ala 200 or 201	Hydrophobic
Gln 193	Asn 197	Hydrogen bond
Asn 197	Gln 193	Hydrogen bond
Ala 200 or 201	Leu 189	Hydrophobic
Asp 205	Lys 182	Ionic
Ala 212	Leu 175	Hydrophobic
Arg 219	Glu 170	Ionic
Arg 219	Arg 166	Hydrogen bond
Gln 224	Gln 163	Hydrogen bond
Glu 235	Arg 155	Ionic
Gln 242	Arg 151	Hydrogen bond
Arg 239	Glu 148	Ionic
Ala 243	Met 145	Hydrophobic

[a]The two ropes are related by a perpendicular diad. Δz(1BU, 1BD) = +2 and the azimuth = $-5°$. The pitch length of the ropes is -14 nm and the maximum permissible distance for residues to be projected onto the contact plane is 0.5 nm (see Fig. 7). The favorable interactions listed above occur between residues less than 0.5 nm apart on the contact plane.

cant contribution to the free energy of the antiparallel system (Sheridan *et al.*, 1982; Shoemaker *et al.*, 1987).

In a study of the *in vitro* assembly of vimentin IF protein, Ip *et al.* (1985) obtained evidence for the existence of an eight-chain particle with a length close to 66 nm, which they concluded had a structure of the type illustrated in Fig. 2e. Although the orientation of the four molecules could not be determined, it seems reasonable to suppose, based on the information available on four-chain particles (Fig. 2b,d), that it has the structure depicted in Fig. 2e. Evidence was also produced that under some conditions, lateral aggregates of these particles form (Fig. 2f) and it was pointed out that if an assembly of the type shown were wrapped around into a tube it possessed the potentiality, by axial repetition, to generate a model for the native IF. Annular structures with outer diameters in the range 18–25 nm have been observed in partially reaggregated IF protein preparations (Renner *et al.*, 1981; Franke *et al.*, 1982; Sauk *et al.*, 1984; Hatzfeld and Franke, 1985; Ip *et al.*, 1985) and it seems likely that they have an interaction pattern similar to that shown in Fig. 2f, although the number of molecular pairs in the ring is likely to be greater than eight.

In general, an aggregate of the type shown in Fig. 2f will not "close" to form a tube unless certain special relationships exist between the staggers of the interacting molecular segments. For the set of contacts shown in Fig. 2f, this is readily shown to be

$$\Delta z(1\text{BU}, 1\text{BD}) - 2\,\Delta z(1\text{BU}, 2\text{D}) + \Delta z(2\text{U}, 2\text{D}) + z(1\text{B}) - z(2) = 0 \quad (1)$$

where $z(1\text{B})$ and $z(2)$ are the lengths of the respective segments measured in the same units (residue lengths) as the intersegment staggers (Fraser *et al.*, 1985). Interaction maxima for vimentin occur at $\Delta z(1\text{BU}, 1\text{BD}) = 3$ and $\Delta z(2\text{U}, 2\text{D}) = 4$ (Fraser *et al.*, 1985), and since $z(1\text{B}) = 101$ and $z(2) = 148$ we obtain

$$\Delta z(1\text{BU}, 2\text{D}) = -20 \quad (2)$$

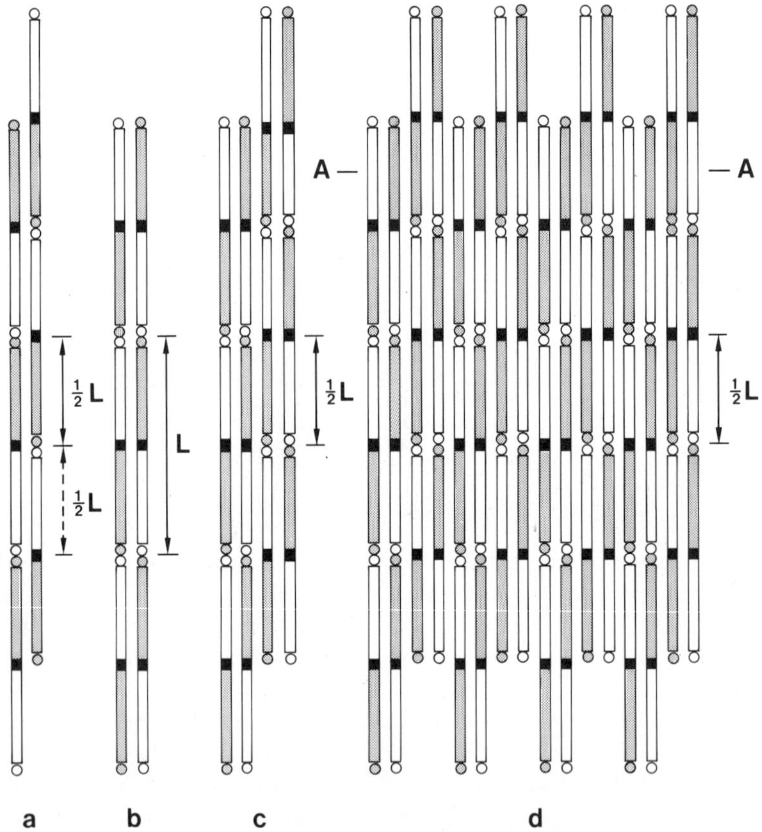

Figure 8. Extensions of the schemes shown in Fig. 2 to give polymeric aggregates: (a) polymer of 2b, (b) polymer of 2d, (c) polymer of 2e, (d) polymer of 2f. If the sheet in (d) is wrapped around to form a tube, a generic model for molecular interactions in IF is obtained. (f) shows how the pattern of interactions in (e) can be subdivided to give four subfilaments with diameters in the range 4–5 nm. Although only (1BU, 1BD) interactions are shown, it is likely that additional interactions, e.g., (2U, 2D), would be formed by appropriate radial and azimuthal displacements of the coiled-coil rope segments. (g) An alternative method of subdivision leading to three subfilaments. In addition to the (2U, 2D) interactions shown, it is likely that (1BU, 1BD) interactions would form by appropriate movements of the coiled-coil rope segments. Note that (f) and (g) could equally well have been drawn with the repeating units rotated 180° about their long axes leading to (2U, 2D) and (1BU, 1BD) interactions, respectively.

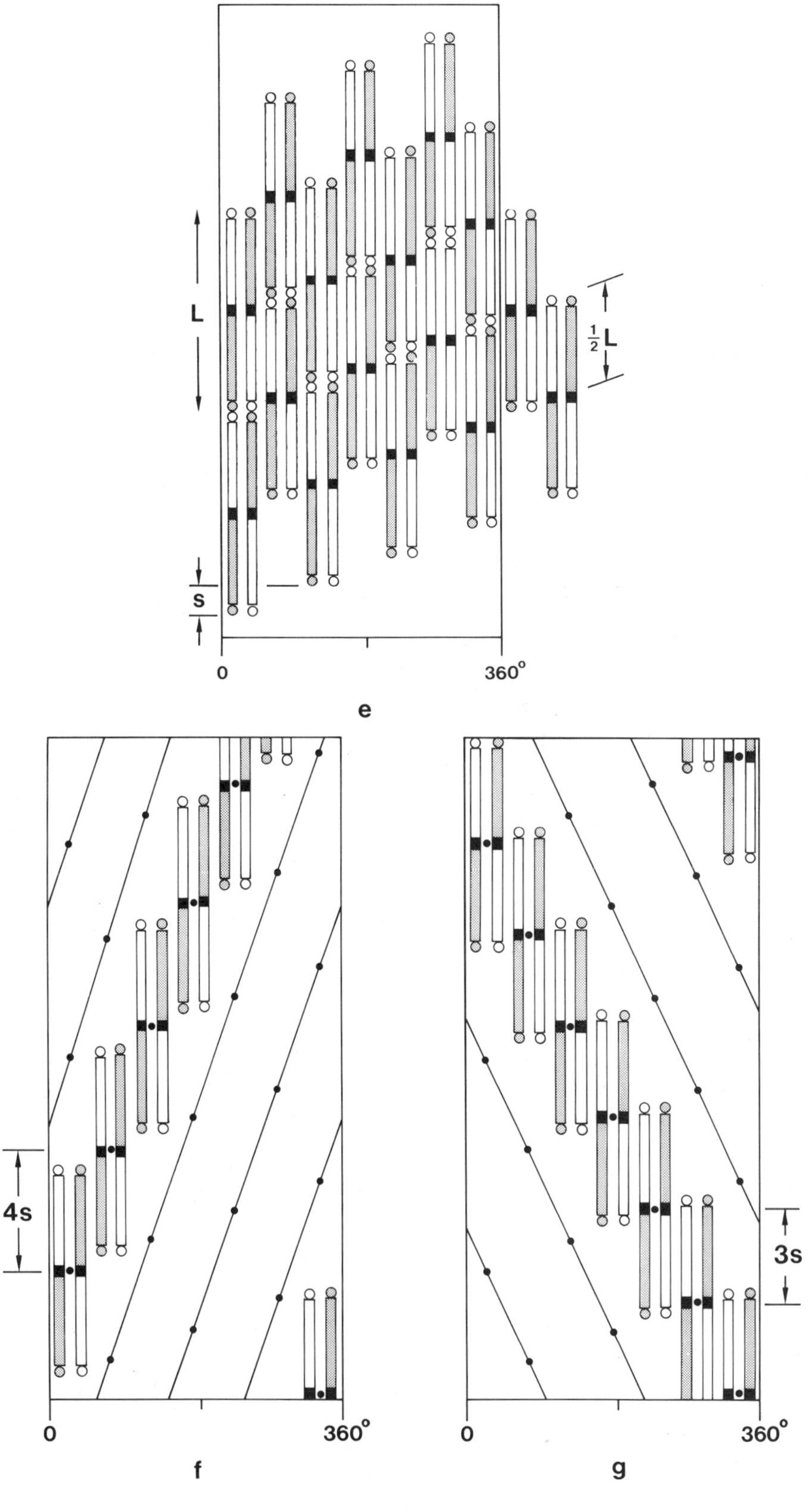
L
½ L
s
0
360°
e
4s
0
360°
f
3s
0
360°
g

This is close in value to a calculated interaction maximum at $\Delta z(1BU, 2D) = -23$ (Fraser *et al.*, 1985), and the relative axial positions of the molecular segments for the lateral aggregate in Fig. 2f, predicted by Eq. (2), are shown in Fig. 3b. No means of deciding the axial translation per residue in the linking segments L1 and L12 is available but it will be seen that the solution in Eq. (2) brings the linking regions into approximate axial register and that the C- and N-terminal domains emerge in the general vicinity of L12. This may be relevant to the suggestion by Traub and Vorgias (1983) that arginine residues in the N-terminal domain of desmin and vimentin interact with L12 during filament assembly. In the schematic representations shown in Figs. 2e, f, and 3b, each segment in the central zone interacts with two neighboring segments. In practice, it is likely that at least one and possibly more additional interactions occur.

3.4. Polymers

In addition to the oligomeric aggregates just described, many examples of polymeric aggregates (subfilaments) have been observed that have diameters less than those of native IF (e.g., see Steinert, 1977; Renner *et al.*, 1981; Franke *et al.*, 1982; Steven *et al.*, 1982; Sauk *et al.*, 1984; Eichner *et al.*, 1985; Engel *et al.*, 1985). Of particular interest are the subfilaments of estimated diameter 2.5–3.5 nm observed in epidermal IF protein preparations by Sauk *et al.* (1984). These subfilaments were beaded and the 6.0- to 8.0-nm-diameter beads were spaced at regular intervals of 21.0 nm. These subfilaments probably involve the pattern of interactions shown in Fig. 8a, which is obtained by the end-to-end aggregation of the oppositely directed molecular pairs in Fig. 2b. The axial period is L but a pair of N-terminal and a pair of C-terminal domains project at approximately equispaced intervals of mean value $L/2 = 22$–23 nm in accord with the observed value. The sum of the volumes of the four-chain segments forming the bead (see Parry and Fraser, 1985) would account satisfactorily for the observed diameter. An alternative possibility for the formation of 2.5- to 3.5-nm subfilaments is shown in Fig. 8b and is obtained by the end-to-end polymerization of molecular pairs interacting as shown in Fig. 2d. In this case, however, the C- and N-terminal domains project at intervals of L, i.e., around 44–45 nm, rather than $L/2$.

Subfilaments of diameter around 4.5 nm are observed in both degraded IF and in partially reconstituted IF (e.g., see Steinert, 1977; Krishnan *et al.*, 1979; Aebi *et al.*, 1983; Eichner *et al.*, 1985) and models for such subfilaments can be formulated by combining two of the subfilaments shown in either Fig. 8a or b with a relative stagger of $L/2$ to give Fig. 8c. The axial period of this type of subfilament is $L/2$; subfilaments with an axial period of L can be formed by omitting the stagger.

In order to obtain a realistic model for the 4.5-nm subfilament, the diagram in Fig. 8c must be wrapped around so that the outer strands of molecules come into contact. In so doing, a helical structure is formed in which molecular pairs of the type shown in Fig. 2d are distributed on a helix with a unit height of $L/2$ and a pitch of L. If the molecules are twisted around each other, this will affect the pitch. Eichner *et al.* (1985) have proposed alternative models of the 4.5-nm subfilament based on optical diffraction patterns obtained from paracrystalline assemblies that are sometimes found in filament preparations reconstituted from keratins of cultured human epidermis.

4. Filament Structure

4.1. A Generic Model for IF

Measurements of mass per unit length for IF obtained using scanning transmission electron microscopy, summarized by Steven (this volume), have yielded a value of 0.67 ± 0.03 chain nm^{-1} (Steven *et al.*, 1983a) and this can be used to estimate the number of subfilaments of the type shown in Fig. 8b that could be accommodated in the IF. The number is simply

$$N = 0.67\ (L/4) \tag{3}$$

giving

L (nm)	N
42	7.0 ± 0.3
44	7.4 ± 0.3
46	7.7 ± 0.3
48	8.0 ± 0.4

If subfilaments of the type shown in Fig. 8a or b are combined with a relative stagger of $L/2$ to form a sheet, we obtain the generic model shown in Fig. 8d for the pattern of interactions in IF. This contains elements of the individual models suggested by Geisler *et al.* (1985), Ip *et al.* (1985), and Fraser *et al.* (1985). If the sheet in Fig. 8d, which has eight subfilaments, were wrapped around to form a tube, the molecules would occupy equivalent positions in the structure and the diameter of the annular ring of molecules would be around 10 nm. The N- and C-terminal domains are aligned in bands that are perpendicular to the filament axis and this would account for the characteristic banding pattern of period $L/2$ around 22 nm, first observed in electron micrographs by Kallman and Wessels (1967) and predicted from an analysis of the x-ray diffraction pattern by Fraser and MacRae (1973b). The stagger between subfilaments is not necessarily $L/2$ as illustrated but could be $L/2$ plus any multiple of $L/8$ for an eight-subfilament model. The mass-per-unit-length data, just discussed, admit the possibility of a seven-subfilament model and in this case the stagger between subfilaments is restricted to $L/2$ plus an odd multiple of $L/14$. In addition, the band of C- and N-terminal domains will necessarily be helical rather than perpendicular to the axis of the filament. In order to calculate the pitch of this helix, it is necessary to know to what extent the tube is distorted by a systematic twist around the filament axis; there is no reason to suppose that the idealized case of straight subfilaments shown in Fig. 8d should be favored over the more general case of a filament with a torsional distortion such that the subfilaments follow helical paths. In the case of α-keratin IF, there is evidence that the subfilaments follow paths that are imperfectly helical (Fraser and MacRae, 1973b, 1983, 1985).

4.2. X-Ray Diffraction Studies

The so-called "hard" α-keratins such as mammalian hair, nail, and claw yield low-angle x-ray diffraction patterns that contain information about the IF in these materials. Particularly detailed patterns can be obtained from porcupine quills (Fig. 1a) and these have been studied in considerable detail (Corey and Wyckoff, 1936; MacArthur, 1943;

Bear, 1944; Lang, 1956; Fraser *et al.*, 1964a, 1965, 1976; Fraser and MacRae, 1973b). The only other material that has yielded patterns of comparable complexity are neurofilaments from *Myxicola* (Wais-Steider *et al.*, 1983) and an example of such a pattern is shown in Fig. 1b. The patterns show a number of similarities including a prominent meridional reflection that has a spacing of 6.7 nm in α-keratin and 6.3 nm in the neurofilament. It is common to encounter such strong meridional reflections in fibrous proteins (Fraser and MacRae, 1973a) and they are generally associated with a uniform axial stagger between molecules, or groups of molecules. Some caution is necessary, however, in interpreting the observation on IF due to the so-called "net-helix ambiguity" (Bear and Selby, 1956; Selby and Bear, 1956), which makes it difficult to decide *a priori* whether the stagger is between repeating units in a helix, i.e., intra-IF, or between IF packed side-by-side on a lattice, i.e., inter-IF (Fraser and MacRae, 1988). The natural breadths of the reflections tend to favor the former, although the latter has been considered as a possibility on at least one occasion (Fraser *et al.*, 1960).

Caution is also required in interpreting the physical significance of the strong meridional reflection even if it is known to originate from within the IF. In a structure such as an IF, in which there is a central framework of coiled-coil segments with nonhelical terminal domains extending outwards, it is possible for the central framework to have a different helical symmetry from the array of terminal domains. An analogous and well-documented situation was found in liquid crystals of poly-γ-benzyl-L-glutamate (Parry and Elliott, 1967; Squire and Elliott, 1969) where the main chain had an α-helical conformation and the side chains were arranged with a different helical symmetry, which gave rise to an intense meridional reflection of spacing 0.506 nm that is not predicted for an α-helical structure. An analysis of the low-angle near-meridional reflections in Fig. 1a (Fraser *et al.*, 1976) suggests that many reflections have their origin in parts of the structure at radii greater than that derived for the central framework and so the possibility of a situation in which the terminal domains pack with a helical symmetry that is related, but different in some way, to the main coiled-coil framework cannot be neglected.

If for the moment we take the simplest interpretation of the observed meridional reflections, i.e., equate them to the stagger between like groups of molecules in the IF, we can attempt to relate this to the observed mass per unit length. An analysis of the x-ray diffraction pattern of α-keratin (Fraser and MacRae, 1983) suggests that the repeating unit in IF contains four polypeptide chains, or two molecules, as in Fig. 2b, and in Table II the axial distance per repeating unit calculated from measurements of mass per unit length are

Table II. Axial Chain Densities Derived from STEM Measurements[a]

Material	Chains per nm	4-chain units per 25 nm	4-chain units per 47 nm	Axial distance per unit (nm)	Reference
Vimentin	0.66	4.1	7.8	6.06	Steven *et al.* (1982)
Epidermal keratin					
Human "light"	0.67	4.2	7.9	5.97	Steven *et al.* (1983a)
Human "heavy"	0.63	3.9	7.4	6.35	Steven *et al.* (1983a)
Bovine	0.71	4.4	8.3	5.63	Steven *et al.* (1983b)

[a]Steven (this volume).

listed. The range of values so obtained includes the values of 6.3 nm obtained for neurofilaments (Wais-Steider *et al.*, 1983) and 6.7 nm for α-keratin (Fraser *et al.*, 1976) although the mean lies closer to the neurofilament value.

4.3. The Surface Lattice

The surface lattice of the generic model for IF (Fig. 8d) is depicted in Fig. 9a (Fraser and MacRae, 1985, 1988). Equivalent points, shown as filled circles, lie on four helices of pitch P at axial intervals of h, the unit height, and the four helices are related by a fourfold symmetry axis. The choice of $P = 4L$ in Fig. 9a is arbitrary. Two molecules, related by a perpendicular diad, are associated with each point so that if all the molecules occupy equivalent positions, the line group (Klug *et al.*, 1958) is sr2 ($n = 4$). The potential for such a high symmetry may not be realized in practice, particularly if the molecules are organized into compact subfilaments.

The low-angle x-ray diffraction pattern produced by hard α-keratin is extremely complex and the meridional reflections can be indexed on a repeat of $h = 47$ nm (Fraser and MacRae, 1971), which is seven times the spacing of the intense 6.7-nm reflection.

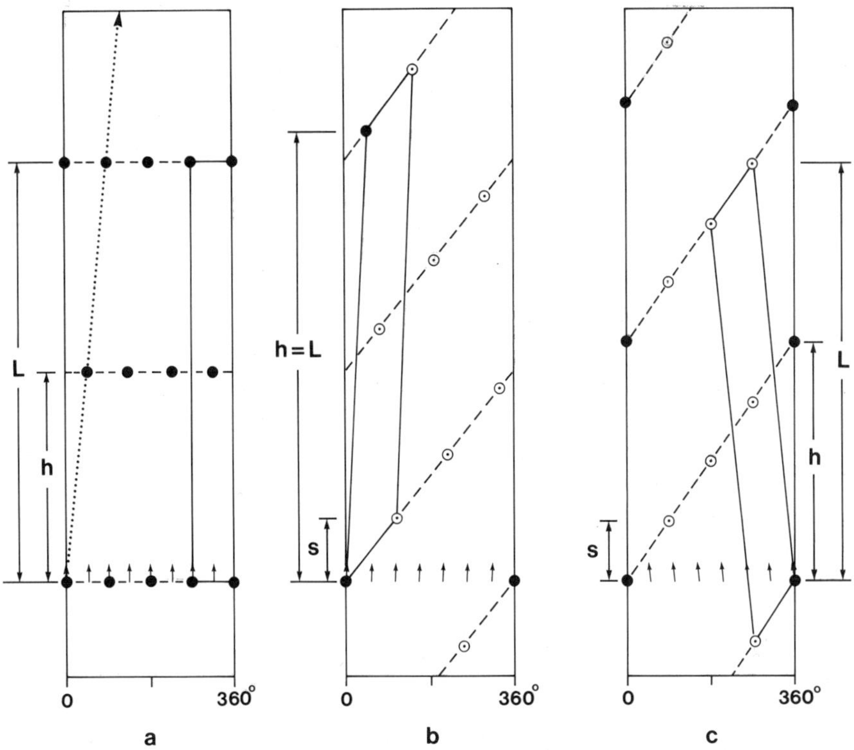

Figure 9. Surface lattices for (a) the generic model shown in Fig. 8d, (b) hard α-keratin, and (c) neurofilaments. The lattices in (b) and (c) are speculations based on the x-ray diffraction data. The topology of molecular interactions in Fig. 8d is preserved as indicated by the transformation of the rectangle in (a) into the trapezoids in (b) and (c). The number of subfilaments of the type shown in Fig. 8a or b is reduced from eight in (a) to seven in (b) and (c).

The simplest interpretation is that there is a mean axial spacing s (Fig. 9b) between equivalent points that is perturbed by a pattern of axial displacements that has a period of 47 nm (Fraser and MacRae, 1973b, 1983, 1985). Similarly, the meridional reflections from *Myxicola* neurofilaments can be indexed on a repeat of $h = 25.2$ nm (Wais-Steider *et al.*, 1983) and a corresponding interpretation is possible for the intense 6.3-nm meridional reflection except that $h = 4s$ in this case (Fig. 9c). The helix is arbitrarily shown as being right-handed and for convenience of illustration has been drawn with a pitch of 25.2 nm, since the true value is unknown. It will be noted that $7s = 44.1$ nm, which is close to values of L determined by electron microscopy. This suggests the possibility that the rectangle in Fig. 9a corresponds to the trapezoid in Fig. 9c. The topological relationship between the molecules and the length L is retained but the intersubfilament displacement is changed from $L/2$ to $(L \pm s)/2$ depending on the convention used for specifying relative axial displacement. Although the interaction topology of Fig. 8d is preserved, there are only seven subfilaments of the type shown in Fig. 8a or 8b per cross section, as illustrated in Fig. 8e. A similar scheme can be devised for the α-keratin surface lattice (Fig. 9b) and again there are only seven subfilaments per cross section (Fraser *et al.*, 1985). In this case the value of L is increased to 47 nm, but this is still well within the range of estimates of molecular length. It is possible in both cases to add an eighth subfilament but this leads to an interruption in the regularity of the helix (Fraser and MacRae, 1985). The surface latttice for hard α-keratin IF, shown in Fig. 9b, predicts a linear density of 0.60 chain nm^{-1}, which is low compared with the mean of the values listed in Table II. Unfortunately, no measurements of linear density for hard α-keratin IF have been reported. One possible explanation for the apparent discrepancy is that hard α-keratin IF have a central subfilament, which would lead to a predicted linear density of 0.66 chain nm^{-1} in better accord with the values given in Table II (Fraser and MacRae, 1983). Alternatively, an eight-subfilament model, without the requirement that one be centrally located, can be constructed and has been described by Fraser and MacRae (1985). Although the lattice mismatch along the locus of the helical dislocation is larger in this case than it is for the seven-subfilament structure, the probability that the rod domains of the IF molecules lie closely parallel (or antiparallel) to this line may indicate that the extent of lattice mismatch may be of little significance *in vivo*. The eight-subfilament model thus remains the most detailed and most likely of any existing at the current time for the structure of an IF.

Fraser and MacRae (1985, 1988) have also been able to show that the meridional (and near-meridional) intensity distribution from the x-ray diffraction pattern of dried α-keratin can be explained quantitatively by a model in which arrays of 7–9 repeating points, each separated from its neighbors by 19.8 nm (or a multiple), are laid down at axial intervals of 47 nm. Such a scheme provides an explanation for an unusual feature of the meridional diffraction pattern to a resolution of about 1 nm, i.e., the absence of about 75% of the orders of the 47-nm repeat. These absences are almost totally accounted for by the sampling of the transform of a finite lattice of period 19.8 nm by the transform of an infinite lattice of period 47 nm. There are, however, a small number of point models that account quantitatively for the low-resolution x-ray data (Fraser and MacRae, 1985, 1988; M. C. Lawrence, unpublished data) and further criteria will be required to distinguish between them.

The surface lattice shown in Fig. 9b can also accommodate a structure composed of subfilaments with diameters in the range 4–5 nm. Fraser and MacRae (1982) pointed out that this lattice can be viewed as consisting of three subfilaments, each having an axial repeat of structure around 20 nm. The relevant interaction topology is shown in Fig. 8g. Although only (2U, 2D) interactions are shown, it is likely that others, e.g., (1BU, 1BD), would be formed by wrapping the overlapping parts of the repeating units around each other. Fraser *et al.* (1985) considered the general problem of the appearance of subfilaments in partially degraded IF and pointed out that in addition to the three-subfilament model just discussed, the surface lattice was consistent with a four-subfilament model having the interaction topology shown in Fig. 8f or even a seven-subfilament model (see Steinert and Parry, 1985). In addition to the (1BU, 1BD) interactions shown, it is again likely that the overlapping segments would associate leading, for example, to (2U, 2D) interactions.

4.4. *Disposition of Terminal Domains*

A major difference between the model depicted in Fig. 9a and those in Fig. 9b and c is the disposition of the N- and C-terminal domains. In the first case they occur in bands that are perpendicular to the axis of the filament and spaced at intervals of $L/2$, whereas in the second they are arranged on a helix with a pitch in the vicinity of $L/2$. In most electron micrographs of IF preparations, the banding does not show any evidence of helical character, and in some instances the banding is fine enough to be strongly suggestive of a lattice of the type shown in Fig. 9a. In other instances, a distinctly helical appearance is observed. Milam and Erickson (1982) have reported that epidermal IF have a helical banding pattern and that the helix is left-handed. This is opposite to that shown in Fig. 9b, which was chosen arbitrarily since the x-ray diffraction patterns of the mirror image structures are indistinguishable.

4.5. *Models Generated Using Interaction Scores*

The calculations of ionic interaction potential, discussed earlier, can be combined with surface lattice information to suggest model structures suitable for further refinement, and the result of such a search (Fraser *et al.*, 1986) is listed in Table III and illustrated in Fig. 10. The pattern of interactions between the molecules for the highest score solution is topologically equivalent to that of Fig. 8d and it is of interest that the score for disulfide bond potential is also a maximum for this solution (see also Fraser *et al.*, 1988, for a discussion of disulfide bonding in α-keratin).

So far we have followed a line of reasoning in which it has been supposed that the building block of the IF is a subfilament of the type shown in Figure 8a or b, but an alternative possibility is that these subfilaments combine in pairs to form a larger subfilament of the type shown in Figure 8c which maintains its identity in the final IF. In this case alternative interpretations of the surface lattice may need to be considered.

4.6. *Radial Distribution of Coiled-Coil Rope Segments*

In conclusion, we consider briefly the evidence for the radial distribution of the coiled-coil rope segments in the IF. Evidence of substructure within the hard α-keratin IF was obtained by Filshie and Rogers (1961) (Fig. 11a) who suggested that a (9 + 2) subfilament structure may be present. However, this was not supported by x-ray diffraction studies (Fraser *et al.*, 1964b, 1972) and later Millward (1970) showed that at close-to-

Table III. Combinations of Ionic Interaction Maxima between Coiled-Coil Rope Segments in Wool IF Protein (Type I Component 8c-1 + Type II Component 7c Heterodimer) That Are Compatible with the Observed Lattice in Hard Keratins[a]

Stagger Δz[b]				Calculated	Ionic		SS bonding[d]		
(1BU, 1BD)	(2U, 2D)	(1BU, 2U)	(1BU, 2D)	Δz_a[b]	score	$\langle h_{12} \rangle$[c]	±1	±2	Ranking
14	13	—	−34	48	182	0.318	6	10	1
15	13	—	−34	49	176	0.323	6	10	
16	13	—	−34	50	176	0.327	6	10	
−5	13	—	−45	51	176	0.230	2	8	2
−5	12	—	−45	50	170	0.234	2	4	
4	4	—	−45	51	172	0.313	0	4	3a
5	4	—	−45	50	170	0.318	0	2	
−5	32	—	−34	49	172	0.142	7	7	3b
−5	−7	—	−54	49	170	0.323	0	1	4
−5	4	−53	—	50	172	0.272	0	0	3c
4	13	−53	—	50	172	0.272	2	8	3d

[a]Fraser *et al.* (1986).
[b]Expressed in residue translations of 0.1485 nm, appropriate to the coiled-coil conformation. The target value of Δz_a is 49.9.
[c]$\langle h_{12} \rangle$ is the mean axial translation per residue in the nonhelical segment linking coiled-coil rope segments 1B and 2.
[d]The potential number of disulfide linkages between the coiled-coil rope segments in a repeating unit calculated on the basis of a tolerance of ±1 and ±2 residues from exact register.

zero defect of focus the subfilament structure was no longer apparent. Instead the image showed evidence of a lightly stained annulus together with a lightly stained core (Fig. 11b, c). The most likely origins of stain exclusion are a high content of regular secondary structure (e.g., coiled-coil ropes) or a high concentration of hydrophobic residues or perhaps disulfide bonds. A similar result was obtained by direct Fourier synthesis from the equatorial reflections in the electron diffraction pattern (Fig. 11d). Eriksson and Thornell (1979) reported a totally different appearance for desmin, which they suggested indicated a fourfold rotation axis.

Information on the radial distribution of electron density in the hard α-keratin IF has also been obtained from x-ray diffraction studies. At low resolution the IF in dry α-keratin scatter like cylinders of diameter around 7.5 nm (Fraser and MacRae, 1958; Fraser *et al.*, 1973). At high resolution the mean diameter on which the coiled-coil ropes are distributed has been estimated to be about 6 nm (Fraser *et al.*, 1965). These dimensions are small in relation to the estimates of 10–12 nm for the diameter obtained from electron microscope studies of other IF, but it must be recalled that the x-ray measurements were performed on dried specimens in which the structure had partially collapsed from its native form (Fraser *et al.*, 1971).

If we assume that the packing diameter of a two-strand coiled-coil rope is around 2 nm, we obtain an estimate of about 10 nm for the diameter of the ring formed when the array of molecules shown in Fig. 8d is wrapped around to give the generic model for IF suggested by *in vitro* electron microscope studies (Fig. 12a). More compact structures with greater potential for intermolecular interactions can be formed by a regular folding of the ring (Fig. 12b–e). In Fig. 12b–d it is possible to allot one type of segment, e.g., 1B,

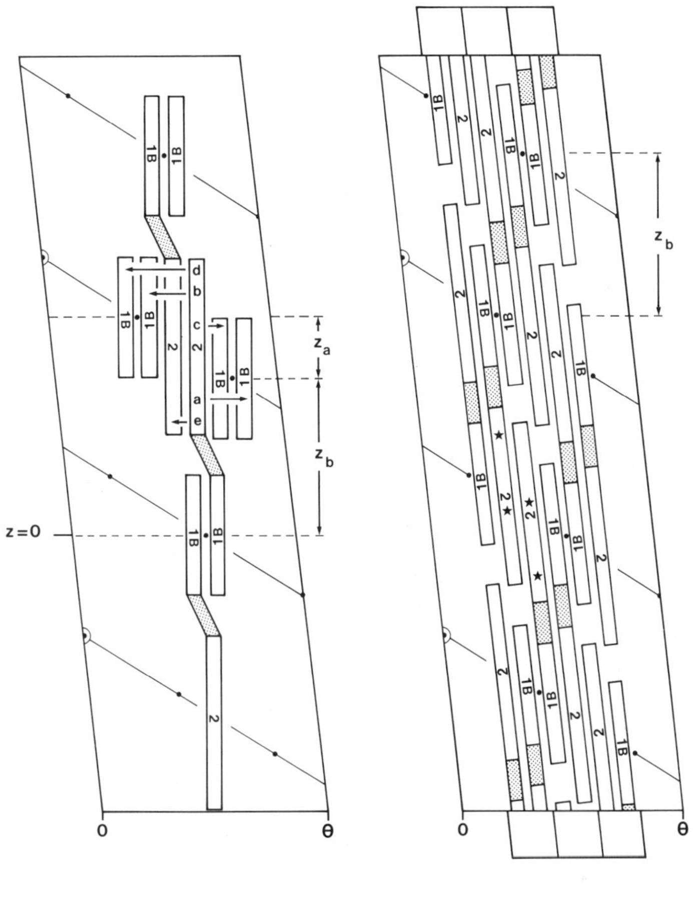

Figure 10. Helical projections of a portion of the surface lattice deduced for hard α-keratin IF (Fraser *et al.*, 1986). The hand of the helix has been changed from right, as in Fig. 9b, to left on the basis of the observations of Milam and Erickson (1982). (a) Five possible types of association (*a–e*) between coiled-coil rope segments for repeating units of the type shown in Fig. 2b. (b) The pattern of interactions that gives the highest interaction score (see Table II). This is topologically similar to the pattern shown in Fig. 8d. The nonhelical linking segment L12 (shown shaded) has an average axial translation per residue close to that observed in the β conformation. The stars indicate breaks in the regularity of the coiled coil and z_a and z_b are axial translations between equivalent points in the surface lattice.

to the inner part of the IF and the other to the outer part of the IF, thus maintaining potential for at least quasi-equivalence of environment. In the most compact form (Fig. 12e) this is not possible and two different types of environment are entailed. Further development of models for the radial distribution of the rope segments requires a knowledge of the extent to which they are clustered into subfilaments and whether these in turn involve a twisting together of the segments.

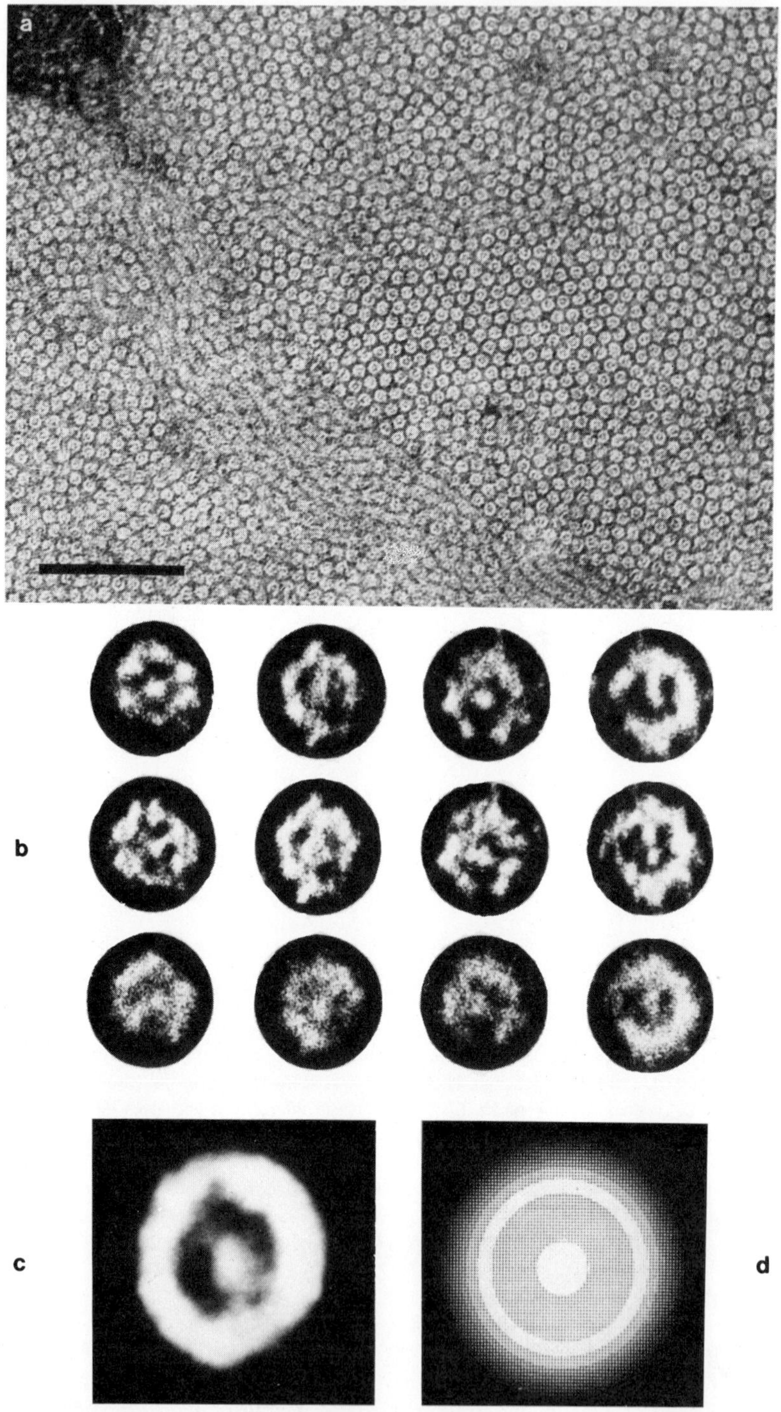

Figure 11. (a) Cross section of a Merino wool fiber showing substructure in the IF. Bar = 100 nm (courtesy of Professor G. E. Rogers). (b) Images of cross sections of individual hard α-keratin IF obtained close to focus (lower set), 0.44 μm underfocus (central set), and 0.66 μm underfocus (upper set). (c) An averaged image (Millward, 1970). (d) A low-resolution Fourier synthesis obtained from an electron diffraction pattern of the specimen used to record the images. The mean diameter of the ring is around 6 nm.

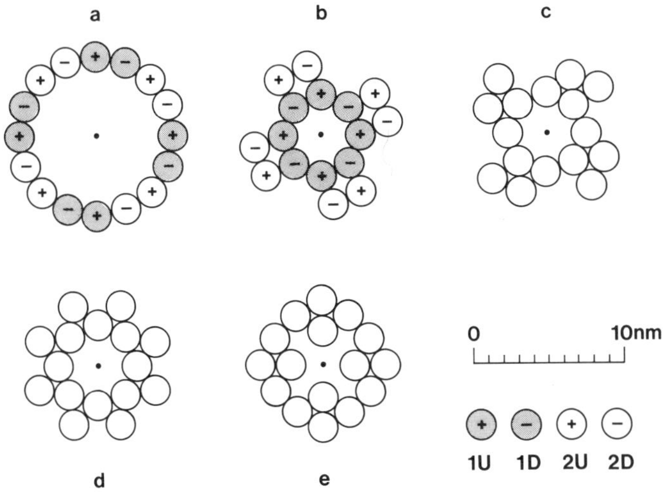

Figure 12. Some basic concepts in coiled-coil rope packing in the framework of the IF. (a) The generic model of Fig. 8d produces a tubular structure of two-strand coiled coils with rope segments arranged on a ring of diameter around 10 nm. (b) More compact structures preserving equivalent, or quasi-equivalent molecular environments can be formulated (Fraser *et al.*, 1985). These would be compatible with the appearance of longitudinal striations and a 4.0- to 4.5-nm subfilament character. In (c), (d), and (e), progressively more compact arrangements are shown.

5. *Concluding Remarks*

In this brief account of present-day knowledge about the three-dimensional structure of IF, we have attempted to bring together some pertinent observations and indicate ways in which they may be used to place constraints on possible models. One weakness of the approach forced upon us by the paucity of precise data on many aspects is the implicit assumption that all IF have a common structure. The high degree of homology between the amino acid sequences of the rod domains (Parry and Fraser, 1985; Steinert and Parry, 1985; Parry, this volume) indicates that the structures of all IF will be closely related, but experience with other fibrous protein assemblies (Fraser and MacRae, 1973a) and particularly muscle proteins (Squire, 1985) and collagen (Doyle *et al.*, 1975; Brodsky *et al.*, 1980) emphasizes the likelihood of polymorphism between different types of IF (Conway and Parry, 1988). One likely origin of polymorphism among IF is the fact that IF of epithelial origin contain two types of chain in equal proportions (Type I and II) whereas desmin and vimentin contain only one (Type III), and neurofilaments contain mixtures of chain types (Type IV).

References

Aebi, U., Fowler, W. E., Rew, P., and Sun, T.-T., 1983, The fibrillar substructure of keratin filaments unraveled, *J. Cell Biol.* **97:**1131–1143.

Ahmadi, B., and Speakman, P. T., 1978, Suberimidate crosslinking shows that a rod-shaped, low-cystine, high-helix protein prepared by limited proteolysis of reduced wool has four protein chains, *FEBS Lett.* **94:**365–367.

Astbury, W. T., 1933, *Fundamentals of Fibre Structure,* Oxford University Press, London, p. 95.

Astbury, W. T., and Street, A., 1931, X-ray studies of the structure of hair, wool and related fibres. I. General, *Philos. Trans. R. Soc. London Ser. A* **230:**75–101.

Astbury, W. T., and Woods, H. J., 1933, X-ray studies of the structure of hair, wool and related fibres. II. The molecular structure and elastic properties of hair keratin, *Philos. Trans. R. Soc. London Ser. A* **232:**333–394.

Bear, R. S., 1944, X-ray diffraction studies on protein fibers. II. Feather rachis, porcupine quill tip and clam muscle, *J. Am. Chem. Soc.* **66:**2043–2050.

Bear, R. S., and Selby, C. C., 1956, The structure of paramyosin fibrils according to X-ray diffraction, *J. Biophys. Biochem. Cytol.* **2:**55–70.

Bragg, W. L., Kendrew, J. C., and Perutz, M. F., 1950, Polypeptide chain configurations in crystalline proteins, *Proc. R. Soc. London Ser. A* **203:**321–357.

Brodsky, B., Eikenberry, E. F., and Cassidy, K., 1980, An unusual collagen periodicity in skin, *Biochim. Biophys. Acta* **621:**162–166.

Conway, J. F., and Parry, D. A. D., 1988, Intermediate filament structure 3: Analysis of sequence homologies, *Int. J. Biol. Macromol.* **10:**79–98.

Corey, R. B., and Wyckoff, R. W. G., 1936, Long spacings in macromolecular solids, *J. Biol. Chem.* **114:**407–414.

Crewther, W. G., and Dowling, L. M., 1971, The preparation and properties of large peptides from the helical regions of the low-sulphur proteins of wool, *Appl. Polym. Symp.* No. 18, 1–20.

Crewther, W. G., Inglis, A. S., and McKern, N. M., 1978, Amino acid sequences of α-helical segments from S-carboxymethylkerateine-A. II. Complete sequence of a type II segment, *Biochem. J.* **173:**365–371.

Crewther, W. G., Dowling, L. M., Steinert, P. M., and Parry, D. A. D., 1983, Structure of intermediate filaments, *Int. J. Biol. Macromol.* **5:**267–274.

Crick, F. H. C., 1952, Is α-keratin a coiled-coil? *Nature* **170:**882–883.

Crick, F. H. C., 1953a, The Fourier transform of a coiled-coil, *Acta Crystallogr.* **6:**685–689.

Crick, F. H. C., 1953b, The packing of α-helices: Simple coiled-coils, *Acta Crystallogr.* **6:**689–697.

Doyle, B. B., Hukins, D. W. L., Hulmes, D. J. S., Miller, A., and Woodhead-Galloway, J., 1975, Collagen polymorphism: Its origins in the amino acid sequence, *J. Mol. Biol.* **91:**79–99.

Eichner, R., Rew, P., Engel, A., and Aebi, U., 1985, Human epidermal keratin filaments: Studies on their structure and assembly, in: *Intermediate Filaments* (E. Wang, D. Fischman, R. K. H. Liem, and T. T. Sun, eds.), New York Academy of Sciences, New York, pp. 381–402.

Engel, A., Eichner, R., and Aebi, U., 1985, Polymorphism of reconstituted human epidermal keratin filaments: Determination of their mass-per-length and width by scanning transmission electron microscopy (STEM), *J. Ultrastruct. Res.* **90:**323–335.

Eriksson, A., and Thornell, L. E., 1979, Intermediate (skeletin) filaments in heart Purkinje fibers: A correlative morphological and biochemical identification with evidence of a cytoskeletal function, *J. Cell Biol.* **80:**231–247.

Filshie, B. K., and Rogers, G. E., 1961, The fine structure of α-keratin, *J. Cell Biol.* **3:**784–786.

Franke, W. W., Schiller, D. L., and Grund, C., 1982, Protofilamentous and annular structures as intermediates during reconstitution of cytokeratin filaments *in vitro, Biol. Cell* **46:**257–268.

Fraser, R. D. B., and MacRae, T. P., 1958, Structural implications of the equatorial X-ray diffraction pattern of α-keratin, *Biochim. Biophys. Acta* **29:**229–240.

Fraser, R. D. B., and MacRae, T. P., 1961, α-Configuration of fibrous proteins, *Nature* **189:**572–573.

Fraser, R. D. B., and MacRae, T. P., 1971, Structure of α-keratin, *Nature* **233:**138–140.

Fraser, R. D. B., and MacRae, T. P., 1973a, *Conformation in Fibrous Proteins and Related Synthetic Polypeptides,* Academic Press, New York.

Fraser, R. D. B., and MacRae, T. P., 1973b, The structure of α-keratin, *Polymer* **14:**61–67.

Fraser, R. D. B., and MacRae, T. P., 1982, The fine structure of keratin fibers, in: *Milton Harris: Chemist, Innovator and Entrepreneur* (M. M. Breuer, ed.), American Chemical Society, Washington, D.C., pp. 109–137.

Fraser, R. D. B., and MacRae, T. P., 1983, The structure of the α-keratin microfibril, *Biosci. Rep.* **3:**517–525.

Fraser, R. D. B., and MacRae, T. P., 1985, Intermediate filament structure, *Biosci. Rep.* **5:**573–579.

Fraser, R. D. B., and MacRae, T. P., 1988, Surface lattice in α-keratin filaments, *Int. J. Biol. Macromol.* **10:**178–184.

Fraser, R. D. B., MacRae, T. P., and Rogers, G. E., 1960, Recent observations on the structure of α-keratin, *J. Text. Inst. Trans.* **51:**T497–T505.

Fraser, R. D. B., MacRae, T. P., and Miller, A., 1964a, The coiled-coil model of α-keratin structure, *J. Mol. Biol.* **10:**147–156.

Fraser, R. D. B., MacRae, T. P., and Miller, A., 1964b, Molecular structure of α-keratin, *Nature* **203:**1231–1233.

Fraser, R. D. B., MacRae, T. P., and Miller, A., 1965, X-ray diffraction patterns of α-fibrous proteins, *J. Mol. Biol.* **14:**432–442.

Fraser, R. D. B., MacRae, T. P., Millward, G. R., Parry, D. A. D., Suzuki, E., and Tulloch, P. A., 1971, The molecular structure of keratins, *Appl. Polym. Symp.* No. 18, 65–83.

Fraser, R. D. B., MacRae, T. P., and Rogers, G. E., 1972, *Keratins: Their Composition, Structure and Biosynthesis,* Thomas, Springfield, Ill.

Fraser, R. D. B., Gillespie, J. M., and MacRae, T. P., 1973, Tyrosine-rich proteins in keratins, *Comp. Biochem. Physiol.* **44B:**943–947.

Fraser, R. D. B., MacRae, T. P., and Suzuki, E., 1976, Structure of the α-keratin microfibril, *J. Mol. Biol.* **108:**435–452.

Fraser, R. D. B., MacRae, T. P., Suzuki, E., and Tulloch, P. A., 1981, Ordered assemblies in fibrous proteins, in: *Structural Aspects of Recognition and Assembly in Biological Macromolecules* (M. Balaban, J. L. Sussman, W. Traub, and A. Yonath, eds.), Balaban ISS, Rehovot, Israel, pp. 327–340.

Fraser, R. D. B., MacRae, T. P., Suzuki, E., and Parry, D. A. D., 1985, Intermediate filament structure: 2. Molecular interactions in the filament, *Int. J. Biol. Macromol.* **7:**258–274.

Fraser, R. D. B., MacRae, T. P., Parry, D. A. D., and Suzuki, E., 1986, Intermediate filaments in α-keratins, *Proc. Natl. Acad. Sci. USA* **83:**1179–1183.

Fraser, R. D. B., MacRae, T. P., Sparrow, L. G., and Parry, D. A. D., 1988, Disulphide bonding in α-keratin, *Int. J. Biol. Macromol.* **10:**106–112.

Fraser, R. D. B., Conway, J. F., and Parry, D. A. D., *Int. J. Biol. Macromol.*, in preparation.

Geisler, N., and Weber, K., 1982, The amino acid sequence of chicken muscle desmin provides a common structural model for intermediate filament proteins, *EMBO J.* **1:**1649–1656.

Geisler, N., Kaufmann, E., and Weber, K., 1985, Antiparallel orientation of the two double-stranded coiled-coils in the tetrameric protofilament unit of intermediate filaments, *J. Mol. Biol.* **182:**173–177.

Gough, K. H., Inglis, A. S., and Crewther, W. G., 1978, Amino acid sequences of α-helical segments from S-carboxymethylkerateine-A. Complete sequence of a type I segment, *Biochem. J.* **173:**373–385.

Gruen, L. C., and Woods, E. F., 1983, Structural studies on the microfibrillar proteins of wool, *Biochem. J.* **209:**587–595.

Hatzfeld, M., and Franke, W. W., 1985, Pair formation and promiscuity of cytokeratins: Formation *in vitro* of heterotypic complexes and intermediate-sized filaments by homologous and heterologous recombinations of purified polypeptides, *J. Cell Biol.* **101:**1826–1841.

Hofmann, H., Fietzek, P. P., and Kuhn, K., 1978, The role of polar and hydrophobic interactions for the molecular packing of type I collagen: A three-dimensional evaluation of the amino acid sequence, *J. Mol. Biol.* **125:**137–165.

Hulmes, D. J. S., Miller, A., Parry, D. A. D., Piez, K. A., and Woodhead-Galloway, J., 1973, Analysis of the primary structure of collagen for the origins of molecular packing, *J. Mol. Biol.* **79:**137–148.

Ip, W., Hartzer, M. K., Pang, Y.-Y. S., and Robson, R. M., 1985, Assembly of vimentin *in vitro* and its implications concerning the structure of intermediate filaments, *J. Mol. Biol.* **183:**365–375.

Kallman, F., and Wessels, N. K., 1967, Periodic repeat units of epithelial cell tonofilaments, *J. Cell Biol.* **32:**227–231.

Kaufmann, E., Weber, K., and Geisler, N., 1985, Intermediate filament forming ability of desmin derivatives lacking either the amino-terminal 67 or the carboxy-terminal 27 residues, *J. Mol. Biol.* **185:**733–742.

Klug, A., Crick, F. H. C., and Wyckoff, H. W., 1958, Diffraction by helical structures, *Acta Crystallogr.* **11:**199–213.

Krishnan, N., Kaiserman-Abramof, I. R., and Raymond, J. L., 1979, Helical structure of neurofilaments isolated from *Myxicola* and giant squid axons, *J. Cell Biol.* **82:**323–335.

Lang, A. R., 1956, An X-ray study of α-keratin. II. Diffractometer measurements of the complete diffraction pattern of Canadian porcupine quill, *Acta Crystallogr.* **9:**446–451.

MacArthur, I., 1943, Structure of α-keratin, *Nature* **152:**38–41.

McLachlan, A. D., and Stewart, M., 1982, Periodic charge distribution in the intermediate filament proteins desmin and vimentin, *J. Mol. Biol.* **162:**693–698.

Milam, L., and Erickson, H. P., 1982, Visualization of a 21 nm axial periodicity in shadowed keratin filaments and neurofilaments, *J. Cell Biol.* **94:**592–596.

Millward, G. R., 1970, The substructure of α-keratin microfibrils, *J. Ultrastruct. Res.* **31:**349–355.

Parry, D. A. D., 1975, Double helix of tropomyosin, *Nature* **256:**346–347.

Parry, D. A. D., and Elliott, A., 1967, The structure of a paracrystalline phase of poly-γ-benzyl-L-glutamate in dimethylformamide, *J. Mol. Biol.* **25:**1–13.

Parry, D. A. D., and Fraser, R. D. B., 1985, Intermediate filament structure: 1. Analysis of IF protein sequence data, *Int. J. Biol. Macromol.* **7:**203–213.

Parry, D. A. D., Crewther, W. G., Fraser, R. D. B., and MacRae, T. P., 1977, Structure of α-keratin: Structural implication of the amino acid sequences of the type I and type II chain segments, *J. Mol. Biol.* **113:**449–454.

Parry, D. A. D., Steven, A. C., and Steinert, P. M., 1985, The coiled-coil molecules of intermediate filaments consist of two parallel chains in exact axial register, *Biochem. Biophys. Res. Commun.* **127:**1012–1018.

Pauling, L., and Corey, R. B., 1951, Atomic coordinates and structure factors for two helical configurations of polypeptide chains, *Proc. Natl. Acad. Sci. USA* **37:**235–240.

Pauling, L., and Corey, R. B., 1953, Compound helical configurations of polypeptide chains: Structure of proteins of the α-keratin type, *Nature* **171:**59–61.

Renner, W., Franke, W. W., Schmid, E., Geisler, N., Weber, K., and Mandelkow, E., 1981, Reconstitution of intermediate-sized filaments from denatured monomeric vimentin, *J. Mol. Biol.* **149:**285–306.

Rudall, K. M., 1952, The proteins of the mammalian epidermis, *Adv. Protein Chem.* **7:**253–290.

Sauk, J. J., Krumweide, M., Cocking-Johnson, D., and White, J. G., 1984, Reconstitution of cytokeratin filaments *in vitro:* Further evidence for the role of nonhelical peptides in filament assembly, *J. Cell Biol.* **99:**1590–1597.

Selby, C. C., and Bear, R. S., 1956, The structure of actin-rich filaments of muscles according to X-ray diffraction, *J. Biophys. Biochem. Cytol.* **2:**71–84.

Sheridan, R. P., Levy, R. M., and Salemme, F. R., 1982, α-Helix dipole moment and electrostatic stabilization of 4-α-helical proteins, *Proc. Natl. Acad. Sci. USA* **79:**4545–4549.

Shoemaker, K. R., Kim, P. S., York, E. J., Stewart, J. M., and Baldwin, R. L., 1987, Tests of the helix dipole model for stabilization for α-helices, *Nature* **326:**563–567.

Skerrow, D., Matoltsy, A. G., and Matoltsy, M. N., 1973, Isolation and characterization of the α-helical regions of epidermal prekeratin, *J. Biol. Chem.* **248:**4820–4826.

Squire, J. M., 1985, Muscle myosin filaments: Internal structure and crossbridge organization, *Comments Mol. Cell. Biophys.* **3:**155–177.

Squire, J. M., and Elliott, A., 1969, Liquid crystalline phases of poly-γ-benzyl-glutamate in solution, *Mol. Cryst. Liquid Cryst.* **7:**457–468.

Steinert, P. M., 1977, The mechanism of assembly of bovine epidermal keratin filaments *in vitro,* in: *Biochemistry of Cutaneous Epidermal Differentiation* (M. Seiji and I. A. Bernstein, eds.), University of Tokyo Press, Tokyo, pp. 444–464.

Steinert, P. M., 1978, Structure of the three-chain unit of the bovine epidermal keratin filament, *J. Mol. Biol.* **123:**49–70.

Steinert, P. M., 1981, Intermediate filaments (IF), in: *Electron Microscopy of Proteins* (J. R. Harris, ed.), Academic Press, New York, Volume 1, pp. 125–166.

Steinert, P. M., and Parry, D. A. D., 1985, Intermediate filaments: Conformity and diversity of expression and structure, *Annu. Rev. Cell Biol.* **1:**41–65.

Steinert, P. M., Idler, W. W., and Goldman, R. D., 1980, Intermediate filaments of baby hamster kidney (BHK-21) cells and bovine epidermal keratinocytes have similar ultrastructures and subunit domain structures, *Proc. Natl. Acad. Sci. USA* **77:**4534–4538.

Steinert, P. M., Idler, W. W., Cabral, F., Gottesman, M. M., and Goldman, R. D., 1981, *In vitro* assembly of homopolymer and copolymer filaments from intermediate filament subunits of muscle and fibroblastic cells, *Proc. Natl. Acad. Sci. USA* **78:**3692–3696.

Steven, A. C., Wall, J., Hainfeld, J., and Steinert, P. M., 1982, Structure of fibroblastic intermediate filaments: Analysis by scanning transmission electron microscopy, *Proc. Natl. Acad. Sci. USA* **79:**3101–3105.

Steven, A. C., Hainfeld, J. F., Trus, B. L., Wall, J. S., and Steinert, P. M., 1983a, Epidermal keratin filaments assembled *in vitro* have masses-per-unit-length that scale according to average subunit mass: Structural basis for homologous packing of subunits in intermediate filaments, *J. Cell Biol.* **97:**1939–1944.

Steven, A. C., Hainfeld, J. F., Trus, B. L., Wall, J. S., and Steinert, P. M., 1983b, The distribution of mass in heteropolymer intermediate filaments assembled *in vitro, J. Biol. Chem.* **258:**8323–8329.

Suzuki, E., Crewther, W. G., Fraser, R. D. B., MacRae, T. P., and McKern, N. M., 1973, X-ray diffraction and infrared studies of an α-helical fragment from α-keratin, *J. Mol. Biol.* **73:**275–278.

Traub, P., and Vorgias, C. E., 1983, Involvement of the N-terminal polypeptide of vimentin in the formation of intermediate filaments, *J. Cell Sci.* **63:**43–67.

Trus, B. L., and Piez, K. A., 1976, Molecular packing of collagen: Three-dimensional analysis of electrostatic interactions, *J. Mol. Biol.* **108:**705–732.

Wais-Steider, C., Eagles, P. A. M., Gilbert, D. S., and Hopkins, J. M., 1983, Structural similarities and differences amongst neurofilaments, *J. Mol. Biol.* **165:**393–400.

Woods, E. F., 1983, The number of polypeptide chains in the rod domain of bovine epidermal keratin, *Biochem. Int.* **7:**769–774.

Woods, E. F., and Inglis, A. S., 1984, Organization of the coiled-coils in the wool microfibril, *Int. J. Biol. Macromol.* **6:**277–283.

9

Intermediate Filament Structure
Diversity, Polymorphism, and Analogy to Myosin

Alasdair C. Steven

1. Introduction

Together with actin microfilaments and microtubules (MT), intermediate filaments (IF, or "intermediate-sized" filaments) constitute one of the three major systems of protein filaments that form the cytoskeletal networks of virtually all eukaryotic cells. However, unlike actin and tubulin, which are highly conserved molecules, IF proteins make up a notably diverse family, whose molecular weights range from ~ 44k to ~ 120k. The helical packing arrangements of protein subunits in actin filaments and MT have long been known, and their quaternary structures have been found (like their primary sequences) to be resistant to evolutionary divergence. In contrast, a conclusive determination of the molecular packing in any kind of IF remains to be achieved, although considerable progress has been made (see Fraser *et al.*, this volume). Consequently, the extent to which different IF are structurally related—apart from their having roughly similar diameters—has been unclear.

Ultimately, the goal of structural studies is to realize an understanding of how the design principles embodied in the polymer's structure equip it to fulfill its biological functions, and in this respect, knowledge of the functional roles assumed by IF remains rather scanty. Nevertheless, information afforded by several different biophysical and biochemical techniques has enabled a general model of IF to be formulated that has specific functional implications (Steven *et al.*, 1985a; Steinert *et al.*, 1985a). This model is based on a filament backbone composed of a bundle of coiled-coil α-helical rods, surrounded by a peripheral shell of non-α-helical domains. The rod domains are conservative moieties common to all IF subunits (polypeptide chains), and their siting in the

Alasdair C. Steven • Laboratory of Physical Biology, National Institute of Arthritis, Musculoskeletal, and Skin Diseases, National Institutes of Health, Bethesda, Maryland 20892.

backbone should allow similar, possibly identical, helical structures to be adopted by IF of all kinds. The highly variable end domains of IF subunits are postulated to occupy exposed peripheral locations where they confer functional specificity: in particular, end-domain deployment is envisaged to be a major regulator of interactions among IF, and between IF and other cytoskeletal or cytoplasmic components. The main purpose of this chapter is to review the lines of evidence on which this hypothesis is based.

Actin filaments and MT are both fibrous polymers of essentially globular subunits. In contrast, IF subunits are highly elongated molecules. In terms of the design hypothesis outlined above, it is apparent that IF structure is more directly comparable to that of another cytoskeletal protein, myosin. The myosin filament is also based on a backbone of close-packed coiled-coil α-helical domains (the myosin "rods"), from which the globular S1 domains protrude. The latter perform the vital role of cross-bridge formation between myosin and actin filaments in the force-generating mechanism of muscular contraction. A second goal of this chapter is to explore the structural analogy between IF and myosin filaments.

2. *How Many Different Kinds of IF Are There?*

Several different classification schemes are used to describe the diversity of IF. These classifications are based on (1) tissue-specific occurrence and expression pattern *in vivo,* (2) degrees of homology exhibited by the amino acid sequences of IF polypeptide chains, (3) capacities of purified subunits to coassemble into filaments *in vitro,* either singly (homopolymer IF) or in combinations (heteropolymer IF), and (4) structural polymorphism, expressed in terms of filament thickness or the number of molecules per cross section. The numbers of different kinds of IF distinguished by these various criteria vary from 3 to over 30 per species, and so the overall situation is somewhat complex. Here, the experimental basis for each classification scheme is outlined, and a summary given of how these schemes interrelate.

2.1. *Tissue-Specific Classification*

Vimentin (54.5k) is the IF protein of fibroblasts and other mesenchymally derived cells. Desmin (54k) is the IF protein characteristic of muscle cells. Glial fibrillary acidic protein (GFAP; 52.5k) is the IF protein associated with astrocytes. Neuronal cells contain the neurofilament proteins NF-L (light), NF-M (medium), and NF-H (heavy). Estimates of their molecular weights by SDS-PAGE—70k, 160k, and 210k, respectively—have been found to be anomalously large in the case of NF-M and NF-H on account of these proteins' unusual amino acid compositions (Kaufmann *et al.*, 1984), but even revised estimates (107k, and 112–138k) make these two polypeptides the most massive by far of all currently recognized IF proteins. Epithelial cells contain a total of about 20 different keratins of 40–70k, and epidermal appendages such as hair and wool may contain about 10 more "hard α-keratins" (Steinert *et al.*, 1985a). However, any given epithelial tissue expresses only a few specific keratins, generally in combinations of one or more particular pairs ("coexpression pairs") (Sun *et al.*, 1985). By this criterion, five kinds of IF are distinguished—vimentin, desmin, GFAP, NF, and the keratins.

2.2. *IF Types Based on Amino Acid Sequence Homologies*

The primary sequences of many IF proteins have been determined in the past few years. All contain a central "rod" domain of closely conserved length (~ 310 residues) and secondary structure, flanked by highly variable end domains (for references and more detailed discussion, see Parry, this volume). The presence of such a rod domain in a sequence may be the ultimate arbiter of whether it should be considered an IF protein. Based primarily on proximities of homology within the rod domain, but also on homologies and subdomainal divisions in the end domains, IF chains have been classified into several sequence-specific types. Early work on peptide sequences from the rod domains of wool keratins divided them into Type I and Type II sequences (Crewther *et al.*, 1978). The first epidermal keratin sequences to be determined were found to conform to this pattern (Hanukoglu and Fuchs, 1982, 1983; Steinert *et al.*, 1983, 1984), and all the other keratin sequences that have since been determined are unequivocally of one or the other Type. Among Type I rod-domain sequences, the degree of homology is typically 70–90% and similarly for Type II, while the corresponding degree of homology is typically 20–40% when Type I and Type II keratins are compared. Moreover, Type II chains contain short, non-α-helical, and mutually homologous subdomains adjacent to both ends of their rod domains. The amino-terminal and carboxy-terminal subdomains are called H1 and H2, respectively. Both H1 and H2 are absent from Type I sequences.

Vimentin, desmin, and GFAP are highly homologous throughout their rod domains and C-terminal end domains. Their sequences are clearly different from those of keratins of Types I and II, although not so remote from them as they are from each other. Accordingly, these three proteins are collectively known as Type III IF (Crewther *et al.*, 1983). As yet, porcine NF-L is the only complete neurofilament protein sequenced (Geisler *et al.*, 1983a). Taken together with partial sequences for NF-M and NF-L, these data suggest that the neurofilament proteins may constitute a fourth class, Type IV, of IF proteins (see Parry, this volume).

2.3. *Classification According to Competency for in Vitro Self-Assembly or Coassembly*

Characteristic conditions required for *in vitro* assembly of IF from purified subunits (ionic strength, pH, critical protein concentration, and presence of compatible copolymerization partners) provide another basis for classification. Under appropriate conditions, any Type III IF protein may either self-assemble into homopolymer IF (e.g., Steinert *et al.*, 1981; Ip *et al.*, 1985a) or coassemble with any other Type III partner into heteropolymer IF (e.g., Steinert *et al.*, 1981). Under somewhat different conditions, NF-L has been found to self-assemble or to coassemble with NF-M and/or NF-H. NF-M may also be competent for homopolymer assembly, but no evidence has been put forward that NF-H possesses this property. Strikingly, no single keratin has been found competent for homopolymer K-IF assembly, although many combinations of at least two subunits coassemble efficiently (Steinert *et al.*, 1976, 1982; Steinert, 1977; Hatzfeld and Franke, 1985). The emerging picture is that the minimal requirement for K-IF assembly is one Type I subunit and one Type II subunit; moreover, virtually any pairing of Type I and Type II subunits appears to be compatible for assembly (Hatzfeld and Franke, 1985). The *in vitro* property of keratins as obligate heteropolymers implies that this requirement may well underlie the *in vivo* phenomenon of "coexpression pairs" whereby specific pairs of Type I and Type II keratin genes are expressed coordinately (Sun *et al.*, 1985). However, a given keratin is not expressed indiscriminately with any keratin of the complementary Type. The basis for this selectivity may well lie in the properties of their respective end domains (excluding H1 and H2), and for some coexpression pairs at least, there is evidence of structural (and presumably also functional) resemblance between corresponding end domains (Steinert *et al.*, 1983, 1985a). However, close homology of end-domain sequences is not a property common to all coexpression partners (B. Knapp *et al.*, 1986).

To summarize, the criterion of compatibility for *in vitro* assembly distinguishes basically three kinds of IF—K-IF, Type III IF, and NF. N-IF are obligate heteropolymers, Type III IF are facultative heteropolymers, and there may be NF subunits of both kinds, viz. facultative heteropolymers (NF-L) and obligate heteropolymers (NF-H) NF subunits. Characteristic conditions have been defined for *in vitro* assembly of each of these three kinds of IF, but assembly *in vivo* may be somewhat different. For instance, the presence of both vimentin and K-IF in tissue culture cells (Connell and Rheinwald, 1983) implies assembly conditions that are acceptable to both. However, there is, as yet, no evidence from either *in vivo* or *in vitro* sources of "hybrid" heteropolymer IF that combine, within the same filament, subunits from more than one of these three classes.

2.4. Polymorphism of IF That Contain the Same Subunits

Direct measurements of mass per unit length performed by scanning transmission electron microscopy (STEM) have provided a sensitive quantitative probe to investigate the structural heterogeneity of IF. Thus, native vimentin IF were found to constitute a homogeneous population apart from a clearly resolved minor fraction of less massive filaments (Steven *et al.*, 1982). The major component's density corresponds to about 16 two-chain molecules in cross section, whereas minor component IF contain about two-thirds of this amount. Both density species were also identified in preparations of reconstituted vimentin IF, as were small amounts of still denser filaments. Subsequent experiments with several different kinds of reassembled K-IF confirmed the observation of polymorphism (Steven *et al.*, 1983a,b; Engel *et al.*, 1985). Since IF structures are thought to consist of bundles of subfilaments, it is plausible that the various density species should correspond to differing numbers of subfilaments. Interpretation of these data in terms of specific configurations of subfilaments will be discussed in greater detail below (see Section 5.4).

In summary, structural polymophism is widespread among reconstituted Type III IF and K-IF. At least three distinct density species have been observed for IF composed of the same kind(s) of subunits. In preparations of native vimentin IF, long "mature" filaments are essentially homogeneous, and the few low-density segments observed, usually rather short IF or end segments, may well represent breakdown products or an intermediate stage of assembly. Technical difficulties in obtaining sufficiently pure, well-dispersed preparations of native K-IF have so far hampered efforts to explore the proposition of structural polymorphism of K-IF in a physiological context.

2.5. Other Differences: Immunological Specificity, Interspecies Variation, and Distinctions Conferred by Associated Proteins

To a substantial degree, the classification of IF subunits according to homology-based sequence Types correlates with their patterns of expression *in vivo* and their predilections for *in vitro* coassembly partners. However, these correlations remain empirical ones in the sense that our understanding of the implications of the now-copious sequence data is not yet sufficiently detailed to explain, for instance, why particular IF proteins will coassemble with some, but not with other, partners.

Immunological specificity has provided another basis for distinguishing IF. Although essentially an operational definition, this criterion has formed the basis for mapping tissue-specific expression properties. Few epitopes have yet been mapped within IF molecules, but it is to be expected that, for the most part, subunit-specific epitopes are likely to reside in variable parts of the molecules and conversely. For instance, end-domain termini have been recognized as regions of extreme sequence variability, and it has proved possible to raise highly specific antisera using synthetic peptides that reproduce these sequences for certain epidermal keratins (Roop *et al.*, 1984). Conversely, the determinant recognized by a monoclonal antibody that reacts with all known IF (Pruss *et al.*, 1981) is thought to reside in a stringently conserved region toward the carboxy-terminus of helix 2 (see Fig. 1) (Geisler *et al.*, 1983b).

A further dimension is introduced by interspecies variation. It appears that there is general correspondence between the complements of IF proteins of the various species of higher animals, but there are also differences among them. For instance, there is substantial variability in electrophoretic mobility as well as in other properties between the NF proteins of different species (Lasek *et al.*, 1985). However, such data as are presently available indicate closer resemblance between the corresponding keratins of different (mammalian) species than between the various keratins of a given species (e.g., Steinert *et al.*, 1985b; A. C. Knapp *et al.*, 1986).

In vivo, a further degree of IF specialization may be accomplished by their association with specific accessory proteins [IF-associated proteins (IFAP)]. Some IFAP have already been characterized (see Section IV of this volume), but it appears likely that variability generated by this mechanism has proliferated markedly less with IF than, for instance, in the actin system, where scores of actin-AP have already been identified (Weeds, 1982; Pollard and Cooper, 1986). In general, AP are thought to participate in regulating the states of assembly and aggregation, and the intracellular locations of their basic filaments, as well as in mediating their interactions with other constituents of the cell. In this vein, it has been suggested (Steven *et al.*, 1985a) that IF end domains—particularly in the keratin system—may be functionally equivalent to AP, with the distinction that, unlike conventional AP, IF end domains are covalently linked to the building blocks of the basic polymer and thus in precise equimolar stoichiometry with them.

Finally, the possibility that further classes of IF proteins may yet emerge is pointed up vividly by the recent discovery that nuclear lamins A and C, major proteins associated with the inner surface of the nuclear envelope, have unmistakably IF-like characteristics (McKeon *et al.*, 1986). These include "classical" IF rod-domain sequences, albeit elongated by the insertion of an additional six heptads in their rod domains.

3. Structure of Oligomers of IF Proteins

When analyzed by secondary structure prediction algorithms, IF rod-domain sequences show strong potential for α-helical structure and coiled-coil formation. These expectations are borne out experimentally by circular dichroism experiments on suspensions of oligomeric IF particles (e.g., Steinert and Idler, 1975) and high-angle x-ray diffraction of intact IF (e.g., Fraser and MacRae, 1973b; Steinert *et al.*, 1976; Renner *et al.*, 1981). However, IF rod domains are not expected to form a single unbroken α helix, but to contain several characteristic and conserved interruptions. The most prominent is a non-α-helical linker, L12, of 16–17 residues that divides each rod domain into two helical segments (called 1 and 2) of approximately equal length (~ 21 nm), both of which contain at least one internal discontinuity (see Parry, this volume). It is now widely accepted that pairs of IF chains combine to form dimeric molecules (Geisler *et al.*, 1982), bonded primarily through a segmented coiled-coil rope formed by the two rod domains. These are associated in a parallel "in-register" configuration, giving polar molecules (Woods and Inglis, 1984; Parry *et al.*, 1985; see Fig. 1).

The next level of organization is the four-chain complex (Ahmadi and Speakman, 1978; Geisler *et al.*, 1982; Ip *et al.*, 1985a), and much current interest is focused on how the two constituent molecules may be associated in such complexes. Assuming that the most stable arrangements are likely to correspond to full or nearly full overlap of the two long helical segments, five distinct possibilities exist (Fig. 1C). These predict particle lengths of ~ 48 and ~ 70 nm. Numerous electron microscopic observations of particles in the former size range (Geisler *et al.*, 1982; Quinlan *et al.*, 1984; Ip *et al.*, 1985a) favor models that predict the shorter length. To distinguish between the alternatives of parallel or antiparallel associations of molecules, the most direct and persuasive evidence to date comes from electron micrographs of desmin particles tagged with the Fab fragments of a monoclonal antibody whose determinant is located in helix 2 (Geisler *et al.*, 1985). Fab-binding at both ends of ~ 50-nm particles was reported to be the predominant tagging

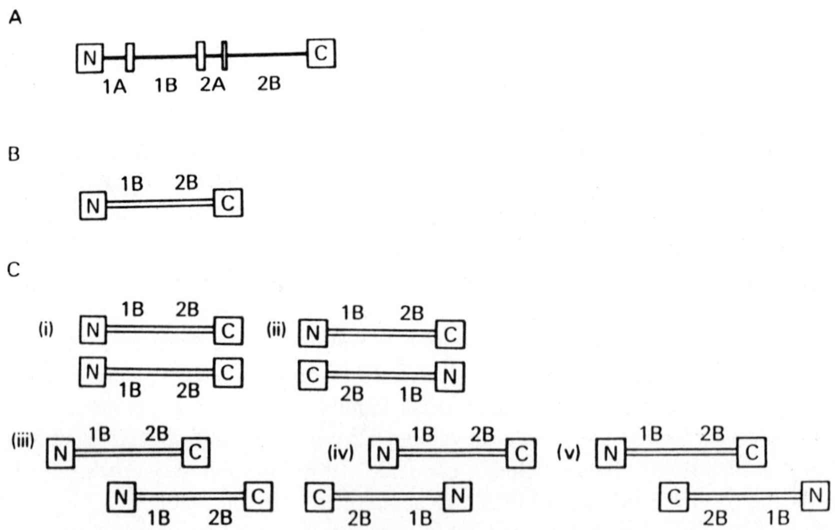

Figure 1. Schematic models of lowest levels of structural organization of IF chains. (A) Domainal model of an individual IF chain distinguishing the non-α-helical end domains (N and C) and the four α-helical tracts (1A, 1B, 2A, 2B) of the rod domain. (B) A two-chain molecule with a parallel, "in-register" association of rod domains forming a two-stranded coiled-coil rope. (C) Five possible modes of association of pairs of molecules into four-chain complexes: the alternatives shown correspond to parallel and antiparallel alignment of molecules, which are either in axial register or half-staggered. From Steinert and Steven (1985) with permission.

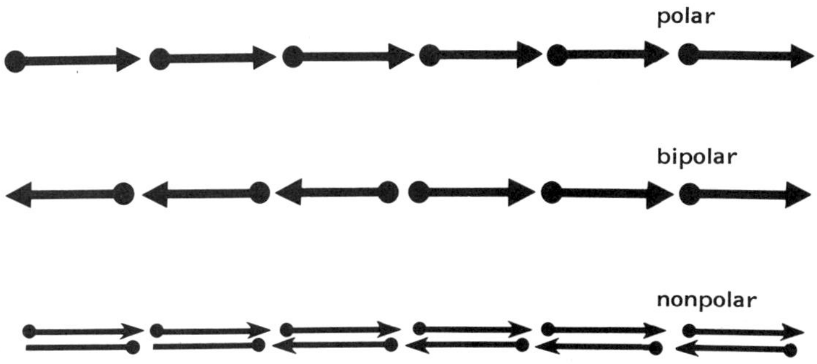

Figure 2. Schematic diagram distinguishing linear polymers that are polar, bipolar, and nonpolar, respectively.

pattern, implying an antiparallel, "in-register" pairing of molecules, provided that the shadowed particles thus visualized were indeed four-chain complexes, and not aligned multimers thereof. The implications of this observation are profound in the sense that any higher-order structure generated from a nonpolar complex is necessarily nonpolar (Fig. 2). If Type III IF do indeed have this property, it sets them apart in a fundamental way from virtually every other naturally occurring protein filament, e.g., F-actin, MT, myosin, flagella, pili, all of which are polar structures. However, it should be noted that the "antiparallel, in-register" pairing of molecules (model Cii of Fig. 1) does not account for the well-characterized tetramer of 1B segments that is produced by proteolysis of K-IF (Woods and Inglis, 1984). To reconcile these observations, it has been suggested that multiple modes of intermolecular association are feasible, of which more than one may be employed within assembled IF (Fraser *et al.*, 1985; Steinert and Steven, 1985).

Finally, we address the subunit composition of these particles with particular reference to K-IF, the obligate heteropolymers. Type I and Type II chains are contained in K-IF in approximately equimolar amounts, and it has been supposed that this stoichiometry is imposed by a heterodimer molecule that contains one chain of each Type (Woods and Inglis, 1984; Quinlan *et al.*, 1984; Parry *et al.*, 1985). However, the alternative possibility has been raised (Hatzfeld and Franke, 1985) that the pertinent level of organization is a four-chain complex containing a Type I homodimer and a Type II homodimer. This issue has not been resolved experimentally, but the latter possibility would have the interesting consequence of giving a keratin four-chain complex that, unlike that of Type III IF, is polar, in which case polarity could persist to the level of fully assembled K-IF.

The foregoing properties have much in common with their counterparts in the myosin system. Myosin also forms a two-chain coiled-coil molecule according to a "parallel, in-register" mode of association, and the next level of organization is also thought to be a four-chain complex (Harrington *et al.*, 1972). Concerning polarity, myosin filaments are distinctively bipolar, i.e., two oppositely oriented polar half-filaments emanate from a common overlap region. Thus, multiple modes of intermolecular interaction must necessarily be operative, since adjacent molecules should be entirely parallel in the filament extremities, whereas at least some antiparallel configurations are required in the central overlap zone. In fact, it is considered likely that myosin molecules associate into several different kinds of dimers (Harrington *et al.*, 1972), as may also be the case in the IF system (see above).

4. Axial Repeats in IF Structure

Evidence for specific axial repeats in IF is based on the detection of periodic distributions of certain kinds of amino acid residues in the rod domains, and of characteristic lengths of rod-domain segments (assuming an axial rise of 0.149 nm per residue in α-helical conformation) that correlate suggestively with axial spacings observed in intact IF. Moreover, both kinds of repeats appear to be essentially common to all IF. In end domains, on the other hand, there are no repetitive sequences common to all IF subunits, and when repeats occur, e.g., in the end domains of some keratin subunits (Steinert *et al.*, 1983, 1985b), these take the form of a few irregularly spaced recurrences rather than a sustained and regular repetition. Electron microscopic measurements put the length of isolated IF molecules at 45–50 nm, which correlates quite well with the expected length of the rod domain (310 residues corresponding to ~ 46 nm total length). Experimental margin of error rules out the possibility of inferring how much the non-α-helical linker regions or the end domains may contribute to the measured lengths, but this correlation does seem to rule out backfolding of the major helical segments to any substantial extent. Furthermore, x-ray diffraction patterns show that the rod domains are oriented in approximate alignment with the filament axes (Fraser and MacRae, 1973a; Steinert *et al.*, 1976; Renner *et al.*, 1981). In contrast, there is no evidence to associate the dimensions or other structural attributes of end domains with intrafilament spacings. Therefore, the correlation that exists between the lengths of rod-domain segments and the most conspicuous axial repeats of IF implies that it is the rod domains that serve as the essential building blocks, no matter what roles may be played by end domains in aiding assembly or in stabilizing IF, once assembled.

4.1. Regular Distribution of Particular Types of Amino Acid Residues in IF Sequences

When analyzed by Fourier transform techniques, the distributions of hydrophobic residues in rod-domain sequences show pronounced repeats of 3.5 and 2.3 residues. These repeats are the second and third harmonics of a seven-residue (heptad) repeat of the form $(a\text{-}b\text{-}c\text{-}d\text{-}e\text{-}f\text{-}g)_n$ wherein the positions *a* and *d* are occupied by hydrophobic residues. Since there are approximately 3.5 residues per turn of an α helix, a hydrophobic stripe is thereby formed on its surface. The interfacing of two such stripes whereby the two chains wind around a common axis provides the primary bonding interaction that stabilizes the coiled coil.

Rod domains have a notably high incidence of charged residues whose distributions also exhibit repeats of 2.3 and 3.5 residues, reflecting in part their exclusion from the *a* and *d* heptad positions. However, the distributions of both acidic and basic residues also exhibit generic repeats of about 9.7 residues (see Parry, this volume). It is thought likely that these regular progressions of alternating charged residues on the outer surfaces of the coiled coils are involved in specifying higher orders of structure, viz. the relative offsets and orientations (parallel or antiparallel) of adjacent molecules. In an α helix, a 9.7-residue repeat corresponds to an axial shift of 1.45 nm (or multiples thereof). Since the progressions of acidic and basic residues are out of phase by ~ 180° (i.e., they alternate), charge-pairing interactions would therefore favor offsets of 0.73, 2.2, 3.6, 5.1 nm, etc. Electron microscopic measurements of the lengths of four-chain complexes suggest modes of association in which the molecules are not far from being in axial register (see Section 3). When two such complexes pair, an offset of about half a molecular length appears to represent one favored configuration (Ip *et al.*, 1985a). However, in both cases, the precision of length measurements is not sufficient to settle whether these complexes correspond to any of the predicted offsets listed above. Nevertheless, we note that a prominent axial repeat of 5.4 nm is exhibited in paracrystals that most likely consist of proteolytic derivatives of rod domains (see Section 4.4), and this spacing is quite close to the 5.1-nm offset in the above list.

A similar heptad repeat is observed in the myosin domain sequence (McLachlan and Karn, 1983), except that the progression of heptads in myosin continues without interruption for much longer than in the segmented IF rod domains. Polar residues in the myosin rod domain also exhibit a nonheptad repeat, but its periodicity, at 9.3 residues (McLachlan and Karn, 1983), is slightly shorter than those of IF proteins, and it has been proposed to represent the third order of a longer, 28-residue, repeat.

4.2. Axial Repeats from Meridional X-Ray Diffraction

From the uniquely well-developed low-angle x-ray diffraction patterns of "hard" α-keratin fibers, an extensive series of meridional reflections has been documented, each of which corresponds to an axial spacing. The meridional reflections can be indexed as orders of an axial repeat of 47 nm (Fraser *et al.*, 1976), and the proposed surface lattice invokes a helical pitch of 22 nm. The 47-nm spacing correlates closely with the length of the IF rod domain, suggesting an end-to-end packing of aligned rod domains. The only other IF system in which low-angle meridional reflections have been recorded are invertebrate NF (Wais-Steider *et al.*, 1983). Although less coherent, these patterns have some features in common with the hard α-keratin data (see Fraser, this volume), although the primary axial repeat is 25 nm in the case of *Myxicola* NF.

4.3. Axial Repeats Detected by Electron Microscopy

When IF are prepared by low-angle rotary shadowing following nebulization onto a mica surface, they exhibit a characteristic beading with a periodicity of about 22 nm (Milam and Erickson, 1982; Henderson *et al.*, 1982; see Fig. 3). This repeat has also been reported to be observable with negative staining (Aebi *et al.*, 1983; Ip *et al.*, 1985a), but it has generally been visualized with greater clarity by the rotary shadowing method. However, the values reported by different investigators vary considerably, from 19.2 nm (Franke *et al.*, 1982) to 23.7 nm (Aebi *et al.*, 1983). To some extent, this dispersion may reflect the use of differing magnification calibration procedures, and substantial experimental uncertainties, with standard deviations typically ≧ 10%. Thus, it has not been clear whether this repeat is closely conserved among different types of IF or whether it is merely a generic feature of IF that, at a quantitative level, exhibits considerable variability.

Precise measurements of periodicities from electron micrographs are most readily effected by optical or digital diffraction (Misell, 1978), but the curvature generally exhibited in micrographs of IF has prevented application of these methods. We have recently developed a computational straightening procedure that renders the particles amenable to diffraction methods (Steven *et al.*, 1985b), and used it to analyze three different kinds of reassembled epidermal K-IF (e.g., Fig. 4), applying a standard magnification calibration

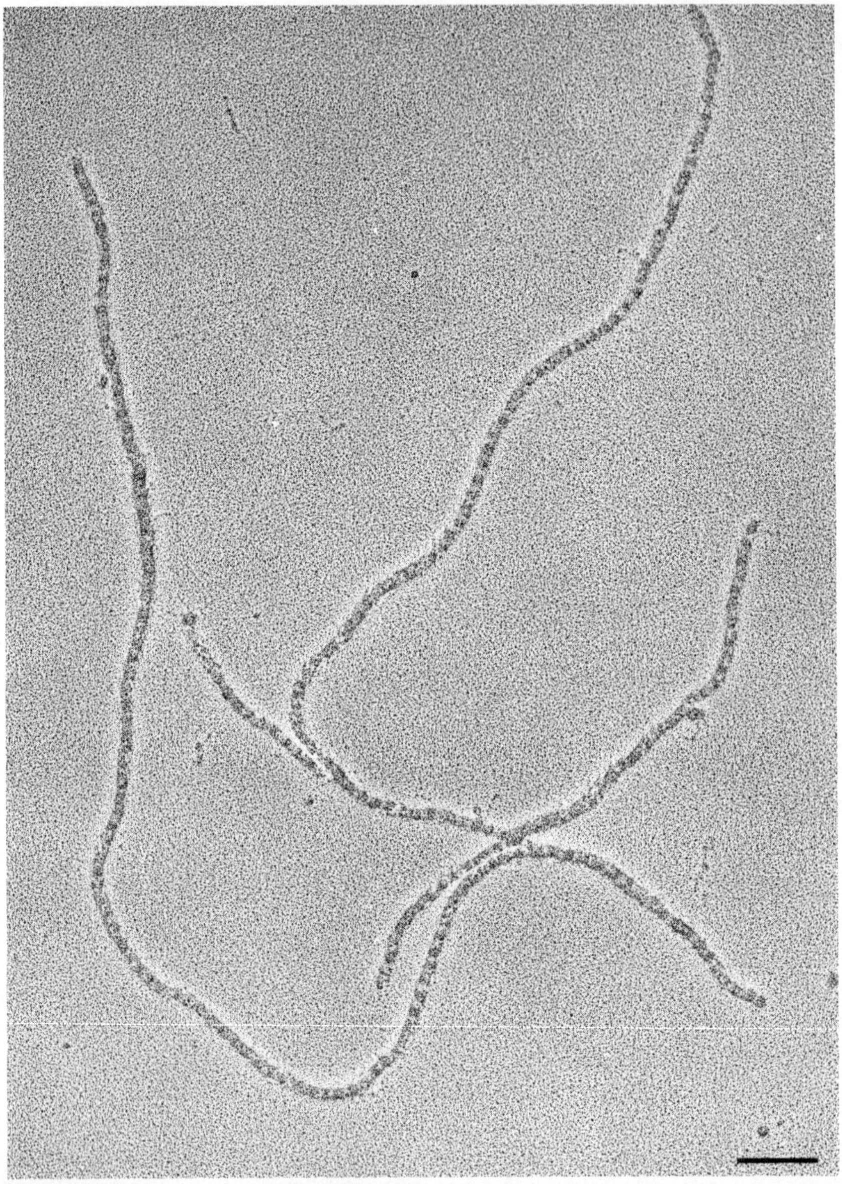

Figure 3. Electron micrograph of epidermal keratin IF, reassembled from proteins isolated from newborn mouse epidermis. The specimen was prepared by low-angle rotary shadowing with platinum after the IF were dried onto a freshly cleaved mica surface from glycerol droplets. Bar = 100 nm.

procedure in each case (A. C. Steven, M. E. Bisher, B. L. Trus, and P. M. Steinert, unpublished results). The standard deviations of the resulting data represent an approximately fourfold improvement in precision over sets of measurements made directly from the micrographs. To within residual experimental uncertainty (0.6 nm, most of which is due to the determination of the calibration spacings), the values that we obtain in all three

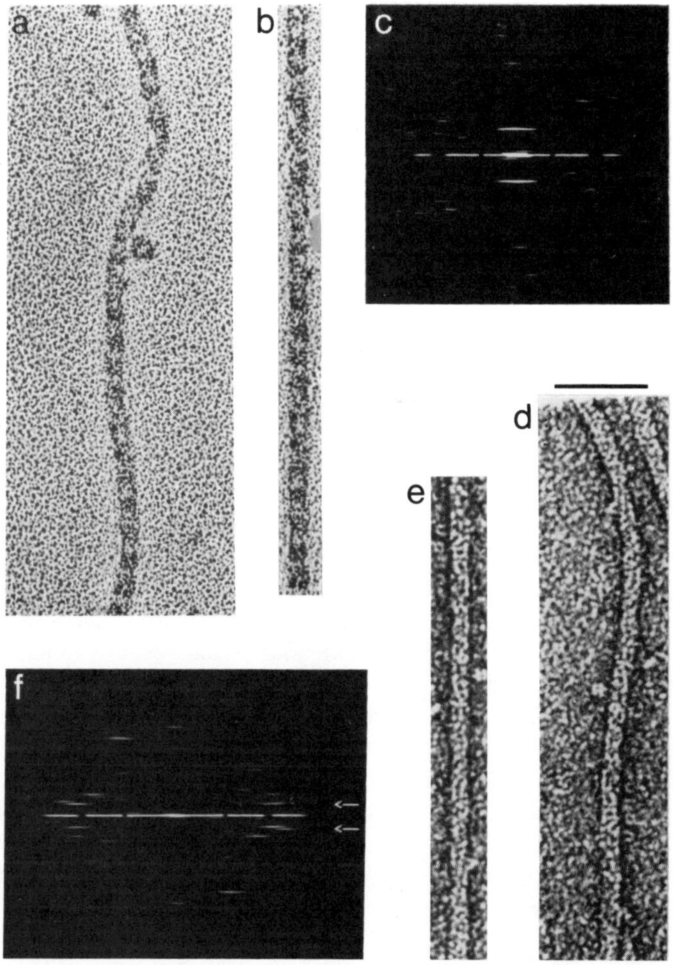

Figure 4. Computational straightening of images of keratin IF. (a, b) A rotary-shadowed murine epidermal keratin IF, before and after straightening; (c) computed diffraction pattern of (b) shows a prominent meridional reflection at $(22.7\ \text{nm})^{-1}$; (d, e) negatively stained bovine epidermal keratin IF, before and after straightening; (f) computed diffraction pattern of (e). Arrows indicate an off-meridional reflection as an axial spacing of $\sim (23\text{nm})^{-1}$. Bars = 50 nm. A. C. Steven *et al.* (unpublished results).

cases are indistinguishable from the global average of 22.7 nm. These data indicate that, for epidermal K-IF at least, this spacing is indeed a closely conserved structural feature.

The observed reflections (e.g., Fig. 4c) generally show little divergence from the meridian, suggesting that this spacing relates to an axial repeat, and not the axial projection of a helical (screw) repeat. On the other hand, it has been reported, on the basis of undirectionally shadowed images of K-IF, that this repeat is, in fact, helical with a left-handed pitch angle of 15° (Milam and Erickson, 1982). However, an alternative explanation of these images is that their apparent handedness may be caused by anisotropy of the shadowing.

For comparison, we note that myosin filaments possess a generic axial repeat of 14.3 nm corresponding to the spacing along the filament of successive clusters ("crowns") of

S1 cross-bridges. To within about 1%, this spacing is conserved among myosin filaments of different kinds. Although this ratio may be just a numerological coincidence, it is intriguing to note that the myosin repeat is almost exactly two-thirds that of IF. We also note that the 14.3-nm myosin repeat is believed to be axial in the sense that all myosin molecules whose S1 domains contribute to a given crown are in axial register, although there is a rotational displacement between successive crowns (Squire, 1981). As noted above, it appears likely that a similar property pertains to the 22.7-nm IF repeat.

4.4. Axial Repeats in Paracrystals

Paracrystalline bundles of aligned filaments have frequently been found to be useful in systems where individual filaments prove refractory to structural analysis. However, little success has yet been achieved with *in vitro* formation of adequate IF paracrystals. Paracrystalline aggregates have been observed in keratin extracts from psoriatic human epidermis, showing a prominent axial repeat of 5.4 nm (Thaler *et al.*, 1980). Diffraction analysis of electron micrographs of similar material (Fig. 5) found this periodicity to be the third order of a longer 16.5-nm repeat, and determined the interfilament spacing to be 5.1 nm (Trachtenberg *et al.*, 1986). Because this width is too small for intact IF, and also because the 16.5-nm spacing correlates with the length of a proteolytically resistant helical segment (1B), it was proposed that these paracrystals are composed of proteolytic derivatives, and not of intact IF. This interpretation is consistent with the detection of a similar 16-nm axial repeat in an x-ray diffraction study of oriented films of material obtained in a chymotryptic digest of keratin (Suzuki *et al.*, 1973), and is further supported by the formation of very similar paracrystals in epidermal keratin preparations treated with chymotrypsin (Aebi *et al.*, 1983), in which fragments in the same M_r range as helical segments 1B and 2 were detected as the major reaction products. Relatively well-ordered paracrystals have also been observed in reassembly experiments with keratins extracted from cultured human epidermal cells, in which endogenous proteolytic activity was suspected to be involved in formation of the paracrystals (Eichner *et al.*, 1985).

Proteolytically derived desmin rod domains (40k) have been found to form paracrystalline tactoids (Geisler *et al.*, 1982), but no information was obtained on their

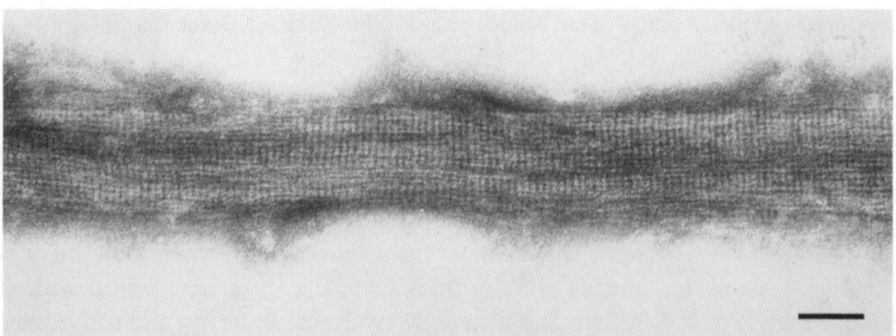

Figure 5. Paracrystal formed *in vitro* from material extracted from psoriatic human epidermis. Such paracrystals contain filaments packed with a center-to-center spacing of ~ 5.5 nm, and are thought to consist of polymerized fragments of keratin molecules (see Section 4.4). Bar = 50 nm. A. C. Steven and P. M. Steinert (unpublished data).

substructure. In summary, while well-ordered IF paracrystals would potentially be of considerable value as aids in IF structure determination, such paracrystals as have been obtained to date appear to contain fragments rather than intact IF. Notwithstanding the intrinsic interest of these aggregates, it remains to be established whether the interactions within them do indeed parallel those that take place in intact IF.

In comparison, we note that the α-helical domains of myosin—the rod, LMM (light meromyosin), and the S2 domain—all form paracrystals quite readily (Squire, 1981). The various types of paracrystals obtained have in common that, unlike the myosin filament itself, they are nonpolar (i.e., they are based on antiparallel molecular packings). Their most prominent axial repeats are usually 14.3 or 43 nm, matching axial periodicities of native myosin filaments. In contrast, no comparably well-ordered paracrystals of intact myosin molecules have been reported, although bipolar "segments" have been obtained (Squire, 1981). These aggregates consist of oppositely oriented layers, in each of which the molecules are packed in parallel: the layers are connected by a partial overlap between the rod domains of molecules in opposing layers. Thus, the relative alacrity with which helical fragments (as opposed to intact molecules) form extensive and reasonably well-ordered paracrystals seems to be a property common to both the myosin and the IF systems. Possibly the presence of protruding cross-bridges or end domains may somehow impede well-ordered paracrystallization or unrestricted radial growth in both cases.

4.5. Proportionality between the Linear Densities of IF and the Molecular Weights of Their Constituent Subunits

As noted above (Section 2.4), STEM dark-field electron microscopy of unstained specimens is capable of determining directly the mass per unit length of filaments. To explore any systematic dependence of this parameter on the kinds of subunits involved, we have analyzed six different kinds of IF by this technique (Steven *et al.*, 1982, 1983a,b; see Fig. 6). The most salient point to emerge from this body of data is that within each polymorphic variant class of IF (see Section 2.4), the linear density of the filament is directly proportional to the average molecular weight of the subunits contained. This relationship has been interpreted in terms of a general model whereby each IF is composed of building blocks that contain conserved and variable elements (Steven *et al.*, 1983b, 1985a). The fixed size and packing arrangement(s) of the conserved elements specify, in essence, the IF structure, whereas the variable elements contribute the additional amounts of mass that account for the observed proportionality between linear density and subunit molecular weight (Fig. 7). The IF rod domains are natural candidates for the postulated conserved elements, and the end domains for the variable elements.

Recently, however, STEM data have been published for several other kinds of reassembled human epidermal K-IF (Engel *et al.*, 1985), and interpreted as being inconsistent with the proportionality observed by our data. A decision between these respective propositions (i.e., proportionality or not) can be made by testing for variability according to subunit type in the characteristic "20–22 nm" axial repeat of IF (Section 4.3). The prediction of our model is that this repeat should be closely conserved, whereas the counterproposal would appear to anticipate variability on a scale comparable to that of the departure from proportionality in the linear density values. We have applied image processing methods to determine this repeat as precisely as possible for several of the IF species already characterized by STEM mass measurements. To within a margin of error

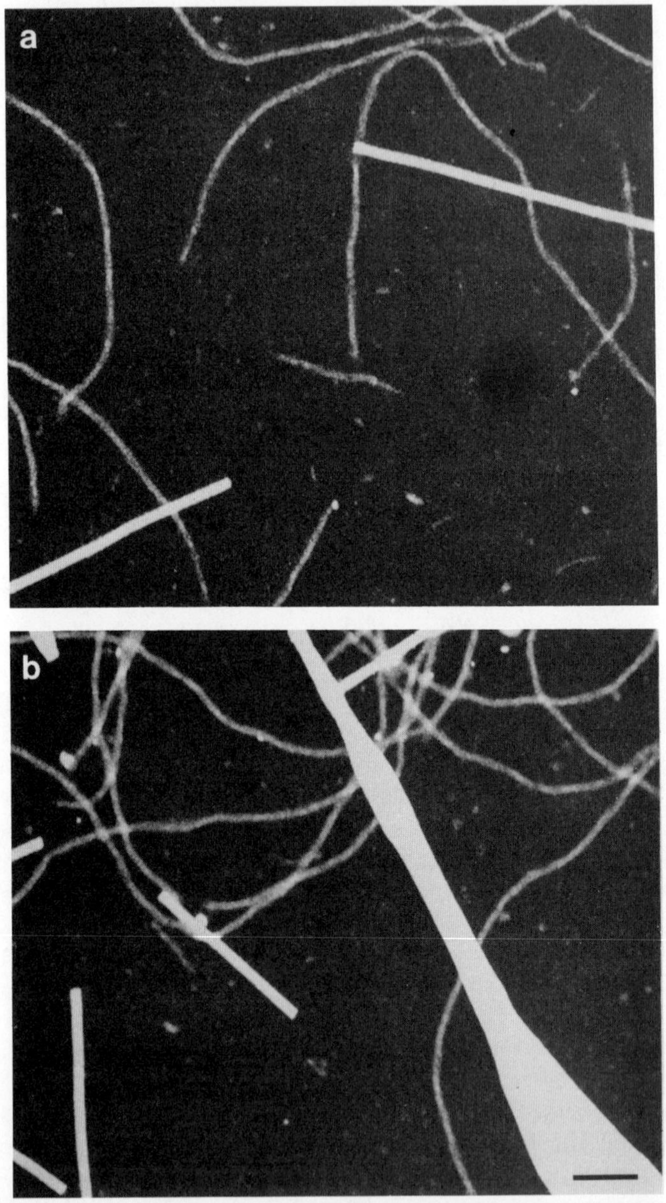

Figure 6. Dark-field scanning transmission electron micrographs of unstained IF prepared by freeze-drying: (a) reassembled bovine epidermal keratin IF; (b) native vimentin IF extracted from Chinese hamster ovary cells. The relatively straight and dense (white) particles present in both fields are TMV particles, included for calibration purposes. Bar = 100 nm.

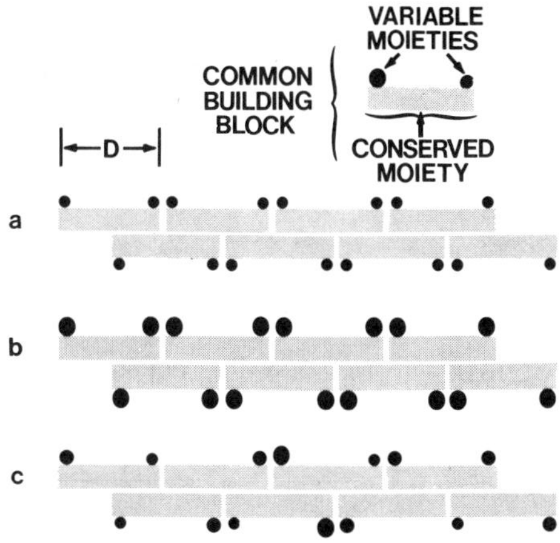

Figure 7. Schematic diagram illustrating how repeating units with conserved elements of fixed size and terminal elements of variable size may be packed according to a common plan. The filament backbones consist of the conserved elements whose size (D) and bonding properties essentially dictate the filament geometry, whereas the variable elements protrude at the filament peripheries; e.g., in (a) and (b), the variable elements are small and large, respectively, and in (c) they are highly heterogeneous. Such filaments have fixed numbers of repeating units per unit length so that their mass-per-unit-length values are directly proportional to the average masses of the repeating units. The numbers of units and packing arrangements shown are entirely illustrative.

of ~ 2.5%, which is comparable to the precision with which we believe our linear densities to be defined, we find indistinguishable values for this axial repeat (Section 4.3). Thus, while it cannot be ruled out that the K-IF analyzed by Engel *et al.* (1985) may differ in some basic way from those that we have studied, there is a more likely explanation for the apparent discrepancy between the respective sets of data. The linear density values are obtained by fitting sets of Gaussian curves to experimental histograms. The positions of the Gaussian maxima—which define the linear density values—may not be stably defined when this procedure is applied to broad distributions whose shapes do not directly suggest that they resolve into several discrete components (see Figs. 6–9 of Engel *et al.*, 1985).

5. *Width and Radial Organization of IF*

5.1. *Traditional Estimates of IF Diameter*

IF were originally recognized as a distinct species of intracellular filament (and so named) on the grounds that their apparent diameters, as perceived in thin-section electron micrographs, i.e., 8–10 nm, were intermediate between those of actin filaments and MT, which are attributed diameters of 5–7 and 22–25 nm, respectively, under these conditions of observation. However, this operational definition of diameter does not take into account vagaries of structural preservation deriving from chemical fixation, dehydration, embedding, and staining. Moreover, few if any of these measurements were strictly quantitative in the sense of applying an objective, numerically defined, edge-detection criterion. It is likely that these considerations contribute to the variability of IF widths reported by different observers, as reviewed by Eriksson and Thornell (1979).

Recent studies of actin filaments and MT analyzed under more native conditions have concluded that their *outer* diameters, viz. 10 nm (Trinick *et al.*, 1986) and 30 nm (Mandelkow *et al.*, 1977), respectively, are considerably greater than the earlier estimates. Assuming that the *outer* diameters of IF are indeed intermediate between those of actin and MT, it follows that they should be considerably greater than the 8–10 nm values traditionally assigned to them. Recent developments in scanning transmission electron microscopy have provided a method for direct characterization of radial distributions of density in filamentous structures that confirms this expectation. This method combines the following advantages: (1) a method of specimen preparation—freeze drying—that is relatively benign; and (2) dark-field STEM imaging allows visualization of unstained specimens with adequate contrast, thereby eliminating ambiguities of interpretation associated with heavy metal staining or shadowing.

Our application of this technique has indicated that K-IF and Type III IF have outer diameters of 14–15 nm (Steven *et al.*, 1982, 1985a, 1987), and that their radial density distributions are distinctly nonuniform, consisting of a nonhollow core, 9–10 nm across, of approximately uniform density, surrounded by a peripheral shell of relatively low (diffuse) average density. In the following sections, these data are summarized and discussed in the light of corresponding information from other types of experiments.

5.2. *Evaluation of Transverse Traces of Projected Density from Electron Micrographs of Negatively Stained, Rotary-Shadowed, and Unstained IF*

In order to perform quantitative determinations of filament widths from electron micrographs, it is essential that these be based on numerical profiles of density derived from digitized images. Because single line traces may be noisy, and also because they give local, possibly nonrepresentative, samples, it is desirable that such profiles should represent axial averages over extended filament segments. However, owing to the characteristic curvature of IF, directly applied axial averaging would introduce lateral smearing. Computational straightening procedures (Steven *et al.*, 1985b) allow one to correct for this form of disorder, and thus obtain axially averaged traces of projected density with favorable signal-to-noise ratios. We have applied this procedure to electron images of IF prepared for electron microscopy by several different techniques (Fig. 10).

5.2.1. *Negatively Stained IF*

Bovine epidermal K-IF and vimentin IF are shown in Figs. 8 and 9 and straightened images of uniform segments are shown in Fig. 4d,e. As is typical for negatively stained IF, these images show few indications of substructure, except for occasional near-axial striations. An axially averaged transverse trace of projected density is shown in Fig. 10b. It is noteworthy that the longitudinal striations discernible in the original images usually show up in the transverse projections as two inner peaks spaced 2.9 ± 0.4 nm apart, lying between two outer peaks. The latter represent stain accumulation around the particles. A central peak in a transverse trace of *projected* density could represent either the penetration of stain into an axial channel or the contrasting of external features. However, in 65% of images analyzed, two distinct inner peaks are resolved. The fact that these peaks are off-centered implies that they do not relate to staining of an axial lumen, which would be centrally and symmetrically located relative to the particle's projection; rather, these peaks represent the accumulation of stain at some nonaxial location, most likely in surface grooves. The outer peaks are spaced 10.7 ± 0.4 nm (S.D.; N = 22) apart, but it is not possible to relate this spacing unambiguously to the outer diameter of these IF without prior knowledge of the depth of the stain layer, the degree of stain infiltration at the outer surface, and whether or not coincidental positive staining of surface domains had taken place.

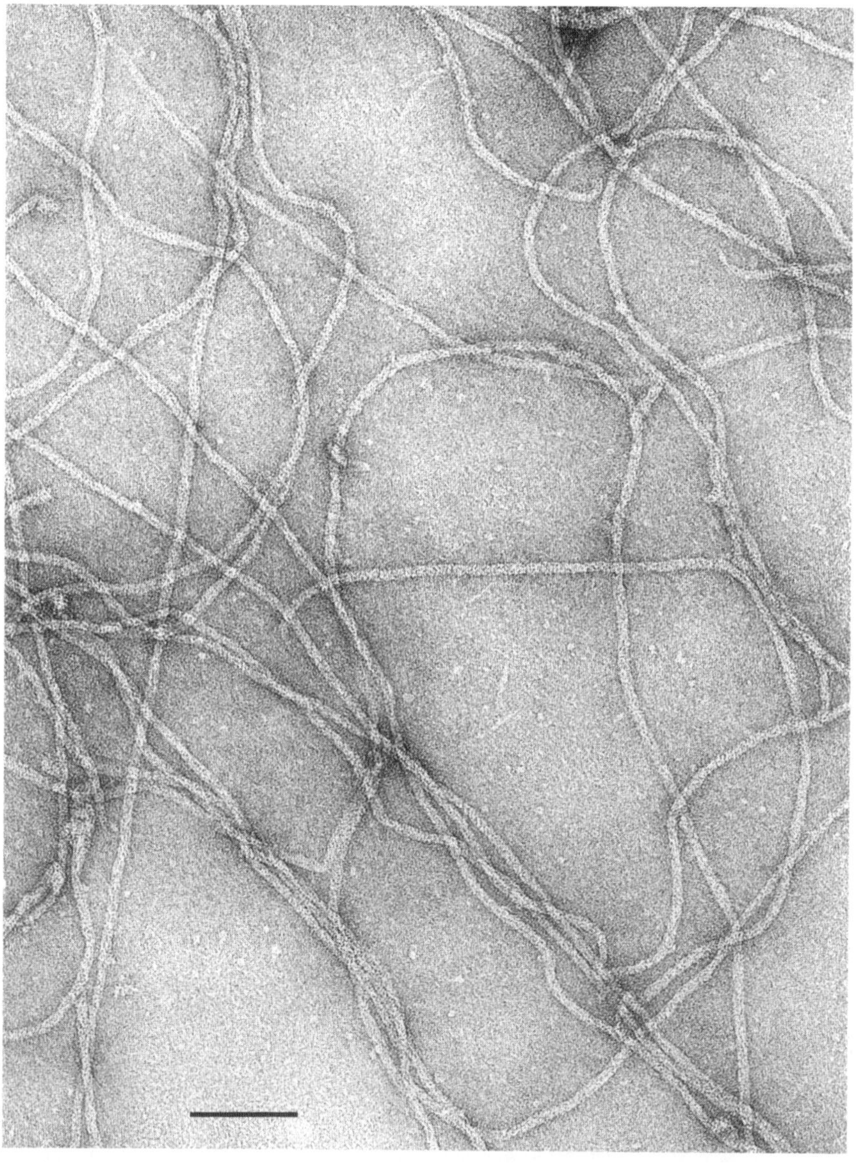

Figure 8. Reassembled bovine epidermal keratin IF negatively stained with uranyl acetate and imaged in bright-field with a conventional transmission electron microscope. Bar = 100 nm. Courtesy of Dr. P. M. Steinert.

5.2.2. IF Prepared by Low-Angle Rotary Shadowing after Glycerol Nebulization

Figure 3 shows a field of bovine epidermal K-IF prepared by this technique, which has been found to reveal (and possibly to accentuate) an intrinsic axial repeat of 22.7 nm (Section 4.3). Averaged transverse traces calculated from straightened particles (Fig. 4a,b) are given in Fig. 10c. The total width spanned by accumulated platinum grains is >

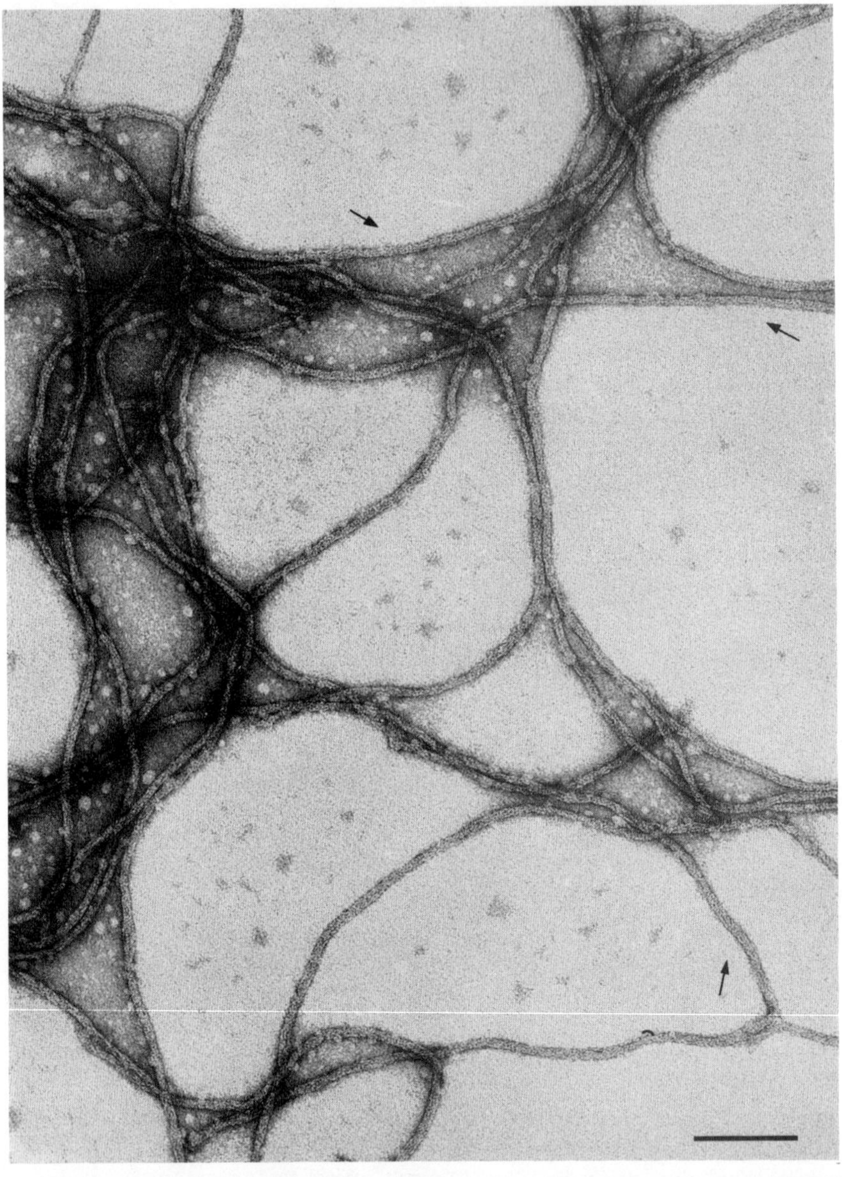

Figure 9. Reassembled vimentin IF negatively stained with uranyl acetate and imaged in bright-field with a conventional transmission electron microscope. Areas of both deep and shallow stain layer are present in this field. Arrows indicate IF segments that are particularly suggestive of subfilamentous structure. Bar = 100 nm.

20 nm. This figure is likely to be an overestimate of the particles' diameters on account of the "decoration" effect caused by the surface mobility of the incident platinum. It is not possible to give an unambiguous correction factor for this effect, but, with lightly shadowed specimens such as these, it does seem unlikely that such widths could be accounted for by particles of 8–10 nm diameter.

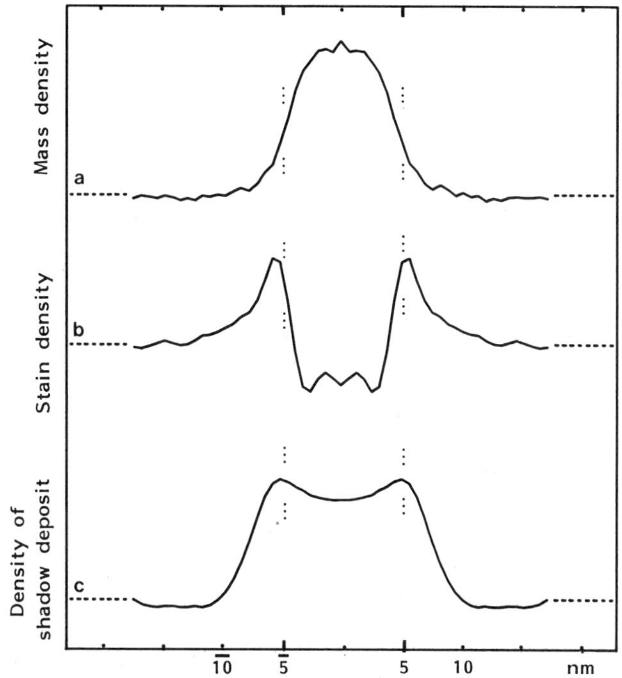

Figure 10. Transverse traces of projected density from epidermal keratin IF prepared for electron microscopy in three different ways. In order to obtain traces with high signal-to-noise ratios, these images were axially averaged after computational straightening (see Section 5.2). (a) Unstained IF prepared by freeze-drying and imaged by dark-field STEM. The projected density is that of the IF protein *per se*. (b) Negatively stained IF imaged in bright-field CTEM. The high outer peaks of projected stain density represent the accumulation of stain around the IF backbone, whereas the two inner peaks most likely represent the penetration of stain into long-pitch helical grooves or crevices in the backbone surface. (c) Low-angle rotary-shadowed IF prepared by glycerol nebulization onto a mica surface followed by vacuum drying. Note that the average projected density is slightly lower than the average background grain density (dashed line) over a region of about 8 nm on either side of the particle. This effect represents surface mobility of the incident platinum, leading to (at least partial) decoration of the IF.

5.2.3. *Unstained Frozen-Dried IF Visualized by Dark-Field STEM*

STEM micrographs of bovine epidermal K-IF are shown in Fig. 6a, and axially averaged traces of projected density in Fig. 10a. In this case, the traces were compiled somewhat differently from those given above, in that (1) the original digital data were resorted to compensate for sparse sampling (while the STEM probe is 0.25 nm in diameter, the specimen is sampled at intervals of 1.0 or 2.0 nm to reduce radiation damage); and (2) rather than straightening long curved IF, data from many short straight segments were combined (Steven *et al.*, 1984). Since these specimens are unstained, the projected density traces relate directly to the IF structure, and these yield outer diameters of 15 nm. To check whether these values might be sensitive to the precise conditions of washing to remove residual salts prior to freeze-drying, these experiments were repeated using several different wash treatments, all of which yielded outer diameters of ~ 15 nm (Steven *et al.*, 1985a). It is conceivable that washing with low-ionic-strength buffers might perturb the structures of vimentin IF (Ip *et al.*, 1985a,b), but indistinguishable results were also obtained with BEK-IF, which are stable at low ionic strengths (Steinert *et al.*, 1976). We also considered the possibility that interaction with the underlying carbon film might perturb the native structure so as to give artifactually high total widths, and so repeated the experiments with substrate-free IF strung across empty holes in a fenestrated carbon film (Fig. 11). However, the diameters obtained for these specimens were also 15 nm, thereby discounting this possibility.

5.3. Radial Density Profiles of IF

It is possible to reconstruct radial density profiles from transverse projections such as that of Fig. 10a (Steven *et al.*, 1984), and such curves are shown in Fig. 12. For "fully assembled" IF corresponding to ~ 16 molecules per cross section, these data consistently reveal an inner region of approximately uniform density extending to a radius of 4.5–5.0 nm, beyond which relatively diffuse density tapers off to an outer limit of ~ 7.5 nm. For polymorphic variants with lower mass-per-unit-length values (see Section 2.4), the core has the same average density as that of the fully assembled form, but extends to a radius of ~ 4 nm and has an outer radius of ~ 6.5 nm (Fig. 12). In both cases, the amounts of mass in the core and in the periphery respectively are about 2 : 1 for native vimentin IF as well as for reassembled vimentin and BEK-IF. The masses of the rod domains and of the end domains combined are also close to 2 : 1 for the subunits in these IF. On these grounds, and for other reasons discussed below, we interpret the radial density profiles in terms of IF backbones composed of laterally packed rod domains surrounded by a peripheral shell of end domains.

The average *dry* density observed in the IF backbone regions is 0.3 kDa/nm^3 (0.5 g/cm^3). This value is lower than the maximum densities that we have observed in other specimens, e.g., 0.6 kDa/nm^3 for tobacco mosaic virus (Steven *et al.*, 1984) suggesting

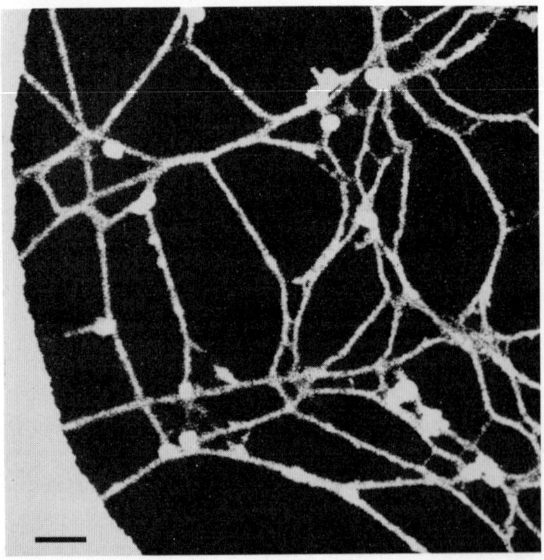

Figure 11. Network of bovine epidermal keratin IF suspended over a hole in a carbon film and imaged without staining by STEM dark-field after freeze-drying. Substrate-free IF are extremely sensitive to radiation damage, which is manifested as gross tearing and disintegration of the filament network. Accordingly, exceptionally low-dose imaging conditions (average dose ~ 20 e^-/nm^2) are required to record such images. Bar = 100 nm. J. Wall, P. M. Steinert, and A. C. Steven (unpublished study).

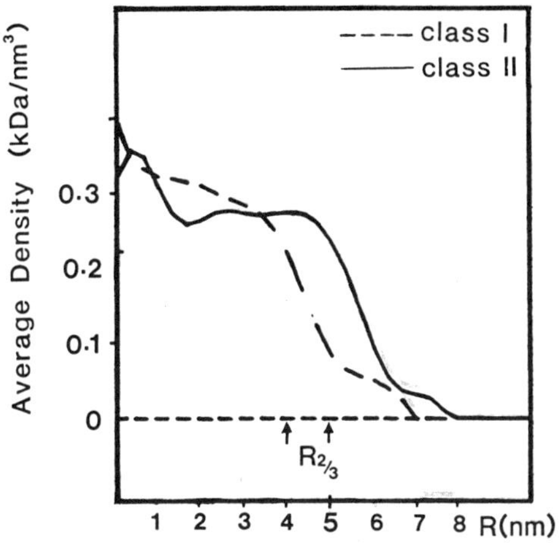

Figure 12. Radial profiles of cylindrically averaged dry density in bovine epidermal keratin IF, calculated from dark-field STEM micrographs of unstained freeze-dried preparations. Data are shown for two polymorphic variants that contain, on average, ~ 16 molecules (class II) and ~11 molecules (class I) in cross section, respectively. Both are found to have solid cores of approximately uniform density surrounded by relatively diffuse peripheral material. Approximately two-thirds of the total mass is contained in the core in each case. The radii that circumscribe two-thirds of the mass ($R_{2/3}$) are marked in both cases (Steven *et al.*, 1987).

that, in solution, this region is heavily hydrated, and occupied by protein that is predominantly polar. This inference is consistent with the high incidence of polar residues ($\geqq$ 70%) in the rod domains of IF subunits. For comparison, we have estimated the spatially averaged dry density for the myosin filament backbone, which is generally accepted to be composed of close-packed α-helical coiled coils. In the case of vertebrate skeletal myosin, taking the backbone to be 15 nm across (Yu *et al.*, 1985; Cantino and Squire, 1986), that it contains the entire rod domains (each 220 kDa), that there are three myosin molecules per 14.3-nm axial repeat, and that the length of the rod is 130 nm (Squire, 1981), an average value of 0.26 kDa/nm^3 dry density is obtained. This value is remarkably close to the density measured directly here for the corresponding region of IF.

5.4. Filaments within Filaments

The proposal has been made on several occasions that IF are organized as bundles of laterally associated ("fasciated") subfilaments (e.g., Krishnan *et al.*, 1979; Lieska *et al.*, 1981; Franke *et al.*, 1982; Aebi *et al.*, 1983). The evidence in support of this idea is mainly the visualization of filamentous substructure in electron micrographs of disrupted or disintegrating IF (e.g., Figs. 13, 14) or of partially assembled specimens. This concept is also compatible with observations that the linear densities of polymorphic variant forms of IF determined by STEM (see Section 2.4) are related by simple integral ratios, which is readily explained if these variants were to contain different numbers of subfilaments. According to our STEM data, the average mass per unit length of low-density variants is about two-thirds that of the "mature" form (characteristic, for instance, of native vimentin IF), which contains about 16 molecules in cross section. This ratio implies complements of two and three subfilaments, respectively (or four and six, etc.). These subfilaments would then contain 5⅓ molecules in cross section (or 2⅔, etc.). Geisler *et al.* (1985) have proposed models according to which mature IF contain seven or eight subfilaments, each a linear string of four-chain complexes (see Section 3). Ip *et al.* (1985b) favor a form of the eight-subfilament version. In terms of this model, our low-density IF with an average relative density of 0.66 would correspond to an unresolved mixture of four-

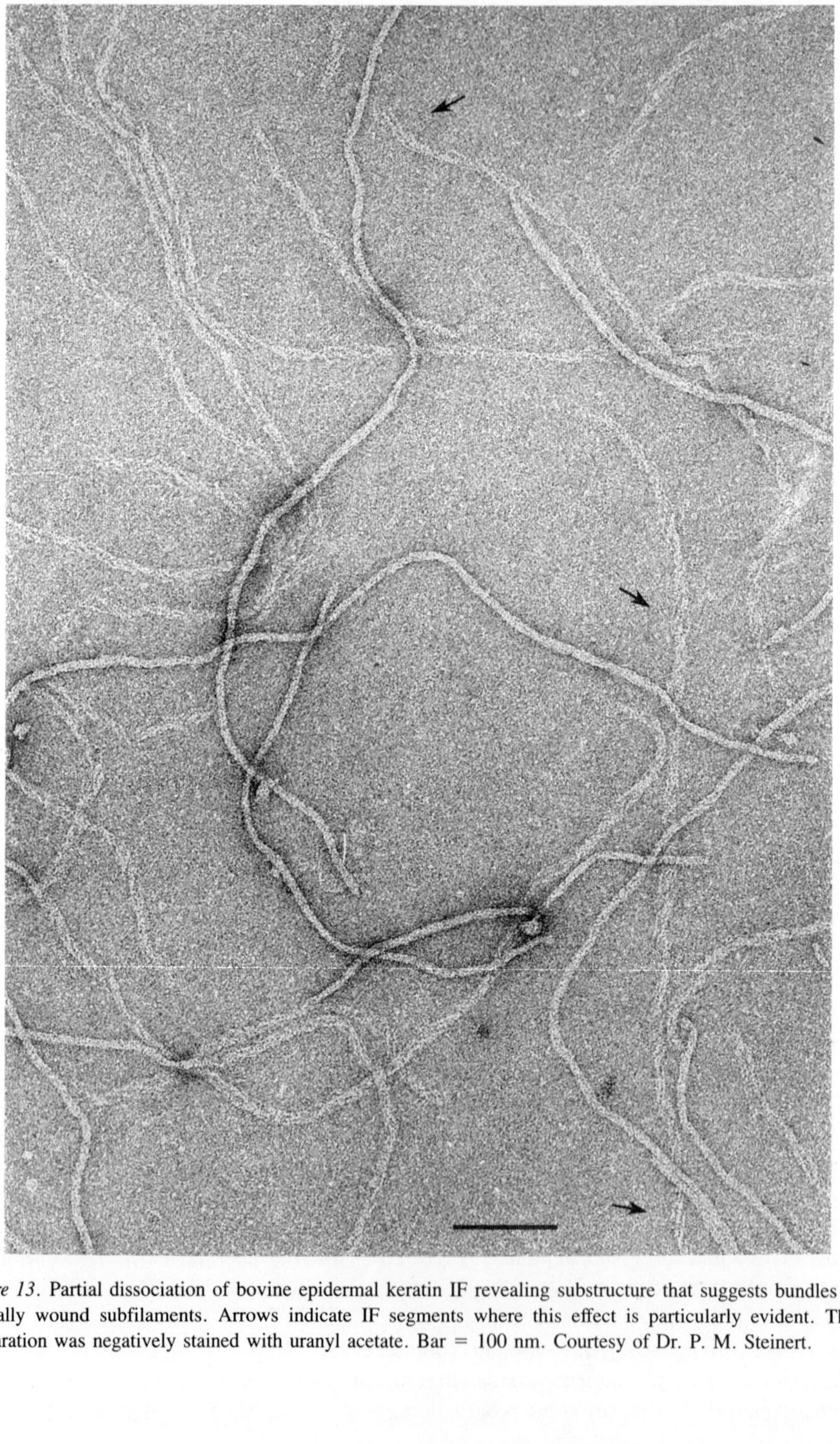

Figure 13. Partial dissociation of bovine epidermal keratin IF revealing substructure that suggests bundles of helically wound subfilaments. Arrows indicate IF segments where this effect is particularly evident. This preparation was negatively stained with uranyl acetate. Bar = 100 nm. Courtesy of Dr. P. M. Steinert.

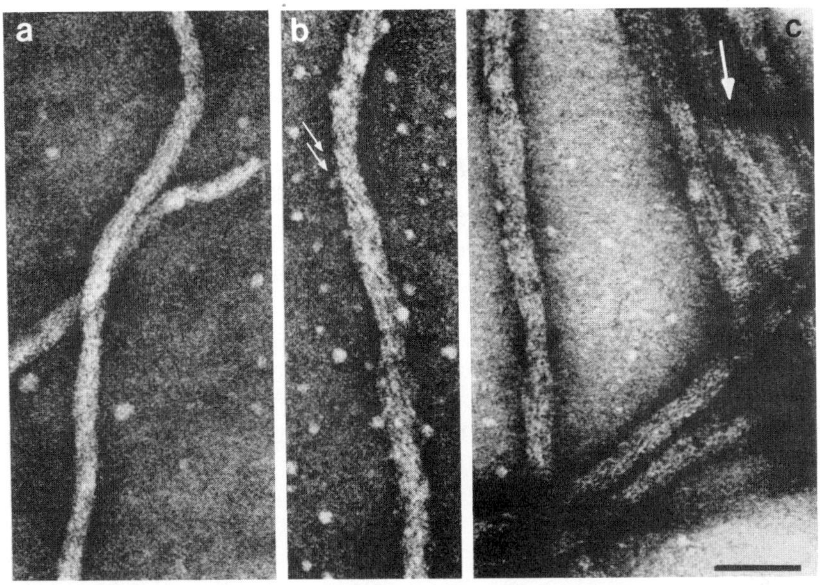

Figure 14. Various degrees of disruption of vimentin IF visualized by STEM after negative staining with uranyl sulfate. (a) The IF backbone has not been penetrated by stain. (b) An IF that is seemingly slightly loosened in structure, permitting the contrasting of long-pitch helically wound striations (arrows) implying a subfilamentous organization. (c) Subfilaments are clearly revealed if an IF in a relatively advanced state of disintegration (arrow). Bar = 50 nm.

subfilament (0.50) and six-subfilament (0.75) species, respectively. Eichner *et al.* (1985, Fig. 11) entertain both four-subfilament and three-subfilament models as viable candidates for mature K-IF, and Krishnan *et al.* (1979) have advocated a four-subfilament configuration for NF. Thus, while there is widespread enthusiasm for the subfilament concept, a consensus has not yet been reached as to the precise numbers of subfilaments involved, their substructures, and mode(s) of packing.

Sauk *et al.* (1984) obtained rotary-shadowed images of what appear to be single K-IF subfilaments (*in vitro* assembly products) that also show conspicuously a "22-nm" axial repeat, as do intact IF. Provided that these polymers do indeed emulate components of intact IF, this observation implies that this periodicity is already imposed at the subfilament level, and the issue of whether the repeat in intact IF is axial or helical (see Section 4.3) relates to whether or not the subfilaments are aligned in register.

With regard to the observed polymorphism of reassembled IF, it is not clear whether this phenomenon has physiological relevance, or whether it represents an artifact of the *in vitro* assembly system. However, we note that an essentially similar form of polymorphism is exhibited *in vivo* by different kinds of myosin filaments. As with polymorphism of IF, its basis appears to reside in the lateral association of differing numbers of subfilaments in closely related surface lattices: these have in common an axial repeat of close to 14.3 nm (Squire, 1981). Similar conclusions as to subfilamentous organization have also been drawn from electron microscopy of isolated myosin filaments in a partially disrupted state (Davey and Graafhuis, 1976; Mau and Rowe, 1980). The least dense native myosin filaments, those of striated vertebrate muscle, contain three subfilaments, those of vertebrate smooth muscle contain four, insect flight muscle have four or six, and increasing numbers are present in the myofilaments of certain crustacea (Wray, 1979; Squire, 1981).

5.5. *Interpretation of Images of Negatively Stained IF with and without Proteolytic Modification*

Although the peaks of projected stain density on either side of a particle provide an *operational* definition of its width, it is not possible to determine from such data how far the particle may, in fact, protrude into the stain layer (see Section 5.2.1). Thus, we believe that the peripherally located end domains of IF are usually not visualized in negatively stained preparations, presumably because they are too diffuse and/or too disordered to exclude stain effectively; and possibly also because they may have a tendency to stain positively, in which case they would not be discriminated from surrounding negative stain. Germane to this point, many IF end domains contain local concentrations of charged residues, particularly basic residues, whose propensity for positive staining has been demonstrated (Tzaphlidou *et al.*, 1982).

It is consistent with the contention that end domains do not, in general, significantly influence the negative staining representation of IF that several studies of proteolytically modified IF whose end domains had been partly or wholly digested found these specimens to be morphologically unchanged. The end domains of the keratins of terminally differentiated epidermal cells contain high (> 50%) concentrations of glycine. Thus, Steinert *et al.* (1983) found that chymotryptic trimming of 60–70% of the Gly residues from these IF did not affect their morphology. Similar observations have been reported for NF (Chin *et al.*, 1983) treated with trypsin (despite the relatively large size of NF end domains), and for desmin IF treated with lysine-specific protease (Kaufmann *et al.*, 1985). These observations imply that, even when present, end domains are not generally visualized by negative staining.

5.6. *In Vitro Assembly Properties of Proteolytically Modified IF*

Although *in situ* proteolysis of end domains does not significantly affect IF morphology, there are cases where IF subunits with proteolytically modified end domains have been found to be impaired in their capacity for self-assembly (e.g., Steinert, 1980; Traub and Vorgias, 1983; Sauk *et al.*, 1984). For instance, desmin rod domains form only tactoids (Geisler *et al.*, 1982), and less radical proteolysis has also been found to diminish or abolish the competence of desmin for assembly (Kaufmann *et al.*, 1985).* It is, however, important to distinguish between a transient regulatory role that may be played by a domain during assembly, and its eventual location (and interactions with other domains) in the fully assembled polymer. Thus, the observed adverse effects of end-

*Albeit indirectly, these observations may afford the best evidence available that when IF subunits are purified under relatively harsh conditions, such as extraction with SDS or in high concentrations of urea or guanidine hydrochloride, the end domains do indeed (largely) renature upon removal of the denaturant. Whereas renaturation of the rod domains is attested by the presence in x-ray diffraction patterns from fibers of reassembled IF, of reflections characteristic of coiled-coil α helices (such as are observed with native IF), no comparably direct data attest to the conformational status of the end domains. Thus, the requirement for desmin self-assembly, of intact and adequately folded end domains, is likely also to apply to the assembly of other kinds of IF.

domain proteolysis on the assembly properties of IF subunits are not incompatible with the postulate of exposed peripheral locations for these domains in intact IF.

These observations have an interesting parallel in the myosin system. Whereas self-assembly of bipolar filaments of the appropriate diameter has been demonstrated with intact myosin molecules (Huxley, 1963; Kaminer *et al.*, 1976), this capability is not possessed by the rod domain, or subfragments thereof, which assemble only into tactoids or paracrystals of indefinite width (see Section 4.4). Thus, while there is little doubt that the myosin S1 domain is fully extraneous in the assembled filament, its presence is somehow required to specify correct assembly.

5.7. Other Pertinent Observations

5.7.1. IF Are Not Hollow

For the IF that we have analyzed (vimentin and BEK-IF), STEM radial density profiles consistently show positive density at the filament centers. For technical reasons (Steven *et al.*, 1984, 1986; Duda *et al.*, 1985), experimental uncertainty in such profiles tends to be higher at near-axial points than farther out. However, for other filamentous polymers known to have axial channels 3–4 nm in diameter, viz. TMV (Steven *et al.*, 1984) and phage T4 tail-tubes treated with guanidine hydrochloride (Duda *et al.*, 1985), these holes are clearly visualized by the STEM technique. Thus, it is highly significant that not even a diminution in density is observed around the axes of these IF, and we conclude that any axial channel must be on a diminutive scale (e.g., $\lesssim 1$ nm in diameter).

We have also argued that the near-axial striations visualized on some negatively stained IF represent penetration of stain into surface grooves and not into a central lumen (Section 5.2.1). In thin sections of IF contrasted by positive staining, there is no consistent pattern to the many experimental observations that have been documented with regard to the staining propensities of IF cores. For instance, transverse thin sections of wool and hair fibers (Filshie and Rogers, 1961; Millward, 1970) suggest a "ring + inner core" structure for these K-IF *in situ*. On the other hand, it has been claimed on the basis of thin-section data that reassembled vimentin IF are hollow (Renner *et al.*, 1981). However, it should be noted that a reduced density of positive staining does not necessarily represent a local absence of protein, but may merely represent inefficient staining. Thus, we conclude that there is no firm evidence that IF of any kind are hollow, but there is direct evidence (see above) that epidermal K-IF and vimentin IF, at least, are not hollow.

To consider the implications of a nonhollow IF core, we note that this property is not easily reconciled with models in which linear rod domains arrayed on a planar surface lattice are simply folded into a cylindrical tube because any such model invokes a substantial axial channel. This property is more readily squared with arrangements whereby rod domains or segments thereof may be positioned at differing radial locations (Fraser *et al.*, 1976; Fraser and MacRae, 1983), or models in which several primary subfilaments entwine to produce secondary subfilaments (e.g., Krishnan *et al.*, 1979; Aebi *et al.*, 1983; Eichner *et al.*, 1985), of which several, in turn, pack together to form the complete IF.

5.7.2. Equatorial X-Ray Diffraction from "Hard" α-Keratin

Using the equatorial intensity distribution from an x-ray fiber diffraction pattern together with a modeled (and to some extent arbitrary) phase set, it is possible to reconstruct the radial distribution of density in the individual filament. Such an analysis has been performed using data from the "hard" K-IF of porcupine quill (Fraser *et al.*, 1976), and also yields a nonhollow core. These results differ slightly from those given above for vimentin and epidermal K-IF (Section 5.3) in that the diameter of the IF backbone, at ~ 7.5 nm, is smaller than the 9.5–10.0 nm, given by the STEM technique. It remains to be determined whether this dimensional discrepancy results from (1) the effects of differing specimen preparation techniques, the STEM specimens being frozen-dried and the quill specimens vacuum-dried, thus undergoing a possibly more traumatic liquid–vapor phase transition; (2) the possibility that native hard K-IF may correspond to the "low-density" polymorphic variant observed in reassembled epidermal K-IF preparations (see Section 2.4); or (3) a more fundamental difference may exist between the respective structures of these different kinds of K-IF. Nevertheless, a recent interpretation of the x-ray fiber diffraction data also places the end domains of these subunits external to the K-IF backbones (Fraser and MacRae, 1982). These K-IF end domains are rich in cysteine residues which are thought to interact, via disulfide bridges, with the "high-sulfur" matrix proteins in which these K-IF are embedded.

5.7.3. *Neurofilament Cross-Bridges*

The carboxy-terminal domains of NF-M and NF-H are by far the largest IF end domains, and several authors have interpreted these domains as constituting the cross-bridge structures that appear to connect adjacent axonal NF at interfilament spacings of about 20 nm, as visualized, for instance, by quick-freeze, deep-etch electron microscopy (Hirokawa *et al.*, 1984). Such an interpretation is certainly in line with the more general hypothesis advanced here of all IF end domains being peripheral protrusions, bearing in mind that these NF end domains, by virtue of their much greater size, are more readily visualized by conventional electron microscopic methods.

5.8. Current Frame of Reference

In this chapter, an attempt has been made to collate the prevailing evidence germane to the structures of IF. Electron microscopic studies have been stressed, and these data assessed in the light of complementary information from other sources. Overall, the data support a general model that is based on IF backbones composed of close packings of the coiled-coil α-helical rod domains, surrounded by peripheral shells containing the non-α-helical end domains. Because of the conservativity of IF rod domains, this scheme allows closely related supramolecular structures to be assumed by IF of all kinds, despite the pronounced diversity of IF proteins. Similarly, peripheral exposure of end domains accommodates a range of structural variation that extends from the vestigial end domains of mesothelial keratins to the massive carboxy-terminal extensions of NF proteins NF-M and NF-H.

A second theme has been to conduct a systematic comparison between the structural properties of IF and those of myosin filaments, prompted by the consideration that myosin filaments are also cytoskeletal protein polymers whose backbones consist of coiled-coil rod domains from which functional globular domains protrude. Their respective properties are compiled in Table I and, disparity in some specific details notwithstanding, this comparison firmly substantiates the basic structural analogy between myosin and IF. It is a

Table I. Structural Properties of Intermediate Filaments and Myosin Filaments Compared

	Intermediate filaments	Myosin filaments	See Section
Polypeptide chains			
Molecular weights	~40,000 to ~135,000	~220,000 (heavy chain)	
Domanial composition	NH_2 end domain/rod/end domain COOH	NH_2 end domain/rod (subfragment 1)	2.2
Rod domains			
Length	~310 residues (~46 nm)	~1117 residues (~1600 nm)	
Secondary structure	α helix (segmented)	α helix	
Repeats of periodically distributed amino acids	2.33, 3.50—polar and nonpolar (heptad repeat) 9.7—acidic and basic	2.33, 3.50—polar and nonpolar (heptad repeat) 9.3—acidic and basic	4.1
Molecules			3
Polypeptide chain content	Two chains	Two chains	3
Mode of association	Interchain α-helical coiled coils formed by rod domains	Interchain α-helical coiled coils formed by rod domains	3
Registration of chains in coiled coils	Parallel, in register	Parallel, in register	3
Filaments			
Prominent axial repeats	22.7 nm	14.3 nm (inter-"crown" spacing) 43 nm (complete axial repeat)	4.3
Polarity	Probably nonpolar (Type III)	Bipolar	3
Polymorphism	Multiple linear density forms reassembled *in vitro*	Multiple linear density forms *in vivo* in different muscles	2.4, 4.5
Location of end domains	External	External	5
Location of rod domains	Backbone	Backbone	5
Average dry density of backbone	~0.3 kDa/nm³	~0.26 kDa/nm³	5.3

plausible notion—albeit one that is not inhibited by much direct experimental evidence—that this analogy may extend to functional attributes. The primary role of myosin is to interact with another set of cytoskeletal filaments, actin filaments, so as to generate mechanical force at the expense of ATP hydrolysis. Both actin-binding and ATP hydrolysis are mediated by the myosin end domain, subfragment 1. Our conjecture does not assign an active role in motility to IF end domains: rather, it envisages that they mediate functionally relevant although still largely obscure interactions between IF and other filaments, or which confine other cytoplasmic components to the cytoskeletal (IF) compartment. It is to be hoped that further innovations in this intriguing area of research will soon test this general structure–function hypothesis, in addition to providing further information as to the molecular substructures of IF.

ACKNOWLEDGMENTS. I thank Drs. P. M. Steinert and R. D. B. Fraser for critical readings, Dr. B. L. Trus for constant collaboration, Ms. M. Bisher for photography, and Ms. A. Hutchings for preparing the manuscript.

References

Aebi, U., Fowler, W. E., Rew, P., and Sun, T.-T., 1983, The fibrillar substructure of keratin filaments unravelled, *J. Cell Biol.* **97:**1131–1143.

Ahmadi, B., and Speakman, P. T., 1978, Suberimidate cross-linking shows that a rod-shaped, low-cystine, high-helix protein prepared by limited proteolysis of reduced wool has four protein chains, *FEBS Lett.* **94:**365–367.

Cantino, M., and Squire, J., 1986, Resting myosin cross-bridge configuration in frog muscle thick filaments, *J. Cell Biol.* **102:**610–618.

Chin, T. K., Eagles, P. A. M., and Maggs, A., 1983, The proteolytic digestion of ox neurofilaments with trypsin and α-chymotrypsin, *Biochem. J.* **215:**239–252.

Connell, N. D., and Rheinwald, J. G., 1983, Regulation of the cytoskeleton in mesothelial cells: Reversible loss of keratin and increase in vimentin during rapid growth in culture, *Cell* **34:**245–253.

Crewther, W. G., Inglis, A. S., and McKern, N. M., 1978, Amino-acid sequence of α-helical segments from S-carboxymethylkerateine A, *Biochem. J.* **173:**365–371.

Crewther, W. G., Dowling, L. M., Steinert, P. M., and Parry, D. A. D., 1983, Structure of intermediate filaments, *Int. J. Biol. Macromol.* **5:**267–274.

Davey, C. L., and Graafhuis, A. E., 1976, Organization of myosin molecules in the muscle thick filament, *Experientia* **32:**32–34.

Duda, R. L., Wall, J. S., Hainfeld, J. F., Sweet, R. M., and Eiserling, F. A., 1985, Mass distribution of a probable tail-length-determining protein in bacteriophage T4, *Proc. Natl. Acad. Sci. USA* **82:**5550–5554.

Eichner, R. A., Rew, P., Engel, A., and Aebi, U., 1985, Human epidermal keratin filaments: Studies on their structure and assembly, *Ann. N.Y. Acad. Sci.* **455:**381–402.

Engel, A., Eichner, R., and Aebi, U., 1985, Polymorphism of reconstituted human epidermal keratin filaments: Determination of their mass-per-unit-length and width by scanning transmission electron microscopy, *J. Ultrastruct. Res.* **90:**323–335.

Eriksson, A., and Thornell, L. E., 1979, Intermediate (skeletin) filaments in heart Purkinje fibers, *J. Cell Biol.* **80:**231–247.

Filshie, B. K., and Rogers, G. E., 1961, The fine structure of α-keratin, *J. Mol. Biol.* **3:**784–786.

Franke, W. W., Schiller, D. L., and Grund, C., 1982, Protofilamentous and annular structures as intermediates during reconstitution of cytokeratin filaments in vitro, *Biol. Cell* **46:**257–268.

Fraser, R. D. B., and MacRae, T. P., 1973a, The structure of α-keratin, *Polymer* **14:**61–67.

Fraser, R. D. B., and MacRae, T. P., 1973b, *Conformation in Fibrous Proteins,* Academic Press, New York.

Fraser, R. D. B., and MacRae, T. P., 1982, The fine structure of keratin fibers, in: *Milton Harris: Chemist, Innovator, and Entrepreneur* (M. M. Breuer, ed.), American Chemical Society, Washington, D.C., pp. 109–137.

Fraser, R. D. B., and MacRae, T. P., 1983, The structure of the α-keratin microfibril, *Biosci. Rep.* **3:**517–525.

Fraser, R. D. B., MacRae, T. P., and Suzuki, E., 1976, Structure of the α-keratin microfibril, *J. Mol. Biol.* **108:**435–452.

Fraser, R. D. B., MacRae, T. P., Suzuki, E., and Parry, D. A. D., 1985, Intermediate filament structure: 2. Molecular interactions in the filament, *Int. J. Biol. Macromol.* **7:**258–274.

Geisler, N., Kaufmann, E., and Weber, K., 1982, Protein-chemical characterization of three structurally distinct domains along the protofilament unit of desmin 10nm filaments, *Cell* **30:**277–286.

Geisler, N., Plessmann, U., and Weber, K., 1983a, The complete amino-acid sequence of the major mammalian neurofilament protein, *FEBS Lett.* **182:**475–478.

Geisler, N., Kaufmann, E., Fischer, S., Plessmann, U., and Weber, K., 1983b, Neurofilament architecture combines structural principles of intermediate filaments with carboxy-terminal extensions increasing in size between triplet proteins, *EMBO J.* **2:**1295–1302.

Geisler, N., Kaufmann, E., and Weber, K., 1985, Antiparallel orientation of the two double-stranded coiled-coils in the tetrameric protofilament unit in intermediate filaments, *J. Mol. Biol.* **182:**173–177.

Hanukoglu, I., and Fuchs, E., 1982, The cDNA sequence of a human epidermal keratin: Divergence of sequence but conservation of structure among intermediate filament proteins, *Cell* **31:**243–252.

Hanukoglu, I., and Fuchs, E., 1983, The cDNA sequence of a Type II cytoskeletal keratin reveals constant and variable structural domains among keratins, *Cell* **33:**915–924.

Harrington, W. F., Burke, M., and Barton, J. S., 1972, Association of myosin to form contractile system, *Cold Spring Harbor Symp. Quant. Biol.* **37:**77–85.

Hatzfeld, M., and Franke, W. W., 1985, Pair formation and promiscuity of cytokeratins: Formation in vitro of heterotypic complexes and intermediate-sized filaments by homologous and heterologous recombinations of purified polypeptides, *J. Cell Biol.* **101:**1826–1841.

Henderson, D., Geisler, N., and Weber, K., 1982, A periodic ultrastructure in intermediate filaments, *J. Mol. Biol.* **155:**173–176.

Hirokawa, N., Glicksman, M. A., and Willard, M. B., 1984, Organization of mammalian neurofilament polypeptides within the neuronal cytoskeleton, *J. Cell Biol.* **98:**1523–1536.

Huxley, H. E., 1963, Electron microscopic studies on the structure of natural and synthetic protein filaments from striated muscle, *J. Mol. Biol.* **7:**281–308.

Ip, W., Hartzer, M. K., Pang, Y.-Y. S., and Robson, R. M., 1985a, Assembly of vimentin in vitro and its implications concerning the structure of intermediate filaments, *J. Mol. Biol.* **183:**365–376.

Ip, W., Hartzer, M. K., Pang, Y.-Y. S., and Robson, R. M., 1985b, Subunit structure of desmin and vimentin protofilaments and how they assemble into intermediate filaments, *Ann. N.Y. Acad. Sci.* **455:**185–199.

Kaminer, B., Szonyi, E., and Belcher, C. D., 1976, "Hybrid" myosin filaments from smooth and striated muscle, *J. Mol. Biol.* **100:**379–386.

Kaufmann, E., Geisler, N., and Weber, K., 1984, SDS-PAGE strongly overestimates the molecular masses of neurofilament proteins, *FEBS Lett.* **170:**81–84.

Kaufmann, E., Geisler, N., and Weber, K., 1985, Intermediate filament forming ability of desmin derivatives lacking either the amino-terminal 67 or the carboxy-terminal 27 residues, *J. Mol. Biol.* **185:**733–742.

Knapp, A. C., Franke, W. W., Heid, H., Hatzfeld, M., Jorcano, J. L., and Moll, R., 1986, Cytokeratin No. 9, an epidermal type I keratin characteristic of a special program of keratinocyte differentiation, *J. Cell Biol.* **103:**657–667.

Knapp, B., Rentrop, M., Schweizer, J., and Winter, H., 1986, Nonepidermal members of the keratin multigene family: cDNA sequence and in situ localization of the mRNAs, *Nucleic Acids Res.* **14:**751–763.

Krishnan, N., Kaiserman-Abramof, I. R., and Lasek, R. J., 1979, Helical substructure of neurofilaments isolated from Myxicola and squid giant axons, *J. Cell Biol.* **82:**323–335.

Lasek, R. J., Phillip, L., Katz, M. J., and Autilio-Gambetti, L., 1985, Function and evolution of neurofilament proteins, *Ann. N.Y. Acad. Sci.* **455:**462–478.

Lieska, N., Maisel, H., and Romero-Herrera, A. E., 1981, Electron microscopy supports a fibrous substructure for lens intermediate filaments, *Curr. Eye Res.* **1:**339–350.

McKeon, F. D., Kirschner, M. W., and Caput, D., 1986, Homologies in both primary and secondary structure between nuclear envelope and intermediate filament proteins, *Nature* **319:**463–468.

McLachlan, A. D., and Karn, J., 1983, Periodic features in the amino-acid sequence of nematode myosin rod, *J. Mol. Biol.* **164:**605–626.

Mandelkow, E., Thomas, J., and Cohen, C., 1977, Microtubule structure at low resolution by x-ray diffraction, *Proc. Natl. Acad. Sci. USA* **74:**3370–3374.

Mau, M. C., and Rowe, A. J., 1980, Fraying of A-filaments into three subfilaments, *Nature* **286:**412–414.

Milam, L. A., and Erickson, H. P., 1982, Visualization of a 21-nm axial periodicity in shadowed keratin filaments and neurofilaments, *J. Cell Biol.* **94:**592–596.

Millward, G. R., 1970, The substructure of α-keratin microfibrils, *J. Ultrastruct. Res.* **31:**349–355.

Misell, D. L., 1978, *Image Analysis, Enhancement and Interpretation,* North-Holland, Amsterdam.

Parry, D. A. D., Steven, A. C., and Steinert, P. M., 1985, The coiled-coil molecules of intermediate filaments consist of two parallel chains in axial register, *Biochem. Biophys. Res. Commun.* **127:**1012–1018.

Pollard, T. D., and Cooper, J. A., 1986, Actin and actin-binding proteins: A critical evaluation of mechanisms and function, *Annu. Rev. Biochem.* **55:**987–1035.

Pruss, R. M., Mirsky, R., Raff, M. C., Thorpe, R., Dowding, A. J., and Anderton, B. H., 1981, All classes of intermediate filaments share a common antigenic determinant defined by a monoclonal antibody, *Cell* **27:**419–428.

Quinlan, R. A., Cohlberg, J. A., Schiller, D. L., Hatzfeld, M., and Franke, W. W., 1984, Heterotypic tetramer (A_2D_2) complexes of non-epidermal keratins isolated from cytoskeletons of rat hepatocytes and hepatoma cells, *J. Mol. Biol.* **178:**365–388.

Renner, W., Franke, W. W., Schmid, E., Geisler, N., Weber, K., and Mandelkow, E., 1981, Reconstitution of intermediate-sized filaments from denatured monomeric vimentin, *J. Mol. Biol.* **149:**285–306.

Roop, D. R., Cheng, C. K., Titterington, L., Meyers, C. A., Stanley, J. R., Steinert, P. M., and Yuspa, S. H., 1984, Synthetic peptides corresponding to keratin subunits elicit highly specific antibodies, *J. Biol. Chem.* **259:**8037–8040.

Sauk, J. J., Krumweide, M., Cocking-Johnson, D., and White, J. G., 1984, Reconstitution of cytokeratin filaments in vitro: Further evidence for the role of nonhelical peptides in filament assembly, *J. Cell Biol.* **99:**1590–1597.

Squire, J., 1981, *The Structural Basis of Muscular Contraction,* Plenum, New York.

Steinert, P. M., 1977, The mechanism of assembly of bovine epidermal keratin filaments in vitro, in: *Biochemistry of Cutaneous Epidermal Differentiation* (M. Seiji and I. A. Bernstein, eds.), University of Tokyo Press, Tokyo, pp. 444–464.

Steinert, P. M., 1980, Structural changes of human epidermal α-keratin in disorders of keratinization, in: *Biochemistry of Normal and Abnormal Epidermal Differentiation* (I. A. Bernstein and M. Seiji, eds.), University of Tokyo Press, Tokyo, pp. 391–405.

Steinert, P. M., and Idler, W. W., 1975, The polypeptide composition of bovine epidermal α-keratin, *Biochem. J.* **151:**603–614.

Steinert, P. M., and Steven, A. C., 1985, Splitting hairs and other intermediate filaments, *Nature* **316:**767.

Steinert, P. M., Idler, W. W., and Zimmerman, S. B., 1976, Self-assembly of bovine epidermal keratin filaments in vitro, *J. Mol. Biol.* **108:**547–567.

Steinert, P. M., Idler, W. W., Cabral, F., Gottesman, M. G., and Goldman, R. D., 1981, In vitro assembly of homopolymer and copolymer filaments from intermediate filament subunits of muscle and fibroblastic cells, *Proc. Natl. Acad. Sci. USA* **78:**3692–3696.

Steinert, P., Idler, W., Aynardi-Whitman, M., Zackroff, R., and Goldman, R. D., 1982, Heterogeneity of intermediate filaments assembled in vitro, *Cold Spring Harbor Symp. Quant. Biol.* **46:**465–474.

Steinert, P. M., Rice, R. H., Roop, D. R., Trus, B. L., and Steven, A. C., 1983, Complete amino-acid sequence of a mouse epidermal keratin subunit and implications for the structure of intermediate filaments, *Nature* **302:**794–800.

Steinert, P. M., Parry, D. A. D., Racoosin, E. L., Idler, W. W., Steven, A. C., Trus, B. L., and Roop, D. R., 1984, The complete cDNA and deduced amino-acid sequence of a Type II mouse epidermal keratin of 60,000 Da: Analysis of sequence differences between Type I and Type II keratins, *Proc. Natl. Acad. Sci. USA* **78:**3692–3696.

Steinert, P. M., Steven, A. C., and Roop, D. R., 1985a, The molecular biology of intermediate filaments, *Cell* **42:**411–419.

Steinert, P. M., Parry, D. A. D., Idler, W. W., Johnson, L. D., Steven, A. C., and Roop, D. R., 1985b, Amino-acid sequences of mouse and human epidermal Type II keratins of Mr 67,000 provide a systematic basis for the structural and functional diversity of the end-domains of keratin intermediate filament subunits, *J. Biol. Chem.* **260:**7142–7149.

Steven, A. C., Wall, J., Hainfeld, J. F., and Steinert, P. M., 1982, Structure of fibroblastic intermediate filaments: Analysis by scanning transmission electron microscopy, *Proc. Natl. Acad. Sci. USA* **79:**3101–3105.

Steven, A. C., Hainfeld, J. F., Trus, B. L., Wall, J. S., and Steinert, P. M., 1983a, The distribution of mass in heteropolymer intermediate filaments assembled in vitro, *J. Biol. Chem.* **258:**8323–8329.

Steven, A. C., Hainfeld, J. F., Trus, B. L., Wall, J. S., and Steinert, P. M., 1983b, Epidermal keratin filaments assembled in vitro have masses per unit length that scale according to average subunit mass, *J. Cell Biol.* **97:**1939–1944.

Steven, A. C., Hainfeld, J. F., Trus, B. L., Wall, J. S., and Steinert, P. M., 1984, Radial distributions of density within macromolecular complexes determined from dark-field electron micrographs, *Proc. Natl. Acad. Sci. USA* **81:**6363–6367.

Steven, A. C., Hainfeld, J. F., Trus, B. L., Wall, J. S., and Steinert, P. M., 1985a, Conformity and diversity in the structures of intermediate filaments, *Ann. N.Y. Acad. Sci.* **455:**371–380.

Steven, A. C., Stall, R., Steinert, P. M., and Trus, B. L., 1985b, Computational straightening of images of curved macromolecular helices by cubic spline interpolation facilitates structural analysis by Fourier methods, *Proc. 43rd Annu. Meet. EMSA,* pp. 738–739.

Steven, A. C., Simpson, T. A., Trus, B. L., Furcinitti, P. S., Hainfeld, J. F., and Wall, J. S., 1986, Radial density profiles of macromolecular filaments determined from dark-field STEM micrographs: Improvements in technique and some applications, *Ann. N.Y. Acad. Sci.,* **483:**188–201.

Sun, T.-T., Tseng, S. C. G., Huang, A. J.-W., Cooper, D., Schermer, A., Lynch, M. H., Weiss, R., and

Eichner, R., 1985, Monoclonal antibody studies of mammalian and epithelial keratins, a review, *Ann. N.Y. Acad. Sci.* **455:**307–329.

Suzuki, E., Crewther, W. G., Fraser, R. D. B., MacRae, T. P., and McKern, N. M., 1973, X-ray diffraction and infrared studies of an α-helical fragment from α-keratin, *J. Mol. Biol.* **73:**275–278.

Thaler, M., Fukuyama, K., Epstein, W. L., and Fisher, K. A., 1980, Comparative studies of keratins isolated from psoriasis and atopic psoriasis, *J. Invest. Dermatol.* **75:**156–158.

Trachtenberg, S., Steinert, P. M., Trus, B. L., and Steven, A. C., 1986, Electron microscopy and image analysis of paracrystalline bundles obtained from psoriatic epidermal keratin, in: *Electron Microscopy and Alzheimer's Disease* (J. Metuzals, ed.), San Francisco Press, San Francisco, pp. 56–59.

Traub, P., and Vorgias, C. E., 1983, Involvement of the N-terminal polypeptides of vimentin in the formation of intermediate filaments, *J. Cell. Sci.* **63:**43–67.

Trinick, J., Cooper, J., Seymour, J., and Egelman, E. H., 1986, Cryo-electron microscopy and three-dimensional reconstruction of actin filaments, *J. Microsc.* **141:**349–360.

Tzaphlidou, M., Chapman, J. A., and Al-Samman, M. H., 1982, A study of positive staining for electron microscopy using collagen as a model system, *Micron* **13:**133–146.

Wais-Steider, C., Eagles, P. A. M., Gilbert, D. S., and Hopkins, J. M., 1983, Structural similarities and differences among neurofilaments, *J. Mol. Biol.* **165:**393–400.

Weeds, A., 1982, Actin-binding proteins—Regulators of cell architecture and motility, *Nature* **296:**811–815.

Woods, E. F., and Inglis, A. S., 1984, Organization of the coiled-coils in the wool microfibril, *Int. J. Biol. Macromol.* **6:**277–283.

Wray, J., 1979, Structure of the backbone in myosin filaments of muscle, *Nature* **277:**37–40.

Yu, L. C., Steven, A. C., Naylor, G. R. S., Gamble, R. C., and Podolsky, R. J., 1985, Distribution of mass in relaxed frog skeletal muscle and its redistribution upon activation, *Biophys. J.* **47:**311–321.

III

Differential Expression of Intermediate Filament Genes

10

Hard Keratin IF and Associated Proteins

Barry C. Powell and George E. Rogers

1. Introduction

The strong and durable nature of the hairs, nails, hooves, horns, and claws of mammals derives from the synthesis and subsequent cross-linked network of keratin proteins. These proteins are traditionally known as hard keratins in distinction to those characteristic of softer, more flexible, keratinized epithelia such as stratum corneum. Hard keratin tissues typically express complex patterns of keratin proteins, yet despite obvious phenotypic variation they all appear to express common sets of keratin genes.

The proteins of hard keratins were first identified from sheep wool and most of our knowledge about them and their genes stems from work on sheep material. They can be grouped into three classes, each comprising two or more protein families, totaling between 30 and 50 proteins in sheep wool (Table I). They were originally classified as the low-sulfur and the high-sulfur and high-glycine/tyrosine keratin proteins owing to their characteristic amino acid composition (Fraser *et al.*, 1972). As terminal differentiation proceeds, the low-sulfur keratins are assembled into microfibrils, forming a filamentous scaffold that becomes enmeshed in a matrix of high-sulfur and high-glycine/tyrosine keratins. The fibrous low-sulfur keratins are now known to belong to the intermediate filament (IF) superfamily of proteins (Hanukoglu and Fuchs, 1982; Weber and Geisler, 1982), the component proteins of the 10-nm-diameter filaments found in the cytoplasm of animal cells. We will refer to them here as hard keratin IF. The information available for other hairs and for nail, hoof, horn, and claw suggests that the expression of genes for their hard keratins represents variations on the theme exemplified by the sheep archetype (Gillespie, 1972; Gillespie and Marshall, 1977; Marshall and Gillespie, 1977; Baden and Kubilus, 1983; Marshall, 1983). This chapter will focus on the patterns of keratin gene

Barry C. Powell and George E. Rogers • Department of Biochemistry, University of Adelaide, Adelaide, South Australia 5000, Australia.

Table I. Proteins of the Hair and Hair Follicle (Hard Keratins)[a]

Protein name	Structural role	Characteristic amino acid composition	Protein family notation	Estimated protein chains in family	M_r
Low-sulfur	Keratin microfibrils, i.e., intermediate filaments (IF)	Cysteine 1–3 mole%	If Type II { SCMKA-5, SCMKA-7	4	56–58,000
		Glutamic acid glutamine ~15 mole% IF	Type I—SCMKA-8	4	38–43,000
High-sulfur	Keratin matrix, i.e., IF-associated protein (IFAP)	Cysteine 16–24 mole%	SCMKB-1/1A	n.d.	23–28,000
			SCMKB-2/1B	7	19,000
			SCMKB-IIIA	11	16,000
			SCMKB-IIIB	4	11,000
Ultrahigh-sulfur (UHS)	Keratin matrix (IFAP) and hair cuticle	Cysteine >30 mole%			>20,000
High-glycine/tyrosine (HGT)	Keratin matrix (IFAP)	Glycine 20–40 mole%	HGT Type I	10	6–9,000
		Tyrosine 12–21 mole%	HGT Type II	5	6–9,000
Trichohyalin	Medulla of hair	Arginine ~16 mole%		>1	190,000
	Inner root sheath of follicle	Glutamic acid/glutamine ~17 mole%			

[a]Data mainly adapted from Powell and Rogers (1986).

expression in sheep wool and will incorporate the data available for the other hard keratin tissues into a broader perspective.

2. The Proteins and Genes of Wool

The keratin proteins are extracted and purified by standard techniques that involve reduction and blocking of the -SH groups to disaggregate the disulfide cross-linked keratins and prevent their reaggregation. Whereas the wool keratin proteins can be extracted with nearly complete recovery, the extractability of human hair keratin is poor, with yields of 5–50% not uncommon (Gillespie and Marshall, 1981). This could be a result of the greater proportion of high-sulfur keratin protein and consequently more extensive cross-linking in human hair (Gillespie and Marshall, 1977; Gillespie, 1983). The three classes of derivatized soluble keratin proteins, often referred to collectively as *S*-carboxymethylkerateines or SCMK, are readily distinguished by two-dimensional gel electrophoresis (Fig. 1). A recognized side effect of blocking the -SH groups by carboxymethylation is an aberrant slower mobility upon SDS gel electrophoresis of proteins with a high SCM-cysteine content (Marshall, 1983). This has led to overestimates in molecular weight of 12–30% for the IF keratin proteins and 75–150% for the high-sulfur keratin proteins (Marshall, 1983).

2.1. Hard Keratin IF

The structure of IF protein has been the subject of recent reviews (Parry and Fraser, 1985; Skerrow, 1985; Steinert and Parry, 1985; Steinert *et al.*, 1985a) and is also covered extensively elsewhere in the present volume. Consequently, this section will concentrate on recent developments regarding hard keratin IF gene structure and organization and will only review IF protein structure insofar as to highlight the similarities and differences between the hard keratin IF and other IF proteins.

2.1.1. *Hard Keratin IF Protein Structure*

The IF keratin proteins of wool are traditionally known as the low-sulfur protein groups, or separately as the component 7 and component 8 protein families. The proteins of the component 8 and component 7 families fall into the molecular weight ranges and the acidic/basic electrophoretic distribution characteristic of the IF Type I and Type II keratin proteins, respectively (Hanukoglu and Fuchs, 1982; Moll *et al.*, 1982b; Weber and Geisler, 1982; Crewther *et al.*, 1983; Dowling *et al.*, 1983).

The heterogeneity of each protein type, their large size, and the presence of N-blocked termini have hampered the determination of the IF keratin protein sequences. Although partial amino acid sequences of Type I and Type II IF keratin proteins from wool were published over 15 years ago (for a review see Crewther, 1976), it was not until several years later (Crewther *et al.*, 1978; Gough *et al.*, 1978) that sufficient contiguous amino acid sequence was established to enable recognition of their helical segments. Currently, only partial sequences of one Type I and two Type II IF keratin proteins from wool have been published (Dowling *et al.*, 1979; Crewther *et al.*, 1980a; Sparrow and Inglis, 1980) but the total Type I protein sequence recently became available (Dowling *et*

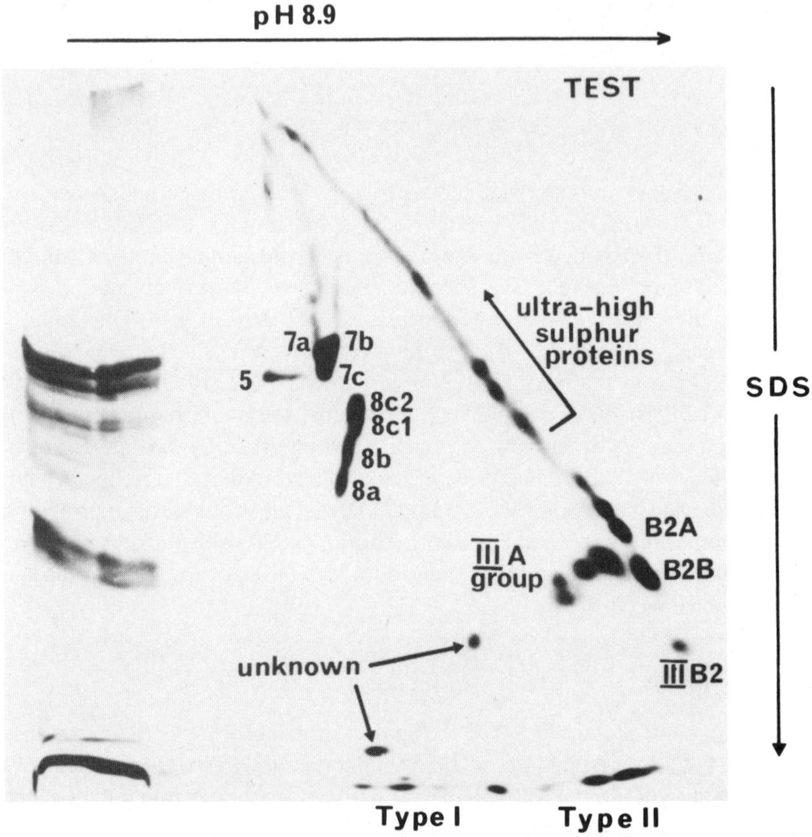

Figure 1. Two-dimensional polyacrylamide gel pattern of the wool keratin proteins. The proteins were *S*-carboxymethylated with [^{14}C]-iodoacetic acid and then separated in the presence of SDS in one direction and at pH 8.9 in the presence of 8 M urea in the other. In addition to the three main protein classes, a subclass of the high-sulfur group, the ultrahigh-sulfur proteins, are shown on the diagonal. Protein positions are revealed by autoradiography. From Powell and Rogers (1986) with permission.

al., 1986) and both Type II sequences (components 5 and 7c) are now nearly complete (L. Sparrow, personal communication). The gene encoding a Type II wool keratin IF recently has been sequenced and the amino acid sequence derived from it is identifiable with that of component 7c (Fig. 2). Unfortunately, no protein sequence data are available for any IF keratin proteins from nail, hoof, horn, or claw. While the hard keratin IF protein sequences of wool have taken several years to complete by protein sequencing techniques, the last 2 years has witnessed the publication of the complete sequences of several epithelial IF keratin proteins, deduced through DNA sequencing of the genes that encode them (Steinert *et al.*, 1984; Johnson *et al.*, 1985; Krieg *et al.*, 1985; Marchuk *et al.*, 1985; Reiger *et al.*, 1985; Tyner *et al.*, 1985). The epithelial IF keratin proteins contain sequences homologous to those of the α-helical regions of the wool proteins (Hanukoglu and Fuchs, 1982; Weber and Geisler, 1982), and when compared they indicate a common plan upon which the IF keratin proteins are built, regardless of tissue distribution. Re-

T C G S G F R G R A F S C V S A C G P R P G R C C I T A A P

Y R G I S C Y R G L T G G F G S R S V C G G F R A G S C G R

S F G Y R S G G V G G L S P P C I T T V S V N E S L L T P L

N L E I D P N A Q C V K Q E E K E Q I K C L N N R F A A F I

D K V R F L E Q Q N K L L E T K L Q F Y Q N R Q C C E S N L

E P L F S G Y I E T L R R E A E C A E A D S G R L S S E L N

S L Q E V L E G Y K K R Y E E E V A L R A T A E N E F V A L

K K D V D C A Y L R K S D L E A N V E A L I Q E T D F L R R

L Y E E E I R V L Q A H I S D T S V I V K M D N S R D L N M

D N I V A E I K A Q Y D D I A S R S R A E A E S W Y R S K C

E E I K A T V I R H G E T L R R T K E E I N E L N R V I Q R

L T A E V E N A K C Q N S K L E A A V T Q A E Q Q G E A A L

A D A K R K L A G L E E A L Q K A K Q D M A C L L K E Y Q E

V M N S K L G L D I E I A T Y R R L L E G Q E Q R L C E G V

G S V N V C V S S S R G G V V C G D L C V S G S R P V T G S

V C S A P C S G N L A V S T G L C A P C G P C N S V T S C G

L G A G G V G S C G I S S Y G V G S C A S V C R K C •

Figure 2. The amino acid sequence of a Type II wool keratin IF protein. It is related to the wool keratin component 7c and was derived from a gene sequence (B. Powell, unpublished). The positions of four α-helical domains generally known as 1A, 1B, 2A, and 2B (in order from the N-terminus) for IF proteins are denoted by boxes and are according to the scheme of Steinert *et al.* (1985b). The positions of seven introns found in the gene sequence are also shown (▼).

markably, this plan is typical of all other IF proteins (Geisler and Weber, 1982; Steinert *et al.*, 1985a) and this suggests the present-day IF protein superfamily has evolved from a primordial protein.

The IF protein superfamily is characterized by a common secondary structure consisting of a conserved central, rodlike region of α helix that is interrupted at conserved locations by three short nonhelical segments. In all IF proteins this α-helical region is about 310 amino acids in length and is flanked by nonhelical amino-terminal and carboxy-terminal domains. The end domains are of variable length and sequence (Steinert *et al.*, 1985a) and confer structural and functional individuality to each IF protein. A comparison of the amino acid sequences of the α-helical regions led to the suggestion that IF proteins can be classified into three main sequence types (Lazarides, 1980; Parry and Fraser, 1985). The IF keratin proteins of epithelial tissue and its appendages are classified into two types, the acidic Type I and the larger, neutral–basic Type II. The Type III IF include desmin, glial fibrillary acidic protein, and vimentin and are commonly found in muscle, astroglial, and mesenchymal tissue, respectively. The neurofilament proteins may comprise a fourth IF sequence type (Steinert *et al.*, 1985a) and it now appears likely that the lamin proteins of the nuclear envelope, newly recognized as IF proteins (McKeon *et al.*, 1986), could constitute a fifth IF sequence type.

Although the IF keratin protein end domains are more variable in length and sequence than the central α-helical region, subdomains have been noted within them (Steinert *et al.*, 1985a,b). End domain substructure was based upon comparisons of epidermal IF keratin protein sequences and may not extend to the hard keratin IF proteins. Preliminary data, derived through gene sequencing (K. Ward, unpublished), suggest that two of the sheep hard keratin Type I proteins have short, highly conserved C-terminal domains (Fig. 3A). However, it is not known if this conservation is a general and distinctive feature of hard keratin IF Type I proteins. With respect to the IF Type II keratin proteins of epidermis, Steinert *et al.* (1985b) have proposed that their C-terminal domain contains three distinct subdomains, designated H2, V2, and C. In this monograph, Steinert and Roop have renamed the C subdomain as the E subdomain. Unlike the proposed variable V2 and C subdomains, the H2 subdomain, comprising 20 amino acids, is thought to be conserved among most Type II proteins. However, the hard keratin Type II proteins appear to be exceptions to this tripartite subdomain arrangement. A comparison of the C-terminal domains of sheep wool protein component 7c and of a related wool protein, deduced through gene sequencing (B. Powell, unpublished), reveals a high degree of homology between these two proteins for 55 amino acids of their C-terminal domains (Fig. 3B). This suggests a bipartite arrangement for the hard keratin IF Type II C-terminal domain in which a long, conserved subdomain is flanked by a variable end domain.

2.1.2. *Hard Keratin IF Tissue Expression*

The amino acid compositions of the IF keratin proteins are very similar, irrespective of species or hard keratin tissue type (Table II). It is not certain whether the few compositional differences reflect actual component variation or result from trace amounts of high-sulfur or high-glycine/tyrosine keratin proteins, which often contaminate these preparations (Gillespie, 1972; Gillespie and Marshall, 1981). Initial one-dimensional electrophoretic analyses suggested that a similar set of IF keratin proteins may be expressed in several hard keratin tissues, albeit in varying proportions (Gillespie and Marshall, 1977). Subsequent two-dimensional electrophoretic analyses of human hair and nail IF keratin proteins from the same individual suggested that both tissues expressed several identical proteins and at least one that was unique to either tissue (Marshall, 1983). Yet, the large number of mammalian IF keratin proteins, coupled with the narrow range of molecular weights and isoelectric points displayed by the members of each IF keratin type preclude a confident assessment of common identity from such electrophoretic analyses. However,

A

```
PCNPCATTNACGKTITPCISSPCAPAAPCTPCVPRSRCGPCNSYVR-COOH        Component 8c1
•••••••••T•••P•G•••N•••————————•S•T••••••TF•H-COOH         Type I gene
```

B

```
CEGVGAVNVCVSSSRGGVVCGDLCVSGSRPVTGSVCSAPCSGNLAVSTGLCAPCG
•••••S••••••••••••••••••••••••••••••••••••••••••••••••••

QLNTTCGGGSCSLGRC-COOH                                       Component 7c
PC•SVTSC•LGAG•VGSCGISSYGVGSCRHVCRKC-COOH                    Type II gene
```

Figure 3. (A) Comparison of amino acid sequences between hard keratin IF Type I C-terminal domains. The amino acid sequences of the C-terminal domains of wool protein 8cl (Dowling *et al.*, 1986) and of a related wool protein, derived through gene sequencing (K. Ward, personal communication), are shown. Identical amino acids are indicated by dots and a deletion, inserted to maximize homology, is represented by a continuous line. (B) Comparison of amino acid sequences between hard keratin IF Type II C-terminal domains. The amino acid sequences of the C-terminal domains of wool protein 7c (L. Sparrow, personal communication) and of a related wool protein, derived through gene sequencing (see Fig. 2), are shown.

Table II. Amino Acid Composition of IF Protein from Various Hard Keratins[a]

	Hair							Nail	Horn		Hoof	
	Monkey[b]	Cat[b]	Raccoon[b]	Horse[b]	Mouse[b]	Human[c]	Sheep[d]	Human[c]	Echidna[c]	Sheep[d]	Sheep[d]	Cow[e]
Lysine	3.6	3.8	3.8	3.8	3.7	3.5	4.1	3.8	4.4	4.1	5.0	4.3
Histidine	0.8	1.2	1.0	0.8	0.9	0.7	0.6	0.8	0.8	0.8	0.8	0.8
Arginine	7.2	6.7	7.3	8.1	7.4	7.1	7.8	6.7	8.7	7.8	7.2	7.1
SCMC[f]	8.1	9.2	8.3	7.4	8.2	7.6	6.0	6.4	4.1	4.7	3.8	4.3
Aspartic acid	8.2	8.3	8.8	8.9	8.4	9.3	9.6	8.5	8.6	8.9	10.1	8.9
Threonine	4.7	5.1	5.0	5.0	4.7	5.4	4.8	5.0	4.7	4.9	4.5	4.6
Serine	8.3	7.7	7.2	8.9	7.3	8.9	8.1	9.6	7.9	8.4	7.9	8.3
Glutamic acid	14.4	14.5	14.4	15.6	14.9	16.5	16.9	14.9	15.2	15.8	17.7	15.8
Proline	3.8	3.9	4.3	3.8	4.2	3.8	3.3	4.3	3.4	3.6	2.4	3.2
Glycine	8.4	8.3	7.0	6.0	8.3	5.1	5.2	8.1	7.0	6.9	6.3	9.4
Alanine	6.5	6.5	7.5	7.2	6.9	6.9	7.7	6.4	8.1	7.4	7.3	7.2
Valine	5.9	5.3	6.3	6.3	5.2	6.1	6.3	5.5	6.0	6.2	5.9	5.8
Methionine	0.6	1.0	1.1	0.4	1.2	0.4	0.6	0.7	0.6	0.7	0.7	0.5
Isoleucine	3.5	3.2	3.4	3.6	3.2	3.6	3.8	3.5	4.4	3.8	3.9	3.9
Leucine	9.5	8.8	9.2	9.8	9.2	10.2	10.2	9.7	10.7	10.1	11.1	10.4
Tyrosine	4.0	3.6	3.0	2.5	4.0	2.5	2.7	3.5	2.7	3.3	3.3	2.9
Phenylalanine	2.7	2.8	2.4	2.2	2.5	2.6	2.0	2.5	2.6	2.5	2.1	2.6

[a]*S*-carboxymethyl derivatives and expressed as mole%.
[b]Gillespie (1983).
[c]Marshall (1985).
[d]Marshall and Gillespie (1977).
[e]Baden and Kubilus (1983).
[f]*S*-carboxymethylcysteine.

another comparative study, also between human hair and nail, using hair monoclonal antibodies that did not cross-react with epithelial keratin IF proteins (Lynch *et al.*, 1985), qualitatively corroborates this dual expression of hard keratin IF proteins in two phenotypically different tissues. Intriguingly, this latter study also suggests that there is a small but significant expression of IF keratin proteins typical of soft epithelia in nail but not in hair, a finding not previously reported.

During the terminal differentiation of the wool fiber, four Type I and four Type II IF keratin proteins appear to be synthesized (Crewther *et al.*, 1980b), typifying the pattern of IF Type I and Type II coexpression observed in other epithelial tissues. The wool IF keratin proteins are richer in half-cystine relative to their epidermal counterparts and comprise a subset in each family. Whereas the epidermal IF keratin proteins contain few half-cystine residues, the wool proteins contain 25–30 half-cystine residues per molecule, which are predominantly located in the end domains (Dowling *et al.*, 1986; L. Sparrow, personal communication). The cystine-rich domains are ideally suited to the formation of the durable hard keratin phenotype through disulfide bond cross-linking. Thus, the cystine-rich IF protein subsets that are characteristic of wool may be prevalent in all hard keratin tissues. Although there are no protein sequence data available for any IF keratin proteins from nail, hoof, horn, or claw, or indeed for other hairs to substantiate this hypothesis, the comparative analyses of the amino acid compositions (Table II) and electrophoretic properties of their IF keratin proteins suggest that the majority of hard keratin IF expressed in mammals are common between the various tissues. The possible conservative expression of a common cystine-rich subset of IF keratin proteins in all hard keratin tissues resembles the wide distribution that seems to be displayed by other epithelial IF keratin proteins (Moll *et al.*, 1982b; Eichner *et al.*, 1984).

2.1.3. Hard Keratin IF Gene Structure and Organization

All the known hard keratin IF nucleotide sequences are derived from sheep cDNA and genomic DNA libraries and there is no equivalent information available for other species. One representative cDNA clone for each hard keratin IF type has been isolated (Ward *et al.*, 1982). The Type I cDNA clone is equivalent to the wool 8c1 protein component and encodes 285 amino acids, about two-thirds of the protein. The Type II cDNA clone is equivalent to the wool 7c protein component and encodes 187 amino acids, about one-third of the protein. Although gene sequences have not been published for any hard keratin IF genes, two are sufficiently near completion (K. Ward, personal communication; B. Powell, unpublished) to indicate that the conserved exon/intron organization characteristic of all known keratin IF genes (Steinert *et al.*, 1985a) not unexpectedly extends to the hard keratin IF genes too. The protein encoded by the sheep IF Type I gene is related to wool component 8c1 and the known sequence, which covers about 75% of the gene, displays the same exon/intron arrangement as a mouse epithelial IF Type I gene (Krieg *et al.*, 1985; K. Ward, unpublished). The protein encoded by the sheep IF Type II gene (see Fig. 2) is related to wool component 7c and the gene has a similar exon/intron arrangement to a human epithelial IF Type II gene (Tyner *et al.*, 1985; B. Powell, unpublished).

The IF keratin protein families comprise 10–20 proteins in each of the Type I and Type II groups which are encoded by a similar number of genes (Fuchs *et al.*, 1981; Lazarides, 1982; Moll *et al.*, 1982b). Close linkage of genes has been observed within, but not between, the IF Type I and Type II gene families in the sheep genome (Powell *et al.*, 1986) and also within the IF Type I family in the human genome (RayChaudhury *et al.*, 1986). Within the sheep IF Type II family, the four coordinately expressed hard keratin IF Type II genes appear to be dispersed among other Type II genes (Powell *et al.*, 1986). An interspersed arrangement of differentially expressed Type II genes suggests that there may be specific factors that promote selective gene activation and/or inhibition. The same factors could coordinate the transcription of the hard keratin IF Type I genes to produce the coexpression that is seen *in vivo*.

It is surprising that the exon that encodes one of the most conserved amino acid regions of IF proteins, the C-terminal end of the α-helical core, has been found in two extragenic locations in sheep DNA (Fig. 4; Powell *et al.*, 1986). In one instance, the orphan exon is located between two normal IF Type II genes, approximately 8 kb 3′ to one of the genes and 12 kb 3′ to the other. In the other case, no IF genes are within at least 8 kb 5′ to the exon or 4.5 kb 3′ to it. The orphan exons could have been artificially produced by DNA rearrangement during library construction or amplification, but there is good evidence to indicate that at least one of them, the one in the cosmid clone (Fig. 4), is not a cloning artifact (Powell *et al.*, 1986). A comparison of the amino acid sequences encoded by these orphan exons with known IF protein sequences suggests that both were derived from IF Type II keratin genes (Powell *et al.*, 1986). The orphan exon found in the λ clone is nearly identical to the equivalent sequence of the cDNA clone encoding the IF Type II wool protein, component 7c, differing in only three amino acids and at 15 nucleotide positions (B. Powell, unpublished). Despite their extragenic locations the two orphan exons, in all respects, appear to represent functional exons. Whether they are recruited into IF genes during transcription remains unknown, yet the location of one exon between two complete IF Type II genes (Fig. 4) suggests that it is not part of the normal transcription unit.

The orphan exons probably arose through unusual recombination events, either during gene duplication or replication and the one found in the cosmid clone was probably

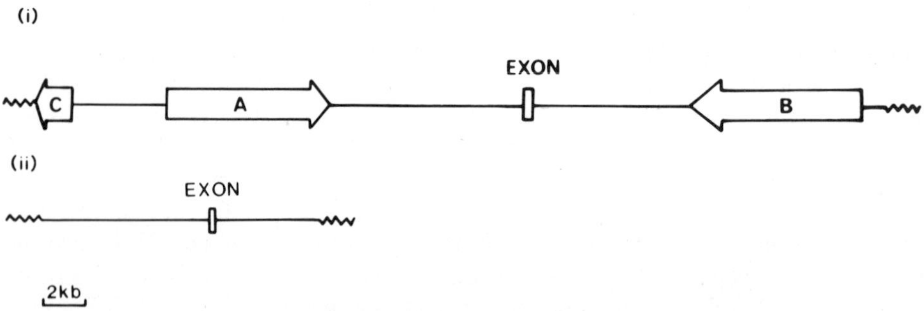

Figure 4. Aberrant genomic arrangement of an IF Type II exon. The location of orphan exons (▯), displaced from IF Type II genes, is shown in two different sheep genomic clones: a cosmid clone (i) and a λ Charon 4A clone (ii). The vector sequences (〰) and the location and direction of transcription (⇧) of IF Type II genes in the cosmid clone are also shown.

Note: As a result of more recent studies, the orientation of gene B has been found to be opposite of that originally published (Powell *et al.*, 1986). Furthermore, the distance separating gene B from gene A is now estimated to be 20 kb, the disposition of the orphan exon between the genes is also different (see text), and the presence of the 5′-end of a third IF Type II gene has been discovered to the left of gene A.

derived from one of the two adjacent IF Type II genes. However, it shows the same level of nucleotide sequence homology to the equivalent exons (68%) and their flanking regions (50%) in those genes (Fig. 5), and so the identity of the parent gene remains unknown. If the orphan exons were generated during the gene duplications that led to the current repertoire of mammalian IF keratin genes, then these arrangements should be detectable in related genomes.

2.2. *Hard Keratin IFAP*

Ever since the recognition that hard keratin microfibrils, known for 30 years, are in fact IF (Hanukoglu and Fuchs, 1982; Weber and Geisler, 1982), it has been realized that the matrix proteins associated with them belong to the general class of IF-associated proteins (IFAP) that stabilize IF aggregates. While the epidermal protein filaggrin is a prime example of this class in epidermal keratinocytes, the hard keratin IFAP were known many years before filaggrin was identified in epidermis and represent the best characterized group of IFAP to date.

It has been mentioned earlier in this section that the matrix of hair keratin is composed of two distinct groups or families of proteins, one glycine/tyrosine rich (high-glycine/tyrosine proteins) and the other rich in half-cystine residues (high-sulfur proteins). In addition to discussing them in some detail below, another protein of the hair system is also included. This is trichohyalin, present in the inner root sheath (IRS) of the hair follicle and in hairs that are medullated. The IRS is prominent in the follicle as a cellular lamina that surrounds the growing hair (see Fig. 10B) and represents about half of the cellular mass differentiating in the growing follicle. It is degraded and lost when the keratinized hair emerges from the follicle at the skin surface. Trichohyalin is a putative IFAP and on present evidence (Rothnagel and Rogers, 1986) appears to disperse between the IF that develop and become densely packed in the IRS cells during their differentiation.

2.2.1. *High-Sulfur Keratin IFAP Structure and Tissue Expression*

The high-sulfur keratins comprise five protein families that are typified by a half-cystine content of at least 16 mole%. Each family is composed of multiple protein components, and the accumulation of protein sequence data (Crewther, 1976; Swart *et al.*, 1976) and more recently gene sequence data (Powell *et al.*, 1983; Frenkel, 1985; B. Powell, unpublished) leave no doubt that the abundance of keratin proteins is a reflection of a similar number of genes.

Of the five families in the high-sulfur keratin class, the B2, BIIIA, and BIIIB protein families are well characterized, whereas relatively little is known about the B1 and the ultrahigh-sulfur (UHS) proteins (Table I). They range in molecular weight from 10k to 30k, but the molecular weights of the B1 and UHS proteins, for which no protein sequence data are available, are necessarily inexact due to the aberrant slower mobility in SDS-PAGE of the high-sulfur proteins after they are *S*-carboxymethylated during the isolation procedure (Marshall, 1983). The precise number of wool high-sulfur keratin proteins is not known but could be in excess of 40 components (Darskus, 1972; Crewther, 1976; Gillespie and Marshall, 1981) of which about half have been directly sequenced (Crewther, 1976). This diversity is often a consequence of only a few amino acid dif-

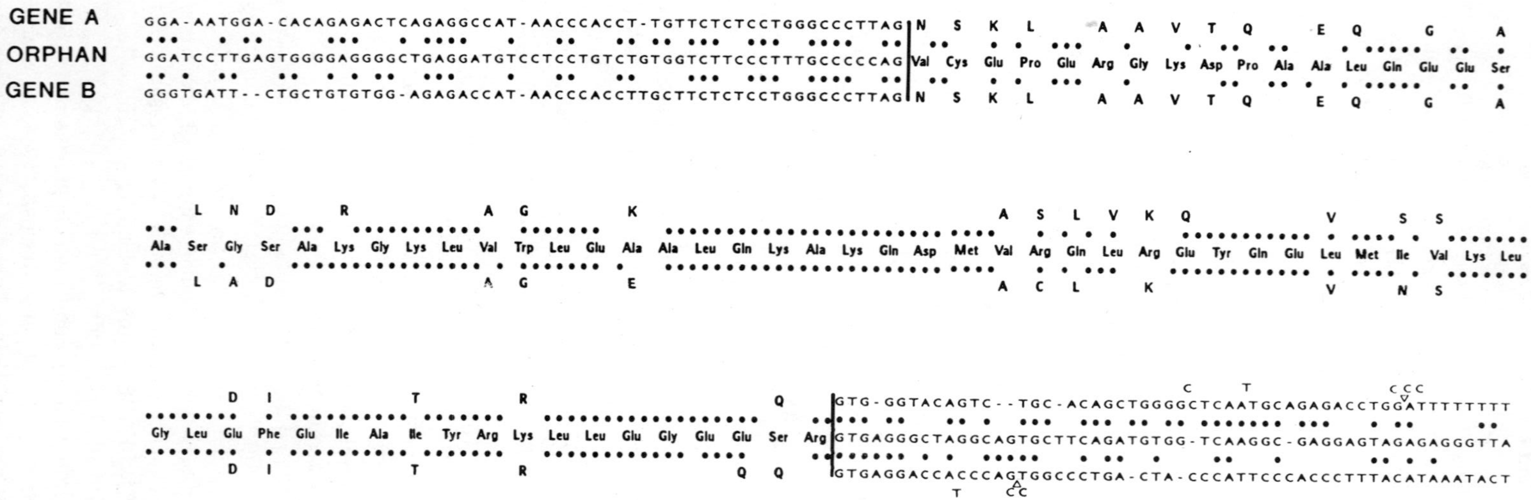

GENE A
ORPHAN
GENE B
GGA-AATGGA-CACAGAGACTCAGAGGCCAT-AACCCACCT-TGTTCTCTCCTGGGCCCTTAG
N S K L A A V T Q E Q G A
GGATCCTTGAGTGGGGAGGGGCTGAGGATGTCCTCCTGTCTGTGGTCTTCCCTTTGCCCCCAG
Val Cys Glu Pro Glu Arg Gly Lys Asp Pro Ala Ala Leu Gln Glu Glu Ser
GGGTGATT--CTGCTGTGTGG-AGAGACCAT-AACCCACCTTGCTTCTCTCCTGGGCCCTTAG
N S K L A A V T Q E Q G A
L N D R A G K A S L V K Q V S S
Ala Ser Gly Ser Ala Lys Gly Lys Leu Val Trp Leu Glu Ala Ala Leu Gln Lys Ala Lys Gln Asp Met Val Arg Gln Leu Arg Glu Tyr Gln Glu Leu Met Ile Val Lys Leu
L A D A G E A C L K V N S
D I T R Q
Gly Leu Glu Phe Glu Ile Ala Ile Tyr Arg Lys Leu Leu Glu Gly Glu Glu Ser Arg
D I T R Q Q
C T CCC
GTG-GGTACAGTC--TGC-ACAGCTGGGGCTCAATGCAGAGACCTGGATTTTTTTT
GTGAGGGCTAGGCAGTGCTTCAGATGTGG-TCAAGGC-GAGGAGTAGAGAGGGTTA
GTGAGGACCACCCAGTGGCCCTGA-CTA-CCCATTCCCACCCTTTACATAAATACT
T CC

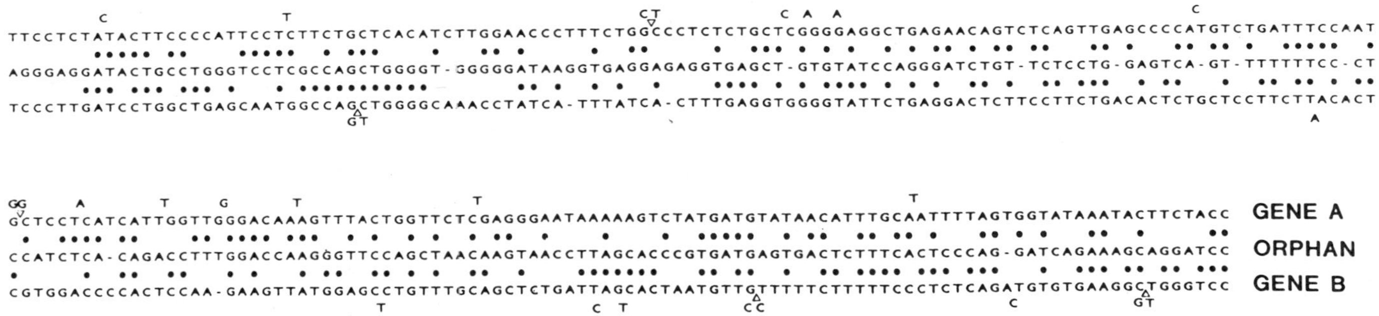

Figure 5. Sequence comparison of an orphan exon with normal, homologous exons from two adjacent IF Type II genes. The nucleotide sequence of an orphan exon and 322 bp of flanking DNA is compared to the equivalent exon and intron sequences in the adjacent IF Type II genes (A and B) from the cosmid clone depicted in Fig. 4. The intron/exon junctions are indicated by vertical lines. The amino acid sequence of the orphan exon is presented in three-letter code and where different amino acids are encoded by the exons from the adjacent genes they are indicated in one-letter code. All dots indicate nucleotide homology of the orphan exon and its flanking sequence with corresponding sequences from either gene A or B. Although the nucleotide sequences of the exons are not given, the dots in the exon region still indicate nucleotide homology between the three exons where the first letter of the three-letter amino acid code is equivalent to the first nucleotide position of the triplet codon. To achieve maximum homology, some nucleotides from the gene A and gene B sequences are shown as insertions relative to the orphan sequence and are presented in uppercase for gene A or lowercase for gene B. Where two or more consecutive nucleotides are treated as an insertion, their point of insertion is indicated by a triangle.

ferences between family members, which precludes a clear identification of all the protein components. A similar complexity has been reported in the high-sulfur keratin proteins of other mammals (Gillespie *et al.*, 1968; Gillespie and Marshall, 1977; Marshall *et al.*, 1977; Marshall, 1980, 1983) and these analyses further suggest that some of the high-sulfur proteins expressed in other species are similar to those of sheep wool. With such complexities, the determination of protein sequences is a prerequisite for unequivocal assessment of identity between species. The only sequence data that are available for such comparisons are from the BIIIA and BIIIB keratin protein families. A similar number of components have been reported in both sheep and goat wool and the homologous proteins vary in only a few amino acids (Swart *et al.*, 1976). However, as sheep and goats only diverged between 6 and 8 million yers ago (Novacek, 1982), this is not unexpected and, unfortunately, there are no sequence data available from more evolutionary distant species.

Comparative electrophoretic analyses of the high-sulfur keratin proteins, between either human hair and nail or between sheep wool, hoof, and horn, suggest that the same complement of proteins is expressed in all hard keratin tissues within a species but differences exist in the relative proportions of some components (Marshall and Gillespie, 1977; Marshall, 1983). A higher expression of high-sulfur keratin protein was observed in both hair and wool compared to the other hard keratin tissues and this was more marked for human hair, but the proteins responsible for the differences have not been identified.

a. B1 Keratin IFAP Family. Less is known about this putative high-sulfur keratin family than any other and since the two early reports (Gillespie, 1962; Haylett *et al.*, 1971) no new data have been published. The B1 protein family has been variously estimated to have a size range of 23–28k (Gillespie, 1962; Haylett *et al.*, 1971). Haylett *et al.* (1971) reported that it has an amino acid composition similar to that of the B2 high-sulfur keratin family and although Gillespie's (1962) values differ in some significant respects, they agree on a half-cystine content of between 16 and 24 mole%, a range typical for a high-sulfur keratin protein. Further, Haylett *et al.* (1971) noted that the first seven amino acids of the B1 proteins were identical to the B2 proteins but no data have ever been published to substantiate that claim as far as we can ascertain. Clearly, unambiguous protein sequence data are required for the putative B1 family to be established as an independent high-sulfur keratin protein family.

b. B2 Keratin IFAP Family. In sheep wool three major and three minor protein components of this family have been identified through protein sequencing (Lindley and Elleman, 1972; Lindley, 1977) and a seventh component was identified through gene sequencing (Powell *et al.*, 1983). They contain about 22 mole% half-cystine and range in size from 151 to 181 amino acids. The major distinguishing feature between them is the repetition frequency of a consecutively repeated ten-amino-acid unit, which is repeated from three to six times in the known B2 proteins (Crewther, 1976; Powell *et al.*, 1983). Apart from this variability in repeat unit frequency, the B2 protein sequences are highly conserved.

Genes encoding two of the known major sheep B2 keratin proteins (B2A and B2C) and a gene encoding a previously unrecognized member of the family (B2D) have been completely sequenced and two of them were found to be tightly linked in the sheep genome (Powell *et al.*, 1983). All genes share a high level of nucleotide sequence homology in the coding region (93%) but their 3′ noncoding regions diverge in both length and sequence, suggesting a distant evolutionary time for their ancestral duplication. However, there is some homology still extant between the 3′ noncoding regions of the tandem B2A/B2D gene pair to suggest they originated from a more recent duplication than the separate B2A and B2C genes (Powell *et al.*, 1983).

c. BIIIA Keratin IFAP Family. The BIIIA protein family of wool is quite complex, comprising 11 proteins, including 8 major and 3 minor components (Swart *et al.*, 1976). All the proteins have been sequenced and are highly homologous. Each contain 130–132 amino acids, including 24 mole% half-cystine. Another notable feature is their high arginine content (about 10 mole%), which gives them an overall basic charge. One BIIIA gene has been completely sequenced and is linked to the B2C high-sulfur keratin gene (B. Powell and L. Crocker, unpublished).

d. BIIIB Keratin IFAP Family. There are four proteins known in this family in sheep wool, one of which is only a minor component (Swart and Haylett, 1971). The major components (BIIIB2, 3, and 4) have been sequenced and contain 97–98 amino acids, of which about 16 mole% are half-cystine (Swart *et al.*, 1976). The BIIIB3 and 4 proteins are more homologous to each other than to the BIIIB2 protein.

Sheep cDNA clones that encode the three major components have been isolated (Ward *et al.*, 1982) and the gene encoding the BIIIB4 component has been completely sequenced (Frenkel, 1985). The combined data indicate that the BIIIB3 and BIIIB4 genes are more closely related to each other than to BIIIB2, suggesting that a primordial BIIIB duplication gave rise to ancestral BIIIB2 and BIIIB3/4 genes and subsequent duplication and divergence produced the current multigene family (Frenkel, 1985).

e. The UHS Keratin IFAP Family. The UHS keratin proteins are expressed at a low level in sheep wool and they are now known to be a more significant component in mouse and human hair (Gillespie, 1983). They were first identified and described after cystine enrichment in experimental sheep 20 years ago (Gillespie and Reis, 1966) but there are still no protein sequence data available for them.

On a two-dimensional pH 8.9/SDS-polyacrylamide gel, the UHS keratin proteins are located on a diagonal, often as discrete spots superimposed on a smear of protein, suggesting a spectrum of molecular weights (Fig. 1). They contain at least 30 mole% half-cystine and although their molecular weights appear to be greater than the other high-sulfur keratin proteins, that feature could be attributable to their higher content of *S*-carboxymethyl groups, promoting anomalous electrophoretic mobility in the presence of SDS (Marshall, 1983). Despite the apparent spectrum of molecular weights, preliminary isolation and sequence analysis of UHS keratin peptides suggests that they may not be such a complex family after all (P. MacKinnon, personal communication).

A wool follicle cDNA clone (K4) that encodes a polypeptide with a half-cystine content of 37 mole% has been isolated (K. Ward *et al.*, unpublished; Fig. 6a) and it may represent a partial UHS clone. Its open reading frame of 75 amino acids is different in sequence from the proteins of the three well-characterized and highly conserved high-sulfur protein families (B2, BIIIA, and BIIIB). Since there are four consecutive decapeptide sequences within the open reading frame of the cDNA clone that exhibit high homology with those of the B2 family (Fig. 6b), it is concluded that the clone K4 and the B2

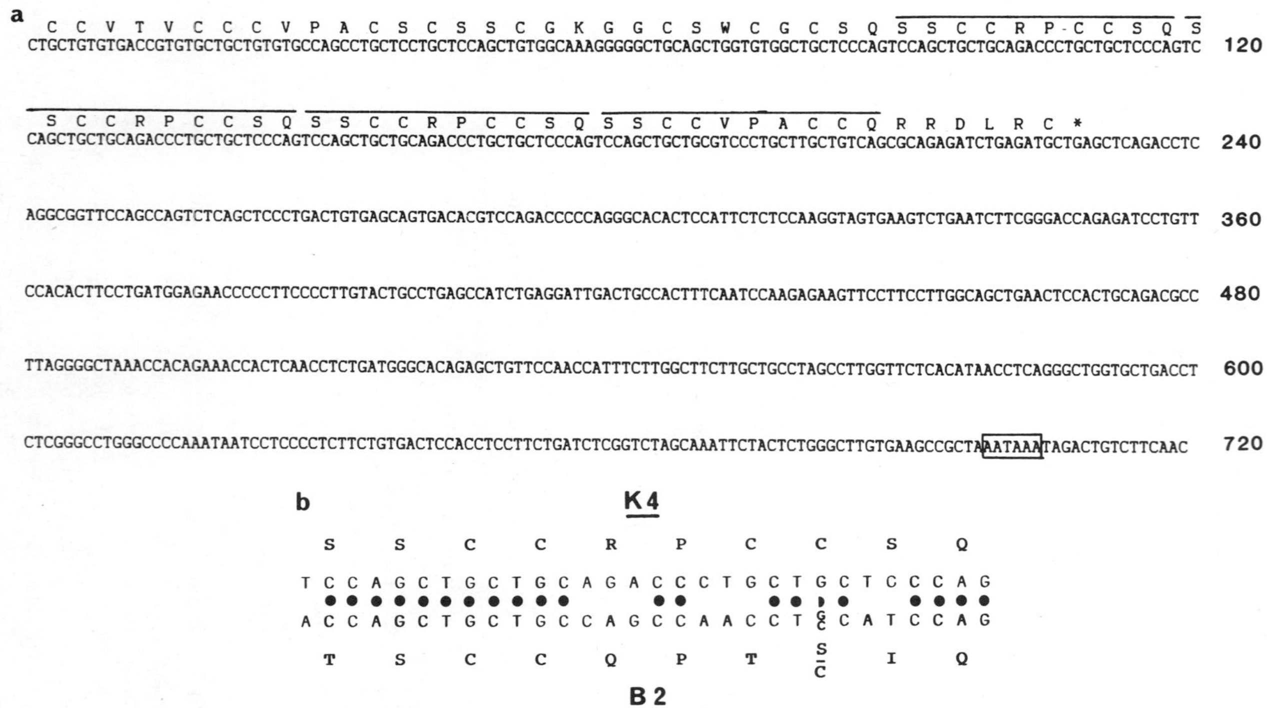

Figure 6. (a) The partial nucleotide sequence of the putative ultrahigh-sulfur keratin cDNA clone, K4. The conceptual amino acid translation of the cysteine-rich (37 mole%) open reading frame is shown in the one-letter code above the nucleotide sequence. Bars indicate the decapeptide repeats and the putative mRNA polyadenylation signal is boxed (courtesy of K. Ward). (b) The consensus amino acid and nucleotide sequences of the decapeptide repeats of the K4 and B2 gene family (Powell *et al.*, 1983). The individual repeats from which each consensus sequence is derived show a minimum of 75% homology to each other at the nucleotide level. Both from Powell and Rogers (1986) with permission.

family are derived from a common ancestral gene. However, for this cDNA clone to be a high-sulfur keratin protein with a typical half-cystine content of 16–24 mole%, the remaining sequence at the N-terminus of the molecule would have to be nearly devoid of half-cystine residues. Such an arrangement would not be typical of the high-sulfur keratin proteins (Crewther, 1976) and hence its tentative identification as a UHS keratin protein.

2.2.2. *High-Glycine/Tyrosine IFAP Structure and Tissue Expression*

At one time (DeDeurwaerder *et al.*, 1964) it was proposed that the high-glycine/tyrosine (HGT) proteins were located in the cuticle of the hair but it is now generally accepted that this is not so. Calculations from x-ray diffraction data indicate that the volume of matrix space available in cortical keratin allows for the location of HGT proteins, together with the high-sulfur proteins, between the keratin IF (Fraser *et al.*, 1973).

The amount of these proteins present in hairs is highly variable between species, varying from as low as 1% for Lincoln wool to as high as 30% for echidna quill. Two families of HGT proteins can be distinguished both in wool (Fig. 1) and in mouse hair; one has a lower half-cystine and a higher phenylalanine content than the other and they are referred to respectively as Type I and Type II (Table I; Gillespie and Frenkel, 1976), to be distinguished of course from the same notation adopted for classifying keratin IF. These proteins have tyrosine contents of 10–20 mole%, which undoubtedly cause their insolubility below pH 10 and affect their behavior in cellulose-based separation procedures. Furthermore, their relatively low molecular weight of 6–9k causes them to travel near the front of SDS-polyacrylamide gels. Notwithstanding these difficulties, both Type I and II proteins have been fractionated, although until recently only Type I sequences were available. Type I proteins when fractionated by QAE-anion exchange cellulose and analyzed by two-dimensional PAGE appear to contain more than 30 components (Gillespie, 1983) and of these only 2 (F and C2) have been sequenced (Dopheide, 1973; Marshall *et al.*, 1980). The reason for the apparent complexity is not understood but the recent characterization of HGT genes now suggests that the number of Type I proteins is much smaller than originally thought, indicating that protein isolation artifacts and/or conformation pecularities might be responsible for anomalous behavior. The Type II HGT proteins are less characterized than the Type I because their higher glycine content (40 mole%) and 40 mole% content of only three other amino acids do not easily allow the separation of closely homologous chains differing by only two or three amino acid residues. Recently, a single Type II chain has been purified by HPLC procedures and almost completely sequenced (A. S. Inglis *et al.*, personal communication). It is characterized by runs of Gly-X sequences that are longer than in Type I proteins, where X can be serine, threonine, tyrosine, or phenylalanine.

The availability of amino acid sequences for Type I proteins, F and C2, enabled the design of synthetic DNA probes corresponding to the amino acid sequences as shown in Fig. 7. They were used to isolate specific cDNA clones from a cDNA library prepared from wool follicle poly-A$^+$ mRNA and both cDNAs were then applied to isolate the genes for F and C2 from a sheep genomic library (E. Kuczek and G. Rogers, unpublished). From extensive protein chemical studies (Gillespie, 1983), it was known that both proteins fractionate together as members of the Type I family and yet a striking aspect of the F and C2 genes is the lack of homology between them at the nucleotide level in both the coding and noncoding regions. Hybridization analysis of genomic DNA with specific

C_2 NH_2-THR ARG PHE PHE CYS CYS GLY SER TYR PHE PRO GLY TYR PRO SER TYR GLY THR ASN PHE HIS ARG THR PHE ARG ALA THR PRO LEU ASN
F NH_2-SER TYR CYS PHE SER SER THR VAL PHE PRO GLY CYS TYR TRP GLY SER TYR GLY TYR PRO LEU GLY TYR SER VAL GLY CYS GLY TYR GLY SER

C_2 CYS VAL VAL PRO LEU GLY SER PRO LEU GLY TYR GLY CYS ASP GLY TYR SER SER LEU GLY TYR GLY PHE GLY GLY SER SER PHE SER ASN
F THR TYR SER PRO VAL GLY TYR GLY PHE GLY TYR GLY TYR ASP GLY GLY SER ALA PHE GLY CYS ARG ARG PHE TRP PRO PHE ALA LEU TYR.COOH

C_2 LEU GLY CYS GLY TYR GLY GLY SER PHE TYR ARG PRO TRP GLY SER GLY SER GLY PHE GLY TYR SER THR TYR.COOH

Figure 7. Comparison of the amino acid sequences for the high-glycine/tyrosine proteins, C2 and F (Dopheide, 1973; Marshall *et al.*, 1980). There are two blocks of homology separated by a region of random sequence. The amino acid sequences upon which the oligonucleotide probes were based are underlined. Adapted from Powell and Rogers (1986) with permission.

probes has demonstrated that the gene for F is unique whereas C2 appears to be represented as a small family of related genes (E. Kuczek, unpublished).

The recent amino acid sequence analysis of an HGT Type II chain mentioned above has enabled the synthesis of a 14-mer DNA fragment in accord with the last five residues at the C-terminus. This oligonucleotide probe will be used to find an equivalent cDNA for selecting genomic clones containing Type II genes, as for the Type I genes. Sequencing of Type II genes will enable confirmation of the Type II protein sequence and comparison with Type I at both protein and nucleotide levels.

2.3. Hard Keratin IFAP Gene Structure and Organization

Current information on hard keratin IFAP genes is only available for sheep keratins. With the possible exception of the high-sulfur B1 IFAP and UHS IFAP families (see Section 2.2.1e), clones representative of most of the sheep keratin IFAP families have been identified and sequenced (Ward *et al.*, 1982; Powell *et al.*, 1983; Frenkel, 1985; Kuczek and Rogers, 1985; E. Kuczek, personal communication). The three genes from each of the high-sulfur B2 and BIIIB IFAP families show a high degree of intrafamily homology in the sequences of the gene coding region. In contrast, little homology is seen in the 3′ noncoding sequences.

Two salient features of the sheep hard keratin IFAP genes are the lack of introns and the conservation of an 18-bp sequence in the 5′ untranslated region. The lack of introns in eukaryotic genes is rare (Breathnach and Chambon, 1981; Blake, 1985) and their absence from seven hard keratin IFAP genes, derived from four gene families, seems remarkable at first sight. However, if all the present-day high-sulfur and HGT keratin genes are descended from a single, intronless primordial gene for each keratin class (see Section 2.4), then the number of keratin IFAP genes without introns is not so exceptional.

The 5′ untranslated regions of the hard keratin IFAP genes are short, typically 50–60 nucleotides in length, with a conserved 18-bp sequence adjacent to the initiation codon (Powell *et al.*, 1983; Powell and Rogers, 1986). This sequence is conserved in five of the six known high-sulfur keratin genes and also in the two sequenced HGT Type I keratin genes (Fig. 8). It is not found in sheep globin or metallothionein genes (Kretschmer *et al.*, 1981; J. Mercer, personal communication) and therefore may be specific for the sheep keratin genes. Unfortunately, a more complete survey cannot be included as the 5′ untranslated sequences of the hard keratin IF genes that are coexpressed with the high-sulfur and HGT IFAP genes are unfinished. In the *Drosophilia* heat shock response, sequences in the mRNA 5′ untranslated region seem to be involved in the high level and preferential translation of those mRNAs (McGarry and Lindquist, 1985) and the conserved sequence in the hard keratin IFAP genes may perhaps function in an analogous manner during follicle differentiation. Future experiments will study this sequence as a possible regulatory element in keratinocyte differentiation.

There is no regular linkage of the sheep hard keratin IFAP genes: some are tightly linked, others are not. For example, in one genomic clone two tandem high-sulfur keratin B2 genes are only 1.8 kb apart (Powell *et al.*, 1983). In another genomic clone, a third gene from that family was identified linked in a transcriptionally divergent orientation to a high-sulfur keratin BIIIA gene (Powell *et al.*, 1983; B. Powell, unpublished) but it is not known whether the two clones are linked. The other three hard keratin IFAP genes that have been sequenced, a high-sulfur keratin BIIIB and two HGT keratin Type I genes, were isolated in individual genomic clones and there is no evidence to indicate linkage.

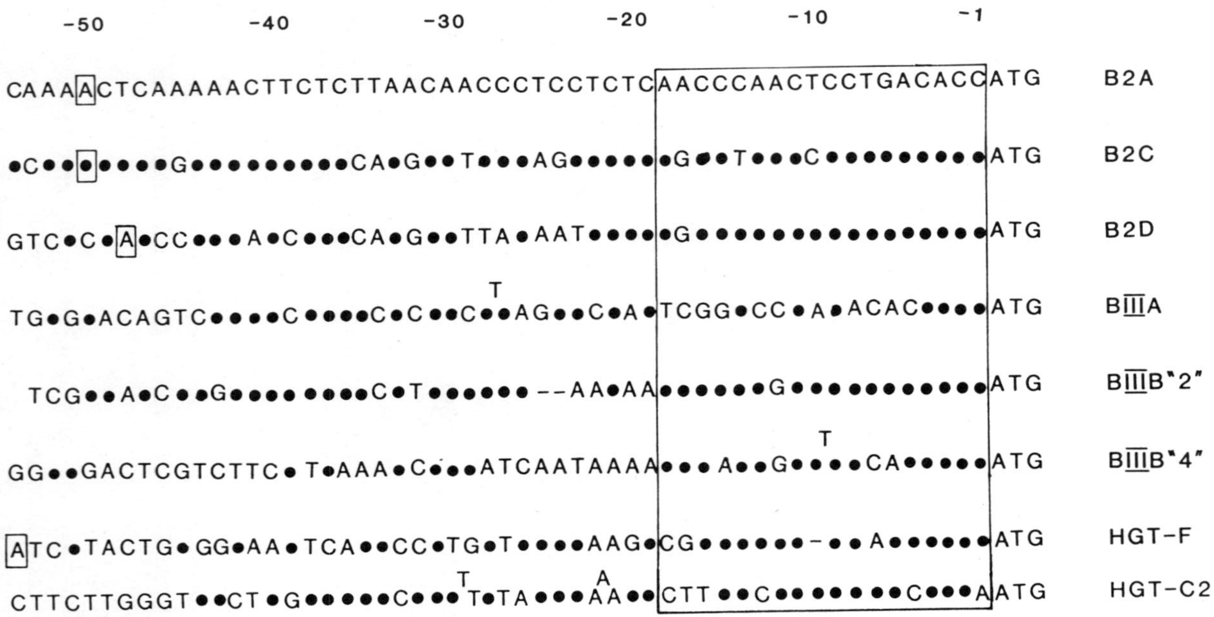

Figure 8. Comparison of the 5′ noncoding regions of six sheep high-sulfur and two high-glycine/tyrosine (HGT) keratin genes. The high-sulfur keratin B2A sequence is presented as the model sequence and the others are aligned with respect to it. Common nucleotide identity is indicated by a dot, and deletions (dash) and insertions (uppercase) have been introduced to maximize homology. Nucleotides are numbered negatively with respect to the methionine initiation codon and boxed residues indicate possible mRNA cap sites. The large box defines a region contiguous with the methionine intiation codon and the sequence of 18 residues is common to seven of the eight genes except for a few base changes. Data from Powell *et al.* (1983), Kuczek and Rogers (1985), Frenkel (1985), B. Powell, K. Ward *et al.* (unpublished).

2.4. Evolutionary Expression of Hard Keratin IF and IFAP

Immunological and sequence hybridization results indicate that IF keratin protein and gene sequences are conserved throughout vertebrate evolution, from fish to man (Fuchs and Marchuk, 1983), covering an evolutionary period of 400–500 million years (Dayhoff, 1978; Goodman *et al.*, 1982; Kimura, 1983). The recent sequencing of an amphibian IF Type I keratin (Jonas *et al.*, 1985) confirms the rigid sequence conservation in the α-helical core across 350 million years. The hard keratin IF appear to be a specialized subset of IF Type I and Type II keratins, richer in half-cystine than epidermal IF keratins, and given the evolutionary appearance of hairs and horny points with monotreme mammals (e.g., echidna) about 200 million years ago (Goodman *et al.*, 1982) they are a relatively recent subset. As the ancestral Type I and Type II IF genes predate the appearance of the hard keratin IF genes, separate changes must have led to the present cystine-rich Type I and Type II IF subsets. Possibly, the acquisition of a cystine-rich character by either type promoted the selection of a similar character in the other.

There is great diversity between the high-sulfur keratin protein families and yet there is evidence to suggest that their interspecies homology dates back at least 100–120 million years to the appearance of eutherian and metatherian mammals (Gillespie and Marshall, 1977; Marshall *et al.*, 1977). Furthermore, the cross-hybridization of a gene probe from the highly conserved sheep B2 high-sulfur keratin family with human and marsupial cat DNA (B. Powell, unpublished) suggests that there has also been conservation of the gene sequences over that time scale. Amino acid composition data and one-dimensional electrophoretic analysis of proteins from echidna hard keratin tissue (Gillespie and Marshall, 1977; Marshall *et al.*, 1977) indicate that the interspecies homology may date back at least 200 million years to the mammalian ancestor. Thus, the ancestral genes for the four or five different high-sulfur keratin IFAP were probably established then and must have originated in the intervening epoch between amphibians and mammals.

Whether the different high-sulfur keratin families evolved from a single primordial gene or whether they are the result of convergent evolution is unknown. A penta- or decapeptide repeat has been proposed as the fundamental unit of the B2 and BIIIA proteins (Ycas, 1972; Elleman *et al.*, 1973; Swart, 1973) and it is also present in the putative UHS cDNA clone (Fig. 6). Limited homology has been noted between the BIIIB proteins and the B2 and BIIIA proteins but their relationship is disputed (Elleman *et al.*, 1973; Swart, 1973). If they were all derived from a common ancestor, then perhaps the ancestral BIIIB sequence diverged long before the B2/BIIIA split. The lack of introns in the different keratin IFAP genes that are expressed in one tissue is surprising if they evolved independently as intronless eukaryotic genes are rare (Breathnach and Chambon, 1981; Blake, 1985). This arrangement would seem more plausible if there had been a single primordial gene for the high-sulfur keratin proteins and this would also be consistent with the ancestral peptide repeat unit proposal. Any progenitor high-sulfur keratin IFAP may have had its origin in the epidermal protein(s) whose present-day descendant(s) could be the cystine-rich protein(s) of keratohyalin granules (Fukuyama and Epstein, 1985).

With little available knowledge of interspecies HGT keratin protein sequence and complexity, or gene content, it is difficult to predict the evolutionary descent of the HGT keratin genes. Nevertheless, the detection of several HGT keratin proteins from the two HGT Type I and Type II gene families in sheep, mouse, and porcupine hard keratin suggests that they have been distinct protein families for 100–120 million years since these therian mammals diverged (Goodman *et al.*, 1982). Interestingly, the quill of the monotreme mammal the echidna, whose evolutionary divergence preceded the therian mammal radiation by 80–100 million years (Goodman *et al.*, 1982), has the highest known percentage of HGT keratin protein (30%), yet only contains HGT Type II protein (Gillespie, 1972). Should echidna hair show a similar restricted expression of HGT protein, then this could indicate that an HGT Type II primordial gene evolved first and subsequent duplication and divergence led to a progenitor HGT Type I gene in the therian mammal ancestor. Clearly, the analysis of more genomes with HGT keratin gene probes is required to clarify whether the species variation in protein content is a reflection of gene content. This should also enable the establishment of a more rigorous evolutionary lineage for this class of keratin genes.

2.5. Trichohyalin in the IRS and Hair Medulla

The hardened cells of the hair medulla and of the mature IRS have been shown to contain insoluble proteins readily distinguishable from the keratin of the hair cortex. Their main features are the presence of citrulline and a high glutamic acid/glutamine content and it is known that the citrulline residues derive from the enzymatic conversion of arginine residues by a peptidylarginine deiminase. Furthermore, these citrulline-containing proteins are highly cross-linked by isopeptide bonds produced by the action of a transglutaminase and thus the insolubility of the proteins in the hardened medulla and IRS cells can be accounted for by that cross-linking and not by disulfide bonds that are of such central importance in the hair cortex (Rogers, 1983).

During the early differentiation of medulla cells and IRS cells, electron-dense granules appear in their cytoplasm and were named trichohyalin by early histologists. From numerous studies using cytochemical and biochemical techniques, it is clear that the granules in both kinds of cells are either identical or very similar in their properties and composition (Rogers, 1983). Recently, a 190k protein was isolated from wool follicles lacking medulla (Rothnagel and Rogers, 1986) and shown by immunogold electron microscopy to originate in the trichohyalin granules of the wool follicle IRS (Fig. 9). The name trichohyalin has been retained for it. It is an arginine-rich protein and antibodies to it have been demonstrated to bind to trichohyalin granules of both the medulla and the IRS and to cross-react with other species (Rothnagel and Rogers, 1986, and unpublished observations). Studies of later stages of differentiation reveal that the trichohyalin granules disappear altogether in the IRS cells and subsequently only condensed IF (8–10 nm type) filaments can be discerned. In the medulla, the granules themselves appear to coalesce and form a solid mass minus any filaments. It has thus been concluded that in the IRS and medulla the trichohyalin of the granule is an arginine-rich precursor protein that, first of all, has some of its arginine residues processed to citrulline residues. Second, the trichohyalin acts as an IFAP in the IRS, whereas in the medulla cells where IF filaments are not present it cannot play that role and forms a cross-linked polymer with itself.

Further studies of trichohyalin must be directed to the characterization of its amino acid sequence. This might be accomplished directly or via the isolation of the gene. Examination of the amino acid sequence should finally prove whether or not trichohyalin is itself an IF protein and converts to an IF, the alternative hypothesis for its function, proposed many years ago (Rogers, 1964). It will also be of interest to isolate the gene even

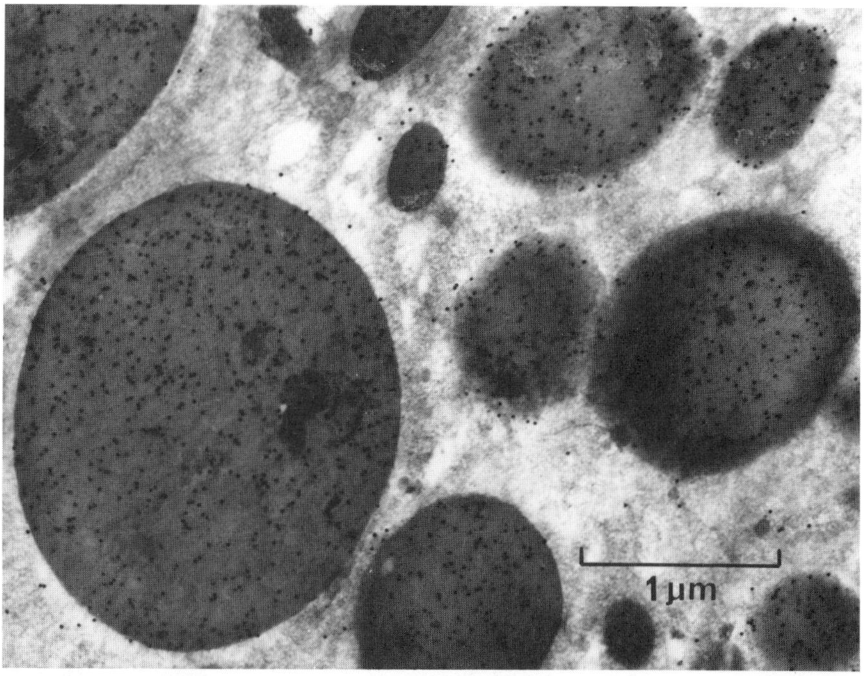

Figure 9. Electron micrograph of trichohyalin granules in the cytoplasm of a hair follicle inner root sheath cell (rat). The section was reacted with antibody, raised in rabbits against trichohyalin protein, followed by protein A–gold. A specific reaction is seen between the granules and the antibody. × 30,000.

if it is not necessary for obtaining thc amino acid sequence. Mapping of the gene and its structure is essential in understanding the expression of trichohyalin and its relationship in the genome to the other major genes that are involved in keratinization. Knowledge of the gene sequence for trichohyalin would also allow the use of specific probes by *in situ* hybridization to establish whether the mRNA for trichohyalin in the IRS is the same as that synthesized in the hair medulla.

3. Differentiation

3.1. The Structure of Hairs

The mature wool or hair fiber has a common structure composed of a cuticle sheath, an inner cortex, and a central medulla, although the relative proportions vary with fiber type. While the cortex is the predominant histological structure of fine fibers, accounting for as much as 90% of the fiber, the amount of medulla generally increases with fiber diameter, representing up to 80% of the fiber in some rodent hairs. The cuticle and cortical cells synthesize keratin whereas the medullary cells embark upon a different developmental program and produce trichohyalin. Within the cortex three different cell types can be recognized, termed the paracortex, orthocortex, and mesocortex (Fig. 10A). Together with the cuticle, they constitute four keratinizing cell types that exhibit distinctive dif-

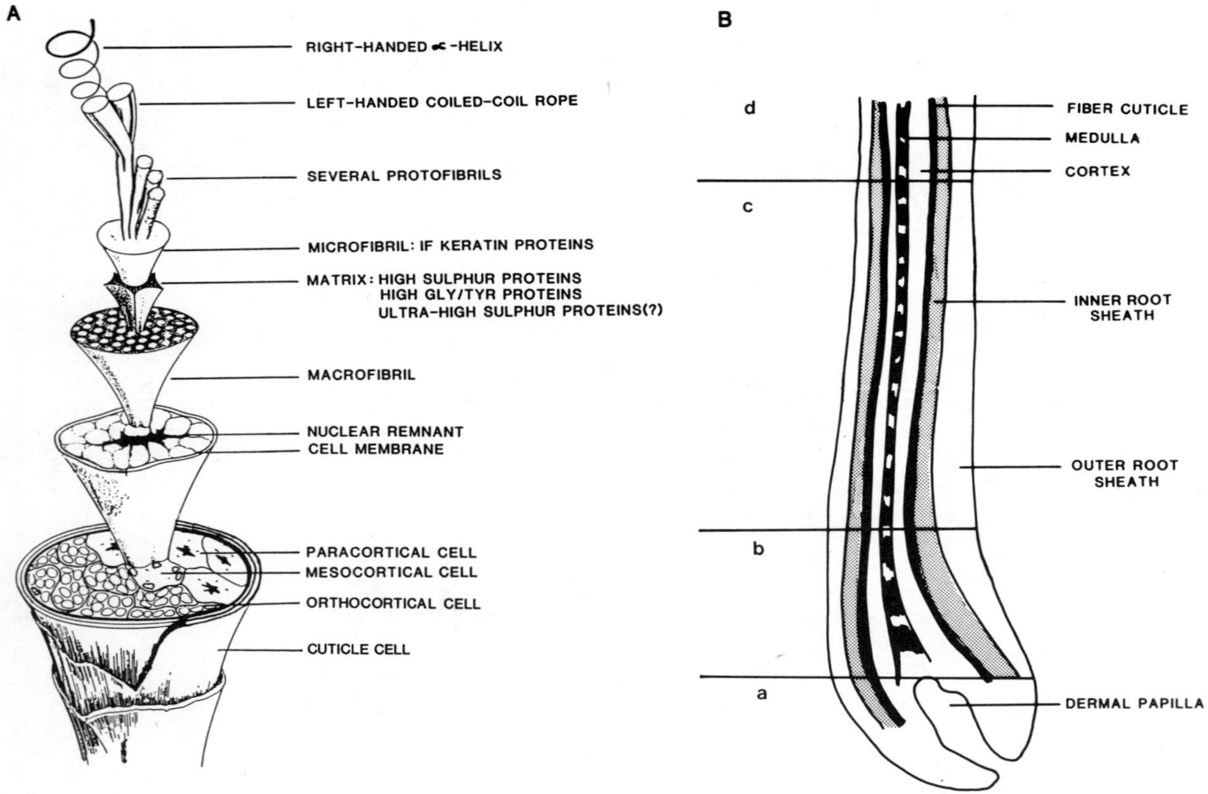

Figure 10. (A) An exploded, diagrammatic representation of a fully formed hair fiber lacking medulla. The constituent components of a paracortical cell are shown in successive enlargements. Adapted from Marshall *et al*. (1985) with permission. (B) A diagrammatic representation of a hair follicle. Different morphological regions within the growing fiber, broadly defined by horizontal lines, are as follows: A, mitotic zone of the follicle bulb; B, cell elongation zone; C, keratogenous zone; D, zone of cross-linking and cell dehydration. The shaded regions are only intended to aid visual distinction of the different cell streams in the hair fiber.

ferences in keratin synthesis, reflecting differential gene activation and expression from the keratin family repertoire.

3.2. Hair Developmental Program

Hair development represents the culmination of a differentiation pathway that is accompanied by striking changes in cell morphology, proceeding from the polygonal stem cell to spindle-shaped cortical cells overlaid by flattened rectangular cuticle cells. The transition from stem cell in the follicle bulb to differentiated cortical and cuticle cells in the emergent follicle occurs in sheep in 2½ to 4 days (Epstein and Maibach, 1969; Chapman, 1971). The initial events that trigger hair biogenesis are at present undefined, although their effects are undoubtedly exerted through the interaction of the dermal papilla with the stem cells (Cohen, 1961; Oliver, 1970; Jahoda *et al.*, 1984). In contrast, the subsequent pathway that leads from these pluripotential stem cells in the follicle bulb to the terminally differentiated cells in the fully formed fiber has been comprehensively detailed by electron microscopic methods. For extensive reviews see Bradbury (1973), Montagna and Parakkal (1974), Swift (1977), and Orwin (1979).

From a morphological viewpoint, fiber development can be conveniently divided into several zones (Fig. 10B). The stem cells proliferate in the follicle bulb around the dermal papilla (zone A) and several cell streams differentiate in the upper part of this mitotic zone (upper zone A/zone B). The location of the stem cells with respect to the dermal papilla appears to regulate the fate of the differentiating cell streams (Epstein and Maibach, 1969; Orwin, 1971). Thus, the central medullary cells differentiate at the apex above the dermal papilla and they are surrounded successively by cortical, cuticle, IRS, and outer root sheath (ORS) cells. The outermost layer, the ORS, is contiguous with the epidermis. Above the mitotic zone, zone B is characterized by changes in cell morphology that indicate the initiation of synthesis of the various terminal differentiation products in all cell types. The first sign of keratin synthesis in the cortical cells is the appearance of microfibrils containing IF protein chains in the cytoplasm. They soon aggregate to form macrofibrils, bundles of microfibrils interspersed with the matrix keratin proteins. Trichohyalin granules appear in the medulla and IRS. In the cuticle cells the onset of keratin synthesis is marked by the appearance of cystine-rich granules in the cytoplasm followed by their ordered deposition around the outermost periphery of the flattening cell. A loose, filamentous network of unknown composition seems to be associated with the granules (Happey and Johnson, 1962; Roth and Helwig, 1964; Woods and Orwin, 1980). Whether this network is composed of IF keratin proteins or other filament-forming proteins is not known. Nevertheless, it is apparent that microfibrils typical of the cortex are not found in the cuticle (Bradbury, 1973; Swift, 1977). Synthesis of fibrillar material also occurs in the ORS and may be the origin of the IF keratin proteins isolated from the ORS of human hair follicle by Moll *et al.* (1982a) but no filament–matrix network is seen at any stage (Rogers, 1964; Orwin, 1971, 1979).

Keratin synthesis in the cortex and cuticle and trichohyalin synthesis in the medulla is maximal in zone C. The IRS cells that surround the growing fiber are the first to fully differentiate and are thought to support and mold the nascent hair fiber. After this period of active biosynthesis, the cells, replete with their differentiation products, undergo extensive cross-linking of the keratin or trichohyalin, destruction of cellular organelles, and finally dehydration (zone D). The final stage of differentiation is the formation of a cross-linked network through oxidation of pairs of cysteine residues to produce disulfide bonds. This process is apparently catalyzed by copper, but it is not known whether the copper catalyst is ionic or is part of a copper protein complex (Gillespie, 1983). The IRS sloughs off into the pilary canal near the epidermal surface and the mature fiber emerges from the follicle.

3.3. Gene Expression during Differentiation in the Hair Follicle

3.3.1. Cuticle Gene Expression

In general, the cuticle accounts for less than 10% of the hair and it has not been studied as extensively as the cortex. In each hair the cuticle forms a cylindrical layer encasing the cortex and it contains a homogeneous population of cells irrespective of whether it has a single layer, as in sheep wool, or is multilayered, as in human hair.

Keratin synthesis in the cuticle appears to be different from that which occurs in the cortex. As cuticle cell differentiation proceeds, cystine-rich granules appear in the cell cytoplasm and their asymmetric deployment leads to distinct layers, aligned parallel to the axis of the fiber and of varying thickness and half-cystine content (Fig. 11). The initial deposition of cystine-rich granules around the cell periphery produces the A layer, which apposes the IRS, and a thinner, less regular layer, the inner layer, apposing the cortex. Next to the A layer is the exocuticle, and while it is also cystine-rich it has been reported to have a lower cystine content than the A layer (Orwin, 1979; Swift, 1981). Beneath the A layer and exocuticle is the endocuticle, which appears to be the result of the condensation and dehydration of the nucleus and cytoplasmic organelles as the cell keratinizes, and, unlike the other layers, it is relatively cystine-poor.

The work of Ley and Crewther (1980) suggested that the UHS keratin proteins are a prominent component of cuticle keratin but a minor one in the cortex. Human hair has a higher half-cystine content (Gillespie and Marshall, 1981) and contains more UHS protein

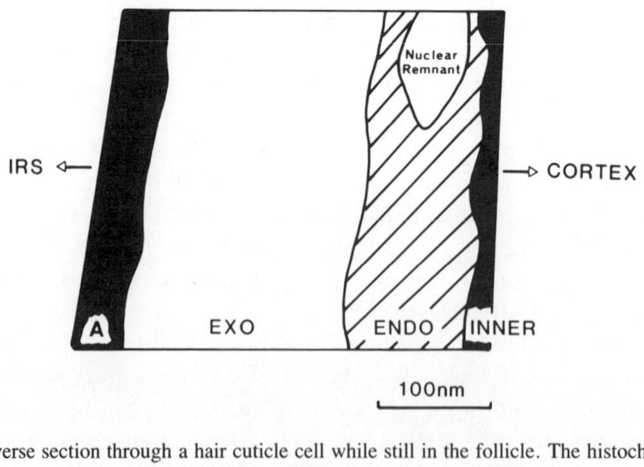

Figure 11. Transverse section through a hair cuticle cell while still in the follicle. The histochemically defined zones (see text) are: A, A layer; EXO, exocuticle; ENDO, endocuticle; and INNER, inner layer. The approximate dimensions shown are those of a human hair cuticle cell (Swift, 1977).

than sheep wool (Gillespie, 1983) and, in view of the suggested location of UHS proteins in the single-layered wool cuticle, it would be interesting to determine their distribution in human hair between the multilayered cuticle and the cortex to see if there is a correlation between UHS protein content and cuticle thickness. Only 30% of the protein could be extracted from wool cuticle, compared to the usual high level (80%) of protein extracted from the cortex, and thus the identity of the bulk of the protein expressed in the cuticle remains uncertain. The intractable cuticle residue contained unmodified cystine, despite the extensive *S*-carboxymethylation reaction incorporated in the extraction procedure, isopeptide cross-links, and, with the exception of glycine (10.4 versus 6.4%), proline (9.4 versus 12.4%), and lysine (1.3 versus 3.4%), it had the same amino acid composition as the extract. No IF or high-sulfur IFAP are detectable in the cuticle using the procedures normally sufficient to extract those components from cortical keratin and if they are present then they must be bound up in an unusual and very stable complex.

3.3.2. Cortical Gene Expression

In the hairs of all species examined, the composition of the cortex is variable. It can be formed from any of three cell types, including two major types, the ortho- and paracortical cells, and a third type, the mesocortical cell. They are recognized by their different proportions and arrangements of microfibrils, whose constituent proteins are the hard keratin IF. The arrangement of the cell types in the cortex is genetically determined and can vary dramatically within a species. Human hair cortex can be all paracortex (straight hair), or a central core of paracortex surrounded by orthocortex (wavy hair), or equal proportions of ortho- and paracortex (curly hair) showing bilateral segmentation (Swift, 1977).

Cortical segmentation is especially evident in sheep wool, and, because of the demands of wool research, the composition of the wool fiber cortex has been extensively studied. The orthocortical cell is composed of about 60% microfibrils arranged in a whorl pattern with small amounts of intermicrofibrillar matrix, while the lower microfibrillar content (about 30%) of the paracortical cell is arranged in a relatively open, often hexagonal array with greater amounts of matrix (Whiteley and Kaplin, 1977; Kaplin and Whiteley, 1978; Orwin, 1979). The mesocortical cell has an intermediate morphology (Rogers, 1959a,b; Brown and Onions, 1960; Bonés and Sikorski, 1967). The variation in microfibril and matrix content of the cells is a direct reflection of the amounts of hard keratin IF and IFAP produced and possibly the synthesis of different keratin proteins too. Although recent immunohistochemical studies suggest that the HGT Type II proteins are located specifically in the orthocortex of the wool fiber (Hewish and French, personal communication), the highly cross-linked nature of the keratinized fiber could hinder antibody penetration. Care should be exercised when interpreting these results as the extent of cross-linking may vary between cell types and the orthocortex, with its lower matrix content, is possibly less extensively cross-linked and therefore more accessible to antibodies than the paracortex.

When several gene products contribute to the terminally differentiated state of a cell, the possibility of coordinate gene expression is raised. This is a central question concerning the synthesis of the different keratin proteins during cortical cell differentiation which has not been unequivocally answered. Early work suggested a two-stage synthesis in which the keratin IF preceded the IFAP (Mercer, 1961; Downes *et al.*, 1963; Rogers,

1964). While this has been essentially confirmed for the paracortex, in which it is now thought the production of IFAP initially lags behind the IF and then increases sharply until there is dual synthesis of both, a simultaneous synthesis was proposed for the orthocortex (Chapman and Gemmell, 1971; Chapman, 1976). The distinctive patterns of keratin synthesis that occur in cortical cells indicate a fine control of keratin gene expression and this is a central feature of differentiation in the follicle. It is not known whether alternative genes from each of the keratin gene families are expressed in the different cortical cell types or whether modulation of transcriptional and/or translational activity is responsible for the observed differences. The immunohistochemical approach to protein localization in the wool fiber is probably not sufficiently sensitive to discriminate between the expression of the closely related proteins of the high-sulfur keratin IFAP families. A better approach might be to employ specific gene probes, derived from unique 3′ noncoding regions of the hard keratin IF and IFAP genes, in *in situ* tissue hybridization of developing hairs. Current DNA sequence data have identified such regions in genes from nearly all the wool keratin protein families. This should permit a more comprehensive investigation of keratin gene expression, not only in the differentiation of the wool fiber, but also in hard keratin tissues in general.

3.3.3. Variation in Keratin Gene Expression

The keratin proteins in the hair fiber are subject to variation through genetic, dietary, and physiological factors (Gillespie and Reis, 1966; Frenkel *et al.*, 1974; Reis, 1979; Gillespie *et al.*, 1982). The multiple components within each family probably fulfill similar roles, and therefore the variable proportions between members of the same species have no obvious phenotypic effect. The least variable are the keratin IF and a proportion of the high-sulfur keratin IFAP, thought to be essential for the development of the basic fiber structure (Gillespie and Reis, 1966; Fraser and MacRae, 1980; Marshall and Gillespie, 1981; Gillespie and Marshall, 1983), while the most dramatic changes occur in the UHS and HGT keratin IFAP families (Frenkel *et al.*, 1974, 1975; Marshall and Gillespie, 1981; Gillespie and Marshall, 1983).

The UHS proteins are one of the most intriguing classes of high-sulfur keratin families because, unlike the others, their expression in sheep is particularly variable, being influenced by nutritional status, administration of epidermal growth factor and chemical defleecing agents, which result in an increase in UHS proteins in the regrowth wool (Gillespie and Reis, 1966; Gillespie *et al.*, 1980, 1982). In wool the UHS proteins are normally expressed at a low level, yet their synthesis can be dramatically induced (Gillespie and Reis, 1966; Reis, 1979; Gillespie, 1983). In contrast, in the hairs of nonruminants they appear to be expressed constitutively, but their synthesis is decreased in some genetic disorders and by protein deficiency (Gillespie, 1983; Gillespie and Marshall, 1983). The reduction in nonruminants is probably due to a decrease in the sulfur-containing amino acids (mainly cysteine) available for keratin synthesis (Pollitt and Stonier, 1971; Gillespie and Marshall, 1980), a situation that mimics the natural state in ruminants where these amino acids are degraded by ruminal flora (Reis, 1979). In view of the location of wool UHS proteins in the fiber cuticle (Ley and Crewther, 1980), it is important to determine their distribution in human hair between its typical multilayered cuticle and cortex and also whether the fluctuation in expression of the UHS proteins, either upward during sulfur enrichment in sheep, or downward during protein deficiency in humans, occurs primarily in the cuticle or in the cortex.

The HGT proteins of wool fluctuate in response to the same conditions that promote UHS synthesis, often in apparent inverse relationship to the UHS protein response (Gillespie *et al.*, 1980, 1982; Gillespie and Marshall, 1981). Thus, there is a downward trend in the content of these proteins in wool produced when sheep are subjected to diets low in aromatic amino acids, or after treatment of the sheep with agents such as cyclophosphamide and epidermal growth factor, which produce total hair loss (Reis, 1979; Gillespie *et al.*, 1980, 1982). Gillespie *et al.* (1980) noted that, in part, these responses reflect a natural phenomenon as the birthcoat wool of sheep has a lower HGT content and a higher sulfur content than wool grown a few months later. A direct comparison of the proteins of birthcoat and sulfur-enriched wool from the same sheep is essential in a further investigation of this effect.

In addition to the physiological changes, there are genetic mutations that are known to affect the expression of the HGT proteins. In sheep there is the felting luster mutant, which lacks detectable HGT proteins in the wool (Gillespie and Darskus, 1971). Mouse hair normally contains higher levels than wool of HGT proteins and with a similar complexity. Tenenhouse and Gold (1976) observed that heterozygotes of the naked (N) mutation produce hair with a content of HGT proteins that is about half that of normal. Recently, use of gene probes for the HGT Type I group (see below) demonstrated that the lowered level arises from a transcriptional defect, probably regulatory, and not explicable by gene dosage (E. Kuczek, personal communication).

The expression patterns of the HGT keratin genes seem to be particularly responsive to environmental factors, as exemplified by the variation in expression from 1 to 12% between breeds of domestic sheep (Gillespie and Darskus, 1971). These sheep breeds have been separate for no more than 10,000 years (Ryder and Stephenson, 1968), an insignificant time in evolutionary terms, and their different contents of HGT keratin protein are unlikely to be due to variation in their HGT keratin gene repertoire. The same conclusion seems to be applicable to human hair, which expresses a low level of HGT protein, because sheep HGT Type I gene probes for the component F and C2 genes each detected two fragments in human genomic DNA (E. Kuczek, personal communication), indicating that those HGT keratin genes are present in the human genome. These variations in HGT expression suggest that the HGT keratin genes are not under a strong selective pressure and it remains possible that very low HGT protein contents in the hard keratin tissue of other mammalian species are, in fact, due to a lack of the genes. The availability of gene probes will make it possible to investigate some aspects of the molecular mechanisms responsible for changes in the expression of HGT proteins in the follicle during wool growth.

ACKNOWLEDGMENT. The writing of this chapter and some of the research work quoted were supported by a grant from the Wool Research Trust Fund on the recommendation of the Australian Wool Corporation.

References

Baden, H. P., and Kubilus, J., 1983, Fibrous proteins of bovine hoof, *J. Invest. Dermatol.* **81**:220–224.

Blake, C. C. F., 1985, Exons and the evolution of proteins, *Int. Rev. Cytol.* **93**:149–185.

Bonés, R. M., and Sikorski, H., 1967, The histological structure of wool fibres and their plasticity, *J. Text. Inst.* **58**:521–532.

Bradbury, J. H., 1973, The structure and chemistry of keratin fibres, *Adv. Protein Chem.* **27:**111–211.

Breathnach, R., and Chambon, P., 1981, Organization and expression of eukaryotic split genes coding for proteins, *Annu. Rev. Biochem.* **50:**349–383.

Brown, T. D., and Onions, W. J., 1960, Anomalies in the microscopic structure of some wools, *Nature* **186:**93–94.

Chapman, R. E., 1971, Cell migration in wool follicles of sheep, *J. Cell Sci.* **9:**791–803.

Chapman, R. E., 1976, Electron microscopic and histochemical features of the formation of the orthocortex and paracortex in wool, in: *Proceedings of the 5th International Wool Textile Research Conference,* Volume 2 (K. Ziegler, ed.), German Wool Research Institute, Aachen, pp. 152–161.

Chapman, R. E., and Gemmell, R. T., 1971, Stages in the formation and keratinization of the cortex of the wool fibre, *J. Ultrastruct. Res.* **36:**342–354.

Cohen, J., 1961, The transplantation of individual rat and guinea pig whisker papillae, *J. Embryol. Exp. Morphol.* **9:**117–127.

Crewther, W. G., 1976, Primary structure and chemical properties of wool, in: *Proceedings of the 5th International Wool Textile Research Conference,* Volume 1 (K. Ziegler, ed.), German Wool Research Institute, Aachen, pp. 1–101.

Crewther, W. G., Inglis, A. S., and McKern, N. M., 1978, Amino acid sequences of α-helical segments from S-carboxymethylkerateine-A: Complete sequence of a Type II segment, *Biochem. J.* **173:**365–371.

Crewther, W. G., Dowling, L. M., and Inglis, A. S., 1980a, Amino acid sequence data from a microfibrillar protein of α-keratin, in: *Proceedings of the 6th International Wool Textile Research Conference,* Volume 2, South African Wool Textile Research Institute, Pretoria, pp. 79–91.

Crewther, W. G., Dowling, L. M., Gough, K. H., Marshall, R. C., and Sparrow, L. G., 1980b, The microfibrillar proteins of α-keratin, in: *Fibrous Proteins: Scientific, Industrial and Medical Aspects,* Volume 2 (D. A. D. Parry and L. K. Creamer, eds.), Academic Press, New York, pp. 151–159.

Crewther, W. G., Dowling, L. M., Steinert, P. M., and Parry, D. A. D., 1983, Structure of intermediate filaments, *Int. J. Biol. Macromol.* **5:**267–274.

Darskus, R. L., 1972, Electrophoretic and chromatographic characterization of sulphur-rich proteins from wool, *J. Chromatogr.* **69:**341–348.

Dayhoff, M. O., 1978, *Atlas of Protein Sequence and Structure,* Volume 5, Suppl. 3, Biomedical Research Foundation, Washington, D.C.

DeDeurwaerder, R. A., Dobb, M. G., and Sweetman, B. H., 1964, Selective extraction of a protein fraction from wool keratin, *Nature* **203:**48–49.

Dopheide, T. A. A., 1973, The primary structure of a protein component 0.62, rich in glycine and aromatic residues obtained from wool keratin, *Eur. J. Biochem.* **34:**120–124.

Dowling, L. M., Gough, K. H.,Inglis, A. S., and Sparrow, L. G., 1979, Comparison of some microfibrillar proteins from wool, *Aust. J. Biol. Sci.* **32:**437–442.

Dowling, L. M., Parry, D. A. D., and Sparrow, L. G., 1983, Structural homology between hard α-keratin and the intermediate filament proteins desmin and vimentin, *Biosci. Rep.* **3:**73–78.

Dowling, L. M., Crewther, W. G., and Inglis, A. S., 1986, The primary structure of component 8c-1, a subunit protein of intermediate filaments in wool keratin, *Biochem. J.* **236:**695–703.

Downes, A. M., Sharry, L. F., and Rogers, G. E., 1963, Separate synthesis of fibrillar and matrix proteins in the formation of keratin, *Nature* **199:**1059–1061.

Eichner, R., Bonitz, P., and Sun, T.-T., 1984, Classification of epidermal keratins according to their immunoreactivity, isoelectric point and mode of expression, *J. Cell Biol.* **98:**1388–1396.

Elleman, T. C., Lindley, H., and Rowlands, R. J., 1973, Periodicity in high sulphur proteins from wool, *Nature* **246:**530–531.

Epstein, W., and Maibach, H. I., 1969, Cell proliferation and movement in human hair bulbs, in: *Advances in Biology of Skin,* Volume 9 (W. Montagna and R. L. Dobson, eds.), Pergamon Press, New York, pp. 83–97.

Fraser, R. D. B., and MacRae, T. P., 1980, Molecular structure and mechanical properties of keratins, in: *The Mechanical Properties of Biological Materials* (J. V. F. Vincent and J. D. Currey, eds.), Cambridge University Press, London, pp. 211–246.

Fraser, R. D. B., MacRae, T. P., and Rogers, G. E., 1972, *Keratins, Their Composition, Structure and Biosynthesis,* Thomas, Springfield, Ill.

Fraser, R. D. B., Gillespie, J. M., and MacRae, T. P., 1973, Tyrosine-rich proteins in keratins, *Comp. Biochem. Physiol.* **44B:**943–947.

Frenkel, M. J., 1985, *Studies on hair keratin genes,* Ph.D. thesis, University of Adelaide, Adelaide, South Australia.

Frenkel, M. J., Gillespie, J. M., and Reis, P. J., 1974, Factors influencing the biosynthesis of the tyrosine-rich proteins of wool, *Aust. J. Biol. Sci.* **27:**31–38.

Frenkel, M. J., Gillespie, J. M., and Reis, P. J., 1975, Studies on the inhibition of synthesis of the tyrosine-rich proteins of wool, *Aust. J. Biol. Sci.* **28:**331–338.

Fuchs, E. V., and Marchuk, D., 1983, Type I and Type II keratins have evolved from lower eukaryotes to form the intermediate filaments in mammalian skin, *Proc. Natl. Acad. Sci. USA* **80:**5857–5861.

Fuchs, E. V., Coppock, S. M., Green, H., and Cleveland, D. W., 1981, Two distinct classes of keratin genes and their evolutionary significance, *Cell* **27:**75–84.

Fukuyama, K., and Epstein, W. L., 1985, Keratohyalin, in: *Biology of the Integument,* Volume 2 (J. Bereiter-Hahn, A. G. Matoltsy, and K. S. Richards, eds.), Springer-Verlag, Berlin, pp. 739–751.

Geisler, N., and Weber, K., 1982, The amino acid sequence of chicken muscle desmin provides a common structural model for intermediate filament proteins, *EMBO J.* **1:**1649–1656.

Gillespie, J. M., 1962, The isolation and properties of some soluble proteins from wool. IV. The isolation of the high sulphur proteins SCMKB1, *Aust. J. Biol. Sci.* **15:**572–588.

Gillespie, J. M., 1972, Proteins rich in glycine and tyrosine from keratins, *Comp. Biochem. Physiol.* **41B:**723–734.

Gillespie, J. M., 1983, The structural proteins of hair: Isolation, characterization and regulation of biosynthesis, in: *Biochemistry and Physiology of the Skin,* Volume 1 (L. A. Goldsmith, ed.), Oxford University Press, London, pp. 475–510.

Gillespie, J. M., and Darskus, R. L., 1971, Relation between the tyrosine content of various wools and their content of a class of protein rich in tyrosine and glycine, *Aust. J. Biol. Sci.* **24:**1189–1197.

Gillespie, J. M., and Frenkel, M. J., 1976, The tyrosine-rich proteins of keratins, *Proceedings of the 5th International Wool Textile Research Conference,* Volume 2 (K. Ziegler, ed.), German Wool Research Institute, Aachen, pp. 265–276.

Gillespie, J. M., and Marshall, R. C., 1977, Proteins of the hard keratins of echidna, hedgehog, rabbit, ox and man, *Aust. J. Biol. Sci.* **30:**401–409.

Gillespie, J. M., and Marshall, R. C., 1980, Variability in the proteins of wool and hair, in: *Proceedings of the 6th International Wool Textile Research Conference,* Volume 2, South African Wool Textile Research Institute, Pretoria, pp. 67–77.

Gillespie, J. M., and Marshall, R. C., 1981, The proteins of normal and aberrant hair keratins, in: *Hair Research* (C. E. Orfanos, W. Montagna, and G. Stüttgen, eds.), Springer-Verlag, Berlin, pp. 76–83.

Gillespie, J. M., and Marshall, R. C., 1983, A comparison of the proteins of normal and trichothiodystrophic human hair, *J. Invest. Dermatol.* **80:**195–202.

Gillespie, J. M., and Reis, P. J., 1966, The dietary-regulated biosynthesis of high sulphur wool proteins, *Biochem. J.* **98:**669–677.

Gillespie, J. M., Haylett, T., and Lindley, H., 1968, Evidence of homology in a high sulphur protein fraction (SCMK-B2) of wool and hair α-keratins, *Biochem. J.* **110:**193–200.

Gillespie, J. M., Frenkel, M. J., and Reis, P. J., 1980, Changes in the matrix proteins of wool and mouse hair following the administration of depilatory components, *Aust. J. Biol. Sci.* **33:**125–136.

Gillespie, J. M., Marshall, R. C., Moore, G. P. M., Panaretto, B. A., and Robertson, D. M., 1982, Changes in the proteins of wool following treatment of sheep with epidermal growth factors, *J. Invest. Dermatol.* **79:**197–200.

Goodman, M., Romero-Herrera, A. E., Dene, H., Czelusniak, J., and Tashian, R. E., 1982, Amino acid sequence evidence on the phylogeny of primates and other eutherians, in: *Macromolecular Sequences in Systematic and Evolutionary Biology* (M. Goodman, ed.), Plenum Press, New York, pp. 115–191.

Gough, K. H., Inglis, A. S., and Crewther, W. G., 1978, Amino acid sequences from S-carboxymethylkerateine-A: Complete sequence of a Type I segment, *Biochem. J.* **173:**373–385.

Hanukoglu, I., and Fuchs, E., 1982, The cDNA sequence of a human epidermal keratin: Divergence of sequence but conservation of structure among intermediate filament proteins, *Cell* **31:**243–252.

Happey, F., and Johnson, A. G., 1962, Some electron microscopic observations on hardening in the human hair follicle, *J. Ultrastruct. Res.* **7:**373–385.

Haylett, T., Swart, L. S., Parris, D., and Joubert, F. J., 1971, The primary structure of some high sulphur proteins of reduced wool, *Appl. Polymer Symp.* **18:**37–44.

Jahoda, C. A. B., Horne, K. A., and Oliver, R. F., 1984, Induction of hair growth by implantation of cultured dermal papilla cells, *Nature* **311:**560–562.

Johnson, L. D., Idler, W. W., Zhou, X.-M., Roop, D. R., and Steinert, P. M., 1985, Structure of a gene for the human epidermal 67 kD keratin, *Proc. Natl. Acad. Sci. USA* **82:**1896–1900.

Jonas, E., Sargent, T. D., and Dawid, I. B., 1985, Epidermal keratin genes expressed in embryos of *Xenopus laevis, Proc. Natl. Acad. Sci. USA* **82:**5413–5417.

Kaplin, I. J., and Whiteley, K. J., 1978, An electron microscopic study of fibril matrix arrangement in high and low-crimp wool fibres, *Aust. J. Biol. Sci.* **31:**231–240.

Kimura, M., 1983, The neutral theory of molecular evolution, in: *Evolution of Genes and Proteins* (M. Nei and R. K. Koehn, eds.), Sinauer Associates, Sunderland, Mass., pp. 208–233.

Kretschmer, P. J., Coon, H. C., Davis, A., Harrison, M., and Nienhuis, A. W., 1981, Hemoglobin switching in sheep: Isolation of the fetal γ-globin gene and demonstration that the fetal γ- and adult β-globin genes lie within eight kilobase segments of homologous DNA, *J. Biol. Chem.* **256:**1975–1982.

Krieg, T. M., Schafer, M. P., Cheng, C. K., Filpula, D., Flaherty, P., Steinert, P. M., and Roop, D. R., 1985, Organization of a Type I keratin gene: Evidence for evolution of intermediate filaments from a common ancestral gene, *J. Biol. Chem.* **260:**5867–5870.

Kuczek, E. S., and Rogers, G. E., 1985, Sheep keratins: Characterization of cDNA clones for the glycine + tyrosine-rich wool proteins using a synthetic probe, *Eur. J. Biochem.* **146:**89–93.

Lazarides, E., 1980, Intermediate filaments as mechanical integrators of cellular space, *Nature* **283:**249–256.

Lazarides, E., 1982, Intermediate filaments: A chemically heterogeneous, developmentally regulated class of proteins, *Annu. Rev. Biochem.* **51:**219–250.

Ley, K. F., and Crewther, W. G., 1980, The proteins of wool cuticle, in: *Proceedings of the 6th International Wool Textile Research Conference,* Volume 2, South African Wool Textile Research Institute, Pretoria, pp. 13–28.

Lindley, H., 1977, The chemical composition and structure of wool, in: *Chemistry of Natural Protein Fibres* (R. S. Asquith, ed.), Plenum Press, New York, pp. 147–191.

Lindley, H., and Elleman, T. C., 1972, The preparation and properties of a group of proteins from the high sulphur fraction of wool, *Biochem. J.* **128:**859–867.

Lynch, M. H., Hardy, C. L., Mak, L., and Sun, T.-T., 1985, Biochemical and immunological characterization of human hair and nail α-keratins, *J. Cell Biol.* **101** (Part 2):21a.

McGarry, T. J., and Lindquist, S., 1985, The preferential translation of Drosophila hsp70 mRNA requires sequences in the untranslated leader, *Cell* **42:**903–911.

McKeon, F. D., Kirschner, M. W., and Caput, D., 1986, Homologies in both primary and secondary structures between nuclear envelope and intermediate filament proteins, *Nature* **319:**463–468.

Marchuk, D., McCrohon, S., and Fuchs, E. V., 1985, Complete sequence of a gene encoding a human Type I keratin: Sequences homologous to enhancer elements in the regulatory region of the gene, *Proc. Natl. Acad. Sci. USA* **82:**1609–1613.

Marshall, R. C., 1980, Genetic variations in the proteins of human nail, *J. Invest. Dermatol.* **75:**264–269.

Marshall, R. C., 1983, Characterization of the proteins of human hair and nail by electrophoresis, *J. Invest. Dermatol.* **80:**519–524.

Marshall, R. C., 1985, Nail, claw, hoof and horn keratin, in: *Biology of the Integument,* Volume 2 (J. Bereiter-Hahn, A. G. Matoltsy, and K. S. Richards, eds.), Springer-Verlag, Berlin, pp. 722–738.

Marshall, R. C., and Gillespie, J. M., 1977, The keratin proteins of wool, horn and hoof from sheep, *Aust. J. Biol. Sci.* **30:**389–400.

Marshall, R. C., and Gillespie, J. M., 1981, Changes in wool protein components following chemical defleecing, in: *Proceedings of the 2nd National Conference on Wool Harvesting Research and Development* (P. R. W. Hudson, ed.), Sydney, pp. 117–121.

Marshall, R. C., Frenkel, M. J., and Gillespie, J. M., 1977, High sulphur proteins in mammalian keratins: A possible aid in classification, *Aust. J. Zool.* **25:**121–132.

Marshall, R. C., Gillespie, J. M., Inglis, A. S., and Frenkel, M. J., 1980, High tyrosine proteins of wool, heterogeneity and biosynthetic regulation, *Proceedings of the 6th International Wool Textile Research Conference,* Volume 2, South African Wool Textile Research Institute, Pretoria, pp. 147–158.

Marshall, R. C., Gillespie, J. M., McGuirk, B. J., Marler, J. W., Reis, P. J., Rogan, I. M., and Whiteley, K. J., 1985, Investigation of the relationship of variation in fibre protein composition upon fleece properties and fabric performance, *Proceedings of the 7th International Wool Textile Research Conference,* Volume 2, Society of Fibre Science and Technology, Japan, pp. 36–44.

Mercer, E. H., 1961, *Keratin and Keratinization,* Pergamon Press, New York.

Moll, R., Franke, W. W., Volc-Platzer, B., and Krepler, R., 1982a, Different keratin polypeptides in epidermis and other epithelia of human skin: A specific cytokeratin of molecular weight 46,000 in epithelia of the pilosebaceous tract and basal cell epitheliomas, *J. Cell Biol.* **95:**285–295.

Moll, R., Franke, W. W., Schiller, D. L., Geiger, B., and Krepler, R., 1982b, The catalogue of human cytokeratins: Patterns of expression in normal epithelia, tumours and cultured cells, *Cell* **31:**11–24.

Montagna, W., and Parakkal, P. F., 1974, *The Structure and Function of Skin,* 3rd ed., Academic Press, New York.

Novacek, M. J., 1982, Information for molecular studies from anatomical and fossil evidence on higher eutherian phylogeny, in: *Macromolecular Sequences in Systematic and Evolutionary Biology* (M. Goodman, ed.) Plenum Press, New York, pp. 3–41.

Oliver, R. F., 1970, The induction of hair follicle formation in the adult hooded rat by vibrissa dermal papillae, *J. Embryol. Exp. Morphol.* **23:**219–236.

Orwin, D. F. G., 1971, Cell differentiation in the lower, outer sheath of the Romney follicle: A companion cell layer, *Aust. J. Biol. Sci.* **24:**989–999.

Orwin, D. F. G., 1979, The cytology and cytochemistry of the wool follicle, *Int. Rev. Cytol.* **60:**331–374.

Parry, D. A. D., and Fraser, R. D. B., 1985, Intermediate filament structure: 1. Analysis of IF protein sequence data, *Int. J. Biol. Macromol.* **7:**203–213.

Pollitt, R. J., and Stonier, P. D., 1971, Proteins of normal hair and of cysteine-deficient hair from mentally retarded siblings, *Biochem. J.* **122:**433–444.

Powell, B. C., and Rogers, G. E., 1986, Hair keratin: Composition, structure and biogenesis, in: *Biology of the Integument,* Volume 2 (J. Bereiter-Hahn, A. G. Matoltsy, and K. S. Richards, eds.), Springer-Verlag, Berlin, pp. 695–721.

Powell, B. C., Sleigh, M. J., Ward, K. A., and Rogers, G. E., 1983, Mammalian keratin gene families: Organization of genes coding for the B2 high sulphur proteins of sheep wool, *Nucleic Acids Res.* **11:**5327–5346.

Powell, B. C., Cam, G. R., Fietz, M. J., and Rogers, G. E., 1986, Clustered arrangement of keratin intermediate filament genes, *Proc. Natl. Acad. Sci. USA* **83:**5048–5052.

RayChaudhury, A., Marchuk, D., Lindhurst, M., and Fuchs, E., 1986, Three tightly linked genes encoding human Type I keratins: Conservation of sequence in the 5′-untranslated leader and 5′-upstream regions of coexpressed keratin genes, *Mol. Cell Biol.* **6:**539–548.

Reiger, M., Jorcano, J. L., and Franke, W. W., 1985, Complete sequence of a bovine Type I cytokeratin gene: Conserved and variable intron positions in genes of polypeptides of the same cytokeratin subfamily, *EMBO J.* **4:**2261–2267.

Reis, P. J., 1979, Effects of amino acids on the growth and properties of wool, in: *Physiological and Environmental Limitations to Wool Growth* (J. L. Black and P. J. Reis, eds.), University of New England Publications Unit, New South Wales, pp. 223–242.

Rogers, G. E., 1959a, Electron microscopic studies of hair and wool, *Ann. N.Y. Acad. Sci.* **83:**378–399.

Rogers, G. E., 1959b, Electron microscopy of wool, *J. Ultrastruct. Res.* **2:**309–330.

Rogers, G. E., 1964, Structural and biochemical features of the hair follicle, in: *The Epidermis* (W. Montagna and W. C. Lobitz, eds.), Academic Press, New York, pp. 179–236.

Rogers, G. E., 1983, The occurrence of citrulline in structural proteins of the hair follicle, in: *Biochemistry and Physiology of the Skin,* Volume 1 (L. A. Goldsmith, ed.), Oxford University Press, London, pp. 511–521.

Roth, S., and Helwig, E. B., 1964, The cytology of the dermal papilla, the bulb and the root sheath of mouse hair, *J. Ultrastruct. Res.* **11:**33–51.

Rothnagel, J. A., and Rogers, G. E., 1986, Trichohyalin and intermediate filament-associated protein of the hair follicle, *J. Cell Biol.* **102:**1419–1429.

Ryder, M. L., and Stephenson, S. K., 1968, *Wool Growth,* Academic Press, New York.

Skerrow, D., 1985, Epidermal α-keratin: Structure and chemical composition, in: *Biology of the Integument,* Volume 2 (J. Bereiter-Hahn, A. G., Matoltsy, and K. S. Richards, eds.), Springer-Verlag, Berlin, pp. 621–643.

Sparrow, L. G., and Inglis, A. S., 1980, Characterization of the cyanogen bromide peptides of component 7c, a major microfibrillar protein from wool, in: *Proceedings of the 6th International Wool Textile Research Conference,* Volume 2, South African Wool Research Institute, Pretoria, pp. 237–246.

Steinert, P. M., and Parry, D. A. D., 1985, Intermediate Filaments, *Annu. Rev. Cell Biol.* **1:**41–65.

Steinert, P. M., Parry, D. A. D., Racoosin, E. L., Idler, W. W., Steven, A. C., Trus, B. L., and Roop, D. R.,

1984, The complete cDNA and deduced amino acid sequence of a Type II mouse epidermal keratin of 60,000 Da: Analysis of sequence differences between Type I and Type II keratins, *Proc. Natl. Acad. Sci. USA* **81:**5709–5713.

Steinert, P. M., Steven, A. C., and Roop, D. R., 1985a, The molecular biology of intermediate filaments, *Cell* **42:**411–419.

Steinert, P. M., Parry, D. A. D., Idler, W. W., Johnson, L. D., Steven, A. C., and Roop, D. R., 1985b, Amino acid sequences of mouse and human epidermal Type II keratins of M_r 67,000 provide a systematic basis for the structural and functional diversity of the end domains of keratin intermediate filament subunits, *J. Biol. Chem.* **260:**7142–7149.

Swart, L. S., 1973, Homology in the amino acid sequences of the high sulphur proteins from wool, *Nature New Biol.* **243:**27–29.

Swart, L. S., and Haylett, T., 1971, Studies on the high sulphur proteins of reduced merino wool: Amino acid sequence of protein SCMKB-IIIB4, *Biochem. J.* **123:**201–210.

Swart, L. S., Jourbert, F. J., and Parris, D., 1976, Homology in the amino acid sequences of the high sulphur proteins from keratings, in: *Proceedings of the 5th International Wool Textile Research Conference,* Volume 2 (K. Ziegler, ed.), German Wool Research Institute, Aachen, pp. 254–263.

Swift, J. A., 1977, The histology of keratin fibres, in: *Chemistry of Natural Protein Fibres* (R. A. Asquith, ed.), Plenum Press, New York, pp. 81–146.

Swift, J. A., 1981, The hair surface, in: *Hair Research* (C. E. Orfanos, W. Montagna, and G. Stüttgen, eds.), Springer-Verlag, Berlin, pp. 65–72.

Tenenhouse, H. S., and Gold, R. J. M., 1976, Loss of a homologous group of proteins in a dominantly inherited ectodermal malformation, *Biochem. J.* **159:**149–160.

Tyner, A. L., Eichman, M. J., and Fuchs, E. V., 1985, The sequence of a Type II keratin gene expressed in human skin: Conservation of structure among all intermediate filament genes, *Proc. Natl. Acad. Sci. USA* **82:**4683–4687.

Ward, K. A., Sleigh, M. J., Powell, B. C., and Rogers, G. E., 1982, The isolation and analysis of the major wool keratin gene families, in: *Proceedings of the 2nd World Congress of Genetics Applied to Livestock Production,* Volume 6, pp. 146–156, Editorial Garsi, Londres 17, Madrid-28, Spain.

Weber, K., and Geisler, N., 1982, The structural relation between intermediate filament proteins in living cells and the α-keratins of sheep wool, *EMBO J.* **1:**1155–1160.

Whiteley, K. J., and Kaplin, I. J., 1977, The comparative arrangement of microfibrils in ortho-, meso- and paracortical cells of merino wool fibres, *J. Text. Inst.* **68:**384–386.

Woods, J. L., and Orwin, D. F. G., 1980, Studies on the surface layers of the wool cuticle, in: *Fibrous Proteins: Scientific, Industrial and Medical Aspects,* Volume 2 (D. A. D. Parry and L. K. Creamer, eds.), Academic Press, New York, pp. 141–149.

Ycas, M., 1972, De novo origin of periodic proteins, *J. Mol. Evol.* **2:**17–27.

11

Differentiation-Specific Expression of Keratin Pairs

W. Michael O'Guin, Alexander Schermer, Marion Lynch, and Tung-Tien Sun

1. Introduction

Keratins are by far the most complex and heterogeneous family of polypeptides that comprise a subclass of the intermediate filaments (IF). A survey of various human epithelia has demonstrated the existence of approximately 25 distinct keratin polypeptides that are expressed as subsets of two to ten polypeptides in any given epithelial cell or tissue. With the goal of understanding the biological significance of keratin heterogeneity, many investigators have studied keratins in detail. Consequently, a tremendous volume of information regarding keratins has been generated during the past 10–12 years. Because of this rapid accumulation of data, the field of keratin research has been a difficult one to follow for those not directly involved. This is due not only to the sheer amount but also to the diversity of the types of information. The keratin literature is filled with seemingly incongruous information derived from a combination of immunological, biochemical, biophysical, and clinical studies with each progressing at a rapid pace. However, general trends in the data have now allowed us to summarize most of the available information into a set of patterns or "rules" of keratin expression. These rules are: (1) Keratin expression is a characteristic of and is mostly restricted to epithelial cells (except a few epithelial tissues, e.g., lens and retinal epithelia, that seem to lack keratins) and their derivatives (Franke *et al.*, 1978, 1979; Sun and Green, 1978b; Sun *et al.*, 1979) (Figs. 1, 2). (2) All known keratin polypeptides, including the "hard" keratins of hair and nail, may be divided into either of two mutually exclusive subfamilies: they are the acidic (type

---

W. Michael O'Guin, Alexander Schermer, Marion Lynch, and Tung-Tien Sun • Epithelial Biology Unit, Departments of Dermatology and Pharmacology, The Kaplan Cancer Center, New York University School of Medicine, New York, New York 10016.

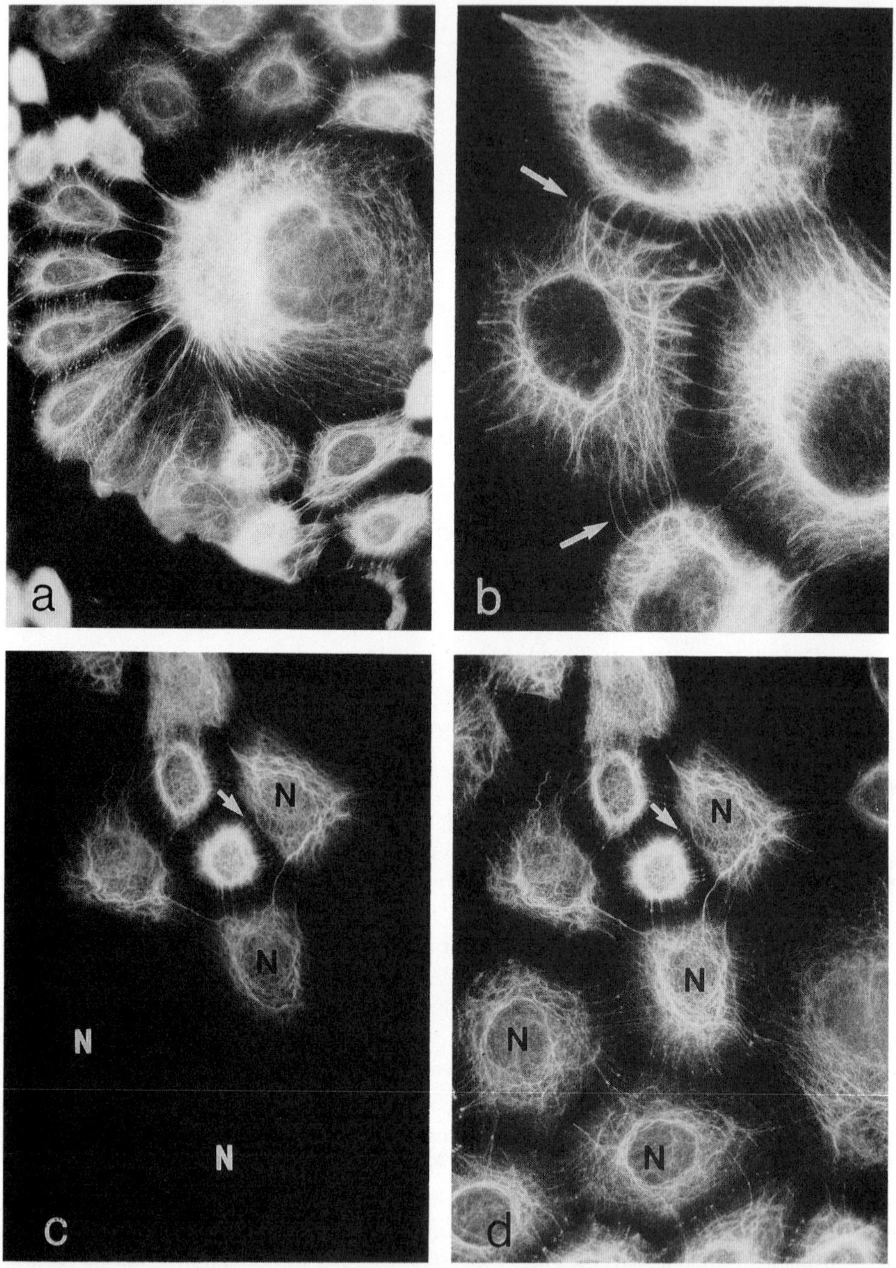

Figure 1. Indirect immunofluorescence staining of keratin intermediate filaments in cultured (A431) human carcinoma cells using (a) a polyclonal rabbit anti-human keratin antiserum and (b) a mouse monoclonal anti-keratin antibody (AE9) to demonstrate the typical pattern of keratin staining seen in epithelial cell cultures. Note the nonstaining nuclei and the extensive network of keratin filaments distributed throughout the cytoplasm. The arrows in (b) identify small discontinuities in the staining between the converging filaments of two apposed cells and represent desmosomal junctions. Double-label immunofluorescence of keratins (c and d) using (c) a monoclonal antibody (AE8) that identifies a subpopulation of A431 cells expressing a unique 51k (No. 13) keratin; (d) is the same field as (c) stained with a broadly reactive rabbit polyclonal anti-human keratin antiserum demonstrating that all of the cells express keratins. The arrows indicate small spots of intense keratin staining at the desmosomal junctions, which are sometimes seen in cultures that have been treated briefly with trypsin. N, nucleus. From O'Guin *et al.* (1985).

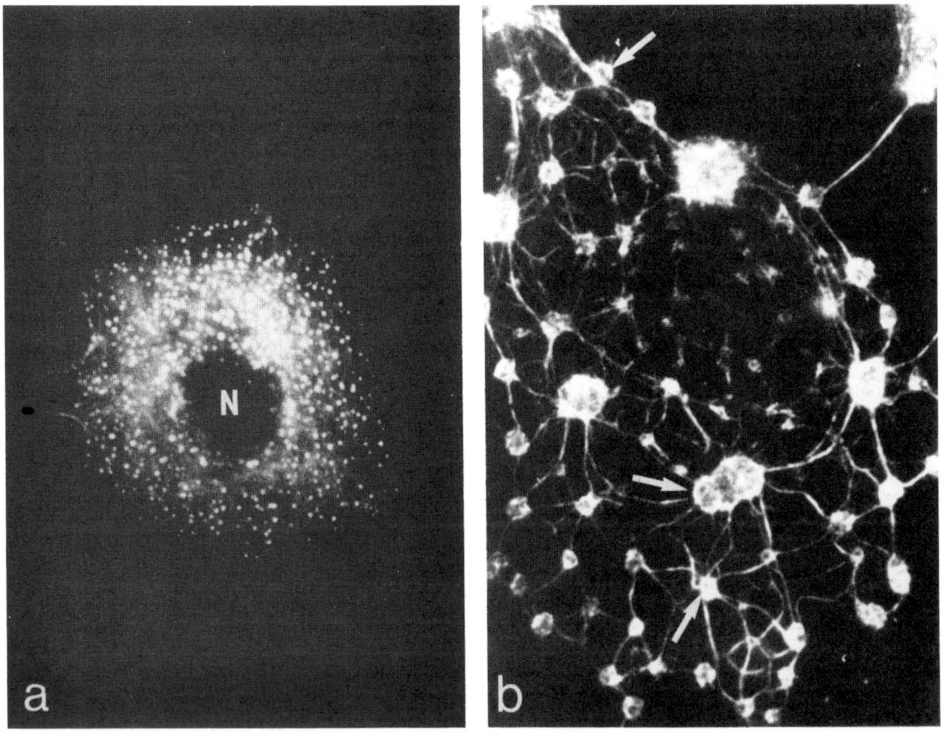

Figure 2. specific reorganizations in keratin cytoskeleton. (a) A mitotic A431 cell in which the keratins are organized into spheroidal aggregates prior to cytokinesis. This commonly occurs in epithelia and is thought to permit equal partitioning of the keratin polypeptides into the resultant daughter cells (Horwitz *et al.*, 1981; Lane *et al.*, 1982; Franke *et al.*, 1982). Indirect immunofluorescence using a monoclonal (CA7) antikeratin. N, nucleus. From O'Guin *et al.*, (1985). (b) A fetal mouse epidermal cell treated with a combination of cytochalasin D and colchicine and then stained with a polyclonal antikeratin demonstrating the effect of antimitotic drugs on cultured epithelia. Note the dramatic reorganization of keratin filaments into a lattice of filaments stabilized at membrane-associated focal centers (arrows) as originally described by Knapp *et al.* (1983a,b). These membrane association sites are thought to involve the redistribution of desmosomal proteins (Knapp *et al.*, 1985). From Knapp *et al.* (1983a).

I) and basic-to-neutral (type II) subfamilies (Eichner *et al.*, 1984; Fuchs *et al.*, 1981; Kim *et al.*, 1983; Lynch *et al.*, 1986; Schiller *et al.*, 1982; Sun *et al.*, 1984; Tseng *et al.*, 1982; reviewed in Cooper *et al.*, 1985; Fuchs *et al.*, 1985; Quinlan *et al.*, 1985; Steinert *et al.*, 1985b). (3) The molar ratio of acidic to basic keratins expressed in many epithelial tissues is approximately 1 : 1 (Kim *et al.*, 1984). (4) Keratin filament formation requires one component from each subfamily (Eichner *et al.*, 1986; Hatzfield and Franke, 1985; Lee and Baden, 1976; Steinert *et al.*, 1976). (5) Each acidic keratin is frequently coexpressed in a more or less coordinated fashion with a particular basic keratin to form what we call a "keratin pair" (Cooper *et al.*, 1985; O'Guin *et al.*, 1987; Sun *et al.*, 1984, 1985). (6) The basic member of a keratin pair is always larger than its acidic partner (frequently, but not necessarily, by a 7–11 kDa margin; Sun *et al.*, 1984). (7) The expression of individual keratin pairs can be correlated with specific programs of epithelial differentiation (Cooper *et al.*, 1985; Cooper and Sun, 1986; O'Guin *et al.*, 1987; Sun *et. al.*, 1984, 1985). (8) The pattern of keratin synthesis associated with a given program of differentiation is developmentally determined and is largely independent of site-specific instruction in normal adult tissues (Doran *et al.*, 1980). (9) The pattern of keratin synthesis is more directly linked to the phenotypic organization of an epithelium than it is to absolute tissue- or site-specificity (reviewed in O'Guin *et al.*, 1987). (10) The patterns of keratin expression found in embryonic, pathological (both benign and neoplastic), and cultured or otherwise experimentally manipulated epithelia are, in general, consistent with the rules deduced from normal tissues (Dale *et al.*, 1985; Eichner *et al.*, 1984; Moll *et al.*, 1983a, b; Weiss *et al.*, 1984). (11) The essence of the pair concept, as defined by coexpression, appears to be valid for many mammalian and nonmammalian species and will probably be found to be common to all keratin-producing species (O'Guin *et al.*, 1987).

These general rules, which obviously are not mutually independent of one another and do have some exceptions, not only provide a unifying concept, but also enhance our understanding of the biological meaning of keratin heterogeneity. It is with this unifying concept in mind that we shall present a summary of recent findings that help to establish many of these rules, and attempt to explain their overall ramifications toward our understanding of the differential expression of keratins by various epithelia.

2. *Keratin Subfamilies*

Through the examination of the keratin composition of an extensive collection of human epithelia, Franke and his co-workers (Franke *et al.*, 1981; Moll *et al.*, 1982a, b) have identified 18–19 distinct polypeptides that comprise the complement of keratins from essentially all soft epithelial tissues. These data were compiled by Moll *et al.* (1982a) into a catalogue of keratins with a standard, numerical system of nomenclature.

Knowledge of the existence of keratin subfamilies results from converging evidence from several laboratories using a diversity of techniques. Working at the gene level, Fuchs *et al.* (1981) and Kim *et al.* (1983) were able to show by positive hybrid-selection with two cDNA probes that the mRNAs coding for keratins 5, 6, 7, and 8 (in the nomenclature of Moll *et al.*, 1982a) were homologous to the type II keratin sequences originally described for wool keratin sequences (Crewther *et al.*, 1976, 1978, 1980; Gough *et al.*, 1978) and the mRNAs for keratins 14, 17, 18, and 19 were related to the type I wool keratin sequences (Fig. 3a). From a protein-chemistry perspective, Schiller *et al.* (1982) were able to demonstrate by two-dimensional peptide mapping that human keratins 1–8 were structurally related to one another, and that keratins 14–16 were distinct from 1–8 forming another subgroup (Fig. 3b). Using a monoclonal keratin antibody, K_G8.14, Gigi and colleagues were also able to demonstrate the immunological relatedness of keratins 5–8 (Gigi *et al.*, 1982) (Fig. 3c). Detailed immunological characterization was greatly facilitated by the AE1 and AE3 monoclonal antibodies, which showed that all of the generally larger and basic keratins (Nos. 1–8) share a common antigenic determinant recognized by AE3 and that most of the smaller, acidic polypeptides (Nos. 10, 13–15, and 19) share an antigenic determinant recognized by AE1 (Eichner *et al.*, 1984; Sun *et al.*, 1983, 1984; Tseng *et al.*, 1982; Woodcock-Mitchell *et al.*, 1982) (Figs. 3d, 4). Based on a compilation of all these data, we came to the conclusion that *all* known human keratins can be classified into two mutually exclusive subfamilies (Sun *et al.*, 1984). These subfamilies are generally known as the basic or type II subfamily comprised of the generally larger, neutral-to-basic keratins 1–8, and the acidic or type I subfamily com-

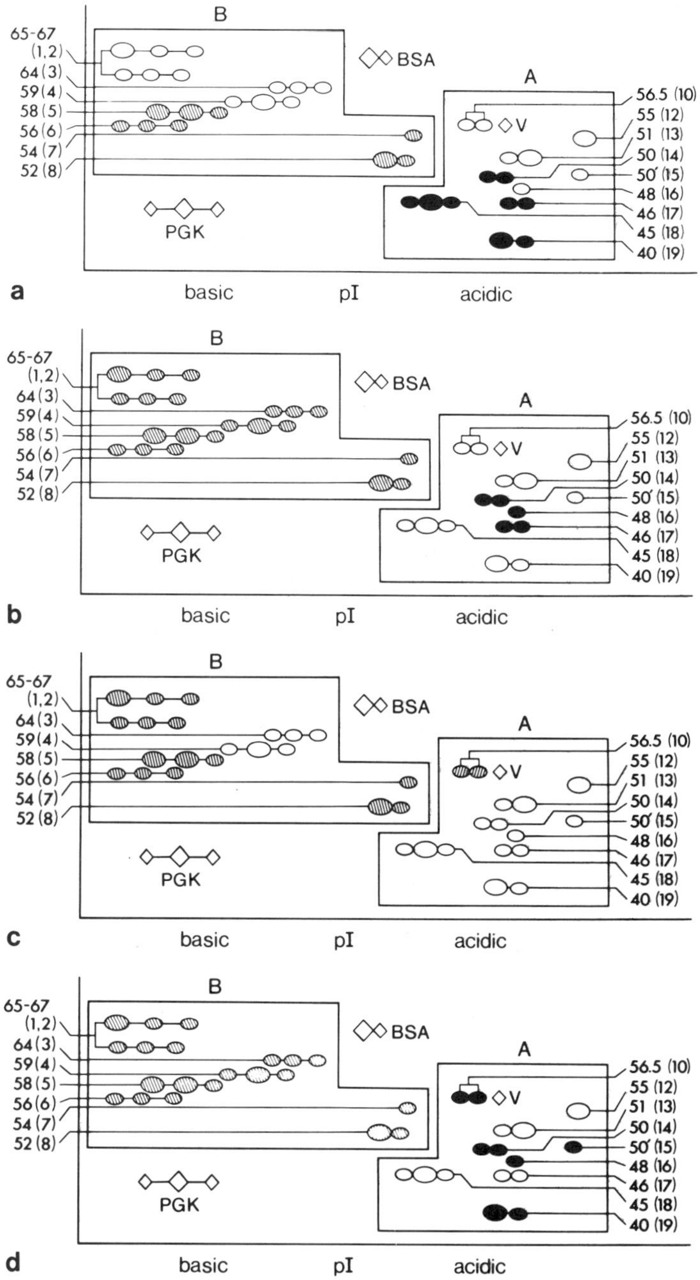

Figure 3. Evidence leading to the identification of two mutually exclusive subfamilies of keratin polypeptides. (a) Hybrid-selection data described by Fuchs *et al.* (1981) and Kim *et al.* (1983) showing relatedness of several acidic (solid ovals) and basic (hatched ovals) keratin mRNAs. Open ovals represent unclassified polypeptides. (b) A summary of peptide mapping data of Schiller *et al.* (1982) showing that all the basic keratins were chemically similar as were some acidic keratins. (c) Data obtained with the $K_G8.14$ monoclonal antibody by Gigi *et al.* (1982) demonstrating immunological relatedness of several basic keratin polypeptides (as well as an acidic 56.5k keratin). (d) The immunological characterization of acidic and basic keratins using the AE1 and AE3 monoclonal antibodies by Woodcock-Mitchell *et al.* (1982), Tseng *et al.* (1982), Sun *et al.* (1983, 1984), and Eichner *et al.* (1984). These data show that all basic keratins share a determinant identified by AE3 (hatched ovals), and most acidic keratins share an epitope recognized by AE1 (solid ovals).

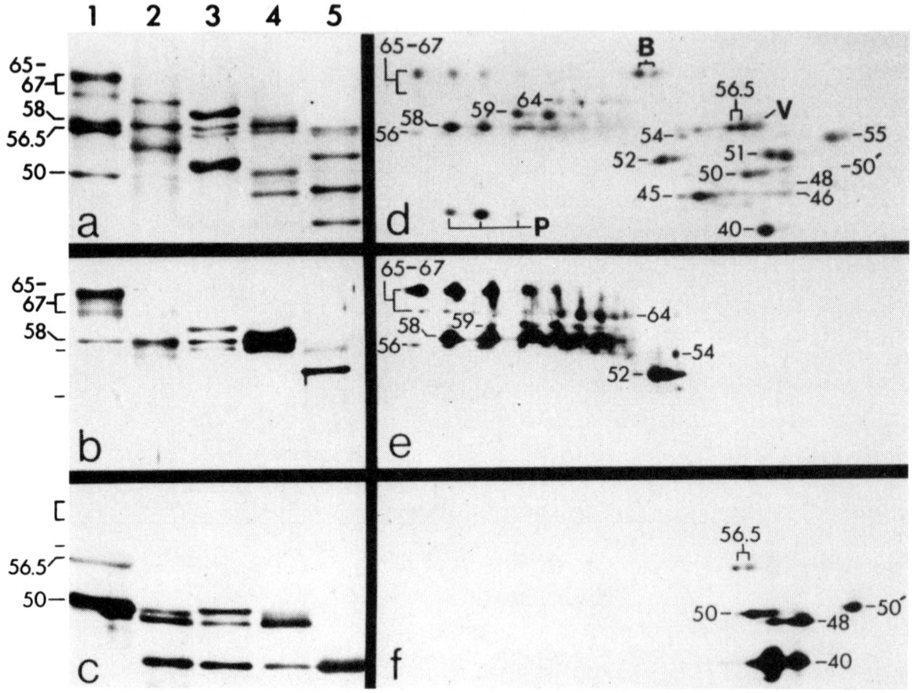

Figure 4. One- and two-dimensional immunoblotting of human keratins with AE1 and AE3 monoclonal anti-keratin antibodies. Keratins extracted from (1) abdominal epidermis, (2) corneal epithelium, (3) esophageal epithelium, (4) cultured epidermal cells, and (5) cultured mesothelial cells were analyzed individually by one-dimensional SDS–PAGE (a–c) or as a total mixture by two-dimensional, nonequilibrium-pH-gradient electrophoresis (d–f). Numbers to the left of SDS gels denote the molecular weights of the four major human epidermal keratins. Numbers on two-dimensional gels are the apparent molecular weights of all keratins. Charge heterogeneity of individual keratins is at least in part due to phosphorylation (Sun and Green, 1978a; Franke *et al.*, 1981; Gilmartin *et al.*, 1980, 1984). Note that AE1 and AE3 recognize two mutually exclusive subsets of keratins and in combination recognize most human keratins. In two-dimensional gels (d–f), basic keratins are on the left and acidic ones of the right. Thus, all AE3-reactive keratins are relatively basic (e) and all AE1-reactive keratins are relatively acidic (f). V denotes vimentin, the mesenchymal-type intermediate filament protein present in mesothelial cells (Wu *et al.*, 1982). B and P denote bovine serum albumin and phosphoglycerate kinase, respectively, which were added as standards. a and d, fast green; b and e, AE3 antibody stain; c and f, AE1 antibody stain. From Cooper *et al.* (1985).

prised of the generally smaller, acidic keratins 9–19. In addition, we and Franke's group have recently described five to eight polypeptides that are the constituents of the so-called "hard" keratins of human hair and nail (Heid *et al.*, 1986; Lynch *et al.*, 1986) (Figs. 5, 6). By a combination of charge properties on two-dimensional polyacrylamide gels and immunoreactivity with the AE1 and AE3 antibodies, this subfamily classification has been extended beyond human keratins to also include cow (Cooper and Sun, 1986; Schiller *et al.*, 1982), rabbit (O'Guin *et al.*, 1987; Schermer *et al.*, 1986), and chicken (O'Guin *et al.*, 1987) keratins (Fig. 7).

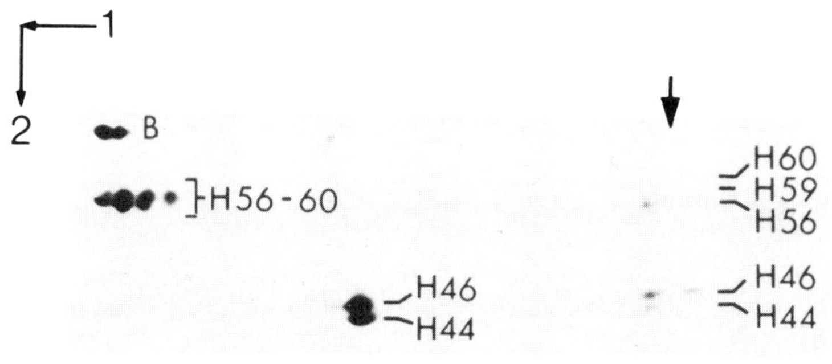

Figure 5. Two-dimensional (IEF/SDS–PAGE) analysis of human hair keratins. 1 and 2 denote the directions of the first dimensional isoelectric focusing and second dimensional SDS–PAGE, respectively. Basic proteins are to the left. For a comparison of the one- and two-dimensional gel patterns, the sample was also applied to a side lane (on the right marked by a downward arrow) during the second dimensional SDS–PAGE. B denotes bovine serum albumin marker. Note the resolution of hair keratins into two distinct groups labeled H56–60 (basic) and H44/46 (acidic) (the most basic H60 component is lost in this IEF gel). From Lynch *et al.* (1986).

3. *Keratin Pairs Defined by Frequent Coexpression*

Careful examination of the keratin composition of a large number of normal and abnormal human epithelia has led to the realization that most acidic human keratins demonstrate preferred coexpression with a particular member of the basic subfamily. Accordingly, the specific combinations of acidic and basic keratins exhibiting coordinate

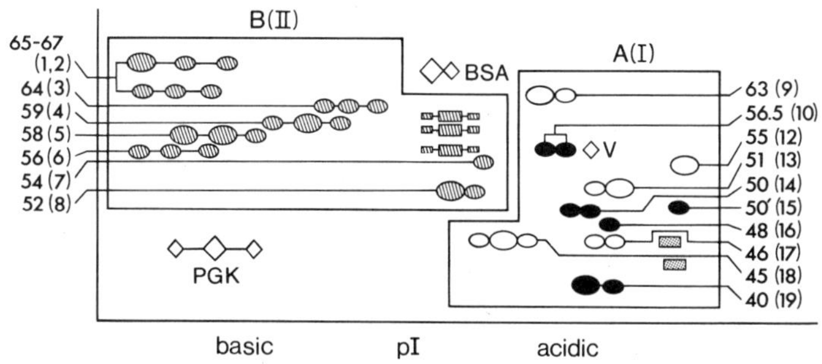

Figure 6. A schematic diagram of the two-dimensional electrophoretic pattern of human soft and hard keratins (a composite of IEF-SDS and NEpHGE-SDS gels). Keratins are identified by molecular weights ($\times\ 10^{-3}$) and catalogue numbers (in parentheses; Moll *et al.*, 1982a). They are divided into the acidic (type I) and basic (type II) subfamilies, according to their relative charges and immunoreactivities. Hard (hair/nail) keratins are represented by rectangles (H56, H59, H60 of the basic subfamily, and H44 and H46 of the acidic subfamily.) Note that all basic keratins are recognized by AE3 (hatched ovals and rectangles), while many but not all acidic keratins are recognized by AE1 (solid ovals). Abbreviations: A(I), acidic Type I keratins; B(II), basic Type II; V, vimentin; BSA, bovine serum albumin; PGK, 3-phosphoglycerate kinase.

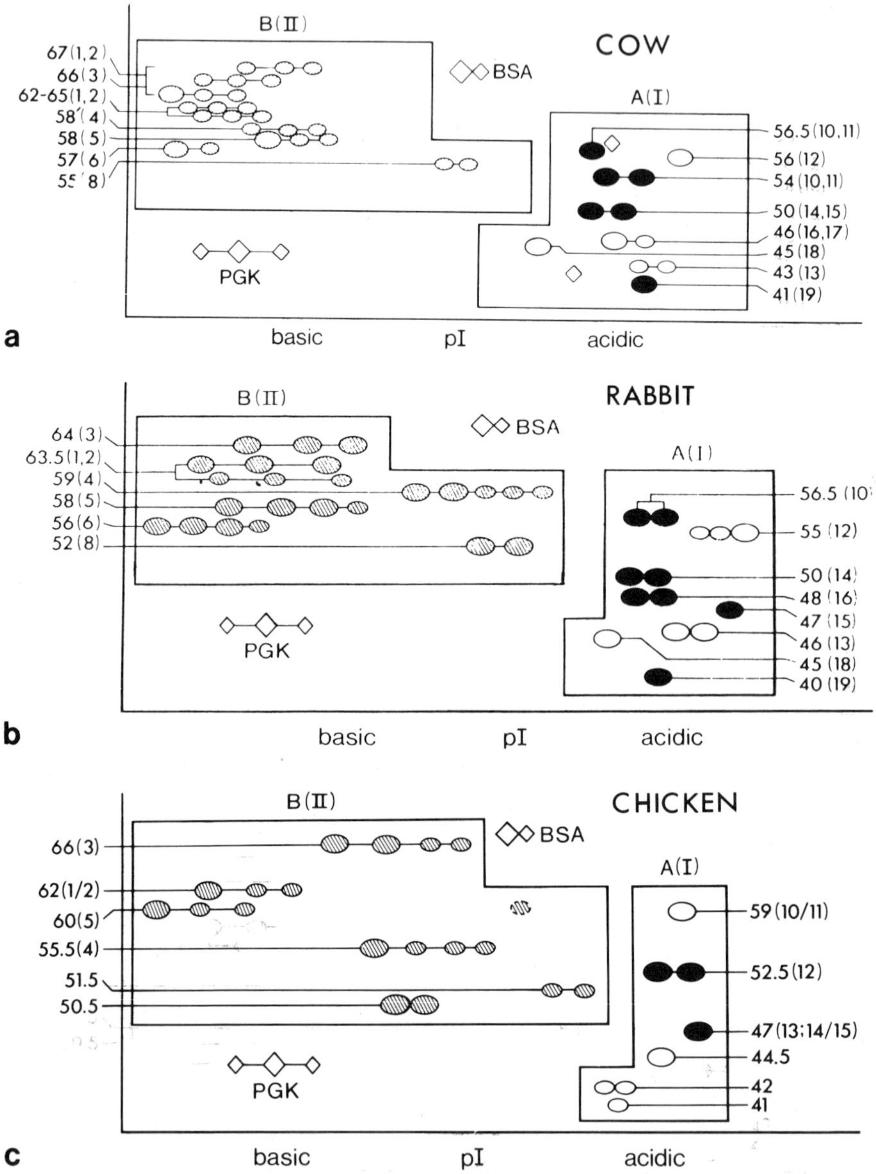

Figure 7. Schematic representation of the two-dimensional gel profiles of keratins from (a) cow, (b) rabbit, and (c) chicken. The approximate molecular weight ($\times\ 10^{-3}$) is given to the left of the catalogue number (Moll *et al.*, 1982a). Ovals that are attached to one another represent major, phosphorylated, isoelectric variants of the same polypeptide. Solid ovals are AE1-positive polypeptides, hatched ovals are AE3-positive polypeptides, and open ovals are polypeptides that do not react with either AE1 or AE3. Note that AE3 recognizes *all* species tested, pointing to the extreme conservation of this epitope among type II keratins. Boxes enclose the acidic (A, type I) and basic (B, type II) subfamilies of polypeptides. The relative locations of bovine serum albumin (BSA) and 3-phosphoglycerate kinase (PGK) are included for reference. From O'Guin *et al.* (1987).

Table 1. Keratin Markers for Different Types of Epithelial Differentiation

Markers	Catalog number	MW	Distribution
Skin-type differentiation[a]	1,2/10	65–67/56.5	Keratinized (cornified) stratified squamous epithelia
Corneal-type differentiation[a]	3/12	64/55	Corneal and snout epithelia
Esophageal-type differentiation[a]	4/13	59/51	All nonkeratinized (noncornified) stratified squamous epithelia except cornea
Palm/sole-type differentiation[a]	1,2/9, 10	65–67/63, 56.6	Thick, glabrous epithelium of palms and soles
Hair/nail-type differentiation[a]	—	H56,H59,H60/H44,H46	Hair shaft and nail plate
Keratinocytes[b]	5/14	58/50	All keratinocytes
Hyperproliferative keratinocytes[c]	6/16, 17	56/48, 46	Embryonic, cultured, and certain diseased keratinocytes (including carcinoma)
Simple epithelia	8/18	52/45	Most simple epithelia and their derivatives

[a]Represent differentiation-specific keratins.
[b]Usually expressed in combination with one or more of the differentiation-specific keratin pairs.
[c]Usually expressed in conjunction with the 5/14 pair and may co-exist within the same suprabasally located keratinocytes with differentiation-specific keratins.

regulation have been described as "keratin pairs" (Sun *et al.*, 1984). These pairs and their general distribution are summarized in Table I. The expression of these various pairs is a direct function of the program of epithelial differentiation shown by a given epithelium and the permissiveness of the environment. Consequently, in most cases the keratin composition can be correlated with the histological organization of the tissue. Thus, most typical simple (nonstratified) epithelia express a 52k/45k (8/18) pair of keratins that is present only as a minor component in stratified epithelia (Debus *et al.*, 1982; Moll *et al.*, 1982a; Ramaekers *et al.*, 1983; Tseng *et al.*, 1982). All stratified epithelia and cultures of their major cell type, the keratinocyte, express various quantities of a 58k/50k (5/14) pair of keratins that has never been found in simple epithelia; we, therefore, refer to this pair as the "keratinocyte marker" for stratified epithelia and related cells (Tseng *et al.*, 1982). Interestingly, pseudostratified epithelia generally demonstrate a combination of these two types of keratins (Moll *et al.*, 1982a; Tseng *et al.*, 1982). Stratified epithelia that exhibit an unusually high rate of turnover and keratinocyte cultures express a 56k/48k, 46k (6/16,17) pair of keratins, which we call a "hyperproliferation marker" (Weiss *et al.*, 1984).

The remaining pairs are expressed mainly by stratified epithelia in a differentiation-dependent manner and demonstrate a highly specific distribution (Franke *et al.*, 1981; Moll *et al.*, 1982a; Sun *et al.*, 1984). Most of these pairs have been found to be suprabasally expressed and are therefore associated with the more advanced stages of epithelial differentiation (thus the term "differentiation-specific keratins"). These differentiation-specific markers are the skin-type 65–67k/56.5k (1,2/10), corneal-type 64k/55k (3/12), esophageal-type 59k/51k (4/13), and palm/sole-type 65–67k/63k, 56.5k

(1,2/9,10) components. These markers are not, however, tissue-specific but are, rather, named for the tissue most frequently associated with their respective programs of differentiation. That is to say, the 1,2/10 pair is characteristic of almost all keratinized, stratified squamous epithelia, the 3/12 pair is mainly found in corneal epithelium, the 4/13 pair is characteristic of virtually all nonkeratinized stratified epithelia (other than cornea), and the expression of the 1,2/10 pair in combination with keratin No. 9 is associated with the thick, hairless skin of palm and sole.

In addition to the differentiation-specific pairs described above, the "hard" keratin polypeptides of human hair and nail have also been shown recently to represent a coexpressed "pair" characteristic of "hair/nail-type differentiation" (Lynch *et al.*, 1986). We have tentatively designated them by molecular weight ($\times 10^{-3}$) preceded by "H" ("hair"). The H56,H59, H60 triplet comprises the basic (type II) members and the H44 and H46 doublet represents the acidic (type I) partners (Fig. 5) (Lynch *et al.*, 1986; for the *in vitro* translation data on these keratins, see Heid *et al.*, 1986). These polypeptides are usually coexpressed as a group, making it difficult to distinguish individual pairs among them. Therefore, they may be considered (at least conceptually) as a single differentiation-specific pair expressed only in hair and nail matrix. Additional differentiation-specific keratin markers have been described for some specialized tissues of nonhuman species such as cow snout (Cooper and Sun, 1986) and avian scales and feathers (β-keratin; see O'Guin *et al.*, 1982; O'Guin and Sawyer, 1982; Sawyer *et al.*, 1984, 1986).

Members of an authentic keratin pair should, by definition, be expected to coexpress under a variety of conditions. Therefore, one must not assume that members of two distinct pairs that happen to *coexist* in a given cell under a particular growth condition are always coregulated. For example, although cultured keratinocytes express both the 5/14 pair and the 6/16 hyperproliferation marker simultaneously, this does not necessarily mean that keratins 6 and 14 represent a "pair" coregulated by a common mechanism.

There are several features that are common to all human keratin pairs. The basic (type II) member of most keratin pairs turns out to be larger than its acidic (type I) partner. Further, this molecular weight difference is frequently by a 7–11 kDa margin (Cooper *et al.*, 1985; O'Guin *et al.*, 1987; Sun *et al.*, 1984). Because of this somewhat constant difference, the various keratin pairs also represent a hierarchy of size-ranks with the largest basic keratin being paired with the largest acidic keratin, and so on (Fig. 8; Sun *et al.*, 1984). However, human hair/nail and palm/sole keratins are clearly exceptions to this pattern of molecular weight divergence within a coexpressed pair. Finally, the existence of keratin pairs is not restricted to humans. Keratin pairs have been defined in a range of mammalian (Cooper and Sun, 1986; Schermer *et al.*, 1986) and at least one nonmammalian (i.e, avian) species (O'Guin *et al.*, 1987); this has allowed for the identification of polypeptides equivalent to individual human keratins (Moll *et al.*, 1982a) in cow, rabbit, and chicken (Fig. 7; O'Guin *et al.*, 1987). Such a unified nomenclature greatly facilitates the discussion of data derived from various species without the confusion caused by multiple sets of nomenclature or variations in molecular weight among species. Although we will primarily restrict our discussion here to human keratins, the equivalent keratins in nonhuman species would have the same designation and exhibit comparable patterns of expression (O'Guin *et al.*, 1987).

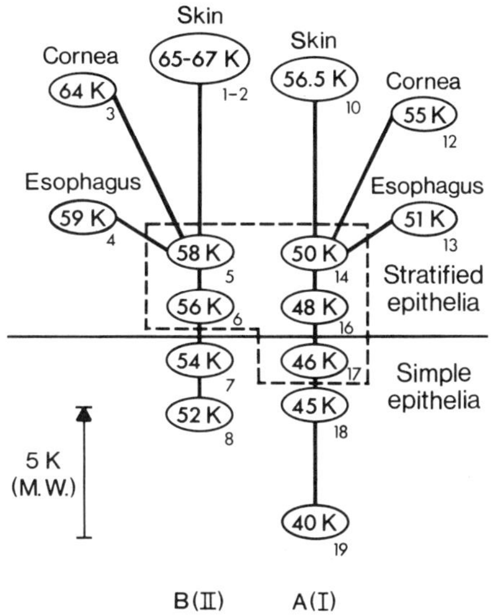

Figure 8. A unifying model of keratin expression. Human keratins of the acidic (type I) and basic (type II) subfamilies are arranged vertically according to their sizes (see the 5000-Da scale). They are also identified by their catalogue numbers (lower right corner; Moll *et al.*, 1982a). Keratins below the horizontal line are mainly expressed by simple epithelia (except the 46-K keratin, which is present in HeLa and hyperproliferative keratinocytes), and those above the line are unique to stratified epithelia. Dashed area encloses the keratins commonly expressed by all stratified epithelia in neoplasms, in hyperproliferative diseases, and in culture. After Sun *et al.* (1984).

4. *Patterns of Keratin Pair Expression*

Consistent patterns of keratin pair expression are established very early during embryogenesis and are closely coordinated with the progression of tissue organization (Banks-Schlegel, 1982; Dale *et al.*, 1976, 1985; Moll *et al.*, 1983b). During early development, most epithelia exist as a single cell layer. Accordingly, they also express only simple epithelial-type keratins. At a time immediately preceding stratification, certain embryonic epithelial cells begin to express the 5/14 keratinocyte markers. The expression of this keratin seems to be coordinated with the acquisition of the potential to stratify (Nelson and Sun, 1983), and marks the divergence of the presumptive stratified epithelial lineages from the various simple epithelial lineages, which usually show little or no additional change in their patterns of keratin expression (Sun *et al.*, 1985). As the stratified epithelial lineages proceed through development, they become progressively multilayered. It is not until stratification has occurred that developmental divergence among the stratified epithelia is marked by the additional expression of the differentiation-specific markers. It must be emphasized, however, that there is no evidence that the pattern of keratin expression directs the morphogenesis of the differentiating epithelium; rather, the keratins appear to be expressed phenotypially in response to mesenchymal induction resulting from a precise sequence of tissue interactions (see Dhouailly *et al.*, 1978; Sawyer *et al.*, 1984). Once the differentiation-specific keratins have been expressed, the overall pattern of keratin synthesis is frequently maintained for the remainder of embryogenesis and carried over into adult life. Therefore, most normal, adult, stratified epithelia stably express the 5/14 keratinocyte marker in combination with their respective differentiation-specific pair (e.g., epidermis expresses the 5/14 pair and the 1,2/10 skin-type differentiation pair).

When the cells from a fully differentiated stratified epithelium are dissociated and maintained *in vitro,* they frequently cease expressing, at least initially, their differentiation-specific keratins and continue or even increase (in terms of relative proportion) their synthesis of the 5/14 keratinocyte pair (Schermer *et al.*, 1986). In addition, a frequent response to tissue culture is the induction of significant amounts of the 6/16,17 hyperproliferation markers (Eichner *et al.*, 1984; Weiss *et al.*, 1984). Sun and Green (1977) compared the patterns of keratin expression by human skin, corneal, and conjunctival keratinocytes grown *in vitro* and found them to be very similar with regard to their keratin composition. Therefore, regardless of their tissue of origin, normal human keratinocytes tend to converge toward a common differentiation pathway when they are placed in cell culture. This presented the problem of establishing the relative importance of intrinsic and extrinsic factors involved in the regulation of keratin expression by adult keratinocytes. To approach this problem, Doran *et al.* (1980) placed keratinocytes cultured from rabbit skin, corneal and esophageal epithelia (Fig. 9) into the "neutral," *in vivo* environment of the athymic (nude) mouse subcutis. After several days, the cellular cysts were removed and found to represent organized epithelia that had reacquired the normal morphological features of their tissue of origin (Figs. 10, 11). In addition, when the keratins from these cysts were analyzed, these keratinocytes had resumed the synthesis of their characteristic differentiation-specific markers (Fig. 12). Thus, it appears that adult epithelia will express their preferred programs of keratin synthesis only when the cells are provided with a permissive environment. These data also demonstrate that, while the preferred programs of keratin synthesis in various epithelia are developmentally determined, in adult tissues the full execution of these programs does not require continued *site-specific* mesenchymal "instructions."

Two frequently cited examples of this kind of "site-specific" expression of keratin are cow snout "epidermis," which synthesizes a 65–67k keratin not found in normal trunk epidermis (Cooper and Sun, 1986; Drochmans *et al.*, 1978; Lee *et al.*, 1979; Steinert *et al.*, 1980), and human palm/sole "epidermis," which possess a 63k (No. 9) acidic keratin that is not detectable in normal trunk epidermis (Moll *et al.*, 1982a; Knapp *et al.*, 1986). If one assumes that snout and palm/sole "epidermis" are intrinsically (developmentally) identical to trunk epidermis, then this means "site-specific" regulation of adult keratin expression has indeed occurred. An equally likely but frequently overlooked possibility, however, is that these epithelia may have diverged from the epidermis during development, and that the 65–67k snout keratin and No. 9 palm/sole keratin simply represent molecular markers for these distinct differentiation pathways. To distinguish between these two possibilities, any proposed "site-specific" expression would have to be proven by conducting studies such as the culturing/transplantation (into a common permissive environment) experiment as performed by Doran *et al.* (1980) or the reciprocal grafting of epithelia to divergent sites.

In retrospect, the failure of many cultured keratinocytes to express their differentiation-specific keratins is perhaps not surprising since most standard tissue culture conditions are designed to promote growth rather than differentiation. Consequently, the keratin pairs associated with the more mature cells in the later stages of differentiation may not be synthesized.

While it is evident that most adult epithelia are committed to a particular pathway of differentiation and will proceed through that program under normal circumstances, it is

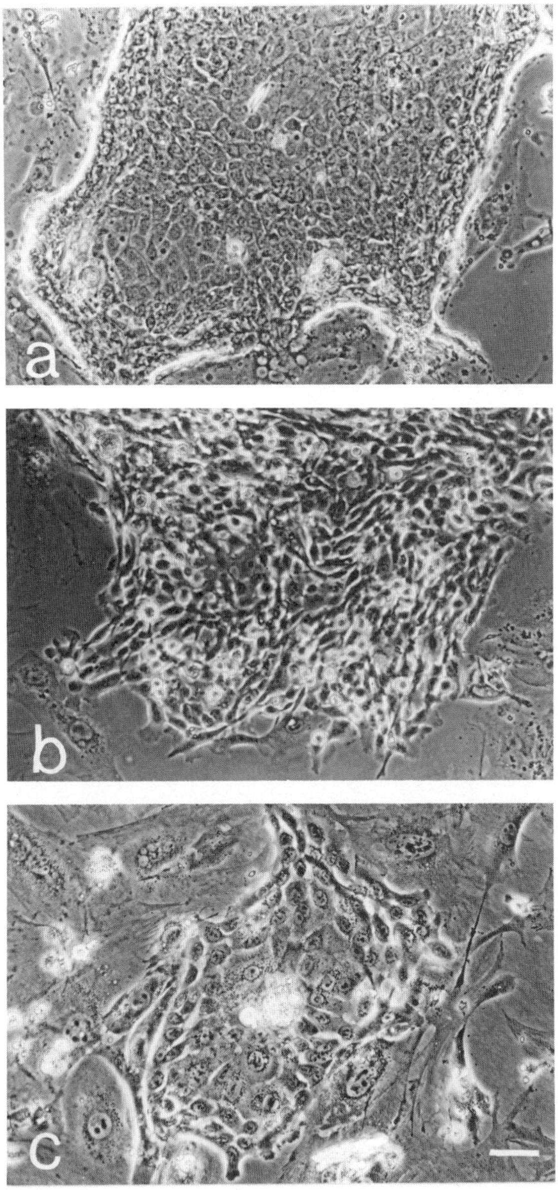

Figure 9. Cultured rabbit keratinocytes from 6- to 9-day-old primary cultures photographed under phase-contrast. (a) An epidermal keratinocyte colony. Cells surrounding the epidermal colony are lethally irradiated 3T3 feeder fibroblasts (Rheinwald and Green, 1975). (b) A corneal keratinocyte colony. Note that the cells are more refractile and "fibroblastic" than epidermal and esophageal cells. (c) An esophageal keratinocyte colony. Under these exponential growth conditions, all three types of keratinocyte express very similar keratins. From Doran *et al.* (1980).

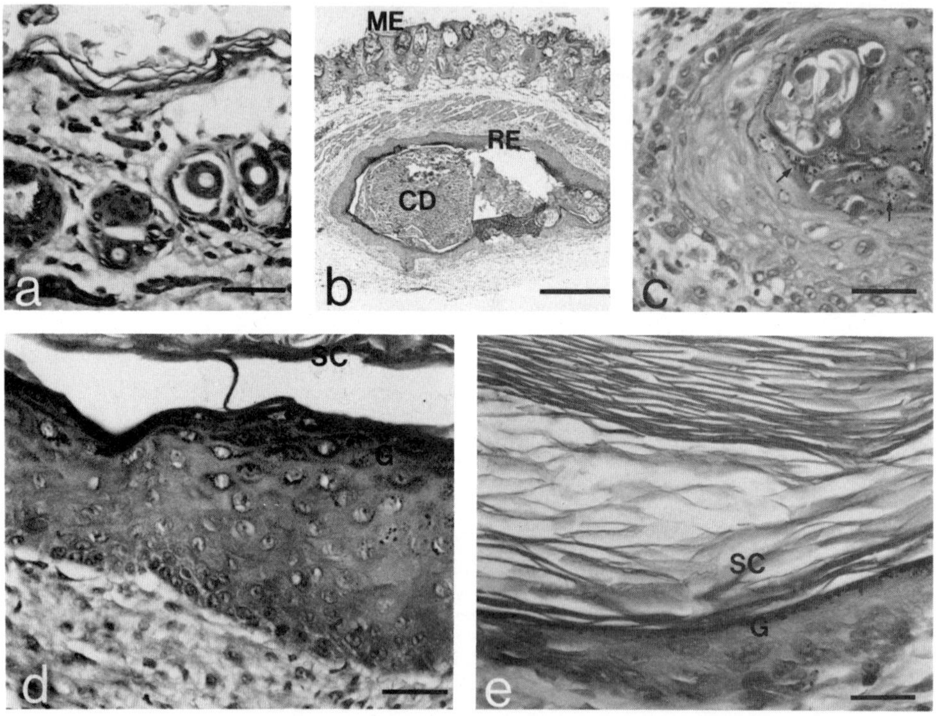

Figure 10. Differentiation of cultured rabbit keratinocytes after subcutaneous injection into athymic (nude) mice. (a) Rabbit trunk skin, *in vivo*. The epidermis is extremely thin, a common feature of furred animals. (b) A low-magnification survey micrograph of a 5-day cyst formed by cultured rabbit epidermal cells. Note the viable epithelium lining the cystic cavity and the cell debris in the center. ME, mouse epidermis; RE, rabbit (cystic) epidermis; CD, cell debris. Bar = 500 μm. (c) An epithelial cyst formed by cultured rabbit epidermal cells 2 days after injection. Note the newly formed keratohyalin granules (arrows). (d) A cyst 3 days after injection. Note the well-organized cell layers and the newly formed keratohyalin granules (G) and enucleated stratum corneum (SC; partial separation of the stratum corneum may have occurred during processing). (e) Cyst 34 days after injection. The main histological features [keratohyalin granules (G), stratum corneum (SC)] are similar to those of the day 3 sample. However, note the extremely thick stratum corneum layers and the relatively thin living cellular layers. From Doran *et al.* (1980).

not true that this is the only possible contingency for these cells. Under various pathological and experimental conditions, adult epithelia may be diverted into a differentiation program that elicits the synthesis of keratins not normally found in those tissues. For example, Tseng *et al.* (1984) provided biochemical evidence that during vitamin A deficiency, corneal, conjunctival, and esophageal epithelia can acquire the 1,2/10 pair. Moreover, immunofluorescent staining data are consistent with the idea that in keratinized corneal epithelium the 1,2/10 pair is expressed suprabasally, most likely coexisting with the previously synthesized 3/11 corneal pair. Weiss *et al.* (1984) have shown that various skin diseases such as psoriasis and actinic keratoses which are associated with an abnormally high rate of cell turnover (i.e., hyperproliferative disorders) present an abnormal pattern of keratin expression. While they continue to express the 5/14 pair along with decreased amounts of the 1,2/10 pair of normal skin, these diseased epithelia also elaborate significant amounts of the 6/16 pair, which is characteristic of hyperproliferative

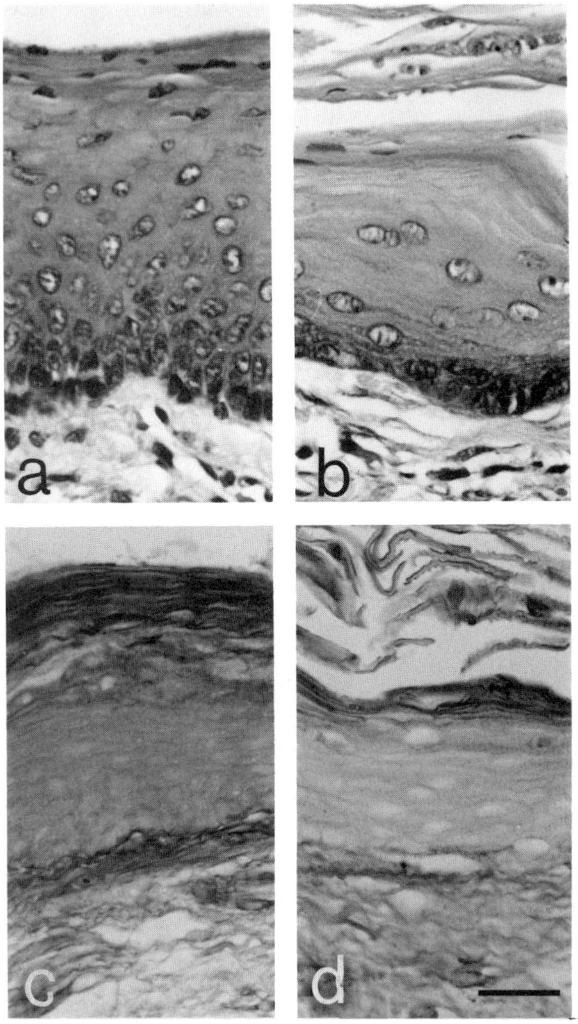

Figure 11. Differentiation of cultured rabbit esophageal keratinocytes after subcutaneous injection into athymic (nude) mice. (a) Esophageal epithelium, *in vivo*. Hematoxylin and eosin (H & E)-stained. (b) Cultured esophageal epithelial cells in an athymic mouse (day 7). Note the formation of parakeratotic (nucleated) stratum corneum and the absence of keratohyalin granules. H & E-stained. (c) Esophageal epithelium, *in vivo*. PAS-stained. (d) Cultured esophageal epithelial cyst. PAS-stained. Note the PAS-staining of the superficial cells. Bar = 25 μm. From Doran *et al.* (1980).

keratinocytes and is usually not produced in normal epidermis (see also McGuire *et al.*, 1984, and Moll *et al.*, 1984). In these tissues there appears to be a reciprocal relationship between the levels of expression of the 1,2/10 pair and the 6/16,17 pair (Weiss *et al.*, 1984). Nonhyperproliferative diseases of the skin did not express the 6/16,17 pair (Weiss *et al.*, 1984).

A large volume of the keratin literature is devoted to patterns of keratin synthesis in various types of epithelial neoplasms (for reviews see Cooper *et al.*, 1985; Moll *et al.*,

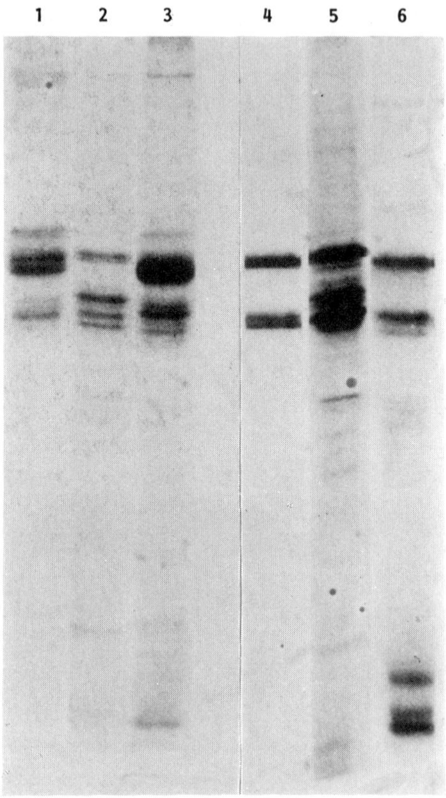

Figure 12. Intrinsic versus extrinsic regulation of keratin expression. The normal keratin composition of adult rabbit epidermis and esophagus *in vivo* (lanes 1 and 4, respectively) is not maintained when these epithelia are dissociated and grown *in vitro* (lanes 2 and 5). However, when these cultured keratinocytes are injected subcutaneously into nude (athymic) mice, both the epidermal (lane 3) and esophageal cells (lane 6) resume their normal, intrinsically regulated patterns of keratin synthesis under this permissive environment. After Doran *et al.* (1980).

1983a; Nelson *et al.*, 1984; Osborn and Weber, 1983). In general, almost all tumors (carcinomas) derived from keratin-positive epithelia, whether of primary or metastatic origin, continue to express keratins but not necessarily in a tumor-type specific manner. For example, some squamous cell carcinomas will continue to express their differentiation-specific keratins, some produce only the keratinocyte and "hyperproliferative keratinocyte" markers, and a very limited number express only simple epithelial keratins despite their presumed stratified-epithelial origin (Moll *et al.*, 1983a; Nelson *et al.*, 1984). This variation in keratin synthesis is probably linked to the relative degree of differentiation found within the individual tumors ranging from the most differentiated ones to the poorly differentiated ones, respectively. Therefore, keratins are valuable tools for pathologists to differentiate carcinomas from other classes of neoplasm and for the identification of many classes of carcinomas, but are of somewhat limited application in the fine-typing of subtle carcinoma variants.

When high levels of retinoic acid are applied to cultured epidermal keratinocytes, the 1,2/10 skin-type differentiation markers are almost totally inhibited while there is a substantial (25- to 40-fold) induction of mRNA for keratins 13 and 19 (simple epithelial-type keratins; Wu *et al.*, 1982; Eckert and Green, 1984). However, these retinoid treatments have many physiological effects on the cells not directly related to keratin expression. Therefore, it is not clear whether the selective inhibition/induction of these keratins is due to direct transcriptional regulation by the retinoids or to some less specific (in terms of keratin expression) physiological effects. Conversely, in cases of severe retinoid deficiency, *in vivo,* both corneal and esophageal epithelia, which are normally nonkeratinized, exhibit patches of frank keratinization (Tseng *et al.*, 1984; for *in vitro* data, see Fuchs and Green, 1981) and produce the 1,2/10 skin-type differentiation pair.

Another important consideration relates to the stability of the program of keratin expression. *In vitro* transformation of human epidermal keratinocytes by SV-40 has shown that it is possible for keratinocytes to lose the potential to express many of the keratins associated with their intrinsic program of differentiation (Hronis *et al.*, 1984). After stable transformation, these T-antigen-positive cells gradually lose the characteristic patterns of keratin synthesis for cultured keratinocytes and express mainly simple epithelial keratins (Fig. 13). This change in keratin expression is accompanied by behavioral changes in the

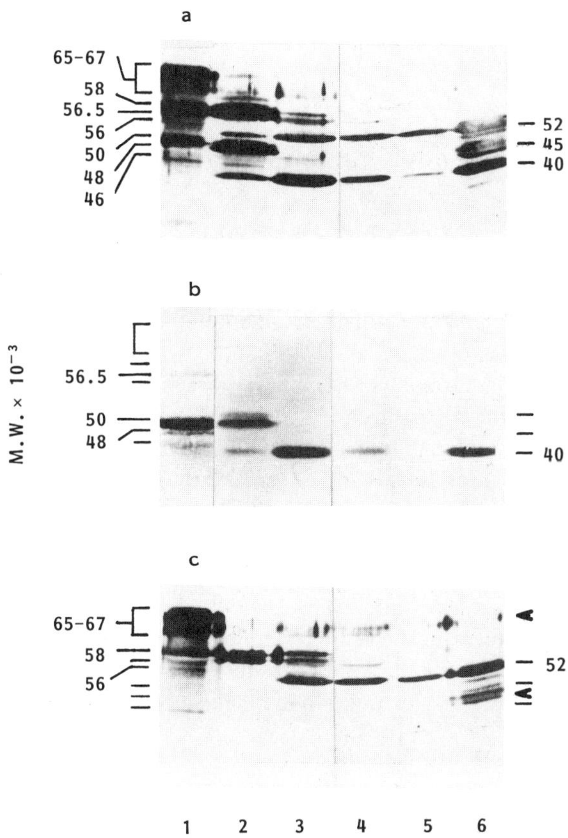

Figure 13. Immunoblot analysis of the cytoskeletal proteins of SV40-transformed human epidermal keratinocytes. (a) AE1 plus AE3 antikeratin antibodies; (b) AE1; (c) AE3. Samples are: lane 1, normal human epidermis; lane 2, cultured normal human epidermal (HE) cells; lane 3, SV40-HE, postcrisis; lane 4, HeLa; lane 5, CRL-1470 pancreatic carcinoma cells; and lane 6, MCF-7 mammary carcinoma cells. In (c) the upper arrowhead on the right denotes a M_r 66,000, AE3-reactive component that is sometimes detected even in lanes without samples; it probably represents human stratum corneum keratin contamination (Ochs, 1983). The lower arrowhead denotes a M_r 45,000, AE3-positive band of MCF-7 cells (lane 6); whether this is a mammary-specific keratin or a degradative product is not known. Note the change in keratin expression resulting from viral transformation. From Hronis *et al.* (1984).

cells. The cells lose the ability to stratify and form cornified envelope precursor molecules (both are normal differentiation functions for cultured skin keratinocytes). The pattern of keratin expression is, therefore, consistent with the phenotypic characteristics of the cells. These findings are unique among studies describing altered patterns of keratin synthesis in that these data describe a gradual, but total and presumably irreversible, change of keratin expression to a pattern that is highly unusual for the parental cell type (Hronis *et al.*, 1984; for hepatocyte data see Venetianer *et al.*, 1983). The most important conclusion to be made from these data is that while patterns of keratin synthesis are generally very stable and highly predictable, they are not irrevocable.

What is also made clear by these data is that the pattern of keratin synthesis in most cases is consistent with the histological organization found in the tissue rather than cell-type or regional specificity (Tseng *et al.*, 1982). While the normal pathway of keratin synthesis is under intrinsic regulation (Doran *et al.*, 1980), this only represents the preferred course of differentiation and does not restrict the potential of epithelia to express other keratins under unusual circumstances.

5. Localization of Various Keratin Pairs

The expression data have described the largely predictable patterns of distribution of the various keratin pairs. The next question is: Where are the individual keratin polypeptides localized within a stratified epithelium? Answers to this question can yield important clues to the functional significance of individual keratins.

Initial attempts at the biochemical identification of polypeptides from different layers of the epidermis relied on horizontal sectioning of tissue and analyzing, by SDS–PAGE, the proteins of each section (Fuchs and Green, 1980). These experiments provided the first biochemical evidence that epidermal differentiation is accompanied by a significant change in keratin composition (Fuchs and Green, 1980). As far as the detailed localization of individual keratins is concerned, however, this approach provides only limited resolution because the undulating profile of epidermal basal layer precludes the isolation of a pure population of basal cells without contamination either by cells from the spinous layer or by underlying ducts and hair follicles. This sort of contamination is probably responsible for the initial assignment of the 6/16 keratin pair to the epidermal basal layer (Fuchs and Green, 1980). These keratins were subsequently found not to be a component of normal epidermis but are, in fact, found in the outer root sheath of hair follicles (Moll *et al.*, 1982b).

To identify the keratins in basal cells, some investigators have turned to analyzing keratins expressed by keratinocytes maintained *in vitro*. When stratified epithelial cells are placed into culture, they cease expressing their differentiation-specific keratins and undergo a period of rapid proliferation. Accordingly, regardless of the tissue of origin, they express mainly the 5/14 keratinocyte markers and the 6/16, 17 hyperproliferation markers (Doran *et al.*, 1980; Sun and Green, 1977, 1978a). The fact that epidermal keratinocyte cultures do not express their respective differentiation-specific keratins has led to the (erroneous) suggestion that the entire culture represents a population of cells equivalent to basal cells found in normal epidermis (Kim *et al.*, 1984). However, cultured keratinocytes express significant amounts of the 6/16, 17 pair, which is not expressed in normal adult epidermis. It is clear that while these cultured cells may resemble basal cells in several aspects, they are distinct from any normal population of cells found *in vivo*. Further, skin keratinocytes tend to stratify relatively rapidly even in small focal centers before the culture becomes confluent. These foci contain cells in various stages of differentiation. Attempts to prevent stratification (e.g., low calcium) do not actually block the progression of some of their differentiation program such as the acquisition of involucrin (Hennings *et al.*, 1980; Watt and Green, 1982). Therefore, the extrapolation of the characteristics of entire cultures of keratinocytes to true basal cells *in vivo* is at best tenuous.

A more sensitive method of keratin localization involved the immunohistochemical localization of the various polypeptides using specific keratin antibodies. This approach provides the highest degree of resolution of all methods of keratin localization. However, the development of antibodies highly specific for a small number of individual keratins can be tedious due to the close immunological relatedness of the various keratin polypeptides, although synthetic peptides have been useful in producing specific antibodies to keratins for which the nucleic acid sequence has been determined (Roop *et al.*, 1984). Another drawback of this approach is that while a positive immunohistochemical reaction proves the presence of at least one antigen defined by the antibody, negative reactivity is essentially uninterpretable in the absence of additional evidence due to the possibility of antigenic masking (Franke *et al.*, 1983b; Woodcock-Mitchell *et al.*, 1982). For these reasons, immunolocalization data should, whenever possible, be supplemented with detailed immunochemical characterization of an antibody with regard to the specific tissue under consideration as well as cell separation studies to eliminate the possibility of antigenic masking.

Using just such an approach, Woodcock-Mitchell *et al.* (1982) deduced from a combination of immunohistochemical staining and horizontal sectioning data that the 5/14 (58k/50k) keratins are the only two IF subunits of the normal epidermal basal layer, whereas the 1,2/10 (65–67k/56.5k) keratins can be localized mainly in suprabasal layers (Fig. 14). This interpretation was strengthened by the tissue-distribution data from the same group showing that the 5/14 keratins were present in various quantities in almost all stratified epithelia, while the 1,2/10 keratins were largely epidermis-specific (Tseng *et al.*, 1982; see also Moll *et al.*, 1982b). Independently, Skerrow and Skerrow (1983) came to a similar conclusion by cell fractionation that human epidermal basal cells contain primarily the 5/14 keratins, while suprabasal cells contain additional No. 1 keratin (in this study keratin 10 was not detected due to the insufficient resolution of one-dimensional SDS gels). These conclusions were later confirmed by several investigators using mouse epidermis as the experimental system (Breitkreutz *et al.*, 1984; Schweizer *et al.*, 1984). These results help to establish the concept that the 1,2/10 keratins are associated with an advanced stage of epidermal differentiation (keratinization).

In order to localize the other two high-molecular-weight, differentiation-related keratin pairs, i.e., the 3/12 corneal keratins and the 4/13 esophageal keratins, Schermer *et al.* (1986) have combined cell fractionation techniques with several newly developed monoclonal antibodies that are highly specific for individual keratins (i.e., AE5 and AE8 antibodies, which are specific for keratins 3 and 13, respectively). These antibodies were obtained by immunizing mice with corneal and esophageal keratins, and screening the hybridoma supernatants against both the original antigen and epidermal keratins. Most antibodies were found to be cross-reactive and were discarded. The few antibodies that reacted specifically or preferentially with corneal or esophageal keratins were cloned and used for these studies (for the specificity of AE5, see Fig. 15).

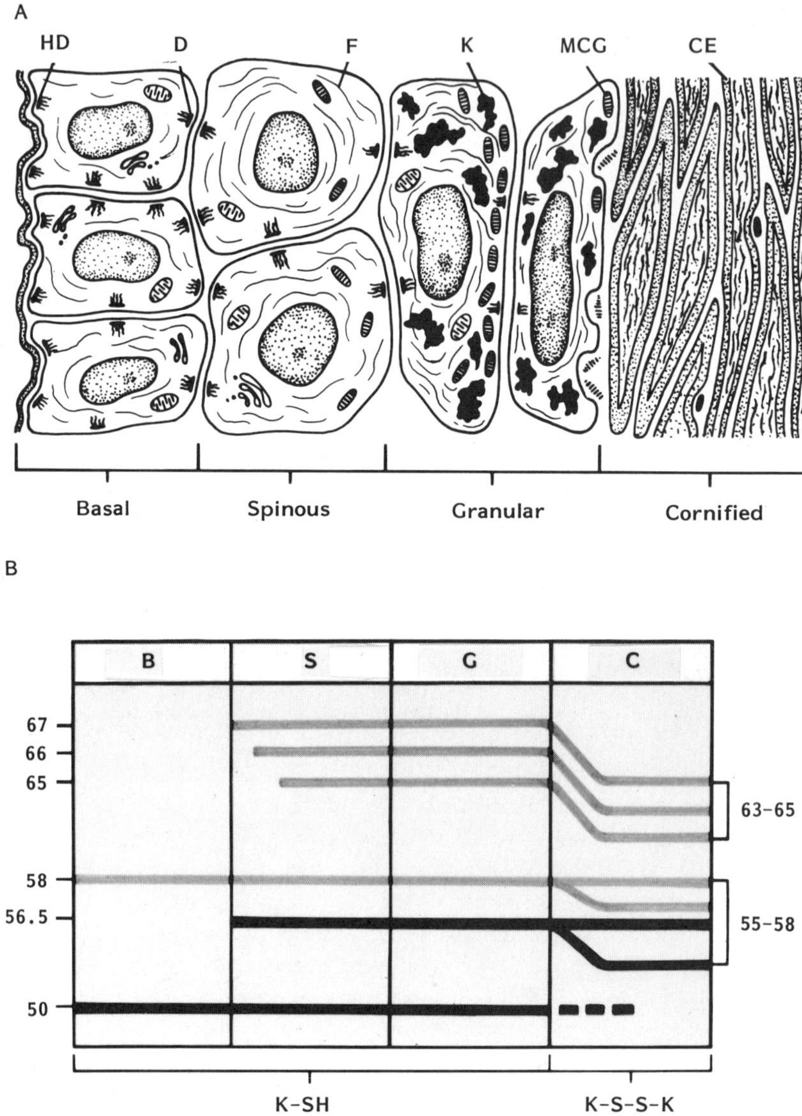

Figure 14. Sequential and coordinate expression of the four major epidermal keratins accompanying morphological differentiation of normal human epidermis. (A) Morphology of the epidermis, which can be divided into four layers: basal (B), spinous (S), granular (G), and cornified (C). Note the presence of hemidesmosomes (HD), desmosomes (D), tonofilaments (F), keratohyalin granules (K), membrane-coating granules (or lamellar granules; MCG), and cornified envelope (CE). (B) A schematic summary of keratin localization data by Woodcock-Mitchell *et al.* (1982) showing that basal cells (B) express mainly an acidic 50k and a basic 58k keratin, while suprabasal cells (S, G, and C) contain, in addition, an acidic 56.5k (No. 10) and a basic 65–67k (No. 1,2) keratin. In normal cornified cells (not callus), most of these keratins are either completely (e.g., the 50k component) or partially (all of the other keratins) degraded, resulting in two diffuse, ill-defined, major bands (one 55–58k, another 63–65k). Keratins of the living cellular layers (B, S, and G) are not cross-linked by intermolecular disulfide bonds (K-SH) and are therefore extractable with urea or SDS alone; whereas keratins in the nonviable cornified layer are known to be cross-linked extensively by intermolecular disulfide bonds (K-S-S-K) and are therefore extractable only after reduction (Sun and Green, 1978a). From Sun *et al.* (1985); adapted from Woodcock-Mitchell *et al.* (1982).

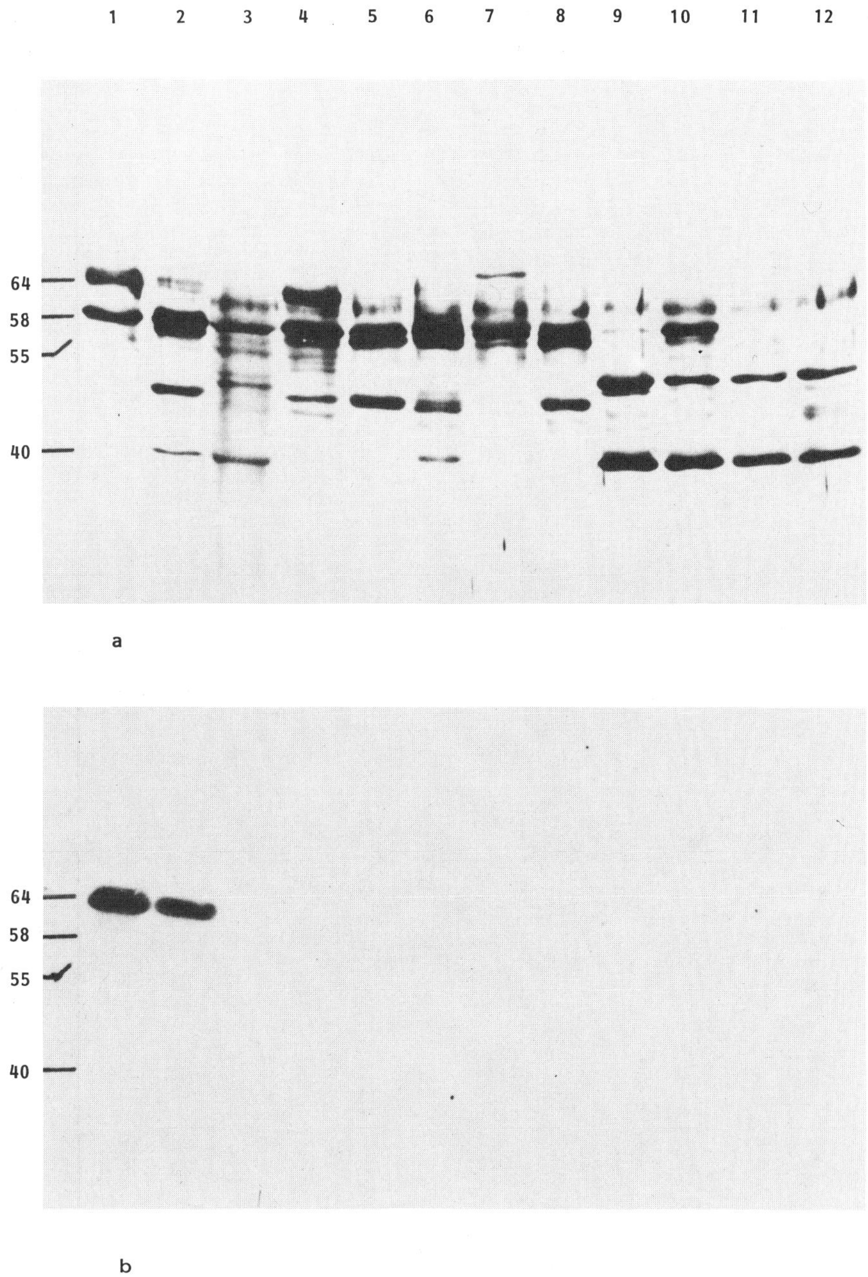

Figure 15. The specificity of AE5 antibody as shown by one-dimensional immunoblot analyses of rabbit epithelial keratins using: (a) a mixture of AE1 and AE3 antibodies; these two antibodies, in combination, allow the detection of most mammalian epithelial keratins (Tseng *et al.*, 1982; Sun *et al.*, 1984). (b) AE5. Rabbit keratins are isolated from: 1, corneal epithelium *in vivo*; 2, cultured corneal epithelial cells; 3, conjunctival epithelium; 4, epidermis; 5, cultured epidermal cells; 6, cultured human epidermal cells included as a reference; 7, esophageal epithelium; 8, cultured esophageal epithelia cells; 9, bladder epithelium; 10, tracheal epithelium; 11, mesothelium; and 12, intestinal epithelium. Note the strong reaction of AE5 with a 64k (No. 3) corneal keratin. From Schermer *et al.* (1986).

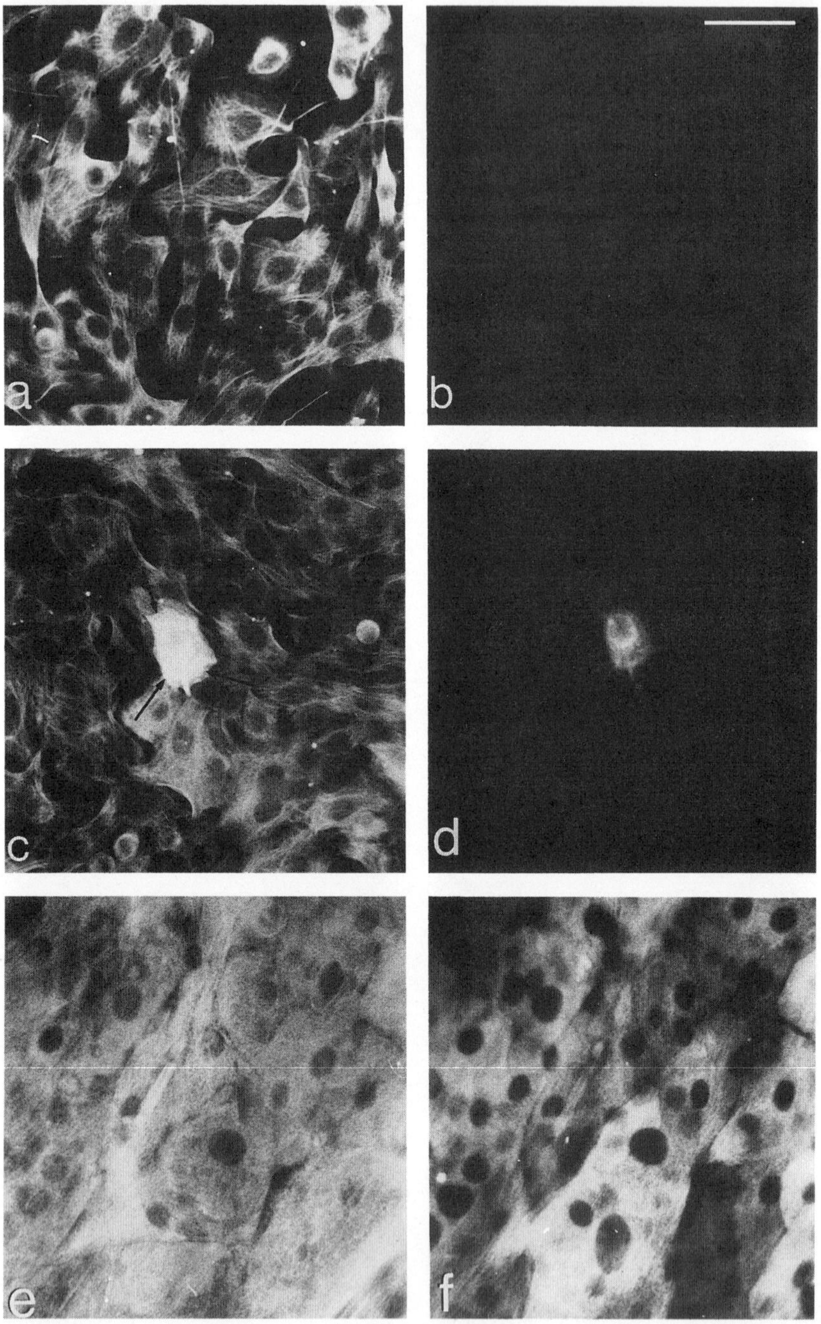

Figure 16. Double immunofluorescence staining of cultured rabbit corneal keratinocytes during successive stages of growth with a broadly reactive rabbit antikeratin antiserum (a, c, e) and the 64k-keratin-specific mouse AE5 antibody (b, d, f). Cell cultures were fixed on day 5 (a, b), 8 (c, d), and 11 (e, f) and processed for double immunofluorescent staining. Note that the rabbit antikeratin antiserum stains almost 100% of the cells at all stages of growth (a, c, e), whereas AE5 stains only suprabasal cells (d, f). The arrows in (c) denote an AE5-positive, suprabasal cell (cf. panel d). Panels (a), (c), and (e) are identical fields as (b), (d), and (f), respectively. Bar = 50 μm. From Schermer *et al.* (1986).

Schermer *et al.* (1986 and unpublished data) found that postconfluent, heavily stratified cultures of rabbit corneal and esophageal epithelial cells synthesize significant quantities of 3/12 and 4/13 keratin pairs, respectively. In such cultures, AE5 and AE8 stain mainly suprabasal cells (for AE5 data see Fig. 16). To rule out the possibility of antigenic masking, these investigators treated the cultures with EGTA for 5 hr and were then able to remove all of the upper cell layers, leaving behind a pure population of basal cells attached to the culture dish. Biochemical analysis confirmed that these basal cells synthesize only the 5/14 keratinocyte pair, while suprabasal cells contain the corneal- or esophageal-type differentiation markers. In conjunction with our earlier epidermal data (Woodcock-Mitchell *et al.*, 1982), these findings suggest that, as a general rule, the differentiation-specific keratin pairs are associated with the relatively mature cells of a stratified epithelium.

With this concept in mind, we were intrigued by the finding that basal cells of adult central corneal epithelium possess keratin No. 3 (Fig. 17a). A corresponding finding is that in older, stratified cultures, the basal cells eventually begin to express this keratin as well. Interestingly, in the limbus, which is the transitional zone between cornea and conjunctiva, the No. 3 (64k) cornea marker is suprabasally expressed (Fig. 17b). This has provided support for the theory that the corneal stem cells might be located in the basal layer of the limbus, and the central corneal epithelium represents the more mature cells that have migrated centripedally from the peripheral limbus (Fig. 18).

The localization of the "hard" keratins of hair and nail has been hampered by a lack

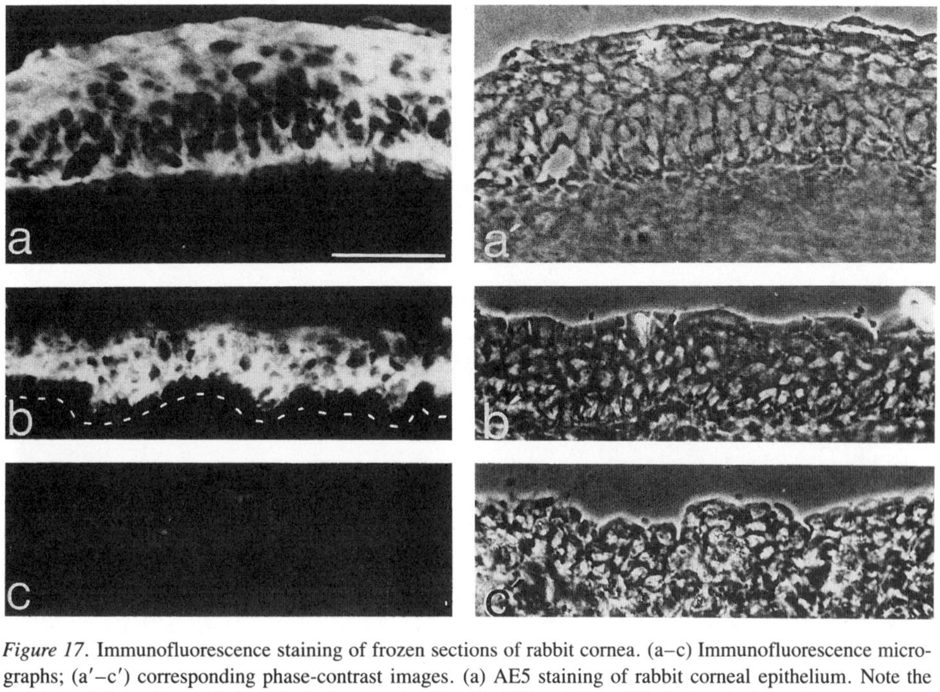

Figure 17. Immunofluorescence staining of frozen sections of rabbit cornea. (a–c) Immunofluorescence micrographs; (a′–c′) corresponding phase-contrast images. (a) AE5 staining of rabbit corneal epithelium. Note the staining of the entire epithelium including basal cells. (b) AE5 staining of limbal epithelium. Note spurabasal staining. (c) AE5 staining of conjunctival epithelium showing no significant reaction. Dashed line in (b) corresponds to the location of epithelial–stromal junction. Bar = 50 μm. From Schermer *et al.* (1986).

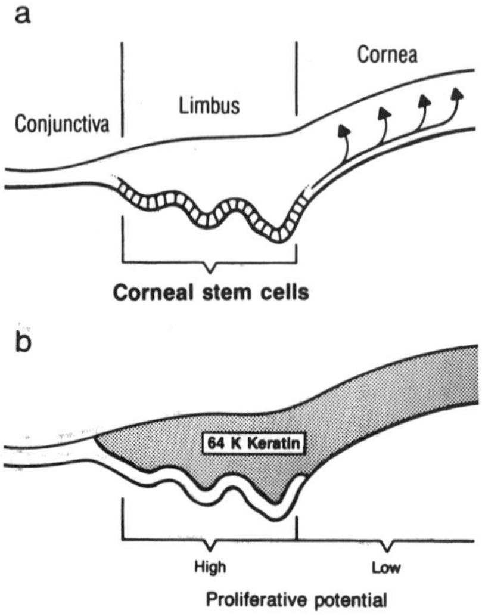

Figure 18. A model of corneal epithelial maturation. (a) The model showing that corneal epithelial stem cells are located in the basal layer of the limbal region. Arrows indicate the direction of cell migration during normal corneal epithelial maturation. (b) Expression of the 64k keratin in corneal and limbal epithelia. The proliferative potential of the 64k-keratin-positive corneal basal cells is presumed to be lower than that of the 64k-keratin-negative limbal basal (stem) cells. From Schermer *et al.* (1986).

of specific antibodies. Recently, however, Lynch *et al.* (1986) have succeeded in producing monoclonal antibodies to various "unmodified" hard keratins from human hair shaft (Fig. 19). Among these are AE12 and AE13, which react with the H44/H46 acidic doublet (Fig. 19, lane 2); our previously described AE3 reacts with the H56/H57/H59 triplet of basic hair keratins (Fig. 19, lane 3). With these antibodies, both the acidic and basic hair/nail-specific keratins have been localized to the suprabasal cells of the hair matrix that eventually give rise to the cortex of the hair shaft and hair cuticle (Fig. 20). Other parts of the hair follicle (i.e., inner and outer root sheath) do not express the hair/nail keratins. In addition to the hard keratin found in hair, human nail plate contains small but significant amounts (ca. 10%) of the 5/14 keratinocyte pair and the 6/16 hyperproliferation markers. These findings explain our previous observations that a poly-

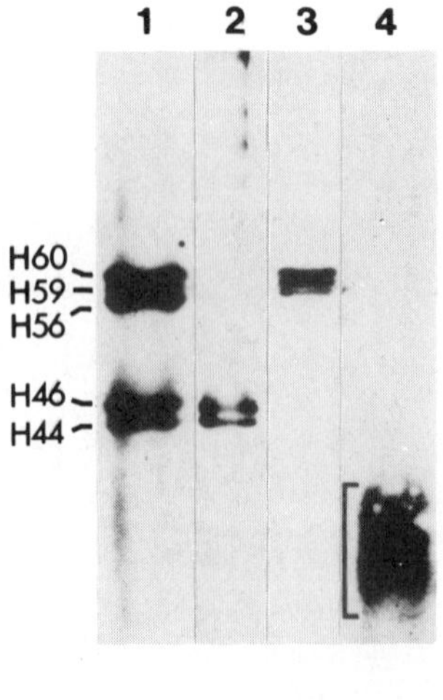

Figure 19. Immunoblot analysis of human hair keratin (lane 1, fast green stain) using AE13 (lane 2) which recognizes only the H44/H46 acidic hair keratins; AE3 (lane 3), which recognizes the H56/H57/H59 basic polypeptides; and AE14 (lane 4), which recognizes the 10–20k, high-sulfur proteins (bracket) associated with hard keratin. After Lynch *et al.* (1986).

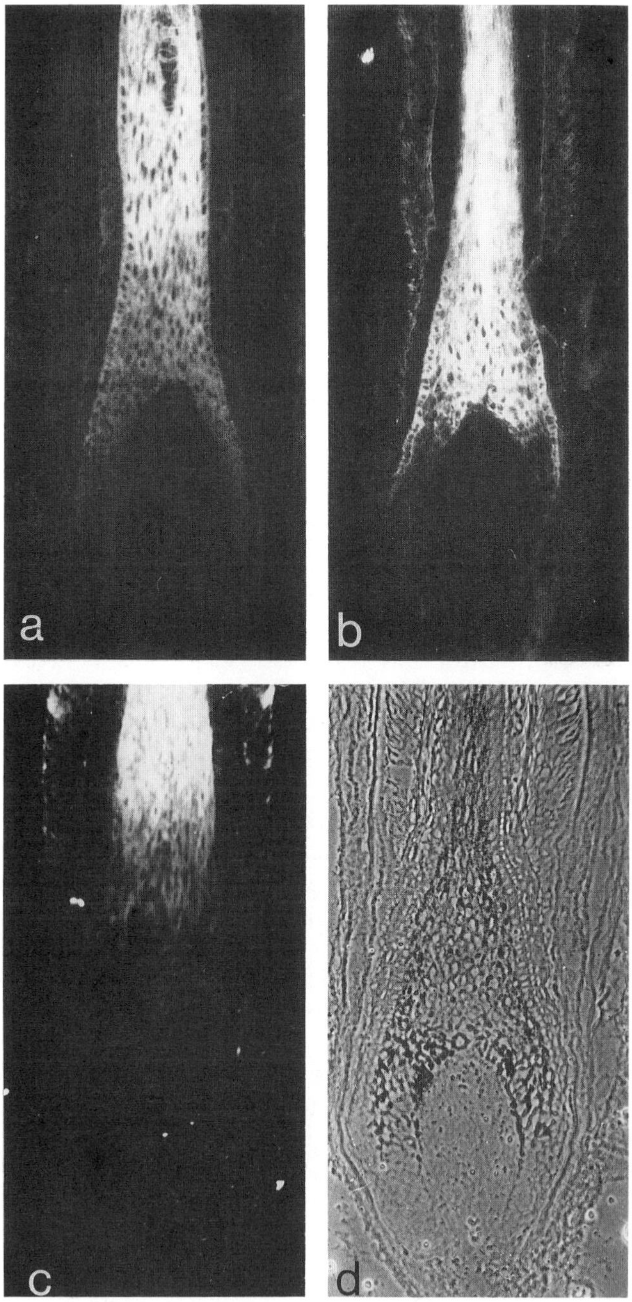

Figure 20. Indirect immunofluorescent localization of human hair keratins using the antibodies described in Fig. 19. Panel (a) is the AE13 staining pattern, which is restricted to the suprabasal hair matrix, hair cortex, and hair cuticle layers. AE3 (b) recognizes the basic hair keratins and gives the same distribution as AE13 in the hair structure; this antibody also cross-reacts with soft keratins in the outer root sheath. The high-sulfur proteins of hair are localized by AE14 (c) in the keratogenous zone of the hair cortex. The staining in the outer root sheath is due to the cross-reactivity of this antibody with the 58k (No. 5) keratin. Panel (d) is the phase-contrast image of a sectioned human hair follicle.

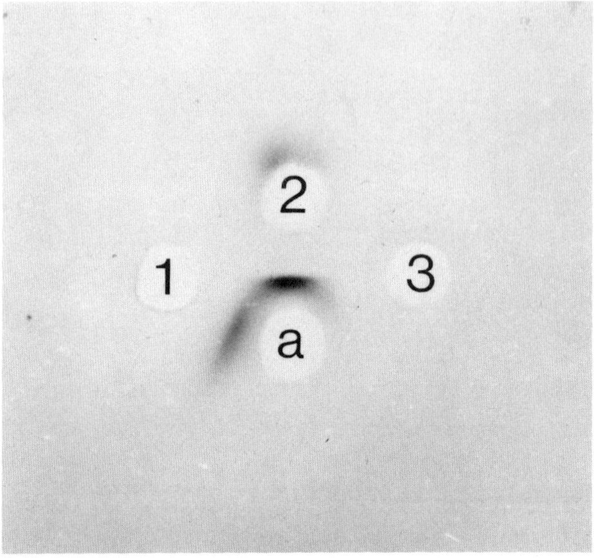

Figure 21. Ouchterlony reaction (double diffusion) of a polyclonal rabbit antiserum to human epidermal keratins (well a) with keratins extracted from (1) abdominal epidermis (Lynch *et al.*, 1986), (2) human nail plate, and (3) human hair shaft. Our recent data indicate that the immunoreactivity of the nail is due to the presence of the 5/15 and 6/16 soft keratin pairs, which are not found in hair.

clonal rabbit antiserum to human epidermal keratins cross-reacts with keratins extracted from human nail plate but not those from human hair shaft (Fig. 21).

Hard-keratin-producing structures also express some IF-associated proteins that have classically been called the "high-sulfur proteins." These 10–25k, cystine-rich molecules are localized by our AE14 antibody in the more mature cells of the keratogenous zone of the hair cortex and are not found in any other hair structures (Fig. 20; Lynch *et al.*, 1986). These proteins are believed to comprise part of the interfilamentous matrix of hard keratinized tissues presumably by establishing extensive sulfhydryl bridges between filaments. It is not currently known what, if any, role these proteins may play in augmenting filament aggregation such as filaggrin does for epidermal keratins (Dale, 1977; Dale *et al.*, 1978; Steinert *et al.*, 1982).

Since hair and nail both express the same major keratins (Baden *et al.*, 1973; Lynch *et al.*, 1986; Marshall, 1983), we suspect that the hard keratins of nail will be localized in the suprabasal nail matrix cells as well as mature nail plate. The association of the coexpressed, hard keratin pairs with the more mature cells of hair and nail structures is consistent with their designation as hair/nail-type differentiation markers (Lynch *et al.*, 1986).

6. *Structural/Functional Significance of Keratin Pairs*

Unlike the other IF subclasses, it was discovered several years ago that a single keratin polypeptide was unable to form a filament *in vitro,* but a mixture of polypeptides taken from a single tissue would readily do so (Lee and Baden, 1976; Steinert *et al.*, 1976). These observations were subsequently refined to the point of recognizing that one member from each subfamily was required to obtain a 10-nm filament (Eichner *et al.*, 1986; Hatzfield and Franke, 1985). Therefore, that all keratin pairs are composed of one acidic and one basic polypeptide reflects the fact that they represent the minimal requirement for 10-nm filament formation. Further, the formation of a keratin filament *in vitro* appears to be entirely nonselective with regard to which acidic and basic components are present. Hatzfield and Franke (1985) have eloquently demonstrated the "promiscuous" proclivities of keratin polypeptides when combined in virtually every conceivable acidic/basic (type I/type II) combination *in vitro*. They also reported that a 1 : 1 molar ratio of the two components was involved and excess polypeptides from one subfamily would be excluded from the filaments. These data explain the observation that the total complement of acidic and basic keratins extractable from any tissue approaches a stable 1 : 1 molar ratio. These data also strongly suggest that the filament-forming properties of keratin polypeptides lie in the highly conserved type I and type II rod domains. These rod domains are highly conserved among all keratins belonging to the same subfamily, and are largely responsible for the relative charge differences between the two subfamilies (Steinert *et al.*, 1982).

The fact that any pair of acidic and basic keratin polypeptides including appropriate xenogenic combinations (Hatzfield and Franke, 1985; Steinert *et al.*, 1982) will form stable filaments *in vitro* raises the question of the fidelity of actual *in vivo* relationships among various polypeptides in cells expressing more than one coexpressed pair. The current data indicate that keratin pairs, as defined by coexpression, may possess characteristics that could potentially direct preferential interactions *in vivo*. It has recently been shown that the 5/14 pair may be specifically isolated as a complex from a mixture of epidermal keratins that also contains the 1,2/10 skin-type differentiation marker. No hybrid filaments consisting of combinations of the two coexpressed pairs have been detected (Eichner *et al.*, 1986), although they will readily form *in vitro* under less stringent conditions (Hatzfield and Franke, 1985). Therefore, despite the fact that any two keratin polypeptides from different subfamilies may interact to form a filament under nonselective, *in vitro* conditions, biologically relevant keratin filaments probably result from highly regulated, sequential expression, or specialized interactions between the two members of various coexpressed pairs of polypeptides. The fact that some coexpressed pairs have demonstrated a preferential affinity for one another (Eichner *et al.*, 1986; Hatzfield and Franke, 1985) and have been isolated as an exclusive complex (Eichner *et al.*, 1986) might reflect a functionally significant stability of coexpressed polypeptides. However, since under experimental conditions some noncoexpressed combinations can form a filament even more stable than that formed by most coexpressed pairs (Hatzfield and Franke, 1985), the correct annealing may never be assured by such a simplistic mechanism as differential affinity. Obviously, little functional advantage would be achieved if an acidic polypeptide highly specialized for epidermal function were to be combined by random cytoplasmic association with a basic polypeptide engineered for very distinct corneal requirements even if they are structurally compatible to form a filament *in vitro*. Therefore, the demonstration of specifically coexpressed keratin pairs strongly implies that this is nature's way of ensuring that the functional requirements of cells are fulfilled in the most efficient manner without permitting inauspicious, random combinations.

Based on the nucleic acid sequences available for several keratin cDNAs and genes, one can now begin to understand the structural basis for properties common to all keratin pairs. The 8–11k difference between members of most coexpressed pairs (Cooper *et al.*, 1985; O'Guin *et al.*, 1987; Sun *et al.*, 1984) is primarily due to the two "H domains" being present only in the basic partner, and the hierarchy of pairing is due to coordinated modifications in the length of the V regions of both members (Steinert *et al.*, 1985a). Therefore, the observed molecular weight difference described for many of the coexpressed pairs is probably a reflection of structural conservation necessary for filament formation and the hierarchy of coexpressed polypeptides is mandated by their functional modifications.

Simple epithelial keratins that presumably have the least specialized functions have greatly diminished or no V_2-region sequences (Bader *et al.*, 1986; Glass *et al.*, 1985; Steinert *et al.*, 1985a). A relative increase in the magnitude of the alterations in the V regions, especially V_2, reflects a corresponding increase in the complexity of the functional specialization of the polypeptide above and beyond that of the generic or baseline filament function that is common to all keratin polypeptides. The acquisition of a more complex modification, either in terms of V-region sequence length or polypeptide mass, is probably closely linked to evolutionary chronology (Sun *et al.*, 1984). However, this is not a strict association, since the skin-type differentiation markers (1,1/10) represent the top pair in the hierarchy of molecular weight in humans, but in rabbits and chickens the top pair is the corneal-type keratins (3/12; O'Guin *et al.*, 1987).

The primordial structure of keratin is probably reflected by the highly conserved type I and type II rod domains which allow the formation of a basic filament structure (Fuchs *et al.*, 1985; Steinert *et al.*, 1985b). The more highly specialized, differentiation-specific keratins are probably the most advanced evolutionarily. There are several indications that the genes coding for a coexpressed pair of polypeptides may have common evolutionary origins. First, the relative degree of modification of coexpressed polypeptides is highly conserved within each subfamily such that polypeptides with similar size ranks are frequently coexpressed (Sun *et al.*, 1984). Second, it has been observed that both members of a coexpressed pair frequently share common antigenic determinants that are not found on any other keratins. For example, our AE2 recognizes the 1,2/10 polypeptides characteristic of skin-type differentiation (Woodcock-Mitchell *et al.*, 1982) and antibody 6.11 recognizes the 8/18 pair of simple epithelial keratins (Brabon *et al.*, 1984; Cooper *et al.*, 1985). This antigenic conservation is probably a result of highly specialized functional domains in the V regions of the coexpressed pair. The fact that both members of a coexpressed pair independently evolved identical protein sequences unique to themselves attests to the strength of the selective pressures driving the evolution of keratin filaments with specific functional properties (Steiner *et al.*, 1985b; Sun *et al.*, 1984).

7. Conclusions and Perspectives

The biological significance of both keratin pairs and the diversity of keratin polypeptides clearly lies in individual specializations of the various polypeptides and their selective expression in a manner that provides for the most effective combination of functional properties suited to the divergent demands of the numerous epithelia. It is because these keratins are so highly specialized to meet the various physiological requirements that the patterns of keratin pair expression are excellent markers for defining and distinguishing several major pathways of epithelial differentiation (Cooper *et al.*, 1985; O'Guin *et al.*, 1987; Sun *et al.*, 1984; for a compilation of human data, see Moll *et al.*, 1982).

We have been able to correlate the expression of specific keratin pairs with definite patterns of epithelial organization. Most simple and stratified epithelia may be distinguished based on their expression of either the 8/18 simple epithelial marker or the 5/14 keratinocyte marker, respectively. As a simple embryonic epithelium develops into a more complex epithelium, the pattern of keratin pair expression usually exhibits a concurrent progression in complexity. Every major pathway of epithelial terminal differentiation is defined by the synthesis of a particular differentiation-specific keratin pair. In humans, we call these pathways skin-type (for all keratinized stratified squamous epithelia), corneal-type (typical of corneal and snout epithelia), esophageal-type (for almost all nonkeratinized stratified squamous epithelia other than cornea), palm/sole-type (for the specialized epithelia of palms and soles), and hair/nail-type (for the "hard" keratin expression of hair and nails; Cooper *et al.*, 1985; Lynch *et al.*, 1986; Sun *et al.*, 1984). Additional programs of epithelial differentiation may exist in some tissues unique to nonhuman species (Cooper and Sun, 1986). The expression of most of the differentiation-specific pairs is primarily restricted to the more mature cells of the epithelium and is associated with advanced stages of keratinocyte differentiation (Schermer *et al.*, 1986; Woodcock-Mitchell *et al.*, 1982). The pattern of differentiation-specific keratin expression is developmentally determined and expressed permissively by adult epithelia, largely independently of site-specific instruction (Doran *et al.*, 1980). The fact that proliferating cultures can be derived from some cells already expressing their differentiation-specific keratins (e.g., basal cells of central corneal epithelium; Schermer *et al.*, 1986) shows that at least in some cases the expression of these markers, *per se,* is not detrimental to cell viability and is separable from "terminal" differentiation (for [^{3}H]thymidine incorporation by some suprabasally located, 67k-keratin-positive, epidermal cells, see Van Neste *et al.*, 1983).

Clearly, in recent years a general sense of accord is emerging from the outwardly overwhelming mélange of keratin data. Research from a number of disciplines is beginning to provide mutual support for observations made from a variety of perspectives, and each piece of the exhaustive puzzle is beginning to fit into an ever-clarifying picture. However, we are still lacking information about very fundamental aspects of the keratins. While we know that it requires both an acidic and a basic polypeptide to form a filament, it is still controversial whether or not this is mediated by a heterodimer or a heterotetramer composed of two homodimers (Ahmadi and Speakman, 1978; Geisler and Weber, 1982; Gruen and Woods, 1983; Quinlan *et al.*, 1984; Parry *et al.*, 1985). While the patterns of keratin expression are extremely stable and highly predictable, it is not known how this is developmentally determined. And although the nucleic acid sequence has been determined for a number of keratin genes, there is virtually no information concerning the mechanism of their regulation. The range of disciplines represented by these few questions serves to emphasize the diversity of interests that has provided the field of keratin research with such a broad base of experimental support. But, more importantly, it demonstrates that no single perspective is adequate to fully understand the comprehensive nature of such a heterogeneous, developmentally regulated multigene family of cytoskeletal elements that accounts for such a broad range of biological properties.

ACKNOWLEDGMENTS. We thank MaryJean Brocia-Foss for typing the manuscript. This investigation was supported in part by grants from the NIH (AM35411 and EY04722). W. M. O. and T.-T. S. are recipients of an NIH postdoctoral training grant (AM07910) and a Monique Weill-Caulier Career Scientist Award, respectively.

References

Ahmadi, B., and Speakman, P. T., 1978, Suberimidate crosslinking shows that a rod shaped, low cystine, high helix protein prepared by limited proteolysis of reduced wool has four chains, FEBS Lett. **94:**365–367.

Baden, H. P., Goldsmith, L. A., and Fleming, B., 1973, A comparative study of the physicochemical properties of human keratinized tissues, *Biochim. Biophys. Acta* **322:**269–278.

Bader, B. L., Magrin, T. M., Hatzfield, M., and Franke, W. W., 1986, Amino acid sequence and gene organization of cytokeratin no. 19, an exceptional tail-less intermediate filament protein, *EMBO J.* **5:**1865–1875.

Banks-Schlegel, S. P., 1982, Keratin alterations during embryonic epidermal differentiation: A presage of adult epidermal maturation, *J. Cell Biol.* **93:**551–559.

Brabon, A. C., Williams, J. F., and Cardiff, R. D., 1984, A monoclonal antibody to a human breast tumor protein released in response to estrogen, *Cancer Res.* **44:**2704–2710.

Breitkreutz, D., Bohnert, A., Herzmann, E., Bowden, P., BouKamp, P., and Fusenig, N.E., 1984, Differentiation specific functions in cultured and transplanted mouse keratinocytes: Environmental influence in ultrastructure and keratin expression, *Differentiation* **26:**154–169.

Cooper, D., and Sun, T. -T., 1986, Monoclonal antibody analysis of bovine epithelial keratins: Specific pairs as defined by coexpression, *J. Biol. Chem.* **261:**4646–4654.

Cooper, D., Schermer, A., and Sun, T. -T., 1985, Classification of human epithelia and their neoplasms using monoclonal antibodies to keratin: Strategies, applications and limitations, *Lab. Invest.* **52:**243–256.

Crewther, W. G., Dowling, L. M., Gough, K. H., Inglis, A. S., McKern, N. M., Sparrow, L. G., and Woods, E. G., 1976, The low-sulfur proteins of wool: Studies on their classification, characterization, primary and secondary structure, *Proc. Fifth Int. Wool Text. Res. Conf. Aachen* **2:**233–242.

Crewther, W. G., Inglish, A. S., and McKern, N. M., 1978, Amino acid sequences of alpha-helical segments from S-carboxymethylkeratein-A: Complete sequence of a type-II segment, *Biochem. J.* **173:**365–371.

Crewther, W. G., Dowling, L. M., Gough, K. H., Marshall, R. C., and Sparrow, L. G., 1980, The microfibrillar proteins of alpha-keratin, in: *Fibrous Proteins: Scientific, Industrial and Medical Aspects,* Volume II (D. A. D. Parry and L. K. Creamer, eds.), Academic Press, London, pp. 151–159.

Dale, B. A., 1977, Purification and characterization of a basic protein from the stratum corneum of mammalian epidermis, *Biochim. Biophys. Acta* **491:**193–204.

Dale, B. A., Stern, I. B., Rabin, M., and Huang, L. -Y., 1976, The identification of fibrous proteins in fetal rat epidermis by electrophoretic and immunologic techniques, *J. Invest. Dermatol.* **66:**230–235.

Dale, B. A., Holbrook, K. A., and Steinert, P. M., 1978, Assembly of stratum corneum basic protein and keratin filaments in macrofibrils, *Nature* **276:**729–731.

Dale, B. A., Holbrook, K. A., Kimball, J. R., Hoff, M., and Sun, T. -T., 1985, Expression of epidermal keratins and filaggrin during human fetal skin development, *J. Cell Biol.* **101:**1257–1269.

Debus, E., Weber, K., and Osborn, M., 1982, Monoclonal cytokeratin antibodies that distinguish simple from stratified squamous epithelia: Characterization on human tissues, *EMBO J.* **1:**1641.

Dhouailly, D., Rogers, G. E., and Sengel, P., 1978, The specification of feather and scale protein synthesis in epidermal–dermal recombinations, *Dev. Biol.* **65:**58–68.

Doran, T. I., Vidrich, A., and Sun, T. -T., 1980, Intrinsic and extrinsic regulation of the differentiation of skin, corneal and esophageal epithelial cells, *Cell* **22:**17–25.

Drochmans, P., Freudenstein, C., Wanson, J., Laurent, L., Keenan, T. W., Stadler, J., Leloup, R., and Franke, W. W., 1978, Structure and biochemical compositin of desmosomes and tonofilaments isolated from calf muzzle epidermis, *J. Cell Biol.* **79:**427–443.

Eckert, R. L., and Green, H., 1984, Cloning of cDNAs specifying vitamin A-responsive human keratins, *Proc. Natl. Acad. Sci. USA* **81:**4321.

Eichner, R., Bonitz, P., and Sun, T. -T., 1984, Classification of epidermal keratins according to their immunoreactivity, isoelectric point, and mode of expression, *J. Cell Biol.* **98:**1388–1396.

Eichner, R., Sun, T. -T., and Aebi, U., 1986, The role of keratin subfamilies and keratin pairs in the formation of human epidermal intermediate filaments, *J. Cell Biol.* **102:**1767–1777.

Franke, W. W., Schmid, E., Osborn, M., and Weber, K., 1978, Different intermediate-sized filaments distinguished by immunofluorescence microscopy, *Proc. Natl. Acad. Sci. USA* **75:**5034–5038.

Franke, W. W., Schmid, E., Weber, K., and Osborn, M., 1979, HeLa cells contain intermediate-sized filaments of the prekeratin type, *Exp. Cell Res.* **118:**95–109.

Franke, W. W., Schiller, D. L., Moll, R., Winter, S., Schmid, E., Engelbrecht, I., Denk, H., Krepler, R., and Platzer, B., 1981, Diversity of cytokeratins: Differentiation-specific expression of cytokeratin polypeptides in epithelial cells and tissues, *J. Mol. Biol.* **153:**933–959.

Franke, W. W., Schmid, E., Grund, C., and Geiger, B., 1982, Intermediate filament proteins in nonfilamentous structures: Transient disintegration and inclusion of subunit protein in granular aggregates, *Cell* **30:**103–113.

Franke, W. W., Schiller, D. L., Hatzfield, M., and Winter, S., 1983a, Protein complexes of intermediate-sized filaments: Melting of cytokeratin complexes in urea reveals different polypeptide separation characteristics, *Proc. Natl. Acad. Sci. USA* **80:**7113–7117.

Franke, W. W., Schmid, E., Wellsteed, J., Grund, C., Gigi, O., and Geiger, B., 1983b, Change of cytokeratin filament organization during the cell cycle: Selective masking of an immunologic determinant in PtK2 cells, *J. Cell Biol.* **97:**1255–1260.

Fuchs, E., and Green, H., 1980, Changes in keratin gene expression during terminal differentiation of the keratinocyte, *Cell* **19:**1033–1042.

Fuchs, E., and Green, H., 1981, Regulation of terminal differentiation of cultured human keratinocytes by vitamin A, *Cell* **25:**617–625.

Fuchs, E., Coppock, H., Green, H., and Cleveland, D. W., 1981, Two distinct classes of keratin genes and their evolutionary significance, *Cell* **17:**75–84.

Fuchs, E., Hanukoglu, I., Marchuk, D., Grace, M., and Kim, K. H., 1985, The nature and significance of differential keratin gene expression, *Ann. N.Y. Acad. Sci.* **455:**436–450.

Geisler, N., and Weber, K., 1982, The amino acid sequence of chicken muscle desmin provides a common structural model for intermediate filament proteins, *EMBO J.* **1:**1649.

Gigi, O., Geiger, B., Eshhar, Z., Moll, R., Schmid, E., Winter, S., Schiller, D. L., and Franke, W. W., 1982, Detection of cytokeratin determinant to diverse epithelial cells by a broadly cross-reacting monoclonal antibody, *EMBO J.* **1:**1429–1437.

Gilmartin, M. E., Culbertson, V. B., and Freedberg, I. M., 1980, Phosphorylation of epidermal keratins, *J. Invest. Dermatol.* **75:**211–216.

Gilmartin, M. E., Mitchell, J., Vidrich, A., and Freedberg, I. A., 1984, Dual regulation of intermediate filament phosphorylation, *J. Cell Biol.* **98:**1144–1149.

Glass, C., Kim, K., and Fuchs, E., 1985, Sequence and expression of a human type II mesothelial keratin, *J. Cell Biol.* **101:**2366–2373.

Gough, K. H., Inglis, A. S., and Crewther, W. G., 1978, Amino acid sequences of alpha-helical segments from S-carboxymethylkeratein-A: Complete sequence of a type-I segment, *Biochem. J.* **173:**373–385.

Gruen, L. D., and Woods, E. F., 1983, Structure studies on the microfibrillar proteins of wool: Interaction between alpha-helical segments and reassembly of a four chain structure, *Biochem. J.* **209:**587–595.

Hatzfield, M., and Franke, W. W., 1985, Pair formation and promiscuity of cytokeratins: Formation in vitro of heterotypic complexes and intermediate filaments by homologous and heterologous recombinations of purified polypeptides, *J. Cell Biol.* **101:**1826–1841.

Heid, H. W., Werner, E., and Franke, W. W., 1986, The complement of native alpha-keratin polypeptides of hair-forming cells: A subset of eight polypeptides different from epithelial cytokeratins, *Differentiation* **32:**101–119.

Hennings, H., Michael, D., Cheng, C., Steinert, P., Holbrook, K., and Yuspa, S., 1980, Calcium regulation of growth and differentiation of mouse epidermal cells in culture, *Cell* **19:**245–254.

Horwitz, B., Kupfer, H., Eshar, Z., and Geiger, B., 1981, Reorganization of arrays of prekeratin filaments during mitosis, *Exp. Cell Res.* **134:**281–290.

Hronis, T. S., Steinberg, M. L., Defendi, V., and Sun, T. -T., 1984, The simple epithelial nature of some simian-virus-40-transformed human epidermal keratinocytes, *Cancer Res.* **44:**5797–5804.

Kim, K. H., Rheinwald, J., and Fuchs, E. V., 1983, Tissue specificity of epithelial keratins: Differential expression of mRNAs from two multigene families, *Mol. Cell Biol.* **3**:495–502.

Kim, K. H., Marchuk, D., and Fuchs, E., 1984, Expression of unusually large keratin during terminal differentiation: Balance of type I and type II keratins is not disrupted, *J. Cell Biol.* **99**:1872–1877.

Knapp, A. E., Franke, W. W., Heid, H., Hatzfield, M., Jorcano, J. L., and Moll, R., 1986, Cytokeratin no. 9, an epidermal type I keratin characteristic of a special program of keratinocyte differentiation displaying body site specificity, *J. Cell Biol.* **103**:657–667.

Knapp, L. W., O'Guin, W. M., and Sawyer, R. H., 1983a, Drug-induced alterations of cytokeratin organization in cultured epithelial cells, *Science* **219**:501–503.

Knapp, L. W., O'Guin, W. M., and Sawyer, R. H., 1983b, Rearrangement of the keratin cytoskeleton after combined treatment with microtubule and microfilament inhibitors, *J. Cell Biol.* **97**:1788–1794.

Knapp, L. W., O'Guin, W. M., and Sawyer, R. H., 1985, Dependence of keratin intermediate filament organization on microtubules, microfilaments and membrane association sites, *Ann. N.Y. Acad. Sci.* **455**:758–761.

Lane, E. B., Goodman, S. L., and Trejdosiewicz, K. K., 1982, Disruption of the keratin filament network during epithelial cell division, *EMBO J.* **1**:1365.

Lee, L. D., and Baden, H. P., 1976, Organization of the polypeptide chains in mammalian keratin, *Nature* **164**:377–379.

Lee, L. D., Kubilus, J., and Baden, H. P., 1979, Intraspecies heterogeneity of epidermal keratins isolated from bovine hoof and snout, *Biochem. J.* **177**:187–196.

Lynch, M. H., O'Guin, W. M., Hardy, C., Mak, L., and Sun, T. -T., 1986, Human hair follicle differentiation: Coordinate expression of acidic and basic hair keratins in upper cortical and cuticle cells and the relationship between "hard" (hair/nail) and "soft" keratin, *J. Cell Biol.* **103**:2593–2606.

McGuire, J., Osber, M., and Lightfoot, L., 1984, Two keratins MW 50,000 and 56,000 are synthesized by psoriatic epidermis, *Br. J. Dermatol.* **111** (Suppl. 27):27–37.

Marshall, R. C., 1983, Characterization of the proteins of human hair and nail by eletrophoresis, J. Invest. Dermatol. **80**:519–524.

Moll, R., Franke, W. W., Schiller, D. L., Geiger, B., and Krepler, R., 1982a, The catalog of human cytokeratins: Patterns of expression in normal epithelia, tumors and cultured cells, *Cell* **31**:11–24.

Moll, R., Franke, W. W., Volc-Platzer, B., and Krepler, R., 1982b, Different keratin polypeptides in epidermis and other epithelia of human skin: A specific cytokeratin of molecular weight 46,000 in epithelia of the pilosebaceous tract and basal cell epitheliomas, *J. Cell Biol.* **95**:285–295.

Moll, R., Krepler, R., and Franke, W. W., 1983a, Complex cytokeratin polypeptide patterns observed in certain human carcinomas, *Differentiation* **23**:256–269.

Moll, R., Moll, I., and Wiest, W., 1983b, Changes in the pattern of cytokeratin polypeptides in epidermis and hair follicle during skin development in human fetuses, *Differentiation* **23**:170–178.

Moll, R., Moll, I., and Franke, W. W., 1984, Differences of expression of cytokeratin polypeptides in various epithelial skin tumors, *Arch. Dermatol. Res.* **276**:349–363.

Nelson, W. G., and Sun, T. -T., 1983, The 50- and 58-kdalton keratin classes as molecular markers for stratified squamous epithelia: Cell culture studies, *J. Cell Biol.* **97**:244–251.

Nelson, W. G., Battifora, H., Santana, H., and Sun, T. -T., 1984, Specific keratins as molecular markers for neoplasms with a stratified epithelial origin, *Cancer Res.* **44**:1600–1603.

Ochs, D., 1983, Protein contaminants of sodium dodecyl sulfate-polyacrylamide gels, *Anal. Biochem.* **135**:470–474.

O'Guin, W. M., and Sawyer, R. H., 1982, Avian scale development. VIII. Relationships between morphogenetic and biosynthetic differentiation, *Dev. Biol.* **89**:485–492.

O'Guin, W. M., Knapp, L. W., and Sawyer, R. H., 1982, Biochemical and immunohistochemical localization of alpha and beta keratins in avian scutate scales, *J. Exp. Zool.* **220**:371–376.

O'Guin, W. M., Schermer, A., and Sun, T. -T., 1985, Immunofluorescence staining of keratin filaments in cultured epithelial cells, *J. Tissue Culture Methods* **9**:123–128.

O'Guin, W. M., Galvin, S., Schermer, A., and Sun, T. -T., 1987, Patterns of keratin expression define distinct pathways of epithelial development and differentiation, *Curr. Top. Dev. Biol.* **22**:97–125.

Osborn, M., and Weber, K., 1983, Tumor diagnosis by intermediate filament typing: A novel tool for surgical pathology, *Lab. Invest.* **48**:372–394.

Parry, D. A. D., Steven, A. C., and Steinert, P. M., 1985, The coiled-coil molecules of intermediate filaments consist of 2 parallel chains in exact axial register, *Biochem. Biophys. Res. Commun.* **127**:1012–1018.

Quinlan, R. A., Cohlberg, J. A., Schiller, D. L., Hatzfield, M., and Franke, W. W., 1984, Heterotypic tetramer (A_2D_2) complex of non-epidermal keratins isolated from cytoskeletons of rat hepatocytes and hepatoma cells, *J. Mol. Biol.* **178:**365–388.

Quinlan, R. A., Schiller, D. L., Hatzfield, M., Achstatter, T., Moll, R., Jorcano, J. L., Magin, T. M., and Franke, W. W., 1985, Patterns of expression and organization of cytokeratin intermediate filaments, *Ann. N.Y. Acad. Sci.* **455:**282–306.

Ramaekers, F. C. S., Huysmans, A., Moesker, O., Kant, A., Jap, P., Herman, C., and Vooijs, P., 1983, Monoclonal antibody to keratin filaments specific for glandular epithelia and their tumors: Use in surgical pathology, *Lab. Invest.* **49:**353–361.

Rheinwald, J. G., and Green, H., 1975, Serial cultivation of strains of human epidermal keratinocytes: The formations of keratinizing colonies from single cells, *Cell* **6:**331.

Roop, D. R., Cheng, C. K., Titterington, L., Meyers, C. A., Stanley, J. R., Steinert, P. M., and Yuspa, S. H., 1984, Synthetic peptides corresponding to keratin subunits elicit highly specific antibodies, *J. Biol. Chem.* **259:**8037–8040.

Sawyer, R. H., O'Guin, W. M., and Knapp, L. W., 1984, Avian scale development. X. Dermal induction of tissue specific keratins in extraembryonic ectoderm, *Dev. Biol.* **101:**8–18.

Sawyer, R. H., Knapp, L. W., and O'Guin, W. M., 1986, The skin of birds—Epidermis, dermis and appendages, in: *Biology of the Integument,* Volume 2 (J. Bereiter-Hahn, A. G. Matoltsy, and K. S. Richards, eds.), Springer-Verlag, Berlin, pp. 194–238.

Schermer, A., Galvin, S., and Sun. T. -T., 1986, Differentiation-related expression of a major 64K corneal keratin in vivo and in culture suggests limbal location of corneal epithelial stem cells, *J. Cell Biol.* **103:**49–62.

Schiller, D. L., Franke, W. W., and Geiger, B., 1982, A subfamily of relatively large and basic cytokeratin polypeptides as defined by one or several polypeptides in epithelial cells, *EMBO J.* **1:**761–769.

Schweizer, J., Kinjo, M., Furstenberger, G., and Winter, H., 1984, Sequential expression of mRNA-encoded keratin sets in neonatal mouse epidermis: Basal cells with properties of terminally differentiating cells, *Cell* **37:**159–170.

Skerrow, D., and Skerrow, C. J., 1983, Tonofilament differentiation in human epidermis: Isolation and polypeptide chain composition of keratinocyte subpopulations, *Exp. Cell Res.* **143:**27–35.

Steinert, P. M., Idler, W. W., and Zimmerman, S. B., 1976, Self assembly of bovine epidermal keratin filaments in vitro, *J. Mol. Biol.* **108:**547–567.

Steinert, P. M., Idler, W. W., and Wantz, M. L., 1980, Characterization of the keratin filament subunits unique to bovine snout epidermis, *Biochem. J.* **187:**913–916.

Steinert, P. M., Idler, W. W., Aynardi-Whitman, M., Zackroff, R. V., and Goldman, R. D., 1982, Heterogeneity of intermediate filaments assembled in vitro, *Cold Spring Harbor Symp. Quant. Biol.* **46:**465–473.

Steinert, P. M., Parry, D. A. D., Idler, W. W., Johnson, L. D., Steven, A. C., and Roop, D. R., 1985a, Amino acid sequences of mouse and human epidermal type II keratins of M_r 67,000 provide a systematic basis for the structural and functional diversity of the end domains of keratin intermediate filament subunits, *J. Biol. Chem* **260:**7142–7149.

Steinert, P. M., Steven, A. C., and Roop, D. R., 1985b, The molecular biology of intermediate filaments, *Cell* **42:**411–419.

Sun, T. -T., and Green, H., 1977, Cultured epithelial cells of cornea, conjunctiva and skin: Absence of marked intrinsic divergence of their differentiated states, *Nature* **269:**489–493.

Sun, T. -T., and Green, H., 1978a, Keratin filaments of cultured human epidermal cells: Formation of intermolecular disulfide bonds during terminal differentiation, *J. Biol. Chem.* **253:**2053–2060.

Sun. T. -T., and Green, H., 1978b, Immunofluorescent staining of keratin fibers in cultured cells, *Cell* **14:**469–476.

Sun, T. -T., Shih, C., and Green, H., 1979, Keratin cytoskeletons in epithelial cells or internal organs, *Proc. Natl. Acad. Sci, USA* **76:**2813–2817.

Sun, T. -T., Eichner, R., Nelson, W. G., Tseng, S. C. G., Weiss, R. A., Jarvinen, M., and Woodcock-Mitchell, J., 1983, Keratin classes: Molecular markers for different types of epithelial differentiation, *J. Invest. Dermatol* **81:**109s–115s.

Sun, T. -T., Eichner, R., Schermer, A., Cooper, D., Nelson, W. G., and Weiss, R. A., 1984, Classification, expression, and possible mechanisms of evolution of mammalian epithelial keratins: A unifying model, in: *The Cancer Cell,* Volume 1 (A. Levine, W. Topp, G. Van de Woude, and J. D. Watson, eds.), Cold Spring Harbor Laboratory, Cold Spring Harbor, N.Y., pp. 169–176.

Sun, T. -T., Tseng, S. C. G., Huang, A. J. -W., Cooper, D., Schermer, A., Lynch, M. H., Weiss, R., and Eichner, R., 1985, Monoclonal antibody studies of mammalian epithelial keratins: A review, *Ann. N.Y. Acad. Sci.* **455:**307–329.

Tseng, S. C. G., Jarvinen, M., Nelson, W. G., Huang, A. J. -W., Woodcock-Mitchell, J., and Sun, T. -T., 1982, Correlation of specific keratins with different types of epithelial differentiation: Monoclonal antibody studies, *Cell* **30:**361–372.

Tseng, S. C. G., Hatchell, D., Tierney, N., Huang, A. J. -W., and Sun, T. -T., 1984, Expression of specific keratin markers by rabbit corneal, conjunctival and esophageal epithelia during vitamin A deficiency, *J. Cell Biol.* **99:**2279–2286.

Tyner, A. L., Eichman, M. J., and Fuch, E., 1985, The sequence of a type II keratin gene expressed in human skin: Conservation of structure among intermediate filament genes, *Proc. Natl. Acad. Sci. USA* **82:**4683–4687.

Van Neste, D., Staquet, M. J., Viac, J., Lachapelle, J. M., and Thivolet, J., 1983, A new way to evaluate the germinative compartment in human epidermis using ^{3}H-thymidine incorporation and immunoperoxidase staining of 67K polypeptide, *Br. J. Dermatol.* **108:**433–439.

Venetianer, A., Schiller, D. L., Magin, T., and Franke, W. W., 1983, Cessation of cytokeratin expression in a rat hepatoma cell line lacking differentiated functions, *Nature* **305:**730–733.

Watt, F. M., and Green, H., 1982, Stratification and terminal differentiation of cultured epidermal cells, *Nature* **295:**434–436.

Weiss, R. A., Eichner, R., and Sun, T. -T., 1984, Monoclonal antibody analysis of keratin expression in epidermal diseases: A 48Kd and a 56Kd keratin as molecular markers for hyperproliferative keratinocytes, *J. Cell Biol.* **98:**1397–1406.

Woodcock-Mitchell, J., Eichner, R., Nelson, W. G., and Sun, T. -T., 1982, Immunolocalization of keratin polypeptides in human epidermis using monoclonal antibodies, *J. Cell Biol.* **95:**580–588.

Wu, Y. -J., Parker, L. M., Binder, N. E., Beckett, M. A., Sinard, J. H., Griffiths, C. T., and Rheinwald, J. G., 1982, The mesothelial keratins: A new family of cytoskeletal proteins identified in cultured mesothelial cells and nonkeratinizing epithelia, *Cell* **31:**693–703.

12

Keratins and Their Genes in Xenopus laevis
Structural and Developmental Aspects

Thomas D. Sargent, Erzsebet Jonas, Milan Jamrich, George S. Michaels, Seiji Miyatani, Jeffrey A. Winkles, and Igor B. Dawid

1. Introduction

Cytokeratins and cytokeratin genes in *Xenopus laevis* have been studied from both a structural and a developmental point of view. The former must be seen in the context of the extensive work on cytokeratins in many vertebrate animals; the information obtained for *Xenopus* provides additional comparative opportunities with important evolutionary implications. Since most other studies on cytokeratins have been done on mammalian material, it is very useful to have extensive data available on a cold-blooded animal as well. The second major impetus for this work came from developmental considerations: cytokeratins are prominent tissue-specific products and the study of their expression proved to be a powerful tool for the characterization of developmental processes.

In this chapter we shall summarize the available information on *Xenopus* cytokeratin genes and compare the sequences among themselves. We shall then review the information about expression of these genes and influences that affect regulation of cytokeratin expression in the frog embryo.

This chapter was originally submitted Feb. 26, 1986.

Thomas D. Sargent, Erzsebet Jonas, Milan Jamrich, George S. Michaels, Seiji Miyatani, Jeffrey A. Winkles, and Igor B. Dawid • Laboratory of Molecular Genetics, National Institute of Child Health and Human Development, National Institutes of Health, Bethesda, Maryland 20892.

2. Cytokeratin Gene Sequences from Xenopus

All sequence information available on *Xenopus* cytokeratins is based on the analysis of cDNA clones and cloned genomic DNA. Two laboratories have carried out these studies: Hoffmann and his colleagues studying cytokeratins expressed in adult skin, and our laboratory focusing on embryonic and larval material. By homology among themselves and to other cytokeratins, these clones can be classified as encoding type I or type II cytokeratins. To date, partial to complete sequence information is available for one adult type I keratin sequence (UF29; Hoffmann and Franz, 1984), three adult type II keratin sequences (Hoffmann *et al.*, 1985), and four type I and two type II embryonic keratins (Jonas *et al.*, 1985; Winkles *et al.*, 1985; and our unpublished results).

2.1. Type I Cytokeratin Sequences

Here we present a detailed amino acid sequence comparison of three embryonic/larval cytokeratins: XK81A1 (originally named XK81; Jonas *et al.*, 1985), XK70 (Winkles *et al.*, 1985), and XK81B2 (unpublished results). The genes encoding these three proteins are all expressed exclusively during premetamorphic development, following two major distinct temporal patterns (see below). All three sequences are clearly identified as type I cytokeratins by their acidic nature, small size, and most clearly, by sequence homology with other type I sequences (see Winkles *et al.*, 1985, for detailed evidence on this point for XK70). Sequence homology between these three proteins are of differing degrees: XK81A1 and XK81B2 are much more closely related to each other than either is to XK70. This is illustrated by an overall comparison of the sequences of the rod domains in Fig. 1A and in the more detailed comparison of Fig. 2A. In contrast to the rod domains, the nonhelical N-terminal regions are essentially nonhomologous between the three compared sequences. At the C-terminus of the proteins listed in Fig. 2A there is a region of about 17 residues that is strongly conserved among most if not all type I sequences of *Xenopus*. As reported earlier, this region also contains some homology to mammalian type I cytokeratins and type III IF protein sequences (Winkles *et al.*, 1985). While XK81A1 and XK81B2 are obviously closely related, they represent clearly distinct genes rather than polymorphic variants of the same gene. XK70 is distantly related to the

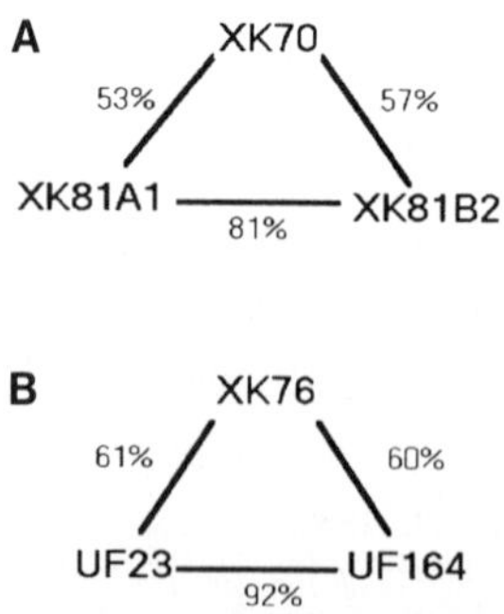

Figure 1. Sequence homology between the rod domains of type I (top) and type II (bottom) cytokeratins from *X. laevis*. Only identities are scored. See Fig. 2 for references to sequence determinations.

XK81 family; in fact, it is somewhat more similar to the human 50k cytokeratin (Marchuk *et al.*, 1984) than to XK81A1 (Winkles *et al.*, 1985).

The major region of homology between cytokeratins is confined to the rod domain. Within this region, homology is distributed unevenly, as illustrated in Fig. 3A, with strong sequence conservation at the extreme ends of the domain. In the interior there is some variation in the homology score (see figure legend for interpretation of the score). Some of the low scoring regions appear to coincide with linker regions (L1, L12, L2), but the correlation is not perfect; a pronounced valley occurs within helical region 2B.

2.2. *Type II Cytokeratin Sequences*

Type I and II cytokeratins, while sharing some homology in sequence and obvious homology in structure, are quite distinct families of proteins. Type II proteins are generally more basic and larger than type I, and intratype homologies clearly exceed intertype homologies (Steinert and Parry, 1985). As shown in Fig. 1B, the two adult type II keratins UF23 and UF164 are very closely related to each other; this relationship extends also through the nonhelical regions (Hoffmann *et al.*, 1985; see Fig. 2B). While the two sequences may have arisen during a genome duplication believed to have occurred during the evolution of *X. laevis* (see below for further discussion), their extreme similarity weighs against such an interpretation. They might more likely result from a recent gene duplication, but it is also conceivable that UF23 and UF164 are in fact polymorphic variants of the same gene. It must be remembered that *Xenopus* material is usually derived from wild-caught animals that are known to harbor extensive polymorphisms (Okada *et al.*, 1985; Winkles *et al.*, 1984). XK76, the larval type II protein, is only moderately homologous to the adult type II sequences in the rod domain and essentially nonhomologous in the C-terminal domain. Furthermore, XK76 is more homologous to a human 56k type II protein (Hanukoglu and Fuchs, 1983) than to the adult *Xenopus* sequences (74% versus 63% identity in the rod domain). A similar situation has been noted with XK70 (see above), and such relationships may be taken as evidence for divergence between members of the cytokeratin gene family early in vertebrate evolution. An inspection of homology distribution within the rod domain (Fig. 3B) shows a similar pattern of peaks and valleys as in the type I sequence comparison in Fig. 3A, except that the valleys are more shallow. This is due to the great similarity of two of the type II sequences (UF164, UF23) that are compared in Fig. 3B.

2.3. *Cytokeratin Gene Structure*

We have isolated genomic DNA fragments containing type I cytokeratin genes (Miyatani *et al.*, unpublished results). The gene encoding XK81A1 has been sequenced together with more than 1 kb of flanking region at each end. The genes encoding 81B1 and B2 have also been obtained and long upstream flanking regions and about one-third of each gene have been sequenced. Figure 4 summarizes these results in the form of an intron/exon map, showing that the placement and approximate size of introns are the same in these genes. These results add to the finding that intron positions in different cytokeratin genes are in generally similar but not precisely conserved positions (Steinert *et al.*, 1985).

```
a               10      20      30      40      50      60      70      80      90      100     110     120

XK81B2  MYFRSSSSYSLQSKGNSGGGGYGA-GFGGGSGAGFGGGSGAGFGGGYGAGFGGGASSGFSLSSA-GGFGAA-AASSSF------------SNFGGNDKQTMQNLNDRLASYLEKVRALE

XK81A1      MTSYRSSSASYYSGSS-SKGGFGSRSLAGSNSYGGSSFG-----AGFSSGVGSGFS-SSG-GNFAMAEAASSSF---------------GGNEKHAMQNLNDRLASYLEKVRALE

XK70          //----SARFGVLASPGVNRARSVAGGASTVRMSS-----ANVTSSAFGG-----SS---------AFAGSSAFAGSPAFNVSVTSNNG----KETMQNLNDRLANYLDRVRSLE

        -1A----------><-----L1---------><-------------------------------------------------1B---------------------------
XK81B2  AANADLELKIREWYEKQKGSGIGAGSKDFSKYFEIISDLRNKILSATIDNSRVVLQIDNAKLAADDFRLKFENELALRQSVETDINGLRRVLDELTLARGDLEMQIESLTEELAYLKK
XK81A1  ATNSDLEGKIRNWYDKQSDAGIGAGSKDYSKYFEIIAELRNKIRAATIDNATVTLQIDNARLAADDFRLKFENELALRQSVEGDSNGLRRVLDELILARGDFELQIESLTEELAYLKK
XK70    QANHELELKIREYLDKKAAVG-----SLDYSGYYNTINLLRSQINDATIDNTRLVLSIDNAKLAADDFKIKYESEMAIRTGAESDIVGLRRVLDELTLNKTDLELEIESLKEELIYLKK

        -------------><------L12----------><--------2A-----------><--L2--><--------------------------------------2B--------
XK81B2  NHEEEMSIAKSSSAGQVNVEMDAAPGIDLNKILSDMRADYETLAEKNRRDAELWFNQKSGELKKEIQTGVEQVQTSKSEINDLRRSLQSLEIELQSQLAMKKSLEDTLAETDGRYGAQ
XK81A1  NHEEEMSHAKSQSAGKVSVEMDAALGVDLTSILNNMRADYEILAEKNRRDAELWFNQKSGELKKEISVGVEQVQASKSEITELKRSLQSLEIELQSQLAMKQSVEGNLNELQGFYSSQ
XK70    NHEEELAVVRSSARGNVDVQVDSAPPVDLAQIMADVRSQYESMMEKNRQELEACYKGQSENLNHEVATNTAALQTSKTAITDLKRTIQSLEIELQSLLSMKGALEGTLAETEAQYGAQ
UF29                                                                  //----------QQKESKSELTELKTTLQSLEIELQSQLAMKKSLEMTLAEVEGSFCMK

        ----------------------------------------------------------------->
XK81B2  QLQTIQFSLRSLEEQLLQIRSDMERQNMEYRQLLDIKTRLEMEIETYRRLLEGEFGSLK--------SSIVQ---A---TEVS-TSQSSSSSKKD----//
XK81A1  QLQQIQNTIGSLEEQLLQIRSDMEHQNTEYKLLLDIKTRLEMEIQTYRRLLEGELGQVT----TVANTSSV-ESK---TESSSTSTTRTRMVKT-IVEEVVDGKVVSSRVE
XK70    QLNHLQAMITQVEMELQNLRSDADHQSLEYKRLLDAKTRLEMEIATYRRLLEGEDTRFSQTETQKAVTIV---SKE-QSSSSIKKVK------KTVI-EEVVDGKVVSSRVEELTETS
UF29    -LSRLQEMIVNVEEQIARLKGESECQTAEYQQLLDIKTRLENEIETYRRLLDGDLSKPKSGGGTSTNTGSTSSKGSTRTVKRRE------------IIEEVVDGKVVSTKVVDM
```

b

```
                10          20          30          40          50          60          70          80          90          100         110         120
                 •           •           •           •           •           •           •           •           •           •           •           •
                                                     <-----------------------1A--------><----L1----><---------------------------------
UF23          //---------------------------------------------FLEQQNKVLETKWKLLQEQGTKGTTKRANLDPLFEKYIADLKKYLDNLISEKGRLQQEMKN

XK76    IPGVSYSGPGIQEVTINQSLLSPLNLEIDPTIQTVRQEEREQIKTLNNKFASFIDKVRFLEQQNRVLETKWELLQNQ-K-SAKASNIGPLFDAYISNLRRQLDSLTSDKGKLEGELRN

UF164         //---------------------------------------------------------------KGSTKRANLDPLFEKYISDLKRYLDNLINEKGRLQQELRT

UF23          //---------------------------------------------FLEQQNKVLETKWKLLQEQGTKGTTKRANLDPLFEKYIADLKKYLDNLISEKGRLQQEMKN

        ----------------1B---------------------------------------------><-----L12-----><--------2A--------><-L2-><----
UF23    LQLLVEDYKKKYEDEINKRTKAENDFVLLKKDVDAAYMVKTELEAKVDTLTSEINFLRTRDAAELSQVHDQVTDTSVVLTMDNNRDLNLDSIIKEVKCQYEQIAQRSKLEAEALYDQKY

XK76    MQDLVEDFKNKYEDEINKRTGAENEFVILKKDVDSAYMNKVELEAKMEGLTDEINFLRALYEAELREMQEQISDTSVVLSMDNNRALDMDSIIAEVKAQYEEIANKSRANAEATYQSKF

UF164   LQDLVEDFKKKYEDEINKRTKAENDFVVLKKDVDAAYMIKTELEAKVDTLTSEINFFETLFAAELSQVHDQVTDTSVVLTMDNNRDLNMDSIIKEVKCQYEQIAQRSKLEAEALYDQRY

UF23    LQLLVEDYKKKYEDEINKRTKAENDFVLLKKDVDAAYMVKTELEAKVDTLTSEINFLRTRDAAELSQVHDQVTDTSVVLTMDNNRDLNLDSIIKEVKCQYEQIAQRSKLEAEALYDQKY

        ------------------------------2B------------------------------------------------------------------------->
UF23    KQLQQTVEGHGDSIKNSKTEISDQNRKIQRLKAEIENVKKQIASSNQSIAGAEERGNLSLKDAEKKLKDLEDAERKLKADMARQLKEYQELMSAKLALNLEISTYRYMLEGEEGRIS-

XK76    QELQATAGQHGNDLRATKTEISELNRAMQRLQAEIESVKAQRAKLEAQIAEAEERGELALKDARNKLAELEAALQKAKQDMNRQLRDYQELMNVKLSLDIEIATYRKLLEGEETRISDG

UF164   KQLQQTVEGHGDSIKNSKTEISDLNLKIQRLKAEIENVKKQIAFLNQSIAGAEERGNLALKDAERKLKDLEDAERKLKADMARQLKEYQELMSAKLALNLEISTYRYMLEGEEGRMS-

UF23    KQLQQTVEGHGDSIKNSKTEISDQNRKIQRLKAEIENVKKQIASSNQSIAGAEERGNLSLKDAEKKLKDLEDAERKLKADMARQLKEYQELMSAKLALNLEISTYRYMLEGEEGRIS-

UF23    -GQIVNKVSISVLSGGSSVYTALGGAAGGMG-----GGGGGGMGS--GHGGGY-GGMVFGGGMVCGMGDGMGGG-GMGGGMGSGHGGGYGHGGAVCFGAGGMGYEGGSMHGGGSSYGHL

XK76    PGPVS--VSVSVVNSTSSSM--------GGSS-----GFGAGYSGSSSGFGAGYSGGSSFGAG---GFGNSSVKF-GSGGSSGVKSYSMSTTSSSRSFRQ

UF164   -GQISNNVSISVIS-------------GGSSVYTALGGAAGGMGGGGMGGGMGGGMGMGGGMGMGGGMGMGGGMGMGGGMGGGMGMGGGMGMGGGMGMGDGMGSGHGGGYGNED--CF

UF23    -GQIVNKVSISVLS-------------GGSSVYTALGGAAGGMGGGGG-GGMGSGHGGGYGGMVFGGGMVCGMGDGMGGGGMGGGM-----------------GSGHGGGYGHGGAVCF

UF23    GGKTSVAIASTTSTTKKTY

UF164   G-GGMGYEGGSMHGGGSSYGHSGGKTSVAIASTTSTTKRSY

UF23    GAGGMGYEGGSMHGGGSSYGHLGGKTSVAIASTTSTTKKTY
```

Figure 2. Sequence comparison between type I (a) and type II (b) cytokeratins from *X. laevis*. The different segments in the rod domains are named according to Steinert and Parry (1985). The sequences are from the following sources. XK81B2: derived from analysis of a cDNA and genomic clone by Miyatani *et al*. (unpublished); XK81A1: Jonas *et al*. (1985); XK70: Winkles *et al*. (1985); UF29: Hoffman and Franz (1984); XK76: Jonas *et al*. (unpublished); UF23 and UF164: Hoffman *et al*. (1985). The proteins were aligned for maximum homology using the k-tuple method of Wilbur and Lipman (1983).

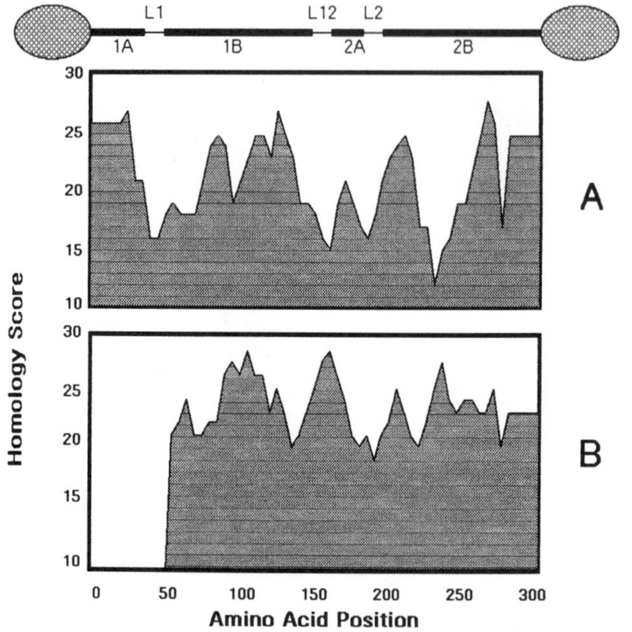

Figure 3. Homology between *X. laevis* cytokeratin rod domains. (A) Type I (XK81A1, XK81B2, XK70); (B) type II (UF164, UF23, XK76). Based on the data in Fig. 2. The homology score was derived from the sum of individual scores at each amino acid position within a 15-residue window. The analysis was then repeated with a 5-residue overlap along the rod domain. Individual scores were assigned as follows: 2 = identity for all sequences compared; 1 = single pair identity; 0 = no homology. These data were then plotted versus the relative amino acid position within the rod domain.

3. Family Relationships among Type I Cytokeratins

In addition to the three fully analyzed cDNAs shown in Fig. 2A, we have cloned and partially characterized an additional cDNA, DG26, which is quite similar to DG70 and encodes a protein homologous, but probably not allelic to XK70 (Sargent and Dawid, 1983; and unpublished).* Information about the XK81 family is based on the cDNA sequences for A1 and B2 components, genomic clone sequences for A1, B1, and B2, and cross-hybridization studies between all of these sequences and the genomic clone for A2. On the basis of these studies, the following conclusions can be drawn: The type I cytokeratin family in *X. laevis* can be subdivided into subfamilies, three of which are known: the XK81, XK70, and UF29 subfamilies, with four, two, and one known members, respectively. In the best analyzed XK81 subfamily, the four members are organized in two pairs of closely related sequences, which we named XK81A1, A2, B1, and B2. This nomenclature follows the model of the *X. laevis* vitellogenins, which are likewise encoded by four distinct genes organized in two pairs of closely related sequences (Wahli

*Upon further analysis of the family represented by DG26 and DG70, we may modify the nomenclature as we have done for the DG81 family.

Figure 4. An intron/exon map for three type I cytokeratin genes (81A1, 81B1, and 81B2) showing their similarities in arrangement and approximate size.

et al., 1981). The analogy is further strengthened by the fact that physical linkage between one member each of the A and B group genes has been demonstrated for both the vitellogenin and cytokeratin families (Wahli *et al.*, 1982; Miyatani *et al.*, unpublished). The presence of pairs of related sequences in *X. laevis* may be due to individual instances of gene duplication, followed by a complete genome duplication during the evolution of the species as suggested by Bisbee *et al.* (1977). In the XK70 subfamily the two known genes are closely related and might have originated from a single copy present before the genome duplication occurred.

These results show that, at least in *X. laevis,* the type I cytokeratin family can be classified into subfamilies whose members are more closely related to each other than to other type I sequences. It would be of interest to know whether such subfamilies also exist in other species.

The probable genome duplication in *X. laevis* evolution together with the duplicative history of the IF gene family has created a rather complex array. To some extent this complexity confounds interpretations of RNA and protein accumulation measurements during development and should be borne in mind in this context. Detailed structural knowledge is required to assign any RNA or protein component to any particular gene.

4. Developmental Aspects

4.1. Distinct Groups of Cytokeratin Genes Are Expressed in Embryonic/Larval and Adult Stages in Xenopus

Five distinct cytokeratin genes, represented by the cDNA clones DG70 (XK70) and its close homologue DG26, DG76 (XK76), DG81 (XK81A1), and DG118 (XK81B2), are known to be active in *Xenopus* embryos; these genes are not detectably expressed in the postmetamorphosis adult (Jonas *et al.*, 1985; Winkles *et al.*, 1985; and unpublished results). Three closely related type II genes (UF23, UF132, and UF164) and one type I gene (UF29) are known to be active in the adult (Hoffman and Franz, 1984; Hoffmann *et al.*, 1985), but it is not known at what time in development these genes are activated. Furthermore, it is clear that not all epidermal cytokeratin genes have been isolated as yet and it is quite possible that some cytokeratins are expressed throughout larval and adult stages in this animal. This situation is in fact suggested by the work of Ellison *et al.* (1985) who studied electrophoretic patterns of cytokeratins in different stages. Several electrophoretic bands were observed that are found in epidermal extracts from all stages from early tadpoles to adults. On the other hand, two-dimensional gel electrophoretic patterns of cytokeratins that had been labeled *in vivo* or *in vitro* and were precipitated with a type I-specific antiserum have provided evidence for qualitative differences between larval and adult proteins (Winkles *et al.*, unpublished results). No firm identification has been made of proteins visualized on gels as products of individual genes represented by cDNA sequences. Such identifications will be required to correlate studies on developmental expression of mRNA and proteins.

In spite of the remaining uncertainties, it is clear that embryonic/larval and adult epidermis in the frog differ substantially in their cytokeratin composition. It has been found generally that skin differentiation is accompanied by an increase in both the diversity and size of cytokeratins (Steinert and Parry, 1985). In the frog, adult skin differs from larval skin in many respects, particularly by being keratinized (Nieuwkoop and Faber, 1967). It is therefore not surprising that different cytokeratins are expressed in the different stages. An increase in the average size of the proteins has in fact been observed (Ellison *et al.*, 1985; Winkles *et al.*, unpublished results), but an increase in the number of expressed components has not been clearly demonstrated.

Why are there distinct embryonic/larval and adult cytokeratins? Two possible explanations have been considered (Jonas *et al.*, 1985). The first invokes functional differences: larval and adult epidermis simply require different proteins for their formation. This hypothesis has much support in the substantial structural and functional differences between larval and adult skin (Nieuwkoop and Faber, 1967). The second explanation is a regulatory one. It is possible that different genes exist because one group has evolved to be coregulated with other embryonic genes, the other to be coregulated with adult genes. Once duplicated genes have arisen, they may be expected to drift apart.

4.2. Correlation between Type I and Type II Cytokeratin Expression

At least one type I and type II cytokeratin are expressed in every cell that expresses either, and IF in epithelial cells are believed to be generated by copolymerization of type I and II molecules (Steinert and Parry, 1985). In this light the frog embryo poses an interesting situation. At least three distinct type I genes have been characterized and shown to be expressed; among these the genes encoding XK81A1 and B2 are quite close, but XK70 is only distantly related (see above). In contrast, only a single type II cDNA has been obtained from embryonic material. While additional type II sequences may simply remain undiscovered, attempts have been made to find such sequences (unpublished work in this laboratory), and the possibility must be considered that XK76 is the only major

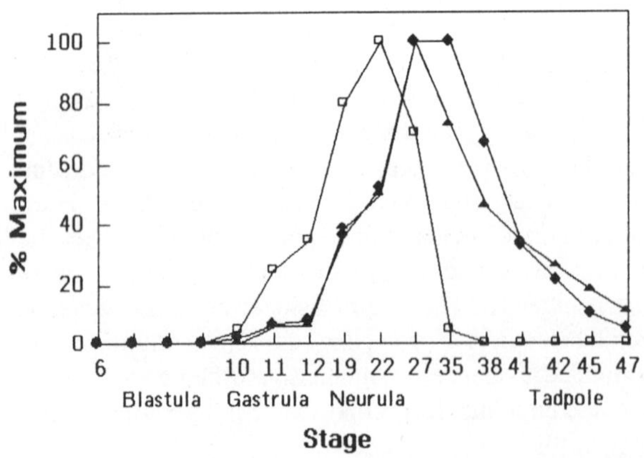

Figure 5. Expression of cytokeratin genes during *X. laevis* development. The accumulation of mRNAs for XK70 (squares), XK81 (all members of the XK81 subfamily were combined in this measurement; triangles), and XK76 (diamonds) is shown. In each case the maximum level was set as 100. Note the abscissa shows developmental stages (Nieuwkoop and Faber, 1967) and is not linear with time.

larval type II cytokeratin. If this were the case, coexpression of type I and II at the transcriptional level would not be upheld strictly in this system. As illustrated in Fig. 5, the mRNAs encoding the type I cytokeratin XK70 accumulates quite early in development and decays after only 2 days. In contrast, the mRNAs for XK81 (type I) and XK76 (type II) accumulate somewhat more slowly in a coordinated fashion and persist much longer. The XK81 variants and XK76 thus form a conventional coordinately expressed set—but does XK70 have a type II partner? An undiscovered gene or possibly stored maternal type II keratin might provide the answer. Alternatively, the different XK81 variants and XK70 might copolymerize with XK76 to form distinct IF in the early epidermis. Abundance levels of these mRNA sets are comparable, and the possibility exists of a surplus of type I over type II cytokeratins with possible consequences for IF structure in the embryo.

4.3. Developmental Regulation of Cytokeratin Gene Expression

Gastrulation leads to the formation of the three germ layers through a series of cell migration and differentiation processes. Molecular markers for these early events are of great utility in unraveling the regulatory phenomena involved. On the basis of such considerations we have sought to identify such markers (Sargent and Dawid, 1983). The embryonic cytokeratin genes proved to be suitable for the study of ectoderm and subsequent epidermal differentiation. Cytokeratin mRNAs are restricted in their expression to the ectoderm of gastrula stage embryos and the developing epidermis during and after neurulation (Jamrich *et al.*, unpublished). The activation of cytokeratin genes occurs at the normal time and mRNAs accumulate to normal levels in fully dispersed cells of dissociated embryos, indicating that this event is programmed into embryonic cells and not dependent on cellular interactions. Nevertheless, cytokeratin gene activation can be influenced by inductive effects: exposing blastula cells to inappropriate cell contacts suppresses cytokeratin mRNA accumulation (Sargent *et al.*, 1986). Likewise, normal neural differentiation in dorsal ectoderm, which is known to be due to induction by underlying chordamesoderm, leads to a rapid suppression of cytokeratin gene activity (Jamrich *et al.*, unpublished). Thus, initial activation of cytokeratin gene expression in the *Xenopus* embryo is cell autonomous, but these genes are responsive to inductive influences during blastula, gastrula, and neurula stages (see also Dawid and Sargent, 1986).

5. Conclusions

Cytokeratins and their genes in *X. laevis* have been studied in some detail. This work adds to the evidence for the structural homology of all vertebrate keratins. The *Xenopus* system, with its accessible early embryonic material, provides a good opportunity to study developmental regulation of cytokeratin gene expression. At the same time, these genes have been most useful in the molecular analysis of developmental processes.

References

Bisbee, C. A., Baker, M. A., Wilson, A. C., Hadji-Adzimi, I., and Fischberg, M., 1977, Albumin phylogeny for clawed frogs (Xenopus), *Science* **195:**785–787.

Dawid, I. B., and Sargent, T. D., 1986, Molecular embryology in amphibians: New approaches to old questions, *Trends Genet.* **2:**47–50.

Ellison, T. R., Mathisen, P. M., and Miller, L., 1985, Developmental changes in keratin patterns during epidermal maturation, *Dev. Biol.* **112:**329–337.

Hanukoglu, I., and Fuchs, E., 1983, The cDNA sequence of a type II cytoskeletal keratin reveals constant and variable structural domains among keratins, *Cell* **33:**915–924.

Hoffmann, W., and Franz, J. K., 1984, Amino acid sequence of the carboxy-terminal part of an acidic type I cytokeratin from Xenopus laevis epidermis as predicted from the cDNA sequence, *EMBO J.* **3:**1301–1306.

Hoffmann, W., Franz, J. K., and Franke, W. W., 1985, Amino acid microheterogeneity of basic (type II) cytokeratins of Xenopus laevis epidermis and evolutionary conservativity of helical and non-helical domains, *J. Mol. Biol.* **184:**713–724.

Jonas, E., Sargent, T. D., and Dawid, I. B., 1985, Epidermal keratin gene expressed in embryos of Xenopus laevis, *Proc. Natl. Acad. Sci. USA* **82:**5413–5417.

Marchuk, D., McCrohon, S., and Fuchs, E., 1984, Remarkable conservation of structure among intermediate filament genes, *Cell* **39:**491–498.

Nieuwkoop, P. D., and Faber, J., 1967, *Normal Table of Xenopus laevis (Daudin),* 2nd ed., North-Holland, Amsterdam.

Okada, A., Shin, T., Dworkin-Rastl, E., Dworkin, M., and Zubay, G., 1985, Constancy of DNA organization of polymorphic and nonpolymorphic genes during development in Xenopus, *Differentiation* **29:**14–19.

Sargent, T. D., and Dawid, I. B., 1983, Differential gene expression in the gastrula of Xenopus laevis, *Science* **222:**135–139.

Sargent, T. D., Jamrich, M., and Dawid, I. B., 1986, Cell interactions and the control of gene activity during early development in Xenopus laevis, *Dev. Biol.* **114:**238–246.

Steinert, P. M., and Parry, D. A. D., 1986, Intermediate filaments: Conformity and diversity of expression and structure, *Annu. Rev. Cell Biol.* **1:**41–65.

Steinert, P. M., Steven, A. C., and Roop, D. R., 1985, The molecular biology of intermediate filaments, *Cell* **42:**411–419.

Wahli, W., Dawid, I. B., Ryffel, G. U., and Weber, R., 1981, Vitellogenesis and the vitellogenin gene family, *Science* **212:**298–304.

Wahli, W., Germond, J.-E., ten Heggeler, B., and May, F. E. B., 1982, Vitellogenin genes A1 and B2 are linked in the Xenopus genome, *Proc. Natl. Acad. Sci. USA* **70:**6832–6836.

Wilbur, W. J., and Lipman, D. J., 1983, Similarity searches of the nucleic acid and protein data banks, *Proc. Natl. Acad. Sci. USA* **80:**726–730.

Winkles, J. A., Jamrich, M., Jonas, E., Kay, B. K., Miyatani, S., Sargent, T. D., and Dawid, I. B., 1984, Gene expression during embryogenesis in Xenopus laevis, in: *Molecular Biology of Development* (E. H. Davidson and R. Firtel, eds.), Liss, New York, pp. 93–108.

Winkles, J. A., Sargent, T. D., Parry, D. A. D., Jonas, E., and Dawid, I. B., 1985, Developmentally regulated cytokeratin gene in Xenopus laevis, *Mol. Cell. Biol.* **5:**2575–2581.

13

Neural Intermediate Filament (Neurofilament) Genes

Linda M. Parysek

In mammalian systems, at least five neural IF proteins (NIFP) have been characterized. These include NF-L, NF-M, and NF-H (68kd, 150kd, and 200kd, respectively, in rat) (Liem *et al.*, 1978; Schlaepfer and Freeman, 1978), a 57kd NIFP (peripherin) (Portier *et al.*, 1984; Parysek and Goldman, 1988), and a 66kd NIFP (Chiu, 1988). These IF polypeptides share the characteristic of neuron-specific expression but there are many distinctions among them.

1. Distinctions among NIFP

A major difference among NIFP is evident from studies on NIFP expression during nervous system (NS) development. The evidence indicates that the expression of NF-H lags far behind that of the other NIFP (Shaw and Weber, 1982; Willard and Simon, 1983; Pachter and Liem, 1984; Carden *et al.*, 1987); the expression of NF-L, NF-M, and the 57kd NIFP in the NS are thought to be linked to the appearance of differentiated neurons (Cochard and Paulin, 1984; Carden *et al.*, 1987; Escurat *et al.*, 1988). In PC12 cells, nerve growth factor (NGF) increases the level of NF-L and NF-M gene transcription (Dickson *et al.*, 1986; Lindenbaum *et al.*, 1988), and of 57kd NIFP mRNA (Leonard *et al.*, 1987), but has no effect on NF-H (Lindenbaum *et al.*, 1988). In another paradigm of development, the axotomy response, there are distinctions among the expression of NIFP genes. Whereas the expression of NF-L, NF-M, and NF-H decreases and remains suppressed for many weeks postaxotomy (Hoffman *et al.*, 1987; Wong and Oblinger, 1987; Oblinger and Lasek, 1988; Tetzlaff *et al.*, 1988; Greenberg and Lasek, 1988), synthesis of the 57kd NIFP is increased and this protein is a major Triton-insoluble protein in the

Linda M. Parysek • Department of Anatomy and Cell Biology, University of Cincinnati Medical School, Cincinnati, Ohio 45267–0521.

growing axonal sprout (Oblinger *et al.*, 1989). The differential expression of these proteins may reflect unique roles for some of the NIFP during differentiation and development.

In IF derived from the adult NS, immunoelectron microscopic studies have indicated that NF-L, NF-M, and NF-H may be found in the same filament (Willard and Simon, 1981; Sharp *et al.*, 1982). Yet, monoclonal antibodies specific for differentially phosphorylated isoforms of these subunits show great distinctions in cellular and subcellular distribution of the isoforms *in vivo* (Goldstein *et al.*, 1983; Lawson *et al.*, 1984; Vitadello *et al.*, 1986, 1987; Lee *et al.*, 1987; Black and Lee, 1988). *In vitro* experiments have indicated the possibility that the 57kd NIFP may be present as a copolymer with one or more of the Type IV neural IF proteins *in vivo* and contribute to the heterogeneity of neural IF (Parysek *et al.*, unpublished data). Indeed, in the adult NS, NF-L, -M, and -H are widely distributed (Shaw *et al.*, 1981; Trojanowski *et al.*, 1986) but the 57kd NIFP is found only in restricted areas of the NS (Portier *et al.*, 1983; Brody *et al.*, 1989). The recently characterized 66kd NIFP, which does form a homopolymer *in vitro,* has a distribution in the NS that complements that of the 57kd NIFP (Chiu, 1988, and personal communication). It is reasonable to consider that differences in neurofilaments in various parts of the NS are linked to differences in function and that neuronal IF may do more than regulate axonal caliber (Lasek *et al.*, 1983; Hoffman *et al.*, 1987).

Recent analyses of NIFP cDNAs have shown that some of the properties of the NIFP, such as potential phosphorylation sites and possible homopolymer-forming ability, are reflected in NIFP amino acid sequences. Genomic analyses have indicated numbers and locations of introns and have provided some clues to the evolutionary relationship of the NIFP genes to other IF genes. Other aspects of the genes, such as nucleotide sequences required for neuron-specific expression, developmental activation, and NGF-inducibility, have yet to be determined. This short review will summarize the recent characterizations of NIFP cDNAs and their respective genes.

2. *NIFP Amino Acid Sequences*

Direct or predicted amino acid sequences have been obtained for NF-L (Geisler et al., 1983, 1985a; Julien *et al.*, 1985, 1986, 1987a; Lewis and Cowan 1985, 1986), NF-M (Geisler *et al.*, 1984; Julien *et al.*, 1986; Napolitano *et al.*, 1987; Levy *et al.*, 1987; Myers *et al.*, 1987; Zopf *et al.*, 1987), NF-H (Geisler *et al.*, 1985b; Robinson *et al.*, 1986; Julien *et al.*, 1986, 1988; Lees *et al.*, 1988), and the 57kd NIFP (Leonard *et al.*, 1988; Parysek *et al.*, 1988). cDNA and genomic clones for the 66kd NIFP have not been reported. In addition, a gene for an NF-M-like protein (Sharpe, 1988) and a cDNA for a novel, low-molecular-weight NIFP (Charnas *et al.*, 1987) have been cloned from *Xenopus laevis* libraries. In general, the predicted molecular structures derived from the sequence information conform to those of the nonneural IF proteins, with the exception that NF-L, -M, and -H contain lengthy, highly acidic carboxy-terminal tails (Geisler *et al.*, 1983). Also, a few anomalies of the α-helical core region of NF-M and -H have been noted. Whereas all other IF proteins except the lamins contain a single, central tryptophan (in coil 2), NF-M and -H contain an additional, conserved tryptophan residue preceding the final 36 amino acids of the IF α-helical core. Unlike most other IF proteins, the rod region of NF-M and -H lack proline residues characteristic of linker regions (Myers *et al.*, 1987; Lees *et al.*, 1988). One other anomaly noted in the rod region of NF-H is the presence of a number of charged amino acid residues positioned such that it is difficult to assign the characteristic heptad repeat motif (Lees *et al.*, 1988). These phenomena have been suggested to be the cause of the inability of NF-M and -H to assemble into homopolymers *in vitro* (Myers *et al.*, 1987; Lees *et al.*, 1988).

3. Sequences Related to the Neural IF Multiphosphorylation Site

Lengthy carboxy-terminal extensions of NF-M and -H distinguish these two proteins from all other IF proteins. These COOH extensions contain an abundance of glutamic acid as does NF-L, but the glutamic acid residues in human, rat, and mouse NF-M and -H are interrupted by repeats of Lys-Ser-Pro-(small side chain hydrophobic amino acid) (Robinson *et al.*, 1986; Myers *et al.*, 1987; Lees *et al.*, 1988; Julien *et al.*, 1988). Monoclonal antibodies to a 13-amino acid synthetic peptide that contains two segments of Lys-Ser-Pro-Val (Ala) or a trimer of the 13-amino-acid peptide cross-reacted with phosphorylated mammalian NF-M and/or NF-H suggesting that the Lys-Ser-Pro-Val (Ala) repeat or an epitope created by this sequence is a phosphorylation site. Furthermore, the monoclonal antibodies to the trimer synthetic peptide identified neural IF subunits from nonmammalian species including chicken, frog, lamprey, and squid (Lee *et al.*, 1988). Examination of the predicted amino acid sequence derived from the chicken NF-M cDNA indicates a hexapeptide repeat of Gly-Lys-Pro/Val/Ala-X-Ser/Thr-Pro that is distinct from the mammalian KSP unit (Zopf *et al.*, 1987). These data suggest evolutionary conservation of at least secondary structures related to the mammalian NF-M and phosphorylation site (Lee *et al.*, 1988). The phosphorylated epitope and the nature of the highly charged tail of NF-L, -M, and -H have allowed immunologic (Shaw *et al.*, 1984; Mencarelli *et al.*, 1987) and silver stain detection (Gambetti *et al.*, 1981; Weber *et al.*, 1983) of neurofilament proteins from nonmammalian species (Phillips *et al.*, 1983; Lasek *et al.*, 1985). The 57kd NIFP does not contain the glutamate-rich carboxy-terminus characteristic of NF-L, -M, and -H. The carboxy-terminus of the 57kd NIFP shares overall similarity to other Type III IF proteins, and contains a polar 10-amino-acid sequence that is common to Type III IF protein carboxy-termini. The region of the 57kd NIFP that contains particularly striking numbers of the phosphorylatable amino acid serine is the amino-terminus. Thus, it is likely that the function of the carboxy-terminus of the 57kd NIFP differs from that of NF-L, -M, and -H.

4. NIFP mRNAs

With the exception of NF-L, Northern blots indicate that apparently single-sized mRNAs are derived from transcription of the neural IF genes. These include a message of 1.8–2.0 kb for the 57kd NIFP (Leonard *et al.*, 1987; Parysek *et al.*, 1988), 3.0 kb for NF-M, and 4.5 kb for NF-H (Julien *et al.*, 1986). In adult NS, a second mRNA species of 3.2 kb was detected for the 57kd NIFP (Parysek *et al.*, 1988). Two mRNA species of 2.5 and 4.0 kb are derived from the single NF-L gene (Lewis and Cowan, 1985) even after transfection into mouse L cells (Julien *et al.*, 1987b). The two NF-L mRNAs are thought to be transcribed from the same gene via a read-through of a polyadenylation site (Lewis and Cowan, 1985); there are two polyadenylation signals located downstream of the NF-L TGA stop codon (Julien *et al.*, 1986). The significance of these two transcripts has not been determined. Both the NF-H (Lees *et al.*, 1988) and 57kd NIFP gene (Parysek *et al.*, unpublished data) contain more than one in-frame ATG translation start site. In the NF-H gene, the ATGs are in tandem (Lees *et al.*, 1988) and in the 57kd NIFP gene, the ATGs are located six amino acids apart. In both genes, the second ATG is in the preferred nucleotide sequence context for efficient translation (Kozak, 1987).

The developmental expression of the NIFP mRNAs has been assayed by Northern blot analyses. In rat, the 57kd NIFP mRNA is present as early as E10 (Parysek *et al.*, 1988b). In the mouse, NF-L and -M mRNAs are detectable by E11 and NF-H message does not accumulate until birth (Julien *et al.*, 1986). The messages for these genes may be present at earlier time points, but there are technical difficulties in obtaining embryos of earlier stages. The NF-L and -M gene are located in tandem in the mouse genome (Julien *et al.*, 1986); this observation may be related to the coordinate expression of these two genes during development.

5. Introns in NIFP Genes

One of the most remarkable features of the NIFP genes is the distinctive number and locations of introns since generally, the IF multigene family contains a remarkable degree of conservation of intron position. Generally seven or eight introns occupy characteristic positions in Type I, II, and III IF genes (Steinert and Roop, 1988). The Type IV neural IF genes are unique in that they contain only two (NF-M) or three (NF-L, -H) introns (Lewis and Cowan, 1986; Julien *et al.*, 1987a, 1988; Levy *et al.*, 1987; Myers *et al.*, 1987; Lees *et al.*, 1988). Each of the Type IV IF genes shares identity in the positions of two of the introns. This suggests that the type IV IF genes were generated from a common precursor neural IF gene (Lees *et al.*, 1988; Julien *et al.*, 1988). The positions of the two introns common among NF-L, -M, and -H occur in coil 2B, the region of most extensive homology among IF of all types. Indeed, one of the two introns interrupts the hallmark sequence TYR ↓ KLLEGEE. NF-L contains a third intron in the glutamic acid-rich carboxy-terminal domain (Lewis and Cowan, 1986; Julien *et al.*, 1987a) while the additional intron of NF-H occurs in rod segment 2A (Lees *et al.*, 1988; Julien *et al.*, 1988). The third intron of NF-L is not shared by NF-M and -H but the latter two genes carry tandem repeat units that interrupt the glutamate-rich carboxy-terminus. Lees *et al.* (1988) have hypothesized that the repeat structure in the NF-M and -H carboxy-terminus was derived either from the intron that interrupted this region in an ancestral NF-L-like gene or by a series of mutations and reiterations of sequences in the glutamate-rich domain (see also Steinert and Roop, 1988). The additional intron of NF-H is located only one amino acid 5′ to the position of a common intron in the Type I and III nonneural IF genes. The position of this first intron in NF-H shows relation to the nonneural IF genes and supports the likelihood that the common progenitor of Type IV IF genes arose by gene duplication rather than an RNA-mediated transposition event (Lees *et al.*, 1988).

Reduced numbers of introns is not a characteristic shared by all neural IF genes. While there is at least one example of intron sliding [corresponding to intron 3 of the vimentin gene (Quax *et al.*, 1983)], the 57 kd NIFP gene appears to be organized like the Type III nonneural IF genes and five of the eight probable intron positions have been discerned (Parysek *et al.*, unpublished data). This suggests that all IF genes expressed exclusively in the NS have not evolved as a group. Although there is no evidence to indicate how the 57kd NIFP gene arose or is related to the other neural IF genes, it is intriguing that the 57kd NIFP, an IF gene with introns like the Type III IF genes, is predominantly found in evolutionarily old areas of the brain.

6. NIFP Gene Promoters

Aspects of NIFP genes that have been examined little, but will likely indicate elements responsible for distinctive expression patterns of the NIFP are the promoter and enhancer elements. The NIFP promoters examined have been found to contain common transcription factor binding sites such as GC boxes (Dynan and Tijan, 1983) and TATAA sequences (Corden *et al.,* 1980), but no CCAAT boxes (Jones *et al.,* 1985) have been delineated. Distinctive sequences such as those that might determine neural-specific expression are not readily apparent. Experiments in transgenic mice, however, have indicated that this determinant lies as close as nucleotide -289 in the NF-L gene (Julien *et al.,* 1988). Indeed, Pieper *et al.* (1987) have shown that sequences responsible for tissue- and stage-specific expression of desmin lie as close as -89 relative to the transcription initiation site. Analysis of NIFP gene transcription regulators, such as the promoter and enhancer elements and the proteins that bind to these sequences, is a major area requiring investigation.

Investigators of IF have not had the advantage of a tool for IF disruption and alteration. The recent availability of molecular probes for the NIFP and methods for their mutation and subsequent introduction into cell lines and mice should tell us how and why the NIFP are important in the development and differentiation of the nervous system.

Note added in proof: Since submission of this manuscript, characterization of the 57-kDa NIFP gene has been published by M. A. Thompson and E. B. Ziff (Thompson and Ziff, 1989).

References

Black, M. M., and Lee, V. M.-Y., 1988, Phosphorylation of neurofilament proteins in intact neurons: demonstration of phosphorylation in cell bodies and axons, *J. Neurosci.* **8:**3296–3305.

Brody, B. A., Ley, C. A., and Parysek, L. M., 1989, Selective distribution of the 57kD neural intermediate filament protein in the rat CNS, *J. Neurosci.,* **9:**2391–2401.

Carden, M. J., Trojanowski, J. Q., Schlaepfer, W. W., and Lee, V. M.-Y., 1987, Two-stage expression of neurofilament polypeptides during rat neurogenesis with early establishment of adult phosphorylation patterns, *J. Neurosci.* **7:**3489–3504.

Charnas, L., Richter, K., Sargent, T., and Dawid, I., 1987, Complementary DNA cloning of a nervous system specific intermediate filament from Xenopus laevis with homology to mammalian neurofilament, *Soc. Neurosci. Abstr.*

Chiu, F.-C., 1988, A 66kDa protein is a putative subunit of mammalian neurofilament, *Soc. Neurosci. Abstr.*

Cochard, P., and Paulin D., 1984, Initial expression of neurofilaments and vimentin in the central and peripheral nervous system of the mouse embryo in vivo, *J. Neurosci.* **4:**2080–2094.

Corden, J., Wasylyk, B., Buchwalder, A., Sassone-Corsi, P., Kedinger, C., and Chambon, P., 1980, Promoter sequences of eukaryotic protein coding regions, *Science* **209:**1406–1414.

Dickson, G., Prentice, H., Julien, J.-P., Ferrari, G., Leon, A., and Walsh, F. S., 1986, Nerve growth factor activates Thy-1 and neurofilament gene transcription in rat PC12 cells, *EMBO J.* **5:**3449–3453.

Dynan, W. S., and Tijan, R., 1983, The promoter-specific transcription factor Sp1 binds to upstream sequences in the SV40 early promoter, *Cell* **35:**79–87.

Escurat, M., Gumpel, M., Lachapelle, F., Gros, F., and Portier, M.-M., 1988, Expression comparee de deux proteines de filaments intermediaires, la peripherine et la proteine de neurofilament 68 kDa, au cours du developpement embryonnaire du rat, *C.R. Acad. Sci.* **306:**447–456.

Gambetti, P., Autilio-Gambetti, L., and Papasozomenos, S. C., 1981, Bodian's silver method stains neurofilament polypeptides, *Science* **213:**1521–1522.

Geisler, N., Kaufmann, E., Fischer, S., Plessmann, U., and Weber, K., 1983, Neurofilament architecture combines structural principles of intermediate filaments with carboxy-terminal extensions increasing in size between triplet proteins, *EMBO J.* **2:**1295–1302.

Geisler, N., Fischer, S., Vandekerckhove, J., Plessmann, U., and Weber, K., 1984, Hybrid character of a large neurofilament protein (NF-M): Intermediate filament type sequence followed by a long and acidic carboxy-terminal extension, *EMBO J.* **3:**2701–2706.

Geisler, N., Plessmann, U., and Weber, K., 1985a, The complete amino acid sequence of the major mammalian neurofilament protein (NF-L). *FEBS Lett.* **182:**475–478.

Geisler, N., Fischer, S., Vandekerckhove, J., Van Damme, J., Plessmann, U., and Weber, K., 1985b, Protein-chemical characterization of NF-H, the largest mammalian neurofilament component: Intermediate filament-type sequences followed by a unique carboxy-terminal extension, *EMBO J.* 4:57–63.

Goldstein, M. E., Sternberger, L. A., and Sternberger, N. H., 1983, Microheterogeneity ("neurotypy") of neurofilament proteins, *Proc. Natl. Acad. Sci. USA* **80:**3101–3105.

Greenberg, S. G., and Lasek, R. J., 1988, Neurofilament protein synthesis in DRG neurons decreases more after peripheral axotomy than after central axotomy, *J. Neurosci.* **8:**1739–1746.

Hoffman, P. N., Cleveland, D. W., Griffin, J. W., Landes, P. W., Cowan, N. J., and Price, D. L., 1987, Neurofilament gene expression: A major determinant of axonal caliber, *Proc. Natl. Acad. Sci. USA* **84:**3472–3476.

Jones, K. A., Yamamoto, K. R., and Tijan, R., 1985, Two distinct transcription factors bind to the HSV thymidine kinase promoter in vitro, *Cell* **42:**559–572.

Julien, J.-P., Ramachandran, K., and Grosveld, F., 1985, Cloning of a cDNA encoding the smallest neurofilament protein from the rat, *Biochim. Biophys. Acta* **825:**398–404.

Julien, J.-P., Meyer, D., Flavell, D., Hurst, J., and Grosveld, F., 1986, Cloning and developmental expression of the murine neurofilament gene family, *Mol. Brain Res.* **1:**243–250.

Julien, J.-P., Grosveld, F., Yazdanbaksh, K., Flavell, D., Meijer, D., and Mushynski, W., 1987a, The structure of a human neurofilament gene (NF-L): A unique exon-intron organization in the intermediate filament gene family, *Biochim. Biophys. Acta* **909:**10–20.

Julien, J.-P., Tretjakoff, I., Beaudet, L., and Peterson, A., 1987b, Expression and assembly of a human neurofilament protein in transgenic mice provide a novel neuronal marking system, *Genes Dev.* **1:**1085–1095.

Julien, J.-P., Cote, F., Beaudet, L., Sidky, M., Flavell, D., Grosveld, F., and Mushynski, W., 1988, Sequence and structure of the mouse gene coding for the largest neurofilament subunit, *Gene* **68:**307–314.

Kozak, M., 1987, An analysis of 5′-noncoding sequences from 699 vertebrate messenger RNAs, *Nucleic Acids Res.* **15:**8125–8148.

Lasek, R. J., Oblinger, M. M., and Drake, P. F., 1983, Molecular biology of neuronal geometry: Expression of neurofilament genes influences axonal diameter, *Cold Spring Harbor Symp. Quant. Biol.* **48:**731–744.

Lasek, R. J., Phillips, L., Katz, M. J., and Autilio-Gambetti, L., 1985, Function and evolution of neurofilament proteins, *Ann. N.Y. Acad. Sci.* **455:**462–478.

Lawson, S. N., Harper, A. A., Harper, E. I., Garson, J. A., and Anderton, B. H., 1984, A monoclonal antibody against neurofilament protein specifically labels a subpopulation of rat sensory neurones, *J. Comp. Neurol.* **228:**263–272.

Lee, V. M.-Y., Carden, M. J., Schlaepfer, W. W., and Trojanowski, J. Q., 1987, Monoclonal antibodies distinguish several differentially phosphorylated states of the two largest rat neurofilament subunits (NF-H and NF-M) and demonstrate their existence in the normal nervous system of adult rats, *J. Neurosci.* **7:**3474–3488.

Lee, V. M.-Y., Otvos, L., Carden, M. J., Hollosi, M., Dietzschold, B., and Lazzarini, R. A., 1988, Identification of the major multiphosphorylation site in mammalian neurofilaments, *Proc. Natl. Acad. Sci. USA* **85:**1998–2002.

Lees, J. F., Shneidman, P. S., Skuntz, S. F., Carden, M. J., and Lazzarini, R. A., 1988, The structure and organization of the human heavy neurofilament subunit (NF-H) and the gene encoding it, *EMBO J.* **7:**1947–1955.

Leonard, D. G. B., Ziff, E. B., and Greene, L. A., 1987, The detection and characterization of messenger RNAs regulated by nerve growth factor in PC12 cells, *Mol. Cell. Biol.* **7:**3156–3167.

Leonard, D. B. G., Gorham, J. D., Cole, P., Greene, L. A., and Ziff, E. B., 1988, A nerve growth factor-related messenger RNA encodes a new intermediate filament protein, *J. Cell Biol.* **106:**181–193.

Levy, Liem, R. K. H., D'Eustachio, P., and Cowan, N. J., 1987, Structure and evolutionary origin of the gene encoding mouse NF-M, the middle-molecular-mass neurofilament protein, *Eur. J. Biochem.* **166:**71–77.

Lewis, S. A., and Cowan, N. J., 1985, Genetics, evolution, and expression of the 68,000-mol-wt neurofilament protein: Isolation of a cloned cDNA probe, *J. Cell Biol.* **100:**843–850.

Lewis, S. A., and Cowan, N. J., 1986, Anomalous placement of introns in a member of the intermediate filament multigene family: An evolutionary conundrum, *Mol. Cell. Biol.* **6:**1529–1534.

Liem, R. K. H., Yen, S.-H., Salomon, G. D., and Shelanski, M. L., 1978, Intermediate filaments in nervous tissue, *J. Cell Biol.* **79:**637–645.

Lindenbaum, M. H., Carbonetto, S., Grosveld, F., Flavell, D., and Mushynski, W. E., 1988, Transcriptional and post-transcriptional effects of nerve growth factor on expression of the three neurofilament subunits in PC-12 cells, *J. Biol. Chem.* **263:**5662–5667.

Mencarelli, C., Bugnoli, M., Contorni, M., Moscatelli, A., Ruggiero, P., and Pallini, V., 1987, Phosphorylated epitopes of neurofilaments have been conserved during chordate evolution, *Biochem. Biophys. Res. Commun.* **149:**807–814.

Myers, M. W., Lazzarini, R. A., Lee, V. M.-Y., Schlaepfer, W. W., and Nelson, D. L., 1987, The human mid-size neurofilament subunit: A repeated protein sequence and the relationship of its gene to the intermediate filament gene family, *EMBO J.* **6:**1617–1626.

Napolitano, E. W., Chin, S. S. M., Colman, D. R., and Liem, R. K. H., 1987, Complete amino acid sequence and in vitro expression of rat NF-M, the middle molecular weight neurofilament protein, *J. Neurosci.* **7:**2590–2599.

Oblinger, M. M., and Lasek, R. J., 1988, Axotomy-induced alterations in the synthesis and transport of neurofilaments and mirotubules in dorsal root ganglion cells, *J. Neurosci.* **8:**1747–1758.

Oblinger, M. M., Wong, J., and Parysek, L. M., 1989, Axotomy-induced changes in the expression of a type III neuronal intermediate filament gene, *J. Neurosci.* (in press).

Pachter, J. S., and Liem, R. K. H., 1984, The differential appearance of neurofilament triplet polypeptides in the developing rat optic nerve, *Dev. Biol.* **103:**200–210.

Parysek, L. M., and Goldman, R. D., 1988, Distribution of a novel, 57kDa intermediate filament protein in the nervous system, *J. Neurosci.* **8:**555–563.

Parysek, L. M., Chisholm, R. L., Ley, C. A., and Goldman, R. D., 1988, A type III intermediate filament gene is expressed in mature neurons, *Neuron* **1:**395–401.

Phillips, L. L., Autilio-Gambetti, L., and Lasek, R. J., 1983, Bodian's silver method reveals molecular variation in the evolution of neurofilament proteins, *Brain Res.* **278:**219–223.

Pieper, F. R., Slobbe, R. L., Ramaekers, F. C. S., Cuypers, H. T., and Bloemendal, H., 1987, Upstream regions of the hamster desmin and vimentin genes regulate expression during in vitro myogenesis, *EMBO J.* **6:**3611–3618.

Portier, M.-M., Croizat, B., de Nechaud, B., Gumpel, M., and Gros, F., 1983, La proteine Y, un nouveau marqueur potentiel dans la diffenciation neuronale du systeme nerveux peripherique, *C. R. Acad. Sci. Paris,* **297:**57–61.

Portier, M.-M., de Nechaud, B., and Gros, F., 1984, Peripherin, a new member of the intermediate filament protein family, *Dev. Neurosci.* **6:**335–344.

Quax, W., Vree Egberts, W., Hendriks, W., Quax-Jeuken, Y., and Bloemendal, H., 1983, The structure of the vimentin gene, *Cell* **35:**215–223.

Robinson, P. A., Wion, D., and Anderton, B. H. 1986, Isolation of a cDNA for the rat heavy neurofilament polypeptide (NF-H). *FEBS Lett.* **209:**203–205.

Schlaepfer, W. W., and Freeman, L. A., 1978, Neurofilament proteins of rat peripheral nerve and spinal cord, *J. Cell Biol.* **78:** 653–662.

Sharp, G., Shaw, G., and Weber, K., 1982, Immunoelectron microscopical localization of the three neurofilament triplet proteins along neurofilaments of cultured dorsal root ganglion neurons, *Exp. Cell Res.* **137:**403–413.

Sharpe, C. R., 1988, Developmental expression of a neurofilament-M and two vimentin-like genes in Xenopus laevis, *Development* **103:**269–277.

Shaw, G., and Weber, K., 1982, Differential expression of neurofilament triplet proteins in brain development, *Nature* **298:**277–279.

Shaw, G., Osborn, M., and Weber, K., 1981, An immunofluorescence microscopial study of the neurofilament triplet proteins, vimentin and glial fibrillary acidic protein within the adult rat brain, *Eur. J. Cell Biol.* **26:**68–82.

Shaw, G., Debus, E., and Weber, K., 1984, The immunological relatedness of neurofilament proteins of higher vertebrates, *Eur. J. Cell Biol.* 34:130–136.

Steinert, P. M., and Roop, D. R., 1988, Molecular and cellular biology of intermediate filaments, *Annu. Rev. Biochem.* **57:**593–625.

Tetzlaff, W. Bisby, M. A., and Kreutzberg, G. W., 1988, Changes in cytoskeletal proteins in the rat facial nucleus following axotomy, *J. Neurosci.* **8:**3181–3189.

Thompson, M. A., and Ziff, E. B., 1989, Structure of the gene encoding peripherin, an NGF-regulated neuronal-specific Type III intermediate filament protein, *Neuron* **2:**1043–1053.

Trojanowski, J. Q., Walkenstein, N., and Lee, V.M.-Y., 1986, Expression of neurofilament subunits in neurons of the central and peripheral nervous system: An immunohistochemical study with monoclonal antibodies, *J. Neurosci.* **6:**650–660.

Vitadello, M., Triban, C., Fabris, M., Gorio, A., and Schiaffino, S., 1986, Heterogeneity of rat neurofilament polypeptides revealed by a monoclonal antibody, *J. Neurochem.* **46:**665–670.

Vitadello, M., Triban, C., Fabris, M., Dona, M., Gorio, A., and Schiaffino, S., 1987, A developmentally regulated isoform of 150,000 molecular weight neurofilament protein specifically expressed in autonomic and small sensory neurons, *Neuroscience* **23:**931–941.

Weber, K., Shaw, G., Osborn, M., Debus, E., and Geisler, N., 1983, Neurofilaments, a subclass of intermediate filaments: Structure and expression, *Cold Spring Harbor Symp. Quant. Biol.* **47:**717–729.

Willard, M., and Simon, C., 1981, Antibody decoration of neurofilaments, *J. Cell Biol.* **89:**198–205.

Willard, M., and Simon, C., 1983, Modulations of neurofilament axonal transport during the development of rabbit retinal ganglion cells, *Cell* **35:**551–559.

Wong, J., and Oblinger, M. M., 1987, Changes in neurofilament gene expression occur after axotomy of dorsal root ganglion neurons: An in situ by hybridization study, *Metab. Brain Dis.* **2:**291–303.

Zopf, D., Hermans-Borgmeyer, I., Gundelfinger, E. D., and Betz, H., 1987, Identification of gene products expressed in the developing chick visual system: Characterization of a middle-molecular-weight neurofilament cDNA, *Genes Dev.* **1:**699–708.

14

The Structure, Complexity, and Evolution of Intermediate Filament Genes

Peter M. Steinert and Dennis R. Roop

1. Introduction

The foregoing chapters in this book have directly or indirectly posed many fundamental questions pertaining to the mechanisms by which the expression of intermediate filament (IF) genes is regulated during development and differentiation. A consensual hypothesis established repeatedly in this book suggests that the morphological similarity of IF resides in the highly conserved secondary structure of the rod domains of all IF chains. The end domains, which vary widely in size and amino acid sequence among IF chains (Table I), are thought to occupy peripheral positions on intact IF where they are intimately involved in the functions of the IF in cells (Steinert *et al.*, 1985a; Steinert and Roop, 1988). Accordingly, the fundamental role of IF gene expression is to generate a particular set of exposed end domains, coupled to specific rod domains, that admit the functions of the IF required by the cell in which they occur.

A first step toward a complete understanding of the role of IF in cells involves characterization of the structure and organization of the IF genes. Such information may provide clues on how the relatively complex yet conserved rod domains evolved into the currently known five sequence types, and how end domain sequences evolved and were recruited onto the various rod domain types to provide the array of modern IF proteins. Moreover, characterization of the genes will ultimately lead to identification of regulatory sequences and factors that control the expression of individual or coexpressed IF genes during development and differentiation.

Peter M. Steinert • Dermatology Branch, National Cancer Institute, National Institutes of Health, Bethesda, Maryland 20892; *present address:* Laboratory of Skin Biology, National Institute of Arthritis and Musculoskeletal and Skin Diseases, National Institutes of Health, Bethesda, Maryland 20205. **Dennis R. Roop** • Departments of Cell Biology and Dermatology, Baylor College of Medicine, Houston, Texas 77030.

To date, our knowledge of these matters is still rather limited, but the study of the molecular biology of IF is a rapidly expanding area of research. In this chapter we will summarize the current data describing gene structure and complexity, and propose possible models that offer explanations for the evolution of IF genes.

2. *Structure of IF Genes*

2.1. *Complexity*

Complete sequence and structure information now exists for many IF genes (Steinert and Roop, 1988). In general, most genes so far characterized appear to exist in one copy per haploid genome, although some exceptions are known to exist. The type II keratin 6 gene system possesses two genes, 6a and 6b (Tyner *et al.*, 1985), that differ significantly only in 3′-noncoding sequences toward the ends of the genes. The chicken vimentin gene can express at least three different mRNA species that differ only in the lengths of their 3′-noncoding regions by virtue of variable termination at one of four possible polyadenylation sites within the single gene (Zehner and Paterson, 1983). In the case of the variants of the keratin 6a/b and chicken vimentin genes, it is not known whether individual members are differentially expressed. The genome appears to possess pseudogenes for some IF gene systems. Human keratins 16 and 14 have one and two pseudogenes, respectively (Savtchenko *et al.*, 1987; Rosenberg *et al.*, 1988) and the mouse endo A (keratin 8) gene has one (Vasseur *et al.*, 1985). However, it should be pointed out that in several studies, the identification of only one copy was based on the finding of a single band on blotting of genomic DNA; experiments using more quantitative methods will be necessary to exclude the possibility of multiple identical copies, or even multiple copies with only minor degrees of polymorphism not readily identifiable by restriction fragment length polymorphisms. In fact, recent protein-chemical data have demonstrated polymorphisms in several keratins (Mischke and Wild, 1987), which could be due to multiple gene copies or extensive allelic variations within a given population. Nevertheless, these studies taken together firmly establish that IF chain diversity originates at the gene level, and is not generated by unusual somatic rearrangement events.

2.2. *IF Gene Linkage and Chromosomal Localization*

Several recent reports have described linkage of IF genes that suggests clustering. There are no substantiated reports of close linkage of genes of different sequence types. Clusters of three type I and two type II sheep wool keratin genes have been located on separate cosmid clones (Powell *et al.*, 1986). Interestingly, in this study, an "orphan" exon encoding the latter half of the 2B rod domain (see Fig. 1) was found between the two type II genes. Three type I human genes are located within 20 kbp (RayChaudhury *et al.*, 1986) and preliminary data from overlapping cosmid clones have extended this cluster to as many as five type I genes within 60 kbp (Bowden *et al.*, 1987). In other studies, a cosmid clone containing keratin 1 and another unspecified type III gene are separated by 15 kbp (Bowden *et al.*, 1987); two type I *Xenopus* genes are separated by 10 kbp (Miyatani *et al.*, 1986); and four type II bovine genes are located within 20 kbp (Blessing *et al.*, 1987).

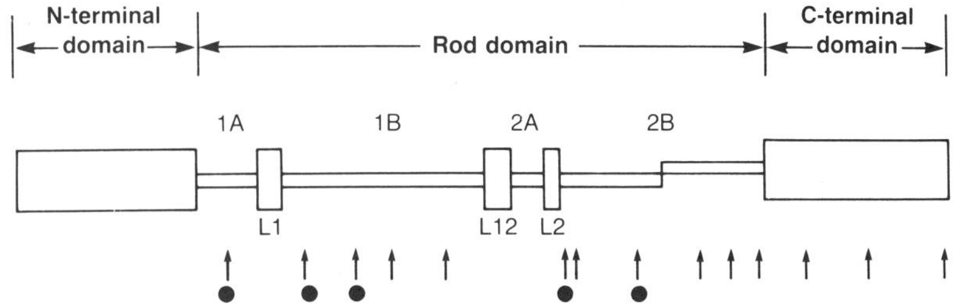

Mouse 59 kD (I)		14	42		94	4		46			119			113
Bovine 54 kD (I)		14	42		94	4		46			119			84
Human K14 52 kD (I)		14	42		94	4		46			120	15		
Human K17 46 kD (I)		14	42		94	4		46			120	15		
Bovine K19 44 kD (I)		14	42		94	4		46						
Human K1 67 kD (II)	17		42	62	94		7	46			119	12		
Human K3 63 kD (II)	17		42	62	94	4		46			119	12		
Human K6b 60 kD (II)	17		42	62	94	4		46			119	11		
Hamster Vimentin 53 kD (III)			42	62	94	4		46			120	14	42	
Hamster Desmin 52 kD (III)			42	62	94	4		46			120	14	41	
Mouse GFAP 50 kD (III)			42	62	94	4		46			120	14	42	
Mouse NF-L 68 kD (IV)									79	111				95
Human NF-L 68 kD (IV)									79	111				95
Human NF-M 105 kD (IV)									79	111				

Figure 1. Locations of introns in selected type I–IV IF genes. The arrows indicate where introns splice the coding regions of a universal model of an IF chain. The numbers refer to the amino acid number along the rod domain or carboxy-terminal end domain. The arrows with solid circles indicate those introns that splice at the first residue of a heptad repeat. The human lamin A and C chains contain an extra six heptads (42 amino acids) in their IB tract, inserted at the conserved intron position 42 in type I–III IF genes (Parry *et al.*, 1986; Weber, 1986). The wool keratin type II "orphan" exon (Powell *et al.*, 1986) spans residues 46–119 of the 2B tract. From Steinert and Roop (1988). References (from the top): Krieg *et al.* (1985), Rieger *et al.* (1985), Marchuk *et al.* (1984), RayChaudhury *et al.* (1986), Bader *et al.* (1986), Johnson *et al.* (1985), Klinge *et al.* (1987), Tyner *et al.* (1985), Quax *et al.* (1983, 1985a), Balcarek and Cowan (1985), Lewis and Cowan (1986), Julien *et al.* (1987), and Myers *et al.* (1987).

By use of human–rodent somatic cell hybrids, the single copy human vimentin and desmin gene are located on chromosome 10 and 2, respectively (Quax *et al.*, 1985b). The human vimentin gene has been further localized to chromosome region 10p13 by *in situ* hybridization (Ferrari *et al.*, 1987). Similarly, the type IV human neurofilament NF-L gene has been localized to chromosome 8 (Julien *et al.*, 1987). The human keratin 3 (type II) and keratin 13 (type 1) gene have been mapped to chromosome 12 and 17, respectively (Romano *et al.*, 1987). Two clusters of human keratin 14 and 16 genes have been mapped on chromosome 17, with functional genes localizing to the 17q12→q21 region and nonfunctional pseudogenes for these two keratins to 17p12 (Rosenberg *et al.*, 1988). Further, the coexpressed human keratins 1 (type II) and 10 (type I) localize to 12q11→q13 and 17q12→q21, respectively (Lessin *et al.*, 1988). Thus, four type I keratins (10, 13, 14, 16) have been mapped to chromosome 17 and two type II keratins (1, 3) have been mapped to chromosome 12. Accordingly, it appears there are at least two clusters of type I keratins, a functional set located at chromosome region 17q12→q21, a smaller nonfunctional set at region 17p12, that perhaps arose due to duplication of part of this type I keratin locus, and a single cluster of type II keratins at chromosomal region 12q11→q13. It remains to be determined whether all keratin genes are localized in these clusters or whether there are multiple loci for each sequence type.

Although the current data base is still incomplete, there is no evidence that keratin genes of either sequence type are arranged in an order related to their expression during development and differentiation, a situation quite different from the organization of globin genes, for example (Proudfoot *et al.*, 1980).

2.3. IF Gene Structure

A comparison of the organization of some IF genes of four sequence types for which complete information on intron/exon structure is available is shown in Fig. 1. Using a schematic model for the organization of IF chain secondary structure, type I IF genes contain seven introns, and type II and III genes possess eight introns, of which most interrupt regions encoding sequences that form the rod domain. The conservation of intron positions among these three types is striking. Three introns are located in identical positions. The positions of two others, one located near the beginning and the other toward the end of the 2B coiled-coil tract, are highly conserved. Type II genes have one unique intron and type II and type II genes share another common intron in their rod domains. Each type possesses one or two introns that splice the regions encoding the carboxyterminal but not amino-terminal end domains. It should also be noted that overlaying the remarkable conservation of intron positions, there is some variation in the exact nucleotide spliced by the intron: for example, that located near the end of segment 2B "wobbles" between different genes by three nucleotides (Rieger *et al.*, 1985). Except for the two introns that splice near the ends of the 2B tract of the rod domain, no others appear to delineate obvious functional domains or subdomains of the IF chains. However, further examination of these genes reveals that several of the introns splice the nucleotides of the codon for the first amino acid of the characteristic heptad repeat of the coiled-coil tracts (Krieg *et al.*, 1985; Roop and Steinert, 1986). This occurs in four of the six introns of type I IF genes, and three of six introns in the type II and III genes (Fig. 1).

The structure of type IV neurofilament genes is, however, markedly different (Fig. 1). There are only two introns in the 2B region of the rod domain, in positions quite different from type I–III IF genes. It is therefore apparent that the neurofilament genes have evolved differently.

3. Models for the Evolution of IF Genes

3.1. Evolution of the Rod Domain

Any model to account for the origin of IF genes must accommodate two facts: first, the striking conservation of the secondary structure of the rod domain of all extant IF chains; and second, type IV IF genes have an intron/exon structure different from that of type I–III IF genes. Current ideas of rod domain evolution are predicated on the tenet that all sequence types evolved from a common precursor. In one hypothesis, Lewis and Cowan (1986) suggested that the intron/exon structure of the mouse NF-L gene may have occurred by the transposition of an mRNA intermediate, by mechanisms previously proposed to account for the origin of processed pseudogenes (Nishoida *et al.*, 1980; Lee *et al.*, 1983). Because all neurofilament genes now appear to contain two introns in the same location in the rod domain, this model would have required the insertion on two introns into a common progenitor of the neurofilament genes before subsequent duplication. A more likely model, illustrated in Fig. 2, is developed below.

Gilbert (1978) postulated that the structure of eukaryote genes reveals their evolutionary history, in that genes evolved by recruitment and combination of smaller sequences, separated by introns, which encode distinct functional protein domains. While this may not be strictly the case for IF genes (see Fig. 1), the nonrandom occurrence of introns between heptad repeats within the coiled-coil tracts of the rod domains of extant type I–III IF genes

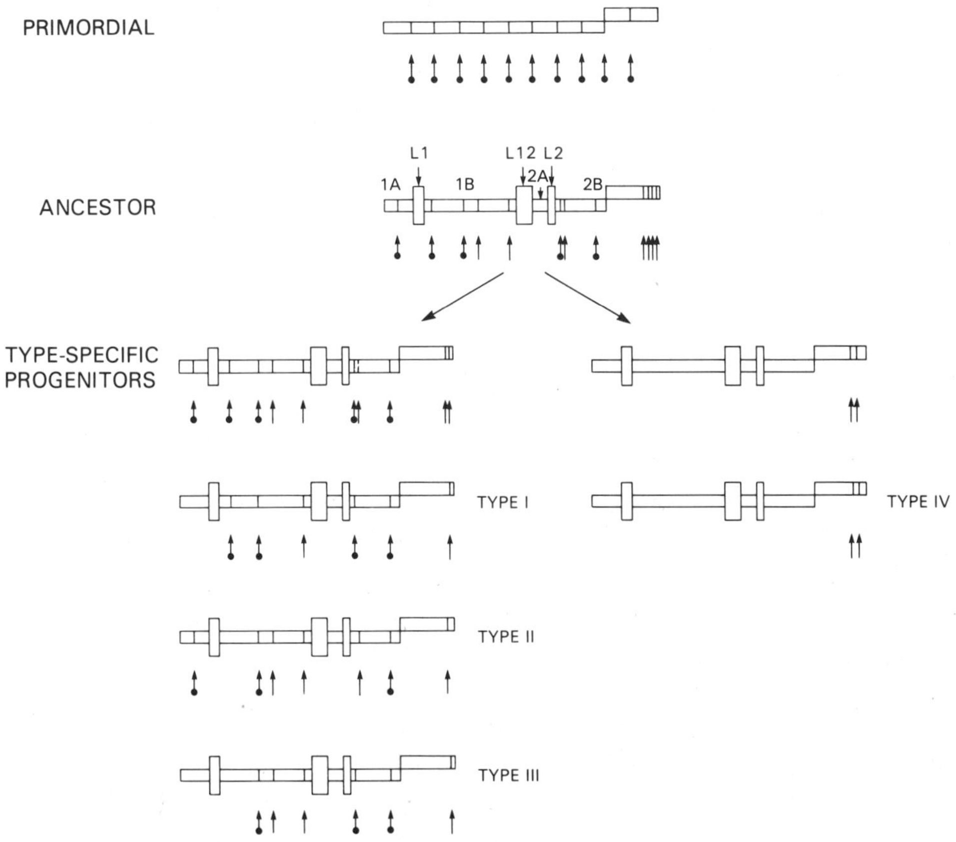

Figure 2. A model for the evolution of the rod domain of IF genes. This envisages several distinct steps: (1) A primordial gene was assembled from peptides containing four (or more) heptad repeats. The heptad discontinuity in the 2B region may have been introduced at this time. (2) Next, an ancestral gene containing three interruptions in the regularity of the heptads was constructed perhaps by sliding of intron/exon boundaries or by mutations of certain residues to form non-coiled-coil linkers. The lamin gene precursor may have originated at or before this time. (3) This ancestral gene duplicated to form two progenitor lineages, one for neuronal IFs and the second for nonneuronal IFs; subsequently, the latter diverged further to form the individual type I, II, and III progenitors. (4) More recent duplication events took place to form the various members of each sequence type. Redrawn from Roop and Steinert (1986).

does suggest a variation of this idea. It seems possible the rod domain of IF genes arose by amplification and duplication of a simple nucleotide sequence encoding multiple heptads that were separated by introns (Roop and Steinert, 1986) (Fig. 2). The length of this original nucleotide sequence is uncertain, but it may have encoded a single heptad (21 bp), or multiples thereof. The smallest exon encoding exact heptad repeats that has been observed in IF genes is four heptads long, encoded by 84 bp, located between amino acid positions 14–42 within the 1B tract of type I genes (Fig. 1). This may represent such a primal nucleotide sequence for two other reasons. First, a 28-amino-acid-residue repeat in the linear distribution of acidic and basic amino acids has been documented in all IF chains (Parry *et al.*, 1977; Crewther *et al.*, 1983; Parry and Fraser, 1985; Conway and Parry, 1988; Parry, this volume). Second, physicochemical studies employing synthetic peptides have revealed that a minimum of four heptads (28 residues) is required to form a stable two-chain coiled coil *in vitro* (Lau *et al.*, 1984). Irrespective of the exact size of this short exon encoding multiple heptad repeats, numerous amplification events would then have been necessary to generate the entire rod domain of about 300 amino acids. Similar mechanisms have been proposed for the formation of progenitors of collagen genes (Yamada *et al.*, 1980), immunoglobulin genes (Ohno *et al.*, 1982), albumin:α-fetoprotein genes (Alexander *et al.*, 1984), and the pyruvate kinase gene (Lonberg and Gilbert, 1985). One other structural feature that has been rigorously conserved among all IF chains is the presence of a heptad polarity reversal near the middle of tract 2B (Parry and Fraser, 1985; Steinert *et al.*, 1985a; Conway and Parry, 1988). This may have been generated by a recombination event during the construction of the primordial gene.

The next stage of evolution may have been the formation of an ancestral gene, in which oligonucleotides encoding the non-coiled-coil amino acid sequences appeared in the gene. The advent of the linker sequences presumably provided some important evolutionary advantage to the cytoskeletal architecture, such as filament formation and stability. The process by which these events may have occurred is unknown, but a mechanism involving sliding of intron/exon boundaries has been suggested for the generation of divergent sequences in other protein family systems (Craik *et al.*, 1983). For example, mutations at a splice junction resulting in the loss of the site and subsequent splicing at an alternate location within an intron could result in the extension of a portion of an exon, perhaps encoding linker amino acid sequences. An alternative possibility is that nucleotide sequences encoding preexisting continuous coiled-coil sequences were altered by deletions and/or rearrangements resulting in the sequence discontinuities observed today.

In addition to this process, another complex series of events occurred during the generation of the ancestral gene to explain the loss of certain of the introns originally thought to be present in the primordial gene. For example, introns that may have delineated the coiled-coil tract 1A, and at the beginning of tracts 1B and 2A are absent in all IF genes (Fig. 1). Interestingly, the precise excision of introns without apparent concomitant loss of exon sequences has been reported for the rat insulin gene (Lomedico *et al.*, 1979) and the chicken pro α-1 collagen gene (Yamada *et al.*, 1980).

Subsequent to the appearance of this ancestral gene, it is thought that another duplication event occurred, with selective loss of certain introns, to generate two progenitor genes, one for the type I–III gene lineage, another for the type IV genes (Roop and Steinert, 1986; Myers *et al.*, 1987; Julien *et al.*, 1987) (Fig. 2). Based on a model for the origin and evolution of neurofilament genes (Lasek *et al.*, 1985), it is estimated that this event took place at least 800–1000 million years ago, during or before the appearance of complex metazoa. Within the first lineage, further duplication events occurred together with additional selective loss of introns and divergence of exon sequences to produce the type I, II, and III-specific progenitors seen today. Variations in the precise splice points of the introns mentioned above may have also occurred at this stage. Finally, additional duplications of these type-specific progenitors would then have generated the various members of each IF type. The driving force for the generation of multiple type-specific genes may have been the evolution of the end domains whose varying sequences were required for the evolution of progressively more complex functions of the IF (see below).

There are at least two predictions of this model of rod domain evolution that seem to have been established by recent work. First, the hypothesis would predict that the structure of the genes encoding chains of the same sequence type should be highly conserved. This now seems to be generally true (Fig. 1). Second, at least some genes of the same type should be linked, as has also now been demonstrated (see Section 2).

3.2. Evolution of the End Domains

3.2.1. Organization of the End Domains

The current view of IF structure states that the hypervariable end domains of IF chains occupy superficial positions on the assembled IF from where they specify the function of the IF. In Fig. 3 and Table I are illustrated the organization and sizes of the various subdomains of a number of type I to IV IF chains based on: (1) basic ends (E1, E2 subdomains); (2) regions of peptide repeats or sequences of special variability (V subdomains); and (3) homologous regions (H subdomains), proposed by Steinert *et al.* (1985b), Conway and Parry (1988), and Zhou *et al.* (1988). It is apparent from Fig. 3 that organization of these end subdomains is highly characteristic of each rod domain sequence type.

Type III chains possess a short homologous region H1 adjacent to the rod domain. The remainder of the amino-terminal end domain consists of very basic (E1) sequences that are hypervariable in size and sequence. The entire carboxyterminal domain has been conserved in size and sequence among the three known type III chains (H2). Type IV neurofilament chains contain short E1, H1, and H2 subdomains and distinctive V2 subdomains containing numerous highly charged peptide repeats. Type II keratins contain homologous H1 and H2 subdomains adjacent to either side of the rod domain that have been conserved within this family. Type I keratins have a short but somewhat variable H1 subdomain (Conway and Parry, 1988; Zhou *et al.*, 1988). The presence of the conserved H1 and H2 subdomains in type II keratins largely accounts for the 6–8 kDa of additional mass between coexpressed type II and type I keratins (Sun *et al.*, 1984; Steinert *et al.*, 1985b). All keratin chains possess end subdomains, E1 and E2, that are typically basic in charge, of random coil conformation and variable in sequence such that they often contain unique epitopes (Roop *et al.*, 1984). The major variation between different keratin chains and neurofilament chains resides in the number of amino acids in their V subdomains (Table I). In keratin chains, the peptide repeats are of the form XY_n, where X = an apolar amino acid (most usually phenylalanine, tyrosine, or isoleucine), Y = glycine or serine, and n = 1–9 (Table II; Steinert *et al.*, 1985b). In keratin chains expressed in simple epithelia, the V2 subdomains tend to be quite short, or entirely absent, and Y of the consensus peptide repeat is most often a serine. In the keratins expressed in more complex epithelia, the V2 subdomains are longer and are enriched in glycines (Table II). The

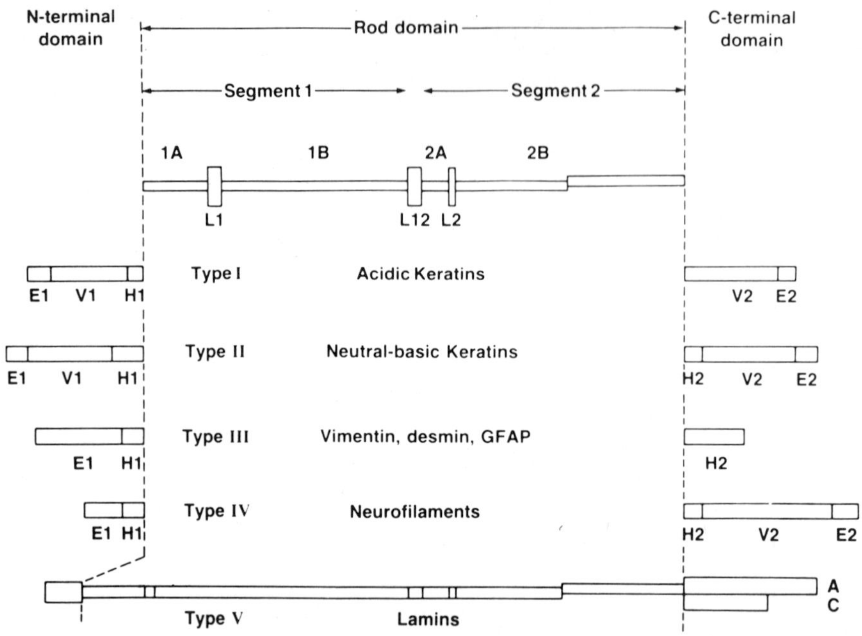

Figure 3. IF chain subdominal organization. All chains contain a central rod domain flanked by end domains. In type I–IV IF chains, the four α-helical tracts are of invariant size (see Table II); the linkers vary slightly in size between the types. In type V lamins, 1B is six heptads longer due to an insertion at position 42 conserved in type I–III genes (Parry *et al.*, 1986; Weber, 1986) and the linkers are α-helical or coiled-coil. Each rod domain type is coupled with end domains that may be divided into specific subdomains based on homologous (H), variable (V), or end (E) sequences. See Table II for amino acid numbers. From Steinert and Roop (1988).

keratin chains expressed in hair and related tissues do not possess V subdomains, but rather elongated cysteine-rich E subdomains instead (Conway and Parry, 1988). In neurofilament chains, the numbers of peptide repeats vary between the different chains and the same chain in different species, but all are variations along a theme of Lys-Ser-Pro-Val-Pro, where the Ser residue is often phosphorylated (Julien *et al.*, 1987; Myers *et al.*, 1987; Napolitano *et al.*, 1987).

3.2.2. End Domains of Keratin Chains

The existence of well-defined subdomains and repeats within the V subdomains of the keratin chains implies an orderly process in their evolution. It is possible these portions evolved about the time of the appearance of the type-specific progenitors of the rod domains. The first step may have been the recruitment of the H1 and H2 subdomains onto the progenitor gene of the type II keratins, since these have been precisely conserved throughout further evolution. Presumably these sequences are critical elements of keratin IF structure or function. A second step was the recruitment of E1 and E2 end subdomains for both type I and type II progenitor genes, which have subsequently diverged widely in size and sequence, but not in overall charge and perhaps conformational properties. In the case of keratin 10 chains, and possibly other keratins as well, the E subdomains have been

Table I. Number of Amino Acids in Various Domains of Selected IF Chains[a]

Protein chain (type)		N-terminal end domain			Rod domain							C-terminal end domain		
		E1	V1	H1	1A	L1	1B	L12	2A	L2	2B	H2	V2	E2
Human K10	I	58	71	9	35	14	101	16	19	8	121	—	89	19
Mouse K10	I	57	75	10	35	14	101	16	19	8	121	—	95	16
Bovine K10	I	57	60	9	35	14	101	16	19	8	121	—	65	19
Human K14	I	14	91	9	35	14	101	16	19	8	121	—	76	9
Human K17	I	14	91	9	35	14	101	16	19	8	121	—	19	27
Bovine K19	I	16	42	8	35	14	101	16	19	8	121 (+9)	—	—	—
Wool 8c	I	25	—	7	35	14	101	16	19	8	121	—	—	47
Human K1	II	37	106	36	35	12	101	17	19	8	121	20	109	22
Mouse K6	II	27	88	36	35	12	101	17	19	8	121	20	54	14
Human K6b	II	25	100	36	35	12	101	17	19	8	121	20	53	15
Human K7	II	26	63	36	35	12	101	17	19	8	121	20	29	17
Hamster desmin	III	86	—	20	35	8	101	16	19	8	121	54	—	—
Hamster vimentin	III	81	—	20	35	8	101	16	19	8	121	55	—	—
Mouse GFAP	III	33	—	20	35	8	101	16	19	8	121	56	—	—
Mouse NF-L	IV	71	—	20	35	10	101	17	19	8	121	97	—	52
Human NF-L	IV	71	—	20	35	10	101	17	19	8	121	88	—	61
Human NF-M		83	—	20	35	10	101	17	19	8	121	88	293	53
Human lamin A	V	—	—	33	35	(12)	143	(17)	19	(8)	121	312	—	—
Human lamin C	V	—	—	33	35	(12)	143	(17)	19	(8)	121	184	—	—

[a]References: type *—Zhou *et al.* (1988), Steinert *et al.* (1983), Rieger *et al.* (1985), Marchuk *et al.* (1984), RayChaudhury *et al.* (1986), Bader *et al.* (1986), Dowling *et al.* (1986); type II—Johnson *et al.* (1985), Steinert *et al.* (1984), Tyner *et al.* (1985), Glass *et al.* (1985); type III—Quax *et al.* (1983, 1985a), Balcarek and Cowan (1985), Lewis and Cowan (1986); type IV—Julien *et al.* (1987), Myers *et al.* (1987); type V—McKeon *et al.* (1986), Fisher *et al.* (1986).

conserved between different species (Zhou *et al.*, 1988). The reason why these subdomains may have been recruited at a later time is that the smallest known keratin 19 (type I) chain does not possess an E2 subdomain at all and may thus represent an early form of type I keratins (Bader *et al.*, 1986). Using the same logic, the third step in end domain evolution involved the formation of the V subdomains. They are very short (or absent) in the smallest keratins of simple epithelia of both sequence types (Table I), but enlarge considerably due to the repetitive consensus peptide in the keratins of more complex epithelia. Thus, it seems likely that as more complex epithelia evolved, the requirement of IF to perform more specialized functions was achieved by progressively enlarging the V subdomains, with subsequent sequence divergence (Table II). The simplest way this may have occurred is by periodic amplification of the consensus peptide repeat. Smaller keratin chains contain V1 subdomains that are somewhat larger than the V2 subdomains (Table I), suggesting that the V1 subdomains evolved first.

In addition to the marked correlation between the complexity of an epithelium and the size of the V subdomains of the keratin chains expressed in it, there is a correlation of the sequence of the V subdomains and the complexity of the cell type (Table II). As the complexity of epithelia increases, the V subdomains change from being serine-rich to glycine-rich. This appears to occur not by addition of glycine-rich peptide repeats to

Table II. Peptide Repeats and Codon Usages in V Subdomains of Keratin IF Chains

Epithelial class/V subdomain	Most common peptide repeats	Codon usage
Simple[a]		
V1	$F_{0.6}Y_{0.3}I_{0.1}(G_{0.2}S_{0.8})_{2-9}$	Glycine: $GGC_{0.4}GGT_{0.2}GGG_{0.2}$ Serine: $AGC_{0.6}TCT_{0.1}TCC_{0.3}$
V2	Apolar$(S_{1.0})_{1-5}$	Serine: $AGC_{0.6}TCT_{0.2}TCC_{0.2}$ Apolars[b]
Stratified squamous[c]		
V1	$F_{0.4}Y_{0.2}I_{0.2}(G_{0.5}S_{0.4})_{2-8}$	Glycine: $GGC_{0.4}GGT_{0.2}GGG_{0.3}$ Serine: $AGC_{0.5}TCT_{0.2}TCC_{0.2}$ Apolars[b]
V2	Apolar$(G_{0.5}S_{0.4})$	Glycine: $GGC_{0.4}GGT_{0.3}GGA_{0.2}$ Serine: $AGC_{0.5}TCT_{0.2}TCC_{0.2}$ Apolars[b]
Terminally differentiating[d]		
V1	$F_{1.0}(G_{1.0})_{1-8}$ $F_{0.8}Y_{0.2}(G_{0.7}S_{0.2})_{2-6}$	Phenylala: $TTT_{0.8}TTC_{0.2}$ Tyrosine: $TAT_{0.4}TAC_{0.6}$ Glycine: $GGT_{0.4}GGC_{0.3}GGG_{0.2}$
V2	$Y_{1.0}(G_{0.6}S_{0.2})_{3-9}$	Tyrosine: $TAT_{0.4}TAC_{0.6}$ Glycine: $GGT_{0.1}GGC_{0.6}GGG_{0.1}$ Serine: $AGC_{0.5}TCT_{0.2}TCC_{0.2}$

[a]Data from human K7 (Glass *et al.*, 1985); K17 (Raychaudhury *et al.*, 1986).
[b]Insufficient data.
[c]Data from human K6a (= 6b) (Tyner *et al.*, 1985); human K14 (Marchuk *et al.*, 1984); mouse 60 kDa (Steinert *et al.*, 1984); mouse 55 kDa and 50 kDa (D. Roop and P. Steinert, unpublished).
[d]Data from human K1 (Johnson *et al.*, 1985); human K10 (Zhou *et al.*, 1988); mouse 59 kDa (Steinert *et al.*, 1983); mouse 67 kDa (Steinert *et al.*, 1985b).

existing serine-rich peptides, but by changing serines to glycines (Table II). This enrichment of glycines contributes to a progressively more insoluble sequence, and thus IF network, in the more specialized epithelia. For example, in terminally differentiating epidermal cells, the V1 and V2 sequences are 60–70% glycine and this is thought to contribute to the insoluble yet flexible structure that is characteristic of the protective barrier of the outer layer of mammalian skin (Steinert *et al.*, 1985a,b). On examination of the nucleotide sequences encoding the V subdomain sequences of keratins, it becomes apparent that such changes from serine to glycine can often be achieved by single base changes in the most commonly used codons (Table II). However, it is not obvious how similar V1 and V2 subdomain exon sequences were recruited onto coexpressed type I and type II keratin genes. Possibly, the V subdomains evolved in a separate location on the genome and were later recruited by complex recombinational events. Analyses of many additional genes may be necessary to resolve these questions.

3.2.3. Evolution in Neurofilament Chains

The type IV NF-L, NF-M, and NF-H neurofilament chains differ primarily in the sizes of their V2 subdomains (Weber and Geisler, 1984; Julien *et al.*, 1987; Lewis and Cowan, 1986; Myers *et al.*, 1987; Napolitano *et al.*, 1987). These are composed of unusual glutamate-lysine-rich peptide repeats that are also highly phosphorylated, presumably form cross-bridges in axons (Hisanaga and Hirokawa, 1988), and bind the Bodian silver stain. Interestingly, there is a progressive increase in the size of the neurofilament chains, i.e., in the size of their V2 subdomains, with increasing complexity of the organism and its nervous system: simple eukaryotes such as *Apylsia* have only low-molecular-weight neurofilament chains, compared to the largest mammalian neurofilament chains (Lasek *et al.*, 1985). This observation may be linked with the fact that developing neurons do not express the neurofilament triplet protein coordinately (Shaw and Weber, 1984), yet mature axons contain predominantly the larger NF-M and NF-H chains. By processes analogous to those proposed for the keratins, the elongation of the V2 subdomain region may have evolved by repeated duplication of an early neurofilament-specific sequence in response to the requirement for more complex neurological systems (Myers *et al.*, 1987). This repeating element may initially have been separated by introns that were later lost (Lasek *et al.*, 1985).

3.3. Evolution of Lamin Genes

The nuclear lamina is a complex fibrous structure located on the inner surface of the nuclear membrane where it is thought to coordinate chromatin structure. Its principal structural components are three or four proteins termed lamins (see Dessev, this volume). The amino acid sequences of lamins A and C have been reported and are quite similar to the cytoplasmic type I–IV IF chains (McKeon *et al.*, 1986; Fisher *et al.*, 1986), yet they display several curious differences. Lamins prefer to form paracrystalline structures *in vitro* of varying widths, depending on the ionic strength (Zackroff *et al.*, 1984; Goldman *et al.*, 1986; Aebi *et al.*, 1986; Parry *et al.*, 1987). Their rod domains differ in the distributions of ionic charges; the linkers are all α-helical or can form a coiled coil; and the exact sequences indicate they constitute a distinct type V class of IF proteins (Fig. 3). Most notably, the 1B tract of their rod domains contains exactly 42 amino acids (six heptads) more than the other IF types, which are inserted precisely at the point of the common intron location of type I–III IF genes (Fig. 1). Thus, type V lamin rod domains are considerably more regular than the other types, more similar to tropomyosin (Parry *et al.*, 1987).

While the structure of the lamin gene is not known, it is likely their evolution is closely related to the cytoplasmic IF types. Lamins presumably emerged very early in eukaryote cell evolution, perhaps 1000 million years ago, well before the appearance of complex metazoa, in order to coordinate chromatin structure and function during cell division and differentiation. Since they possess a considerably more regular rod domain than the cytoplasmic IF types, it is possible they are closely related to or are even the antecedents of the ancestral IF gene prior to its divergence into the distinct neuronal and nonneuronal lineages. Characterization of the complexity and structure of lamin genes of different phyla will advance our understanding of these events.

4. Conclusions and Future Perspectives

A model for the evolution of IF genes has been proposed and is based on the recent advances in our understanding of their structures, complexity, and organization in the genome. The central rod domain common to all IF may have arisen in a primordial gene by amplification of a simple DNA sequence encoding multiple heptad repeats to form an extended coiled-coil chain. Subsequently, this was reorganized to form an ancestral gene encoding a protein capable of forming an IF structure. This may be closely related to modern type V lamins. During the evolution of complex metazoa and the development of specialized cell types, this ancestral gene was duplicated to form two progenitor lineages, one for type IV neuronal IF and a second for nonneuronal IF that further diverged to generate the type I–III type-specific progenitor genes. Finally, each progenitor line diverged to produce the various members seen in modern IF proteins. The end domains that define IF function in cells may also have evolved by amplification and then divergence of simple sequences at about the same time as the type-specific progenitors. This model should prove to be very useful for further analyses of the complexity of the IF gene system.

New work will now be directed toward identification and characterization of the sequences that regulate the expression and coexpression of the IF genes during development and differentiation. In this way, the function of IF in the regulation of cytoskeletal structure and cellular activity will be clarified.

References

Aebi, U., Cohn, J., Buhle, L., and Gerace, L., 1986, The nuclear lamina is a meshwork of intermediate-type filaments, *Nature* **323:**560–564.

Alexander, F., Young, P. R., and Tilghman, S. M., 1984, Evolution of the albumin α-fetoprotein ancestral gene from the amplification of a 27 nucleotide sequence, *J. Mol. Biol.* **173:**159–176.

Bader, B. L., Magin, T. M., Hatzfeld, M., and Franke, W. W., 1986, Amino acid sequences and gene organization of cytokeratin no. 19, an exceptional tail-less intermediate filament protein, *EMBO J.* **5:**1865–1875.

Balcarek, J., and Cowan, N. J., 1985, Structure of the mouse glial fibrillary acidic protein gene: Implications for the evolution of the intermediate filament multigene family, *Nucleic Acids Res.* **13:**5527–5543.

Blessing, M., Zentgraf, H., and Jorcano, J. L., 1987, Differentially expressed bovine cytokeratin genes: Analysis of gene linkage and evolutionary conservation of 5′-upstream sequences, *EMBO J.* **6:**567–575.

Bowden, P. E., Blanchard, A., and Steinert, P. M., 1987, Linkage of epidermal keratin genes isolated on large cosmid clones, *J. Invest. Dermatol.* **88:**478.

Conway, J. F., and Parry, D. A. D., 1988, Intermediate filament structure. 3. Analysis of sequence homologies, *Int. J. Biol. Macromol.* **10:**79–98.

Craik, C. S., Rutter, W. J., and Fletterick, R., 1983, Splice junctions: Association with variation in protein structure, *Science* **220:**1125–1129.

Crewther, W. G., Dowling, L. M., Steinert, P. M., and Parry, D. A. D., 1983, The structure of intermediate filaments, *Int. J. Biol. Macromol.* **5:**267–274.

Dowling, L. M., Crewther, W. G., and Ingles, A. S., 1986, The primary structure of component 8c-1, a subunit protein of intermediate filaments in wool keratin. Relationships with proteins from other intermediate filaments, *Biochem. J.* **236:**695–703.

Ferrari, S., Cannizzaro, L. A., Battini, R., Huebner, K., and Baserga, R., 1987, The gene encoding human vimentin is located on the short arm of chromosome 10, *Am. J. Human Genet.* **41:**616–626.

Fisher, D. Z., Chaudhary, N., and Blobel, G., 1986, cDNA sequence of nuclear lamins A and C reveals primary and secondary structural homology to intermediate filament proteins, *Proc. Natl. Acad. Sci. USA* **83:**6450–6455.

Gilbert, W., 1978, Why genes in pieces? *Nature* **271:**501.

Glass, C., Kim, K. H., and Fuchs, E., 1985, Sequence and expression of a human type II mesothelial keratin, *J. Cell Biol.* **101:**2366–2373.

Goldman, A. E., Maul, G., Steinert, P. M., Yang, H.-Y., and Goldman, R. D., 1986, Keratin-like proteins that co-isolate with intermediate filaments of BHK-21 cells are nuclear lamins, *Proc. Natl. Acad. Sci. USA* **83:**3839–3843.

Hisanaga, S., and Hirokawa, N., 1988, Structure of the peripheral domains of neurofilaments revealed by low angle rotary shadowing, *J. Mol. Biol.* **202:**297–302.

Johnson, L. D., Idler, W. W., Zhou, X.-M., Roop, D. R., and Steinert, P. M., 1985, The structure of the gene for the human epidermal keratin of 67,000 molecular weight, *Proc. Natl. Acad. Sci. USA* **82:**1896–1900.

Julien, J.-P., Grosveld, F., Yasdanbaksh, K., Flavell, D., Meijer, D., and Mushynski, W., 1987, The structure of a human neurofilament gene (NF-L): A unique exon–intron organization in the intermediate filament gene family, *Biochim. Biophys. Acta* **909:**10–21.

Klinge, E. M., Sylvestre, Y., Freedeburg, I. M., and Blumenberg, M., 1978, Evolution of keratin genes: different protein domains evolve by different pathways, *J. Mol. Evol.* **24:**319–329.

Krieg, T. M., Schafer, M. P., Cheng, C. K., Filpula, D., Flaherty, P., Steinert, P. M., and Roop, D. R., 1985, Organization of a type I keratin gene: Evidence for evolution of intermediate filaments from a common ancestral gene, *J. Biol. Chem.* **260:**5867–5870.

Lasek, R. J., Philips, L., Katz, M. J., and Autilio-Gambetti, L., 1985, Function and evolution of neurofilament proteins, *Ann. N.Y. Acad. Sci.* **455:**462–478.

Lau, S. Y. M., Taneja, A. K., and Hodges, R. S., 1984, Synthesis of a model protein of defined secondary and quaternary structure. Effects of chain length on the stabilization and formation of two-stranded α-helical coiled-coils, *J. Biol. Chem.* **259:**13253–13264.

Lee, M. G. S., Lewis, S. A., Wilde, C. D., and Cowan, N. J., 1983, Evolutionary history of a multigene family: An expressed human β-tubulin gene and three processed pseudogenes, *Cell* **33:**477–487.

Lessin, S. R., Huebner, K., Isobe, M., Croce, C. M., and Steinert, P. M., 1988, Chromosomal mapping of human keratin genes: Evidence of non-linkage, *J. Invest. Dermatol.* **91:**572–578.

Lewis, S. A., and Cowan, N. J., 1986, Anomalous placement of introns in a member of the intermediate filament multigene family: An evolutionary conundrum, *Mol. Cell. Biol.* **6:**1529–1534.

Lomedico, P., Rosenthal, N., Efstratiadis, S., Gilbert, W., Kolodner, R., and Tizard, R., 1979, The structure and evolution of the two nonallelic rat preproinsulin genes, *Cell* **18:**545–558.

Lonberg, N., and Gilbert, W., 1985, Intron/exon structure of the chicken pyruvate kinase gene, *Cell* **40:**81–90.

McKeon, F. D., Kirschner, M. W., and Caput, D., 1986, Homologies in both primary and secondary structure between nuclear envelope and intermediate filament proteins, *Nature* **319:**463–468.

Marchuk, D., McCrohon, S., and Fuchs, E., 1984, Remarkable conservation of structure among intermediate filament genes, *Cell* **39:**491–498.

Mischke, D., and Wild, G., 1987, Polymorphic keratins in human epidermis, *J. Invest. Dermatol.* **88:**191–197.

Miyatani, S., Winkles, J. A., Sargent, T. D., and Dawid, I. B., 1986, Stage-specific keratins of *Xenopus laevis* embryos and tadpoles: The XK81 family, *J. Cell Biol.* **103:**1957–1965.

Myers, M. W., Lazzarini, R. A., Lee, V. M.-Y., Schlaepfer, W. W., and Nelson, D. L., 1987, The human mid-size neurofilament subunit: A repeated protein sequence and relationship of its gene to the intermediate filament family, *EMBO J.* **6:**1617–1626.

Napolitano, E. N., Chin, S. S. M., Colman, D. R., and Liem, R. K. H., 1987, Complete amino acid sequence and in vitro expression of rat NF-M, the middle molecular weight neurofilament protein, *J. Neurosci.* **7:**2590–2595.

Nishoida, Y., Leder, A., and Leder, P., 1980, An unusual α-globin-like gene that has cleanly lost both globin intervening sequences, *Proc. Natl. Acad. Sci. USA* **77:**2806–2809.

Ohno, S., Matsunaga, T., and Wallace, R. B., 1982, Identification of the 48-base-long primordial building block sequence of mouse immunoglobulin variable region genes, *Proc. Natl. Acad. Sci. USA* **79:**1999–2002.

Parry, D. A. D., and Fraser, R. D. B., 1985, Intermediate filament structure. I. Analysis of IF protein sequence data, *Int. J. Biol. Macromol.* **7:**203–213.

Parry, D. A. D., Crewther, W. G., Fraser, R. D. B., and MacRae, T. P., 1977, Structure of alpha-keratin: Structural implications of the amino acid sequences of the type I and type II chain segments, *J. Mol. Biol.* **113:**448–454.

Parry, D. A. D., Conway, J. F., and Steinert, P. M., 1986, Structural studies on lamin: Similarities and differences between lamin and intermediate filament proteins, *Biochem. J.* **238:**305–308.

Parry, D. A. D., Conway, J. F., Goldman, A. E., Goldman, R. D., and Steinert, P. M., 1987, Nuclear lamin proteins: Common structures for paracrystalline, filamentous and lattice forms, *Int. J. Biol. Macromol.* **9:**137–145.

Powell, B. C., Cam, G. R., Fietz, M. J., and Rogers, G. E., 1986, Clustered arrangement of keratin intermediate filament genes, *Proc. Natl. Acad. Sci. USA* **83:**5048–5052.

Proudfoot, N. J., Shander, M. H. M., Manley, J. L., Gefter, M. L., and Maniatis, T., 1980, Structure and in vitro transcription of human globin genes, *Science* **209:**1329–1336.

Quax, W., Egberts, W. V., Hendricks, W., Quax-Jeuken, Y., and Bloemendal, H., 1983, The structure of the vimentin gene, *Cell* **35:**215–223.

Quax, W., van der Brock, L., Egberts, W. V., Ramaekers, F., and Bloemendal, H., 1985a, Characterization of the hamster desmin gene: Expression and formation of desmin filaments in nonmuscle cells after gene transfer, *Cell* **43:**327–338.

Quax, W., Meera-Khan, P., Quax-Jeuken, Y., and Bloemendal, H., 1985b, The human desmin and vimentin genes are located on different chromosomes, *Gene* **38:**189–196.

RayChaudhury, A., Marchuk, D., Lindhurst, M., and Fuchs, E., 1986, Three tightly linked genes encoding human type I keratins: Conservation of sequences in the 5′-untranslated leader and 5′-upstream regions of co-expressed keratins, *Mol. Cell Biol.* **6:**539–548.

Rieger, M., Jorcano, J. L., and Franke, W. W., 1985, Complete sequence of a bovine type I cytokeratin gene: Conserved and variable intron positions in genes of polypeptides of the same cytokeratin family. *EMBO J.* **4:**2261–2267.

Romano, V., Bosco, P., Costa, G., Leube, R., Franke, W. W., Rochi, M., and Romeo, G., 1987, Chromosome assignment of cytokeratin genes, *Cytogenet. Cell Genet.* **46:**683.

Roop, D. R., and Steinert, P. M., 1986, The structure and evolution of intermediate filament genes, in: *Cell and Molecular Biology of the Cytoskeleton* (J. W. Shay, ed.), Plenum Press, New York, pp. 69–83.

Roop, D. R., Cheng, C. K., Titterington, L., Meyers, C. A., Stanley, J. R., Steinert, P. M., and Yuspa, S. H., 1984, Synthetic peptides corresponding to keratin subunits elicit highly specific antibodies, *J. Biol. Chem.* **259:**8037–8040.

Rosenberg, M., RayChaudhury, A., Shows, T. B., LeBeau, M. M., and Fuchs, E., 1988, A group of type I keratins on human chromosome 17; characterization and expression, *Mol. Cell Biol.* **8:**722–736.

Savtchenko, E. S., Freedberg, I. M., and Blumenberg, M., 1987, Human keratin genes are linked, *J. Invest. Dermatol.* **88:**516.

Shaw, G., and Weber, K., 1984, Differential expression of neurofilament triplet proteins in brain development, *Nature* **298:**277–279.

Steinert, P. M., and Roop, D. R., 1988, Molecular and cellular biology of intermediate filaments, *Annu. Rev. Biochem.* **57:**593–625.

Steinert, P. M., Rice, R. H., Roop, D. R., Trus, B. L., and Steven, A. C., 1983, Complete amino acid sequence of a mouse epidermal keratin subunit: Implications for the structure of intermediate filaments, *Nature* **302:**794–800.

Steinert, P. M., Parry, D. A. D., Racoosin, E. L., Idler, W. W., Steven, A. C., Trus, B. L., and Roop, D. R., 1984, The complete cDNA and deduced amino acid sequence of a type II mouse epidermal keratin of 60,000 molecular weight: Analysis of sequence differences between type I and type II keratins, *Proc. Natl. Acad. Sci. USA* **81:**5709–5713.

Steinert, P. M., Steven, A. C., and Roop, D. R., 1985a, The molecular biology of intermediate filaments, *Cell* **42:**411–419.

Steinert, P. M., Parry, D. A. D., Idler, W. W., Johnson, L. D., Steven, A. C., and Roop, D. R., 1985b, Amino acid sequences of mouse and human epidermal type II keratins of M_r 67,000 provide a systematic basis for the structural and functional diversity of the end domains of keratin intermediate filament subunits, *J. Biol. Chem.* **260:**7442–7449.

Sun, T.-T., Eichner, R., Schermer, A., Cooper, D., Nelson, W. G., and Weiss, R. A., 1984, Classification, expression and possible mechanisms of evolution of mammalian epithelial keratins: A unifying model, in: *Cancer Cells 1: The Transformed Phenotype,* Cold Spring Harbor Laboratory, Cold Spring Harbor, N.Y., pp. 169–176.

Tyner, A. L., Eichman, M. J., and Fuchs, E., 1985, The sequence of a type II keratin gene expressed in human skin: Conservation of structure among all intermediate filament genes, *Proc. Natl. Acad. Sci. USA* **82:**4683–4687.

Vasseur, M., Duprey, P., Brulet, P., Jacob, F., 1985, One gene and one pseudogene for the cytokeratin endo A, *Proc. Natl. Acad. Sci. USA* **82:**1155–1159.

Weber, K., 1986, Link between lamins and intermediate filaments, *Nature* **320:**402.

Weber, K., and Geisler, N., 1984, Intermediate filaments—from wool α-keratin to neurofilaments: A structural overview, in: *Cancer Cells 1: The Transformed Phenotype,* Cold Spring Harbor Laboratory, Cold Spring Harbor, N.Y., pp. 153–159.

Yamada, K., Avvedimento, V. E., Mudryji, M., Ohkubo, H., Vogeli, G., Irani, M., Pastan, I., and De Crombrugghe, B., 1980, The collagen gene: Evidence for its evolutionary assembly by amplification of a DNA segment containing an exon of 54 bp, *Cell* **22:**887–892.

Zackroff, R. V., Goldman, A. E., Jones, J. C. R., Steinert, P. M., and Goldman, R. D., 1984, Isolation and characterization of keratin-like proteins from cultured cells with fibroblastic morphology, *J. Cell Biol.* **98:**1231–1237.

Zehner, Z. E., and Paterson, B. M., 1983, Characterization of the chicken vimentin gene: Single copy gene producing multiple mRNAs, *Proc. Natl. Acad. Sci. USA* **80:**911–915.

Zhou, X.-M., Idler, W. W., Steven, A. C., Roop, D. R., and Steinert, P. M., 1988, The complete sequence of the human intermediate filament chain keratin 10; subdomainal divisions and model for folding of end domain sequences, *J. Biol. Chem.* **263:**15584–15589.

IV

Accessory Proteins Involved in Regulating the Organization of Intermediate Filaments

15

Intermediate Filament-Associated Proteins

Hsi-Yuan Yang, Norman Lieska, and Robert D. Goldman

1. Introduction

Intermediate filaments (IF) comprise a set of ~10-nm-diameter cytoskeletal filaments that are distributed throughout the cytoplasm from the level of the nuclear surface to the region of the plasma membrane in most eukaryotic cells (Ishikawa *et al.*, 1968; Traub, 1985a; Wang *et al.*, 1985; Goldman *et al.*, 1986). While IF from the various vertebrate cells share certain structural characteristics and constitute a unique fibrous organelle (Pruss *et al.*, 1981; Fuchs and Hanukoglu, 1983; Steinert and Parry, 1985), the structural proteins comprising them can be classified into subgroups whose distribution in most cases is tissue-type specific (Goldman *et al.*, 1979; Lazarides, 1980; Steinert and Roop, 1988). Due to this tissue-type specificity, IF have been used as markers to determine the origin of cells in studies of differentiation and in clinical pathology, e.g., neoplasia (Osborn and Weber, 1982).

As has been shown to be the situation with the other major cytoskeletal systems, i.e., microtubules and microfilaments, IF also have associated proteins (IFAP) that may play important roles in the regulation of their supramolecular organization and the interrelationships between IF and other cellular components (Starger *et al.*, 1978). One of the earliest IFAP candidates to be described was a species of approximately 300,000 molecular weight (300k) that coisolated with IF in IF-enriched cytoskeletal preparations (Triton/high-salt-insoluble cellular residues) obtained from baby hamster kidney (BHK-21) cells. This high-molecular-weight protein cocycled with the major IF structural subunits (vimentin and desmin) during *in vitro* IF disassembly/assembly (Starger *et al.*,

Hsi-Yuang Yang • Department of Anatomy and Cell Biology, College of Medicine, University of Illinois, Chicago, Illinois 60612. **Norman Lieska** • Department of Anatomy, Texas College of Osteopathic Medicine, Fort Worth, Texas 76107. **Robert D. Goldman** • Department of Cell, Molecular, and Structural Biology, Northwestern University Medical School, Chicago, Illinois 60611.

1978; Zackroff and Goldman, 1979). Since that time, there have been reports of a number of molecules exhibiting close association with IF in a variety of cell types. These include: filaggrin (Dale, 1977; Dale *et al.*, this volume); plectin (Pytela and Wiche, 1980), synemin (Granger and Lazarides, 1980); p95 (Lin and Feramisco, 1981); paranemin (Breckler and Lazarides, 1982); epinemin (Lawson, 1983); p50 (Wang *et al.*, 1983); IFAP-300k (Yang *et al.*, 1985a); IFAP-70-280k (Yang *et al.*, 1985b), NAPA-73 (Ciment *et al.*, 1986); and nuclear lamin B (Georgatos and Blobel, 1987; Georgatos *et al.*, 1987). Several other proteins, which are known to be associated with other cellular elements, may also be IF-associated. These are nonerythrocyte spectrin or ankyrin, two of the major components of the membrane skeleton (Lehto and Virtanen, 1983; Mangeat and Burridge 1984; Georgatos and Marchesi, 1985; Georgatos *et al.*, 1985); microtubule-associated proteins (MAP-2, Bloom and Vallee, 1983; a MAP-1A-like protein, Brown and Binder, 1988; Map-tau, Miyata *et al.*, 1986); and desmoplakins, desmosome-associated proteins (Jones and Goldman, 1985).

What is an IFAP? It is impossible to give a universally acceptable definition for an IFAP at the present time since the study of IFAP is relatively recent, and our knowledge about them is very limited. In our opinion, an IFAP may be tentatively defined as a protein that appears to be morphologically and functionally related to IF, but does not represent a structural component of IF, i.e., does not form IF by itself. Eventually, the identification of an IFAP should be based on more stringent biochemical and morphological characterizations of the binding parameters between a putative IFAP and IF, similar to those described for ligand–receptor interactions. Most of the IFAP mentioned have been identified by criteria such as coisolation, cocycling, and/or colocalization with IF. At the present time, it seems reasonable to use these less stringent criteria to initiate the preliminary identification of IFAP. It should also be mentioned that, even with such a nonstringent approach, some of the potential IFAP may be very difficult to identify. For example, IFAP bound only to the ends of IF, such as the proteins putatively linking IF and the nuclear surface (Goldman *et al.*, 1985), would not be expected to show a codistribution pattern with IF by immunofluorescence microscopy, and IFAP present in very small amounts, or with low binding affinities, may not be detected in coisolation and/or cocycling studies. Indeed, most IFAP identified so far appear to be IF–IF linking proteins because an IF-like pattern is usually demonstrated by immunofluorescence using antibodies directed against these molecules (see below). In spite of these problems, we feel that *in vitro* recombination studies using purified putative IFAP and purified IF proteins are necessary for further defining the specificity and functions of associated polypeptides. Before reviewing the work of other laboratories, we will first describe our approach to defining IFAP in cultured BHK-21 cells.

2. IFAP in BHK-21 Cells

In addition to the IF structural proteins vimentin and desmin present in large quantities, IF preparations from BHK-21 cell cultures contain a number of polypeptides present in relatively small amounts, as determined by SDS–PAGE (Starger *et al.*, 1978; Zackroff and Goldman, 1979). For the past 6 years, we have been determining whether some of these coisolating components are IFAP. The monoclonal antibody technique has been employed to great success in this undertaking. This technique is especially advantageous in cases in which the specific antigens are difficult to identify and to purify in quantities sufficient for immunization. Using the total IF preparation as an antigen mixture for immunization, a variety of monoclonal antibodies have been produced by the mouse hybridoma technique. These monoclonal antibodies are then used as immunological probes for the identification and characterization of IFAP in BHK-21 cells and for monitoring their purification.

2.1. Criteria for Identification of IFAP in BHK-21 Cells

To identify an IFAP, the putative species must first be proven to be distinct from the structural subunit proteins by various biochemical analyses. In addition, we have developed an experimentally defined set of four morphological and biochemical criteria for the preliminary identification of an IFAP (Yang *et al.*, 1985a; Lieska *et al.*, 1985a). These criteria are: (1) colocalization of the IFAP with IF *in situ;* (2) coisolation of the IFAP with IF in IF-enriched cytoskeletal preparations, i.e., in Triton/high-salt-insoluble cellular residues; (3) cocycling of the IFAP with IF through disassembly/assembly cycling of IF; and (4) *in vitro recombination* of the purified IFAP with IF assembled from purified IF subunit protein(s).

Morphologically, this experimental protocol involves the production of antibodies and the subsequent immunofluorescence localization of the associated protein under a variety of normal (replating/spreading; cell cycle/mitosis) and drug-altered cell states. Double-label immunofluorescence studies have been employed to elucidate potential complementary distribution and/or codistribution with a wide spectrum of cytoskeletal, nuclear, plasma membrane, and extracellular matrix proteins. Ultrastructural immunogold localization gives higher-resolution information on the above states and provides some data on the structural nature of the association with IF, as well as on the secondary association of the IF/IFAP complex with other organelles. Ultrastructural localization is especially important for IFAP bound to the ends of IF (such as those intervening between IF and the cell or nuclear surface) since such species will not exhibit an IF codistribution pattern detectable by immunofluorescence microscopy (Lehto and Virtanen, 1983).

The antibodies are also used in parallel studies of identifying the appropriate polypeptide species by immunoblot analysis of SDS-PAGE-separated proteins comprising IF-enriched cytoskeletal preparations. Preparative purification of the interacting species (including the respective IF subunit proteins) is necessary for the further characterization and *in vitro* recombination studies. The latter analyses confirm the association seen *in situ* and allow ultrastructural immunogold and biochemical investigations of the nature of the interactions under more controlled conditions. It is important to note that the criteria based upon cosedimentation may constitute definitive evidence with certain IFAP, but may be inadequate for other IFAP, e.g., those present in very small quantities or with low binding affinities, because such IFAP may not be detected under these circumstances. Using these methods and criteria, two IFAP (IFAP-300k and IFAP-70-280k) have been identified in IF preparations from BHK cell cultures. The results of these studies are discussed below.

2.2. IFAP-300k

2.2.1. *Identification, Characterization, and in Vitro Recombination of IFAP-300k*

A monoclonal antibody (anti-IFAP-300k) has been prepared that reacts exclusively with a 300k high-molecular-weight polypeptide doublet (IFAP-300k) present in native BHK IF preparations (Yang *et al.*, 1985a; Lieska *et al.*, 1985a). Immunological techniques and comparative one-dimensional peptide mapping have indicated that IFAP-300k is biochemically distinct from the IF structural subunits (vimentin and desmin) of the BHK-21 cell line. Double-label immunofluorescence investigations using anti-IFAP-300k and rabbit anti-vimentin/desmin serum show that IFAP-300k codistributes with IF in a fibrous network in fully spread BHK cells (Fig. 1). Double-labeling with these antibodies further reveals coexistence of the respective antigens in the juxtanuclear IF cap, which characterizes cells treated with colchicine and those in the early stages of spreading following trypsinization/replating (Fig. 1). Following the removal of colchicine and in the late stages of cell spreading, IFAP-300k and the IF structural subunits become reorganized to form the highly coincident, cytoplasmic fibrous patterns characteristic of normal spread cells.

Ultrastructural localization by the immunogold technique using this anti-IFAP-300k confirms that IFAP-300k is associated with IF in various physiological and morphological states (Fig. 2). The distribution of gold particles is nonuniform along IF, being present primarily at crossover points between IF or in areas where IF are closely associated laterally with other IF. As such, the pattern is focal in nature and appears to be associated only with IF that are interacting with other IF in various organizational states, especially in the dispersed networks that are characteristic of spread cells. Gold particles are not seen in association with microtubules, microfilaments, or other cellular organelles (Fig. 2). These observations suggest that IFAP-300k may play an important role in cross-linking IF into a dynamic, complex network.

It has been determined by immunoblot analyses IFAP-300k is retained along with the vimentin/desmin IF subunits through two *in vitro* disassembly/assembly cycles of BHK IF preparations in a stoichiometry comparable to that seen in freshly prepared IF-enriched cytoskeletal preparations. This result further suggests a functional association. IFAP-300k, vimentin, and desmin have been purified using chromatographic procedures. The molecule's isoelectric point has been determined to be 5.35 by chromatofocusing. *In vitro* recombination experiments analyzed by biochemical and morphological methods have demonstrated that purified IFAP-300k recombines at sites of close IF–IF association in preparations reassembled from purified vimentin and desmin. In contrast to the focal localization of IFAP-300k along IF, anti-vimentin/desmin decorates these IF along their entire lengths.

2.2.2. *Distribution of IFAP-300k Immunoreactivity in Various Cultured Mammalian Cell Lines*

Anti-IFAP-300k immunoreactivity, as demonstrated by immunofluorescence microscopy, has been revealed in a number of other cultured mammalian cell lines, including mouse (3T3) and human (ENSON) fibroblasts; marsupial (PtK-2) epithelial cells; and mouse neuroblastoma (Nb2a) cells (Yang *et al.*, 1984). No significant immunofluorescence can be detected in primary cultures of mouse epidermal (PME) cells. Double-label immunofluorescence studies using mouse monoclonal anti-IFAP-300k and rabbit antivimentin have confirmed the codistribution of IFAP-300k labeling and vimentin IF in normal and colchicine-treated fibroblasts (3T3, ENSON) and in Nb2a cells, whose IF have also been reported to be of the vimentin type (Lazarides, 1982).

Both vimentin and keratin types of IF have been demonstrated in PtK-2 epithelial cells (for review, see Jones *et al.*, 1985). IFAP-300k immunoreactivity has been shown to be associated only with the vimentin IF (both in the presence and absence of colchicine)

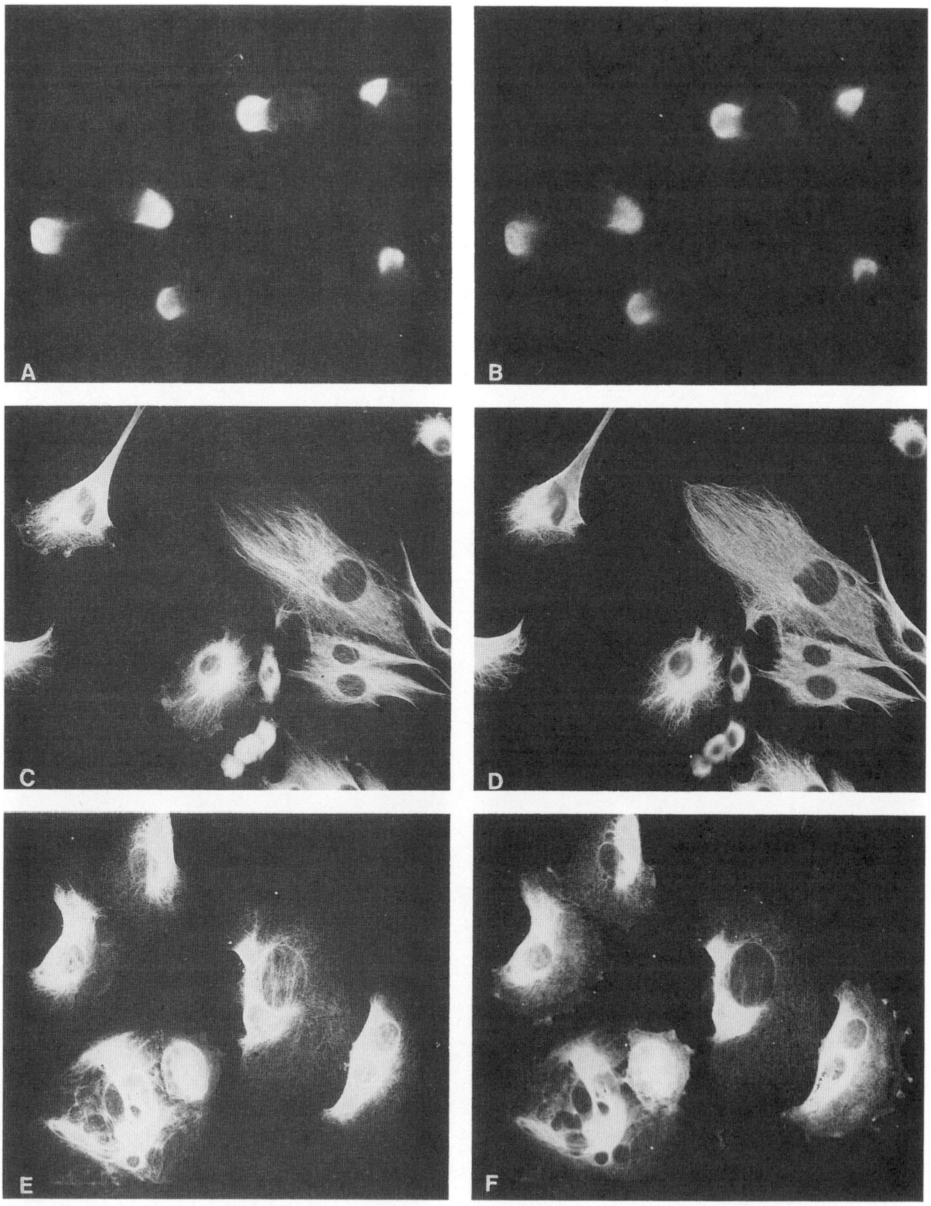

Figure 1. Double-label immunofluorescence localization of vimentin/desmin-containing intermediate filaments (A, C, E) and IFAP-300k (B, D, F) in baby hamster kidney (BHK-21) cells. A rabbit anti-vimentin/desmin and a mouse monoclonal anti-IFAP-300k were used for the double labeling. Note that the two staining patterns were essentially coincident. (A, B) Early cell spreading 20 min after trypsinization and replating. Fluorescence was concentrated in the juxtanuclear IF cap. (C, D) Spread cells 24 hr after replating. The staining was seen in the dispersed IF networks. (E, F) BHK cells 24 hr after colchicine treatment. The majority of the immunoreactivity of both antibodies was codistributed in the juxtanuclear cap of retracted intermediate filaments. × 1320.

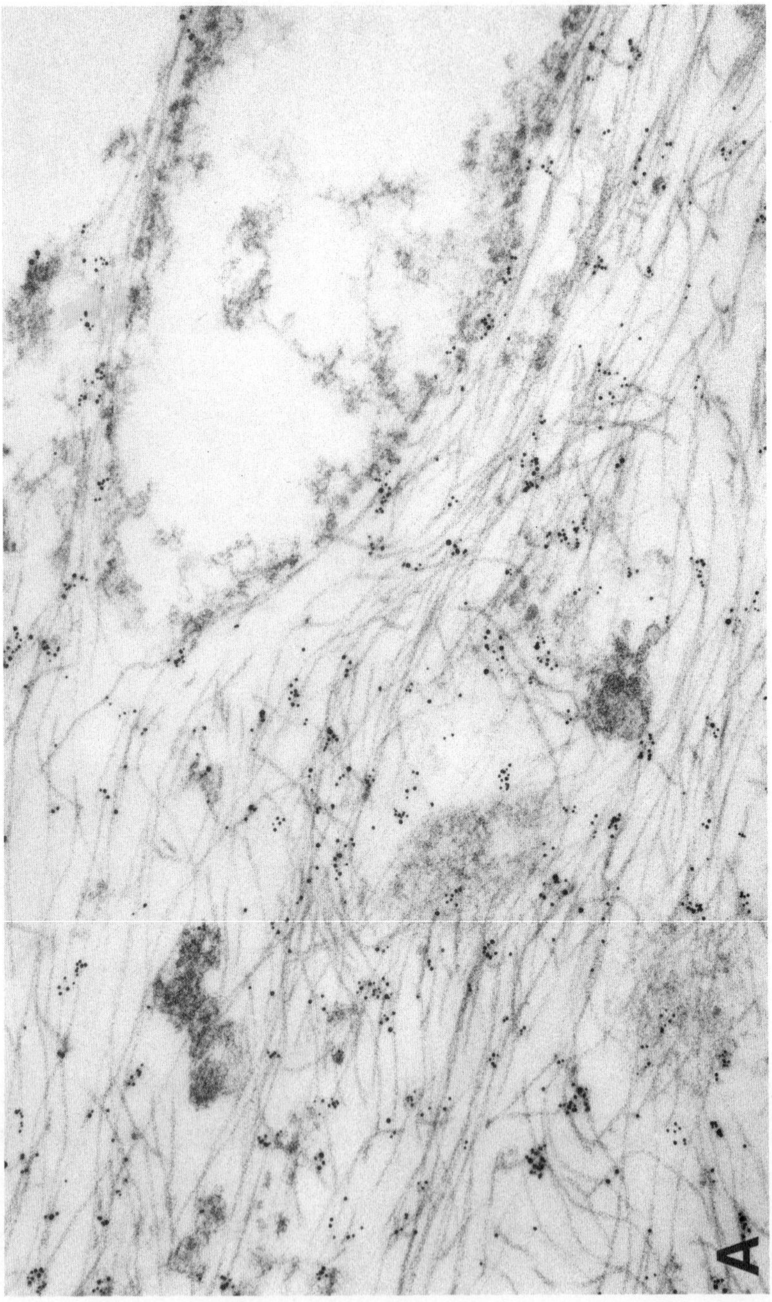
A

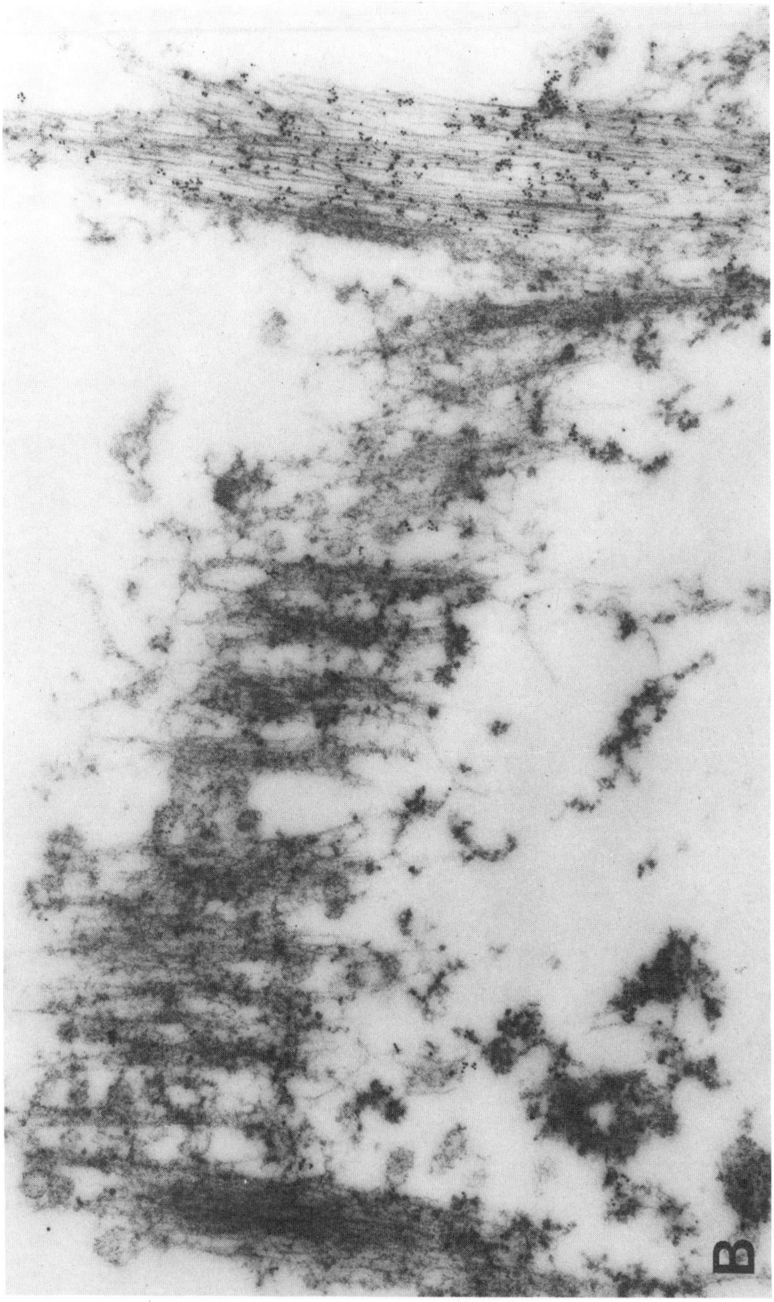

Figure 2. Ultrastructural immunogold localization of IFAP-300k in BHK-21 cells. Indirect immunolocalization of this IFAP was carried out using monoclonal anti-IFAP-300k and 6-nm colloidal gold-conjugated secondary antibodies. (A) Association of IFAP-300k with dispersed IF networks. (B) Association of IFAP-300k with an IF bundle. Note that immunoreactivity of this IFAP was not found to be associated with microfilaments (B) or microtubules (C). (A) × 93,800; (B) × 46,900; (C) × 50,600.

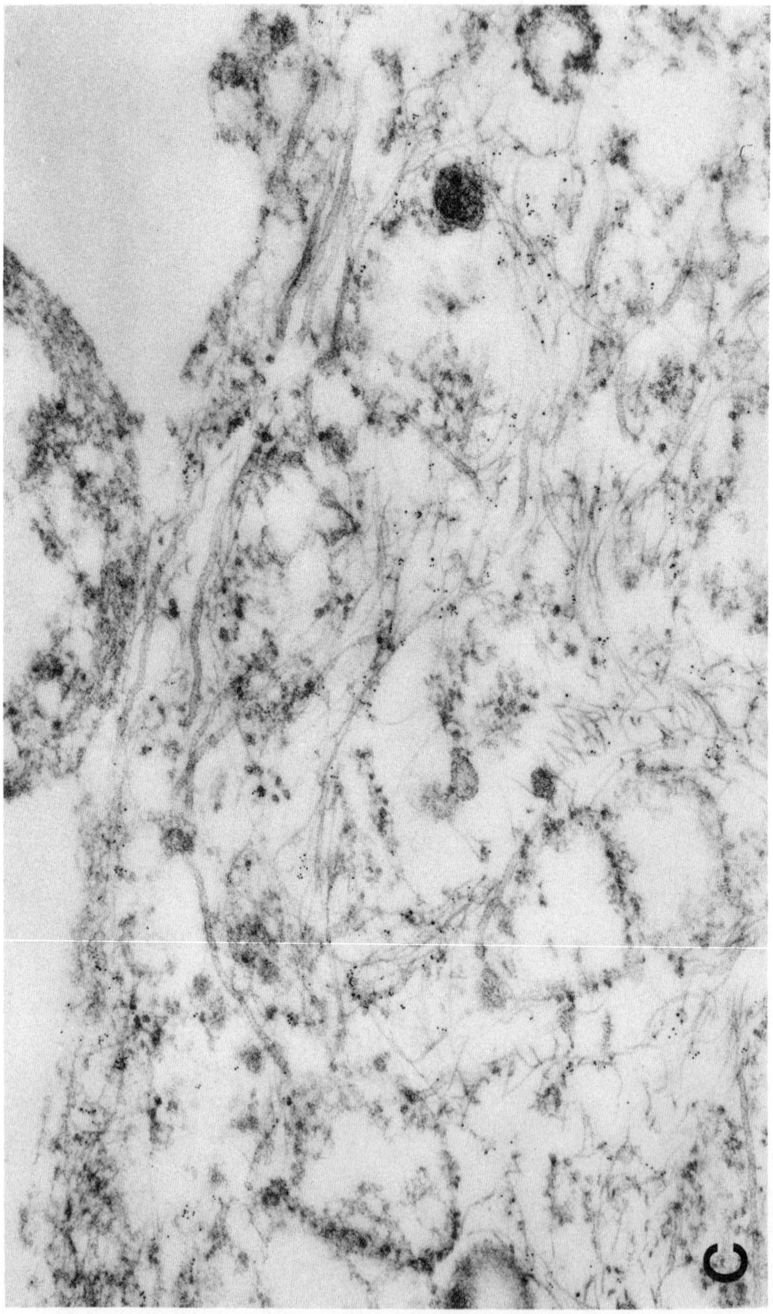

Figure 2. (continued)

by double-labeling using anti-IFAP-300k with either antivimentin or antikeratin sera. These immunofluorescence results demonstrate that anti-IFAP-300k immunoreactivity is present in vimentin-containing cells of diverse evolutionary origins, ranging from marsupial (PtK-2) to human (ENSON).

Immunoblotting analyses using the monoclonal anti-IFAP-300k have been performed on BHK-21, ENSON, PtK-2, and PAM cells. The anti-IFAP-300k reacts with a 300k polypeptide in native IF preparations and whole cell extracts from BHK-21 and ENSON cells. No immunoreactivity is detected in PtK-2 IF-enriched cytoskeletal preparations, but anti-IFAP-300k does react with a 100k doublet in whole cell extracts from PtK-2 cells. No immunoreactivity is detected in either native IF preparations or whole cell extracts obtained from mouse epidermal (PAM) cells. These results indicate that there is heterogeneity with respect to the molecular weight and binding affinity of IFAP-300k-related molecules in cultured cell lines.

2.3. IFAP-70-280k

2.3.1. *Identification and Characterization of IFAP-70-280k*

Using the hybridoma technique, a monoclonal antibody (anti-IFAP-70-280k) has been produced that has been characterized by SDS-PAGE, immunoblotting, immunofluorescence screening to react specifically with a group of 70–280k polypeptides present in BHK IF preparations (Yang *et al.*, 1985b, 1990; Lieska *et al.*, 1985b). Anti-IFAP-70-280k has been used to probe the relationship between these polypeptides and IF *in situ* and *in vitro*. On the basis of SDS-PAGE and immunoblotting analyses these polypeptides are retained along with the vimentin/desmin IF subunits through two *in vitro* disassembly/assembly cycles of the BHK IF preparations as is the case with IFAP-300k. Results obtained from double-labeled immunofluorescence and ultrastructural immunogold localization demonstrate that this group of immunologically related polypeptides is associated with IF in BHK cells.

2.3.2. *Different Distribution of IFAP-300k and IFAP-70-280k in BHK-21 Cells*

Double-label immunofluorescence studies using anti-IFAP-300k or anti-IFAP-70-280k, in conjunction with rabbit anti-vimentin/desmin, reveal that IFAP-300k appears to be associated with the total IF network in BHK cells, while anti-IFAP-70-280k reacts with only a part of the network (Yang *et al.*, 1986, 1990). In cells treated with colchicine, most of the IF withdraw from their peripheral distribution to form a juxtanuclear IF cap as cytoplasmic microtubules depolymerize, while the remaining IF maintain associations with the cell surface and the plasma membrane-associated stress fibers. Double-label immunofluorescence studies reveal the codistribution of IFAP-300k with both the IF in juxtanuclear caps and those remaining associated with the cell surface in colchicine-treated cells. However, anti-IFAP-70-280k immunoreactivity can only be demonstrated in the colchicine-induced juxtanuclear cap of IF in BHK cells. The selective association of IFAP-70-280k with colchicine-sensitive IF indicates that this IFAP may be more specifically associated with the microtubule-related (i.e., colchicine-sensitive) IF.

3. Other IFAP

In this section we review the literature on other IFAP, with the exception of filaggrin, the IF–IF bundling protein of keratinocytes. This IFAP is discussed in detail in the chapter by Dale and co-workers.

3.1. Plectin

Plectin is a 300kD protein originally identified in IF-enriched preparations of several cultured cell lines, including rat glioma C6, HeLa, CHO, and SV-40-transformed hamster cells (Pytela and Wiche, 1980). In CHO, it accounts for up to 1% of the cellular protein (Herrmann and Wiche, 1983). Plectin has been found to be phosphorylated by both cAMP-independent and -dependent protein kinases (Herrmann and Wiche, 1983). Peptide mapping reveals that the two kinases phosphorylate different sites.

Plectin has been described as a general "cytoplasmic element cross-linker" (Wiche and Baker, 1982; Wiche *et al.*, 1983), as well as a possible factor involved in the formation of cell junctions (Wiche *et al.*, 1983). These conclusions are based upon several observations: (1) the similarity between plectin and high-molecular-weight MAP and the former's possible role as a mediator of microtubule–IF interaction (Pytela and Wiche, 1980; Wiche *et al.*, 1980); (2) plectin's codistribution with vimentin-containing IF in colcemid-treated cells as determined by immunofluorescence (Wiche and Baker, 1982; Wiche *et al.*, 1982); and (3) a wide spectrum of cellular localizations, (e.g., desmosomes and other junctional complexes, muscle Z lines, cardiac intercalated disks) stained by antiplectin serum during indirect immunofluorescence (Wiche *et al.*, 1983). Recently, plectin has been found to be enriched in the surface membranes of epithelial basal cells from frozen sections of human cornea and skin (Wiche *et al.*, 1984). Plectin has also been localized in various cultured cell types, including fibroblasts, epithelial cells, and glial cells (Pytela and Wiche, 1980; Wiche and Baker, 1982; Wiche *et al.*, 1984) and has a broad distribution in frozen sections of various rat and human tissues (Wiche *et al.*, 1983). Antiplectin immunoreactivity has been demonstrated in fibroblasts, endothelial cells, neurons, epithelial cells, epidermal cells, and all three types of muscle cells. In all of these, the staining pattern varies from one cell type to another. Many of these observations must be interpreted very cautiously, since they are mainly based on immunofluorescence observations using a rabbit polyclonal antiserum. Specific biochemical analyses and ultrastructural immunolocalization (with satisfactory preservation of the necessary ultrastructure) should help to resolve the very complex interaction pattern that this molecule exhibits. Finally, recent studies using a monoclonal plectin antibody have suggested that plectin and IFAP-300k (see above) may possess some degree of homology (Herrmann and Wiche, 1987; Foisner *et al.*, 1988). However, our antibody preparations against the 300k component reveal very different staining patterns in cultured cells (Yang *et al.*, 1984) and in tissues (unpublished observations) using immunofluorescence techniques. The relationship between the 300kD IFAP and plectin can only be established through direct biochemical comparison of purified preparations of the two proteins.

3.2. Synemin

Synemin is a 230kD polypeptide that cocycles with desmin and vimentin through disassembly/assembly cycles of IF isolated from avian smooth muscle, where it was originally identified (Granger and Lazarides, 1980). Synemin purified from chicken gizzard smooth muscle is phosphorylated and has a pI of 5.34 (Sandoval *et al.*, 1983). This IFAP has also been found in skeletal muscle, wherein it codistributes with desmin- and vimentin-containing IF during *in vitro* myogenesis (Granger and Lazarides, 1980). During this process, a cytoplasmic filament pattern is observed for all three antigens, which appear to move to the periphery of the myofibril Z disks as development proceeds. Colcemid causes the filamentous forms of these three proteins to coaggregate into cytoplasmic bundles in cultured myotubes, but has little effect on these antigens after they have become associated with the Z disks. Extraction of cells with nonionic detergents and high salt yields cytoskeletons that contain all three proteins whose respective immunofluorescence patterns are similar. These data suggest that synemin is associated with desmin- and vimentin-containing IF in skeletal muscle cells. However, as with plectin, these conclusions must be treated with caution, since they are based solely upon immunofluorescence microscopy. Evidence from specific biochemical studies (e.g., reconstitution) and ultrastructural immunolocalization are necessary to further confirm the association between synemin and IF. This is especially true concerning skeletal muscle cells because details of the organization of IF in these cells are not available (for relevant discussion, see Tokuyasu *et al.*, 1984).

Synemin, as well as vimentin, have been identified and colocalized with IF in mature avian erythrocytes that have been sonicated (Granger *et al.*, 1982). The structure and distribution of IF in sonicated chicken erythrocytes and embryonic erythroid cells have been studied by antibody decoration and low-angle shadowing (Granger and Lazarides, 1982). Incubation of sonicated cells with antivimentin results in a uniform decoration of the IF, while incubation with antisynemin results in a pattern of periodically spaced foci along the filaments. Based on these light microscopic observations, it has been suggested that synemin also mediates the cross-linking of IF. However, there have been no comparable *in situ* ultrastructural studies to demonstrate the presence and organization of either the IF cytoskeleton or the localization of synemin in erythrocytes. Subsequent pulse-labeling studies in avian erythrocytes have indicated that, in the assembly of newly synthesized vimentin and synemin from soluble forms into IF, the rate of vimentin assembly limits the rate of synemin assembly (Moon and Lazarides, 1983). It has been suggested that the elongation of vimentin-containing IF generates synemin binding sites.

In immunofluorescence observations, synemin has been shown to associate with desmin filaments in smooth muscle cells, with desmin- and vimentin-containing IF in skeletal muscle cells, and with vimentin IF in avian erythrocytes and lens cells (Granger *et al.*, 1982; Price and Lazarides, 1983; Granger and Lazarides, 1980, 1984). Since synemin has not been demonstrated in neurons, epithelial cells, and most glial cells, it appears that it is expressed only in conjunction with desmin and/or vimentin. Conversely, synemin does not appear to be present in all cells that contain desmin and/or vimentin.

3.3. Paranemin

Paranemin is a 280kD polypeptide originally isolated from IF preparations of 14-day-old chicken embryonic skeletal muscle by extraction with a physiological salt solution followed by gel filtration (Breckler and Lazarides, 1982). On isoelectric focusing gels, it appears as two closely spaced variants with an average pI of 4.5. This protein differs from synemin by its isoelectric point, molecular weight, and immunological reactivity. In cultured myogenic cells, immunofluorescence microscopy using paranemin antiserum has revealed that this protein exhibits a spatial distribution similar to that seen with desmin, vimentin, and synemin antibodies, i.e., associated with cytoplasmic filamentous structure at the early myotube stage or with myofibril Z lines in later stage myotubes. In putative chicken embryonic fibroblasts grown in culture, only a subpopulation of the cells show antiparanemin immunoreactivity, although all of them are vimentin-positive. Double-label immunofluorescence has suggested the codistribution of paranemin and vimentin in these latter cell types both in the presence and absence of colchicine.

Paranemin immunofluorescence has been shown in each of the three types (skeletal, cardiac, and smooth) of developing chicken muscle cells (Breckler and Lazarides, 1982; Price and Lazarides, 1983). However, the expression of this protein appears to vary among these types of muscle cells during differentiation. It has been suggested that expression of paranemin during muscle differentiation does not parallel that of synemin, desmin, or vimentin, as demonstrated by immunofluorescence microscopy and immunoautoradiography (Price and Lazarides, 1983). Interestingly, paranemin immunofluorescence has been shown in certain vimentin-containing, nonmuscle cells such as Schwann cells and some endothelial cells. Paranemin does not appear to be present in neurons, epithelial cells, or most types of glial cells, as determined by immunofluorescence microscopy (Price and Lazarides, 1983). This suggests that paranemin is only expressed in certain types of vimentin- and/or desmin-containing cells.

3.4. Epinemin

Epinemin has been identified by use of a mouse monoclonal antibody prepared by immunization with extracts from fresh chicken gizzard (Lawson, 1983). It has a molecular weight of 44.5 k and an isoelectric point of 5.0. This protein appears to be different from vimentin and actin, as suggested by the results of various biochemical, immunological, and immunolocalization techniques. In double-label immunofluorescence, antiepinemin displays a uniform fibrous pattern that is similar to that of antivimentin but that appears different from that of antitubulin. Immunoferritin electron microscopic observations resolve the antiepinemin pattern into an intermittent distribution along vimentin-containing IF. Antiepinemin immunoreactivity has been demonstrated in various cultured cells and in skeletal muscle. By immunofluorescence criteria, it is absent from smooth muscle, from cells without vimentin, and from neural cell containing vimentin. This protein is resistant to nonionic detergent extraction, but can be solubilized from vimentin-containing IF by extraction with high salt solutions. Furthermore, it "collapses" with IF in colchicine-treated cells.

Recently, the ultrastructural distribution of epinemin has been demonstrated by the immunogold method in quick-frozen, deep-etched cytoskeletons (Lawson, 1984). The results reveal that epinemin is expressed asymmetrically at discrete sites on the vimentin filaments and that one to three antiepinemin molecules are found at each of these discrete foci. This study also suggests that epinemin does not cross-link adjacent IF.

3.5. p50

The p50 protein, an IFAP with a molecular weight of 50kD, has been identified by mouse monoclonal antibodies generated by using native BHK IF preparations as antigen

(Wang *et al.*, 1983). Codistribution of this IFAP and IF has been confirmed by double-label immunofluorescence microscopy of BHK-21 cells. The anti-p50 immunofluorescence is associated with the BHK IF in a dotted pattern. Ultrastructural immunogold localization using monoclonal anti-p50 has further shown that the gold particles are found in clusters or aggregates around proteinaceous material associated with the BHK IF. Recently, this IFAP has been shown to be increased in senescent BHK-21 cells and suggested to be involved in age-related IF bundle formation by way of increased interfilament cross-bridges (Wang, 1985). The presence of these IF bundles may be a physical impediment to the locomotion of these nondividing cells.

Immunoreactivity of p50 has also been demonstrated by immunofluorescence microscopy in a variety of cultured cells, including epithelial cells (PtK-1 and HeLa) and cells of astroglial origin, and in many types of tissues, including epidermis, smooth muscle, and subdermal pericytes, but not in the white or gray matter of the cerebellum and spinal cord (Wang *et al.*, 1983).

3.6. NAPA-73

NAPA-73 (an acronym for neurofilament-associated protein, avian specific, 73kD) has been identified by a monoclonal antibody that reacts with a variety of avian embryonic cells containing neurofilaments (Ciment *et al.*, 1986). When compared with the neurofilament triplet proteins by immunological reactivity and electrophoretic mobility, this protein appears to be distinct. Whole mount immunogold electron microscopy has demonstrated the association of NAPA-73 with the neurofilament bundles in primary cultures of neurons prepared from chicken embryo dorsal root ganglia. Furthermore, other studies with the monoclonal anti-NAPA-73 have shown that this IFAP is associated with vimentin IF during early development stages and with neurofilaments in later developmental stages. These results suggest that NAPA-73 may bind to different IF types during the development of the chicken nervous system.

3.7. p230 and Spectrin-Related Proteins

A 230kD polypeptide (p230) has been isolated from cytoskeletal preparations of bovine lenses (Lehto and Virtanen, 1983). In various cultured mammalian cells and in some human tissues (kidney and liver), p230 has been localized by immunofluorescence in the peripheral cytoplasm of the cells in a layer distinct from that of the major cytoskeletal structures. While this peripheral distribution is not dependent on the presence of intact stress fibers or microtubules, it is strikingly affected by colchicine treatment in that it appears to codistribute with the induced perinuclear aggregates of IF. This implies an association with IF. These results suggest that p230 may be part of a cell surface/cytoskeleton-associated, subplasmalemmal structure, perhaps important in cell surface–cytoskeletal interactions. This speculation is supported by the evidence that IF interact with the plasma membrane or with subplasmalemmal material in some cultured fibroblasts, lens-forming cells, and nucleated erythrocytes (Granger and Lazarides, 1982; Ramaekers *et al.*, 1982; Lehto, 1983; Green and Goldman, 1986), although definitive proof in support of this possibility is not available.

Based on the detergent resistance and electrophoretic mobility of p230, it has been suggested that p230 may represent α-spectrin (Lehto and Virtanen, 1983). Although no data have been provided to indicate the biochemical identity between p230 and spectrin-related molecules, the possibility of such an identity has been highlighted by the fact that several bona fide spectrinlike molecules have been shown to be associated with IF. By immunoelectron microscopy, a close association between the spectrin-related molecule TW260/240 and IF has been demonstrated in the terminal web of the brush border of the intestinal epithelium (Hirokawa *et al.*, 1983). Probably the most convincing data available in support of an association between IF and spectrin are provided by microinjection studies of Mangeat and Burridge (1984). Intracellular precipitation of nonerythrocyte spectrin follows the microinjection of either a monoclonal antibody directed against the α-chain of nonerythrocyte spectrin or an affinity-purified polyclonal antibody against bovine brain spectrin (fodrin). This intracellular immunoprecipitation induces no alteration in the distribution of actinlike microfilament bundles, microtubules, or coated vesicles, nor is cell shape affected; however, the distribution of IF is affected. The IF become more distorted and condensed, and the clumped IF are frequently observed to coincide with regions of aggregated spectrin. This aggregation of IFs is not induced by microinjection of irrelevant antibodies, nor is it induced by monoclonal antispectrin in cells unreactive with the antibody. These results suggest an association between vimentin filaments and noerythrocyte spectrin and raise the possibility that the latter is involved in linking IF to the plasma membrane (Mangeat and Burridge, 1984).

3.8. MAP-2

The existence of a structural and functional association between IF and microtubules in a variety of cells has been well documented (for review, see Goldman and Knipe, 1972; Goldman *et al.*, 1980; Shelanski *et al.*, 1981). This association is especially prominent in neurons, wherein IF (neurofilaments) and microtubules (neurotubules) are the major cytoplasmic components in neuronal processes. These elements are interconnected by a distinct cross-bridging system to form a three-dimensional network in the axoplasm (Metuzals *et al.*, 1981; Schnapp and Reese, 1982; Tsukita *et al.*, 1982; Hirokawa, 1982; Hirokawa *et al.*, 1984). Evidence suggests that this network plays an important role in axonal transport (Lasek and Hoffman, 1976; Baitinger *et al.*, 1983; Willard, 1983). Some biochemical evidence also supports an association between IF and microtubules, e.g., structural subunits of neural IF are often found in tubulin preparations (Berkowitz *et al.*, 1977; Thorpe *et al.*, 1979).

Although the precise biochemical nature of the IF–microtubule association is not clear, several lines of evidence indicate that some of the MAP may be implicated. High-molecular-weight MAP (MAP-1 and MAP-2; for review, see Vallee and Bloom, 1984; Vallee *et al.*, 1984) have been shown to bind not only to tubulin, but also to neural IFs (Runge *et al.*, 1981; Shelanski *et al.*, 1981; Minami *et al.*, 1982). Other studies have further shown that mammalian neural IF can partially inhibit the *in vitro* assembly of microtubules and that this inhibition is due to the association of a complex of MAP-1, MAP-2, and tubulin with neurofilaments (LeTerrier *et al.*, 1982). Morphologically, the appearance of the high-molecular-weight MAP as arms on the microtubule surface also favors the suggestion that these proteins are involved in the association between microtubules and other cellular elements (Dentler *et al.*, 1975; Murphy and Borisy, 1975; Zingsheim *et al.*, 1979; Voter and Erickson, 1982).

The possibility that a particular MAP may mediate IF–microtubule association has been further studied by immunofluorescence microscopy of primary cultures of rat brain cells (Bloom and Vallee, 1983). Double-label immunofluorescence demonstrates codistribution of MAP-2 with both microtubules and IF in flat cells present in newborn rat brain cultures. Pretreatment of the cultures with vinblastine induces the redistribution of the MAP-2 into perinuclear cables, wherein the coexistence of MAP-2 and vimentin has primarily been shown by double-label immunofluorescence. Tubulin paracrystals in vinblastine-treated cells are not stained by anti-MAP-2. In cells extracted with digitonin, which destroys microtubules but does not affect IF, a normal fibrillar distribution of MAP-2 is maintained. Similar experiments have been conducted with monoclonal anti-MAP-1A; however, no evidence for an interaction of this MAP with IF cables has been obtained (Vallee *et al.*, 1984). These data are in contrast to the results of previous *in vitro* reconstitution experiments in which it was found that all high-molecular-weight brain MAP are associated with purified neural IF at relatively low stoichiometries (LeTerrier *et al.*, 1982). The reason for this contradiction is not clear.

These results suggest that MAP-2 may bind to IF in some types of cells, and suggest that one of the functions of this protein is to mediate the IF–microtubule association. Since MAP-2 is only found in neurons (for review, see Vallee *et al.*, 1984), it may be one of the cross-linking proteins between neurofilaments and microtubules. However, rigorous proof of this possibility awaits further detailed biochemical characterization of neural IF.

3.9. Desmoplakin 1 and 2

The possibility that the two major desmosomal plaque proteins, desmoplakins 1 and 2 (Mueller and Franke, 1982), are associated with IF is based upon studies of desmosome assembly in primary mouse epidermal (PME) cell cultures (Jones and Goldman, 1985). In this culture system, desmosome formation is induced by switching the Ca^{2+} content of the medium from a lower (0.1 mM) to a higher (1.2 mM) concentration (Hennings *et al.*, 1980). The results have shown that under lower levels of Ca^{2+}, antidesmoplakin immunofluorescence exists in a pattern of discrete spots concentrated mainly in the perinuclear region where the cytoplasmic network of keratin IF bundles is also localized. After the switch to higher levels of Ca^{2+}, time sequence studies show that the desmoplakin spots appear to move, along with the keratin IF network, toward the cell surface at areas of cell–cell contact. Desmosomes appear to form after the desmoplakin spots have reached the cell surface. These immunofluorescence observations have been further confirmed by ultrastructural immunogold localization. The data indicate that these preformed desmosomal components are intimately associated with the keratin IF network in the PME cells, and suggest that the movement of both the desmoplakin proteins and the keratin IF bundles to the cell surface is an important initial event in desmosome morphogenesis (see also Green and Jones, this volume).

3.10. Cellular Expression of the IFAP: A Summary

While there is virtually no information available with respect to the specific functions of any of the IFAP (except perhaps filaggrin), a number of their interesting features can be generalized: (1) IF-type specificity has been demonstrated by most of the known IFAP. (2) The expression of most of the IFAP is cell-type specific, i.e., an IFAP associated with a certain type of IF is usually found in some, but not all, of the cells that contain that type of IF. (3) Different IFAP are present in a single cell, and they have been shown to be associated with various IF organizational states (e.g., see the discussions of the three IFAP found in BHK-21 cells: IFAP-300k, IFAP-70-280k, and p50). (4) Within a single cell, the expression of some of its IFAP has been shown to be related to the functional and developmental status of the cell [e.g., functional filaggrin is present only in terminally differentiated, cornified epidermal cells (see Dale *et al.*, this volume), and p50 is specifically increased in senescent BHK-21 cells].

4. Speculations Regarding the Possible Functional Significance of IFAP

As a major cytoskeletal element, IF have been assigned numerous functions, generally based on their collective intracellular architecture (Lasek and Hoffman, 1976; Duffy *et al.*, 1982; Ben-Ze'ev, 1983, 1984; Goldman *et al.*, 1984, 1985, 1986). IF form a very complex network in the cytoplasm through extensive interactions among themselves. Thus, a variety of supramolecular organizational states of IF (e.g., dispersed individual filaments, complex networks, tight IF bundles, and compact juxtanuclear caps of IF) can be found in a single BHK-21 cell (Goldman and Follett, 1969; Yang *et al.*, 1985a). Furthermore, IF have been shown to be closely associated with other cytoplasmic elements, including microfilaments, microtubules, and some membrane-bound organelles such as mitochondria (see Goldman *et al.*, this volume; see also Celis *et al.*, 1984; Wolf and Mullins, 1985; Green *et al.*, 1986, 1987; Wang and Goldman, 1978; Chen *et al.*, 1982; Mose-Larsen *et al.*, 1982). IF also appear to be closely associated with the nuclear surface (Lehto *et al.*, 1978; Jones *et al.*, 1982; Fey *et al.*, 1984; Goldman *et al.*, 1985, 1986) and with the cell surface via such structures as desmosomes and hemidesmosomes (Arnn and Staehelin, 1981; Ellison and Garrod, 1984; Franke *et al.*, 1984; Jones and Goldman, 1985). Recently, evidence for a possible IF–cell surface interaction via the membrane skeleton has also been reported (Granger and Lazarides, 1982; Lazarides, 1984).

Among the many factors that may contribute functionally to the distinct organizational states of the IF cytoskeleton as well as to the interactions of IF with other cytoplasmic and cell surface components, IFAP are the most likely candidates. IFAP may also control dynamic changes in IF arrangements *in vivo* by regulating assembly/ disassembly of IF *in vivo*. Furthermore, IFAP may be involved in the cellular signal-transduction–effector control system thereby permitting the IF cytoskeleton to be functionally linked with other cellular activities (Goldman *et al.*, 1984, 1985, 1986; Traub, 1985a,b). For example, some of the IFAP may be phosphoproteins, or may even contain certain phosphokinase activities. In addition to the known IFAP, which are associated with the cytoplasmic IF system, others as yet undetected may be associated with the nuclear IF network (i.e., the nuclear lamina) thereby constituting a possible new category of IFAP. Such nuclear IFAP could be significant with respect to regulation of gene expression.

ACKNOWLEDGMENTS. We thank Laura J. Davis and Diane Johnson-Seaton for their excellent assistance. This work has been supported by NIH grants to R.D.G. (NCI and NIGMS), H.-Y.Y. (NINCDS), and N.L. (NEI).

References

Arnn, J., and Staehelin, L. A., 1981, The structure and function of spot desmosomes, *Dermatology* **20:**330–339.

Baitinger, C., Cheney, R., Clements, D., Glickman, M., Hirokawa, N., Levine, J., Meiri, K., Simon, C., Skene, P., and Willard M., 1983, Axonally transported proteins in axon development, maintenance and regeneration, *Cold Spring Harbor Symp. Quant. Biol.* **47:**791–802.

Ben-Ze'ev, A., (1983, Cell configuration-related control of vimentin biosynthesis and phosphorylation in cultured mammalian cells, *J. Cell Biol.* **97:**858–865.

Ben-Ze'ev, A., 1984, Control of intermediate filament protein synthesis by cell–cell interaction and cell configuration, *FEBS Lett.* **17:**107–110.

Berkowitz, S., Katagiri, J., Binder, H., and Williams, R., 1977, Separation and characterization of microtubule proteins from calf brain, *Biochemistry* **16:**5610–5617.

Bloom, G. S., and Vallee, R. B., 1983, Association of microtubule associated protein 2 (MAP-2) with microtubules and intermediate filaments in cultured brain cells, *J. Cell Biol.* **96:**1523–1531.

Breckler, J., and Lazarides, E., 1982, Isolation of a new high molecular weight protein associated with desmin and vimentin filaments from avian embryonic skeletal muscle, *J. Cell Biol.* **92:**795–806.

Brown, K. D., and Binder, L. I., 1988, A monoclonal antibody to a high molecular weight brain MAP recognizes intermediate filaments in HeLa and CHO cells, *J. Cell Biol.* **107:**459a.

Celis, J. E., Small, J. V., Larsen, P. M., Fey, S. J., De Mey, J., and Celis, A., 1984, Intermediate filaments in monkey kidney TC7 cells: focal center and interrelationship with other cytoskeletal system, *Proc. Natl. Acad. Sci. USA* **81:**1117–1121.

Chen, L. B., Summerhayes, I. C., Johnson, L. V., Walsh, M. L., Bernal, S. D., and Lampidis, T. J., 1982, Probing mitochondria in living cells with rhodamine 123, *Cold Spring Harbor Symp. Quant. Biol.* **46:**141–155.

Ciment, G., Resler, A., Letourneau, P. C., and Weston, J. A., 1986, A novel intermediate filament-associated protein, NAPA-73, that binds to different filament types at different stages of nervous system development, *J. Cell Biol.* **102:**246–251.

Dale, B. A., 1977, Purification and characterization of a basic protein from the stratum corneum of mammalian epidermis, *Biochim. Biophys. Acta* **491:**193–204.

Dentler, W. L., Granette, S., and Rosenbaum, J. L., 1975, Ultrastructural localization of the high molecular weight proteins associated with *in vitro* assembled brain microtubules, *J. Cell Biol.* **65:**237–241.

Duffy, P. E., Huang, Y. Y., and Rapport, M. M., 1982, The relationship of glial fibrillary acidic protein to the shape, motility, and differentiation of human astrocytoma cells, *Exp. Cell Res.* **139:**145–157.

Ellison, J., and Garrod, D. R., 1984, Anchoring filaments of the amphibian epidermal–dermal junction traverse the basal lamina entirely from the plasma membrane of the hemidesmosomes to the dermis, *J. Cell Sci.* **72:**163–172.

Fey, E. G., Wan, K. M., and Penman, S., 1984, The epithelial cytoskeletal framework and nuclear matrix–intermediate filament scaffold: Three dimensional organization and protein composition, *J. Cell Biol.* **100:**93–102.

Foisner, R., Leichtfield, F. E., Herrmann, H., Small, J. V., Lawson, D., and Wiche, E., 1988, Cytoskeleton-associated plectin: *In situ* localization, *in vitro* reconstitution, and binding to immobilized intermediate filament proteins, *J. Cell Biol.* **106:**723–733.

Franke, W. W., Schiller, D. L., Hatzfield, M., Magin, T. M., Jorcano, J. L., Mittnacht, S., Schmid, E., Cohlberg, J. A., and Quinlan, R. A., 1984, Cytokeratins: Complex formation biosynthesis, and interactions with desmosomes, in: *Cancer Cell,* Volume 1 (A. Levine, W. Topp, G. V. Wonde, and J. D. Watson, eds.), Cold Spring Harbor Laboratory, Cold Spring Harbor, N.Y., pp. 177–190.

Fuchs, E., and Hanukoglu, I., 1983, Unraveling the structure of the intermediate filaments, *Cell* **34:**332–334.

Georgatos, S. D., and Blobel, G., 1987, Lamin B constitutes an intermediate filament attachment site at the nuclear envelope, *J. Cell Biol.* **105:**117–125.

Georgatos, S. D., and Marchesi, V. T., 1985, The binding of vimentin of human erythrocyte membranes: A model system for the study of intermediate filament–membrane interactions, *J. Cell Biol.* **100:**1955–1961.

Georgatos, S. D., Weaver, D. C., and Marchesi, V. T., 1985, Site specificity in vimentin–membrane interactions: Intermediate filament subunits associate with the plasma membrane via their head domains, *J. Cell Biol.* **100:**1962–1967.

Georgatos, S. D., Weber, K., Geisler, N., and Blobel, G., 1987, Binding of two desmin derivatives of the plasma membrane and the nuclear envelope of avian erythrocytes: Evidence for a conserved site-specificity in intermediate filament–membrane interactions, *Proc. Natl. Acad. Sci. USA* **84:**6780–6784.

Goldman, R. D., and Follett, E. A. C., 1969, The structure of the major cell processes of isolated BHK-21 fibroblasts, *Exp. Cell Res.* **57:**263–276.

Goldman, R. D., and Knipe, D. M., 1972, Functions of cytoplasmic fibers in nonmuscle cell motility, *Cold Spring Harbor Symp. Quant. Biol.* **37:**534.

Goldman, R. D., Milsted, A., Schloss, J. A., Starger, J., and Yerna, M. J., 1979, Cytoplasmic fibers in mammalian cells: Cytoskeletal and contractile elements, *Annu. Rev. Physiol.* **41:**703–722.

Goldman, R. D., Hill, B. F., Steinert, P., Aynardi-Whitman, M., and Zackroff, R. V., 1980, Intermediate filament–micortubule interactions: Evidence in support of a common organization center, in: *Microtubules and Microtubule Inhibitors* (M. DeBrabander and J. DeMey, eds.), Elsevier/North-Holland, Amsterdam. pp. 91–102.

Goldman, R. D., Goldman, A. E., Green, K. J., Jones, J. C. R., Lieska, N., Talian, J., and Yang, H.-Y., 1984, Intermediate filaments: Their interactions with various cell organelles and their associated proteins, *J. Submicrosc. Cytol.* **16:**73–74.

Goldman, R. D., Goldman, A. E., Green, K. J., Jones, J. C. R., Lieska, N., and Yang, H.-Y., 1985, Intermediate filaments: Possible functions as cytoskeletal connecting links between the nucleus and the cell surface, *Ann. N.Y. Acad. Sci.* **455:**1–17.

Goldman, R. D., Goldman, A. E., Green, K., Jones, J. C. R., Jones, S. M., and Yang, H.-Y., 1986, Intermediate filament networks: Organization and possible functions of a diverse group of cytoskeletal elements, *J. Cell Sci. Suppl.* 5:69–97.

Granger, B. L., and Lazarides, E., 1980, Synemin: A new high molecular weight protein associated with desmin and vimentin filaments in muscle, *Cell* **22:**727–738.

Granger, B. L., and Lazarides, E., 1982, Structural associations of synemin and vimentin filaments in avian erythrocytes revealed by immunoelectron microscopy, *Cell* **30:**263–275.

Granger, B. L., and Lazarides, E., 1984, Expression of the intermediate-filament-associated protein synemin in chicken lens cells, *Mol. Cell Biol.* **4:**1943–1950.

Granger, B. L., Repasky, E. A., and Lazarides, E., 1982, Synemin and vimentin are components of intermediate filaments in avian erythrocyte, *J. Cell Biol.* **92:**299–312.

Green, K. J., and Goldman, R. D., 1986, Evidence for an interaction between the cell surface and the intermediate filaments in cultured fibroblasts, *Cell Motil. Cytoskel.* **6:**389–405.

Green, K. J., Talian, J. C., and Goldman, R. S., 1986, Relationship between intermediate filaments and microfilaments in cultured fibroblasts: Evidence for common foci during cell spreading, *Cell Motil. Cytoskel.* **6:**604–618.

Green, K. J., Geiger, B., Jones, J. C. R., and Talian, J. C., 1987, The relationship between intermediate filaments and microfilaments before and during the formation of desmosomes and adherens-type junctions in mouse epidermal keratinocytes, *J. Cell Biol.* **104:**1389–1402.

Hennings, U., Michael, D., Cheng, C., Steinert, P., Holbrook, K., and Yuspa, S. H., 1980, Calcium regulation of growth and differentiation of mouse epidermal cells in culture, *Cell* **19:**245–254.

Herrmann, H., and Wiche, G., 1983, Specific *in situ* phosphorylation of plectin in detergent-resistant cytoskeletons from cultured Chinese hamster ovary cells, *J. Biol. Chem.* **258:**14610–14618.

Herrmann, H., and Wiche, G., 1987, Plectin and IFAP-300K are homologous proteins binding to microtubule-associated proteins 1 and 2 and to the 240 kilodalton subunit of spectrin, *J. Biol. Chem.* **262:**1320–1325.

Hirokawa, N., 1982, Cross-linker system between neurofilaments, microtubules and membranous organelles in frog axons revealed by quick-freeze, deep-etching method, *J. Cell Biol.* **94:**129–142.

Hirokawa, N., Cheney, R. E., and Willard, M., 1983, Location of a protein of the fodrin/spectrin TW 260/240 family in the mouse intestinal brush border, *Cell* **32:**953–965.

Hirokawa, N., Glicksman, M. A., and Willard, M. B., 1984, Organization of mammalian neurofilament polypeptides within the neuronal cytoskeleton, *J. Cell Biol.* **98:**1523–1536.

Ishikawa, H., Bischoff, R., and Holtzer, H., 1968, Mitosis and intermediate sized filaments in developing skeletal muscle, *J. Cell Biol.* **38:**538–555.

Jones, J. C. R., and Goldman, R. D., 1985, Intermediate filaments and the initiation of desmosome assembly, *J. Cell Biol.* **101:**506–517.

Jones, J. C. R., Goldman, A. E., Steinert, P. M., Yuspa, S., and Goldman, R. D., 1982, Dynamic aspects of the supramolecular organization of intermediate filament networks in cultured epidermal cells, *Cell Motil.* **2:**197–213.

Jones, J. C. R., Goldman, A. E., Yang, H.-Y., and Goldman, R. D., 1985, The organizational fate of intermediate filament networks in two epithelial cell types during mitosis, *J. Cell Biol.* **100:**93–102.

Lasek, R. J., and Hoffman, P. N., 1976, The neuronal cytoskeleton, axonal transport and axonal growth, in: *Cell Motility* (R. Goldman, T. Pollard, and J. Rosenbaum. eds.), Cold Spring Harbor Laboratory, Cold Spring Harbor, N.Y., pp. 1021–1049.

Lawson, D., 1983, Epinemin: A new protein associated with vimentin filaments in non-neural cells, *J. Cell Biol.* **97:**1891–1905.

Lawson, D., 1984, Distribution of epinemin in colloidal gold-labeled, quick-frozen deep-etched cytoskeletons, *J. Cell Biol.* **99:**1451–1460.

Lazarides, E., 1980, Intermediate filaments as mechanical integrators of cellular space, *Nature* **283:**249–253.

Lazarides, E., 1982, Intermediate filaments: A chemically heterogeneous, developmentally regulated class of proteins, *Annu. Rev. Biochem.* **51:**219–250.

Lazarides, E., 1984, Assembly and morphogenesis of the avian erythrocyte cytoskeleton, in: *Molecular Biology of the Cytoskeleton* (G. G. Borisy, D. W. Cleveland, and D. B. Murphy, eds.), Cold Spring Harbor Laboratory, Cold Spring Harbor, N.Y., pp. 131–150.

Lehto, V.-P., 1983, 140,000 dalton surface glycoprotein. A plasma membrane component of the detergent-resistant cytoskeletal preparation of cultured human fibroblasts, *Exp. Cell Res.* **143:**271–286.

Lehto, V.-P., and Virtanen, I., 1983, Immunolocalization of a novel, cytoskeleton-associated polypeptide of Mr 230,000 daltons (p230), *J. Cell Biol.* **96:**703–716.

Lehto, V.-P., Virtanen, I., and Kurki, P., 1978, Intermediate filaments anchor the nuclei in nuclear monolayers of cultured human fibroblasts, *Nature* **272:**175–177.

Le Terrier, J. F., Liem, R. K., and Shelanski, M. L., 1982, Interactions between neurofilaments and microtubule-associated proteins: A possible mechanism for intraorganellar bridging, *J. Cell. Biol.* **95:** 982–986.

Lieska, N., Yang, H.-Y., and Goldman, R. D., 1985a, Purification of the 300,000-mol-wt intermediate filament (IF) associated protein and its *in vitro* recombination with IF, *J. Cell Biol.* **101:**802–813.

Lieska, N., Yang, Y.-Y., and Goldman, R. D., 1985b, Purification and some *in vitro* properties of BHK-21 cell intermediate filament-associated proteins 70/280kD, *J. Cell Biol.* **101:**15a.

Lin, J. J.-C., and Feramisco, J. R., 1981, Disruption of the *in vivo* distribution of the intermediate filaments in fibroblasts through the microinjection of a specific monoclonal antibody, *Cell* **24:**185–193.

Mangeat, P. H., and Burridge, K., 1984, Immunoprecipitation of non-erythrocyte spectrin within live cells following microinjection of specific antibodies: Relation of cytoskeletal structures, *j. Cell Biol.* **98:**1363–1377.

Metuzals, J., Montpetit, V., and Clapin, D.-F., 1981, Organization of the neurofilamentous network, *Cell Tissue Res.* **214:**455–482.

Minami, Y., Murofishi, V., and Sakai, H., 1982, Interaction of tubulin with neurofilaments: Formation of networks by neurofilament dependent tubulin polymerization, *J. Biochem.* **92:**889–898.

Miyata, Y., Hoshi, M., Nishida, E., Minami, Y., and Sakai, H., 1986, Binding of microtubule-associated protein 2 and tau to the intermediate filament reassembled from neurofilament 70-kDa subunit protein, *J. Biol. Chem.* **261:**13026–13030.

Moon, R. T., and Lazarides, E., 1983, Synthesis and post-translational assembly of intermediate filaments in avian erythroid cells: Vimentin assembly limits the rate of synemin assembly, *Proc. Natl. Acad. Sci. USA* **80:**5495–5499.

Mose-Larsen, P., Bravo, R., Fey, S. J., Small, J. V., and Celis, J. E., 1982, Putative association of mitochondria with a subpopulation of intermediate-sized filaments in cultured human skin fibroblasts, *Cell* **31:**681–692.

Mueller, M., and Franke, W. W., 1982, Biochemical and immunological characterization of desmoplakins I and II, the major polypeptides of the desmosomal plaque, *Differentiation* **23:**189–205.

Murphy, D. B., and Borisy, G. G., 1975, Association of high-molecular-weight proteins with microtubules and their role in microtubule assembly *in vitro, Proc. Natl. Acad. Sci. USA* **72:**2696–2700.

Osborn, M., and Weber, K., 1982, Intermediate filaments: Cell type specific markers in differentiation and pathology, *Cell* **31:**303–306.

Price, M. G., and Lazarides, E., 1983, Expression of intermediate filament-associated protein paranemin and synemin in chicken development, *J. Cell Biol.* **97:**1860–1874.

Pruss, R. M., Mirsky, R., Raff, M. C., Thorpe, R., Dowding, A. J., and Anderson, B. H., 1981, All classes of intermediate filaments share a common antigenic determinant defined by a monoclonal antibody, *Cell* **27:**419–428.

Pytela, R., and Wiche, G., 1980, High molecular weight polypeptides (270,000–340,000) from cultured cells are related to hog brain microtubule-associated proteins but copurify with intermediate filaments, *Proc. Natl. Acad. Sci. USA* **77:**4808–4812.

Ramaekers, F. C. S., Dunia, S. I., Dodemont, H. J., Benedetti, E. L., and Bloemendal, H., 1982, Lenticular intermediate-sized filaments: Biosynthesis and interactions with plasma membrane, *Proc. Natl. Acad. Sci. USA* **79:**3208–3212.

Runge, M. S., Lau, T. M., Yphantis, D., Lifiscs, M., Saito, A., Altin, M., Reinke, K., and Williams, R. C., Jr., 1981, ATP-induced formation of an associated complex between microtubules and neurofilaments, *Proc. Natl. Acad. Sci. USA* **78:**1431–1435.

Sandoval, I. V., Colaco, C. A., and Lazarides, E., 1983, Purification of the intermediate filament-associated protein, synemin, from chicken smooth muscle. Studies on its physiochemical properties, interaction with desmin and phosphorylation, *J. Biol. Chem.* **258:**2568–2576.

Schnapp, B. J., and Reese, T. S., 1982, Cytoplasmic structure in rapid frozen axons, *J. Cell Biol.* **94:**667–669.

Shelanski, M. L., LeTerrier, J.-F., and Liem, R. K. H., 1981, Evidence for interaction between neurofilaments and microtubules, *Neurosci. Res. Progr. Bull.* **19:**32–43.

Starger, J. M., Brown, W. E., Goldman, A. E., and Goldman, R. D., 1978, Biochemical and immunological analysis of rapidly purified 10nm filaments from baby hamster kidney (BHK-21) cells, *J. Cell Biol.* **78:**93–109.

Steinert, P. M., and Parry, D. A. D., 1985, Intermediate filaments, *Annu. Rev. Cell Biol.* **1:**41–65.

Steinert, P. M., and Roop, D. R., 1988, Molecular and cellular biology of intermediate filaments, *Annu. Rev. Biochem.* **57:**593–625.

Thorpe, R., Delacourte, A., Ayers, M., Bullock, C., and Anderson, B. H., 1979, The polypeptides of isolated brain 10nm filaments and their association with polymerized tubulin, *Biochem. J.* **181:**275–84.

Tokuyasu, K. T., Maher, P. A., and Singer, S. J., 1984, Distribution of vimentin and desmin in developing chick myotubes *in vivo*. I. Immunofluorescence study, *J. Cell Biol.* **98:**1961–1972.

Traub, P., 1985a, *Intermediate Filaments: A Review*, Springer-Verlag, Berlin.

Traub, P., 1985b, Are intermediate filament proteins involved in gene expression? *Ann. N.Y. Acad. Sci.* **455:**68–78.

Tsukita, S., Usukura, J., Tsukita, S., and Ishikawa, H., 1982, The cytoskeleton in myelinated axons: A freeze-etch replica study, *Neuroscience* **7:**2135–2147.

Vallee, R. B., and Bloom, G. S., 1984, High molecular weight microtubule-associated proteins (MAPs), *Modern Cell Biology 3*, Liss, New York, pp. 21–75.

Vallee, R. B., Bloom, G. S., and Theurkauf, W. E., 1984, Microtubule-associated proteins: Subunits of the cytomatrix, *J. Cell Biol.* **99:**38s–44s.

Voter, W. A., and Erickson, H. P., 1982, Electron microscopy of MAP 2 (microtubule-associated protein 2). *J. Ultrastruct. Res.* **80:**374–382.

Wang, E., and Goldman, R. D., 1978, Functions of cytoplasmic fibers in intracellular movement in BHK-21 cells, *J. Cell Biol.* **79:**708–726.

Wang, E., Ciarncross, E. J., Yung, W. K. S., Garber, E. R. and Liem, R. K. H., 1983, An intermediate filament-associated protein, p50, recognized by monoclonal antibodies, *J. Cell Biol.* 97:1507–1514.

Wang, E., Fischman, D., Liem, R. K. H., and Sun, T.-T. (eds.), 1985, *Intermediate Filaments, Ann. N.Y. Acad. Sci.* **455.**

Wang, W., 1985, Are cross-bridge structures involved in the bundle formation of intermediate filaments and the decrease in locomotion that accompany cell aging? *J. Cell Biol.* **100:**1466–1473.

Wiche, G., and Baker, A. A., 1982, Cytoplasmic network arrays demonstrated by immunolocalization using antibodies to a high molecular weight protein present in cytoskeletal preparations from cultured cells, *Exp. Cell Res.* **138:**15–29.

Wiche, G., Baker, M. A., Kindas-Mugge, I., Leichtfried, F., and Pytela, R., 1980, High molecular weight polypeptides (around 300,000) from cultured cells and their possible role as mediators of microtubule–intermediate filament interaction, in: *Microtubules and Microtubule Inhibitors* (M. DeBrabander and J. DeMey, eds.), Elsevier/North-Holland, Amsterdam, pp. 189–200.

Wiche, G., Herrmann, H., Leichtfried, F., and Pytela, R., 1982, Plectin: A high molecular weight cytoskeletal polypeptide component that copurifies with intermediate filaments of the vimentin types, *Cold Spring Harbor Symp. Quant. Biol.* **46:**475–482.

Wiche, G., Krepler, R., Antlieb, U., Pytela, R., and Denk, H., 1983, Occurrence and immunolocalization of plectin in tissues, *J. Cell Biol.* **97:**887–901.

Wiche, G., Krepler, R., Antlieb, U., Pytela, R., and Aberer, W., 1984, Identification of plectin in different human cell types and immunolocalization at epithelial basal cell surface membranes, *Exp. Cell Res.* **155:**43–49.

Willard, M., 1983, Neurofilaments and axonal transport, in: *Neurofilaments* (C. A. Marotta, ed.), University of Minnesota Press, Minneapolis, pp. 86–116.

Wolf, K. M., and Mullins, J. M., 1985, Cytochalasin-induced redistribution of cytokeratin filaments in PtK-1 cells, *J. Cell Biol.* **101:**14a.

Yang, H.-Y., Lieska, N., and Goldman, R. D., 1984, The distribution of a 300K molecular weight intermediate filament associated protein in different cell types, *J. Cell Biol.* **99:**324a.

Yang, H.-Y., Lieska, N., Goldman, A. E., and Goldman, R. D., 1985a, A 300,000-mol-wt. intermediate filament associated protein in baby hamster kidney (BHK-21) cells, *J. Cell Biol.* **100:**620–631.

Yang, H.-Y., Lieska, N., and Goldman, R. D., 1985b, Identification and localization of a group of immunologically-related, intermediate filament-associated polypeptides in BHK-21 cells, *J. Cell Biol.* **101:**15a.

Yang, H.-Y., Lieska, N., and Goldman, R. D., 1986, Different distribution of two intermediate filament-associated proteins, IFAP-300kD and IFAP-70-280kD, in BHK-21 cells, *J. Cell Biol.* **103:**419a.

Yang, H.-Y., Lieska, N., and Goldman, R. D., 1990, A group of immunologically-related, intermediate filament-associated polypeptides (70–280kD) in BHK-21 cells, submitted for publication.

Zackroff, R. V., and Goldman, R. D., 1979, *In vitro* assembly of intermediate filaments from baby hamster kidney (BHK-21) cells, *Proc. Natl. Acad. Sci. USA* **76:**6226–6230.

Zingsheim, H.-P., Hertzog, W., and Weber, K., 1979, Difference in surface morphology of microtubules reconstituted from pure brain tubulin using two different microtubule-associated proteins in high molecular weight MAP-2 proteins and tau proteins, *Eur. J. Cell Biol.* **19:**175–183.

16

Filaggrins

Beverly A. Dale, Katheryn A. Resing, and Paul V. Haydock

1. Introduction

Filaggrins are a group of cationic, species-distinct proteins named for their ability to cause intermediate filament (IF) aggregation *in vitro*. Several types of intermediate filament associated proteins (IFAP) are now recognized and have been classified in a recent review (Steinert and Roop, 1988, and described elsewhere in this volume). Filaggrin was the first epidermal IFAP to be isolated, and is a readily studied example of this class of proteins. Various investigators have made significant progress on the protein structure and function. More recently, the cDNA has been used for studies of profilaggrin expression and gene structure.

A large phosphorylated precursor (profilaggrin) is synthesized in the granular cell layer of the epidermis (Fig. 1) and other keratinizing epithelia. It is converted to filaggrin late in the process of keratinization. Profilaggrin does not associate with the abundant keratin IF of the epidermis, while filaggrin forms large fibrous aggregates (macrofibrils) similar in appearance to those in keratinized cells; thus, filaggrin may function *in vivo* as a keratin filament aggregating protein to aid in the alignment and dense packing of these filaments in the differentiated cells of the stratum corneum.

The synthesis of profilaggrin and its conversion to filaggrin are apparently tightly controlled and occur only in those tissues that synthesize high-molecular-weight keratins that are unique to keratinized epithelia (56.5k and 65–67k in humans; for a review see Sun *et al.*, 1983). Filaggrin, as well as these keratins, are differentiation-specific markers for keratinization. Thus, the occurrence, function, and fate of filaggrin are linked to the IF proteins that are the major structural proteins of keratinized epithelia.

Beverly A. Dale • Departments of Oral Biology, Periodontics, Medicine/Dermatology, and Biochemistry, University of Washington, Seattle, Washington 98195. ***Katheryn A. Resing*** • Department of Biochemistry, University of Washington, Seattle, Washington 98195. ***Paul V. Haydock*** • Departments of Periodontics and Medicine/Dermatology, University of Washington, Seattle, Washington 98195.

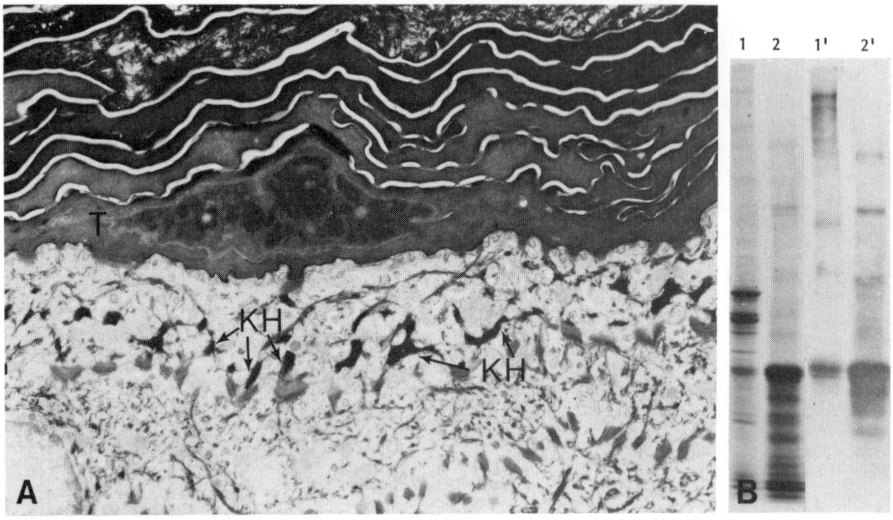

Figure 1. (A) Morphology of human epidermis in the region of profilaggrin expression, posttranslational modification to filaggrin, and degradation. Note the irregular-shaped keratohyalin granules (KH) within cells of the granular layer, the transition cell (T) with degenerating nucleus, the lower four to five cell layers within the stratum corneum that contain densely packed keratin filaments embedded in a matrix material, and the upper cell layers of the stratum corneum in which the matrix material has disappeared and the filaments are more dispersed. From K. Holbrook. × 13,000. Bar = 1 μm. (B) Detection of human epidermal filaggrin and profilaggrin by Western blotting. 8 M urea crude epidermal extract (lanes 1 and 1′) and cationic protein fraction that does not bind to DE52 at pH 8 (lanes 2 and 2′). Lanes 1 and 2 show the Coomassie blue-stained gels. Lanes 1′ and 2′ show the PAP-stained blot using a monoclonal antibody to filaggrin. Profilaggrin (P) is present in the crude extract (lane 1′), but is partially proteolyzed. Filaggrin (F) is present in both the crude extract and cationic fraction. Profilaggrin binds to DE52 and is absent from lane 2′. Several cross-reactive cationic bands are probable dephosphorylated intermediates (I) in the conversion of profilaggrin to filaggrin, and are distinguished from artifactual proteolytic fragments of profilaggrin by their DE52-binding properties. The keratins (stained in lane 1) do not react with antibody to filaggrin.

The size of filaggrin varies in different mammalian species, but its function, biochemical properties, composition, synthesis from a large phosphorylated precursor, localization within developing and differentiating epithelia, and degradation within the stratum corneum are common features in all species studied. These features will be reviewed in this chapter. In addition, we will briefly review several genetic disorders of epidermal keratinization now known to be associated with filaggrin alterations.

2. *Isolation and Properties of Filaggrin and Profilaggrin*

2.1. *Extraction and Purification*

Filaggrin and profilaggrin are preferentially extracted from epidermis by buffers containing high concentrations of salt (Ugel, 1971; Sibrack *et al.*, 1974; Dale and Ling, 1979a; Resing *et al.*, 1984), while filaggrin is preferentially extracted with formic acid or

4 M urea. Buffered 8 M urea extracts of epidermis contain both profilaggrin and filaggrin as well as the keratins (Fig. 1B).

Because filaggrin is extremely cationic, it is readily isolated by ion-exchange chromatography; it is a major component of the cationic protein fraction which does not bind to DEAE-cellulose (Fig. 1B). It can be further purified by gel filtration, HPLC, or preparative SDS–PAGE (Dale, 1977; Steinert *et al.*, 1981; Resing *et al.*, 1985). The purified protein is soluble in dilute buffers at neutral pH and is relatively stable. In contrast, profilaggrin binds to DEAE-cellulose due the presence of large amounts of phosphate (Lonsdale-Eccles *et al.*, 1980). Profilaggrin is soluble only in solutions with urea or high salt concentrations.

2.2. *Properties of Profilaggrin and Filaggrin*

Properties of profilaggrin and filaggrin are summarized in Table I. Filaggrin ranges in size from 26k in mouse to 64k in guinea pig. Profilaggrin has a high molecular weight (≥ 400k) in all species studied. Antibodies to either profilaggrin or filaggrin react with both forms of the protein on immunoblots (e.g., Fig. 1B) and with the granular and cornified layers of the epidermis (see Fig. 10) (Dale and Ling, 1979b; Murozuka *et al.*, 1979).

The filaggrins were originally called "histidine-rich proteins" (Hoober and Bernstein, 1966) because they contain approximately 7–8% histidine. Both profilaggrin and filaggrin have an unusual amino acid composition characterized by a predominance of Arg, Glu, Gly, Ser, and His, but are virtually lacking in nonpolar residues. Pulse–chase labeling of epidermal proteins and autoradiography of tissue sections convincingly demonstrate a precursor–product relationship between profilaggrin and filaggrin. Profilaggrin in the granular cell layer is rapidly labeled with [^{3}H]histidine, but filaggrin becomes labeled only after a lag of more than 6 hr (Ball *et al.*, 1978; Dale and Ling, 1979a) corresponding to the time of labeling of transition cells of the stratum corneum (Fukuyama and Epstein, 1966).

Table I. Characterization of Filaggrin and Profilaggrin

Property	Profilaggrin	Filaggrin
Localization in epidermis	Granular layer	Cornified
Molecular weight by SDS–PAGE[a]		
Rat	~600k	45–48k
Human	~400k	38k
Mouse	~500k	27k
Guinea pig	>300k	64k
Isoelectric point[b]	6.5	>10
Phosphate (P-ser)	8–20 moles/filaggrin unit	0
Reaction with antibody to filaggrin	+	+
Reaction with keratins to form macrofibrils	−	+

[a]Molecular weights from Dale (1977), Lonsdale-Eccles *et al.* (1982), Harding and Scott (1983), Lynley and Dale (1983), Steinert *et al.* (1981), Ramsden *et al.* (1983), Fleckman *et al.* (1985), Resing *et al.* (1984) for various mammalian species. Additional size variants of filaggrin (not indicated) are also notable in mouse and guinea pig.
[b]Values for rat filaggrin.

2.3. Profilaggrin Protein Structure

The structure of the high-molecular-weight profilaggrin has been elucidated by peptide mapping and cDNA sequencing. Lonsdale-Eccles *et al.* (1984) suggested that profilaggrin has a repeating structure in which filaggrin units are arranged in a linear sequence separated by linker peptides that are lost during the conversion of profilaggrin to filaggrin (Fig. 2). Evidence leading to this model includes (1) the observation that profilaggrin is synthesized from a very large mRNA, 30 S and 34 S in mouse and rat, respectively (Meek *et al.*, 1983), (2) peptide mapping studies demonstrated a striking similarity between profilaggrin and filaggrin (Bernstein *et al.*, 1983; Lonsdale-Eccles *et al.*, 1982; Resing *et al.*, 1985), and (3) profilaggrin is readily proteolyzed to a unit ϕ, slightly larger than filaggrin (Scott and Harding, 1981; Lonsdale-Eccles *et al.*, 1984), via a polymeric series of related proteins. Support for the tandem-repeat model was subsequently obtained by detailed tryptic mapping of filaggrin, a two-domain intermediate, and profilaggrin from C57B16 mice. Results showed that (1) profilaggrin is composed of 16 to 18 repeated copies of filaggrin, (2) several peptides that are common to both profilaggrin

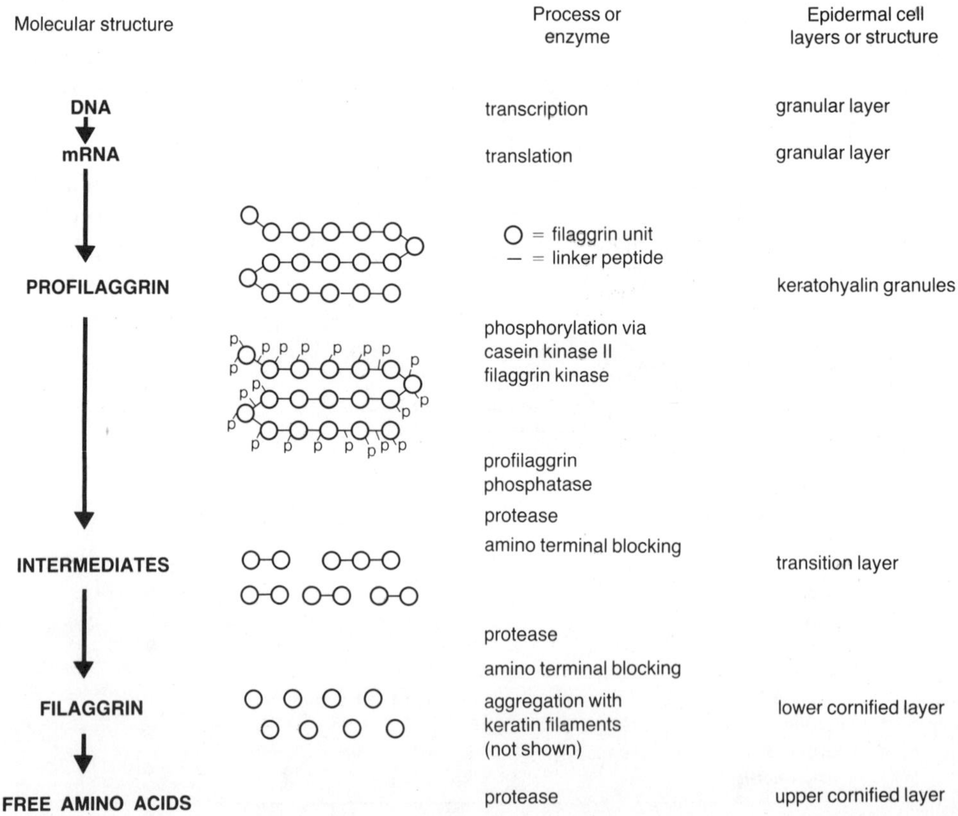

Figure 2. Diagrammatic scheme of filaggrin synthesis showing correlation of expression and posttranslational modification steps with epidermal cell differentiation.

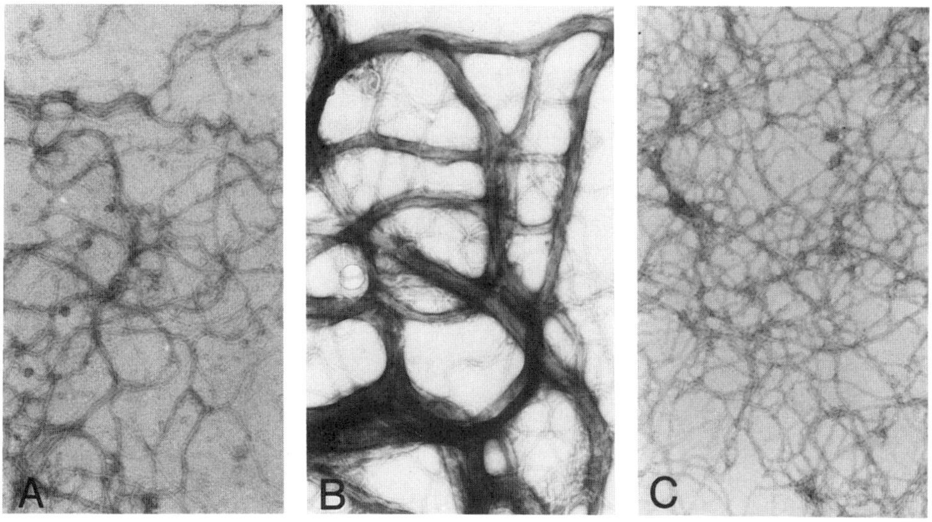

Figure 3. (A) Association of rat epidermal filaggrin and keratin filaments to form macrofibrils. Keratin filaments isolated and formed by the method of Steinert *et al.* (1976). (B) Macrofibrils formed from filaggrin plus keratin filaments at a ratio of 0.25 : 1 (w/w). (C) Profilaggrin (ϕ unit) plus keratin filaments at a ratio of 0.25 : 1 (w/w). Modified from Lonsdale-Eccles *et al.* (1982).

and filaggrin are phosphorylated in profilaggrin, and (3) additional tyrosine-containing peptides are unique to profilaggrin (Resing *et al.*, 1985; Dale *et al.*, 1985a). The latter peptides would be very susceptible to proteolysis via chymotrypsin-like enzymes, and thus may explain the ease of proteolytic breakdown. These peptide sequences have been confirmed in sequences of a cDNA clone for mouse profilaggrin (Rothnagel *et al.*, 1987). Homologous sequences are present in the rat cDNA sequence (see Section 3.5 and Fig. 6; Haydock and Dale, 1989).

2.4. Aggregation with Keratin IF

Filaggrin and cationic intermediates (see Fig. 2) aggregate with keratin (Harding and Scott, 1983), forming fibers in which the filaments are closely aligned and parallel to each other (Fig. 3) (Dale *et al.*, 1978; Steinert *et al.*, 1981; Lonsdale-Eccles *et al.*, 1982). The resulting macrofibrils contain filaggrin and the intermediates as a structural component that can be recovered upon solubilization (Fig. 4). The stoichiometry of the filaggrin–filament association is approximately 2 moles filaggrin to 3 moles filament protein. An association plot of the binding of labeled filaggrin to a constant quantity of keratin filaments gives a value of 0.83 (approximately 1.0) per keratin filament polypeptide with a dissociation constant of 10^{-6} (Dale *et al.*, 1985a). It is not yet clear whether the strength of the filaggrin–keratin interaction is related to the keratin composition of the filaments. Filaggrin is also capable of association with other classes of IF, although specific parameters of the association are not known (Steinert *et al.*, 1981).

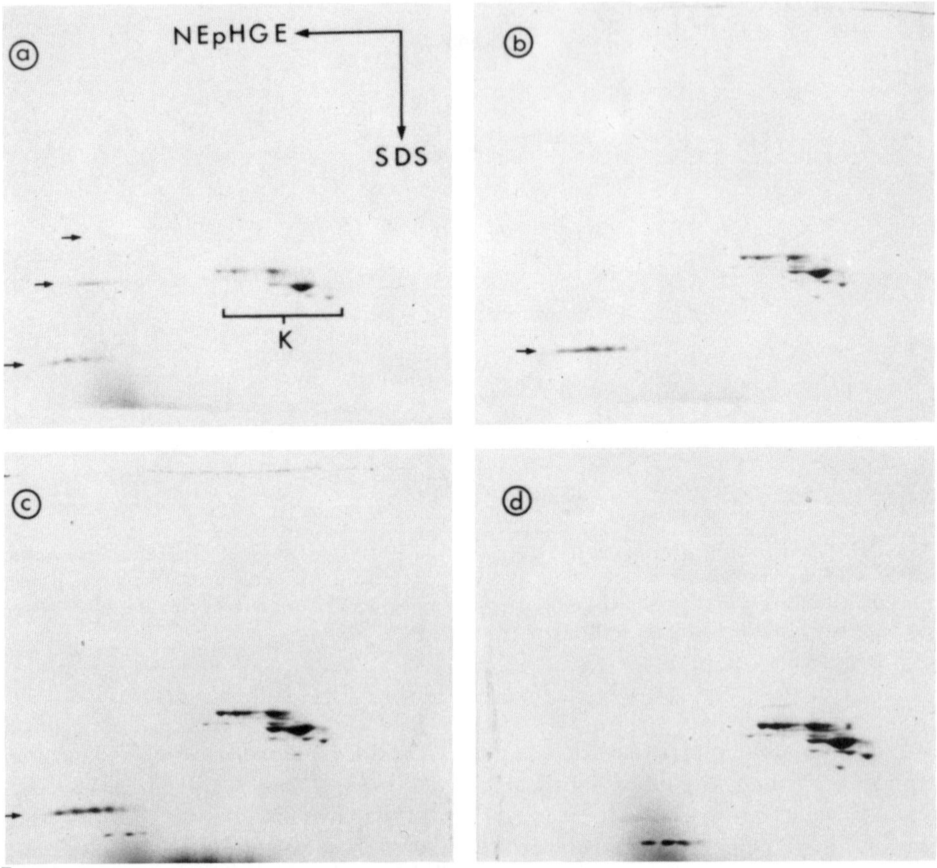

Figure 4. Two-dimensional analysis of *in vitro* macrofibril formation with newborn mouse filaggrin. Aliquots of guinea pig keratin were sequentially added to a solution of partially purified mouse filaggrins until macrofibrils no longer formed. The macrofibrils were recovered after each addition and analyzed by NEpHGE/SDS gel electrophoresis. (a) First macrofibril pellet, (b) second pellet, (c) last pellet, (d) soluble nonaggregating proteins. K denotes the major keratins, arrow denotes filaggrin and intermediates that can be recovered from macrofibril pellets. From Harding and Scott (1983).

3. Profilaggrin mRNA and Gene Structure

3.1. Use of cDNA Clones

The unique features of the model for profilaggrin structure cannot be adequately tested via peptide and protein sequencing studies. The proposed repeating structure of profilaggrin has been confirmed by the use of cDNA clones. Profilaggrin cDNA clones have also been used to determine the size of the mRNA, to obtain the sequence of filaggrin, and to serve as a source of "recombinant filaggrin" produced by *E. coli* via an expression vector, and to begin to understand the filaggrin gene structure.

3.2. Selection and Characterization of Profilaggrin cDNA Clones

Oligonucleotide probes encoding unique peptide sequences were used to screen mouse and rat epidermal cDNA libraries prepared from epidermal RNA enriched for profilaggrin mRNA by sucrose gradient centrifugation. Haydock and Dale (1986) used an oligoprobe coding for the tyrosine-containing linker region and obtained a rat cDNA clone with a 2314-bp insert. Mouse profilaggrin clones were similarly screened using a different oligoprobe; a clone of 1479 bp was characterized and sequenced (Rothnagel *et al.*, 1987).

The cDNA clones derived from rat and mouse have several interesting features in common. Both cDNA clones have the same overall structure, encoding most of a filaggrin domain, a linker segment, a filaggrin domain that is broken by the stop codon, and the 3′ noncoding region. The second filaggrin domain (the last on the large profilaggrin molecule) diverges in sequence in the last 23 amino acids; the divergent sequence and the 3′ untranslated region are extremely AT rich (70%) in contrast to the remainder of the clone (30% AT). Clones from both rat and mouse have large regions of essentially identical sequence within the filaggrin domain which define the size of the profilaggrin repeat unit in that species; 1218 bp for rat and 750 bp for mouse. This was initially shown by restriction endonuclease mapping for the rat cDNA and later by complete sequencing for mouse (Rothnagel *et al.*, 1987) and rat (Haydock and Dale, 1989). Together, these data strongly suggested that the tandem-repeat model for profilaggrin structure was correct. The model was proven at the genomic level by the demonstration of a repetitive structure in the gene by Southern blotting of genomic DNA and by characterization of a cosmid clone containing the mouse profilaggrin gene (Haydock and Dale, 1986; Rothnagel *et al.*, 1987).

3.3. Size and Expression of Profilaggrin mRNA

Profilaggrin mRNA is extremely large; the rat and mouse mRNAs are 23 and 17kb, respectively, by Northern blotting (Haydock and Dale, 1986; Rothnagel *et al.*, 1987) (for a representative blot for rat, see Fig. 8), confirming earlier results by sucrose density gradient centrifugation (Meek *et al.*, 1983). Hybridization signals are smeared, indicating that these mRNAs either are labile during purification, or are present in epidermal cells as a series of incomplete transcripts. In fact, the 3′ noncoding region of both rat and mouse profilaggrin mRNAs have several characteristics in common with mRNAs that have been shown to be unstable.

In situ hybridization using cRNA clones shows expression of profilaggrin in the granular layer of newborn rat (Fisher *et al.*, 1987) and mouse epidermis (Rothnagel *et al.*, 1987), the location of keratohyalin granules. This suggests that regulation of profilaggrin expression occurs at the transcriptional level.

3.4. Repetitive Structure of the Profilaggrin Gene

The profilaggrin gene structure has been examined in both rat and mouse (Haydock and Dale, 1986; Rothnagel *et al.*, 1987). Hybridization patterns of a variety of restriction enzyme digests of genomic DNA closely match those obtained with the cDNA. For example, when rat genomic DNA was digested with *Pst*I, a single genomic fragment of approximately 1200 bp hybridized with the 1218-bp *Pst*I fragment of the rat cDNA clone (Fig. 5, lane 5). Partial *Pst*I digests of rat genomic DNA yield a series of hybridizing bands that are apparent multiples of the 1218-bp fragment (Fig. 5, lanes 1–4) and indicate that the rat profilaggrin gene is periodic with at least 13 (and possibly as many as 20 or more) tandemly arranged repeating units of 1218 bp. Because the cDNA and the gene have the same size repeating unit, the gene must lack introns within the coding region. These results show that the entire profilaggrin protein is synthesized as a single translation product, which in the rat is at least 700 kDa and may be as high as 1000 kDa. Similar findings have been observed with the mouse profilaggrin gene, which has at least 20 repeats of the 750-bp unit (Rothnagel and Steinert, 1988). The size of the expected repeat unit for both mouse and rat filaggrin as determined from the size of the protein agrees well with that found using these cDNA techniques.

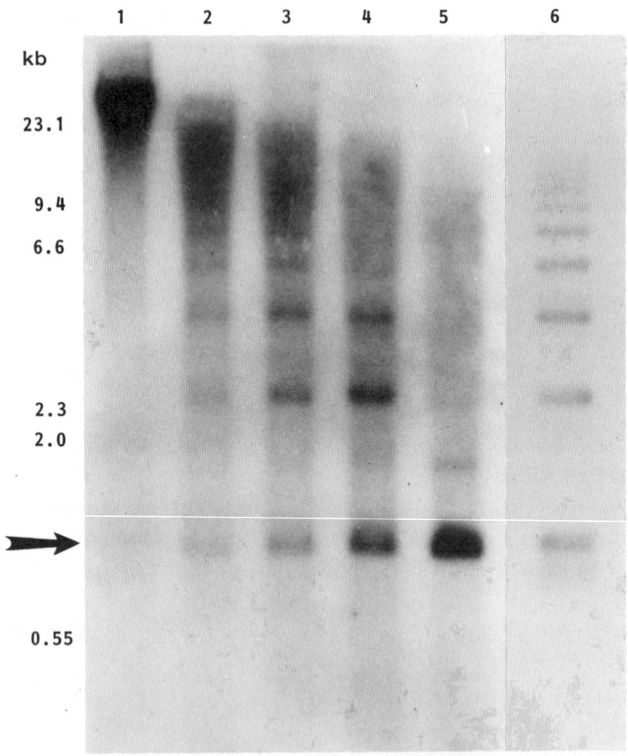

Figure 5. Repetitive structure of the rat profilaggrin gene. Hybridization of the 1218-bp *Pst*I fragment with genomic DNA digested with increasing amounts of *Pst*I. Lanes 1–5 contain digests using 0, 0.5, 1, 1.5, and 2 U *Pst*I/μg rat genomic DNA. Lane 6 shows a similar hybridization of the 1218-bp cDNA concatamerized with 2 U T4 ligase overnight at 4°C. The sample in lane 6 was briefly digested with 1 U *Pst*I to ensure that intermediate-sized fragments were represented. Note that a polymeric series of bands based on a monomeric fragment of 1218 bp (arrow) is detected in both the genomic DNA and the concatamerized cDNA. Mobilities of concatamerized 1218-bp fragment are identical to those of the partially hybridizing genomic DNA bands. The approximate mobility of standards is shown on the left.

```
R          Q Q Q R Q R Q Q H E Q E R E H E Q Q Q P E S S Q R Q Q G S      29
M        G H Q H Q H Q H Q H E H E - - - - - - - Q P E S G H R Q Q Q S    24
m   P17  G H Q H Q H Q H E H

R        S G R T Q R A A R H E Q D S D S T R Q R G S H Q A H S S A R      59
M        S G R G H Q G A H Q E Q G R D S A R P R G S N Q G H S S S R      54
m         F14 G H Q G A H Q E Q G R

R        T Q E E I A R G R S G A T A S E G P G P Q R E A A R D S S E      89
M        P Q - - - - - - - - - - - - - - - - - - - - - - - - - - - -      56
m

R        Q A Q S R R T E T S S R G R S G H S T G R A H E D R H Q Q A      119
M        - - - - - - - - - - - - - - - - - - - - - - - - - - - - - -
m

R        T D R S A R S G S R G G Q A G S H S E S E A S G G Q A G R R      149
M        - - - - - - - - - - - - - - - - - - - - - - - - - - - - - -
m

R        G T A A T R H T S R P E Q S P D T A G R T G S S R G Q Q S A      179
M        - - - - - - - - - - - - - - - - - - - - - - - - - - - - - -
m

R        Q R H G D S T P G S T R T G S R G R G E S P A G Q Q S P D R      209
M        - - - A D S P R V S A R S G S G G R G - - - - - - - Q S P D G/A  77
m                                 F16* G - - - - - - - Q S P D A

R        A R H I E S R R G R T R E A S A S Q S S D S E G H S G A H A      239
M        S G R S S N R R D R P R Q P/L S P S Q S S/G D S Q H H S G V Q V  107
m        S G R                F18* Q P S P S Q S S D S Q V H S G V Q V

R        G I G Q G Q T S T I H R R A G S S S G S - - Q R A S A G G Q      267
M        E G R R G H S S S A N R R A G S S S G S G V Q G A S A G G L      137
m        E G R                 F19 A G S S S G S G V Q G A S A G G L

R        A A D T S S R S G A S Q G Q A S G Q G R A G S R S R E A Q G      297
M        A A D A S R R S G A R Q G Q A S A Q G R A G S Q - - - - - - -    161
m        A A D A S R

R        I A R H G Q S T D S H R H S G A G Q A Q A H T A T S I D S R      327
M        - - - - - - - - - - - - - - - - G Q A Q G R V S S S A D R Q      175
m                                       F11* V G S S A D R

R        R P R G S S E S Q A S D S E G H S D Y S E A H T Q G A H G G      357
M        G R R G V S E S R A S D S E G H S D F S E G Q A V G A H R Q      205
m         F20* G V S E S Q A S

R        I Q T S Q R H E Q R P S R G Q Q G S G H P Q V | Y Y Y G V E E    387
M        S G A G Q R H E Q R S S R G Q H G S G - - - - | Y Y Y E Q E H    231
m                                             P33* | Y Y Y E Q E H

R        T E D E S D A Q Q G H Q Q Q Q Q Q R                              406
M        S E E E S D S Q H Q H G H Q H E Q Q R                            250
m        S
```

Figure 6. Amino acid sequence of one repeating unit of rat (R) and mouse (M) profilaggrin based on the cDNA sequence (Rothnagel *et al.*, 1987; Haydock and Dale, 1989). Identical residues are indicated by ♦; residues absent in the shorter mouse sequence are shown by –. The linker region (boxed) with multiple tyrosine residues is present in profilaggrin but absent from filaggrin. The sites of proteolysis are defined by the ends of the linker. The placement of peptides from mouse profilaggrin (P) or filaggrin (F) sequenced by automated Edmund degradation is shown in line m. Peptide numbering is as given in Resing *et al.* (1985). An asterisk indicates the peptide is phosphorylated in profilaggrin labeled *in vivo* with ^{32}P-inorganic phosphate. Serine residues near acidic residues or other phosphoserines are likely phosphorylation sites.

3.5. Predicted Sequence of the Profilaggrin Repeat Unit

To clarify structural and functional features of profilaggrin and filaggrin, the sequences predicted for the two partial domains of filaggrin from the cDNA have been rearranged to form one complete profilaggrin repeat unit (Fig. 6). The arrangement is based on the known amino-terminus of mouse filaggrin (residue 1) and the strong similarities between the mouse and rat sequences. The rat protein consists of 380 amino acids with a linker of 26 (Haydock and Dale, 1989); the mouse has 224 amino acids with a linker of 26 residues (Rothnagel *et al.*, 1987). An obvious feature of this alignment is the segment of 121 amino acids missing from the amino-terminal portion of mouse filaggrin, and a second shorter deletion of 22 residues in the carboxy-terminal third. Although there are other short missing segments, these two account for most of the size difference between rat and mouse filaggrin. The aligned portions show 50–60% identical residues, with the greatest similarity at regions of posttranslational modifications. Both species have a linker composed of multiple tyrosines, a highly acidic segment, then a Gln/His-rich segment.

Peptide sequences previously determined from mouse filaggrin (Resing *et al.*, 1985) can be readily positioned on the mouse sequence and by homology on the rat sequence. Those peptides that are phosphorylated in profilaggrin are located in the carboxy-terminal half of rat filaggrin, and in the linker region (Fig. 6). Arginine residues, thought to be important to the keratin filament aggregating activity of filaggrin, frequently occur in pairs, but are not concentrated in any one region. They are distributed throughout the protein, suggesting that charge–charge interactions may be important for IF aggregation. In fact, secondary structure predictions according to Chou and Fasman (1978) suggest that the bulk of the protein lacks a rigid structure, and thus filaggrin may be free to fit the charge distribution on the keratins.

Secondary structure calculations of rat profilaggrin indicate extended secondary structure around and in the linker region. A β sheet is centered around residue 380. This is followed by an α helix beginning at approximately residue 387 extending through the many glutamines of the linker and into the amino-terminal portion of filaggrin. In addition, just preceding the linker is a sequence with four histidines which could form a zinc binding site (reviewed in Vallee and Galdes, 1984). The possibility of zinc binding is in agreement with previous observations that epidermal tissue sections bind zinc in the region of the granular and lower cornified cell layers (Reaven and Cox, 1963); in fact, this was the initial reason for a search for a histidine-rich epidermal protein.

4. Filaggrin Synthesis and Relation to Epidermal Cell Differentiation

4.1. Introduction

In vivo labeling experiments with [^{3}H]histidine and ^{32}P-inorganic phosphate demonstrated the multiple posttranslational modifications involved in the synthesis of profilaggrin and its conversion to filaggrin (Lonsdale-Eccles *et al.*, 1982, 1984; Harding and Scott, 1983; Resing *et al.*, 1984, 1985). Figure 2 shows the present state of our understanding of this processing and the epidermal cell type in which specific events occur.

4.2. Profilaggrin and Phosphorylation

The nascent profilaggrin is synthesized and phosphorylated in the granular layer and deposited in the keratohyalin granules. Phosphate is located on serine residues in both filaggrin and linker peptides. The phosphopeptides of mouse profilaggrin (Resing *et al.*, 1985) are not similar in sequence to known cAMP-dependent kinase sites, but rather have serines located near acidic residues. Mamrack (1984) has demonstrated phosphorylation of filaggrin by casein kinase II (CKII), which phosphorylates threonine and serine near acidic residues (Pinna *et al.*, 1979). Using filaggrin as a substrate analogous to profilaggrin, several kinases that phosphorylate filaggrin efficiently have been partially purified from rat epidermis (Resing *et al.*, 1987). These enzymes are not regulated by cAMP or Ca^{2+}/calmodulin. One was identified as CKII by phosphorylation of a specific peptide, Arg3-Glu3-Thr-Glu3 (Kuenzel and Krebs, 1985), use of both ATP and GTP as phosphate donor, subunit structure of 38 and 25 kDa, and autophosphorylation of the 25-kDa subunit (Hathaway and Traugh, 1982).

Other kinases have some properties in common with CKII, but phosphorylate filaggrin more efficiently than casein. They are temporarily referred to as filaggrin kinases. Peptide mapping indicates these six kinases phosphorylate distinct sites in filaggrin or the linker segment, except for one peptide which was phosphorylated by several kinases (Resing *et al.*, 1988).

4.3. Profilaggrin Phosphatase

Processing of profilaggrin to filaggrin occurs as the cells undergo remodeling in the conversion of granular to cornified cells and involves complete dephosphorylation and specific limited proteolysis (Lonsdale-Eccles *et al.*, 1984; Resing *et al.*, 1984). The timing of the dephosphorylation and the first proteolytic step is probably correlated with the dispersal of the keratohyalin in the transition cell.

Epidermal acid phosphatases that hydrolyze *p*-nitrophenyl phosphate (e.g., Mäkinen and Mäkinen, 1981) were localized to the granular layer by histochemical and electron microscopic studies (Eisen *et al.*, 1964), suggesting a potential role in the conversion of granular to cornified cells (Freinkel and Traczyk, 1983). However, acid phosphatase has little effect on profilaggrin (Lonsdale-Eccles *et al.*, 1982) or phosphofilaggrin; thus, more specific phosphatases may be involved in this pathway. One candidate is an Fe^{2+}-activated protein phosphatase that can be extracted by high salt buffers from the epidermis. This enzyme hydrolyzes *p*-nitrophenyl phosphate as well as casein and phosvitin (Hara *et al.*, 1985). Using filaggrin phosphorylated with CKII as a model substrate, a neutral protein phosphatase was isolated that dephosphorylates phosphofilaggrin and profilaggrin, but not casein (Haugen-Scofield *et al.*, 1988) or *p*-nitrophenyl phosphate. Preliminary evidence suggests that this enzyme may be regulated by the interplay of calcium, pH, and ionic strength. It has a narrow substrate specificity consistent with known protein phosphatases that act on proteins phosphorylated by CKII-like kinases (Ullman and Perlman, 1975; Donella Deana *et al.*, 1981), in contrast to the broad substrate specificity shown by most protein phosphatases, which act on proteins phosphorylated by cyclic nucleotide-dependent kinases (as reviewed in Ingebritsen and Cohen, 1983).

4.4. Proteolytic Processing

The amino- and carboxy-terminal sequences of mouse filaggrin have been determined (both termini are glycine), thus identifying the sites of proteolytic processing in the conversion of profilaggrin to filaggrin (see Figs. 2 and 6). Because the sequences are similar in this region, the rat probably has homologous sites. These sites define a linker segment between each filaggrin domain that is removed during processing. Although the proteases involved are not well characterized, pulse–chase studies (Resing *et al.*, 1984; Scott and Harding, 1982) suggest that there are two distinct proteolytic steps that are separated by several hours. These two steps can be correlated with movement of [^{3}H]histidine in the epidermis in studies by Fukuyama and Epstein (1966) (Fig. 2). The first step corresponds to breakdown of keratohyalin and conversion of the granular cell to a transition cell; it releases intermediates containing two or more domains of filaggrin [i.e., two-domain intermediate (2DI)]. The second proteolytic event occurs at conversion to the cornified cell and yields filaggrin. The structural basis for this two-stage processing lies in the fact that profilaggrin has two types of linker segments, but only one is present in 2DI, suggesting that two types of linkers are removed sequentially. One linker sequence (in 2DI) is predicted by the mouse cDNA (Fig. 6), while the other has an insert of Phe-Tyr-Pro-Val between the filaggrin carboxy-terminal glycine and the multiple tyrosine residues. The two linkers must be present in a nearly alternating distribution as shown schematically in Fig. 7 (Resing *et al.*, 1989). This distribution has also been shown by restriction mapping of a mouse filaggrin genomic clone in which the two types of linkers could be distinguished (Rothnagel and Steinert, 1988). These results suggest a physical basis for the two-stage processing in which linker A, containing Gly-Phe, is cleaved in the first stage, and linker B, containing Gly-Tyr, is cleaved in the second stage. This model is supported by the identification of an epidermal protease cleaving profilaggrin to intermediate-sized fragments (Resing *et al.*, 1989). The

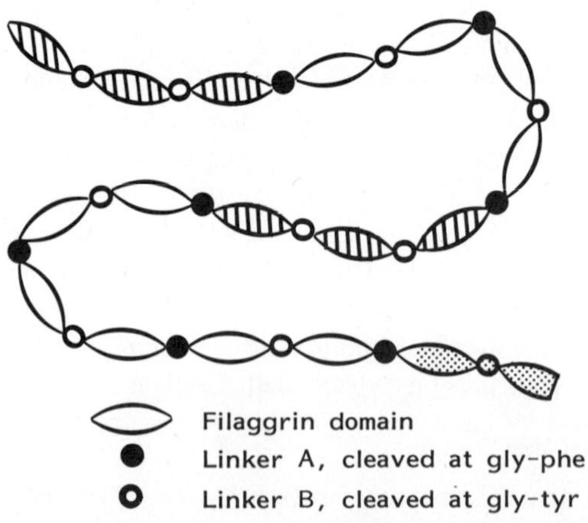

Figure 7. Hypothetical structure of profilaggrin with 18 filaggrin domains and two types of linker segments. Linkers are cleaved at their amino-termini. Linker A (solid ●) is cleaved at a Gly-Phe bond and linker B (○) at a Gly-Tyr bond and these sites alternate irregularly along the profilaggrin polymer. Cleavage of linker A yields dimers and trimers; subsequent cleavage of linker B yields transient monomers (filaggrin with a linker segment still attached to its amino-terminus). The dotted region represents that portion of profilaggrin that is encoded by the cDNA. Modified from Resing *et al.* (1989).

second protease has not been identified but may be similar to proteases that produce filaggrin-sized fragments (Kashima *et al.*, 1988).

The amino-terminal tryptic peptide of mouse filaggrin was identified by its blocked amino-terminus, compared to the unblocked profilaggrin peptide with an amino-terminal glycine (Resing *et al.*, 1985). The glycine is preceded by arginine in the sequence predicted from the mouse cDNA sequence (Rothnagel *et al.*, 1987); thus, a proteolytic cleavage between arginine and glycine completes removal of the linker segment. The newly exposed amino-terminus is then blocked, probably by an acetyl group.

Within the stratum corneum, another proteolytic step degrades filaggrin to free amino acids (Scott *et al.*, 1982). The resulting high concentration of amino acids may be important for normal hydration of the stratum corneum. Scott and Harding (1986) showed that the protease that degrades filaggrin is activated soon after birth as a result of dehydration of the stratum corneum; its activity in the adult is a function of the state of hydration of the tissue. This protease is unusual in that it completely degrades filaggrin, but not the keratin filaments. Filaggrin degradation may be correlated with morphologic observations of loosening of the keratin fibrils in the midcornified layer (see Fig. 1) suggesting that the structural role of filaggrin as an IFAP is a transient one associated with the morphologic occurrence of densely packed keratin filaments in the lower stratum corneum.

5. Synthesis of Profilaggrin and Filaggrin during Fetal Development

5.1. Expression in Rodent

The expression of profilaggrin and filaggrin is strongly correlated with morphologic changes during epidermal development. These proteins are first detectable at the time of morphologic evidence of keratinization indicated by formation of the granular layer (with its keratohyalin granules) and the initial cornified layers (Fukuyama *et al.*, 1981). In the rat this occurs at day 19–20 of gestation. Epidermal extracts from 16- and 18-day fetal rat epidermis give no immunoreaction on tissue sections (Fig. 8) or immunoblots (not shown) using monospecific antifilaggrin antibody; no hybridizable mRNA can be detected with profilaggrin cDNA (Fig. 8). In contrast, 20-day and newborn animals show strong immunoreactivity in the granular and cornified layers, and readily detectable profilaggrin mRNA and protein. The fact that there is little or no lag between the mRNA synthesis and the appearance of immunologically detectable protein suggests that profilaggrin expression is under transcriptional control as are the keratins (Roop *et al.*, 1983, 1985).

5.2. Expression in Human Fetal Skin Development

During human fetal development the epidermis has four distinguishable stages of development: embryonic [< 9 weeks estimated gestational age (EGA)], stratification (9–14 weeks), follicular keratinization (14–23 weeks), and interfollicular keratinization (> 23 weeks). A correlation between expression of keratins and filaggrin with morphologic changes during this long developmental period is very clear (Moll *et al.*, 1983; Dale *et al.*, 1985b). The high-molecular-weight keratins that are associated with differentiation (see Sun *et al.*, 1983) are expressed in cells of the intermediate layer during the period of stratification in interfollicular regions. However, filaggrin is first detectable in the devel-

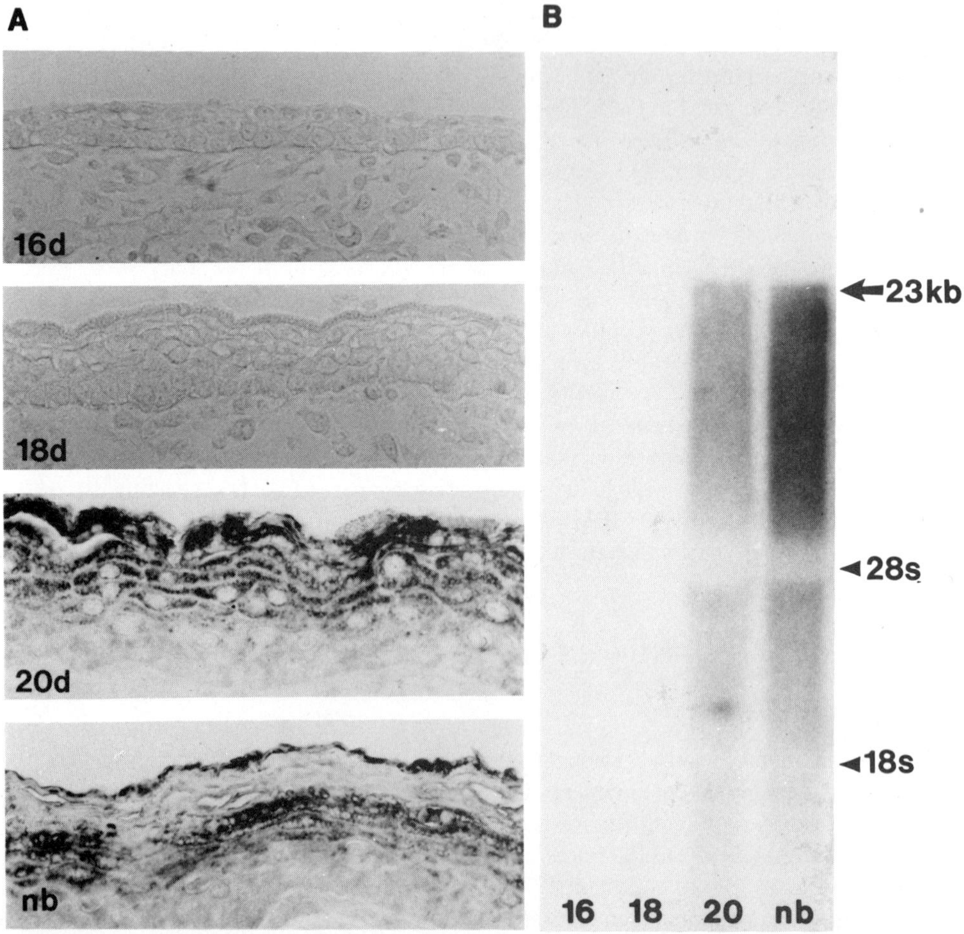

Figure 8. Expression of profilaggrin/filaggrin during development in fetal rat epidermis. Skin samples were obtained from fetuses from time-mated rats at the gestational days indicated for immunohistochemistry and RNA extraction. (A) Expression of profilaggrin/filaggrin protein. Tissue sections of fetal skin were stained with a monospecific polyclonal antibody for rat filaggrin by the avidin–biotin–peroxidase complex method (Hsu *et al.*, 1981). At days 16 and 18, the epidermis is relatively simple in structure and is not stained with filaggrin antibody. In contrast, the 20-day and newborn samples react strongly with antibody in the newly formed stratum corneum and granular layers. (B) Expression of profilaggrin mRNA. Total skin RNA was purified and analyzed for profilaggrin mRNA by hybridization with labeled rat cDNA pR4D6 under stringent conditions. No hybridization is detected at days 16 and 18, but a reaction is seen in the 20-day and newborn samples. Hybridization signals are smeared (see text) with a maximum size of 23 kb. The light spaces in the hybridization pattern are due to the presence of 18 and 28 S ribosomal RNA.

oping hair follicles at about 15 weeks EGA as keratinization occurs in this region, and only later (about 22 weeks EGA) in the interfollicular epidermis, again correlated with morphologic evidence of keratinization (Fig. 9). Immunoreaction is confined to granular and cornified cells. Filaggrin can be detected on immunoblots at corresponding times (Fig. 9).

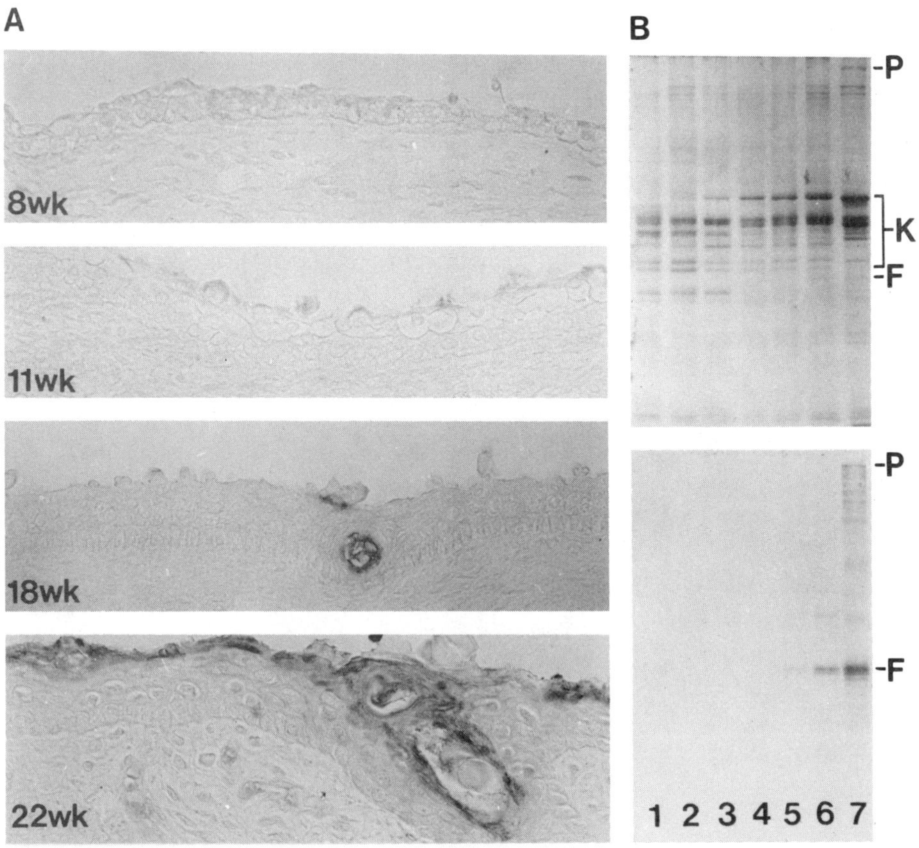

Figure 9. Expression of filaggrin in human fetal epidermis correlated with morphology. (A) Immunohistochemistry using monoclonal antibody AKH1 to profilaggrin/filaggrin. Note the lack of reaction in embryonic stage of development (8 weeks EGA), and fetal stratified period (11 weeks EGA), but positive reaction within the keratinizing cells of the developing hair follicle (18 weeks EGA), and the interfollicular epidermis at 22 weeks EGA. (B) SDS–PAGE and immunoblot using AKH1 of fetal and neonatal epidermal extracts (EGA: lane 1, 10 weeks; lane 2, 11 weeks; lane 3, 13 weeks; lane 4, 15 weeks; lane 5, 16 weeks; lane 6, 20 weeks; lane 7, neonatal). Note that filaggrin is detectable in samples of greater than 15 weeks EGA. Profilaggrin was not detected in fetal samples, probably due to its small quantity in these samples.

6. *Filaggrin Alterations Associated with Disorders of Keratinization*

Filaggrin is an excellent marker of differentiation in normal and pathologic skin. Its expression is altered in some disorders of keratinization (e.g., Voorhees *et al.*, 1968; Baden *et al.*, 1974; Kanitakis *et al.*, 1988). This may occur via effects on profilaggrin synthesis or posttranslational modification. Alteration in the localization and quantity of filaggrin is shown in Fig. 10 for several types of ichthyosis. For example, in ichthyosis vulgaris the amount of filaggrin and profilaggrin is reduced; the relative reduction is correlated with the amount of keratohyalin and with the clinical severity of the case (Sybert *et al.*, 1985). In lamellar ichthyosis (or nonbullous congenital ichthyosiform erythroderma) and in bullous congenital ichthyosiform erythroderma (epidermolytic hy-

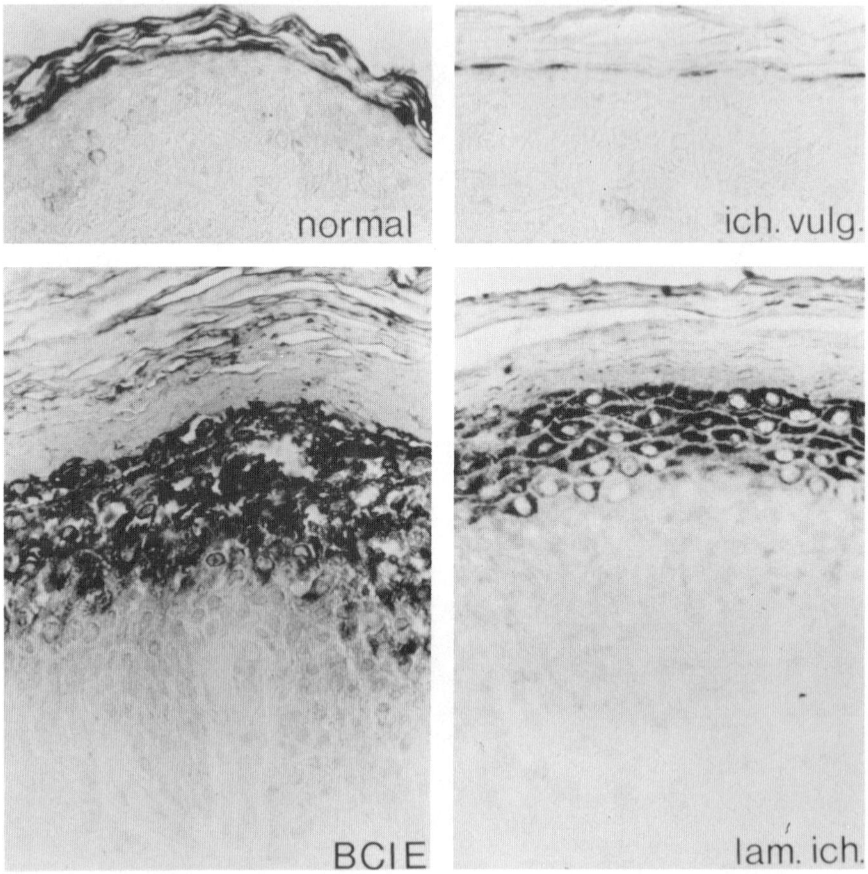

Figure 10. Comparison of profilaggrin/filaggrin localization by immunohistochemistry in normal human epidermis and three disorders of keratinization using AKH1. Reaction is minimal in ichthyosis vulgaris, but the thickened granular layer is reactive in both bullous congenital ichthyosiform erythroderma (epidermolytic hyperkeratosis) and lamellar ichthyosis.

perkeratosis) (BCIE, EH), an increase in the number of granular layers is seen with a strong antifilaggrin reactivity in these cells but lack of reaction in the stratum corneum. In Harlequin ichthyosis (not shown), profilaggrin remains phosphorylated in the thickened stratum corneum and is not converted to filaggrin, suggesting a defect in dephosphorylation (Dale *et al.*, 1990).

Abnormalities in filaggrin function also could be due to keratin alterations; thus, abnormalities in the structural genes for keratins may contribute to severe disorders. For instance, in BCIE keratin filaments form clumps in cells of the spinous layer. Although filaggrin seems to be overproduced in this disorder and is detectable by immunohistochemistry in the upper spinous cells, its localization is not strictly correlated with the keratin clumps, which may be due to an abnormal keratin protein (Ogawa *et al.*, 1979; Holbrook *et al.*, 1983).

It is our hope that studies of disorders of keratinization will help us understand more about biosynthetic steps of the filaggrin pathway, control of epidermal differentiation, and the relationship between expression of specific keratins and filaggrin function.

7. *Concluding Remarks and Future Directions*

The protein filaggrin and its precursor form, profilaggrin, are specifically associated with keratinization and provide a model system for the study of IFAP. The synthesis and accumulation of profilaggrin as an inactive, polymeric, and insoluble protein ensures that premature association and aggregation of keratin filaments does not occur during normal epidermal differentiation. The extensive posttranslational processing of profilaggrin to filaggrin must accompany differentiation of a granular cell to a cornified cell in which the keratin filaments are densely packed in arrays parallel to the skin surface. As these cells move toward the surface and eventual desquamation, filaggrin breaks down in the stratum corneum and its pool of amino acids then contributes to the osmotic properties of these cells. Thus, filaggrin functions as a transient filament cross-linking protein that facilitates the formation of permanent disulfide bonds by inducing and maintaining the proper alignment of keratin filaments. The filaggrin–keratin IF association is a useful model system for investigations of biochemical, cell biological, and functional aspects of IFAP.

ACKNOWLEDGMENT. These studies were supported by USPHS Grants DEO4660, AM21557, GM15731, 5T32-DEO7063 (K.A.R.), HDO6673 (P.V.H.).

References

Baden, H. P., Roths, S. I., Goldsmith, L. A., Baden, S. B., and Lee, L. D., 1974, Keratohyalin protein in disorders of keratinization, *J. Invest. Dermatol.* **62:**411–414.

Ball, R. C., Walker, G. K., and Bernstein, I. A., 1978, Histidine-rich proteins as molecular markers of epidermal differentiation, *J. Biol. Chem.* **253:**5861–5868.

Chou, P. Y., and Fasman, G. D., 1978, Prediction of the secondary structure of proteins from their amino acid sequence, *Adv. Enzymol.* **47:**45–148.

Dale, B. A., 1977, Purification and characterization of a basic protein from the stratum corneum of mammalian epidermis, *Biochim. Biophys. Acta* **491:**193–204.

Dale, B. A., and Ling, S. Y., 1979a, Evidence of a precursor form of stratum corneum basic protein in rat epidermis, *Biochemistry* **18:**3539–3546.

Dale, B. A., and Ling, S. Y., 1979b, Immunologic cross-reaction of stratum corneum basic protein and a keratohyalin granule protein, *J. Invest. Dermatol.* **72:**257–261.

Dale, B. A., Holbrook, K. A., and Steinert, P. M., 1978, Assembly of stratum corneum basic protein and keratin filaments in macrofibrils, *Nature* **276:**729–731.

Dale, B. A., Resing, K. A., and Lonsdale-Eccles, J. D., 1985a, Filaggrin: a keratin filament associated protein, *Ann. N.Y. Acad. Sci.* **455:**330–342.

Dale, B. A., Holbrook, K. A., Kimball, J. R., Hoff, M., and Sun, T.-T., 1985b, Expression of epidermal keratins and filaggrin during human fetal skin development, *J. Cell Biol.* **101:**1257–1269.

Dale, B. A., Holbrook, K. A., Fleckman, P., Kimball, J. R., Brumbaugh, S., and Sybert, V. P., 1990, Heterogeneity in Harlequin ichthyosis, an inborn error of epidermal keratinization: variable morphology and structural protein expression and a defect in lamellar granules, *J. Invest. Dermatol.* **94:** (in press).

Donella Deana, A., Meggio, F., and Pinna, L. A., 1981, Specificity of a rat liver protein phosphatase active on caseins phosphorylated by two cAMP-independent protein kinases, *FEBS Lett.* **125:**77–82.

Eisen, A. Z., Arndt, K. A., and Clark, W. A., 1964, The ultrastructural localization of acid phosphatase in human epidermis, *J. Invest. Dermatol.* **43:**319–326.

Fisher, C., Haydock, P. V., and Dale, B. A., 1987, Localization of profilaggrin mRNA in newborn rat skin by in situ hybridization, *J. Invest. Dermatol.* **88:**661–664.

Fleckman, P., Dale, B. A., and Holbrook, K. A., 1985, Profilaggrin, a high-molecular-weight precursor of filaggrin in human epidermis and cultured keratinocytes, *J. Invest. Dermatol.* **85:**507–512.

Freinkel, R. K., and Traczyk, T. N., 1983, Acid hydrolases of the epidermis: Subcellular localization and relationship to cornification, *J. Invest. Dermatol.* **80:**441–446.

Fukuyama, K., and Epstein, W. L., 1966, Epidermal keratinization: Localization of isotopically labeled amino acids, *J. Invest. Dermatol.* **47:**551–560.

Fukuyama, K., Marshburn, I., and Epstein, W. L., 1981, Histidine-rich protein in developing rat epidermis, *Dev. Biol.* **81:**201–207.

Hara, A., Kato, T., Sawada, H., Fukuyama, K., and Epstein, W. L., 1985, Characterization of Fe^{2+}-activated acid phosphatase in rat epidermis, *Comp. Biochem. Physiol.* **82B:**269–274.

Harding, C. R., and Scott, I. R., 1983, Histidine-rich proteins (filaggrins): Structural and functional heterogeneity during epidermal differentiation, *J. Mol. Biol.* **170:**651–673.

Hathaway, G. M., and Traugh, J. A., 1982, Casein kinases: Multipotential protein kinases, *Curr. Top. Cell. Regul.* **21:**101–127.

Haugen-Scofield, J. A., Resing, K. A., and Dale, B. A., 1988, Characterization of an epidermal phosphatase specific for filaggrin phosphorylated by casein kinase II, *J. Invest. Dermatol.*, **91:**553–559.

Haydock, P. V., and Dale, B. A., 1986, Repetitive structure of the profilaggrin gene as demonstrated using epidermal profilaggrin cDNA, *J. Biol. Chem.* **261:**12520–12525.

Haydock, P. V., and Dale, B. A., 1989, Filaggrin, an intermediate filament associated protein: Sequence of a cDNA from rat and implications for species specific divergence, submitted for publication.

Holbrook, K. A., Dale, B. A., Sybert, V. P., and Sagegiel, R. W., 1983, Epidermolytic hyperkeratosis: Ultrastructure and biochemistry of skin and amniotic fluid cells from two affected fetuses and a newborn infant, *J. Invest. Dermatol.* **80:**222–227.

Hoober, J. K., and Bernstein, I. A., 1966, Protein synthesis related to epidermal differentiation, *Proc. Natl. Acad. Sci. USA* **56:**594–601.

Hsu, S. M., Raine, L., and Fanger, H., 1981, Use of avidin–biotin–peroxidase complex (ABC) in immunoperioxidase techniques: A comparison between ABC and unlabeled antibody (PAP) procedures, *J. Histochem. Cytochem.* **29:**577–580.

Ingebritsen, T. S., and Cohen, P., 1983, Protein phosphatases: Properties and role in cellular regulation, *Science* **221:**331–338.

Kanitakis, J., Ramirez-Bosca, A., Reano, A., Viac, J., Roche, P., and Thivolet, J., 1988, Filaggrin expression in normal and pathological skin, *Virchows Arch. A* **412:**375–382.

Kashima, M., Fukuyama, K., Mamoru, K., and Epstein, W. L., 1988, Limited proteolysis of high molecular weight histidine-rich protein of rat epidermis by epidermal proteinases, *J. Invest. Dermatol.* **90:**829–833.

Kuenzel, E. A., and Krebs, E. G., 1985, A synthetic peptide substrate specific for casein kinase II, *Proc. Natl. Acad. Sci. USA* **82:**737–741.

Lonsdale-Eccles, J. D., Haugen, J. A., and Dale, B. A., 1980, A phosphorylated keratohyalin-derived precursor of epidermal stratum corneum basic protein, *J. Biol. Chem.* **255:**2235–2238.

Lonsdale-Eccles, J. D., Teller, D. C., and Dale, B. A., 1982, Characterization of a phosphorylated form of the intermediate filament-aggregating protein filaggrin, *Biochemistry* **21:**5940–5948.

Lonsdale-Eccles, J. D., Resing, K. A., Meek, R. L., and Dale, B. A., 1984, High-molecular-weight precursor of epidermal filaggrin and hypothesis for its tandem repeating structure, *Biochemistry* **23:**1239–1245.

Lynley, A. M., and Dale, B. A., 1983, Characterization of human epidermal filaggrin, a histidine-rich, keratin filament-aggregating protein, *Biochim. Biophys. Acta* **744:**28–35.

Mäkinen, P.-L., and Mäkinen, K. K., 1981, Purification and properties of rat skin acid phosphatases, *Int. J. Peptide Protein Res.* **18:**352–369.

Mamrack, M. D., 1984, Protein phosphorylation and tumor promotion, in: *Mechanisms of Tumor Promotion,* Volume 2 (T. J. Slaga, ed.), CRC Press, Boca Raton, pp. 143–158.

Meek, R. L., Lonsdale-Eccles, J. D., and Dale, B. A., 1983, Epidermal filaggrin is synthesized on a large messenger ribonucleic acid as a high-molecular-weight precursor, *Biochemistry* **22:**4867–4871.

Moll, R., Moll, I., and Wiest, W., 1983, Changes in the pattern of cytokeratin polypeptides in epidermis and hair follicles during skin development in human fetuses, *Differentiation* **23:**170–178.

Murozuka, T., Fukuyama, K., and Epstein, W. L., 1979, Immunochemical comparison of histidine-rich protein in keratohyalin granules and cornified cells, *Biochim. Biophys. Acta* **579:**334–345.

Ogawa, H., Hattori, M., and Ishibushi, Y., 1979, Abnormal fibrous protein isolated from the stratum corneum of a patient with bullous congenital ichthyosiform erythroderma (BCIE), *Arch. Dermatol. Res.* **266:**109–116.

Ramsden, M., Loehren, D., and Balmain, A., 1983, Identification of rapidly labeled 350k histidine-rich protein in neonatal mouse epidermis, *Differentiation* **23:**243–249.

Reaven, E. P., and Cox, A. J., Jr., 1963, The histochemical localization of histidine in the human epidermis and its relationship to zinc binding, *Cytochemistry* **11:**782–790.

Resing, K. A., Walsh, K. A., and Dale, B. A., 1984, Identification of two intermediates during processing of profilaggrin to filaggrin in neonatal mouse epidermis, *J. Cell Biol.* **99:**1372–1378.

Resing, K. A., Dale, B. A., and Walsh, K. A., 1985, Multiple copies of phosphorylated filaggrin in epidermal profilaggrin demonstrated by analysis of tryptic peptides, *Biochemistry* **24:**4167–4175.

Resing, K. A., Haugen-Scofield, J. A., Mamrack, M. C., and Dale, B. A., 1987, Use of filaggrin as a model substrate for epidermal kinases and phosphatases, in: *Processes in Cutaneous Epidermal Differentiation* (I. A. Bernstein and T. Hirone, eds.), Praeger Scientific, New York, pp. 165–185.

Resing, K. A., Al-Alawi, N., and Dale, B. A., 1988, Purification of six nucleotide-independent kinases which phosphorylate an intermediate filament associated protein, filaggrin, *Faseb J.* **2:**A1759.

Resing, K. A., Walsh, K. A., and Dale, B. A., 1989, Identification of proteolytic cleavage sites in the conversion of profilaggrin to filaggrin in mammalian epidermis, *J. Biol. Chem.* **264:**1837–1846.

Roop, D. R., Hawley-Nelson, P., Cheng, C. K., and Yuspa, S. H., 1983, Keratin gene expression in mouse epidermis and cultured epidermal cells, *Proc. Natl. Acad. Sci. USA* **80:**716–720.

Roop, D. R., Cheng, C. K., Toftgard, R., Stanley, J. R., Steinert, P. M., and Yuspa, S. H., 1985, The use of cDNA clones and monospecific antibodies as probes to monitor keratin gene expression, *Ann. N.Y. Acad. Sci.* **455:**426–435.

Rothnagel, J. A., and Steinert, P. M., 1988, Structure of the mouse filaggrin gene: characterization of its repeating structure and a comparison of the gene from different strains of mice, *J. Cell Biol.* **107:** 250a.

Rothnagel, J. A., Mehrel, T., Idler, W. W., Roop, D. R., and Steinert, P. M., 1987, The gene for mouse epidermal filaggrin precursor: Its partial characterization, expression, and sequence of a repeating filaggrin unit, *J. Biol. Chem.* **262:**15643–15648.

Scott, I. R., and Harding, C. R., 1981, Studies on the synthesis and degradation of a high molecular weight, histidine-rich phosphoprotein from mammalian epidermis, *Biochim. Biophys. Acta* **669:**65–78.

Scott, I. R., and Harding, C. R., 1982, Histidine-rich proteins: Synthesis and post-translational modifications during epidermal differentiation, *Biochem. Soc. Trans.* **9:**1981–1982.

Scott, I. R., and Harding, C. R., 1986, Filaggrin breakdown to water binding compounds during development of the rat stratum corneum is controlled by the water activity of the environment, *Dev. Biol.*, **115:**84–92.

Scott, I. R., Harding, C. R., and Barrett, J. G., 1982, Histidine rich protein of the keratohyaline granules: Source of the free amino acids, urocanic acid and pyrrolidone carboxylic acid in the stratum corneum, *Biochim. Biophys. Acta* **719:**110–117.

Sibrack, L. A., Gray, R. H., and Bernstein, I. A., 1974, Localization of the histidine-rich protein in keratohyalin: A morphologic and macromolecular marker in epidermal differentiation, *J. Invest. Dermatol.* **62:**394–405.

Steinert, P. M., and Roop, D. R., 1988, Molecular and cellular biology of intermediate filaments, *Annu. Rev. Biochem.* **57:**593–625.

Steinert, P. M., Idler, W. W., and Zimmerman, S. B., 1976, Self-assembly of bovine epidermal keratin filaments in vitro, *J. Mol. Biol.* **108:**547–567.

Steinert, P. M., Cantieri, J. S., Teller, D. C., Lonsdale-Eccles, J. D., and Dale, B. A., 1981, Characterization of a class of cationic proteins that specifically interact with intermediate filaments, *Proc. Natl. Acad. Sci. USA* **78:**4097–4101.

Sun, T.-T., Eichner, R., Nelson, W. G., Tseng, S. C. G., Weiss, R. A., Jarvinen, M., and Woodcock-Mitchell, J., 1983, Keratin classes: Molecular markers for different types of epithelial differentiation, *J. Invest. Dermatol.* **81:**109s–115s.

Sybert, V. P., Dale, B. A., and Holbrook, K. A., 1985, Ichthyosis vulgaris: Identification of a defect in synthesis of filaggrin correlated with an absence of keratohyaline granules, *J. Invest. Dermatol.* **84:**191–194.

Ugel, A. R., 1971, Studies on isolated aggregating oligoribonucleoproteins of the epidermis with histochemical and morphological characteristics of keratohyalin, *J. Cell Biol.* **49:**405–422.
Ullman, B., and Perlman, R. L., 1975, Purification and characterization of a phosphoprotein phosphatase from bovine adrenal cortex, *Biochim. Biophys. Acta* **403:**393–411.
Vallee, B. L., and Galdes, A., 1984, The metallobiochemistry of zinc enzymes, *Adv. Enzymol.* **56:**283–430.
Voorhees, J. J., Chakrabarti, S. G., and Bernstein, I. A., 1968, The metabolism of "histidine-rich" protein in normal and psoriatic keratinization, *J. Invest. Dermatol.* **51:**344–354.

V

Pathological Conditions Relating to Intermediate Filaments

17

Expression, Organization, and Involvement of Intermediate Filaments in Disease Processes

Omar Skalli and Giulio Gabbiani

1. Introduction

Cytoskeletal and cytocontractile elements are involved in many cell physiological processes such as maintenance of cell shape, locomotion, mitosis, and secretion (for review see Alberts *et al.*, 1989). Moreover, the study of these structures may contribute to the understanding of cellular alteration during pathological situations (for review see Rungger-Brändle and Gabbiani, 1983). Different types of cytoskeletal changes may cause pathological situations. In some instances, a disease is due to the presence of an abnormal cytoskeletal protein. A well-documented example of this is the immotile cilia syndrome, characterized by an impaired mucociliary transport leading to chronic infections of airways (Camner *et al.*, 1975; Eliasson *et al.*, 1977); here, the basic defect is due to abnormal or missing dynein arms (Afzelius, 1985; Neustein *et al.*, 1980; Pedersen and Mygind, 1976). In other situations the primary defect affects a noncytoskeletal cellular component but leads to typical changes of the cytoarchitecture, as is the case for sickle-cell anemia where the primary defect is a mutant β-globin molecule leading to an altered organization of the cytoskeleton during the sickling process (Lux, 1979). In most instances, however, changes in the organization of cytoskeletal elements associated with pathological conditions reflect the degree of cellular adaptation to pathological stimuli (Rungger-Brändle and Gabbiani, 1983). Thus, cytoskeletal organization can be profoundly modified by pathological processes affecting cell activities such as migration or proliferation. Moreover, certain important pathological processes, such as tumor inva-

Omar Skalli and Giulio Gabbiani • Department of Pathology, University of Geneva, 1211 Geneva 4, Switzerland.

sion, may at least in part depend on altered cytoskeletal organization (for review see Boschek, 1982).

In this review, we will summarize the most known pathological conditions in which a direct or indirect alteration of intermediate filament (IF) function takes place. As we shall see, the mechanisms leading to IF as well as to cytoskeletal dysfunction are only rarely understood.

2. IF as Differentiation Markers of Neoplastic and Nonneoplastic Cells

Since the recognition of five classes of IF proteins and the availability of antibodies specific for each IF type (Bennett *et al.*, 1978; Franke *et al.*, 1978; Lazarides and Balzer, 1978), a considerable amount of work has been done to determine which subunit is expressed in various cells, during physiological and pathological conditions. Some "rules" governing the expression of IF subunits have emerged from these immunohistochemical and biochemical studies. These rules have been summarized in many reviews (for recent reviews see: Cooper *et al.*, 1985; Gown and Gabbiani, 1984; Miettinen *et al.*, 1984a; Moll *et al.*, 1982; Osborn and Weber, 1983; Ramaekers *et al.*, 1983a; Rungger-Brändle and Gabbiani, 1983; Steinert *et al.*, 1984; Virtanen *et al.*, 1984).

It was rapidly clear that each of the five classes of IF protein is specific for a distinct group of cells or tissues and that most cells contain only a single IF class (Table I).

Table I. Tissue Specificity of IF Expression in Normal Mammalian Tissues

Cell type	Vimentin	Desmin	Keratin
Nonmuscular mesenchymal cells			
Fibroblasts	+	−	−
Endothelial cells	+	−	−
Chondroblasts	+	−	−
Melanocytes	+	−	−
Lymphocytes	+	−	−
Monocytes/macrophages	+	−	−
Sertoli cells	+	−	−
Muscular mesenchymal cells			
Skeletal muscle	−	+	−
Cardiac muscle	−	+	−
Parenchymal smooth muscle	−	+	−
Vascular smooth muscle	+	+	−
Epithelial and mesothelial cells			
Squamous epithelium	−	−	+
Urothelium	−	−	+
Simple epithelia of various tracts	−	−	+
Glandular epithelia	−	−	+
Pneumocytes	−	−	+
Hepatocytes	−	−	+
Renal tubular epithelium	−	−	+
Thyroid follicular epithelium	−	−	+
Adrenal cortical cells	−	−	+
Mesothelium	+	−	−

Mesenchymal cells contain vimentin and most muscle cells desmin (for the expression of desmin and vimentin in smooth muscle cells, see Section 3.3). Epithelial cell IF are made up of keratins, a family of 19 polypeptides (Franke *et al.*, 1979a; Moll *et al.*, 1982). The keratin composition of epithelia varies according to the type, differentiation, and disease state (see Cooper *et al.*, 1985; Moll *et al.*, 1982; and Section IV of this volume). The other two classes of IF are found in neural and glial cells and are reviewed in Section 7 of this volume.

As a general rule, the IF protein composition of a given cell type is the same in related mammalian species (Gown and Vogel, 1984; Ramaekers *et al.*, 1983a). However, some differences are found between related species: e.g., human and rat mesothelial cells (Czernobilsky *et al.*, 1985) and human and rat mesangial cells (Stamenkovic *et al.*, 1986; see also Section 3.3 for species differences in smooth muscle cells).

The tissue-specific expression of IF subunits, and the finding that, in general, tumor cells (either from the primary tumor of from its metastases) retain the IF proteins contained in the tissue from which they originate have led to important practical applications in the differential diagnosis of human tumors.

Sarcomas can be easily distinguished from carcinomas by their positivity to vimentin and/or desmin antibodies and their negativity to prekeratin antibodies (Table II). Use of these criteria has allowed the diagnosis of mesodermal mixed tumors in a case in which this diagnosis had been missed by light and electron microscopic studies (Ramaekers *et al.*, 1983b) and has allowed distinguishing between the sarcomatous and carcinomatous cells present within carcinosarcoma (Huszar *et al.*, 1984). Analysis of soft tissue tumors with IF antibodies has raised questions regarding "traditional" histogenetic interpretation of some sarcomas. Thus, synovial sarcoma cells can be distinguished from the vimentin-positive, prekeratin-negative synovial lining cells (Gown and Vogel, 1984; Miettinen and Virtanen, 1984; Osung *et al.*, 1982a) by their staining with prekeratin antibodies (Miettinen *et al.*, 1983a). This suggests that synovial sarcoma has no histogenetic relationships to synovial membrane cells.

Antibodies against vimentin and desmin are also useful for the differential diagnosis between sarcomatous tumors, though their use has some limitations. The first of these limitations is that positive staining to vimentin and negative staining to prekeratin does not allow the distinction of sarcomas from lymphomas or leukemia (Altmannsberger *et al.*, 1982a; Gabbiani *et al.*, 1981a). This distinction can, however, be accomplished by means of other techniques. Another limitation is that positivity to vimentin and negativity to desmin does not rule out a muscular origin since many vascular smooth muscle cells do not express desmin. Although replicating nontumoral, e.g., granulation, tissue (Skalli *et al.*, 1986a) (Fig. 1) and tumoral fibroblastic tissues (Schürch *et al.*, 1984) contain only vimentin-positive cells, the possibility still exists that such cells are derived from those vascular smooth muscle cells that do not contain desmin. In addition, desmin staining cannot discriminate between striated and smooth muscle tumors (Table II). The expression of desmin and vimentin has been extensively studied in rhabdomyosarcomas where it is related to cytological differentiation (Altmannsberger *et al.*, 1982a, 1984, 1985; Denk *et al.*, 1983; Gabbiani *et al.*, 1981a; Miettinen *et al.*, 1982a,b; Molenaar *et al.*, 1985) (Fig. 2).

Extensive studies have proven that, like normal epithelial cells, carcinoma cells contain in general prekeratin. Thus, carcinomas can be distinguished from other tumors by their positivity to broad-specificity antibodies against cytokeratins (Altmannsberger *et al.*, 1981, 1982b; Bannasch *et al.*, 1980; Gabbiani *et al.*, 1981a; Nagle *et al.*, 1983) (Fig. 3).

Table II. IF Expression in Soft Tissue Neoplasms

Tumor	Vimentin	Desmin	Cytokeratin	References
Angiosarcoma	+	−	−	Altmannsberger *et al.* (1982a), Miettinen *et al.* (1982a)
Fibrosarcoma	+	−	−	Miettinen *et al.* (1982a)
Malignant fibrous histiocytoma	+	−	−	Denk *et al.* (1983), Gabbiani *et al.* (1981a), Miettinen *et al.* (1982a, 1984b)
Liposarcoma	+	−	−	Altmannsberger *et al.* (1982a), Gabbiani *et al.* (1981a), Miettinen *et al.* (1982a)
Dermatofibrosarcoma protuberans	+	−	−	Altmannsberger *et al.* (1982a)
Atrial myxoma	+	−	−	Gabbiani *et al.* (1981a)
Chondrosarcoma	+	−	−	Miettinen *et al.* (1983b)
Melanoma	+	−	−	Huszar *et al.* (1983), Ramaekers *et al.* (1983d)
Ewing's sarcoma	+	−	−	Miettinen *et al.* (1982d)
Epithelioid sarcoma	+	−	−	Miettinen *et al.* (1982a)
Liver embryonal sarcoma	+	−	−	Keating and Taylor (1985)
Seminomas	+	−	−	Ramaekers *et al.* (1985a)
Alveolar soft part sarcoma	+	+	−	Denk *et al.* (1983)
Leiomyoma	+	+	−	Denk *et al.* (1983), Gabbiani *et al.* (1981a)
Leiomyosarcoma	+	+	−	Denk *et al.* (1983), Miettinen *et al.* (1982a, 1984c)
Rhabdomyosarcoma	+	+	−	Altmannsberger *et al.* (1982a, 1984, 1985), Denk *et al.* (1983), Gabbiani *et al.* (1981a), Miettinen *et al.* (1982a,b), Molenaar *et al.* (1985)

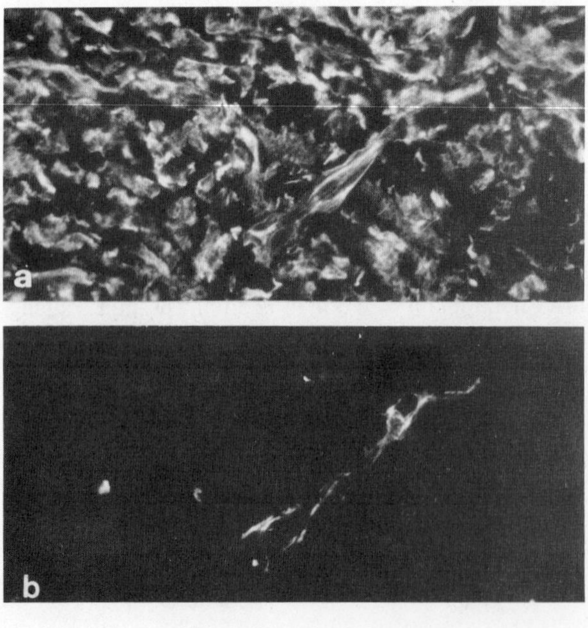

Figure 1. Double immunofluorescent staining for vimentin (a) and desmin (b) of rat granulation tissue 2 weeks after removal of a 4-cm^2 skin fragment. All myofibroblasts are vimentin-positive, desmin-negative. Only some cells in the blood vessel wall are also positive for desmin. × 400.

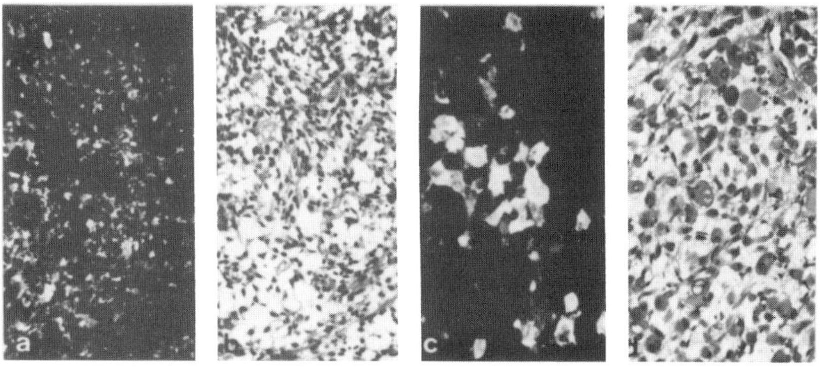

Figure 2. Rhabdomyosarcoma of the orbit. (a) Area containing many vimentin-positive cells. (b) Similar area stained with hematoxylin and eosin shows the presence of undifferentiated small cells. (c) Large cellular elements are positive after incubation with antidesmin antibodies. (d) Similar area stained with hematoxylin and eosin shows several large eosinophilic, relatively well-differentiated cells. × 350. From Gabbiani *et al.*, *Am. J. Pathol.* **104:**206–216, 1981, with the kind permission of the publisher.

Osborn and Weber (1983) reviewed a series of 131 carcinomas of various sites and degrees of differentiation which were all positive to cytokeratin. Immunohistochemical analysis of IF has proven to be a valuable aid in problematic cases such as differential diagnosis between thymomas and lymphoepitheliomas (vimentin-negative, prekeratin-positive) from lymphomas (vimentin-positive, prekeratin-negative) (Battifora *et al.*, 1980; Miettinen *et al.*, 1982c).

Normal epithelial cells can be classified by the set of keratin polypeptides they express; this classification can be achieved by two-dimensional gel electrophoresis or by the use of monoclonal antibodies recognizing subsets of keratins (Cooper *et al.*, 1985; Moll *et al.*, 1982). Carcinomas can also be subtyped using similar methods (Cooper *et al.*,

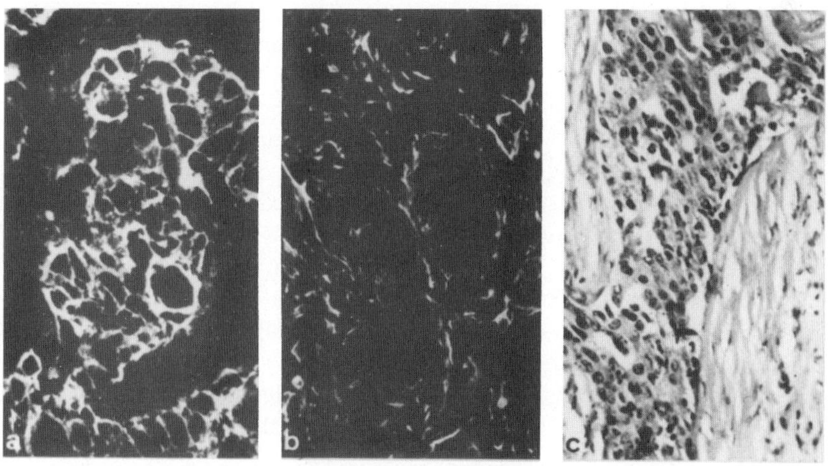

Figure 3. Invasive duct carcinoma of the breast. (a) Positive staining of tumor cells after incubation with prekeratin antibodies. (b) Adjacent section: stromal cells surrounding tumor cells are stained with antivimentin antibodies. (c) Section stained with hematoxylin and eosin. × 500. From Gabbiani *et al.*, *Am. J. Pathol.* **104:**206–216, 1981, with the kind permission of the publisher.

1985; Moll *et al.*, 1982). The keratin pattern of simple epithelia undergoes only minor, if any, changes during malignancy (Denk *et al.*, 1982; Moll *et al.*, 1982). Stratified squamous epithelium neoplasms generally express keratin of lower molecular weight than their normal counterparts (Moll *et al.*, 1982, 1983, 1984b; Nelson *et al.*, 1984). Similar modifications of the prekeratin pattern are observed during hyperproliferative diseases such as psoriasis (McGuire *et al.*, 1984; Weiss *et al.*, 1984). Changes in the pattern of expression of cytokeratins have also been found *in vitro* during viral transformation (Darmon *et al.*, 1984; Hronis *et al.*, 1984; Morris *et al.*, 1985).

The use of IF typing for tumor diagnosis has become a routine procedure due to the availability of commercial IF antibodies which can be used on frozen and paraffin-embedded material. However, one should be aware of the possibility of some technical artifacts (for discussion of this problem, see Gown and Gabbiani, 1984; Miettinen *et al.*, 1984a; Osborn and Weber, 1983) which could complicate the interpretation of the results.

The use of IF in tumor diagnosis may suffer a number of limitations as suggested by recent reports showing that the expression of IF proteins in various tissues is far more complex than has been previously realized. Exceptions to the rule of tissue-specific expression of IF proteins have already been described in earlier works. For instance, desmin is absent from a number of differentiated smooth muscle cells (see Section 3.3), and has recently been identified in a number of stromal cells (Franke and Moll, 1987; Glasser and Julian, 1986; Skalli *et al.*, 1986a; Yokoi *et al.*, 1984) and endothelial cells (Fujimoto and Singer, 1986; Toccanier-Pelte *et al.*, 1987). The finding that nonmesenchymal cells may in some cases coexpress vimentin together with their characteristic IF protein (Blobel *et al.*, 1985; Caselitz *et al.*, 1981; Cremer *et al.*, 1981; Czernobilsky *et al.*, 1985; Franke *et al.*, 1979b, 1983; Holthöfer *et al.*, 1983, 1984; Krepler *et al.*, 1982; Lane *et al.*, 1983; LaRocca and Rheinwald, 1984; Lehtonen *et al.*, 1983b; Miettinen *et al.*, 1984d; Osborn *et al.*, 1982; Ramaekers *et al.*, 1983c; Regauer *et al.*, 1985; Virtanen *et al.*, 1981; Vogel *et al.*, 1984; Waldherr and Schwechheimer, 1985) was considered to have minor consequence on the diagnostic value of IF typing which relied on the nonvimentin IF subunit expressed by these cells. However, numerous recent findings have demonstrated that IF subunits can display complex coexpression patterns in normal cells as well as during development, pathological and experimental conditions. Thus, cytokeratin can be coexpressed in nonepithelial cells together with other IF subunits such as neurofilament proteins (Doglioni *et al.*, 1987; Kasper *et al.*, 1986; Lehto *et al.*, 1985; Miettinen *et al.*, 1985b), glial fibrillary acidic protein (Franko *et al.*, 1987), and desmin (Franke and Moll, 1987; Jahn *et al.*, 1987; Van Muijen *et al.*, 1987). A striking example of IF protein expression complexity is represented by smooth muscle cells, which can be subdivided into six types according to their immunoreactivity for desmin, vimentin, and keratin antibodies (Jahn *et al.*, 1987). Interestingly, a positivity for keratin antibodies was also found in some leiomyosarcomas and rhabdomyosarcomas (Brown *et al.*, 1987; Coindre *et al.*, 1988; for review see Jahn *et al.*, 1987). Finally, it is worth noting that some normal and tumor cells appear to be devoid of cytoplasmic IF protein (Czernobilsky *et al.*, 1985; Dellagi *et al.*, 1983; Jackson *et al.*, 1980, 1981; Miettinen *et al.*, 1985a; Traub *et al.*, 1983; Venetianer *et al.*, 1983), although some of these results are controversial (Lehtonen *et al.*, 1983a,b).

Attempts to unravel the histogenesis of cells of uncertain origin have been made by determining their IF composition (Franke *et al.*, 1980; Gould *et al.*, 1985; Holthöfer *et al.*, 1984; Mahrle *et al.*, 1983; Moll *et al.*, 1984a; Saurat *et al.*, 1984; Stamenkovic *et al.*,

1986). However, the use of IF proteins as histogenetic markers should be considered critically since in a few cases differentiated cells do not express the IF proteins characteristic of the germ layer from which they stem. Thus, the ectoderm-derived lens cells express vimentin (Barritault *et al.*, 1980; Lieska *et al.*, 1980; Ramaekers *et al.*, 1980) and the mesoderm-derived kidney tubule cells contain prekeratin (Holthöfer *et al.*, 1983).

The recent data on the complex expression of IF proteins suggest that normal and neoplastic tissues should be characterized more systematically than what has been done up to now by antibodies against all classes of IF proteins. Additional cytoskeletal markers such as actin isoforms may prove to be useful for diagnostic purposes (Babaï *et al.*, 1988; Miettinen, 1988; Schmidt *et al.*, 1988; Schürch *et al.*, 1987; Skalli *et al.*, 1987, 1988; Tsukuda *et al.*, 1987; Vanderkerckhove *et al.*, 1987). Finally, the complex pattern of IF expression (and/or of other cytoskeletal proteins) may be helpful to evaluate the degree of differentiation of some neoplasms, thereby furnishing new criteria for their classification and possibly for the definition of their clinical behavior.

3. Pathological Situations Characterized by Altered IF Expression and/or Organization

3.1. Myopathies and Cardiomyopathies

Numerous ultrastructural and immunofluorescence data are available on IF organization in embryonic and adult skeletal and cardiac muscle (for review see Stromer, this volume; Price and Sanger, 1983; Thornell *et al.*, 1983). Abnormal accumulations of IF have been observed in various muscle diseases. Inclusion bodies stained with desmin but not with vimentin antibodies were noted in myocardial cells of patients with a familial cardiomyopathy (Stoeckel *et al.*, 1981) and in muscle fibers of patients with different hereditary congenital myopathy (Fidzianska *et al.*, 1983; Osborn and Goebel, 1983). In various myopathies, inclusion bodies made up of desmin were associated with increased amounts of desmin (Rappaport *et al.*, 1988) and posttranslational modifications of this protein, such as phosphorylation (Rappaport *et al.*, 1988) and ubiquitination (Lowe *et al.*, 1988). Whether these posttranslational modifications play a role in the formation of desmin inclusions remains to be elucidated. Numerous IF were associated with sarcoplasmic bodies in most muscle fibers in a type of hereditary distal myopathy starting in the hand flexor muscles (Edström *et al.*, 1980). Numerous IF were also located at the level of the Z disk, between myofibrils and beneath the plasmalemma. The sarcoplasmic bodies exhibited an unspecific orange fluorescence and were not stained with desmin antibodies; they were interpreted as residual bodies unrelated to IF.

In nemaline myopathy, a congenital, nonprogressive disease, muscle cells contain typical electron-dense rods that consist of massive accumulations of α-actinin (Jockusch *et al.*, 1980), desmin being only peripherally associated, as in normal Z lines (Jockusch *et al.*, 1980; Thornell *et al.*, 1980). Thus, nemaline bodies may be seen as a kind of hypertrophic Z lines.

Antidesmin may provide helpful information in the evaluation of muscle regeneration versus atrophy, as suggested for Duchenne muscular dystrophy (Thornell *et al.*, 1980) and eosinophilic fasciitis (Thornell and Bjelle, 1981). In these diseases, regenerating muscle fibers can be distinguished by their strong and diffuse staining with antidesmin antibodies, due to their large content of IF.

IF accumulation is associated with cardiac hypertrophy, a common functional modification of this organ. In hypertrophic human heart, myocardial cells contain focal accumulations of IF together with degenerated myofibrils and proliferation of sarcotubular elements (Ferrans and Roberts, 1973a; Porte *et al.*, 1980; Thiedemann and Ferrans, 1976). Similar changes occur in the rat myocardium after treatment with an anabolic steroid (Behrendt, 1977). It was suggested that increased synthesis or impaired degradation may be responsible for the IF accumulation occurring during heart hypertrophy (Edström *et al.*, 1980; Stoeckel *et al.*, 1981).

3.2. Alcoholic Liver Disease

Mallory bodies are one of the well-documented examples of an association between an IF protein accumulation and a pathologic condition. Mallory bodies, alcoholic hyalin, and Mallory's hyalin are synonyms used to describe coarse, eosinophilic cytoplasmic inclusions found in hepatocytes during alcoholic hepatitis and cirrhosis (for review see Denk et al., 1981a; French, 1983; Hall, 1985a; Phillips, 1982) and, although less frequently, in other liver diseases (for review see Denk *et al.*, 1979a; French, 1981a) such as hepatocellular carcinoma (Keeley *et al.*, 1972) and primary biliary cirrhosis (Monroe *et al.*, 1973). Cytoplasmic inclusions morphologically and ultrastructurally similar to Mallory bodies were also reported in alveolar epithelial cells in several lung diseases (reviewed by Hall, 1985b) such as asbestosis (Kuhn and Kuo, 1973) and radiation pneumonitis (Warnock *et al.*, 1980). Mallory bodies can be experimentally induced in mice fed for 2–4 months with a diet supplemented with the fungistatic antibiotic griseofulvin (Denk *et al.*, 1976). These disappear 4 weeks after interruption of the griseofulvin-supplemented diet; however, they reappear almost immediately after refeeding the mice with griseofulvin or with the microtubule-disrupting drug colchicine; interestingly, colchicine alone is not able to induce Mallory bodies even when ingested for a prolonged period of time (Denk and Eckerstorfer, 1977; Denk *et al.*, 1976). Mallory bodies are also present in cells of hepatocarcinoma induced by carcinogens such as diethylnitrosamine (Borenfreud and Bendich, 1978) or dieldrin (Meierhenry *et al.*, 1981, 1983). When the hepatocarcinoma cells were grown in culture, their Mallory bodies persisted after several passages (Borenfreund and Bendich, 1978), representing a heritable phenotypic feature (Borenfreund *et al.*, 1979, 1980a).

Mallory body ultrastructure and resistance to solubilization by detergents led to the suggestion that these structures are made up of abnormally assembled IF (French and Davies, 1975; French *et al.*, 1977). The filamentous nature of Mallory bodies was observed *in situ* and after isolation by biochemical procedures (Okamura *et al.*, 1975; Wiggers *et al.*, 1973; Yokoo *et al.*, 1972). In alcoholic cirrhosis, Mallory bodies appear as three ultrastructurally distinguishable types of inclusions (Yokoo *et al.*, 1972). Type 1 inclusions are bundles of fibrils aligned in parallel arrays or in a concentric whorled formation, type 2 inclusions consist of clusters of randomly oriented fibrils, and type 3 inclusions appear as a granular or amorphous substance containing only scattered remains of fibrils. Type 1 Mallory bodies were not observed in griseofulvin-fed mice (Denk *et al.*, 1979a); in this animal model, Mallory bodies differ from the human ones in that they are covered by a fine, fimbriated fuzzy coat (Franke *et al.*, 1979c).

Hepatocytes contain IF of the prekeratin type that are related to but not identical with epidermal prekeratins (Denk *et al.*, 1981b; Franke *et al.*, 1981a). Immunofluorescence studies have clearly established that Mallory bodies found in various pathological and experimental conditions are related to prekeratins since they share antigenic determinants with prekeratins of bovine hoof or muzzle (Borenfreund *et al.*, 1981; Denk *et al.*, 1979b; Franke *et al.*, 1979c; Tinberg and Mednick, 1981) and of liver cells (Denk *et al.*, 1981b; Franke *et al.*, 1981a; Hazan *et al.*, 1986) (Fig. 4). The existence of a close relationship between IF and Mallory bodies is also supported by the binding of antibodies raised against human Mallory bodies on IF of liver and epidermal cells (Kimoff and Huang, 1981). However, Mallory bodies contain antigenic determinants (Fleming *et al.*, 1981;

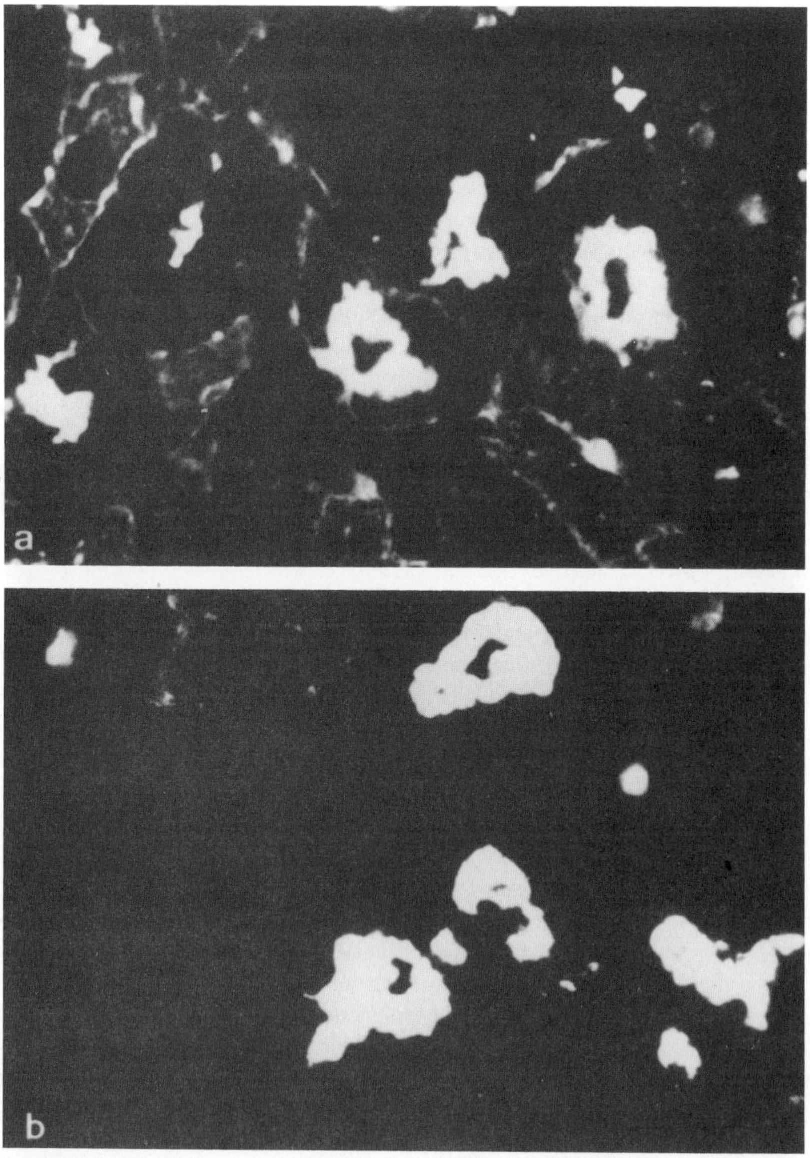

Figure 4. Indirect immunofluorscence microscopy on cryostat sections of human liver biopsy in a case of alcoholic hepatitis. Mallory bodies are decorated by antibodies to mouse liver cytokeratin component D (a, × 400) and by antibodies to bovine hoof prekeratin (b, × 640). Courtesy of Dr. H. Denk, Department of Pathology, Graz, Austria.

Hazan *et al.*, 1986; Morton *et al.*, 1980, 1981) and sugar residues (Borenfreund *et al.*, 1979, 1980b; Luisanda *et al.*, 1977; Tinberg *et al.*, 1978) not present in prekeratins. Mallory bodies of cultured hepatoma cells have the unique feature of binding both prekeratin and vimentin antibodies (Borenfreund *et al.*, 1980c). This result is not unexpected since cultured hepatocytes and hepatoma cells (with few exceptions) express vimentin and cytokeratin (Franke *et al.*, 1981b).

Relatively pure preparations of Mallory bodies can be obtained by different biochemical methods (French *et al.*, 1972; Okamura *et al.*, 1975; Tinberg *et al.*, 1978). When analyzed by SDS polyacrylamide gel electrophoresis, these preparations show a polypeptide composition that somehow varies according to the species, experimental or pathological conditions, and method of purification (Franke *et al.*, 1979c; Tinberg *et al.*, 1978; Tinberg, 1981a). For example, in griseofulvin (Franke *et al.*, 1979c)- and diethylnitrosamine (Borenfreund *et al.*, 1980a)-induced Mallory bodies, the molecular weight of these polypeptides ranged from 48,000 to 66,000. This polypeptide pattern differs from that of hepatocyte IF because of the presence of high-molecular-weight components similar to those present in epidermal cytokeratins (Denk *et al.*, 1981b; Franke *et al.*, 1979c). The major polypeptides of human Mallory bodies are probably related, since they share a common antigenic determinant recognized by an antibody raised against their 56kD component (Tinberg, 1981b).

The development of Mallory bodies from hepatocyte IF has been examined by electron microscopy and immunofluorescence. IF antigenic determinants visualized in normal hepatocytes by immunofluorescence or immunoelectron microscopy are decreased or absent in hepatocytes containing Mallory bodies; this suggests that Mallory bodies are formed at the expense of IF (Denk *et al.*, 1981a; Franke *et al.*, 1981a; Kimoff and Huang, 1981) and that during this process there are changes in IF organization and conformation (Hazan *et al.*, 1986; Katsuma *et al.*, 1987). This view was also supported by scanning electron microscopy studies showing that in the hepatocytes of griseofulvin-fed mice, the number of IF was decreased, and that Mallory bodies were formed as the result of condensation or collapse of IF (Okanoue *et al.*, 1985). However, Irie *et al.* (1982) found by morphometric measurement that the amount of IF was similar in hepatocytes of control and griseofulvin-fed mice. The discrepancy between this result and those of immunofluorescence, transmission and scanning electron microscopy may be due to the time at which Mallory body formation was examined and/or to loss or masking of IF antigenic determinants present in normal hepatocytes (Katsuma *et al.*, 1987). Interestingly, an increase of IF may sometimes precede Mallory body formation (Kimoff and Huang, 1981; Okanoue *et al.*, 1985; Yokoo *et al.*, 1972).

The development of Mallory bodies in hepatocytes of griseofulvin-fed mice is a potentially reversible phenomenon (Denk *et al.*, 1976). After the withdrawal of griseofulvin from the diet, Mallory bodies are dispersed into small granules accumulating at the hepatocyte periphery and often showing conspicuous association with desmosomes (Denk and Franke, 1981). This suggests that Mallory body material may be, at least in part, utilized for the rearrangement of the hepatocyte cytoskeleton and especially for the constitution of the tonofilament–desmosome complexes (Denk and Franke, 1981). The dispersal of Mallory bodies by Ca^{2+}-activated proteases specific for IF (Nelson and Traub, 1981, 1982) is unlikely since Mallory bodies, unlike hepatocyte tonofilaments, seem to be resistant to Ca^{2+}-dependent proteolysis (Irie *et al.*, 1984). The resistance of Mallory bodies to proteolytic degradation is puzzling since they have been shown to be ubiquitinated (Lowe *et al.*, 1988).

Three hypotheses have been proposed regarding the pathogenesis of Mallory bodies. The "microtubular failure" hypothesis postulates that Mallory bodies represent a pathological aggregation of IF due to drugs or conditions favoring microtubule disassembly (French and Davies, 1975), since it is known that microtubule disassembly by various drugs causes IF aggregation (Goldman, 1971; Ishikawa *et al.*, 1968; Weber *et al.*, 1976). However, this hypothesis seems unlikely, at least for the experimental model of mice fed with griseofulvin since in their hepatocytes the number of microtubules is not decreased (Irie *et al.*, 1982). The "vitamin A deficiency" hypothesis came from the proposition that Mallory bodies may be regarded as a pathological and largely hepatocyte-specific form of keratinization (Denk *et al.*, 1979a). The rationale for this hypothesis is that vitamin A deficiency is associated with squamous metaplasia of various epithelia and that Mallory bodies occur in conditions in which vitamin A deficiency is produced by malabsorption of lipid-soluble substances or lack of vitamin A activation. Moreover, the amount of vitamin A is reduced in the liver of griseofulvin-fed mice (Denk *et al.*, 1979a). This "vitamin A deficiency" hypothesis gained support from the finding that vitamin A mediates changes in keratin expression in situations such as malignant transformation (Kim *et al.*, 1984). Finally, the last hypothesis considers Mallory body appearance as one of the multiple steps involved in the conversion of normal cells to neoplastic cells. Clinical evidence for this "preneoplasia" hypothesis is the occurrence of Mallory bodies not only in some hepatic neoplastic cells (Keeley *et al.*, 1972; Norkin and Campagna-Pinto, 1968; Purtilo and Gottlieb, 1973; Wetzel and Alexander, 1979) but also in pulmonary adenocarcinomas (Michel *et al.*, 1982). Moreover, evidence has been provided suggesting that human hepatocarcinoma cells express Mallory bodies more frequently in early stages (Nakanuma and Ohta, 1986) than in advanced stages. Studies of experimental hepatic carcinogenesis have demonstrated an association between Mallory bodies and γ-glutamyl-transpeptidase, an oncofetal marker (Borenfreund *et al.*, 1980b). This marker of neoplastic transformation was found in diethylnitrosamine-induced rat hepatocarcinoma cells containing Mallory bodies (Borenfreund *et al.*, 1980b) and in the hepatocytes of mice fed griseofulvin (Ihrig *et al.*, 1980; Tazawa *et al.*, 1983). Mallory bodies disappeared from cultured rat hepatoma cells when sodium butyrate was added in the medium (Borenfreund *et al.*, 1980c) and this was accompanied by a sharp decrease in γ-glutamyl-transferase activity, by the induction of alkaline phosphatase activity, and by the suppression of *in vitro* growth properties characteristic of tumor cells (Borenfreund *et al.*, 1980b,c). From these observations, Borenfreund *et al.*, (1980b) postulated that Mallory bodies are part of the cytoskeletal disorganization caused by oncogenic transformation. They also suggested that one of the effects of the transformation process could be an increased phosphorylation of IF, resulting in the formation of Mallory bodies and that this could be reversed after the induction of alkaline phosphatase by butyrate. One problem with the preneoplasia hypothesis is that certain diseases in which Mallory bodies are commonly seen, such as Wilson's disease, are not associated with hepatoma formation (French, 1981b).

Considering that Mallory bodies are produced by various agents such as drugs, injury, cirrhosis, and neoplasia, French (1983) suggested that the common factor underlying all these disease processes may be cellular injury followed by hepatocyte proliferation.

3.3. *Atheromatosis*

Atherosclerosis is the most important cause of death in the Western world. Two main events lead to the formation of an atheromatous plaque: (1) migration of smooth muscle cells from the media to the intima where they replicate and accumulate, and (2) intra- and extracellular accumulation of lipids (mainly cholesterol and its esters).

Studies on IF of medial and atherosclerotic smooth muscle cells were initiated by the finding that contrary to what is known for other smooth muscles, aortic smooth muscle cells of mammals contain abundant amounts of vimentin and relatively little desmin (Berner *et al.*, 1981a; Frank and Warren, 1981; Gabbiani *et al.*, 1981b). Using an affinity-purified antidesmin antibody, Schmid *et al.* (1982) undertook a systematic survey of the desmin and vimentin content in the aorta of various vertebrates. They showed that mammalian and avian aortas contain two main cell types; (1) cells containing only vimentin, and (2) cells containing both vimentin and desmin; cells containing only desmin were occasionally observed. The finding of a heterogeneity in the IF content of mammalian medial aortic smooth muscle cells provided clear evidence of arterial smooth muscle cell heterogeneity contrasting with the observation that in mammals these cells are ultrastructurally identical (Karrer, 1961; Pease and Paule, 1960). This is not the case for the avian media in which two cell types have been described by means of electron microscopy (Moss and Benditt, 1970), one having a typical smooth muscle morphology and the other a fibroblastlike morphology; the first type of cell is strongly positive for desmin antibodies and may also stain with vimentin antibodies, whereas the second type of cells stains only with vimentin antibodies (Schmid *et al.*, 1982). These results also proved that two IF proteins can be expressed in a differentiated cell type in adult animals.

Primary culture of rat aortic smooth muscle cells confirmed the heterogeneity of their IF content and showed that the percentage of cells containing vimentin plus desmin decreases with the length of time in culture (Travo *et al.*, 1982). A heterogeneity in the expression of vimentin and desmin was also observed at different levels of the vascular tree: a low number of desmin-positive cells is characteristic of elastic arteries, whereas muscular arteries and other blood vessels contain large number of desmin-positive cells (Osborn *et al.*, 1981). From these observations, the hypothesis was raised that desmin-positive and desmin-negative vascular smooth muscle cells may perform different functions and that the expression of desmin may coincide with contractility. This hypothesis is supported by the fact that in the thoracic aorta the number of desmin- plus vimentin-containing cells and the amount of desmin per cell increase regularly from fetal to adult rats (Kocher *et al.*, 1985). That the heterogeneity of vascular smooth muscle cells (as far as their content of IF is concerned) may be relevant for the pathogenesis of atheromatosis was suggested using an experimental model of intimal smooth muscle accumulation (intimal thickening) (Gabbiani *et al.*, 1982; Kocher *et al.*, 1984). In this model, the endothelium of the thoracic aorta is removed with an embolectomy catheter (Baumgartner and Studer, 1966) leading to an intimal proliferation of smooth muscle cells which ceases when the endothelium is regenerated. Proliferating cells of the intimal thickening are almost exclusively positive for vimentin and only a few are positive for vimentin and desmin (Kocher *et al.*, 1984). Biochemical analysis showed that these cells contain increased amounts of vimentin compared to cells of the normal aortic media. This was also shown in the thoracic aorta of rabbits fed cholesterol (Molony *et al.*, 1986). Interestingly, this situation is reversible since after endothelial regeneration the intimal thickening and the underlying media are similar in their number of vimentin- plus desmin-containing cells and in the amount of vimentin and desmin per cell (Kocher *et al.*, 1984). Subsequent studies showed that in early human fibrous plaques, desmin-containing cells are practically absent, but reappear in complicated lesions (Kocher and Gabbiani, 1986;

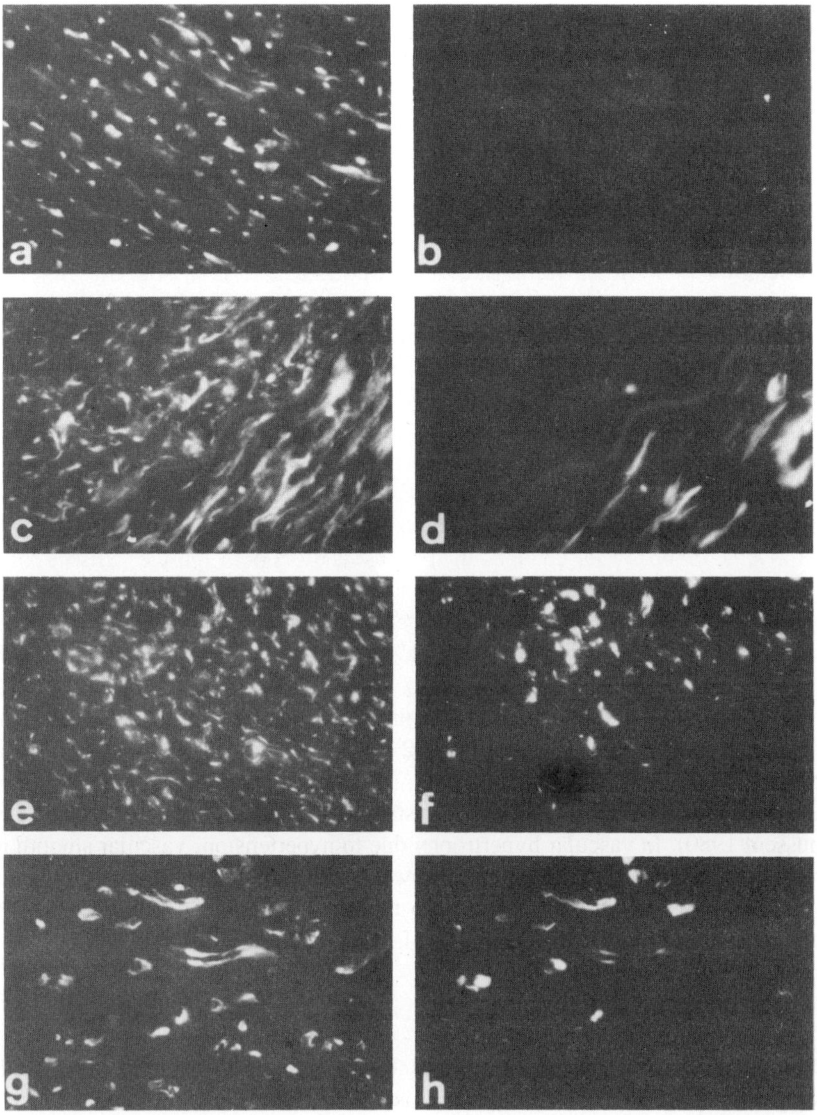

Figure 5. Double immunofluorescent staining for vimentin (a, c, e, g) and desmin (b, d, f, h) of representative fields of human femoral artery fibrous plaque (a, b), coronary artery fibrous plaque with some adjacent media (c, d), complicated femoral artery plaque (e, f), and complicated aortic plaque (g, h). Desmin-positive cells are practically absent in femoral artery fibrous plaque (b). In (d), desmin-positive cells appear well represented in the coronary artery media but practically absent in the fibrous plaque clearly separated from the media by the internal elastic lamina. Numerous desmin-positive cells can be seen in complicated plaques from the femoral artery (f), and the aorta (h). SMC of plaques are oriented differently than are those of the media. This explains why in our sections SMC of plaques appear generally smaller than SMC of the media. All × 400. From Kocher and Gabbiani, *Hum. Pathol,* **17:** 875–880, 1986, with the kind permission of the publisher.

Osborn *et al.*, 1987) (Fig. 5). These changes may thus represent markers of the phenotypic features acquired by smooth muscle cells during atheroma formation. Interestingly, desmin antibodies, used together with monoclonal antibodies against cell lineage-specific antigens, have been also useful to learn the cellular composition of human atherosclerotic plaques (Jonasson *et al.*, 1986) and to show that smooth muscle cells of atheromatous plaques express HLA-DR antigens in contrast to smooth muscle cells of normal human arteries (Jonasson *et al.*, 1985).

It was suggested that culture of aortic smooth muscle cells may represent a useful model for the understanding of mechanisms leading to smooth muscle cell differentiation and atheroma formation, since during culture rat aortic smooth muscle cells develop cytoskeletal features similar to those observed in normal fetal smooth muscle cells or to cells of the atheromatous plaque, that is: increased content of vimentin per cell, increased number of cells containing only vimentin, and decreased number of vimentin- plus desmin-containing cells (Skalli *et al.*, 1986b). These immunohistochemical and biochemical findings are in agreement with ultrastructural observations showing an increase in the number of IF during cell culture (Chamley-Campbell *et al.*, 1979; Palmberg *et al.*, 1985). The selection of vimentin-containing smooth muscle cells during cell culture may be due, at least in part, to a higher replicative activity of these cells compared to vimentin- plus desmin-containing cells, as shown by pulse–chase experiments with tritiated thymidine (Skalli *et al.*, 1986b). That changes in IF content of smooth muscle cells in culture go together with their replicative activity is also suggested by the fact that aortic smooth muscle cells entering for the first time in S phase after plating, increase their vimentin content per cell (Fig. 6, Table III) and that when these cells are plated in a serum-free medium, their IF content remains stable (Skalli *et al.*, 1986b).

Hypertension, cigarette smoking, and diabetes, three well-established risk factors for atherosclerosis (for review see Woolf, 1982), appear to modify the organization and/or number of IF in aortic smooth muscle cells. Thus, compact bundles of IF were observed by electron microscopy in human vascular smooth muscle cells of smokers and diabetics (Asmussen, 1980). In vascular hypertrophy due to hypertension, vascular smooth muscle cells have a massive increase of IF as shown by a morphometric study (Berner *et al.*, 1981b). A similar increase of IF was also reported in hypertrophied intestinal smooth muscle (Gabella, 1979) and hypertrophied heart (see Section 3.1).

3.4. Sarcoidosis

Sarcoidosis is a systemic disorder of unknown etiology characterized by the formation of granulomas without caseous necrosis (for review see James and Neville, 1977). Granulomas are characterized by macrophage-derived giant cells containing vimentin IF (Cain and Kraus, 1983). These are concentrated around each nucleus and extend into the cytoplasm in a pattern similar to that found in mononucleated cells. In approximately 10% of granulomas, the giant cells contain stellate inclusions called asteroid bodies. As seen by transmission electron microscopy, asteroid bodies arise within the area of the cytocenter while their arms extend toward the Golgi area (Cain and Kraus, 1977). Asteroid bodies in pulmonary sarcoidosis are made up of IF organized as dense bundles and curllike whorls; they brightly stain with monospecific antivimentin antibodies (Cain *et al.*, 1983) and sometimes include some microtubules (Cain and Kraus, 1977).

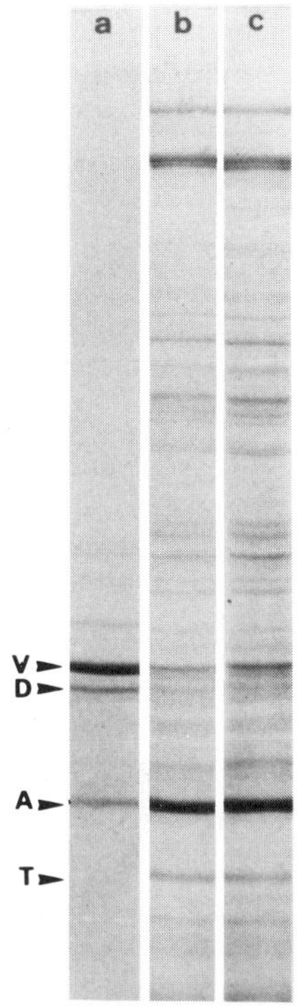

Figure 6. 8% SDS–PAGE of total extracts of SMC during different phases of the cell cycle. SMC were sorted by cytofluorometry 36 hr after plating as G1 cells (b) and S–G2 cells (c). A cytoskeletal preparation of BHK-21 cells was added as marker (a) for vimentin (V), desmin (D), and actin (A). T, tropomyosin. Note the increase of vimentin in S–G2 cells (c) compared to G1 cells (b). From Skalli *et al.*, *J. Submicrosc. Cytol.* **18:**481–493, 1986, with the kind permission of the publisher.

3.5. Tumors and Transformed Cells

The expression of IF proteins in tumor cells has been investigated by several laboratories (see Section 2), although relatively few studies have focused on the ultrastructural organization of IF within these cells. Some works have shown that IF organization can be profoundly modified in neoplastic cells compared to their normal counterparts. Thus, when compared to normal epithelial cells, colonic carcinoma cells exhibited a striking disorganization of their IF network: large accumulations of IF were seen in the cytoplasm, particularly in the perinuclear region, and desmosomes were distributed at random, some of

Table III. Vimentin, Actin, and Tropomyosin Content in Rat Aortic Smooth Muscle Cells during Different Phases of the Cell Cycle[a,b]

	Vimentin	Actin	Tropomyosin
G1	3.1 ± 0.2	20.7 ± 0.2	3.2 ± 0.1
S–G2	4.7 ± 0.6[c]	19.3 ± 0.6	2.8 ± 0.3

[a]From Skalli *et al.*, *J. Submicrosc. Cytol.* **18:**481–493, 1986, with the kind permission of the publisher.
[b]Percentage of total proteins; data are expressed as mean ± S.E.M. of three independent experiments.
[c]Significantly different ($p < 0.005$) from values of G1 cells.

them being intracytoplasmic (Brown *et al.*, 1983a,b). Bundles of "filaments" were also described in different tumor cells but the exact nature of such filaments was not always established (e.g., see Carstens and Broghamer, 1978; Zucker-Franklin, 1974). Some reports have described increased amounts of IF in tumor cells; these were organized in whorls, fascicles, and sometimes perinuclear rings. Examples for these changes include cardiac myxoma (Ferrans and Roberts, 1973b), malignant mesothelioma (Davis, 1974; Lehto and Virtanen, 1978), giant cell carcinoma of the lung (N. S. Wang *et al.*, 1976), T-lymphocyte leukemia (Sezary cells) (Zucker-Franklin, 1974), bronchial squamous cell carcinoma (Inoue and Dionne, 1977), duodenal carcinoid (Carstens and Broghamer, 1978), and bronchial cells of animals treated with nitrosamines (Reznick-Schüller, 1977). Interestingly, an increase in IF protein synthesis or gene transcription is observed in cultured cells transformed with viruses (Keski-Oja *et al.*, 1983; Rubin *et al.*, 1979; Zehner and Paterson, 1983) or treated with the tumor promoter phorbol myristate (Laszlo and Bissell, 1983; Siebert and Fukuda, 1985). *In vitro* viral transformation also causes a rearrangement of the IF network similar to that observed in human tumor cells: IF retract around the nucleus (Ball and Singer, 1981; Bourgeade *et al.*, 1981; Hynes and Destree, 1978), may form tight bundles (Lenk and Penman, 1979), or even disappear (Murti and Goorha, 1983; Sharpe *et al.*, 1980).

IF accumulations sometimes constitute inclusions present in tumor cells (see also Section 3.2). In a case of bronchial carcinoid, argyrophilic inclusions were made up of aggregates of 8- to 10-nm-diameter filaments which were strongly stained by prekeratin antibodies (Berger *et al.*, 1984). Similar inclusions were described ultrastructurally in other cases of bronchial (Alvarez, 1980; An, 1978) and thymic carcinoid (Fetissof *et al.*, 1982). IF also composed the cytoplasmic inclusions found in a case of meningioma (Goldman *et al.*, 1980) and the so-called fibrous bodies of growth-hormone-secreting pituitary adenomas (Horvath and Kovacs, 1978; Roy, 1978). Fibrous bodies are spherical inclusions measuring up to 5 μm which are decorated with an anticytokeratin serum (Neumann *et al.*, 1985); they were proposed to be morphological markers in the differential diagnosis of pituitary adenomas (Horvath and Kovacs, 1978). Kidney rhabdoid tumor cells contain cytoplasmic globules composed of 10-nm filaments (Fung *et al.*, 1981; Haas *et al.*, 1981); these globules are stained with vimentin antibodies and are retained in cell culture (Vogel *et al.*, 1984).

In epithelial cells, bundles of prekeratin filaments are associated with the plasma membrane in desmosomes. IF probably play an important role in desmosome assembly (Jones *et al.*, 1985) but, at least in hepatocytes, mature desmosome integrity and position seem to be independent of the anchorage of IF (Denk *et al.*, 1985). A gradual loss of desmosomes has been correlated to the degree of malignancy in certain human carcinoma cells (Fisher and Sharkey, 1962; Kocher *et al.*, 1981; Schindler *et al.*, 1982); it has also been reported that the number of desmosomes per cell increases during dimethylformamide-induced differentiation of cultured human colon carcinoma cells (Christensen *et al.*, 1985). Other reports have shown that the number of desmosomes was lower in invasive tumors of transitional cell carcinoma of the urinary bladder than in normal transitional epithelial cells (Alroy *et al.*, 1981). However, the role of reduced intracellular adhesion in tumor invasiveness and metastasis is still debated (Pauli *et al.*, 1978; Pitelka *et al.*, 1980; Weinstein *et al.*, 1976). Intracytoplasmic desmosomes have been described in different carcinoma cells (Brown *et al.*, 1983a; Bülow and Klingmüller, 1971; Caputo and Prandi, 1972; Li *et al.*, 1980; Takaki *et al.*, 1971). Intracytoplasmic desmosomes originate from vesicular formations, probably derived from the cell membrane (Caputo and Prandi, 1972). IF are sometimes arranged tangentially to these desmosomes or alternatively are independent of them. Desmosome internalization can be induced in cultured cells treated with EGTA (Kartenbeck *et al.*, 1982). This endocytic resumption of desmosomal material perhaps may explain reversible cell–cell interactions involved in embryogenesis, mitosis, as well as epithelial cell detachment and locomotion during metastasis formation (Kartenbeck *et al.*, 1982).

3.6. Aging

When compared to young fibroblasts, senescent fibroblasts show: decreased DNA synthesis and proliferative capacity, as well as a lengthened intermitotic interval (for review see Choe and Rose, 1976). Other characteristics of senescent cells such as reduced rate of locomotion, increase in size, and change of morphology may be at least partially due to altered cytoskeletal element organization. Thus, in cultured fibroblasts, senescence is associated with an increase in the number of IF, microtubules, and microfilaments (Wang and Gundersen, 1984). In addition, numerous prominent actin stress fibers traverse the cytoplasm and IF are organized as tight sheets or bundles. The formation of IF bundles in senescent fibroblasts is probably due to an increased cross-linking of IF rather than accumulation of vimentin subunits (Wang, 1985). A protein called p50 or epinemin (Lawson, 1983; Liem *et al.*, 1985) may be responsible for this cross-linking. It was suggested that the cytoskeletal organization of senescent cells may contribute to their inability to replicate and that the presence of IF bundles in the cytoplasm may physically impede cell locomotion (Wang, 1985; Wang and Gundersen, 1984).

The cytoskeletal modifications described in aged cultured fibroblasts may not occur in all cell types since in lens cells aging is accompanied by an increase in F-actin but by a decrease in vimentin (Ramaekers *et al.*, 1981).

4. Autoantibodies Against IF

Autoantibodies are present in the serum of a large number of individuals apparently free of autoimmune diseases (Guilbert *et al.*, 1982; Serre *et al.*, 1987). Some of them are directed against cytoskeletal structures and in particular IF (for review see Kurki and Virtanen, 1984; Toh, 1979). Autoantibodies have been the first antibodies available against IF of epithelial and mesenchymal cells (Gordon *et al.*, 1978; Kurki *et al.*, 1977; Osborn *et al.*, 1977). Some of the autoantibodies against IF (Yildiz *et al.*, 1980) present in rabbits recognize antigenic determinants common to smooth muscle cells, fibroblasts, neuroblastoma, glial, and epithelial cells.

Autoantibodies against IF are usually detected by immunofluorescence on cultured cells, or on cryostat sections of various tissues (for review see Kurki and Virtanen, 1984, 1985). Identification of the autoantigen at the molecular level by means of immunoblotting has unfortunately been rarely performed.

Autoantibodies against all types of IF have been reported in association with many diseases. Some of these were classified as smooth muscle antibodies according to their staining pattern on tissue sections (Farrow *et al.*, 1971; Johnson *et al.*, 1965) and indeed are directed against smooth muscle IF (Kurki *et al.*, 1977, 1978). However, smooth muscle autoantibodies (for review see Holborow, 1979; Toh, 1979) include antibodies directed against non-IF cytoskeletal proteins such as actin (Chaponnier *et al.*, 1977; Gabbiani *et al.*, 1973) and microtubules (Whitehouse *et al.*, 1974).

Much attention has been paid to autoantibodies present in the sera of patients with liver diseases. Sera of patients with chronic aggressive hepatitis contain autoantibodies against actin (Chaponnier *et al.*, 1977; Gabbiani *et al.*, 1973) or IF (Kurki *et al.*, 1978; McMillan and Haire, 1979). The specificity of these antibodies seems to correlate with their immunoglobulin class: IgG are directed against actin and IgM against IF (McMillan and Haire, 1979). Chronic aggressive hepatitis includes two major types: the autoimmune type and the viral type associated with hepatitis B virus chronic infection (Mackay, 1978); interestingly, actin autoantibodies occur more frequently in patients with the autoimmune type of the disease, whereas anti-IF antibodies are more frequently associated with the viral type (Pedersen *et al.*, 1982). Anti-IF IgM also occur frequently during the acute phase of hepatitis A (Pedersen *et al.*, 1981; Toh *et al.*, 1979), but are infrequent during the acute phase of hepatitis B and during acute non-A non-B hepatitis (Pedersen *et al.*, 1981). IgM autoantibodies against vimentin were also found during primary biliary cirrhosis and alcoholic liver disease (Kurki *et al.*, 1983a).

IgM antibodies reacting with IF of nonmuscular mesenchymal cells have been associated with the acute phase of various infectious diseases such as chicken pox, measles, and mumps (Toh *et al.*, 1979), mononucleosis (Bretherton and Toh, 1981; Linder *et al.*, 1979a), *Mycoplasma pneumoniae* infection (Bretherton *et al.*, 1981), malaria (Mortazavi-Milani *et al.*, 1984), and leishmaniasis (Böhme *et al.*, 1986).

Autoantibodies reacting with various classes of IF were shown by means of immunoblotting in sera of patients with Waldenström macroglobulinemia (Dellagi *et al.*, 1982; Virtanen *et al.*, 1981) and in 15 of 20 serum samples from patients with angioimmunoblastic lymphadenopathy (Dellagi *et al.*, 1984). These antibodies may bind to an epitope similar or identical to that recognized by a monoclonal antibody reacting with the five classes of IF (Pruss *et al.*, 1981).

On the basis of immunofluorescent staining on rat esophagus cryostat sections, "keratin antibodies" were described in patients having various inflammatory diseases (Johnson *et al.*, 1981; Quismorio *et al.*, 1983; Scott *et al.*, 1981; Young *et al.*, 1979). However, since no other control of antibody specificity was done, the exact nature of the antigen(s) recognized by these antibodies remains to be shown. Immunofluorescence on cultured cells established that there is a high incidence of anti-IF IgM in sera of patients with rheumatoid arthritis (Kurki *et al.*, 1983b; Osung *et al.*, 1982b). Immunoabsorption experiments with pure vimentin or keratin revealed that most of these antibodies reacted with vimentin alone and the remaining with both vimentin and keratin (Osung *et al.*, 1982b). Antibodies with similar reactivities but of the IgG or IgA class have also been found in the sera of patients with Sjögren's syndrome (Kurki *et al.*, 1983b). Autoantibodies directed against keratins of the upper layers of the epidermis were described in sera of patients with cutaneous diseases such as bullous dermatoses, basal cell carcinoma, and melanoma but also in the sera of patients with noncutaneous diseases, such as pulmonary neoplasms (Hintner *et al.*, 1983a). Finally, antibodies to IF were associated with chronic graft-versus-host disease (Tazzari *et al.*, 1987).

The association of IF autoantibodies with various diseases has not been shown to be of diagnostic or prognostic value, with the possible exception of the antibodies found in angioimmunoblastic lymphadenopathy (Dellagi *et al.*, 1984).

The mechanism of IF autoantibody production is not known. Some authors have suggested that they may result from cell destruction (Fudenberg, 1971). That such a mechanism may exist is supported by the finding that antigens related to alcoholic hyalin are found in sera of patients during the early phase of alcoholic hepatitis, during which many hepatocytes die; this phase is followed by the appearance of alcoholic hyalin antibodies (Kanagasundaram *et al.*, 1977). The occurrence of IF antibodies during the acute phase of viral infection may also be explained by this mechanism; moreover, some viruses increase synthesis of IF protein *in vitro* (see Section 2). This association of IF autoantibodies with the acute phase of some infectious diseases may also be due to other causes such as occurrence of common antigenic determinants between infectious agents and IF proteins. Common determinants between cytoskeletal proteins and infectious agents were described for vimentin and various viruses (Fagraeus *et al.*, 1983; Fujinami *et al.*, 1983; Singh *et al.*, 1987) and for myosin and *Streptococcus pyogenes* (Krisher and Cunningham, 1985). Moreover, infectious agents may express proteins having structural and functional properties common with cytoskeletal proteins. *Mycoplasma pneumoniae* contains an actinlike protein which interacts with vertebrate heavy meromyosin (Neimark, 1977). Another example is that of the v-frg gene product of the Gardner–Rasheed feline sarcoma virus which is the result of a recombination event between the genes of actin and of a tyrosin kinase (Naharro *et al.*, 1984). Alternatively, cytoskeletal proteins such as actin may be incorporated in enveloped viruses during the process of budding on the plasma membrane (Damsky *et al.*, 1977; Naito and Matsumoto, 1978; Stanislawsky *et al.*, 1984; Tyrrell and Norrby, 1978; E. Wang *et al.*, 1976). It was suggested that the monoclonal antibodies against cytoskeletal proteins produced by plasmocytic cells of patients with Waldenström macroglobulinemia may represent the expansion of a clone producing an autoantibody, due to an altered immunoregulation (Dighiero *et al.*, 1982). An alternative explanation is that viruses act as polyclonal B-cell activators (Bird and Britton, 1979; Butchko *et al.*, 1978; Cafruny and Plagemann, 1982). The relevance of this mechanism for the production of some IF autoantibodies has been suggested experimentally for Epstein–Barr virus infection (Mortazavi-Milani *et al.*, 1982) and leishmaniasis (Böhme *et al.*, 1986).

5. IF as Activators of Inflammatory Reactions

Whether IF autoantibodies, and more generally cytoskeletal autoantibodies, have a physiological role is unclear. However, some observations suggest that they may be involved in the clearance of damaged cells (Hintner *et al.*, 1987). IF by themselves may also trigger inflammatory reactions. The first report of this IF activity came from Linder *et al.* (1979b) who showed that the complement fragments C1q, C3, and C4 bind to the IF network in human embryo lung fibroblasts permeabilized by nonionic detergents. C1q binding to IF was antibody-independent and initiated complement activation by the classical pathway. Complement components also bind to the desmin-containing Z line of bovine heart muscle (Linder *et al.*, 1980), to endothelial cells (Linder, 1981), and to various human skin cells such as keratinocytes of the upper epidermal layers, melanocytes, Schwann cells, nerve axons, dermal fibroblasts, and vascular smooth muscle cells (Hintner *et al.*, 1982; Linder, 1982; Schuler *et al.*, 1982). Immunoelectron microscopy revealed that IF were invariably responsible for this binding (Schuler *et al.*, 1982). Since

IF of various cell types are able to bind complement factors, it was proposed that a domain common to all IF proteins could be the binding site (Schuler *et al.*, 1982). The failure of some normal cells, such as basal keratinocytes, to bind complement factors was attributed to masking of complement binding sites by IF-associated proteins peculiar to these cell types. Other cell types lacking complement binding properties include melanoma cells, basal carcinoma cells, and keratinocytes of psoriatic lesions (Hintner *et al.*, 1983b). The relevance of these findings is still unclear, but it is worth noting that complement-mediated binding of monocytes to IF has been shown to occur *in vitro,* suggesting that this may be a mechanism for the removal of damaged or dead cells (Linder *et al.*, 1983).

IgG, another category of molecules playing a role during inflammatory processes, have also been reported to bind IF of fibroblasts and endothelial cells (Hansson *et al.*, 1984). This occurs through the $C_{\gamma}2$ domain of their Fc fragment in a saturable, high-affinity fashion (Fig. 7). IgG binding to IF activates the complement cascade (Hansson *et al.*, 1987), further suggesting that this binding plays a role in scavenging of dead cells. In addition, these data suggest that it is important to reevaluate the reports showing the presence of low-titer autoantibodies against IF in the serum of various patients (Hansson *et al.*, 1984). A development of these findings is an immunohistochemical technique for the detection of dead cells in the aortic endothelium by staining the cytoplasmic IgG bound to IF (Hansson and Schwartz, 1983).

Taken together, the observations of Linder *et al.* (1979b) and of Hansson *et al.* (1984) may be relevant for pathological processes *in vivo,* and in particular for the pathology of the vessel wall, since *in vivo* injured endothelial cells accumulate fibrinogen, IgG (Hansson *et al.*, 1979, 1980), and complement (Linder, 1981). Hintner *et al.* (1983b) also noticed that complement binding by IF of various skin cells may contribute to the initiation of repair mechanisms after cellular injury and that lack of complement binding on certain skin cells in various proliferative skin disorders may lead to an impairment of these mechanisms. Autoantibodies to IF could play a role in this scavenging process by opsonizing IF aggregates, thereby promoting their phagocytosis (Hintner *et al.*, 1987).

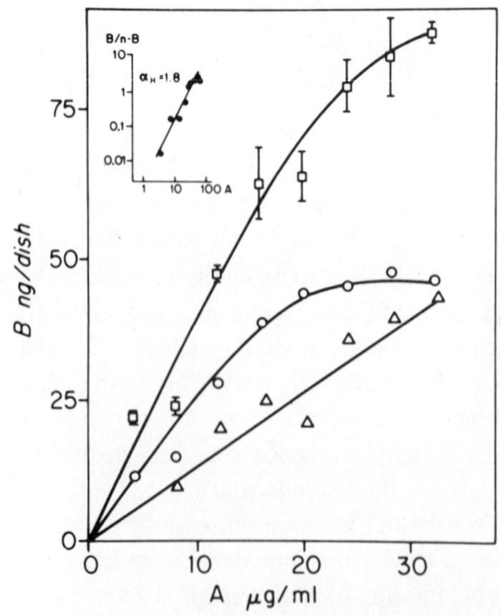

Figure 7. Binding of IgG to vimentin-enriched cytoskeletons. Binding (B) of [^{125}I]-IgG1 to vimentin-enriched cytoskeletons. Total binding (□, ng/dish) and specific binding (○, ng/dish) after subtraction of binding to serum-coated dishes (△, ng/dish) are plotted versus concentration (A) of IgG1 added (ng/μl). In a typical experiment, addition of 20 ng of [^{125}I]-IgG1 per ml gave a total binding of 2000 cpm. (Inset) Hill plot of the binding data shown. A value of $n = 2 \times 10^5$ per cell was used for this analysis. From Hansson *et al.*, *Proc. Natl. Acad. Sci. USA* **81**:3103–3107, 1984, with the kind permission of Dr. Hansson.

Some reports suggested that IF may be involved in the development of certain deposits of amyloid. Immunofluorescence and immunoelectron microscopy with anti-serum amyloid A showed a staining of IF; this staining was inhibited by preincubation of the antibody with purified antiserum amyloid A (Linder *et al.*, 1977). However, no control examined whether the antibody used had a cross-reactivity with vimentin. Surprisingly, IF of cultured fibroblasts display a green birefringence in polarization microscopy after staining with Congo red, a property characteristic of amyloid (Linder *et al.*, 1979c). This probably shows that there is some structural similarity between amyloid and IF but not necessarily that they are related. However, in some cases, amyloid deposits associated with tumors resemble IF ultrastructurally (David and Buchner, 1978; Page *et al.*, 1975) and react with anti-IF antibodies (Eto *et al.*, 1984; Gomes *et al.*, 1981; Kobayashi and Hashimoto, 1983; Masu *et al.*, 1980).

6. Conclusions

In the last few years, our general knowledge of cytoskeletal structures and particular of IF biology has furnished new information concerning the mechanisms and characteristics of several pathological situations (for a review see Rungger-Brändle and Gabbiani, 1983). This has resulted in practical applications such as the use of immunohistochemical techniques with antibodies directed against IF proteins for the differential diagnosis of human tumors. Moreover, the study of IF subunit expression has been instrumental for the understanding of the biology of important diseases such as cancer and atheromatosis. In general, it appears that quantitative and/or qualitative changes in IF subunit expression during various diseases represent useful indicators of cellular adaptation taking place during these situations (e.g., see Ben-Ze'ev *et al.*, 1986; Ramaekers *et al.*, 1983c, 1985b; Skalli *et al.*, 1986b; Summerhayes *et al.*, 1981).

IF probably play an important role in inflammatory reactions and in the scavenging of dead cells. Their insolubility after cell death may be related to the production of IF autoantibodies which are found in patients suffering from various chronic diseases, and in particular inflammatory diseases. These autoantibodies may only be indicators of cell death or may actively participate, through complement binding or opsonization, in the scavenging of dead cells.

A reorganization of the cytoplasmic IF network is frequently found in pathological cells irrespective of their embryological origin. The most frequent morphological features of this reorganization are an increase in the number of IF and the formation of tight bundles, which often assume a perinuclear position. In addition, many intracytoplasmic bodies, some of which have been classically described during various diseases, appear to result from IF accumulation. In certain situations it was suggested that the increased number of IF and their bundling may be the first step toward their aggregation (Kimoff and Huang, 1981; Okanoue *et al.*, 1985; Yokoo *et al.*, 1972). In various cell lines, mitosis is also characterized by an IF reorganization which may assume many of the forms observed in pathological cells (Jones *et al.*, 1985). It is thus tempting to hypothesize that the IF reorganization observed in cells during pathological situations may be linked, at least in part, to their growth control, either by affecting it, or as a consequence of growth alteration. Phosphorylation and proteolysis by specific Ca^{2+}-activated proteases have been proposed to be regulatory mechanisms of IF organization (for a review see Traub,

1985). The finding that the IF aggregates present in many cells during various diseases are ubiquitinated may be of interest to the understanding of the pathogenesis of these structures (Lowe *et al.*, 1988). Finally, the IF reorganization found during pathological conditions may affect some cellular processes for which these structures are required; in that respect, pathological situations may help cell biologists to understand IF function in normal cells.

ACKNOWLEDGMENTS. This work has been supported by the Swiss National Science Foundation, Grant 3.107-0.85. We thank Drs. H. Denk (University of Graz, Graz, Austria) and G. K. Hansson (University of Göteborg, Göteborg, Sweden) for allowing reproduction and providing prints of Figs. 4 and 7, *The American Journal of Pathology* for allowing reproduction of Fig. 2 and 3, *Human Pathology* for allowing reproduction of Fig. 5, and *The Journal of Submicroscopic Cytology* for allowing reproduction of Fig. 6 and Table III. We thank Mrs. M.-M. Rossire for typing the manuscript.

References

Afzelius, B. A., 1985, The immotile-cilia syndrome: A microtubule-associated defect, *CRC Crit. Rev. Biochem.* **19:**63–87.

Alberts, B., Bray, D., Lewis, J., Raff, M., Roberts, K., and Watson, J. D., 1989, *Molecular Biology of the Cell*, Garland Publishing, New York, Chapter 11, pp. 613–681.

Alroy, J., Pauli, B. U., and Weinstein, R. S., 1981, Correlation between numbers of desmosomes and the aggressiveness of transitional cell carcinoma in human urinary bladder, *Cancer* **47:**104–112.

Altmannsberger, M., Osborn, M., Hölscher, A., Schauer, A., and Weber, K., 1981, The distribution of keratin type intermediate filaments in human breast cancer. An immunohistological study, *Virchows Arch. B* **37:**277–284.

Altmannsberger, M., Osborn, M., Treuner, J., Hölscher, A., Weber, K., and Schauer, A., 1982a, Diagnosis of human childhood rhabdomyosarcoma by antibodies to desmin, the structural protein of muscle specific intermediate filaments, *Virchows Arch. B* **39:**203–215.

Altmannsberger, M., Weber, K., Hölscher, A., Schauer, A., and Osborn, M., 1982b, Antibodies to intermediate filaments as diagnostic tools. Human gastrointestinal carcinomas express prekeratin, *Lab. Invest.* **46:**520–526.

Altmannsberger, M., Osborn, M., Schäfer, H., Schauer, A., and Weber, K., 1984, Distinction of nephroblastomas from other childhood tumors using antibodies to intermediate filaments, *Virchows Arch. B* **45:**113–124.

Altmannsberger, M., Weber, K., Droste, R., and Osborn, M., 1985, Desmin is a specific marker for rhabdomyosarcomas of human and rat origin, *Am. J. Pathol.* **118:**85–95.

Alvarez, F. E., 1980, Intracytoplasmic fibrillary inclusions in bronchial carcinoid, *Cancer* **46:**144–151.

An, T., 1978, Cytoplasmic inclusions in bronchial carcinoid, *Hum. Pathol.* **9:**241–242.

Asmussen, I., 1980, Intermediate filaments arranged in bundles in human vascular smooth muscle cells, *J. Submicrosc. Cytol.* **12:**673–680.

Babaï, F., Skalli, O., Schürch, W., Seemayer, T. A., and Gabbiani, G., 1988, Chemically induced rhabdomyosarcomas in rats. Ultrastructural, immunohistochemical, biochemical features and expression of α-actin isoforms, *Virchows Arch. B* **55:**263–277.

Ball, E. H., and Singer, J., 1981, Association of microtubules and intermediate filaments in normal fibroblasts and its disruption upon transformation by a temperature-sensitive mutant of Rous sarcoma virus, *Proc. Natl. Acad. Sci. USA* **78:**6986–6990.

Bannasch, P., Zerban, H., Schmid, E., and Franke, W. W., 1980, Liver tumors distinguished by immunofluorescence microscopy with antibodies to proteins of intermediate-sized filaments, *Proc. Natl. Acad. Sci. USA* **77:**4948–4952.

Barritault, D., Courtois, Y., and Paulin, D., 1980, Biochemical evidence that vimentin is the only in vivo constituent of the intermediate-sized filaments in adult bovine epithelial lens cells, *Biol. Cell* **39:**335–338.

Battifora, H., Sun, T. T., Bahu, R. M., and Rao, S., 1980, The use of antikeratin antiserum as a diagnostic tool: Thymoma versus lymphoma, *Hum. Pathol.* **11:**635–640.

Baumgartner, H. R., and Studer, A., 1966, Folgen des Gefässkatheterismus am normo- und hypercholesterinaemischen Kaninchen, *Pathol. Microbiol.* **29:**393–405.

Behrendt, H., 1977, Effect of anabolic steroids on rat heart muscle cells. I. Intermediate filaments, *Cell Tissue Res.* **180:**303–315.

Bennett, G. S., Fellini, S. A., Croop, J. M., Otto, J. J., Bryan, J., and Holtzer, H., 1978, Differences among 100 Å filament subunits from different cell types, *Proc. Natl. Acad. Sci. USA* **75:**4364–4368.

Ben-Ze'ev, A., Zöller, M., and Raz, A., 1986, Differential expression of intermediate filament proteins in metastatic and nonmetastatic variants of the BSp73 tumor, *Cancer Res.* **46:**785–790.

Berger, G., Berger, F., Bejui, F., Bouvier, R., Rochet, M., and Feroldi, J., 1984, Bronchial carcinoid with fibrillary inclusions related to cytokeratins: An immunohistochemical and ultrastructural study with subsequent investigation of 12 foregut APUDomas, *Histopathology* **8:**245–257.

Berner, P. F., Frank, E., Holtzer, H., and Somlyo, A. P., 1981a, The intermediate filament proteins of rabbit vascular smooth muscle: Immunofluorescent studies of desmin and vimentin, *J. Muscle Res. Cell Motil.* **2:**439–452.

Berner, P. F., Somlyo, A. V., and Somlyo, A. P., 1981b, Hypertrophy-induced increase of intermediate filaments in vascular smooth muscle, *J. Cell Biol.* **88:**96–101.

Bird, A. G., and Britton, S., 1979, A new approach to the study of human B lymphocyte function using an indirect plaque assay and a direct B cell activator, *Immunol. Rev.* **45:**41–67.

Blobel, G. A., Moll, R., Franke, W. W., Kayser, K. W., and Gould, V. E., 1985, The intermediate filament cytoskeleton of malignant mesotheliomas and its diagnostic significance, *Am. J. Pathol.* **121:**235–247.

Böhme, M. W. J., Evans, D. A., Miles, M. A., and Holborow, E. J., 1986, Occurrence of autoantibodies to intermediate filament proteins in human visceral leishmaniasis and their induction by experimental polyclonal B-cell activation, *Immunology* **59:**583–588.

Borenfreund, E., and Bendich, A., 1978, In vitro demonstration of Mallory body formation in liver cells from rats fed diethylnitrosamine, *Lab. Invest.* **38:**295–303.

Borenfreund, E., Higgins, P. J., Steinglass, M., and Bendich, A., 1979, Carcinogen-induced abnormalities in rat liver cells and their modification by chemical agents, *Cancer Res.* **39:**800–807.

Borenfreund, E., Higgins, P. J., and Peterson, E., 1980a, Intermediate-sized filaments in cultured rat liver tumor cells with Mallory body-like cytoplasm abnormalities, *J. Natl. Cancer Inst.* **64:**323–333.

Borenfreund, E., Higgins, P. J., and Bendich, A., 1980b, In vivo–in vitro rat liver carcinogenesis: Modification in protein synthesis and ultrastructure, *Ann. N.Y. Acad. Sci.* **349:**357–372.

Borenfreund, E., Schmid, E., Bendich, A., and Franke, W. W., 1980c, Constitutive aggregates of intermediate-sized filaments of the vimentin and cytokeratin type in cultured hepatoma cells and their dispersal by butyrate, *Exp. Cell Res.* **127:**215–235.

Borenfreund, E., DeHarven, E., and Garra, L., 1981, Mallory body-like abnormalities in carcinomas induced by cultured transformed rat liver cells, *Hepatology* **1:**408–415.

Boschek, C. B., 1982, Organizational changes of cytoskeletal proteins during cell transformation, in: *Advances in Viral Oncology,* Volume 1 (G. Klein, ed.), Raven Press, New York, pp. 173–187.

Bourgeade, M. M., Rousset, S., Paulin, D., and Chany, C., 1981, Reorganization of the cytoskeleton by interferon in MSV-transformed cells, *J. Interferon Res.* **1:**323–332.

Bretherton, L., and Toh, B. H., 1981, IgM autoantibody to intermediate filaments in infectious mononucleosis, *J. Clin. Lab. Immunol.* **5:**7–10.

Bretherton, L., Toh, B. H., and Jack, I., 1981, IgM autoantibody to intermediate filaments in Mycoplasma pneumoniae infections, *Clin. Immunol. Immunopathol.* **18:**425–430.

Brown, D. T., Anderton, B. H., and Wylie, C. C., 1983a, The organization of intermediate filaments in normal human colonic epithelium and colonic carcinoma cells, *Int. J. Cancer* **32:**163–169.

Brown, D. T., Anderton, B. H., and Wylie, C. C., 1983b, Alterations in the organization of cytokeratin filaments in normal and malignant human colonic epithelial cells during mitosis, *Cell Tissue Res.* **233:**619–628.

Brown, D. C., Theaker, J. M., Banks, P. M., Gatter, K. C., and Mason, D. Y., 1987, Cytokeratin expression in smooth muscle and smooth muscle tumors, *Histopathology* **11:**477–486.

Bülow, M. V., and Klingmüller, G., 1971, Elektronenmikroskopische Untersuchungen des Keratoakanthoms. Vorkommen intracytoplasmatischer Desmosomen, *Arch. Dermatol. Forsch.* **241:**292–304.

Butchko, G. M., Armstrong, R. B., Martin, W. J., and Ennis, F. A., 1978, Influenza A viruses of the H2N2 subtype are lymphocyte mitogens, *Nature* **271:**66–67.

Cafruny, W. A., and Plagemann, P. G. W., 1982, Immune response to lactate dehydrogenase-elevating virus: Serologically specific rabbit neutralizing antibody to the virus, *Infect. Immun.* **37:**1007–1012.

Cain, H., and Kraus, B., 1977, Asteroid bodies: Derivatives of the cytosphere. An electron microscopic contribution to the pathology of the cytocentre, *Virchows Arch. B* **26:**119–132.

Cain, H., and Kraus, B., 1983, Immunofluorescence microscopic demonstration of vimentin filaments in asteroid bodies of sarcoidosis. A comparison with electron microscopic findings, *Virchows Arch. B* **42:**213–226.

Cain, H., Kraus, B., Krauspe, R., Osborn, M., and Weber, K., 1983, Vimentin filaments in peritoneal macrophages at various stages of differentiation and with altered function, *Virchows Arch. B* **42:**65–81.

Camner, P., Mossberg, B., and Afzelius, B. A., 1975, Evidence for congenitally nonfunctioning cilia in the tracheobronchial tract in two subjects, *Am. Rev. Resp. Dis.* **112:**807–809.

Caputo, R., and Prandi, G., 1972, Intracytoplasmic desmosomes, *J. Ultrastruct. Res.* **41:**358–368.

Carstens, P. H. B., and Broghamer, W. L., Jr., 1978, Duodenal carcinoid with cytoplasmic whorls of microfilaments, *J. Pathol.* **124:**235–238.

Caselitz, J., Osborn, M., Seifert, G., and Weber, K., 1981, Intermediate-sized filament proteins (prekeratin, vimentin, desmin) in the normal parotid gland and parotid gland tumours. Immunofluorescence study, *Virchows Arch. A* **393:**273–286.

Chamley-Campbell, J., Campbell, G. R., and Ross, R., 1979, The smooth muscle cell in culture, *Physiol. Rev.* **59:**1–61.

Chaponnier, C., Kohler, L., and Gabbiani, G., 1977, Fixation of human anti-actin autoantibodies on skeletal muscle fibres, *Clin. Exp. Immunol.* **27:**278–284.

Choe, B. K., and Rose, N. R., 1976, In vitro senescence of mammalian cells, *Gerontology* **22:**89–108.

Christensen, T. G., Burke, B., Dexter, D. L., and Zamcheck, N., 1985, Ultrastructural evidence of dimethylformamide-induced differentiation of cultured human colon carcinoma cells, *Cancer* **56:**1559–1565.

Coindre, J. M., De Mascarel, A., Trojani, M., De Mascarel, I., and Pages, A., 1988, Immunohistochemical study of rhabdomyosarcoma. Unexpected staining with S100 protein and cytokeratin, *J. Pathol.* **155:**127–132.

Cooper, D., Schermer, A., and Sun, T. T., 1985, Classification of human epithelia and their neoplasms using monoclonal antibodies to keratins: Strategies, applications, and limitations, *Lab. Invest.* **52:**243–256.

Cremer, M., Treiss, I., Cremer, T., Hager, D., and Franke, W. W., 1981, Characterization of cells of amniotic fluids by immunological identification of intermediate-sized filaments: Presence of cells of different tissue origin, *Hum. Genet.* **59:**373–379.

Czernobilsky, B., Moll, R., Levy, R., and Franke, W. W., 1985, Co-expression of cytokeratin and vimentin filaments in mesothelial, granulosa and rete ovarii cells of the human ovary, *Eur. J. Cell Biol.* **37:**175–190.

Damsky, C. H., Sheffield, J. B., Tuszynski, G. P., and Warren, L., 1977, Is there a role for actin in virus budding? *J. Cell Biol.* **75:**593–605.

Darmon, M., Delescluse, C., Semat, A., Bernard, B., Bailly, J., and Prunieras, M., 1984, A keratin of fetal skin is reexpressed in human keratinocytes transformed by SV40 virus or treated with the tumor promoter TPA, *Exp. Cell Res.* **154:**315–319.

David, R., and Buchner, A., 1978, Amyloid stroma in a tubular carcinoma of palatal salivary gland. A histochemical and ultrastructural study, *Cancer* **41:**1836–1844.

Davis, J. M. G., 1974, Ultrastructure of human mesotheliomas, *J. Natl. Cancer Inst.* **52:**1715–1725.

Dellagi, K., Brouet, J. C., Perreau, J., and Paulin, D., 1982, Human monoclonal IgM with autoantibody activity against intermediate filaments, *Proc. Natl. Acad. Sci. USA* **79:**446–450.

Dellagi, K., Vainchenker, W., Vinci, G., Paulin, D., and Brouet, J. C., 1983, Alteration of vimentin intermediate filament expression during differentiation of human hemopoietic cells, *EMBO J.* **2:**1509–1514.

Dellagi, K., Brouet, J. C., and Seligmann, M., 1984, Antivimentin autoantibodies in angioimmunoblastic lymphadenopathy, *N. Engl. J. Med.* **310:**215–218.

Denk, H., and Eckerstorfer, R., 1977, Colchicine-induced Mallory body formation in the mouse, *Lab. Invest.* **36:**563–565.

Denk, H., and Franke, W. W., 1981, Rearrangement of the hepatocyte cytoskeleton after toxic damage:

Involution, dispersal and peripheral accumulation of Mallory body material after drug withdrawal, *Eur. J. Cell Biol.* **23:**241–249.

Denk, H., Eckerstorfer, R., Gschnait, F., Konrad, K., and Wolff, K., 1976, Experimental induction of hepatocellular hyalin (Mallory bodies) in mice by griseofulvin treatment. 1. Light microscopic observations, *Lab. Invest.* **35:**377–382.

Denk, H., Franke, W. W., Kerjaschki, D., and Eckerstorfer, R., 1979a, Mallory bodies in experimental animals and man, *Int. Rev. Exp. Pathol.* **20:**77–121.

Denk, H., Franke, W. W., Eckerstorfer, R., Schmid, E., and Kerjaschki, D., 1979b, Formation and involution of Mallory bodies ("alcoholic hyalin") in murine and human liver revealed by immunofluorescence microscopy with antibodies to prekeratin, *Proc. Natl. Acad. Sci. USA* **76:**4112–4116.

Denk, H., Franke, W. W., and Kerjaschki, D., 1981a, Mallory bodies: New facts and findings, in: *Frontiers in Liver Disease* (P. D. Berk and T. C. Chalmers, eds.), Thieme Stratton, Stuttgart, pp. 93–105.

Denk, H., Franke, W. W., Dragosics, B., and Zeiler, I., 1981b, Pathology of cytoskeleton of liver cells: Demonstration of Mallory bodies (alcoholic hyalin) in murine and human hepatocytes by immunofluorescence microscopy using antibodies to cytokeratin polypeptides from hepatocytes, *Hepatology* **1:**9–20.

Denk, H., Krepler, R., Lackinger, E., Artlieb, U., and Franke, W. W., 1982, Biochemical and immunocytochemical analysis of the intermediate filament cytoskeleton in human hepatocellular carcinomas and in hepatic neoplastic nodules of mice, *Lab. Invest.* **46:**584–596.

Denk, H., Krepler, R., Artlieb, U., Gabbiani, G., Rungger-Brändle, E., Leoncini, P., and Franke, W. W., 1983, Proteins of intermediate filaments. An immunohistochemical and biochemical approach to the classification of soft tissue tumors, *Am. J. Pathol.* **110:**193–208.

Denk, H., Lackinger, E., Cowin, P., and Franke, W. W., 1985, Maintenance of desmosomes in mouse hepatocytes after drug-induced rearrangement of cytokeratin filament material, *Exp. Cell Res.* **161:**161–171.

Dighiero, G., Guilbert, B., and Avrameas, S., 1982, Naturally occurring antibodies against nine common antigens in human sera. II. High incidence of monoclonal Ig exhibiting antibody activity against actin and tubulin and sharing antibody specificities with natural antibodies, *J. Immunol.* **128:**2788–2792.

Doglioni, C., Dell'Orto, P., Coggi, G., Iuzzolino, P., Bontempini, L., and Viale, G., 1987, Choroid plexus tumors. An immunocytochemical study with particular reference to the coexpression of intermediate filament proteins, *Am. J. Pathol.* **127:**519–529.

Edström, L., Thornell, L.-E., and Eriksson, A., 1980, A new type of hereditary distal myopathy with characteristic sarcoplasmic bodies and intermediate (skeletin) filaments, *J. Neurol. Sci.* **47:**171–190.

Eliasson, R., Mossberg, B., Camner, P., and Afzelius, B. A., 1977, The immotile-cilia syndrome: A congenital ciliary abnormality as an etiologic factor in chronic airway infections and male sterility, *N. Engl. J. Med.* **297:**1–6.

Eto, H., Hashimoto, K., Kobayashi, H., Fukaya, T., Matsumoto, M., and Sun, T. T., 1984, Differential staining of cytoid bodies and skin-limited amyloids with monoclonal anti-keratin antibodies, *Am. J. Pathol.* **116:**473–481.

Fagraeus, A., Örvell, C., Norberg, R., and Norrby, E., 1983, Monoclonal antibodies to epitopes shared by actin and vimentin obtained by paramyxovirus immunization, *Exp. Cell Res.* **145:**425–432.

Farrow, L. J., Holborow, E. J., and Brighton, W. D., 1971, Reaction of human smooth muscle antibody with liver cells, *Nature* **232:**186–187.

Ferrans, V. J., and Roberts, W. C., 1973a, Intermyofibrillar and nuclear–myofibrillar connections in human and canine myocardium. An ultrastructural study, *J. Mol. Cell. Cardiol.* **5:**247–257.

Ferrans, V. J., and Roberts, W. C., 1973b, Structural features of cardiac myxomas. Histology, histochemistry, and electron microscopy, *Hum. Pathol.* **4:**111–146.

Fetissof, F., Boivin, F., Jobard, P., Arbeille-Brassard, B., Romet, J. L., and Maillet, M. Y., 1982, Microfilamentous carcinoid of the thymus: Correlation of ultrastructural study with Grimelius stain, *Ultrastruct. Pathol.* **3:**9–15.

Fidzianska, A., Goebel, H. H., Osborn, M., Lenard, H. G., Osse, G., and Langenbeck, U., 1983, Mallory body-like inclusions in a hereditary congenital neuromuscular disease, *Muscle Nerve* **6:**195–200.

Fisher, E. R., and Sharkey, D. A., 1962, The ultrastructure of colonic polyps and cancer with special reference to the epithelial inclusion bodies of Leuchtenberger, *Cancer* **15:**160–170.

Fleming, K. A., Morton, J. A., Barbatis, C., Burns, J., Canning, S., and McGee J. O., 1981, Mallory bodies in alcoholic and non-alcoholic liver disease contain a common antigenic determinant, *Gut* **22:**341–344.

Frank, E. D., and Warren, L., 1981, Aortic smooth muscle cells contain vimentin instead of desmin, *Proc. Natl. Acad. Sci. USA* **78:**3020–3024.

Franke, W. W., and Moll, R., 1987, Cytoskeletal components of lymphoid organs. I. Synthesis of cytokeratins 8 and 18 and desmin in subpopulations of extrafollicular reticulum cells of human lymph nodes, tonsils, and spleen, *Differentiation* **36:**145–163.

Franke, W. W., Schmid, E., Osborn, M., and Weber, K., 1978, Different intermediate-sized filaments distinguished by immunofluorescence microscopy, *Proc. Natl. Acad. Sci. USA* **75:**5034–5038.

Franke, W. W., Appelhans, B., Schmid, E., Freudenstein, C., Osborn, M., and Weber, K., 1979a, Identification and characterization of epithelial cells in mammalian tissues by immunofluorescence microscopy using antibodies to prekeratin, *Differentiation* **15:**7–25.

Franke, W. W., Schmid, E., Breitkreutz, D., Lüderr, M., Boukamp, P., Fusenig, N. E., Osborn, M., and Weber, K., 1979b, Simultaneous expression of the two different types of intermediate sized filaments in mouse keratinocytes proliferating in vitro, *Differentiation* **14:**35–50.

Franke, W. W., Denk, H., Schmid, E., Osborn, M., and Weber, K., 1979c, Ultrastructural, biochemical, and immunologic characterization of Mallory bodies in livers of griseofulvin-treated mice. Fimbriated rods of filaments containing prekeratin-like polypeptides, *Lab. Invest.* **40:**207–220.

Franke, W. W., Schmid, E., Freudenstein, C., Appelhans, B., Osborn, M., Weber, K., and Keenan, T. W., 1980, Intermediate-sized filaments of the prekeratin type in myoepithelial cells, *J. Cell Biol.* **84:**633–654.

Franke, W. W., Denk, H., Kalt, R., and Schmid, E., 1981a, Biochemical and immunological identification of cytokeratin proteins present in hepatocytes of mammalian liver tissue, *Exp. Cell Res.* **131:**299–318.

Franke, W. W., Mayer, D., Schmid, E., Denk, H., and Borenfreund, E., 1981b, Differences of expression of cytoskeletal proteins in cultured rat hepatocytes and hepatoma cells, *Exp. Cell Res.* **134:**345–365.

Franke, W. W., Grund, C., Jackson, B. W., and Illmensee, K., 1983, Formation of cytoskeletal elements during mouse embryogenesis. IV. Ultrastructure of primary mesenchymal cells and their cell–cell interactions, *Differentiation* **25:**121–141.

Franko, M. C., Gibbs, C. J., Jr., Rhoades, D. A., and Gajdusek, D. C., 1987, Monoclonal antibody analysis of keratin expression in the central nervous system, *Proc. Natl. Acad. Sci. USA* **84:**3482–3485.

French, S. W., 1981a, Nature, pathogenesis and significance of the Mallory body, *Semin. Liver Dis.* **1:**217–231.

French, S. W., 1981b, The Mallory body: Structure, composition, and pathogenesis, *Hepatology* **1:**76–83.

French, S. W., 1983, Present understanding of the development of Mallory's body, *Arch. Pathol. Lab. Med.* **107:**445–450.

French, S. W., and Davies, P. L., 1975, The Mallory body in the pathogenesis of alcoholic liver disease, in: *Alcoholic Liver Pathology* (J. M. Khanna, Y. Israel, and H. Kalant, eds.), Addiction Research Foundation of Ontario, Toronto, pp. 113–143.

French, S. W., Ihrig, T. J., and Norum, M. L., 1972, A method of isolation of Mallory bodies in a purified fraction, *Lab. Invest.* **26:**240–244.

French, S. W., Sim, J. S., Franks, K. E., Burbige, E. J., Denton, T., and Caldwell, M. G., 1977, Alcoholic hepatitis, in: *Alcohol and the Liver* (M. M. Fisher and J. G. Rankin, eds.), Plenum Press, New York, pp. 261–286.

Fudenberg, H. H., 1971, Genetically determined immune deficiency as the predisposing cause of "autoimmunity" and lymphoid neoplasias, *Am. J. Med.* **51:**295–298.

Fujimoto, T., and Singer, S. J., 1986, Immunocytochemical studies of endothelial cells in vivo. I. The presence of desmin only, or of desmin plus vimentin, or vimentin only, in the endothelial cells of different capillaries of the adult chicken, *J. Cell Biol.* **103:**2775–2786.

Fujinami, R. S., Oldstone, M. B. A., Wroblewska, Z., Frankel, M. E., and Koprowski, H., 1983, Molecular mimicry in virus infection: Crossreaction of measles virus phosphoprotein or of herpes simplex virus protein with human intermediate filaments, *Proc. Natl. Acad. Sci. USA* **80:**2346–2350.

Fung, C. H. K., Gonzalez-Crussi, F., Yonan, T. N., and Martinez, N., 1981, "Rhabdoid" Wilms' tumor: An ultrastructural study, *Arch. Pathol. Lab. Med.* **105:**521–523.

Gabbiani, G., Ryan, G. B., Lamelin, J. P., Vassalli, P., Majno, G., Bouvier, C. A., Cruchaud, A., and Lüscher, E. F., 1973, Human smooth muscle autoantibody. Its identification as antiactin antibody and a study of its binding to "nonmuscular" cells, *Am. J. Pathol.* **72:**473–488.

Gabbiani, G., Kapanci, Y., Barazzone, P., and Franke, W. W., 1981a, Immunochemical identification of intermediate-sized filaments in human neoplastic cells. A diagnostic aid for the surgical pathologist, *Am. J. Pathol.* **104:**206–216.

Gabbiani, G., Schmid, E., Winter, S., Chaponnier, C., de Chastonay, C., Vandekerckhove, J., Weber, K., and

Franke, W. W., 1981b, Vascular smooth muscle cells differ from other smooth muscle cells: Predominance of vimentin filaments and a specific α-type actin, *Proc. Natl. Acad. Sci. USA* **78:**298–302.

Gabbiani, G., Rungger-Brändle, E., de Chastonay, C., and Franke, W. W., 1982, Vimentin-containing smooth muscle cells in aortic intimal thickening after endothelial injury, *Lab. Invest.* **47:**265–269.

Gabella, G., 1979, Hypertrophic smooth muscle. IV. Myofilaments, intermediate filaments and some mechanical properties, *Cell Tissue Res.* **201:**277–288.

Glasser, S. R., and Julian J., 1986, Intermediate filament protein as a marker of uterine stromal cell decidualization, *Biol. Reprod.* **35:**463–474.

Goldman, J. E., Horoupian, D. S., and Johnson, A. B., 1980, Granulofilamentous inclusions in a meningioma, *Cancer* **46:**156–161.

Goldman, R. D., 1971, The role of three cytoplasmic fibers in BHK-21 cell motility. I. Microtubules and the effects of colchicine, *J. Cell Biol.* **51:**752–762.

Gomes, M. A., Staquet, M. J., and Thivolet, J., 1981, Staining of colloid bodies by keratin antisera in lichen planus, *Am. J. Dermatopathol.* **3:**341–347.

Gordon, W. E., III, Bushnell, A., and Burridge, K., 1978, Characterization of the intermediate (10 nm) filaments of cultured cells using an autoimmune rabbit antiserum, *Cell* **13:**249–261.

Gould, V. E., Moll, R., Moll, I., Lee, I., and Franke, W. W., 1985, Neuroendocrine (Merkel) cells of the skin: Hyperplasias, dysplasias, and neoplasms, *Lab. Invest.* **52:**334–353.

Gown, A. M., and Gabbiani, G., 1984, Intermediate-sized (10-nm) filaments in human tumors, in: *Advances in Immunohistochemistry* (R. A. DeLellis, ed.), Masson, Paris, York, pp. 89–109.

Gown, A. M., and Vogel, A. M., 1984, Monoclonal antibodies to human intermediate filament proteins. II. Distribution of filament proteins in normal human tissues, *Am. J. Pathol.* **114:**309–321.

Guilbert, B., Dighiero, G., and Avrameas, S., 1982, Naturally occurring antibodies against nine common antigens in human sera. I. Detection, isolation, and characterization, *J. Immunol.* **128:**2779–2787.

Haas, J. E., Palmer, N. F., Weinberg, A. G., and Beckwith, J. B., 1981, Ultrastructure of malignant rhabdoid tumor of the kidney: A distinctive renal tumor of children, *Hum. Pathol.* **12:**646–657.

Hall, P., 1985a, The pathological spectrum of alcoholic liver disease, *Pathology* **17:**209–218.

Hall, P., 1985b, Pathology and pathogenesis of alcoholic liver disease, in: *Alcoholic Liver Disease* (P. Hall, ed.), Arnold, London, pp. 41–68.

Hansson, G. K., and Schwartz, S. M., 1983, Evidence for cell death in the vascular endothelium in vivo and in vitro, *Am. J. Pathol.* **112:**278–286.

Hansson, G. K., Bondjers, G., and Nilsson, L. A., 1979, Plasma protein accumulation in injured endothelial cells. Immunofluorescent localization of IgG and fibrinogen in the rabbit aortic endothelium, *Exp. Mol. Pathol.* **30:**12–26.

Hansson, G. K., Bondjers, G., Bylock, A., and Hjalmarsson, L., 1980, Ultrastructural studies on the localization of IgG in the aortic endothelium and subendothelial intima of atherosclerotic and nonatherosclerotic rabbits, *Exp. Mol. Pathol.* **33:**302–315.

Hansson, G. K., Starkebaum, G. A., Benditt, E. P., and Schwartz, S. M., 1984, Fc-mediated binding of IgG to vimentin-type intermediate filaments in vascular endothelial cells, *Proc. Natl. Acad. Sci. USA* **81:**3103–3107.

Hansson, G. K., Lagerstedt, E., Bengtsson, A., and Heideman, M., 1987, IgG binding to cytoskeletal intermediate filaments activates the complement cascade, *Exp. Cell Res.* **170:**338–350.

Hazan, R., Denk, H., Franke, W. W., Lackinger, E., and Schiller, D. L., 1986, Change of cytokeratin organization during development of Mallory bodies as revealed by a monoclonal antibody, *Lab. Invest.* **54:**543–553.

Hintner, H., Stingl, G., Schuler, G., and Wolff, K., 1982, In vitro complement-binding on cytoplasmic structures in normal human skin: I. Immunofluorescence studies, *J. Invest. Dermatol.* **79:**119–124.

Hintner, H., Steinert, P. M., and Lawley, T. J., 1983a, Human upper epidermal cytoplasmic antibodies are directed against keratin intermediate filament protein, *J. Clin. Invest.* **72:**1344–1351.

Hintner, H., Stanzl, U., Schuler, G., Klein, G., Fritsch, P., and Stingl, G., 1983b, In vitro complement binding in human skin cells with altered differentiation, *J. Invest. Dermatol.* **80:**490–493.

Hintner, H., Romani, N., Stanzl, U., Grubauer, G., Fritsch, P., and Lawley, T. J., 1987, Phagocytosis of keratin filament aggregates following opsonization with IgG-anti-keratin filament autoantibodies, *Differentiation* **36:**234–254.

Holborow, E. J., 1979, Contractile proteins as autoantigens, *Methods Achiev. Exp. Pathol.* **9:**244–260.

Holthöfer, H., Miettinen, A., Paasivuo, R., Lehto, V. P., Linder, E., Alfthan, O., and Virtanen, I., 1983,

Cellular origin and differentiation of renal carcinomas. A fluorescence microscopic study with kidney-specific antibodies, antiintermediate filament antibodies, and lectins, *Lab. Invest.* **49:**317–326.

Holthöfer, H. Miettinen, A., Lehto, V. P., Lehtonen, E., and Virtanen, I., 1984, Expression of vimentin and cytokeratin types of intermediate filament proteins in developing and adult human kidneys, *Lab. Invest.* **50:**552–559.

Horvath, E., and Kovacs, K., 1978, Morphogenesis and significance of fibrous bodies in human pituitary adenomas, *Virchows Arch. B* **27:**69–78.

Hronis, T. S., Steinberg, M. L., Defendi, V., and Sun, T. T., 1984, The simple epithelial nature of some simian-virus-40-transformed human epidermal keratinocytes, *Cancer Res.* **44:**5797–5804.

Huszar, M., Halkin, H., Herczeg, E., Bubis, J., and Geiger, B., 1983, Use of antibodies to intermediate filaments in the diagnosis of metastatic amelanotic malignant melanoma, *Hum. Pathol.* **14:**1006–1008.

Huszar, M., Herczeg, E., Lieberman, Y., and Geiger, B., 1984, Distinctive immunofluorescent labeling of epithelial and mesenchymal elements of carcinosarcoma with antibodies specific for different intermediate filaments, *Hum. Pathol.* **15:**532–538.

Hynes, R. O., and Destree, A. T., 1978, 10 nm filaments in normal and transformed cells, *Cell* **13:**151–163.

Ihrig, T. J., Pettit, N. B., Akeda, S., Benson, N. C., and French, S. W., 1980, Liver injury due to griseofulvin: Gamma glutamyl transpeptidase induction correlated with Mallory bodies, *Fed. Proc.* **39:**876.

Inoue, S., and Dionne, G. P., 1977, Tonofilaments in normal and human bronchial epithelium in squamous cell carcinoma, *Am. J. Pathol.* **88:**345–354.

Irie, T., Benson, N. C., and French, S. W., 1982, Relationship of Mallory bodies to the cytoskeleton of hepatocytes in griseofulvin-treated mice, I Lab. Invest. **47:**336–345.

Irie, T., Benson, N. C., and French, S. W., 1984, Electron microscopic study of the in vitro calcium-dependent degradation of Mallory bodies and intermediate filaments in hepatocytes, *Lab. Invest.* **50:**303–312.

Ishikawa, H. Bischoff, R., and Holtzer, H., 1968, Mitoses and intermediate-sized filaments in developing skeletal muscle, *J. Cell Biol.* **38:**538–555.

Jackson, B. W., Grund, C., Schmid, E., Bürki, K., Franke, W. W., and Illmensee, K., 1980, Formation of cytoskeletal elements during mouse embryogenesis. Intermediate filaments of the cytokeratin type and desmosomes in preimplantation embryos, *Differentiation* **17:**161–179.

Jackson, B. W., Grund, C., Winter, S., Franke, W. W., and Illmensee, K., 1981, Formation of cytoskeletal elements during mouse embryogenesis. II. Epithelial differentiation and intermediate-sized filaments in early postimplantation embryos, *Differentiation* **20:**203–216.

Jahn, L., Fouquet, B., Rohe, K., and Franke, W. W., 1987, Cytokeratins in certain endothelial and smooth muscle cells of two taxonomically distant vertebrate species, Xenopus laevis and man, *Differentiation* **36:**234–254.

James, D. G., and Neville, E., 1977, Pathobiology of sarcoidosis, *Pathobiol. Annu.* **7:**31–61.

Jockusch, B. M., Veldman, H., Griffiths, G. W., Van Oost, B. A., and Jennekens, F. G. I., 1980, Immunofluorescence microscopy of a myopathy. α-actin is a major constituent of nemaline rods, *Exp. Cell Res.* **127:**409–420.

Johnson, G. D., Holborow, E. J., and Glynn, L. E., 1965, Antibody to smooth muscle in patients with liver disease, *Lancet* **2:**878–879.

Johnson, G. D., Carvalho, A., Holborow, E. J., Goddard, D. H., and Russell, G., 1981, Antiperinuclear factor and keratin antibodies in rheumatoid arthritis, *Ann. Rheum. Dis.* **40:**263–266.

Jonasson, L., Holm, J., Skalli, O., Gabbiani, G., and Hansson, G. K., 1985, Expression of class II transplantation antigen on vascular smooth muscle cells in human atherosclerosis, *J. Clin. Invest.* **76:**125–131.

Jonasson, L., Holm, J., Skalli, O., Bondjers, G., and Hansson, G. K., 1986, Regional accumulations of T cells, macrophages, and smooth muscle cells in the human atherosclerotic plaques, *Arteriosclerosis* **6:**131–138.

Jones, J. C. R., Goldman, A. E., Yang, H. Y., and Goldman, R. D., 1985, The organizational fate of intermediate filament networks in two epithelial cell types during mitosis, *J. Cell Biol.* **100:**93–102.

Kanagasundaram, N., Kakumu, S., Chen, T., and Leevy, C. M., 1977, Alcoholic hyalin antigen (AHAg) and antibody (AHAb) in alcoholic hepatitis, *Gastroenterology* **73:**1368–1373.

Karrer, H. E., 1961, An electron microscope study of the aorta in young and aging mice, *J. Ultrastruct. Res.* **5:**1–27.

Kartenbeck, J., Schmid, E., Franke, W. W., and Geiger, B., 1982, Different modes of internalization of proteins associated with adhaerens junctions and desmosomes: Experimental separation of lateral contacts induces endocytosis of desmosomal plaque material, *EMBO J.* **1:**725–732.

Kasper, M., Goertchen, R., Stosiek, P., Perry, G., and Karsten, U., 1986, Coexistence of cytokeratin, vimentin and neurofilament protein in human choroid plexus. An immunohistochemical study of intermediate filaments in neuroepithelial tissues, *Virchows Arch. A* **410:**173–177.

Katsuma, Y., Swierenga, S. H. H., Khettry, U., Marceau, N., and French, S. W., 1987, Changes in the cytokeratin intermediate filament cytoskeleton associated with Mallory body formation in mouse and human liver, *Hepatology* **7:**1215–1233.

Keating, S., and Taylor, G. P., 1985, Undifferentiated (embryonal) sarcoma of the liver: Ultrastructural and immunohistochemical similarities with malignant fibrous histiocytoma, *Hum. Pathol.* **16:**693–699.

Keeley, A. F., Iseri, O. A., and Gottlieb, L. S., 1972, Ultrastructure of hyaline cytoplasmic inclusions in a human hepatoma: Relationship to Mallory's alcoholic hyalin, *Gastroenterology* **62:**280–293.

Keski-Oja, J., Lehto, V. P., Vartio, T., Badley, R. A., and Virtanen, I., 1983, Microfilaments and intermediate filaments in epithelial cells transformed by murine sarcoma or leukemia viruses, *Eur. J. Cell Biol.* **30:**191–199.

Kim, K. H., Schwartz, F., and Fuchs, E., 1984, Differences in keratin synthesis between normal epithelial cells and squamous cell carcinomas are mediated by vitamin A, *Proc. Natl. Acad. Sci. USA* **81:**4280–4284.

Kimoff, R. J., and Huang, S.-N., 1981, Immunocytochemical and immunoelectron microscopic studies on Mallory bodies, *Lab. Invest.* **45:**491–503.

Kobayashi, H., and Hashimoto, K., 1983, Amyloidogenesis in organ-limited cutaneous amyloidosis: An antigenic identity between epidermal keratin and skin amyloid, *J. Invest. Dermatol.* **80:**66–72.

Kocher, O., and Gabbiani, G., 1986, Cytoskeletal features of human normal and atheromatous arterial smooth muscle cells, *Hum. Pathol.* **17:**875–880.

Kocher, O., Amaudruz, M., Schindler, A. M., and Gabbiani, G., 1981, Desmosomes and gap junctions in precarcinomatous and carcinomatous conditions of squamous epithelia. An electron microscopic and morphometrical study, *J. Submicrosc. Cytol.* **13:**267–281.

Kocher, O., Skalli, O., Bloom, W. S., and Gabbiani, G., 1984, Cytoskeleton of rat aortic smooth muscle cells. Normal conditions and experimental intimal thickening, *Lab. Invest.* **50:**645–652.

Kocher, O., Skalli, O., Cerutti, D., Gabbiani, F., and Gabbiani, G., 1985, Cytoskeletal features of rat aortic cells during development. An electron microscopic, immunohistochemical, and biochemical study, *Circ. Res.* **56:**829–838.

Krepler, R., Denk, H., Artlieb, U., and Moll, R., 1982, Immunocytochemistry of intermediate filament proteins present in pleomorphic adenomas of the human parotid gland: *Characterization of different cell type in the same tumor, Differentiation* **21:**191–199.

Krisher, K., and Cunningham, M. W., 1985, Myosin: A link between streptococci and heart, *Science* **227:**413–415.

Kuhn, C., and Kuo, T. T., 1973, Cytoplasmic hyaline in asbestosis. A reaction of injured alveolar epithelium, *Arch. Pathol. Lab. Med.* **95:**190–194.

Kurki, P., and Virtanen, I., 1984, The detection of human antibodies against cytoskeletal components, *J. Immunol. Methods* **67:**209–223.

Kurki, P., and Virtanen, I., 1985, The detection of smooth muscle antibodies reacting with intermediate filaments of desmin type, *J. Immunol. Methods* **76:**329–335.

Kurki, P., Linder, E., Virtanen, I., and Stenman, S., 1977, Human smooth muscle autoantibodies reacting with intermediate (100 Å) filaments, *Nature* **268:**240–241.

Kurki, P., Virtanen, I., Stenman, S., and Linder, E., 1978, Characterization of human smooth muscle autoantibodies reacting with cytoplasmic intermediate filaments, *Clin. Immunol. Immunopathol.* **11:**379–387.

Kurki, P., Miettinen, M., Salaspuro, M., Virtanen, I., and Stenman, S., 1983a, Cytoskeleton antibodies in chronic active hepatitis, primary biliary cirrhosis, and alcoholic liver disease, *Hepatology* **3:**297–302.

Kurki, P., Helve, T., and Virtanen, I., 1983b, Antibodies to cytoplasmic intermediate filaments in rheumatic diseases, *J. Rheumatol.* **10:**558–562.

Lane, E. B., Hogan, B. L. M., Kurkinen, M., and Garrels, J. I., 1983, Co-expression of vimentin and cytokeratins in parietal endoderm cells of early mouse embryo, *Nature* **303:**701–704.

LaRocca, P. J., and Rheinwald, J. G., 1984, Coexpression of simple epithelial keratins and vimentin by human mesothelium and mesothelioma in vivo and in culture, *Cancer Res.* **44:**2991–2999.

Laszlo, A., and Bissell, M. J., 1983, TPA induces simultaneous alterations in the synthesis and organization of vimentin, *Exp. Cell Res.* **148:**221–234.

Lawson, D., 1983, Epinemin: A new protein associated with vimentin filaments in non-neural cells, *J. Cell Biol.* **97:**1891–1905.

Lazarides, E., and Balzer, D. R., Jr., 1978, Specificity of desmin to avian and mammalian muscle cells, *Cell* **14:**429–438.

Lehto, V. P., and Virtanen, I., 1978, Intermediate (10 nm) filaments in human malignant mesothelioma, *Virchows Arch. B* **28:**229–234.

Lehto, V. P., Miettinen, M., and Virtanen, I., 1985, A dual expression of cytokeratin and neurofilaments in bronchial carcinoid cells, *Int. J. Cancer* **35:**421–425.

Lehtonen, E., Lehto, V. P., Vartio, T., Badley, R. A., and Virtanen, I., 1983a, Expression of cytokeratin polypeptides in mouse oocytes and preimplantation embryos, *Dev. Biol.* **100:**158–165.

Lehtonen, E., Lehto, V. P., and Paasivuo, R., and Virtanen, I., 1983b, Parietal and visceral endoderm differ in their expression of intermediate filaments, *EMBO J.* **2:**1023–1028.

Lenk, R., and Penman, S., 1979, The cytoskeletal framework and poliovirus metabolism, *Cell* **16:**289–301.

Li, C. L., Chew, E. C., Huang, D. P., and Ho, H. C., 1980, Intracytoplasmic desmosomes in tumor cells, *Cell Biol. Int. Rep.* **4:**593–597.

Liem, R. K. H., Moraru, E., Chin, S., and Wang, E., 1985, Immunoelectron microscopic localization of antibodies to the neurofilament triplet, *Exp. Cell Res.* **156:**419–428.

Lieska, N., Chen, J., Maisel, H., and Romero-Herrera, A. E., 1980, Subunit characterization of lens intermediate filaments, *Biochim, Biophys. Acta* **626:**136–153.

Linder, E., 1981, Binding of C1q and complement activation by vascular endothelium, *J. Immunol.* **126:**648–658.

Linder, E., 1982, Antibody-independent binding of C1q and activation of serum complement by human skin in vitro, *J. Invest. Dermatol.* **78:**116–150.

Linder, E., Lehto, V. P., Virtanen, I., Stenman, S., and Natvig, J. B., 1977, Localization of amyloid-related serum protein SAA-like material to intermediate (10 nm) filaments of cultured human embryonal fibroblasts, *J. Exp. Med.* **146:**1158–1163.

Linder, E., Kurki, P., and Andersson, L. C., 1979a, Autoantibody to "intermediate filaments" in infectious mononucleosis, *Clin. Immunol. Immunopathol.* **14:**411–417.

Linder, E., Lehto, V. P., and Stenman, S., 1979b, Activation of complement by cytoskeletal intermediate filaments, *Nature* **278:**176–178.

Linder, E., Lehto, V. P., and Virtanen, I., 1979c, Amyloid-like green birefringence in cytoskeletal 10 nm filaments after staining with Congo red, *Acta Pathol. Microbiol. Scand. Sect. A* **87:**299–306.

Linder, E., Hormia, M. I., and Eriksson, A., 1980, Binding and activation of complement by intermediate filaments of both "vimentin" and "desmin" type, *Eur. J. Cell Biol.* **22:**374.

Linder, E., Helin, H., Chang, C. M., and Edgington, T. S., 1983, Complement-mediated binding of monocytes to intermediate filaments in vitro, *Am. J. Pathol.* **112:**267–277.

Lowe, J. A., Blanchard, A., Morrell, K., Lennox, G., Reynolds, L., Billett, M., Landon, M., and Mayer, R. J., 1988, Ubiquitin is a common factor in intermediate filament inclusion bodies of diverse type in man, including those of Parkinson's disease, Pick's disease, and Alzheimer's disease, as well as Rosenthal fibres in cerebellar astrocytomas, cytoplasmic bodies in muscle, and Mallory bodies in alcoholic liver disease, *J. Pathol.* **155:**9–15.

Luisanda, A. V., Kanagasundaram, N., and Leevy, C. M., 1977, Chemical nature of alcoholic hyalin, *Gastroenterology* **73:**1374–1376.

Lux, S. E., 1979, Spectrin–actin membrane skeleton of normal and abnormal red blood cells, *Semin. Hematol.* **16:**21–51.

McGuire, J., Osber, M., and Lightfoot, L., 1984, Two keratins MW 50,000 and 56,000 are synthesized by psoriatic epidermis, *Br. J. Dermatol.* **111**(Suppl. 27):27–37.

Mackay, I. R., 1978, Chronic active hepatitis, cirrhosis and other diseases of the liver, in: *Immunological Diseases,* Volume II (M. Samter, ed.), Little, Brown, Boston, pp. 1454–1477.

McMillan, S. A., and Haire, M., 1979, The specificity of IgG- and IgM-class smooth muscle antibody in the sera of patients with multiple sclerosis and active chronic hepatitis, *Clin. Immunol. Immunopathol.* **14:**256–263.

Mahrle, G., Bolling, R., Osborn, M., and Weber, K., 1983, Intermediate filaments of the vimentin and prekeratin type in human epidermis, *J. Invest. Dermatol.* **81:**46–48.

Masu, S., Hosokowa, M., and Seiji, M., 1980, Immunofluorescence studies on cutaneous amyloidosis with anti-keratin antibody, *Tohoku J. Exp. Med.* **131:**121–122.

Meierhenry, E. F., Ruebner, B. H., Gershwin, M. E., Hsieh, L. S., and Grench, S. W., 1981, Mallory body formation in hepatic nodules of mice ingesting dieldrin, *Lab. Invest.* **44:**392–396.

Meierhenry, E. G., Ruebner, B. H., Gershwin, M. E., Hsieh, L. S., and French, S. W., 1983, Dieldrin-induced Mallory bodies in hepatic tumors of mice of different strains, *Hepatology* **3:**90–95.

Michel, R. P., Limacher, J. J., and Kimoff, R. J., 1982, Mallory bodies in scar adenocarcinoma of the lung, *Hum. Pathol.* **13:**81–85.

Miettinen, M., 1988, Antibody specific to muscle actins in the diagnosis and classification of soft tissue tumors, *Am. J. Pathol.* **130:**205–215.

Miettinen, M., and Virtanen, I., 1984, Synovial sarcoma—a misnomer, *Am. J. Pathol.* **117:**18–25.

Miettinen, M., Lehto, V. P., Badley, R. A., and Virtanen, I., 1982a, Expression of intermediate filaments in soft-tissue sarcomas, *Int. J. Cancer* **30:**541–546.

Miettinen, M., Lehto, V. P., Badley, R. A., and Virtanen, I., 1982b, Alveolar rhabdomyosarcoma. Demonstration of the muscle type of intermediate filament protein, desmin, as a diagnostic aid, *Am. J. Pathol.* **108:**246–251.

Miettinen, M., Lehto, V. P., and Virtanen, I., 1982c, Nasopharyngeal lymphoepithelioma: Histological diagnosis as aided by immunohistochemical demonstration of keratin, *Virchows Arch. B* **40:**163–169.

Miettinen, M., Lehto, V. P., and Virtanen, I., 1982d, Histogenesis of Ewing's sarcoma. An evaluation of intermediate filaments and endothelial cell markers, *Virchows Arch. B* **41:**277–284.

Miettinen, M., Lehto, V. P., and Virtanen, I., 1983a, Monophasic synovial sarcoma of spindle-cell type: Epithelial differentiation as revealed by ultrastructural features, content of prekeratin and binding of peanut agglutinin, *Virchows Arch. B* **44:**187–199.

Miettinen, M., Lehto, V. P., Dahl, D., and Virtanen, I., 1983b, Differential diagnosis of chordoma, chondroid, and ependymal tumors as aided by anti-intermediate filament antibodies, *Am. J. Pathol.* **112;**160–169.

Miettinen, M., Lehto, V. P., and Virtanen, I., 1984a, Antibodies to intermediate filament proteins in the diagnosis and classification of human tumors, *Ultrastruct. Pathol.* **7:**83–107.

Miettinen, M., Lehto, V. P., and Virtanen, I., 1984b, Malignant fibrous histiocytoma within a recurrent chordoma, *Am. J. Clin. Pathol.* **82:**738–743.

Miettinen, M., Lehto, V. P., Ekblom, P., Tasanen, A., and Virtanen, I., 1984c, Leiomyosarcoma of the mandible: Diagnosis as aided by immunohistochemical demonstration of desmin and laminin, *J. Oral Pathol.* **13:**373–381.

Miettinen, M., Franssila, K., Lehto, V. P., Paasivuo, R., and Virtanen, I., 1984d, Expression of intermediate filament proteins in thyroid gland and thyroid tumors, *Lab. Invest.* **50:**262–270.

Mietinen, M., Lehto, V. P., and Virtanen, I., 1985a, Immunofluorescence microscopic evaluation of the intermediate filament expression of the adrenal cortex and medulla and their tumors, *Am. J. Pathol.* **118:**360–366.

Miettinen, M., Lehto, V. P., Dahl, D., and Virtanen, I., 1985b, Varying expression of cytokeratin and neurofilaments in neuroendocrine tumors of human gastrointestinal tract, *Lab. Invest.* **52:**429–436.

Molenaar, W. M., Oosterhuis, J. W., Oosterhuis, A. M., and Ramaekers, F. C. S., 1985, Mesenchymal and muscle-specific intermediate filaments (vimentin and desmin) in relation to differentiation in childhood rhabdomyosarcomas, *Hum. Pathol.* **16:**838–843.

Moll, R., Franke, W. W., Schiller, D. L., Geiger, B., and Krepler, R., 1982, The catalog of human cytokeratins: Patterns of expression in normal epithelia, tumors and cultured cells, *Cell* **31:**11–24.

Moll, R., Krepler, R., and Franke, W. W., 1983, Complex cytokeratin polypeptide patterns observed in certain human carcinomas, *Differentiation* **23:**256–269.

Moll, R., Moll, I., and Franke, W. W., 1984a, Identification of Merkel cells in human skin by specific cytokeratin antibodies: Changes of cell density and distribution in fetal and adult plantar epidermis, *Differentiation* **28:**136–154.

Moll, R., Moll, I., and Franke, W. W., 1984b, Differences of expression of cytokeratin polypeptides in various epithelial skin tumors, *Arch. Dermatol. Res.* **276:**349–363.

Molony, L., Hagen, P. O., and Schachat, F. H., 1986, Intermediate filament heterogeneity in normal and hypercholesterolemic rabbit vascular smooth muscle cells, *Exp. Cell Res.* **163:**78–86.

Monroe, S., French, S. W., and Zamboni, L., 1973, Mallory bodies in a case of primary biliary cirrhosis: An ultrastructural and morphogenetic study, *Am. J. Clin. Pathol.* **59:**254–262.

Morris, A., Steinberg, M. L., and Defendi, V., 1985, Keratin gene expression in simian virus 40-transformed human keratinocytes, *Proc. Natl. Acad. Sci. USA* **82:**8498–8502.

Mortazavi-Milani, S. M., Stierle, H. E., and Holborow, E. J., 1982, In vitro induction of anti-intermediate filament antibody in lymphocyte cultures by Epstein–barr virus, *Immunol. Lett.* **5:**203–205.

Mortazavi-Milani, S. M., Badakere, S. S., and Holborow, E. J., 1984, Antibody to intermediate filaments of the cytoskeleton in the sera of patients with acute malaria, *Clin. Exp. Immunol.* **55:**177–182.

Morton, J. A., Fleming, K. A., Trowell, J. M., and McGee, J. O., 1980, Mallory bodies—immunohistochemical detection by antisera to unique non-prekeratin components, *Gut* **21:**727–733.

Morton, J. A., Bastin, J., Fleming, K. A., McMichael, A., Burns, J., and McGee, J. O., 1981, Mallory bodies in alcoholic liver disease: Identification of cytoplasmic filament/cell membrane and unique antigenic determinants by monoclonal antibodies, *Gut* **22:**1–7.

Moss, N. S., and Benditt, E. P., 1970, Spontaneous and experimentally induced arterial lesions. I. An ultrastructural survey of the normal chicken aorta, *Lab. Invest.* **22:**166–183.

Murti, K. G., and Goorha, R., 1983, Interaction of frog virus-3 with the cytoskeleton. I. Altered organization of microtubules, intermediate filaments, and microfilaments, *J. Cell Biol.* **96:**1248–1257.

Nagle, R. B., McDaniel, K. M., Clark, V. A., and Payne, C. M., 1983, The use of antikeratin antibodies in the diagnosis of human neoplasms, *Am. J. Clin. Pathol.* **79:**458–466.

Naharro, G., Robbins, K. C., and Reddy, E. P., 1984, Gene product of v-fgr onc: Hybrid protein containing a portion of actin and a tyrosine-specific protein kinase, *Science* **223:**63–66.

Naito, S., and Matsumoto, S., 1978, Identification of cellular actin within the rabies virus, *Virology* **91:**151–163.

Nakanuma, Y., and Ohta, G., 1986, Expression of Mallory bodies in hepatocellular carcinoma in man and its significance, *Cancer* **57:**81–86.

Neimark, H. C., 1977, Extraction of an actin-like protein from the prokaryote Mycoplasma pneumoniae, *Proc. Natl. Acad. Sci. USA* **74:**4041–4045.

Nelson, W. G., Battifora, H., Santana, H., and Sun, T. T., 1984, Specific keratins as molecular markers for neoplasms with a stratified epithelial origin, *Cancer Res.* **44:**1600–1603.

Nelson, W. J., and Traub, P., 1981, Properties of a Ca^{2+}-activated protease specific for the intermediate-sized filament protein vimentin in Ehrlich ascites tumour cells, *Eur. J. Biochem.* **116;**51–57.

Nelson, W. J., and Traub, P., 1982, Intermediate (10 nm) filament proteins and the Ca^{2+}-activated proteinase specific for vimentin and desmin in the cells from fish to man: An example of evolutionary conservation, *J. Cell Sci.* **57:**25–49.

Neumann, P. E., Goldman, J. E., Horoupian, D. S., and Hess, M. A., 1985, Fibrous bodies in growth hormone-secreting adenomas contain cytokeratin filaments, *Arch. Pathol. Lab. Med.* **109:**505–508.

Neustein, H. B., Nickerson, B., and O'Neal, M., 1980, Kartagener's syndrome with absence of inner dynein arms of respiratory cilia, *Am. Rev. Resp. Dis.* **122:**979–981.

Norkin, S. A., and Campagna-Pinto, D., 1968, Cytoplasmic hyaline inclusions in hepatoma, *Arch. Pathol.* **86:**25–32.

Okamura, K., Harwood, T. R., and Yokoo, H., 1975, Isolation and electrophoretic study on Mallory bodies from the livers of alcoholic cirrhosis, *Lab. Invest.* **33:**193–199.

Okanoue, T., Ohta, M., Ou, O., Kachi, K., Kagawa, K., Yuki, T., Okuno, T., Takino, T., and French, S. W., 1985, Relationship of Mallory bodies to intermediate filaments in hepatocytes. A scanning electron microscopic study, *Lab. Invest.* **53:**534–540.

Osborn, M., and Goebel, H. H., 1983, The cytoplasmic bodies in a congenital myopathy can be stained with antibodies to desmin, the muscle-specific intermediate filament protein, *Acta Neuropathol.* **62:**149–152.

Osborn, M., and Weber, K., 1983, Tumor diagnosis by intermediate filament typing: A novel tool for surgical pathology, *Lab. Invest.* **48:**372–394.

Osborn, M., Franke, W. W., and Weber, K., 1977, Visualization of a system of filaments 7–10 nm thick in cultured cells of an epithelioid line (Pt K2) by immunofluorescence microscopy, *Proc. Natl. Acad. Sci. USA* **74:**2490–2494.

Osborn, M., Caselitz, J., and Weber, K., 1981, Heterogeneity of intermediate filament expression in vascular smooth muscle: A gradient in desmin positive cells from the rat aortic arch to the level of the arteria iliaca communis, *Differentiation* **20:**196–202.

Osborn, M., Geisler, N., Shaw, G., Sharp, G., and Weber, K., 1982, Intermediate filaments, *Cold Spring Harbor Symp. Quant. Biol.* **46:**413–429.

Osborn, M., Caselitz, J., Püschel, K., and Weber, K., 1987, Intermediate filament expression in human vascular smooth muscle and in arteriosclerotic plaques, *Virchows Arch. A.* **411:**449–458.

Osung, O. A., Chandra, M., and Holborow, E. J., 1982a, Intermediate filaments in synovial lining cells in rheumatoid arthritis and other arthritides are of vimentin type, *Ann. Rheum. Dis.* **41:**74–77.

Osung, O. A., Chandra, M., and Holborow, E. J., 1982b, Antibody to intermediate filaments of the cytoskeleton in rheumatoid arthritis, *Ann. Rheum. Dis.* **41:**69–73.

Page, D. L., Whelan Weiss, S., and Eggleston, J. C., 1975, Ultrastructural study of amyloid material in the calcifying epithelial ondontogenic tumor, *Cancer* **36:**1426–1435.

Palmberg, L., Sjölund, M., and Thyberg, J., 1985, Phenotype modulation in primary cultures of arterial smooth-muscle cells: Reorganization of the cytoskeleton and activation of synthetic activities, *Differentiation* **29:**275–283.

Pauli, B. U., Cohen, S. M., Alroy, J., and Weinstein, R. S., 1978, Desmosome ultrastructure and the biological behavior of chemical carcinogen-induced urinary bladder carcinomas, *Cancer Res.* **38:**3276–3285.

Pease, D. C., and Paule, W. J., 1960, Electron microscopy of elastic arteries. The thoracic aorta of the rat, *J. Ultrastruct. Res.* **3:**469–483.

Pedersen, H., and Mygind, N., 1976, Absence of axonemal arms in nasal mucosa cilia in Kartagener's syndrome, *Nature* **262:**494–495.

Pedersen, J. S., Toh, B. H., Locarnini, S. A., Gust, I. D., and Shyamala, G. N. S., 1981, Autoantibody to intermediate filaments in viral hepatitis, *Clin. Immunol. Immunopathol.* **21:**154–161.

Pedersen, J. S., Toh, B. H., Mackay, I. R., Tait, B. D., Gust, I. D., Kastelan, A., and Hadzic, N., 1982, Segregation of autoantibody to cytoskeletal filaments, actin and intermediate filaments with two types of chronic active hepatitis, *Clin. Exp. Immunol.* **48:**527–532.

Phillips, M. J., 1982, Mallory bodies and the liver, *Lab. Invest.* **47:**311–313.

Pitelka, D. R., Hamamoto, S. T., and Taggart, B. N., 1980, Epithelial cell junctions in primary and metastatic mammary tumors of mice, *Cancer Res.* **40:**1588–1599.

Porte, A., Stoeckel, M.-E., Sacrez, A., and Batzenschlager, A., 1980, Unusual familial cardiomyopathy with storage of intermediate filaments in the cardiac muscular cells, *Virchows Arch. (Pathol Anat)* **386:**43–58.

Price, M. G., and Sanger, J. W., 1983, Intermediate filaments in striated muscle. A review of structural studies in embryonic and adult skeletal and cardiac muscle, in: *Cell and Muscle Motility,* Volume 3 (R. M. Dowben and J. W. Shaw, eds.), Plenum Press, New York, pp. 1–40.

Pruss, R. M., Mirsky, R., Raff, M. C., Thorpe, R., Dowding, A. J., and Anderton, B. H., 1981, All classes of intermediate filaments share a common antigenic determinant defined by a monoclonal antibody, *J. Cell Biol.* **27:**419–428.

Purtilo, D. T., and Gottlieb, L. S., 1973, Cirrhosis and hepatoma occurring at Boston City Hospital (1917–1968), *Cancer* **32:**458–462.

Quismorio, F. P., Kaufman, R. L., Beardmore, T., and Mongan, E. S., 1983, Reactivity of serum antibodies to the keratin layer of rat esophagus in patients with rheumatoid arthritis, *Arthritis Rheum.* **26:**494–499.

Ramaekers, F. C. S., Osborn, M., Schmid, E., Weber, K., Bloemendal, H., and Franke, W. W., 1980, Identification of the cytoskeletal proteins in lens-forming cells, a special epithelioid cell type, *Exp. Cell Res.* **127:**309–327.

Ramaekers, F. C. S., Boom-Kens, T. R., and Bloemendal, H., 1981, Cytoskeletal and contractile structures in bovine lens cell differentiation, *Exp. Cell Res.* **135:**454–461.

Ramaekers, F. C. S., Puts, J. J. G., Moesker, O., Kant, A., Huysmans, A., Haag, D., Jap, P. H. K., Herman, C. J., and Vooijs, G. P., 1983a, Antibodies to intermediate filament proteins in the immunohistochemical identification of human tumours: An overview, *Histochem. J.* **15:**691–713.

Ramaekers, F. C. S., Verheijen, R. H. M., Moesker, O., Kant, A., Vooijs, G. P., and Herman, C. J., 1983b, Mesodermal mixed tumor. Diagnosis by analysis of intermediate filament proteins, *Am. J. Surg. Pathol.* **7:**381–385.

Ramaekers, F. C. S., Haag, D., Kant, A., Moesker, O., Jap, P. H. K., and Vooijs, G. P., 1983c, Coexpression of keratin- and vimentin-type intermediate filaments in human metastatic carcinoma cells, *Proc. Natl. Acad. Sci. USA* **80:**2618–2622.

Ramaekers, F. C. S., Puts, J. J. G., Moesker, O., Kant, A., Vooijs, G. P., and Jap, P. H. K., 1983d, Intermediate filaments in malignant melanomas. Identification and use as marker in surgical pathology, *J. Clin. Invest.* **71:**635–643.

Ramaekers, F., Feitz, W., Moesker, O., Schaart, G., Herman, C., Debruyne, F., and Vooijs, P., 1985a, Antibodies to cytokeratin and vimentin in testicular tumour diagnosis, *Virchows Arch. A* **408:**127–142.

Ramaekers, F., Huysmans, A., Moesker, O., Schaart, G., Herman, C., and Vooijs, P., 1985b, Cytokeratin expression during neoplastic progression of human transitional cell carcinomas as detected by a monoclonal and a polyclonal antibody, *Lab. Invest.* **52:**31–38.

Rappaport, L., Contard, F., Samuel, J. L., Delcayre, C., Marotte, F., Tomé, F., and Fardeau, M., 1988, Storage of phosphorylated desmin in a familial myopathy, *FEBS Lett.* **231**:421–425.

Regauer, S., Franke, W. W., and Virtanen, I., 1985, Intermediate filament cytoskeleton of amnion epithelium and cultured amnion epithelial cells: Expression of epidermal cytokeratins in cells of a simple epithelium, *J. Cell Biol.* **100**:997–1009.

Reznick-Schüller, H., 1977, Ultrastructural alterations of APUD cells during nitrosamine-induced lung carcinogenesis, *J. Pathol.* **121**:79–82.

Roy, S., 1978, Cytoplasmic filamentous masses in chromophobe adenoma of the human pituitary gland, *J. Pathol.* **125**:151–154.

Rubin, R. W., Howard, J., and Leonardi, C., 1979, A biochemical and ultrastructural comparison of Triton X-100 models of normal and transformed cells, *Tissue Cell* **11**:413–423.

Rungger-Brändle, E., and Gabbiani, G., 1983, The role of cytoskeletal and cytocontractile elements in pathologic processes, *Am. J. Pathol.* **110**:359–392.

Saurat, J. H., Didierjean, L., Skalli, O., Siegenthaler, G., and Gabbiani, G., 1984, The intermediate filament proteins of rabbit normal epidermal Merkel cells are cytokeratins, *J. Invest. Dermatol.* **83**:431–435.

Schindler, A. M., Amaudruz, M. A., Kocher, O., Riotton, G., and Gabbiani, G., 1982, Desmosomes and gap-junctions in various epidermoid preneoplastic and neoplastic lesions of the cervix uteri, *Acta Cytol.* **26**:466–470.

Schmid, E., Osborn, M., Rungger-Brändle, E., Gabbiani, G., Weber, K., and Franke, W. W., 1982, Distribution of vimentin and desmin filaments in smooth muscle tissue of mammalian and avian aorta, *Exp. Cell Res.* **137**:329–340.

Schmidt, R. A., Cone, R., Haas, J. E., and Gown, A. M., 1988, Diagnosis of rhabdomyosarcomas with HHF35, a monoclonal antibody directed against muscle actins, *Am. J. Pathol.* **131**:19–28.

Schuler, G., Hintner, H., Wolff, K., Fritsch, P., and Stingl, G., 1982, In vitro complement binding on cytoplasmic structures in normal human skin: Immunoelectronmicroscopic studies, *J. Cell Biol.* **95**:543–551.

Schürch, W., Seemayer, T. A., Lagacé, R., and Gabbiani, G., 1984, The intermediate filament cytoskeleton of myofibroblasts: An immunofluorescence and ultrastructural study, *Virchows Arch. A* **403**:323–336.

Schürch, W., Skalli, O., Seemayer, T. A., and Gabbiani, G., 1987, Intermediate filament proteins and actin isoforms as markers for soft tissue tumor differentiation and origin. I. Smooth muscle tumors, *Am. J. Pathol.* **128**:91–103.

Scott, D. L., Delamere, J. P., Jones, L. J., and Walton, K. W., 1981, Significance of laminar antikeratin antibodies to rat oesophagus in rheumatoid arthritis, *Ann. Rheum. Dis.* **40**:267–271.

Serre, G., Vincent, C., Viraben, R., and Soleilhavoup, J. P., 1987, Natural IgM and IgG autoantibodies to epidermal keratins in normal human sera. I: ELISA-titration, immunofluorescence study, *J. Invest. Dermatol.* **88**:21–27.

Sharpe, A. H., Chen, L. B., Murphy, J. R., and Fields, B. N., 1980, Specific disruption of vimentin filament organization in monkey kidney CV-1 cells by diphteria toxin, exotoxin A, and cycloheximide, *Proc. Natl. Acad. Sci. USA* **77**:7267–7271.

Siebert, P. D., and Fukuda, M., 1985, Induction of cytoskeletal vimentin and actin gene expression by a tumor-promoting phorbol ester in the human leukemic cell line K562, *J. Biol. Chem.* **260**:3868–3874.

Singh, B., Goldman, R., Hutton, L., Herzog, N. K., and Arlinghaus, R. B., 1987, The P55 protein affected by v-mos expression is vimentin, *J. Virol.* **61**:3625–3629.

Skalli, O., Ropraz, P., Trzeciak, A., Benzonana, G., Gillessen, G., and Gabbiani, G., 1986a, A monoclonal antibody against α-smooth muscle actin: A new probe for smooth muscle differentiation, *J. Cell Biol.* **103**:2787–2796.

Skalli, O., Bloom, W. S., Ropraz, P., Azzarone, B., and Gabbiani, G., 1986b, Cytoskeletal remodeling of rat aortic smooth muscle cells "in vitro": Relationships to culture conditions and analogies to "in vivo" situations, *J. Submicrosc. Cytol.* **18**:481–493.

Skalli, O., Vandekerckhove, J., and Gabbiani, G., 1987, Actin isoform pattern as a marker of normal or pathological smooth-muscle and fibroblastic tissues, *Differentiation* **33**:232–238.

Skalli, O., Gabbiani, G., Babaï, F., Seemayer, T. A., Pizzolato, G., and Schürch, W., 1988, Intermediate filament proteins and actin isoforms as markers for soft tissue tumor differentiation and origin. II. Rhabdomyosarcomas, *Am. J. Pathol.* **130**:515–531.

Stamenkovic, I., Skalli, O., and Gabbiani, G., 1986, Distribution of intermediate filament proteins in normal and diseased human glomeruli, *Am. J. Pathol.* **125**:465–475.

Stanislawsky, L., Mongiat, F., and Moura Neto, V., 1984, Presence of actin in oncornaviruses, *Biochem. Biophys. Res. Commun.* **118:**580–586.

Steinert, P. M., Jones, J. C. R., and Goldman, R. D., 1984, Intermediate filaments, *J. Cell Biol.* **99:**22s–27s.

Stoeckel, M.-E., Osborn, M., Porte, A., Sacrez, A., Batzenschlager, A., and Weber, K., 1981, An unusual familial cardiomyopathy characterized by aberrant accumulations of desmin-type intermediate filaments, *Virchows Arch. A* **393:**53–60.

Summerhayes, I. C., Cheng, Y. S. E., Sun, T. T., and Chen, L. B., 1981, Expression of keratin and vimentin intermediate filaments in rabbit bladder epithelial cells at different stages of benzo(a)pyrene-induced neoplastic progression, *J. Cell Biol.* **90:**63–69.

Takaki, Y., Masutani, M., and Kawada, A., 1971, Electron microscopic study of keratoacanthoma, *Acta Derm. Venereol.* **51:**21–26.

Tazawa, J., Irie, T., and French, S. W., 1983, Mallory body formation runs parallel to γ-glutamyl transferase induction in hepatocytes of griseofulvin-fed mice, *Hepatology* **3:**989–1001.

Tazzari, P. L., Gobbi, M., Zauli, D., Tassinari, A., Crespi, C., Miserocchi, F., Dinota, A., Bandini, G., Ricci, P., and Tura, S., 1987, Close association between antibodies to cytoskeletal intermediate filaments, and chronic graft-versus-host disease, *Transplantation* **44:**234–236.

Thiedemann, K.-U., and Ferrans, V. J., 1976, Ultrastructure of sarcoplasmic reticulum in atrial myocardium of patients with mitral valvular disease, *Am. J. Pathol.* **83:**1–38.

Thornell, L.-E., and Bjelle, A., 1981, Eosinophilic fasciitis: An ultrastructural and immunohistochemical study of the intermediate filament protein skeletin in regenerating muscle fibres, *Neuropathol. Appl. Neurobiol.* **7:**435–449.

Thornell, L.-E., Edström, L., Eriksson, A., Henriksson, K.-G., and Angqvist, K.-A., 1980, The distribution of intermediate filament protein (skeletin) in normal and diseased human skeletal muscle, *J. Neurol. Sci.* **47:**153–170.

Thornell, L.-E., Eriksson, A., and Edström, L., 1983, Intermediate filaments in human myopathies, in: *Cell and Muscle Motility,* Volume 4 (R. M. Dowben and J. W. Shay, eds.), Plenum Press, New York, pp. 85–136.

Tinberg, H. M., 1981a, Intermediate filaments: Analysis of filamentous aggregates induced by griseofulvin, an antitubulin agent, *Biochem. Biophys. Res. Commun.* **99:**458–465.

Tinberg, H. M., 1981b, Intermediate filaments: Immunochemical comparison of major polypeptides of alcoholic hyalin, *FEBS Lett.* **125:**53–59.

Tinberg, H. M., and Mednick, D. L., 1981, Immunochemical comparison of prekeratin and alcoholic hyalin intermediate filaments, *Biochem. Biophys. Res. Commun.* **102:**867–876.

Tinberg, H. M., Regan, R. J., Geier, E. A., Peterson, G. E., and French, S. W., 1978, Isolation of hepatocellular hyalin and electrophoretic resolution of polypeptide components, *Lab. Invest.* **39:**483–490.

Toccanier-Pelte, M.-F., Skalli, O., Kapanci, Y., and Gabbiani, G., 1987, Characterization of stromal cells with myoid features in lymph nodes and spleen in normal and pathologic conditions, *Am. J. Pathol.* **129:**109–118.

Toh, B. H., 1979, Smooth muscle autoantibodies and autoantigens, *Clin. Exp. Immunol.* **38:**621–628.

Toh, B. H., Yildiz, A., Sotelo, J., Osung, O., Holborow, E. J., Kanakoudi, F., and Small, J. V., 1979, Viral infections and IgM autoantibodies to cytoplasmic intermediate filaments, *Clin. Exp. Immunol.* **37:**76–82.

Traub, P., 1985, *Intermediate Filaments: A Review,* Springer-Verlag, Berlin, pp. 150–163.

Traub, U. E., Nelson, W. J., and Traub, P., 1983, Polyacrylamide gel electrophoretic screening of mammalian cells cultured in vitro for the presence of the intermediate filament protein vimentin, *J. Cell Sci.* **62:**129–147.

Travo, P., Weber, K., and Osborn, M., 1982, Co-existence of vimentin and desmin type intermediate filaments in a subpopulation of adult rat vascular smooth muscle cells growing in primary culture, *Exp. Cell Res.* **139:**87–94.

Tsukuda, T., McNutt, M. A., Ross, R., and Gown, A. M., 1987, HHF35, a muscle actin-specific monoclonal antibody. II. Reactivity in normal, reactive, and neoplastic human tissues, *Am. J. Pathol.* **127:**389–402.

Tyrrell, D. L. J., and Norrby, E., 1978, Structural polypeptides of measles virus, *J. Gen. Virol.* **39:**219–229.

Vandekerckhove, J., Osborn, M., Altmannsberger, M., and Weber, K., 1987, Actin typing of rhabdomyosarcomas shows the presence of the fetal and adult forms of sarcomeric muscle actin, *Differentiation* **35:**126–131.

Van Muijen, G. N. P., Ruiter, D. J., and Warnaar, S. O., 1987, Coexpression of intermediate filament polypeptides in human fetal and adult tissues, *Lab. Invest.* **57:**359–369.

Venetianer, A., Schiller, D. L., Magin, T., and Franke, W. W., 1983, Cessation of cytokeratin expression in a rat hepatoma cell line lacking differentiated functions, *Nature* **305:**730–733.

Virtanen, I., Lehto, V. P., Lehtonen, E., Vartio, T., Stenman, S., Kurki, P., Wager, O., Small, J. V., Dahl, D., and Badley, R. A., 1981, Expression of intermediate filaments in cultured cells, *J. Cell Sci.* **50:**45–63.

Virtanen, I., Miettinen, M., and Lehto, V. P., 1984, Cytoskeletal proteins as tissue markers for cancer, *Cancer Bull.* **36:**174–178.

Vogel, A. M., Gown, A. M., Caughlan, J., Haas, J. E., and Beckwith, J. B., 1984, Rhabdoid tumors of the kidney contain mesenchymal specific and epithelial specific intermediate filament proteins, *Lab. Invest.* **50:**232–238.

Waldherr, R., and Schwechheimer, K., 1985, Co-expression of cytokeratin and vimentin intermediate-sized filaments in renal cell carcinomas. Comparative study of the intermediate-sized filament distribution in renal cell carcinomas and normal human kidney, *Virchows Arch. A* **408:**15–27.

Wang, E., 1985, Are cross-bridging structures involved in the bundle formation of intermediate filaments and the decrease in locomotion that accompany cell aging? *J. Cell Biol.* **100:**1466–1473.

Wang, E., and Gundersen, D., 1984, Increased organization of cytoskeleton accompanying the aging of human fibroblasts in vitro, *Exp. Cell Res.* **154:**191–202.

Wang, E., Wolf, B. A., Lamb, R. A., Choppin, P. W., and Goldberg, A. R., 1976, The presence of actin in enveloped viruses, in: *Cell Mobility* (R. Goldman, T. Pollard, and Y. Rosenbaum, eds.), Cold Spring Harbor Laboratory, Cold Spring Harbor, N.Y., pp. 589–600.

Wang, N. S., Seemayer, T. A., Ahmed, M. N., and Knaack, J., 1976, Giant cell carcinoma of the lung. A light and electron microscopic study, *Hum. Pathol.* **7:**3–16.

Warnock, M. L., Press, M., and Churg, A., 1980, Further observations on cytoplasmic hyaline in the lung, *Hum. Pathol.* **11:**59–65.

Weber, K., Wehland, J., and Herzog, W., 1976, Griseofulvin interacts with microtubules both in vivo and in vitro, *J. Mol. Biol.* **102:**817–829.

Weinstein, R. S., Merk, F. B., and Alroy, J., 1976, The structure and function of intercellular junctions in cancer, *Adv. Cancer Res.* **23:**23–89.

Weiss, R. A., Eichner, R., and Sun, T. T., 1984, Monoclonal antibody analysis of keratin expression in epidermal diseases: A 48 kd and a 56 kd keratin as molecular markers for hyperproliferative keratinocytes, *J. Cell Biol.* **98:**1397–1406.

Wetzel, W. J., and Alexander, R. W., 1979, Focal nodular hyperplasia of the liver with alcoholic hyalin bodies and cytologic atypia, *Cancer* **44:**1322–1326.

Whitehouse, J. M. A., Ferguson, N., and Currie, G. A., 1974, Antibody to microtubules in infectious mononucleosis, *Clin. Exp. Immunol.* **17:**227–235.

Wiggers, K. D., French, S. W., French, B. A., and Carr, B. N., 1973, The ultrastructure of Mallory body filaments, *Lab. Invest.* **29:**652–658.

Woolf, N., 1982, Factors that may affect the extent and severity of atherosclerosis, in: *Pathology of Atherosclerosis* (N. Woolf, ed.), Butterworth Scientific, London, pp. 153–186.

Yildiz, A., Toh, B. H., Sotelo, J., Osung, O., Holborow, E. J., and Small, J. V., 1980, Smooth muscle autoantibodies reacting with cytoplasmic intermediate filaments in sera from normal, nonimmunized rabbits, *Clin. Immunol. Immunopathol.* **16:**279–286.

Yokoi, Y., Namihisa, T., Kuroda, H., Komatsu, I., Miyazaki, A., Watanabe, S., and Usui, K., 1984, Immunocytochemical detection of desmin in fat-storing cells (Ito cells), *Hepatology* **4:**709–714.

Yokoo, H., Minick, O. T., Batti, F., and Kent, G., 1972, Morphologic variants of alcoholic hyalin, *Am. J. Pathol.* **69:**25–40.

Young, B. J. J., Mallya, R. K., Leslie, R. D., Clark, C. J. M., and Hamblin, T. J., 1979, Anti-keratin antibodies in rheumatoid arthritis, *Br. Med. J.* **2:**97–99.

Zehner, Z. E., and Paterson, B. M., 1983, Vimentin gene expression during myogenesis: Two functional transcripts from a single copy gene, *Nucleic Acids Res.* **11:**8317–8332.

Zucker-Franklin, D., 1974, Properties of the Sezary lymphoid cell. An ultrastructural analysis, *Mayo Clin. Proc.* **49:**567–574.

18

The Cytoskeleton in Neurological Disease

Michael L. Shelanski and Carol M. Troy

1. Introduction

The history of the microscopic study of diseases of the nervous system is little older than the appreciation that these maladies are often accompanied by alterations in fibrillar or reticular elements within nerve cells. The most common of these alterations is known as neurofibrillary degeneration. This change, marked by an increase in the apparent number and thickness of the argentophilic fibrillar elements of the neuronal cytoplasm, is seen in Alzheimer's disease, amyotrophic lateral sclerosis, lathyrism, sporadic motor neuron disease, and a number of drug- and toxin-induced neuropathies.

2. Neurofibrillary Degeneration

The introduction of silver impregnation methods for staining neurons at the end of the 19th century revealed a delicate fibrillar network in the neuronal cell body which extended into its processes, especially the axon. This confirmed earlier work that described similar networks and had even postulated that such fibrils might run from one neuron to the other and subserve connectivity. It was not long before these staining methods were applied to tissue from patients who had succumbed to neurological diseases. One of the earliest to do this was the German neuropathologist Alois Alzheimer, who reported characteristic lesions in patients who had died of a rapidly progressive presenile dementia (Alzheimer, 1907). This entity, which we now know as *Alzheimer's disease* (AD), showed silver-positive intraneuronal inclusions or "tangles" in the cell body and the proximal axon and dendrites. These tangles are also found in clusters of

Michael L. Shelanski and Carol M. Troy • Department of Pathology and Center for Neurobiology and Behavior, College of Physicians and Surgeons, Columbia University, New York, New York 10032.

degenerating nerve terminals, where they form part of the senile plaque. The plaques and tangles are the histopathological hallmarks of AD. With time it has become apparent that a variety of idiopathic, toxic, and iatrogenic diseases of the nervous system have neurofibrillary degeneration as a major morphological feature (Wisniewski *et al.*, 1970).

The advent of electron microscopy revealed that these neurofibrillary degenerations fell into two major categories. In the first, comprising all of the diseases listed above with the exception of AD and certain related conditions, the "tangle" was found to be made up of bundles of 10-nm filaments indistinguishable in appearance from normal neurofilaments. In the second group, which includes AD, the Guam parkinsonism–dementia complex, some cases of Down's syndrome (mongolism), and dementia pugilistica, the "tangle" is composed of filaments that differ in morphology from the normal cytoskeletal elements. These structures were originally referred to as "twisted tubules" (Terry, 1963; Kidd, 1963) but are now commonly called paired helical filaments (PHF). The PHF are made up of two helically wound filaments of 9- to 11-nm diameter that cross each other every 70 to 80 nm. The PHF are approximately 20 nm wide at their widest and 10 nm at the crossover point.

In addition to normal intermediate filaments (IF) and PHF, straight filaments of 15-nm diameter are also seen in some cases of AD as are specialized inclusions known as Hirano bodies.

In this chapter we will review existing evidence on the chemical composition of tangles, on their mode of formation, and on their effect on the economy of the neuron. We will start with the diseases in which the tangle is composed of filaments of normal morphology and then discuss the PHF and AD against this background. Finally, we will discuss briefly other intraneuronal inclusions in which cytoskeletal elements are found.

2.1. *Neurofibrillary Proliferations Composed of 10-nm IF*

The lack of a suitable model for AD has led to extensive study of experimental conditions in which argentophilic neurofibrillary tangles can be induced in neurons. While these tangles differ in their ultrastructure from AD, there are certain similarities in their morphology at the light microscopic level. They also are remarkably similar at both the light and electron microscopic level to human diseases such as amyotrophic lateral sclerosis (Hirano *et al.*, 1984) and sporadic motor neuron disease (Schochet *et al.*, 1969), in which the tangles are composed of 10-nm filaments. Among the agents that induce this type of change are aluminum (Klatzo *et al.*, 1965), mitotic spindle poisons (Wisniewski *et al.*, 1968; Ghetti and Gambetti, 1983) (colchicine, vinca alkaloids, podophyllotoxin, nocodazole, and maytansine), IDPN (Chou and Hartmann, 1965), a toxin that induces lathyrsin in animals, acrylamide (Jones and Cavanagh, 1984), and a number of other agents.

2.2. *Biochemical Composition of Neurofibrillary Tangles*

At the time neurofibrillary tangles were first described, the chemical composition of the normal neurofilament was not known. It was clear from electron micrographs that neurofilaments differed morphologically from the IF of astroglia (Wuerker, 1970), but they did not appear much different from IF in a variety of cultured nonneural cells. The composition of neurofilaments was first inferred by Hoffman and Lasek (1975), who found three radioactively labeled, axonally transported proteins with apparent molecular weights of 68,000, 150,000, and 200,000 in peripheral nerve. This identification was confirmed by direct biochemical separation and analyses (Liem *et al.*, 1978; Schlaepfer, 1977). The filaments are composed of a core of a 70k protein (NF-L), which in turn is surrounded by two additional proteins with apparent molecular weights on gel electrophoresis of 150k (NF-M) and 200k (NF-H) (Liem and Hutchison, 1982).

Neurofibrillary proliferation was induced in neurons by the intrathecal administration of aluminum salts. When these neurons were isolated and compared electrophoretically to isolated neurons from control animals, an enrichment in each of the neurofilament triplet proteins was found. With the exception of an additional peptide at about 56–58k, the neurofilament proteins were the only ones increased in these cells (Selkoe *et al.*, 1975). This selective increase in neurofilament proteins correlates well with the increased protein content of these neurons as compared to controls (Embree *et al.*, 1967). In a variety of other studies, antibodies against one or more of the neurofilament components have been used to demonstrate that the "tangles" are composed of apparently normal neurofilament proteins (Goldman and Yen, 1986).

The existence of an apparently normal composition does not rule out a variety of posttranslational modifications in the neurofilaments in these conditions. All three of the triplet subunits can be phosphorylated, with NF-M and NF-H being more highly phosphorylated than NF-L (Julien and Mushynski, 1983). Even minor alterations in phosphorylation could alter interactions of the neurofilaments with other cytoskeletal elements and alter the integrity of the cytoskeleton. These factors will be considered in the sections that follow, but it is broadly correct to say that the tangles in these conditions are composed of normal IF—that is to say, they are not the result of aggregation of a new gene product nor the result of the turning on of another member of the IF class.

2.3. Mechanism of Formation of Neurofibrillary Tangles

The accumulation of neurofilaments into tangles could be the result of one or more of the following: (1) the collapse of filaments, which are normally widely dispersed, into readily visible bundles; (2) an increase in the synthesis of filaments; (3) a decrease in the transport of newly synthesized filaments from the cell body; or (4) a decrease in the rate of degradation of filaments.

Several lines of evidence argue for a role for the "collapse" mechanism. To begin with, mature neurofibrillary tangles contain few microtubules or other cytoskeletal elements in contrast to the normal cytoplasm in which microtubules and neurofilaments appear to have very precise spacing between them, which is regulated by fine bridges or arms connecting filaments to each other and to microtubules. Among the candidates for these bridges are the high-molecular-weight, microtubule-associated proteins (MAP), which bind to the core of the neurofilament (NF-L) (Leterrier *et al.*, 1982, 1984; Heimann *et al.*, 1984; Runge *et al.*, 1981).

Treatments that depolymerize microtubules (i.e., colchicine or colcemid) result in the formation of filamentous bundles (Wisniewski *et al.*, 1968), and this formation will occur in the presence of protein synthesis inhibitors. This treatment not only uncouples microtubules from neurofilaments but also disrupts their links to mitochondria (Raine *et al.*, 1971) and inhibits those parts of axoplasmic transport that depend on microtubule-associated motors such as kinesin (Vale *et al.*, 1985; Brady, 1985). This bundling of IF after uncoupling of the three-dimensional lattice of the cytoskeleton by microtubule depolymerization is not limited to the neuron. Similar effects are seen in a variety of cell types that have IF of composition other than that of the neurofilament. This suggests that the conserved regions of the IF contain a site that recognizes "linkers" to microtubules or microtubule-associated structures.

The uncoupling need not be the result of alterations of either the neurofilament or the microtubule. In neuropathies induced by IDPN, the tubules remain intact but segregate into a cluster in the central part of the axoplasm (Papasozomenos *et al.*, 1982). The neurofilaments appear to dissociate from the tubules and lie peripherally. At least some of the MAP are associated with the filaments in the condition, suggesting a failure in a linking function in the cytoplasm (Griffin *et al.*, 1983). In this case the effect of the toxin appears to be to dissociate the MAP from its microtubule binding site. This is functionally equivalent to the dissociation of MAP from microtubules that occurs when microtubules are depolymerized. All of these mechanisms could lead to the segregation or "collapse" of neurofilaments into the bundles observed under the electron microscope.

Since the MAP consist not only of linkers but also of kinesin and other "motor" proteins (Vale *et al.*, 1985; Brady, 1985; Paschal *et al.*, 1987), one would expect that the depolymerization of microtubules or alteration of normal cytoplasmic interactions might lead to an alteration of axoplasmic transport.

A second possible mechanism for the induction of neurofibrillary tangles is the overproduction of neurofilaments. In the case of neurofilament overproduction, it is possible that the number of filaments could exceed the available supply of linkage sites to microtubules, resulting in a segregation of the filaments into areas in which no microtubules were present. This same scenario would occur in the case where synthesis of filaments was unaltered but axoplasmic transport was blocked, resulting in an excess of neurofilaments in the cell body. Existing data argue against the overproduction mechanism. In addition to the aforementioned ability of colchicine to induce bundling of neurofilaments and other IF in the presence of protein synthesis inhibitors, newer studies have used cDNAs complementary to the mRNAs for neurofilaments to measure the mRNA levels in axotomy and in aluminum-induced neurofibrillary degeneration. In each of these cases, the specific mRNAs are seen to fall rather than rise during the time when neurofilament bundling is taking place (Hoffman *et al.*, 1987). This happens in spite of evidence that the mass of individual neuron cell bodies (Embree *et al.*, 1967) and the amount of neurofilament triplet protein in the cell body (Selkoe *et al.*, 1975) are both increased in experimental aluminum encephalopathy, an increase that must depend on abnormalities in either export from the cell body or the degradation of filaments—or both. It is risky to generalize to all causes of neurofibrillary degeneration from these two cases, so it remains possible that cases of overproduction do exist. However, even if they do, it is clear that this is not a requirement for formation of neurofibrillary tangles.

This leads to the third mechanism, failure to transport the filaments from the cell body. In the normal neuron, materials are transported down the axon in two major rate classes; the first moves rapidly, $>$ 100 mm/day, while the second, which includes the principal cytoskeletal elements, moves far more slowly—at approximately 1 mm/day (Lasek and Hoffman, 1976). To the best of our knowledge, most of the elements of the neuronal cytoskeleton depend on this slow transport to move them from the perikaryon down the axon where they are degraded (Roots, 1983). The slow component consists of neurofilaments and microtubules together with a variety of proteins that are likely to serve in the interactions between these filaments. In experimental conditions such as aluminum encephalopathy and the IDPN neuropathy, there is a drastic slowing and possibly even a complete arrest of the slow transport process (Roots, 1983; Bizzi *et al.*, 1984; Griffin *et al.*, 1985). The proteins are synthesized but fail to leave the cell body and the proximal axon—the principal areas in which neurofilament bundling is observed in these conditions. Though the mRNAs for neurofilaments are reduced in these conditions, it is not clear that this is reflected in the synthetic rate of the peptides—at least not at short times after induction of the disease process. This in turn depends on whether the major control of neurofilament synthesis is at the transcriptional or translational level, information that is still lacking for neurofilaments in both normal and diseased tissue. At the same time that neurofilament mRNA levels are depressed, mRNAs for tubulin are normal to elevated. Since tubulin is also "blocked" from transport in these conditions and since tubulin synthesis has been shown to be inhibited by excess tubulin subunits (Cleveland *et al.*, 1983), tubulin must either be rapidly degraded or a second control mechanism favoring synthesis must be called into play. Such a mechanism might involve the failure of a normally retrogradely transported repressor to reach the cell body. This, in turn, would cause the synthetic machinery to respond by producing more tubulin to be used in axonal repair or regeneration. In the regenerative process, synthesis of neurofilament mRNA is also increased (Tesser *et al.*, 1986).

Perhaps the reciprocal regulation of message levels for neurofilaments and tubulin argues for different temporal roles for these two organelles in attempts at axonal regeneration. In certain developing systems, such as the rodent optic nerve, the microtubules are present prior to the neurofilaments (Peters and Vaughn, 1967). The neurofilaments then appear perhaps to provide the structural and organizational framework for a larger axonal diameter. Understanding of this process is limited by lack of knowledge of the physiological role of neurofilaments. It is clear that neurofilaments are important elements in the determination of the three-dimensional structure of axoplasm and their numbers correlate strongly with axonal diameter, suggesting that they are the critical determinant of diameter (Hoffman *et al.*, 1984). Whether they serve other functions as well remains to be determined.

The development of antibodies that selectively recognized the phosphorylated and dephosphorylated forms of the neurofilament yielded the unexpected result that neurofilaments in the axon were normally phosphorylated while those in the cell body were not (Sternberger and Sternberger, 1983). When these antibodies were used to examine experimental neurofibrillary degenerations, it was found that the filamentous accumulations in the cell body were also phosphorylated. Taken together with the axoplasmic transport data, this suggests that filaments were phosphorylated in preparation for transport, but since this system was blocked, they accumulated in the cell body and proximal processes. The function of phosphorylation of neurofilaments is unclear. It could serve to modulate the cross bridging between filaments and tubules (Shelanski *et al.*, 1981) or filaments and other cytoskeletal elements. It has also been suggested that it protects against proteolytic degradation of the filaments (Goldstein *et al.*, 1987). Whatever the case, a common hallmark of a variety of toxic neuropathies as well as animal and human idiopathic motor neuron disease is the accumulation of phosphorylated neurofilament epitopes in the cell body and proximal processes (Muñoz *et al.*, 1988; Wiley *et al.*, 1987; Bizzi and Gambetti, 1986). Presumably this reflects the failure of the machinery necessary for their export. However, it is impossible to rule out a primary role in which phosphorylation itself might block a filament from entering into transport, perhaps by prematurely associating the filaments with one another rather than with the transport mechanism.

For example, in *n*-hexane-induced neuropathy the toxin induces covalent cross-linking of the neurofilaments (Graham *et al.*, 1985). This, in turn, may render the filaments incapable of transport. The tangles in aluminum-induced encephalopathy have aluminum associated with them, raising the possibility that the key step is participation of the phosphate groups on the filaments in the formation of nondissociable aluminum phosphates. These in turn could impede the proper interactions of filaments with other cytoplasmic organelles and the transport mechanism. A polyvalent ion such as aluminum could also serve to cross-link adjacent filaments via their phosphate groups in analogy to the cross-linking seen in *n*-hexane neuropathies.

Failure to degrade filaments at the appropriate points, whether at the nerve ending or, as has recently been suggested, along the extent of the process (Nixon *et al.*, 1982), could also lead to a buildup of filaments in the cell body and a mechanical inhibition of slow axoplasmic transport. This cannot be the case in Wallerian degeneration where a key feature is distal degradation of neurofilaments, but it does remain a possibility in other conditions. Since the enzymes implicated in this process are calcium-activated proteases of the calpain type (Zimmerman and Schlaepfer, 1982; Vitto and Nixon, 1986; Hamakubo *et al.*, 1986), alterations in calcium metabolism might also be implicated in these processes.

The possibilities considered above deal only with events leading directly to the formation of neurofibrillary tangles and only deal with part of the picture, since alterations in cytoskeletal elements other than neurofilaments, microtubules and MAP have been omitted. The formation of neurofibrillary tangles need not be a direct result of the pathogenetic insult but could reflect the end product of a cascade of receptor-mediated, metabolic, and other events. In turn, the formation of neurofibrillary tangles and the failure of axonal transport can have profound implications on neuronal survival, neurotransmission, and repair. The data on neurofibrillary tangles also provide a valuable reference against which to judge the more complex and perplexing data on AD.

3. Alzheimer's Disease

AD, a progressive dementing illness, is the cause of serious disability requiring long-term care of over 2 million people in the United States today. As people live longer, this disease has become more prominent and its social and economic costs more significant. Biological research on AD has moved more slowly than research on the experimental neuropathies described above largely because there is no satisfactory animal model. This has restricted neurobiological research to postmortem brains and to a limited number of peripheral tissues. In spite of these limitations, research has progressed on four major fronts: (1) the study of neurofibrillary tangles in AD; (2) the study of the senile plaque and amyloid in AD; (3) the study of the levels of neurotransmitters and their synthetic enzymes in AD; (4) the study of genetic defects in familial AD. While neurotransmitter studies have pointed to deficits in several transmitter and modulator systems, most notably in the cholinergic system and in choline acetyltransferase activity (Terry and Davies, 1980), and genetic studies show a genetic deficit on chromosome 21 in familial cases (St. George-Hyslop *et al.*, 1987), these studies are beyond the scope of this chapter. We will restrict ourselves to the neuronal cytoskeleton in AD and to amyloid to the extent that its discussion impinges on the cytoskeleton.

3.1. The Alzheimer Neurofibrillary Tangle

Given the structural differences between the normal neurofilament and the paired helical filament (PHF) in AD, we will refer to the neurofibrillary tangle in AD as an Alzheimer neurofibrillary tangle (ANT). The ANT are formed only in CNS neurons with a predilection for the larger neurons of the temporal and frontal cortices, and hippocampus (Price *et al.*, 1986). At the light microscopic level they are quite similar in appearance to neurofibrillary tangles in the conditions we have discussed before and stain with Bodian's silver stain. This fact is interesting because in normal brain extracts run on SDS–polyacrylamide gels, the Bodian stain is specific for the neurofilament triplet (Gambetti *et al.*, 1981).

PHF and ANT can be isolated from Alzheimer brain by a variety of subcellular fractionation methods (Iqbal *et al.*, 1974, 1984; Selkoe *et al.*, 1982). The highest yield is obtained by utilizing the insolubility of these structures in detergents such as SDS to remove contaminants from the preparations (Selkoe *et al.*, 1982). Unfortunately, the structures obtained in this manner appear to be highly cross-linked and resist further solubilization even in strong denaturants such as 6 M guanidine hydrochloride. Other procedures have been devised that claim that a population of PHF are soluble in SDS and yield proteins in the 60–80k range (Grundke-Iqbal *et al.*, 1984). It is still not clear whether these soluble proteins are actually constituents of the PHF or simply associated with them. These difficulties have led to an almost complete reliance on immunohistochemical methods to try to unravel the biochemical composition of the PHF. Numerous studies have been done using antibodies against known cytoskeletal proteins and against partially purified fractions enriched in PHF. The first antibody staining of ANT was with an antibody raised against a preparation of microtubules. This antibody gave strong staining of ANT but failed to stain microtubule-rich areas of normal brain, suggesting that it was against a minor component in the microtubule preparation (Grundke-Iqbal *et al.*, 1979). Antibodies against neurofilaments have given conflicting results with some antibodies strongly staining the ANT (Gambetti *et al.*, 1983; Rasool and Selkoe, 1984; Miller *et al.*, 1986) but others not reacting at all. In general, it appeared that the best results were obtained with antibodies that have activity against NF-M and/or NF-H. Antibodies that are selective for phosphorylated neurofilament proteins stain ANT strongly. Antibodies that did not differentiate between the two forms appear to be the ones that failed to stain. This suggests that much of the variability of staining with antibodies to neurofilaments is related to whether the antibodies recognize the phosphorylated forms of the protein (Cork *et al.*, 1986). Those antibodies that did stain reacted against ANT in both cell bodies and processes, with the presence of phosphorylated epitopes in the perikaryon being reminiscent of the accumulation of these epitopes in the experimental neuropathies previously discussed (Dickson *et al.*, 1987).

Antibodies to the MAP tau (Weingarten *et al.*, 1975) also have been reported to stain ANT (Brion *et al.*, 1985; Wood *et al.*, 1986; Kosik *et al.*, 1986). This result is consistent, with several different antibodies giving similar results. In addition, a number of antibodies raised against purified PHF have been shown to react specifically with purified tau proteins (Yen *et al.*, 1987). However, other anti-PHF antibodies do not recognize tau and react either solely with ANT or with ANT and the high-molecular-weight MAP (Yen *et al.*, 1987; Ksiezek-Reding and Yen, 1987). In addition, there are ANT in brain sections that fail to stain with antitau. Some neurofilament antibodies that stain ANT have been shown to cross-react with tau, raising the possibility that what has been called neurofilament in ANT is really the phosphorylation site on tau. In addition to this array of epitopes in the ANT, ubiquitin has also been identified (Mori *et al.*, 1987; Perry *et al.*, 1987a).

A critical analysis of the literature leads one to suppose that the ANT are composed of derivatives from tau and possibly other MAP together with another component that is recognized by the unique anti-PHF antibodies. The presence of ubiquitin suggests that the cell has attempted to target this material for degradation but that this has been abortive. The unidentified component in the PHF could either be another protein or a unique epitope formed by one of the identified components as a result of a conformational change. It appears unlikely at this point that neurofilament proteins form a significant part of the ANT. In addition to the immunological considerations, it is hard to reconcile the highly α-helical structure of neurofilaments with the β-sheet conformation of PHF (Kirschner *et al.*, 1986), though such transitions are possible under extreme conditions.

Since these studies have had to rely on immunohistology, it is not possible to tell which activities are part of the PHF themselves and which might be part of the ANT by binding to the PHF. There has been one report that PHF are composed of material that is biochemically similar to the β protein of Alzheimer amyloid (Masters *et al.*, 1985). However, attempts to stain ANT with antibodies to β protein have been unsuccessful and this report must be treated with skepticism.

Thus, the composition of ANT and PHF still awaits the complete purification and analysis of these structures. In spite of the differences between AD and the experimental models, certain important similarities remain. These are the involvement of cytoplasmic "linkers"—in the case of AD, tau and possibly MAP—the accumulation of phosphorylated epitopes in "tangles" in the cell body and finally a failure in degradation or transport of cytoskeletal components from the cytoplasm. While axoplasmic transport cannot be measured in postmortem material, its blockade has long been suggested as playing a role in this disease.

In addition to PHF, cases of AD also show straight filaments of narrow diameter, which have been studied immunochemically (Perry *et al.*, 1987b). Recently, an antibody (Alz-50) has been reported to react specifically with AD brain recognizing an antigen with an apparent molecular weight of 68k (Wolozin *et al.*, 1986). More extensive work has shown this protein to also be present in developing brain and it appears to be a variant of the tau protein (Nukina *et al.*, 1988), although Alz-50 does not stain ANT. Another indication of tau protein modification in AD is the finding by Kosik *et al.*, that nerve fibers in AD brain stain strongly with tau antibodies even in the absence of obvious tangles, while normal brain is only faintly reactive (Kowall and Kosik, 1987). Both these findings raise the possibility that a modified tau protein may play a role in the pathogenesis of AD separate from the role that tau plays in ANT formation.

3.2. Nonneural Changes in Alzheimer's Disease

In AD the major symptoms and the major pathology are focused in the CNS. There are, however, indications that abnormalities of cytoskeletal function and calcium metabolism exist in nonneural tissues in this disease.

For example, polymorphonuclear neutrophils (PMN) derived from Alzheimer patients have been reported to fail to migrate along a thermal gradient and to behave like PMN that have been exposed to colchicine (Fu *et al.*, 1986). Other studies show a decrease in membrane fluidity in neuronal membranes from AD brain (Zubenko *et al.*, 1987a,b). These results suggest an alteration in the cytoskeleton but studies that have focused directly on the cytoskeleton in these cells are lacking.

Fibroblasts isolated from patients with the familial form of AD have elevated total cellular calcium content when compared both to age-matched controls and to young adults (Peterson and Goldman, 1986). When cytoplasmic free calcium $[Ca^{2+}]_i$ was measured in these cells using the calcium reporter dyes Quin-2 and Fura-2 and a microspectrofluorometer, the ionic calcium was found to be greatly altered in the AD cells with a reduction of 70% below age-matched controls and 81% below young adult controls (Peterson *et al.*, 1986a). The changes were not due to a failure to be able to elevate $[Ca^{2+}]_i$ since the AD cells responded to bradykinin or f-Met-Leu-Phe with a rise in $[Ca^{2+}]_i$, though this was always somewhat less than in control cells (Peterson *et al.*, 1988). The Alzheimer cells also spread far more slowly on the tissue culture substrata than did the age-matched controls. The depression levels of $[Ca^{2+}]_i$ could impair many physiological functions including activation of calcium-activated proteases, which would function in filament degradation, and calmodulin, which would act to depolymerize microtubules. Whether these provocative findings are generally valid in AD or are limited to familial cases, or even a subset of familial cases, awaits further research.

4. Other Fibrillar Structures

In addition to normal neurofilaments and PHF, patients with progressive supranuclear palsy show neurofibrillary tangles composed of 15-mm straight filaments which may be present in the same cells as PHF that display a longer periodicity (Steele *et al.*, 1964). These straight structures cross-react with antibodies against ANT (Yen *et al.*, 1983; Dickson *et al.*, 1985).

5. Inclusions Containing Cytoskeletal Elements

A variety of neurological diseases contain characteristic cytoplasmic inclusions that have recently been studied using immunocytochemical techniques. Some of the most important are discussed below.

5.1. Pick Bodies

These round structures are found in the cytoplasm of neurons in the central cortex and basal ganglia in cases of Pick's disease, another form of dementia. They are composed of a heterogeneous-appearing collection of filaments with diameters between 10 and 20 nm as well as granular material and, in some cases, PHF of longer periodicity (Rewcastle and Ball, 1968). A number of workers have reported the presence of neurofilament antigens in Pick bodies (Dickson *et al.*, 1985; Gambetti *et al.*, 1983; Muñoz-Garcia and Ludwin,

1984; Probst *et al.*, 1983). Antigenic determinants shared only with ANT have also been found in these bodies (Dickson *et al.*, 1985; Rasool and Selkoe, 1985). These studies do not settle the question of the composition of Pick bodies. More precise localization studies at the electron microscopic level and eventual purification and analysis of these structures are desirable. The relative rarity of Pick's disease makes this latter goal difficult to achieve.

5.2. *Lewy Bodies*

These spheroidal cytoplasmic inclusions are a hallmark of neuronal pathology in idiopathic Parkinson's disease. They occur most prominently in the substantia nigra and in other aminergic neurons (Den Hartog Jager and Bethlem, 1980; Forno, 1969; Lewy, 1912; Lipkin, 1959; Ohama and Ikuta, 1976). These bodies are composed of filaments differing little in diameter from normal IF. As in the case of Pick bodies, these inclusions are recognized by antibodies directed against the components of the neurofilament triplet (Forno *et al.*, 1983; Goldman *et al.*, 1983). Goldman and his co-workers have found no actin reactivity in these structures nor did antibodies directed against ANT react.

Thus, it appears that Lewy bodies may, like non-ANT be composed of neurofilaments. However, one would imagine that they are formed by a route totally different from the route by which the neurofibrillary tangle is generated.

5.3. *Hirano Bodies*

These rod-shaped inclusions are present in a variety of neurological diseases, including Guam parkinsonism–dementia complex, AD, Pick's disease and in small numbers in apparently normal brain as well as in the neurologically mutant mouse, brindled (Mo^{br}) (Nagara *et al.*, 1980; Peterson *et al.*, 1986b). They are found both in the CNS and PNS and in glial cells as well as neurons.

These structures are composed of a highly ordered array of thin filaments that are arranged in parallel sheets. These sheets cross each other at an angle with the elements of one sheet contacting the elements of the adjacent sheets (Hirano *et al.*, 1968; Schochet *et al.*, 1968; Schochet and McCormick, 1972). The individual filaments are about 6 nm in diameter. Depending on the angle of section through these structures, they may appear cross-hatched, beaded, or as arrays of parallel filaments. Immunocytochemical studies (Goldman, 1983; Galloway *et al.*, 1987) show that the structures react with antibodies against actin but not with antibodies against myosin, neurofilaments, or tubulin. The manner in which these structures form is still unknown, but the brindled mouse model will allow study of the dynamics of this process.

5.4. *Other Inclusions*

In addition to the inclusions discussed above, there exist a number of rarer cytoplasmic and intranuclear inclusions of poorly understood chemistry and etiology, which are beyond the scope of this chapter.

6. Conclusion

This review has attempted to use examples from the literature to illustrate the range of alterations in cytoskeletal organelles that are seen in human and animal neurological diseases. The current state of our knowledge has allowed identification of many of the proteins involved and enabled the formulation of hypotheses for the pathophysiology of these processes. However, the task is far from complete and a full understanding ultimately depends on an understanding of the normal functions of the cytoskeleton.

References

Alzheimer, A., 1907, Über ein eigenartige Erkrankung der Hirnrinde, *Allg. Z. Psychiatr.* **64:**146–148.

Bizzi, A., and Gambetti, P., 1986, Phosphorylation of neurofilaments is altered in aluminum intoxication, *Acta Neuropathol.* **71:**154–158.

Bizzi, A., Crane, R. C., Autilio-Gambetti, L., and Gambetti, P., 1984, Aluminum effect on slow axonal transport: A novel impairment of neurofilament transport, *J. Neurosci.* **4:**722–731.

Brady, S. T., 1985, A novel brain ATPase with properties expected for the fast axonal transport motor, *Nature* **317:**73–75.

Brion, J. P., van den Bosch de Aquilar, P., and Flament-Durand, J., 1985, Senile dementia of the Alzheimer type: Morphological and immunocytochemical studies in senile dementia of Alzheimer type, in: *Advances in Applied Neurological Science,* Volume 2 (J. Traben and W. H. Gispen, eds.), Springer Verlag, Berlin, pp. 164–174.

Chou, S. M., and Hartmann, H. A., 1965, Electron microscopy of focal neuroaxonal lesions produced by β,β′-iminodipropionitrite (IDPN) in rats. I. The advanced lesions, *Acta Neuropathol.* **4:**598–603.

Cleveland, D. W., Pittenger, M. F., and Lopatz, M. A., 1983, Autoregulatory control of the expression of α and β tubulin, *J. Submicrosc. Cytol.* **15:**353–358.

Cork, L. C., Sternberger, N. H., Sternberger, L. A., Casanova, M. F., Struble, R. G., and Price, D. L., 1986, Phosphorylated neurofilament antigens in neurofibrillary tangles in Alzheimer's disease, *J. Neuropathol. Exp. Neurol.* **45:**59–64.

Den Hartog Jager, W. A., and Bethlem, J., 1980, The distribution of Lewy bodies in the central and autonomic nervous systems in idiopathic paralysis agitans, *J. Neurol. Neurosurg. Psychiatry* **23:**283–290.

Dickson, D. W., Kress, Y., Crowe, A., and Yen, S.-H., 1985, Monoclonal antibodies to Alzheimer neurofibrillary tangles (ANT). 2. Demonstration between ANT and neurofibrillary degeneration in progressive supranuclear palsy, *Am. J. Pathol.* **120:**292–303.

Dickson, D. W., Ksiezak-Reding, H., Davies, P., and Yen, S.-H., 1987, A monoclonal antibody that recognizes a phosphorylated epitope in Alzheimer neurofibrillary tangles, neurofilaments and tau proteins immunostains granulovacuolar degeneration, *Acta Neuropathol.* **73:**254–258.

Embree, L. J., Hamberger, A., and Sjostrand, J., 1967, Quantitative cytochemical studies and histochemistry in experimental neurofibrillary degeneration, *J. Neuropathol. Exp. Neurol.* **26:**427–434.

Forno, L. S., 1969, Concentric hyaline intraneuronal inclusions of Lewy tape in the brains of elderly persons (50 incidental cases): Relationships to parkinsonism, *J. Am. Geriatr. Soc.* **17:**557–575.

Forno, L. S., Strefling, A. M. Sternberger, L. A., and Eng, L. F., 1983, Immunocytochemical staining of neurofibrillary tangles and the periphery of Lewy bodies with a monoclonal antibody to neurofilaments, *J. Neuropathol. Exp. Neurol.* **42:**342 (abstract).

Fu, T. K., Matsuyama, S. S., Kellser, J. O., and Jarvik, L. F., 1986, Philothermal response, microtubules and dementia, *Neurobiol. Aging* **7:**41–43.

Galloway, P. G., Perry, G., and Gambetti, P., 1987, Hirano body filaments contain actin and actin-associated proteins, *J. Neuropathol. Exp. Neurol.* **46:**185–199.

Gambetti, P., Autilio-Gambetti, L., and Papasozomenos, S. C., 1981, Bodian's silver method stains neurofilament polypeptides, *Science* **213:**1521–1522.

Gambetti, P., Autilio-Gambetti, L., Perry, G., Shecket, G., and Crane, R. C., 1983, Antibodies to neurofibrillary tangles of Alzheimer's disease raised from human and animal neurofilament fractions, *Lab. Invest.* **49**:430–435.

Ghetti, B., and Gambetti, P., 1983, Comparative immunocytochemical characterization of neurofibrillary tangles in experimental maytansine and aluminum encephalopathies, *Brain Res.* **276**:388–393.

Goldman, J. E., 1983, The association of actin with Hirano bodies, *J. Neuropathol. Exp. Neurol.* **42**:146–152.

Goldman, J. E., and Yen, S.-H., 1986, Cytoskeletal protein abnormalities in neurodegenerative diseases, *Ann. Neurol.* **19**:209–223.

Goldman, J. E., Yen, S.-H., Chiu, F.-C., and Peress, N., 1983, Lewy bodies of Parkinson's disease contain neurofilament antigens, *Science* **221**:1082–1084.

Goldstein, M. E., Sternberger, N. H., and Sternberger, L. H., 1987, Phosphorylation protects neurofilaments against proteolysis, *J. Neuroimmunol.* **14**:149–160.

Graham, D. G., Anthony, D. C., Szakal-Quin, G., Gottfried, M. R., and Boekelheide, K., 1985, Covalent crosslinking of neurofilaments in the pathogenesis of n-hexane neuropathy, *Neurotoxicology* **6**:55–63.

Griffin, J. W., Fahnestock, K. E., Price, D. L., and Hoffman, P. N., 1983, Microtubule–neurofilament segregation produced by β,β′-iminodipropionitrile: Evidence for the association of fast axoplasmic transport with microtubules, *J. Neurosci.* **3**:557–566.

Griffin, J. W., Parhad, I., Gold, B., Price, D. L., Hoffman, P. N., and Fahnestock, K., 1985, Axonal transport of neurofilament proteins in IDPN neurotoxicity, *Neurotoxicology* **6**:43–53.

Grundke-Iqbal, I., Johnson, A. B., Wisniewski, H. M., Terry, R. D., and Iqbal, K., 1979, Evidence that Alzheimer neurofibrillary tangles originate from neurotubules, *Lancet* **1**:578–580.

Grundke-Iqbal, I., Iqbal, K., Tung, Y. C., and Wisniewski, H. M., 1984, Alzheimer paired helical filaments: Immunochemical identification of polypeptides, *Acta Neuropathol.* **62**:259–267.

Hamakubo, T., Kannagi, R., Murachi, T., and Matus, A., 1986, Distribution of calpains I and II in rat brain, *J. Neurosci.* **6**:3103–3111.

Heimann, R., Shelanski, M. L., and Liem, R. K. H., 1984, Microtubule-associated proteins bind specifically to the 70 kDa neurofilament protein, *J. Biol. Chem.* **254**:12160–12166.

Hirano, A., Dembitzer, H. M., Kurland, L. T., and Zimmerman, H. M., 1968, The fine structure of some intraganglionic alterations, *J. Neuropathol.* **28**:365–366.

Hirano, A., Donnenfeld, H., Sasaki, S. and Nakano, I., 1984, Fine structured observation of neurofilamentous changes in amyotrophic lateral sclerosis, *J. Neuropathol. Exp. Neurol.* **43**:461–470.

Hoffman, P. N., and Lasek, R., 1975, The slow component of axoplasmic transport. Identification of major structural polypeptides of the axon and their generality among mammalian neurons, *J. Cell Biol.* **66**:351–366.

Hoffman, P. N., Griffin, J. W., and Price, D. L., 1984, Control of axonal caliber by neurofilament transport, *J. Cell Biol.* **99**:705–714.

Hoffman, P. N., Cleveland, D. W., Griffin, J. W., Landes, P. W., Cowan, N. J., and Price, D. L., 1987, Neurofilament gene expression: A major determinant of axonal caliber, *Proc. Natl. Acad. Sci. USA* **84**:3472–3476.

Iqbal, K., Wisniewski, H., Shelanski, M. L., Brostoff, S., Liwnicz, B. H., and Terry, R. D., 1974, Protein changes in senile dementia, *Brain Res.* **77**:337–343.

Iqbal, K., Zaidi, T., Thompson, C. H., Merz, P., and Wisniewski, H. M., 1984, Alzheimer paired helical filaments: Bulk isolation, solubility and protein composition, *Acta Neuropathol.* **62**:167–177.

Jones, H. B., and Cavanagh, J. B., 1984, The evolution of intracellular responses to acrylamide in rat spinal ganglion neurons, *Neuropathol. Appl. Neurobiol.* **10**:101–121.

Julien, J. P., and Mushynski, W. E., 1983, The distribution of phosphorylation sites among identified proteolytic fragments of mammalian neurofilaments, *J. Biol. Chem.* **258**:4019–4025.

Kidd, M., 1963, Paired filaments in electron microscopy of Alzheimer's disease, *Nature* **197**:192–193.

Kirschner, D. A., Abraham, C., and Selkoe, D. J., 1986, X-ray diffraction from intraneuronal paired helical filaments and extraneuronal amyloid fibers in Alzheimer's disease indicates cross-beta conformation, *Proc. Natl. Acad. Sci. USA* **83**:503–507.

Klatzo, I., Wisniewski, H. M., and Streicher, E., 1965, Experimental production of neurofibrillary degeneration, *J. Neuropathol. Exp. Neurol.* **24**:187.

Kosik, V. S., Joachim, C. L., and Selkoe, D. J., 1986, Microtubule-associated protein tau is a major antigenic component of paired helical filaments in Alzheimer's disease, *Proc. Natl. Acad. Sci. USA* **83**:4044–4048.

Kowall, N. W., and Kosik, K. S., 1987, The cytoskeletal pathology of Alzheimer's disease is characterized by aberrant tau distribution, *Ann. Neurol.* **22**:639–643.

Ksiezak-Reding, H., and Yen, S.-H., 1987, Two monoclonal antibodies recognize Alzheimer's neurofibrillary tangles, neurofilament, and microtubule-associated proteins, *J. Neurochem.* **48:**455–462.

Lasek, R. J., and Hoffman, P. N., 1976, The neuronal cytoskeleton, axonal transport and axonal growth, in: *Cell Motility* (R. Goldman, T. Pollard, and J. Rosenbaum, eds.), Cold Spring Harbor Laboratory, Cold Spring Harbor, NY., pp. 1021–1050.

LeTerrier, J.-F., Liem, R. K. H., and Shelanski, M. L., 1982, Interactions between neurofilaments and microtubule-associated proteins: Possible mechanism for intraorganellar bridging, *J. Cell Biol.* **95:**982–986.

LeTerrier, J.-F., Wong, J., Liem, R. K. H., and Shelanski, M. L., 1984, Promotion of microtubule assembly by neurofilament-associated, microtubule-associated proteins, *J. Neurochem.* **43:**1385–1391.

Lewy, F. H., 1912, Paralysis agitans. I. Pathologische Anatomie, in: *Handbuch der Neurologie* (M. Lewandowsky, ed.), Springer, Berlin, pp. 920–933.

Liem, R. K. H., and Hutchison, S., 1982, Purification of the individual components of the neurofilament triplet. Filament assembly from the 70,000 dalton subunit, *Biochemistry* **21:**3221–3226.

Liem, R. K. H., Yen, S.-H., Salomon, G. D., and Shelanski, M. L., 1978, Intermediate filaments in nervous tissues, *J. Cell Biol.* **78:**637–645.

Lipkin, L. E., 1959, Cytoplasmic inclusions in ganglion cells associated with parkinsonian states, *Am. J. Pathol.* **35:**1117–1133.

Masters, C. L., Multhaup, G., Simms, G., Pottgiesser, J., Martins, R. N., and Beyreuther, K., 1985, Neuronal origin of a cerebral amyloid: Neurofibrillary tangles of Alzheimer's disease contain the same protein as the amyloid of plaque cores and blood vessels, *EMBO J.* **4:**2757–2763.

Miller, C. C., Brion, J. P., Calvert, R., Chin, T. K., Eagles, P. A., Downes, M. J., Flament-Durand, J., Haugh, M., Kahn, J., Probst, A., Olrich, J., and Anderton, B. H., 1986, Alzheimer's paired helical filaments share epitopes with neurofilaments side arms, *EMBO J.* **5:**269–276.

Mori, H., Kondo, J., and Ihara, Y., 1987, Ubiquitin is a component of paired helical filaments in Alzheimer's disease, *Science* **235:**1641–1644.

Muñoz, D. G., Greene, C., Peel, D. P., and Selkoe, D. J., 1988, Accumulation of phosphorylated neurofilaments in anterior horn motorneurons of amyotrophic lateral sclerosis patients, *J. Neuropathol. Exp. Neurol.* **47:**9–18.

Muñoz-Garcia, D., and Ludwin, S. K., 1984, Classic and generalized variants of Pick's disease: A clinicopathological, ultrastructural and immunocytochemical comparative study, *Ann. Neurol.* **16:**467–480.

Nagara, H., Yakajima, K., and Suzuki, K., 1980, An ultrastructural study of the cerebellum of the brindled mouse, *Acta Neuropathol.* **52:**41–50.

Nixon, R. A., Brown, B. A., and Marotta, C. A., 1982, Posttranslational modification of a neurofilament protein during axoplasmi transport: Implications for regional specialization of CNS axons, *J. Cell Biol.* **94:**150–158.

Nukina, N., Kosik, K. S., and Selkoe, D. J., 1988, The monoclonal antibody, Alz 50, recognizes tau proteins in Alzheimer's disease brain, *Neurosci. Lett.*, **87:**240–246.

Ohama, E., and Ikuta, F., 1976, Parkinson's disease: Distribution of Lewy bodies and monoamine neuron systems, *Acta Neuropathol.* **34:**311–319.

Papasozomenos, S. C., Yoon, M., Crane, R., Autilio-Gambetti, L., and Gambetti, P., 1982, Redistribution of proteins of fast axonal transport following administration of β,β′-iminodipropionitrile: A quantitative autoradiographic study, *J. Cell Biol.* **95:**672–675.

Paschal, B. M., Shpetner, H. S., and Vallee, R. B., 1987, Map 1C is a microtubule-activated ATPase which translates microtubules *in vitro* and has dynein-like properties, *J. Cell Biol.* **105:**1273–1285.

Perry, G., Friedman, R., Shaw, G., and Chau, V., 1987a, Ubiquitin is detected in neurofibrillary tangles and senile plaque neurites of Alzheimer's disease brains, *Proc. Natl. Acad. Sci. USA* **84:**3033–3036.

Perry, G., Mulvihill, P., Manetto, V., Autilio-Gambetti, L., and Gambetti, P., 1987b, Immunocytochemical properties of Alzheimer straight filaments, *J. Neurosci.* **7:**3736–3738.

Peters, A., and Vaughn, J. E., 1967, Microtubules and filaments in the axons and astrocytes of early postnatal optic nerve, *J. Cell Biol.* **32:**113–119.

Peterson, C., and Goldman, J. E., 1986, Alterations in calcium content and biochemical processes during aging and Alzheimer's disease, *Proc. Natl. Acad. Sci. USA* **83:**2758–2762.

Peterson, C., Ratan, R. R., Shelanski, M. L., and Goldman, J. E., 1986a, Cytosolic free calcium and cell spreading decrease in fibroblasts from aged and Alzheimer's donors, *Proc. Natl. Acad. Sci. USA* **83:**7999–8001.

Peterson, C., Suzuki, K., Kress, Y., and Goldman, J. E., 1986b, Abnormalities of dendritic actin organization in the brindled mouse, *Brain Res.* **382:**205–212.

Peterson, C., Ratan, R. R., Shelanski, M. L., and Goldman, J. E., 1988, Altered response of fibroblasts from aged and Alzheimer donors to drugs that elevate cytoplasmic free calcium, *Neurobiol. Aging* **9:**261–266.

Price, D. L., Whitehouse, P. J., and Struble, R. G., 1986, Cellular pathology in Alzheimer's and Parkinson's disease, *Trends Neurosci.* **9:**29–33.

Probst, A., Anderson, B. H., Ulrich, J., Kohler, R., Karen, J., and Hertz, P. U., 1983, Pick's disease: An immunocytochemical study of neuronal changes: Monoclonal antibodies show that Pick bodies share antigenic determinants with neurofibrillary tangles and neurofilaments, *Acta Neuropathol.* **60:**175–182.

Raine, C. S., Ghetti, B., and Shelanski, M. L., 1971, On the association between microtubules and mitochondria in axons, *Brain Res.* **34:**389–393.

Rasool, C. S., and Selkoe, D. J., 1984, Alzheimer's disease: Exposure of neurofilament immunoreactivity in SDS-insoluble paired helical filaments, *Brain Res.* **322:**194–198.

Rasool, C. S., and Selkoe, D. J., 1985, Recognition of Pick bodies by antibodies to neurofibrillary tangles in Alzheimer's disease, *N. Engl. J. Med.* **312:**700–705.

Rewcastle, N. B., and Ball, M. J., 1968, Electron microscopy of the inclusion bodies in Pick's disease, *Neurology* **18:**1205–1213.

Roots, B. I., 1983, Neurofilament accumulation induced in synapses by leupeptin, *Science* **221:**971–972.

Runge, M. S., Lane, T. M., Yphantis, D. A., Lifsics, M. R., Saito, A., Altin, M., Reinke, K., and Williams, R. C., Jr., 1981, ATP-induced formation of a complex between microtubules and neurofilaments, *Proc. Natl. Acad. Sci. USA* **78:**1431–1435.

St. George-Hyslop, P. H., Tanzi, R. E., Polinsky, R. J., Haines, J. L., Nee, L., Watkins, P. C., Myers, R. H., Feldman, R. G., Pollen, D., Drachman, D., Growdon, J., Bruni, A., Foncin, J.-F., Salmon, D., Frommelt, P., Amaducci, L., Sorbi, S., Placentini, S., Stewart, G. D., Hobbs, W. J., Conneally, P. M., and Gusella, J. F., 1987, The genetic defect causing familial Alzheimer's disease maps on chromosome 21, *Science* **235:**885–890.

Schlaepfer, W. W., 1977, Immunological and ultrastructural studies of neurofilaments isolated from rat peripheral nerve, *J. Cell Biol.* **74:**226–240.

Schochet, S. S., Jr., and McCormick, W. F., 1972, Ultrastructure of Hirano bodies, *Acta Neuropathol.* **21:**50–60.

Schochet, S. S., Jr., Lampert, P. W., and Lindenberg, R., 1968, Fine structure of the Pick and Hirano bodies in a case of Pick's disease, *Acta Neuropathol.* **11:**330–337.

Schochet, S. S., Jr., Hartman, J. M., Ladewig, P. P., and Eakle, K. M., 1969, Intraneuronal conglomerates in sporadic motor neuron disease, *Arch. Neurol.* **20:**548–553.

Selkoe, D. J., Liem, R. K. H., Yen, S.-H., and Shelanski, M. L., 1979, Biochemical and immunological characterization of neurofilaments in experimental neurofibrillary degeneration induced by aluminum, *Brain Res.* **163:**235–252.

Selkoe, D. J., Ihara, Y., and Salazar, F. J., 1982, Alzheimer's disease: Insolubility of partially purified helical filaments in sodium dodecyl sulfate and urea, *Science* **215:**1243–1245.

Shelanski, M. L., Liem, R. K. H., LeTerrier, J.-F., and Keith, C. H., 1981, The cytoskeleton and neuronal disease, in: *International Cell Biology* (H. G. Schweiger, ed.), Springer, Berlin, pp. 428–439.

Steele, J. C., Richardson, E. P., and Olszewski, J., 1964, Progressive supranuclear palsy: A heterogeneous degeneration involving the brain stem, basal ganglia and cerebellum with vertical gaze and pseudobulbar palsy, nuctal dyptonia and dementia, *Arch. Neurol.* **10:**333–359.

Sternberger, L. A., and Sternberger, N. H., 1983, Monoclonal antibodies distinguish phosphorylated and nonphosphorylated forms of neurofilament *in situ, Proc. Natl. Acad. Sci. USA* **80:**6126–6130.

Terry, R. D., 1963, The fine structure of neurofibrillary tangles in Alzheimer's disease, *J. Neuropathol. Exp. Neurol.* **22:**629–642.

Terry, R. D., and Davies, P., 1980, Dementia of the Alzheimer type, *Annu. Rev. Neurosci.* **3:**77–95.

Tesser, P., Jones, P. S., and Schechter, N., 1986, Elevated levels of retinal neurofilament mRNA accompany optic nerve regeneration, *J. Neurochem.* **47:**1235–1243.

Vale, R. D., Reese, T. S., and Sheetz, M. P., 1985, Identification of a novel force-generating protein, kinesin, involved in microtubule based motility, *Cell* **42:**39–50.

Vitto, A., and Nixon, R. A., 1986, Calcium-activated neutral proteinase of human brain: Subunit structure and enzymatic properties of multiple molecular forms, *J. Neurochem.* **47:**1039–1051.

Weingarten, M. D., Lockwood, A. H. Hwo, S. Y., and Kirschner, M. W., 1975, A protein factor essential for microtubule assembly, *Proc. Natl. Acad. Sci. USA* **72:**1858–1862.

Wiley, C. A., Love, S., Skoglund, R. R., and Lampert, P. W., 1987, Infantile neurodegenerative disease with neuronal accumulation of phosphorylated neurofilaments, *Acta Neuropathol.* **72:**369–376.

Wisniewski, H., Shelanski, M., and Terry, R. D., 1968, Effects of mitotic spindle inhibitions on neurotubules and neurofilaments of anterior horn cells, *J. Cell Biol.* **38:**224–229.

Wisniewski, H., Terry, R. D., and Hirano, A., 1970, Neurofibrillary pathology, *J. Neuropathol. Exp. Neurol.* **29:**163–176.

Wolozin, B. L., Pruchnicki, A., Dickson, D. W., and Davies, P., 1986, A neuronal antigen in the brain of Alzheimer patients, *Science* **232:**648–650.

Wood, J. G., Mirra, S. S., Pollock, N. J., and Bindex, L. I., 1986, Neurofibrillary tangles of Alzheimer's disease share antigenic determinants with the axonal microtubule-associated protein tau, *Proc. Natl. Acad. Sci. USA* **83:**4040–4043.

Wuerker, R., 1970, Neurofilaments and glial filaments, *Tissue Cell* **2:**1–9.

Yen, S.-H., Horoupian, D., and Terry, R. D. 1983, Immunocytochemical comparison of neurofibrillary tangles in senile dementia of Alzheimer type, progressive supranuclear palsy and postencephalitic parkinsonism, *Ann. Neurol.* **13:**172–175.

Yen, S.-H., Dickson, D. W., Crowe, A., Butler, M., and Shelanski, M. L., 1987, Alzheimer's neurofibrillary tangles contain unique epitopes and epitopes in common with the heat-stable microtubule-associated proteins tau and MAP_2, *Am. J. Pathol.* **126:**63–73.

Zimmerman, U. J., and Schlaepfer, W. W., 1982, Characterization of a brain calcium-activated protease that degrades neurofilament proteins, *Biochemistry* **21:**3977–3982.

Zubenko, G. S., Cohen, B. M., Boller, F., Malmakova, I., Keefe, N., and Chojnacki, B., 1987a, Platelet membrane abnormality in Alzheimer's disease, *Ann. Neurol.* **22:**237–244.

Zubenko, G. S., Wusylko, M., Cohen, B. M., Boller, F., and Teply, I., 1987b, Family study of platelet fluidity in Alzheimer's disease, *Science* **238:**539–542.

Index

Acidic residues, in protein chains, 183
  structural periodicity, 186
Acrylamide, inducing neurofibrillary proliferation, 452
Actin
  IgG against, 432
  in smooth muscle cells, 428, 429
Actin-containing structures
  cortical, IF associated with, 158–159
  IF associated with, 7, 12
AD: *see* Alzheimer's disease (AD)
Aggregation; *see also* Molecular aggregation
  with keratin IF, 397–398
  oligomeric, 213, 215–216, 218
  polymeric, 216–217, 218
Aging, altered IF expression and, 431
Albumin: α-fetoprotein genes, in rod domain evolution, 358
Alcoholic liver disease, altered IF expression and, 422–425
α-helix; *see also* Central α-helix-rich domain
  protein chain secondary structure, 177–179
  rod domain structure and, 187
α-keratin: *see* Hard α-keratins
α-pattern, of wool particles, 4
Aluminum, inducing neurofibrillary proliferation, 452, 453
Aluminum encephalopathy, filament transport and, 455
Alzheimer neurofibrillary tangle (ANT), 457–458
Alzheimer's disease (AD), 456–457
  neurofibrillary tangle, 451–452, 457–458
  nonneural changes in, 458–459
Amino acid sequences
  of desmins and vimentin, 22–24, 26
  high-glycine/tyrosine proteins, 283–285
  of high-sulfur proteins, 111
  of high-tyrosine proteins, 115–117
  of IF protein chains, 176–177, 179–180
    structural periodicities, 185–187
Amino acid sequences (*cont.*)
  IF types based on, 235
  Myxicola 155k neurofilament polypeptide, 57
  of neurofilament triplet polypeptides, 45–48
  NIFP, 346–347
  profilaggrin repeat unit, 401
  of wool keratin IF protein, 271
    C-terminals, 273
Amino acids
  of hard keratin IF proteins, 274
  in high-sulfur proteins
    from hard α-keratins, 107
    from wool and mouse hair, 109
  of high-tyrosine protein fractions, 115, 116
  in IF chain domains, 361
  normal versus mutant mouse hair, 121
  residue distribution in IF sequences, 240–241
cAMP, neurofilament phosphorylation and, 71–72
Amyloid deposits, IF involvement in, 435
Anagen, 101
Ancestral gene, in rod domain evolution, 358
Angioimmunoblastic lymphadenopathy, antibodies to IF in, 432
ANT: *see* Alzheimer neurofibrillary tangle (ANT)
Anti-IFA antibody, 48, 72–80
  in *Myxicola*, 54
  in squid, 62–63
Antibody(ies)
  decoration studies, in neurofilament isolation, 43
  for diagnostic histopathology, 72–80; *see also* Anti-IFA antibody
  against IF, 431–433
  localization techniques, for cardiac muscle 10-nm filaments, 30–31
  monoclonal: *see* Monoclonal antibodies (MAbs)
Aplysia spp., neurofilament function and evolution, 81
Aromatic residues, in protein chains, 184
  conservation of, 184, 185

Atheromatosis, altered IF expression and, 425–428, 429
Attachment plaques, 157
Autoantibodies, against IF, 431–433
Axial repeats, 240
  amino acid residue distribution and, 240–241
  detected by electron microscopy, 241–244
  from meridional X-ray diffraction, 241
  in paracrystals, 244–245
Axolemma, and neurofilament relationships, 66
Axons
  neurofilament content and, 38
  neurofilament transport by, 80–81
Axoplasm
  liquification of, CANP studies and, 67
  Myxicola, 53–54, 57
  neurofilaments interactions in, 63–66
  squid, 59–61

B1 keratin IFAP family, 280
B2 keratin IFAP family, 280–281
"Band 6," 155
Basal cell carcinoma, antibodies to IF in, 432
Basal cells, keratin identification in, 318
Basic residues, in protein chains, 183
  structural periodicity, 186
β-keratin, four-chain aggregate structure and, 194
BHK-21 cells, IFAP in, 372–379; *see also under* Intermediate filament-associated proteins (IFAP)
  identification criteria for, 373
BIIIA keratin IFAP family, 281
BIIIB keratin IFAP family, 281
Biosynthesis, of lamins, 133
Bodian silver method, 72–73
Boutons terminaux, 38
Breast, invasive duct carcinoma of, 419
Bullous dermatoses, antibodies to IF in, 432

C-terminal domain
  acidic and basic residues in, 183
  amino acid sequences compared, 273
  amino acids in, 361
  IF morphology and, 258
  in IF protein chains, 181
  structure of, 191–194
C-terminal region, peripheral, 51–52
Calcium-activated neutral proteases (CANPs), in neurofilament modification, 66–70
Calcium-activated proteolysis, neurofilament modification by, 66–70
Calcium-dependent neutral proteinases: *see* Calcium-activated neutral proteases (CANPs)
Calpains: *see* Calcium-activated neutral proteases (CANPs)
CANPs: *see* Calcium-activated neutral proteases (CANPs)
Carcinoma; *see also* Tumors
  basal cell, antibodies to IF in, 432
  invasive duct, of breast, 419
Cardiac hypertrophy, thin filaments and, 32
Cardiac muscle
  IF-cell surface association in, 157–158
  10-nm filaments in, 30–32
Cardiomyopathies, altered IF expression and, 421–422
Catagen-telogen periods, 101
Cations
  desmin filament formation and, 26–27
  filaggrin and, 397–398
Cell cycle dynamics
  intermediate filaments and, 140–141
  of nuclear lamina, 137–140
    dissassembly regulation, 141
Cell fractionation
  for keratin pair localization, 319
  NL and NPC associations, 134–135
Cell membrane complex, IF interaction with, 160–164
Cell surface, IF interactions with, 147–148; *see also* Desmosome; Hemidesmosome
  in muscle cells, 157–158
Cells
  functions, CANP implication in, 69
  IF role in, 353
  neoplastic and nonneoplastic differentiation, 416–421
  10-nm filament location in
    cardiac muscle, 30–31
    skeletal muscle, 28–29
    smooth muscle, 19–21
Central α-helix-rich domain
  in IF protein chains, 182–187
  rod domain and, 190
Central nervous system, neurofilament isolation from, 41–42
Chemical fractionation, of wool and hard α-keratin proteins, 97, 98
Chemotherapy, hair growth affected by, 101
Chicken pox, IgM antibodies reaction with IF in, 432
Chromatin, nuclear lamina association with, 136–137
Chromosomes
  localization, IF gene linkage and, 354–356
  NL association with, 137
Cigarette smoking, 428
Circular dichroism, neurofilaments and, 42
Codons, in subdominal keratin IF chains, 362
Coiled-coil rope structure, 206–208
  favorable interactions between, 212, 215
  IF protein molecules, 188–189
  packing concepts, 227
  radial and contact plane projections, 210–211, 213

Coiled-coil rope structure (*cont.*)
  radial distribution of, 223–227
Colchicine, inducing neurofibrillary proliferation, 452
Cold Spring Harbor Cell Motility (1978) meeting, 4
Collagen genes, in rod domain evolution, 358
Constituent proteins
  of intermediate filaments, 104–106
  of keratins: *see* Keratin proteins
Cornea
  epithelial maturation model, 324
  immunofluorescence staining techniques, 322–323
Cortex
  cell types in, 289–290
  gene expression, 293–294
  keratin synthesis in, 291
Cortical actin-containing structures, IF association with, 158–159
Cortical cells, nutrition and, 123
Cross-bridge, neurofilament, 258
Cross-linking
  fiber protein, 103–104
  intrachain disulfide, high-sulfur proteins and, 112
  plectin and, 380
  triplet polypeptide arrangement, 49, 51
  zero-length intermolecular, 137
Cuticle gene expression, 292–293
Cystine, hair content
  dietary manipulation, 122–124
  TTD and, 119
Cytokeratin genes, in *Xenopus laevis*: *see Xenopus laevis*
Cytokeratin polypeptide chains, 177
Cytokeratins, in *Xenopus laevis*: *see Xenopus laevis*
Cytoplasm
  IF in, 3
    in vitro procedures and, 13–15
    nuclear lamina association with, 136
  inclusions, in neurological diseases, 459–460
  organization, IF research and, 6
Cytoskeletal networks, 147–148
Cytoskeleton
  keratin, reorganization in, 303
  neurofilaments in vertebrates, 38–41
  nuclear lamina association with, 136
  protein immunoblot analysis, 317
  10-nm filament role in, 28
Cytoskeleton-cell membrane complex, molecular architecture of, 147–148

Defleecing, wool protein composition after, 122–123
Dementia pugilistica, neurofibrillary degeneration and, 452
Dense bodies, and 10-nm filaments relationship, 20–21
Dephosphorylation
  lamin, 139
  neurofilament modification and, 71
Desmin
  antibodies against, 417
  filament formation concept, 26
  IB chain segments, 212, 214, 215
  IF molecule chain composition in, 189–190
  immunofluorescence staining for, 418
  isolation and preparation of
    in cardiac muscle, 31
    in skeletal muscle, 29
    in smooth muscle, 21–22
  in mitosis, 140–141
  molecular properties of, 22–24
  primary structures of, 177
Desmocalmin, 154
Desmoplakin I, 155–156
  IFAP and, 385
Desmoplakin II, 155–156
  IFAP and, 385
Desmoplakin IV, 155
Desmosome
  biochemical composition of, 151–152, 153–154
  IF attachment to, 7, 13
  IF interaction with, 148
    morphological analysis, 148–151
  putative linkers in, 152, 154–156
Diabetes, 428
Diagnostic histopathology, antibodies as tool in, 72–80
Diet, keratin composition and, 122–124
Differentiation
  gene expression in hair follicle, 292–295
  hair developmental program, 291–292
  hair structure, 289–291
  of medulla and IRS cells, 288
  of wool fiber terminals, 275
"Differentiation-specific keratins," 309
Diffraction studies: *see* X-ray diffraction
cDNA
  CANPs nature and, 69
  clones, profilaggrin structure and, 398–399
  NIFP, 346
  ultrahigh-sulfur keratin clone, 281–283
DNA sequences, NL association with, 137
Domain
  C-terminal: *see* C-terminal domain
  central α-helix-rich, 182–187
  end: *see* End domains
  N-terminal: *see* N-terminal domain
  protein chain and lamins compared, 197
  rod: *see* Rod domain
  size in IF protein chains, 181
  subdomainal: *see* Subdomain
Dosidicus gigas, neurofilament investigations in, 59–63; *see also* Squid

Drosophila spp.
  IF genes in, 285
  NL association with, 137

ECM: *see* Extracellular matrix (ECM)
EGF: *see* Epidermal growth factor (EGF)
Ehrlich ascites tumor cells, NL structures in, 131
Electron microscopy (EM); *see also* Immunoelectron microscopy
  axial repeats detected by, 241–244
  cardiac muscle cells (rabbit)
  of desmin and vimentin filaments, 25
  desmosomes and hemidesmosome, 149
  of hard α-keratins, 95, 99–100
  interphase cell, 129, 130
  Myxicola axoplasm, 53, 55
  neurofibrillary degeneration, 452
  neurofilament density and structure, 38–41, 43
  scanning transmission: *see* Scanning transmission electron microscopy (STEM)
  squid neurofilaments, 60
  transverse traces of IF projected density
    negatively stained, 248–249, 250
    rotary-shadowed, 249–250
    tracing methods compared, 252
    unstained, 251–252
  of trichohyalin granules, 289
Electrophoresis
  high-sulfur wool proteins, 108
  high-tyrosine proteins, 114
  low-sulfur wool proteins, 105
  two-dimensional, protein species specificity, 119, 120
  of wool and hard α-keratin proteins, gel, 97, 98
EM: *see* Electron microscopy (EM)
Embryogenesis
  cytokeratin gene expression in *X. laevis*, 341–342
  keratin pair expression established during, 311
  profilaggrin and filaggrin expression in, 405–407
Encephalopathy, aluminum, 455
End domains; *see also* C-terminal domain; N-terminal domain
  amino acids in, 361
  of keratin chains, 360–362
  organization of, 359–360
Endocrine status, hair growth regulation and, 101
Epidermal growth factor (EGF), defleecing with, wool protein composition and, 122–123, 124
Epidermis
  appendages described, 95
  cell differentiation, filaggrin synthesis and, 402–405
  fetal development and, 405–407
    profilaggrin and filaggrin expression during, 405–407
  and hair follicle morphogenesis, 100–101
  keratin expression in, 320
Epidermis (*cont.*)
  morphology, profilaggrin expression and, 394
Epinemin, 382
Epithelial cells, IF systems in, 7, 8
Epithelium
  corneal maturation model, 324
  neoplasms of, keratin synthesis in, 315–316
  prekeratin filament bundles in, 430–431
Erythrocytes
  IF-cell membrane complex interaction and, 160–161
  in vitro reconstitution of membrane, 136
  spectrin in, 10-nm filaments and, 27
Escherichia coli, cDNA clones and, 398
Eukaryotic cells, IF in, 3
Evolution, IF keratin protein conservation and, 287–288
Exons
  aberrant genomic arrangement of, 276–277
  cytokeratin genes, 341
Extended β, protein chain secondary structure, 177–179
Extracellular matrix (ECM), 161, 163

Fast Fourier transform (FTT) technique, IF protein sequence, 184–185
Felting lustre mutant wool, high-tyrosine protein content in, 118
Fetus, development of
  filaggrin synthesis and, 405–407
  human, 405–407
Fiber proteins
  assembly and cross-linking of, 103–104
  development of, 102–103
Fibroblasts
  actin-containing structures and IF association in, 158–159
  IF-cell membrane complex interaction and, 161–163
  IF systems in, 7, 8
  senescent, 431
Fibronexus: *see* Extracellular matrix (ECM)
15-nm filaments, 459
55-kDa monomer, and 10-nm filament relationship, 24–27
Filaggrin, 393–394; *see also* Profilaggrin
  aggregation with keratin IF, 397–398
  extraction and purification, 394–395
  keratinization disorders and, 407–409
  macrofibrils and, 397
  properties of, 395
  synthesis of: *see* Filaggrin synthesis
  10-nm filaments and, 27
Filaggrin synthesis, 396
  epidermal cell differentiation and, 402–405
  during fetal development, 405–407

Filaments; *see also* Protofilaments
  filaments within, IF morphology and, 253–256
  IF research and, 6
  paired helical: *see* Paired helical filament (PHF)
  10-nm: *see* 10-nm filaments
  traversing, desmosome preparation and, 150
Fodrin, neurofilament interaction with, 66
Folding, protein, characteristics of, 194
Follicle bulb, hair development and, 291; *see also* Hair follicle
Food intake, wool growth rate and, 123; *see also* Nutrition, role in keratin composition
Four-chain aggregates, structure of, 194–196
Fractionation
  cell: *see* Cell fractionation
  chemical, 97, 98
Fragments, enzymatically derived, 209
Freeze drying, of IF, 248, 251–252
FTT: *see* Fast Fourier transform (FTT) technique, IF protein sequence

Ganglion cells, antibody labeling of, 74–80
Gardner–Rasheed feline sarcoma virus, 433
Gene expression
  cortical, 293–294
  cuticle, 292–293
  hard keratin IF protein, 269–289; *see also under* Hard keratin
  IF research and, 6
  keratin, variation in, 294–295
Gene regulation: *see* Transcription, neurofilaments and
Generic model, for IF structure, 219
  surface lattice of, 221–223
Genes
  cytokeratin: *see Xenopus laevis*
  IF: *see* Intermediate filament (IF) genes
  NIFP
    introns in, 348–349
    promotors, 349
  of wool, 275–277, 285–286; *see also under* Hard keratin
Genetics, role in keratin composition, 118
  mutations, 119, 121–122
  species specificity, 118–119, 120
Giant axon
  in CANP studies, 67–68
  ganglion, antibody labeling of, 74–75
  in *Myxicola infundibulum*, 53
  in squid, antibody labeling and, 77–79
  squid studies, 59
Glial fibrillary acidic protein
  IF molecule chain composition in, 189–190
  primary structures of, 177
Graft-versus-host disease, chronic, antibodies to IF in, 432
Granulomas, sarcoidosis and, 428

Guam parkinsonism-dementia complex, neurofibrillary degeneration and, 452

Hair follicle; *see also* Wool follicle
  fiber protein and: *see* Fiber proteins
  gene expression in, 292–295
  hair growth rate regulation, 101
  morphogenesis of, 100–101
Hairs
  development of, 291–292
  growth regulation factors, 101
  structure of, 289–291
Hard α-keratins, 95–96
  amino acid composition of, 96
  composition regulation, 117–124
    genetic, 118–122
    molecular basis of variability in, 117–118
    nutritional, 122–124
    physiological, 124
  fine structure of, 99–100
  helical projections of surface lattice, 225
  high-tyrosine protein content of, 113; *see also* High-tyrosine proteins
  protein composition of, 96
    constituent, 96–99; *see also* Keratin proteins
  solubility, 96–97
Hard keratin
  α: *see* Hard α-keratins
  IF protein, 269
    evolutionary expression of, 287–288
    gene structure and organization, 275–277
    structure, 269–272
    tissue expression, 272–275
  IFAP family
    evolutionary expression of, 287–288
    high-glycine/tyrosine, 283–285
    high-sulfur, 277–283
  IFAP gene, structure and organization, 285–288
  protein classification of, 267–268
Helical projections, of hard α-keratin surface lattice, 225
Hemidesmosome
  biochemical composition of, 151–152, 153–154
  IF interaction with, 148
    morphological analysis, 148–151
Hepatocytes, 422–423
  Mallory bodies development in, 424
Heptad polarity, in rod domain evolution, 358
HGT: *see* High-glycine/tyrosine (HGT) protein
High-glycine/tyrosine (HGT) protein; *see also* High-tyrosine proteins
  evolutionary expression of, 287–288
  gene expression variations and, 295
  gene noncoding regions compared, 285–286
  hard keratin IFAP family, 283–285
High-sulfur proteins
  chemical composition of, 107–108

High-sulfer proteins (*cont.*)
  evolutionary expression of, 287
  gene noncoding regions compared, 285–286
  heterogeneity of, 108–110
  keratin IFAP family, 277–283
  occurrence and isolation, 107
  size and amino acid sequences of, 111
  structure of, 111–112
  and UHS proteins: *see* Ultrahigh-sulfur (UHS) proteins
High-tyrosine proteins
  amino acid sequence and structure of, 115–117
  hetereogeneity of, 114–115
  occurrence and isolation of, 113
Hirano bodies, 460
Histidine-rich proteins: *see* Filaggrin
Heterodimers, 189–190
Homodimers, 189–190
Hyperproliferation marker, 309
  keratin pair expression patterns and, 312
Hypertension, 428

IDPN
  filament transport and, 455
  inducing neurofibrillary proliferation, 452
  neurofibrillary tangles and, 455
IF: *see* Intermediate filaments (IF)
IFAP: *see* Intermediate filament-associated proteins (IFAP)
IFB: *see* Intermediate filament bundles (IFB)
IFFC: *see* Intermediate filament focal centers (IFFC)
Immunoblot analysis
  anti-IFAP-300k immunoreactivity, 379
  autoantibody reaction with IF, 432
  of human epidermal keratinocytes, 317
  of human keratins, 306
    hair, 324
Immunoelectron microscopy, for cardiac muscle 10-nm filaments, 31
Immunofluorescence staining
  corneal keratinocytes, 322, 323
  epinemin, 382
  IFAP-300k, 374
  keratin IF expression, 301, 302
  p50 protein, 383
  paranemin, 382
  plectin, 380
  synemin, 381
  for vimentin and desmin, 418, 427
Immunoglobulins
  genes in rod domain evolution, 358
  IgG
    and autoantibody correlation, 432
    inflammatory reactions and, 434
  IgM, and autoantibody correlation, 432
Immunohistochemical staining
  human hair keratin localization, 325
Immunohistochemical staining (*cont.*)
  for keratin pair localization, 319
  profilaggrin/filaggrin localization by, 408
Immunological probes, of neurofilaments, 72–80
Immunological specificity, IF classification and, 236–237
Immunoreactivity: *see* Immunofluorescence staining
in vitro procedures
  chymotryptic proteolysis of desmin, 23
  IF inflammatory reactions, 434
  IF isolation and characterization by, 13–15
  IFAP-300k recombination, 373–374, 375–378
  keratin pair expression and, 327
  keratinocyte cultures, 312–315, 317–318
    basal cells, 318–319
  macrofibril formation, 398
  molecular aggregation, 208
  self assembly/coassembly, IF classification based on, 235–236
  spectrin-IF association, 163–164
  vimentin and erythrocyte membrane reconstitution, 136
  vimentin IF protein assembly, 215
in vivo procedures
  filaggrin synthesis, 402
  IF-nuclear association, 136
  keratin pair expression and, 327
  keratinocyte cultures, 312–315
Infectious diseases, IgM antibodies reaction with IF in, 432
Inflammation
  IF as activators, 433–435
  "keratin antibodies" and, 432
Inner root sheath (IRS)
  hair development and, 291
  trichohyalin in, 277, 288–289
Interaction score: *see* Ionic interaction score
Intercellular plasma bridges, 3
Interchain ionic interactions, 188–189
Intermediate filament-associated proteins (IFAP), 107–117, 371–372; *see also* High-sulfur proteins; High-tyrosine proteins
  in BHK-21 cells, 372–379
  cellular expression of, 385–386
  desmoplakin 1 and 2, 385
  epinemin, 382
  fiber protein assembly and cross-linking, 103–104
  functional significance speculations, 386
  hard keratin, 277; *see also under* Hard keratin, IFAP family
  IF classification and, 236–237
  IF research and, 6
  IFAP-70-280k, 379
    and IFAP-300k distribution compared, 379
  IFAP-300k, 373–374, 375–378
    and IFAP-70-280k distribution compared, 379

Intermediate filament-associated proteins (*cont.*)
  IFAP-300k (*cont.*)
    immunoreactivity in mammalian cell lines, 374–379
  MAP-2, 384–385
  NAPA-73, 383
  p50, 382–383
  p230 and spectrin-related proteins, 383–384
  paranemin, 381–382
  plectin, 380
  in smooth muscle, 27
  synemin, 380–381
Intermediate filament bundles (IFB), 7, 10
Intermediate filament focal centers (IFFC), 158
Intermediate filament (IF) genes
  complex nature of, 354
  evolutionary models for, 356–363
  linkage and chromosomal localization, 354–356
  structure, 356
  of wool, 269–277
Intermediate filaments (IF)
  classification of, 175–176; *see also* Protein chains, classification of
    schemes for, 5, 234–237
  complexity of, 6
  generic model for, 219
  historical overview, 3–5
  molecular structure of, 206–208; *see also* Molecular aggregation
  neural: *see* Neurofilaments
  nomenclature, 4, 5
  -plaque linkers, putative, 152, 154–156
  subcellular organization of, 7–15
  10-nm: *see* 10-nm filaments
  three-dimensional structure of, 205, 219–227
  width and radial organization of, 247–259; *see also* Morphology, of IF
Interphase, 129, 130
  M phase and, 141
Interspecies variation, IF classification and, 236–237
Introns
  cytokeratin genes, 341
  IF genes, 355
  NIFP genes, 348–349
  rod domain evolution and, 357–358
Invasive duct carcinoma, of breast, 419
Ionic interaction score
  IF structure model generation using, 223
  molecular aggregation and, 209–213
IRS: *see* Inner root sheath (IRS)

Juxtanuclear region, IF in, 7, 9–10

Keratin; *see also* Hard α-keratins; Hard keratin
  constituent proteins of, 97–99; *see also* Keratin proteins
Keratin (*cont.*)
  expression rules, 301, 303–304
  filaggrin aggregation with, 397–398
  gel profiles (cow, rabbit, and chicken), 308
  gene expression variations, 294–295
  and protein chain classification, 179–182
  subfamilies, 304–307
  synthesis of, 291
"Keratin antibodies," 432
Keratin chains
  early physiochemical studies, 188
  end domains of, 360–362
  sequence homology, 179, 180
  terminal domains of, 192
Keratin expression
  keratin pairs and, 307–311
  rules of, 301–304
Keratin pairs
  coexpression, defined by, 307, 309–311
  common features, 307, 309–311
  expression
    patterns, 311–318
    rules, 301–304
  localization of, 318–326
  structural/functional significance of, 326–328
Keratin proteins, 97, 98
  amino acid composition, 99
  genetic control over, 118
  morphological location of, 97, 99
  mutations affecting, 119–122
  primary structures of, 176–177
  species specificity, 118–119, 120
Keratinization disorders, filaggrin alterations and, 407–409
Keratinocytes, 13
  keratin pair expression patterns and, 311–318

Lamin genes, evolution of, 363
Lamins; *see also* Nuclear lamina (NL)
  biosynthesis of, 133
  classified, 130–131
  dephosphorylation of, 139
  IF-like proteins as, 196–197
  molecular structure and polymer assembly, 133–134
  and protein chains compared, 197
Larva, cytokeratin gene expression in *X. laevis*, 341–342
Leiomyosarcomas, 420
Leishmaniasis, IgM antibodies reaction with IF in, 432
Lewy bodies, 460
Lincoln wool, high-tyrosine protein content in, 118
Linear density, molecular weight and, proportionality between, 245–247
Liver, alcoholic, altered IF expression and, 422–425

Loligo spp.; *see also* Squid
neurofilament investigations in, 59–63
neurofilament polypeptide structure in, 64
Low-sulfur proteins, preparation and properties of, 104–105
Lowicryl K4M-embedded bovine tongue mucosa, 152, 153–154

M-line site, 10-nm filaments at, 29
M phase, and interphase, 141
M phase-promoting factor (MPF), 141
Macrofibrils
filaggrin and, 397
in vitro formation, 398
Macroglobulinemia, Waldenström, 432
Malaria, IgM antibodies reaction with IF in, 432
Mallory bodies, 422, 424
pathogenesis hypotheses, 425
Mammals
neurofilament polypeptides, 43–45; *see also* Triplet polypeptides
neurofilaments in, 38–41
MAP-2, 384–385
MAPs: *see* Microtubule-associated proteins (MAPs)
Marine fan worm: *see Myxicola infundibulum*
Maytansine, inducing neurofibrillary proliferation, 452
Measles, IgM antibodies reaction with IF in, 432
Medulla
hair fiber lacking, 290
trichohyalin in, 288–289
synthesis of, 291
Meiotic cells, nuclear lamina-chromation association and, 136–137
Melanoma, antibodies to IF in, 432
Merino wool
fibre cross section, 226
follicle size, 102–103
high-tyrosine protein content in, 118
Meso cells, nutrition and, 123
Mesocortex, 289–290
Microfilaments, actin-containing, 7, 12
Microtubular failure hypothesis, of Mallory bodies pathogenesis, 425
Microtubule-associated proteins (MAPs), and neurofilament relationship, 65
Microtubules, 7, 11
IF versus, 4
and neurofilament relationships
in invertebrates, 65–66
in vertebrates, 38
Mitosis, vimentin and desmin in, 140–141
Mitotic spindle poisons, inducing neurofibrillary proliferation, 452
Molecular aggregation, 208
enzymatically derived fragments, 209
ionic interaction score, 209–213
oligomers, 213–218
Molecular analysis, of desmins and vimentin, 22–24
Molecular structure, of intermediate filaments, 206–208
Molecular weight
IF primary structures, 176–177
squid neurofilaments, 61
triplet polypeptides, 44
Molecule
changes during NL disassembly, 139–140
IF protein, schematic representation of, 207
lamin, structure and polymer assembly, 133–134
two-chain coiled-coil, IF research and, 6
Mongolism, neurofibrillary degeneration and, 452
Monoclonal antibodies (MAbs)
in diagnostic histopathology, 72–80
for keratin pair localization, 319
Monoclonal antibody, against IFAP-300k, 373–374
Mononucleosis, IgM antibodies reaction with IF in, 432
Morphogenesis, of hair follicle, 100–101
Morphology, of IF
diameter, traditional estimates, 247–248
equatorial intensity distribution, 257–258
and myosin filaments compared, 259
neurofilament cross-bridges and, 258
positive density, 257
proteolytic modification of
image interpretation, 256
in vitro assembly, 256–257
radial density profiles, 252–253
subfilament bundles, 253–256
width, quantitative determination by EM, 248–252
MPF: *see* M phase-promoting factor (MPF)
Multiphosphorylation site, NIFP, 347
Mumps, IgM antibodies reaction with IF in, 432
Muscles, IF-cell surface association in, 157–158; *see also* Cardiac muscle; Skeletal muscle; Smooth muscle
Mutations, affecting keratin proteins, 119, 119–122, 121–122
Mycoplasma pneumoniae infection
IF autoantibody production and, 433
IgM antibodies reaction with IF in, 432
Myocardium, thin filaments rearrangement in, 31
Myofibril axis, skeletal muscle 10-nm filaments and, 29
Myopathies, altered IF expression and, 421–422
Myosin filaments, and IF properties compared, 259
Myxicola infundibulum
calcium-activated neutral proteases and, 67

Myxicola infundibulum (*cont.*)
  neurofilaments in
    function and evolution, 81
    investigations on, 53–59
    polypeptide structure in, 64
  proteolytic digestion and, 50
  x-ray diffraction studies, 220

N-terminal domain
  acidic and basic residues in, 183
  amino acids in, 361
  structure of, 191–194
NAPA-73, 383
Neoplastic cells
  epithelial, keratin synthesis in, 315–316
  IF as differentiation marker of, 416–421
Nerve cells: *see* Neurons
Nerve degeneration, neurofilament destruction mediated by CANP, 69–70
Nerve growth factor (NGF), 345
Nerve terminals, neurofilament turnover in, CANP and, 69–70
Nervous system, NIFP expression during development of, 345–346
Nervous tissue, CANP purification from, 68–69
"Net-helix ambiguity," 220
Neural IF proteins (NIFP)
  amino acid sequences of, 346–347
  distinctions among, 345–346
  gene promotors, 349
  intron number and location, 348–349
  multiphosphorylation site, 347
  mRNAs, 347–348
Neural intermediate filaments: *see* Neurofilaments
Neurofibrillary degeneration, 451–456
Neurofibrillary proliferation, agents inducing, 452, 453
Neurofibrillary tangles
  in Alzheimer's disease, 457–458
  composition of, 452–453
  formation mechanism, 453–456
  progressive supranuclear palsy and, 459
Neurofibrils, 3
Neurofilament-associated protein, avian specific, 73k: *see* NAPA-73
Neurofilament chains, evolution in, 362–363
Neurofilament protein
  C-terminal domains in, 192
  primary structures of, 177
Neurofilaments, 37
  amino acid composition and sequence studies, 45–48
  axonal transport of, 80–81
  cross-bridges, 258
  immunological probes of, 72–80
Neurofilaments (*cont.*)
  interactions of, 63–66
  from invertebrates, 52–53
    Myxicola studies, 53–59
    squid studies, 59–63
  isolation of, 41–42
  ontogeny and phylogeny of, 81–82
  polypeptides
    mammalian, 43–45
    triplet type, arrangement of, 48–52
  posttranslational modifications of, 66–72
  proteins: *see* Neural IF proteins (NIFP)
  reassembly properties of, 45
  solubility and structure of, 42–43
  from vertebrates, 38–52
Neurological disease
  Alzheimer's disease, 456–459
  cytoplasmic inclusions and, 459–460
  neurofibrillary degeneration, 451–456
  progressive supranuclear palsy, 459
Neurons
  IF systems in, 7, 8
  neruofilamentous skeleton in, 37
    vertebrates, 38–41
NGF: *see* Nerve growth factor (NGF)
NIFP: *see* Neural IF proteins (NIFP)
NL: *see* Nuclear lamina (NL)
Nocodazole, inducing neurofibrillary proliferation, 452
Nomenclature
  cytokeratin polypeptide chains, 5
  IF classification, 5
  IF descriptive terms, 4
Noncoding regions, keratin genes compared, 285–286
Nonneoplastic cells, IF as differentiation marker of, 416–421
NPC: *see* Nuclear pore complexes (NPC)
Nuclear lamina (NL), 129–130; *see also* Lamins
  association with nuclear and cytoplasmic structural elements, 134–137
  cell cycle dynamics of, 137–140
    intermediate filaments and, 140–141
  composition and developmental changes, 132–133
  defined, 130–131
  dissassembly
    cell cycle regulation of, 141
    molecular changes in, 139–140
  isolation of, 131–132
Nuclear membrane; *see also* Nuclear lamina (NL)
  fragmentation factors, 140
  nuclear lamina association with, 135
Nuclear pore complexes (NPC), 9–10
  nuclear lamina association with, 134–135
Nutrition, role in keratin composition, 122–124

Oligomers
IF research and, 6
molecular aggregation and, 213–218
structure of, 237–239
Ontogeny, of neurofilaments, 81–82
Oocytes, 137
NL dsassembly, 138–139
Orbit, rhabdomysarcoma of, 419
Organelles, 7
and neurofilament relationships, 66
Orphan exons, 276–277
nucleotide sequence of, 278–279
ORS: *see* Outer root sheath (ORS), hair development and
Ortho cells, nutrition and, 123
Orthocortex, 289–290
Ouchterlony reaction, 326
Outer root sheath (ORS), hair development and, 291

p50, 382–383
p230, 383–384
Paired helical filament (PHF)
in Alzheimer's disease, 457–459
neurofibrillary degeneration and, 452
Pick bodies and, 459–460
in progressive supranuclear palsy, 459
Para cells, nutrition and, 123
Paracortex, 289–290
Paracrystals, axial repeats in, 244–245
Paranemin, 27, 381–382
Pentapeptides, folded, thiol-disulfide system of, 112
Peptide repeats, in subdomainal keratin IF chains, 362
Peripheral nervous system, neurofilament isolation from, 41–42
Phenylalanine, conservation in protein chains, 184, 185
PHF: *see* Paired helical filament (PHF)
Phosphopeptide mapping, nuclear lamina and, 140
Phosphorylation
IF assembly and disassembly by, 15
Myxicola polypeptides, 55
neurofilament modification by, 70–72
neurofilament triplet polypeptides and, 48, 70–71
and NL disassembly relationship, 139
profilaggrin and, 403
Phylogeny, of neurofilaments, 81–82
Physiology, keratin composition and, 124
Pick bodies, 459–460
Pick's disease, 459–460
Plaque
atheromatous, formation of, 425–426
attachment, 157
desmosome-IF complex, 150–151
putative linkers in, 152, 154–156
hemidesmosome-IF complex, 151
putative linkers in, 152, 154–156
Plaque protein, and IF interaction, 152, 154–156
Plasma membrane, 7, 13
Plectin, 380
PMN: *see* Polymorphonuclear neutrophils (PMN), from Alzheimer patients
Podophyllotoxin, inducing neurofibrillary proliferation, 452
Polarity
heptad, in rod domain evolution, 358
of thin filaments, 21
Poly-γ-benzyl-L-glutamate, 220
Polymers
aggregation of, 216–217, 218
lamin molecules and, 133–134
Polymorphism, IF classification and, 236
Polymorphonuclear neutrophils (PMN), from Alzheimer patients, 459
Polypeptides; *see also* Keratin
230k, 383–384
mammalian neurofilament, 43–45; *see also* Triplet polypeptides
in *Myxicola* neurofilaments, 54–55, 56
subunit: *see* Subunit polypeptides
Porcupine quill, hard α-keratins structural studies, 95–96
X-ray diffraction, 219–220
Posttranslational mechanisms, in neurofilament modification, 66–72
Preneoplasia hypothesis, of Mallory bodies pathogenesis, 425
Primary structures, of IF protein chains, 176–177
Profilaggrin, 393–394; *see also* Filaggrin
cDNA clones, 398–399
extraction and purification, 394–395
and filaggrin synthesis, 402–405
gene, repetitive structure of, 399–401
hypothetical structure, 404
properties of, 395
protein structure, 396–397
mRNA, site and expression of, 399
synthesis during fetal development, 405–407
Profilaggrin phosphatase, 403
Profilaggrin repeat unit, 402
amino acid sequence of, 401
Progenitor genes, in rod domain evolution, 358–359
Progressive supranuclear palsy, neurofibrillary tangles in, 459
Protein chains
central α-helix-rich domain in, 182–187
classification of, 179–182
four-chain aggregates, 194–196
IF research and, 6
and lamins A and C compared, 197
molecular structure
rod domain, 187–191
terminal domain, 191–194
nuclear lamins and, 196–197

Protein chains (*cont.*)
schematic representation of, 196
type I–IV, 176
primary structures, 176–177
secondary structures, 177–179
Protein synthetic apparatus, wool growth rate and, 123
Proteins; *see also individually named proteins*
in fiber, 102–104
and filament properties, in skeletal muscle, 30
high-sulfur: *see* High-sulfur proteins
high-tyrosine: *see* High-tyrosine proteins
IF-associated: *see* Intermediate filament-associated proteins (IFAP)
low-sulfur, preparation and properties of, 104–105
microtubule-associated (MAPs), 65
neural IF: *see* Neural IF proteins (NIFP)
neurofilament-associated, 66
normal versus mutant mouse hair, 121
of wool, 269–289; *see also under* Hard keratin
Proteolysis, calcium-activated, neurofilament modification by, 66–70
Proteolytic processing
IF image interpretation, 256
IF *in vitro* assembly properties, 256–257
neurofilaments and, 49, 50
profilaggrin synthesis and, 404–405
Protofilaments
electron microscopy and, 25
in *Myxicola* neurofilaments, 57–59
packing of, 24–25
subunit organization in, 24
Putative IF-plaque linkers, 152, 154–156
Pyruvate kinase gene, in rod domain evolution, 358

Quill: *see* Porcupine quill

Radial density profiles, of IF, 252–253
Random-coil, protein chain secondary structure, 177–179
Recombinant filaggrin, 398
Repeat unit; *see also* Axial repeats
IF size and linear density, 402
profilaggrin, 402
Residues
azimuthal distribution of, 210
pattern in protein chains, 183
patterns in rod domain, 190–191
position projections in 1B rope segment, 212, 214, 215
in protein chains, structural periodicity, 186
Retinoic acid, cultured epidermal keratinocytes and, 316–317
Rhabdomyosarcomas, 420
of the orbit, 419
Rheumatoid arthritis, antibodies to IF in, 432

mRNA
neurofibrillary tangles, 454, 455
NIFP, 347–348
profilaggrin, 399
sequences in *Drosphila* heat shock response, 285
Rod domain
amino acids in, 361
cytokeratin sequence from *X. laevis*, 336, 338–339, 340
degree of homology in, 182
IF genes, evolution of, 356–359
of IF protein molecules, 187–191
long types, possible interactions between, 210
residues in
acidic and basic, 183
patterns, 190–191

SAM: *see* Substrate-attached material (SAM)
Sarcoidosis, altered IF expression and, 428
Sarcolemma, IF-cell surface association and, 158
Sarcoma, 417
Gardner–Rasheed feline virus, 433
Scanning transmission electron microscopy (STEM)
advantages of, 248
axial chain densities derived from, 220–221
dark-field, IF visualization by, 251–252
desmin and vimentin filaments, 26
SDS-PAGE
human hair keratin analysis, 307
mammalian neurofilament polypeptides, 44–45
smooth muscle cell extracts, 428, 429
squid axoplasm, 61, 62
Secondary structures
of IF protein chains, 177–179
of terminal domains, 192–193
Sedimentation equilibrium techniques, 44
Segment
protein chain and lamins compared, 197
size in IF protein chains, 181
Sheep, hard keratin IFAP gene structure, 285–286; *see also* Merino wool; Wool
Sidearms
of neurofilaments, 40–41
triplet polypeptide identity and, 52
Sjögren's syndrome, antibodies to IF in, 432
Skeletal muscle, 10-nm filaments in, 28–30
Skeletin: *see* Desmin
Smooth muscle
atherosclerotic, 426
IF-cell surface association in, 158
SDS-PAGE of cell extracts, 428, 429
10-nm filaments in, 19–28
vimentin, actin and tropomyosin in, 428, 429
Soft tissue neoplasms, IF expression in, 418
Solubility
of hard α-keratins, 96–97
hard α-keratins and, 95

Solubility (*cont.*)
of neurofilaments, 42
Spectrin
IF association with, 163–164
related proteins, 383–384
Spinal nerve root, neurofilaments in, 38–41
Spisula solidissima, 137
NL disassembly, 138–139
Squid
calcium-activated neutral proteases and, 67
neurofilament investigations in, 59–63
neurofilament polypeptide structure in, 64
Squid stellate ganglion
antibody labeling of, 74, 76
neurofilament topographical regulation model, 80
STEM: *see* Scanning transmission electron microscopy (STEM)
Streptococcus pyogenes, IF autoantibody production and, 433
Subdomain
IF chain, 360
keratin chain end domains, 360–362
in neurofilament chains, 362–363
peptide repeats and codon usage in, 362
structure in terminal domains, 193
Subfilaments: *see* Subunits
Substrate-attached material (SAM), 161
Subunit polypeptides
amino acid sequences of, 106
assembly of, 106
properties of, 105
Subunits
IF morphology and, 253–256
of neurofilaments, 43
CANP studies and, 68–69
gene encoding and, 81–82
organization in protofilament, 24
polymeric, 216–217, 218
Supranuclear palsy, fibrillar structures in, 459
Surface lattice
helical projections of, 225
IF structure and, 221–223
Synemin, 380–381
Synovial sarcoma cells, 417

Temperature, desmin filament formation and, 26–27
10-nm filaments
in cardiac muscle, 30–32
55-kDa monomer relationship with, 24–27
neurofibrillary proliferations composed of, 452
in skeletal muscle, 28–30
in smooth muscle, 19–28
Terminal differentiation, of wool fiber, 275
Terminal domains; *see also* C-terminal domain; N-terminal domain
disposition of, 223
of IF protein molecules, 191–194
Thin filaments: *see* 10-nm filaments
Thiol-disulfide system, of folded pentapeptides, 112
Tissue espression
high-glycine/tyrosine keratin IFAP structure and, 283–285
high-sulfur keratin IFAP structure and, 277–283
Tissue organization, IF research and, 6
Tissue-specific classification, of IF structure, 234
Tissue specificity, of IF expression in mammalian tissue, 416
Tonofibrillae, 3
Tonofilaments: *see* Intermediate filament bundles (IFB)
Transcription, neurofilaments and, 82
Transformed cells, altered IF expression and, 429–431
Transverse traces, of IF projected density, 248–252; *see also under* Electron microscopy (EM)
Traversing filaments, desmosome preparation and, 150
Trichohyalin, in IRS and hair medulla, 277, 288–289
Trichothiodystrophy (TTD), 119
UHS fraction loss in, 121
Triplet polypeptides
amino acid sequences of, 45–48
anomalous behavior in SDS, 44
arrangement in neurofilament, 48–52
isolation of, 41
molecular weights of, 44
reassembly properties of, 45
Tropomyosin, in smooth muscle cells, 428, 429
TTD: *see* Trichothiodystrophy (TTD)
Tumors; *see also individually named tumors*
altered IF expression and, 429–431
IF use in diagnosis of, 420
Turn, protein chain secondary structure, 177–179
Twisted tubules: *see* Paired helical filament (PHF)
Two-chian coiled-coil molecule, IF research and, 6
Tyrosine, conservation in protein chains, 184, 185; *see also* High-tyrosine proteins

UHS: *see* Ultrahigh-sulfur (UHS) proteins
Ultrahigh-sulfur (UHS) proteins
follicle bulbs and, 103
gene expression variations and, 294
high-sulfur proteins and, 110–111
keratin IFAP family, 277, 280, 281–283

Vesicles, and neurofilament relationships, 66
Vimentin
antibodies against, 417
depolymerized-biotinylated, 14–15
filament formation concept, 26
IF molecule chain composition in, 189–190
immunofluorescence staining for, 418
in vitro assembly study, 215

Vimentin (*cont.*)
  in vitro assembly study (*cont.*)
    polymeric aggregation and, 216–217
  in vitro reconstitution of membrane, 136
  isolation and preparation of, 21–22
  in mitosis, 140–141
  molecular properties of, 22–24
  primary structures of, 177
  in smooth muscle cells, 428, 429
Vinca alkaloids, inducing neurofibrillary proliferation, 452
Vitamin A deficiency hypothesis, of Mallory bodies pathogenesis, 425

Waldenström macroglobulinemia, autoantibody reaction with IF in, 432
Wool; *see also under* Hair; Merino wool
  early studies on, 3–4
  growth regulation factors, 101
  hard α-keratins biochemistry, 95–96
  helix-rich particles from, 209
  IF genes of, 269–277
  ionic interaction score and, 224
  terminal differentiation, 275
Wool follicle
  cDNA clone, 281–283
  Merino sheep, 102–103

X-ray diffraction
  equatorial, from hard α-keratin, 257–258
  four-chain aggregate structure and, 194, 196
  of hard α-keratins, 95, 99–100
  high angle patterns, coiled-coil rope structure and, 206–208
  IF structure studies, 219–221
  meridional reflection, 220–221
    axial repeats from, 241
    surface lattice, 222–223
  Myxicola axoplasm, 57, 58
  surface lattice, 221–223
Xenopus laevis
  cytokeratin gene sequences from, 337
    developmental aspects, 341–342
    type I, 336–337, 340–341
    type II, 342
  cytokeratin gene structure from, 337–340
  lamin structure in, 134
  NL composition changes in, 132

Z line
  10-nm filaments and
    cardiac muscle, 30–31
    skeletal muscle, 28–29
    smooth muscle, 21
  IF-cell surface association and, 158